Cognitive Radio Technology

Cognitive Radio
Technology
Second Edition

Bruce A. Fette

Editor

AMSTERDAM • BOSTON • HEIDELBERG • LONDON
NEW YORK • OXFORD • PARIS • SAN DIEGO
SAN FRANCISCO • SINGAPORE • SYDNEY • TOKYO
Academic Press is an imprint of Elsevier

Academic Press is an imprint of Elsevier
30 Corporate Drive, Suite 400
Burlington, MA 01803

Library of Congress Cataloging-in-Publication Data
Application submitted.

ISBN 13: 978-0-12-374535-4

For information on all Academic Press publications,
visit our Website at *www.books.elsevier.com*

Printed in the United States
09 10 11 12 13   10 9 8 7 6 5 4 3 2 1

Contents

Preface

Dr. Joseph Mitola III
Stevens Institute of Technology
Castle Point on the Hudson, New Jersey

This preface[1] takes a visionary look at ideal cognitive radios (iCRs) that integrate advanced software-defined radios (SDRs) with CR techniques to arrive at radios that learn to help their user using computer vision, high-performance speech understanding, GPS navigation, sophisticated adaptive networking, adaptive physical layer radio waveforms, and a wide range of machine learning processes.

CRS KNOW RADIO LIKE TELLME KNOWS 800 NUMBERS

When you dial 1-800-555-1212, a speech synthesis algorithm may say, "Toll Free Directory Assistance powered by TellMe®. Please say the name of the listing you want." If you mumble, it says, "OK, United Airlines. If that is not what you wanted press 9, otherwise wait while I look up the number." Reportedly, some 99 percent of the time TellMe gets it right, replacing the equivalent of thousands of directory assistance operators of yore. TellMe, a speech-understanding system, achieves a high degree of success by its focus on just one task: finding a toll-free telephone number. Narrow task focus is one key to algorithm successes.

The cognitive radio architecture (CRA) is the building block from which to build cognitive wireless networks (CWN), the wireless mobile offspring of TellMe. CRs and networks are emerging as practical, real-time, highly focused applications of computational intelligence technology. CRs differ from the more general artificial intelligence (AI) based services (e.g., intelligent agents, computer speech, and computer vision) in degree of focus. Like TellMe, ideal cognitive radios (iCRs) focus on very narrow tasks. For iCRs, the task is to adapt radio-enabled information services to the specific needs of a specific user. TellMe, a network service, requires substantial network computing resources to serve thousands of users at once. CWNs, on the other hand, may start with a radio in your purse or on your belt—a cell phone on steroids—focused on the narrow task of creating from myriad available wireless information networks and resources just what is needed by one user: you. Each CR fanatically serves the needs and protects the personal information of just one owner via the CRA using its audio and visual sensory perception and autonomous machine learning.

---

[1]Adapted from J. Mitola III, *Cognitive Radio Architecture: The Engineering Foundations of Radio XML*, Wiley, 2006.

TellMe is here and now, while iCRs are emerging in global wireless research centers and industry forums such as the Software-Defined Radio Forum and Wireless World Research Forum (WWRF). This book introduces the technologies to evolve SDR to dynamic spectrum access (DSA) and towards iCR systems. It introduces technical challenges and approaches, emphasizing DSA and iCR as a technology enabler for rapidly emerging commercial CWN services.

FUTURE ICRS SEE WHAT YOU SEE, DISCOVERING RF USES, NEEDS, AND PREFERENCES

Although the common cell phone may have a camera, it lacks vision algorithms, so it does not see what it is imaging. It can send a video clip, but it has no perception of the visual scene in the clip. With vision processing algorithms, it could perceive and categorize the visual scene to cue more effective radio behavior. It could tell whether it were at home, in the car, at work, shopping, or driving up the driveway at home. If vision algorithms show you are entering your driveway in your car, an iCR could learn to open the garage door for you wirelessly. Thus, you would not need to fish for the garage door opener, yet another wireless gadget. In fact, you would not need a garage door opener anymore, once CRs enter the market. To open the car door, you will not need a key fob either. As you approach your car, your iCR perceives this common scene and, as trained, synthesizes the fob radio frequency (RF) transmission to open the car door for you.

CRs do not attempt everything. They learn about your radio use patterns leveraging a-priori knowledge of radio, generic users, and legitimate uses of radios expressed in a behavioral policy language. Such iCRs detect opportunities to assist you with your use of the radio spectrum, accurately delivering that assistance with minimum tedium.

Products realizing the visual perception of this vignette are demonstrated on laptop computers today. Reinforcement learning (RL) and case-based reasoning (CBR) are mature machine learning technologies with radio network applications now being demonstrated in academic and industrial research settings as technology pathfinders for iCR[2] and CWN.[3] Two or three Moore's law cycles, or three to five years from now, these vision and learning algorithms will fit into your cell phone. In the interim, CWNs will begin to offer such services, presenting consumers with new trade-offs between privacy and ultrapersonalized convenience.

CRS HEAR WHAT YOU HEAR, AUGMENTING YOUR PERSONAL SKILLS

The cell phone you carry is deaf. Although this device has a microphone, it lacks embedded speech-understanding technology, so it does not perceive what it hears. It can let you talk to your daughter, but it has no perception of your daughter, nor of your

---

[2] J. Mitola III, *Cognitive Radio Architecture*, 2006.
[3] M. Katz and S. Fitzek, *Cooperation in Wireless Networks*, Elsevier, 2007.

conversation's content. If it had speech-understanding technology, it could perceive your dialog. It could detect that you and your daughter are talking about a common subjects such as a favorite song. With iCR, speech algorithms detect your daughter telling you by cell phone that your favorite song is now playing on WDUV. As an SDR, not just a cell phone, your iCR determines that she and you both are in the WDUV broadcast footprint and tunes its broadcast receiver chipset to FM 105.5 so that you can hear "The Rose." With your iCR, you no longer need a transistor radio in your pocket, purse, or backpack. In fact, you may not need an MP3 player, electronic game, and similar products as high-end CR's enter the market (the CR may become the single pocket pal instead). While today's personal electronics value propositions entail product optimization, iCR's value proposition is service integration to simplify and streamline your daily life. The iCR learns your radio listening and information use patterns, accessing songs, downloading games, snipping broadcast news, sports, and stock quotes you like as the CR reprograms its internal SDR to better serve your needs and preferences. Combining vision and speech perception, as you approach your car, your iCR perceives this common scene and, as you had the morning before, tunes the car radio to WTOP for your favorite "traffic and weather together on the eights."

For effective machine learning, iCRs save speech, RF, and visual cues, all of which may be recalled by the radio or the user, acting as an information prosthetic to expand the user's ability to remember details of conversations, and snapshots of scenes, augmenting the skills of the ⟨Owner/⟩.[4] Because of the brittleness of speech and vision technologies, CRs may also try to "remember everything" like a continuously running camcorder. Since CRs detect content (e.g., speakers' names and keywords such as "radio" and "song"), they may retrieve content requested by the user, expanding the user's memory in a sense. CRs thus could enhance the personal skills of their users (e.g., memory for detail).

IDEAL CRS LEARN TO DIFFERENTIATE SPEAKERS TO REDUCE CONFUSION

To further limit combinatorial explosion in speech, CR may form speaker models—statistical summaries of speech patterns—particularly of the ⟨Owner/⟩. Speaker modeling is particularly reliable when the ⟨Owner/⟩ uses the iCR as a cell phone to place a call. Contemporary speaker classification algorithms differentiate male from female

---

[4]Semantic Web: Researchers formulate CRs as sufficiently speech-capable to answer questions about ⟨Self/⟩ and the ⟨Self/⟩ use of ⟨Radio/⟩ in support of its ⟨Owner/⟩. When an ordinary concept, such as "owner," has been translated into a comprehensive ontological structure of computational primitives (e.g., via Semantic Web technology), the concept becomes a computational primitive for autonomous reasoning and information exchange. Radio XML, an emerging CR derivative of the eXtensible Markup Language (XML) offers to standardize such radio-scene perception primitives. They are highlighted in this brief treatment by ⟨Angle-brackets/⟩. All CR have a ⟨Self/⟩, a ⟨Name/⟩, and an ⟨Owner/⟩. The ⟨Self/⟩ has capabilities such as ⟨GSM/⟩ and ⟨SDR/⟩, a self-referential computing architecture, which is guaranteed to crash unless its computing ability is limited to real-time response tasks; this is appropriate for a CR but may be too limiting for general-purpose computing.

speakers with a high level of accuracy. With a few different speakers to be recognized (i.e., fewer than 10 in a family) and with reliable side information (e.g., the speaker's telephone number), today's state-of-the-art algorithms recognize individual speakers with better than 95 percent accuracy.

Over time, each iCR can learn the speech patterns of its ⟨Owner/⟩ in order to learn from the ⟨Owner/⟩ and not be confused by other speakers. The iCR may thus leverage experience incrementally to achieve increasingly sophisticated dialogs. Today, a 3-GHz laptop supports this level of speech understanding and dialog synthesis in real time, making it likely to be available in a cell phone in 3 to 5 years.

The CR must both know a lot about radio and learn a lot about you, the ⟨Owner/⟩, recording and analyzing personal information, and the related aggregation of personal information places a premium on trustworthy privacy technologies. Therefore, the CRA incorporates ⟨Owner/⟩ speaker recognition as one of multiple soft biometrics in a biometric cryptology framework to protect the ⟨Owner/⟩'s personal information with greater assurance and convenience than password protection.

MORE FLEXIBLE SECONDARY USE OF THE RADIO SPECTRUM

In 2008, the US Federal Communications Commission (FCC) issued its second Report and Order (R&O) that radio spectrum allocated to TV, but unused in a particular broadcast market (e.g., because of the transition from analog to digital TV) could be used by CRs as secondary users under Part 15 rules for low-power devices—for example, to create ad hoc networks. SDR Forum member companies have demonstrated CR products with these elementary spectrum-perception and use capabilities. Wireless products, both military and commercial, already implement the FCC vignettes.

Integrated visual- and speech-perception capabilities needed to evolve the DSA CR to the situation-aware iCR are not many years distant. Productization is underway. Thus, many chapters of Bruce's outstanding book emphasize CR spectrum agility, suggesting pathways toward enhanced perception technologies, with new long-term growth paths for the wireless industry. Those who have contributed to this book hope that it will help you understand and create new opportunities for CR technologies.

Acknowledgments

This Second Edition of *Cognitive RadioTechnology* has been a collaborative effort of many leading researchers in the field of cognitive radio with whom I have had the pleasure of interacting over the last 10 years through participation in the Software Defined Radio Forum, and in some cases, a few of whom I have worked with over nearly my entire career. To each of these contributors, I owe great thanks, as well as to all the other participants in the SDR Forum who have contributed their energy to advance the state of the art. In addition to the authors, each contributor or contributor's team in turn, has also been supported by their staffs and we appreciate their contributions as well.

I owe much to my family, Elizabeth, Alexandra, and Nicholas, who suffered my long distractions with their patience, love, understanding, and substantial help in editing and reviewing. I also owe many thanks to my editor, Sandy Rush, who has patiently guided me through this difficult but very creative process. I dedicate this book to my mother, who provided the perfect mixture of guidance and responsibility; to my grandfather; to my father; and Aunt Margaret, whose early guidance into the many aspects of science led me to this career.

I also acknowledge the support from General Dynamics C4 Systems for the support to work in this exciting new field.

Bruce A. Fette

Chapter 2

This chapter is dedicated to the regulatory community that struggles tirelessly to balance technical rigor with good policy making.

Pail Kolodzy

Chapters 4 and 8

The chapters are dedicated to Mona and Ashley. Thank you both for your love and friendship, and thank you for the time I needed to work on this chapter.

John Polson

Chapter 7

The authors of this chapter wish to thank all of the researchers, colleagues, and friends who have contributed to our work. Specifically, we are pleased to recognize the members of the Virginia Tech research group, including Ph.D. students Bin Le, David Maldonado, and Adam Ferguson; master's students David Scaperoth and Akilah Hugine; and faculty members Allen MacKenzie and Michael Hsiao. Finally, a very big thank you goes to three former colleagues who helped start this research: Christian Rieser, Tim Gallagher, and Walling Cyre.

Thomas W. Rondeau, Charles W. Bostian

Chapter 9

Ronald Brachman, Barbara Yoon, and J. Christopher Ramming helped to refine my understanding of cognition and cognitive networking. Joseph Mitola III and Preston Marshall greatly enhanced my knowledge of radio systems, and Mitola interested me in the intersection of radios and robotics. Harry Lee and Marc Olivieri helped me to understand fine-scale variations in RF reception. Larry Jackel and Thomas Wagner helped me to understand the challenges of decentralized control of robots. In addition, the author is indebted to Joseph Mitola III, Daniel Koditschek, and Bruce Fette for their kindness in reviewing and critiquing draft versions of this chapter.

Jonathan M. Smith

Chapter 10

The work for this chapter was sponsored by the Department of Defense under Air Force contract FA8721-05-C-0002. Opinions, interpretations, conclusions, and recommendations are those of the authors and are not necessarily endorsed by the US government. The authors are grateful to Joe Mitola for creating the DARPA seedling effort that supported this work.

Joseph P. Campbell, William M. Campbell, Scott M. Lewandowski, Alan V. McCree, Clifford J. Weinstein

Chapter 11

Youping Zhao was supported through funding from Cisco, Electronics and Telecommunications Research Institute (ETRI), and Texas Instruments, and advised by Jeffrey H. Reed. Bin Le was supported by the National Science Foundation (NSF) under Grant No. CNS-0519959 and advised by Charles W. Bostian. Special thanks to Bruce Fette, Jody Neel, David Maldonado, Joseph Gaeddert, Lizdabel Morales, Kyung K. Bae, Shiwen Mao, and David Raymond for their helpful discussions and comments. Any opinions, findings, and conclusions or recommendations expressed in this chapter are those of the author(s) and do not necessarily reflect the views of the sponsor(s).

Youping Zhao, Bin Le, Jeffrey H. Reed

Chapter 13

The work presented in this chapter was partially supported by NSF Grant No. 0225442.

Mieczyslaw M. Kokar, David Brady, Kenneth Baclawski

Chapter 14

This chapter is dedicated to Lynné, Barb, Max, and Madeline Sophia.

Although the views expressed are exclusively my own, I would like to express appreciation to The MITRE Corporation's commitment to technical excellence in the public interest through which one can step back and study the evolution of cognitive radio architecture from a variety of perspectives—US DoD, military, emergency services, aviation, commercial, and global.

Joseph Mitola III

Chapter 16

Work done for this chapter was supported by HY-SDR Research Center at Hanyang University, Seoul, Korea, under the ITRC program of Ministry of Knowledge Economy, and by the Korea Science and Engineering Foundation (KOSEF) grant funded by the Korean government to INHA-WiTLAB as a National Research Laboratory.

Jae Moung Kim, Seungwon Choi, Yusuk Yun, Sung Hwan Sohn, Ning Han, Gyeonghua Hong, Chiyoung Ahn

Chapter 17

Research for this chapter was supported by DARPA's neXt Generation Communications Program under Contract Nos. FA8750-05-C-0230 and FA8750-05-C-0150. SRI's XG project web page can be found at *http://xg.csl.sri.com.*

Grit Denker, Daniel Elenius, David Wilkins

Chapter 19

The work for this chapter was partially supported by DARPA through Air Force Research Laboratory (AFRL) Contract FA8750-07-C-0169. The views and conclusions contained in it are those of the authors and should not be interpreted as representing the official policies, either expressed or implied, of the Defense Advanced Research Projects Agency or the US government.

Luiz A. Dasilva, Ryan W. Thomas

Chapter 21

The preparation of this chapter was supported by Grant N00014-04-1-0563 from the US Office of Naval Research. Thomas Royster was also supported by a fellowship from the US National Science Foundation. The authors thank Steven Boyd for many beneficial suggestions during the preparation of the chapter.

Michael B. Pursley, Thomas C. Royster IV

Chapter 23

The authors would like to thank the following participants in IEEE activities with whom the direct interactions have been most valuable. Special recognition and acknowledgment for reviewing and commenting: James M. Baker, BAE Systems, Apuvra Mody, BAE Systems, AS&T, IEEE 802.22 voting member. For their contributions, we acknowledgments Douglas Sicker and James Hoffmeyer. Recognition is due also to Matt Sherman, Christian Rodriquez, Jacob Wood, David Putnam, Paul Kolodzy, and Vic Hsiao for topical discussions.

Ralph Martinez, Donya He

History and Background of Cognitive Radio Technology

Bruce A. Fette
General Dynamics C4 Systems
Scottsdale, Arizona

1.1 THE VISION OF COGNITIVE RADIO

Just imagine if your cellular telephone, personal digital assistant (PDA), laptop, automobile, and television were as smart as "Radar" O'Reilly from the popular TV series *M*A*S*H*.[1] They would know your daily routine as well as you do. They would have things ready for you as soon as you ask, almost in anticipation of your needs. They would help you find people, things, and opportunities; translate languages; and complete tasks on time. Similarly, if a radio were smart, it could learn services available in locally accessible wireless computer networks, and could interact with those networks in their preferred protocols, so you would have no confusion in finding the right wireless network for a video download or a printout. Additionally, it could use the frequencies and choose waveforms that minimize and avoid interference with existing radio communication systems. It might be like having a friend in everything that's important to your daily life, or like you were a movie director with hundreds of specialists running around to help you with each task, or like you were an executive with a hundred assistants to find documents, summarize them into reports, and then synopsize the reports into an integrated picture. A cognitive radio (CR) is the convergence of the many pagers, PDAs, cell phones, and array of other single-purpose gadgets we use today. They will come together over the next decade to surprise us with services previously available to only a small select group of people, all made easier by wireless connectivity and the Internet.

1.2 HISTORY AND BACKGROUND LEADING TO COGNITIVE RADIO

The sophistication possible in a Software Defined Radio (SDR) has now reached the level where each radio can conceivably perform beneficial tasks that help the user, help the network, and help minimize spectral congestion. Some radios are able to

---

[1] "Radar" O'Reilly is a character in the popular TV series *M*A*S*H*, which ran from 1972 to 1983. He always knew what the Colonel needed before the Colonel knew he needed it.

demonstrate one or more of these capabilities in limited ways. A simple example is the adaptive Digital European Cordless Telephone (DECT) wireless phone, which finds and uses a frequency within its allowed plan with the least noise and interference on that channel and time slot. Of these capabilities, conservation of spectrum is already a national priority in international regulatory planning. This book leads the reader through the regulatory considerations, the technologies, and the implementation details to support three major applications that raise an SDR's capabilities to make it a CR:

1. Spectrum management and optimizations.
2. Interface with a wide variety of wireless networks, leading to management and optimization of network resources.
3. Interface with a human, providing electromagnetic resources to aid the human in his or her activities.

Many technologies have come together to result in the spectrum efficiency and CR technologies that are described in this book. This chapter gives the reader the background context of the remaining chapters of this book. These technologies represent a wide swath of contributions from many leaders in the field. These cognitive technologies may be considered as an application on top of a basic SDR platform.

To truly recognize how many technologies have come together to drive CR techniques, we begin with a few of the major contributions that have led up to today's CR developments. The development of digital signal processing (DSP) techniques arose due to the efforts of leaders such as Alan Oppenheim [1], Lawrence Rabiner [2, 3] and Ronald Schaefer, Ben Gold and Thomas Parks [4], James McClellen [4], James Flanagan [5], fred harris [6], and James Kaiser. These pioneers[2] recognized the potential for digital filtering and DSP, and prepared the seminal textbooks, innovative papers, and breakthrough signal-processing techniques to teach an entire industry how to convert analog signal processes to digital processes. They guided the industry in implementing new processes that were entirely impractical in analog signal processing.

Somewhat independently, Cleve Moler, Jack Little, John Markel, Augustine Gray, and others began to develop software tools that would eventually converge with the DSP industry to enable efficient representation of the DSP techniques and would provide rapid and efficient modeling of these complex algorithms [7, 8].

Meanwhile, the semiconductor industry, continuing to follow Moore's Law [9], evolved to the point where the computational performance required to implement digital signal processes used in radio modulation and demodulation were not only practical, but resulted in improved radio communication performance, reliability, flexibility, and increased value to the customer. This meant that analog functions implemented with large discrete components were replaced with digital functions implemented in silicon, and consequently were more producible, less expensive, more reliable, smaller, and lower power [10].

During this same period, researchers all over the globe explored various techniques to achieve machine learning and related methods for improved machine behavior.

---

[2]This list of contributors is only a partial representative listing of the pioneers with whom the author is personally familiar, and not an exhaustive one.

Among these were analog threshold logic, which led to fuzzy logic and neural networks, a field founded by Frank Rosenblatt [11]. Similarly, languages to express knowledge and to understand knowledge databases evolved from list processing (LISP) and Smalltalk and from massive databases with associated probability statistics. Under funding from the Defense Advanced Research Projects Agency (DARPA), many researchers worked diligently on natural language understanding and understanding spoken speech. Among the most successful speech-understanding systems were those developed by Janet and Jim Baker (who subsequently founded Dragon Systems) [12] and Kai Fu Lee et al. [13]. Both of these systems were developed under the mentoring of Raj Reddy at Carnegie Mellon. Today, we see Internet search engines reflecting the advanced state of artificial intelligence (AI).

In networking, DARPA and industrial developers at Xerox, BBN Technologies, IBM, ATT, and Cisco each developed computer networking techniques, which evolved into the standard Ethernet and Internet we all benefit from today. The Internet Engineering Task Force (IETF), and many wireless networking researchers, continue to evolve networking technologies with a specific focus on making radio networking as ubiquitous as our wired Internet. These researchers are exploring wireless networks that range from access directly via a radio access point to more advanced techniques in which intermediate radio nodes serve as repeaters to forward data packets toward their eventual destination in an ad hoc network topology.

All of these threads come together as we arrive today at the cognitive radio era (see Figure 1.1). Cognitive radios are nearly always applications that sit on top of a software defined radio, which in turn is implemented largely from digital signal processors and general-purpose processors (GPPs) built with silicon. In many cases, the spectral efficiency and other intelligent support to the user arises by sophisticated networking of many radios to achieve the end behavior, which provides added capability and other benefits to the user.

1.3 A BRIEF HISTORY OF SOFTWARE DEFINED RADIO

A software defined radio is a radio in which the properties of carrier frequency, signal bandwidth, modulation, and network access are defined by software. Modern SDR also implements any necessary cryptography, forward error correction coding, and source coding of voice, video, or data in software as well. As shown in the timeline of Figure 1.2, the roots of SDR design go back to 1987, when Air Force Rome Labs (AFRL) funded the development of a programmable modem as an evolutionary step beyond the architecture of the integrated communications, navigation, and identification architecture (ICNIA). ICNIA was a federated design of multiple radios—that is, a collection of several single-purpose radios used as one piece of equipment.

Today's SDR, in contrast, is a general-purpose device in which the same radio tuner and processors are used to implement many waveforms at many frequencies. The advantage of this approach is that the equipment is more versatile and cost effective. Additionally, it can be upgraded with new software for new waveforms and new applications after sale, delivery, and installation. Following the programmable modem, AFRL and DARPA joined forces to fund the SPEAKeasy-I and SPEAKeasy-II programs.

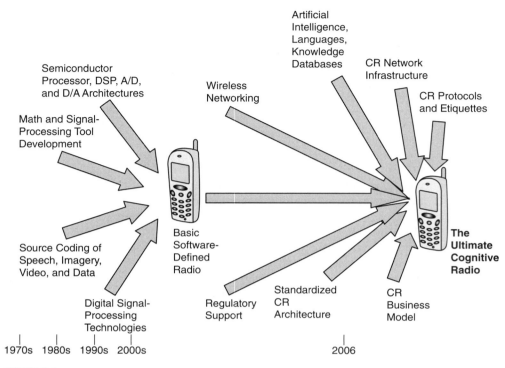

FIGURE 1.1

Technology timeline. Synergy among many technologies converge to enable the SDR. In turn, the SDR becomes the platform of choice for the CR.

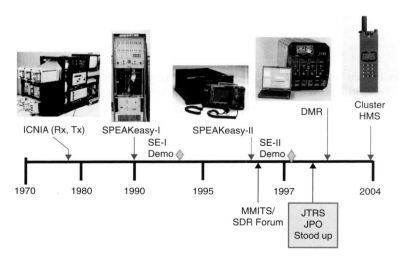

FIGURE 1.2

SDR timeline. Images of ICNIA, SPEAKeasy-I, SPEAKeasy-II, and DMR on their contract award timelines and corresponding demonstrations. These radios are the evolutionary steps that led to today's SDRs.

SPEAKeasy-I was a six-foot-tall rack of equipment (not easily portable), but it did demonstrate that a completely software programmable radio could be built, and included a software programmable crytography chip called Cypress, developed by Motorola Government Electronics Group (subsequently purchased by General Dynamics). SPEAKeasy-II was a complete radio, packaged in a practical radio size (the size of a stack of two pizza boxes), and was the first SDR to include programmable voice coder (vocoder), and sufficient analog and digital signal-processing resources to handle many different kinds of waveforms. It was subsequently tested in field conditions at Ft. Irwin, California, where its ability to handle many waveforms underlined its extreme utility, and its construction from standardized commercial off-the-shelf (COTS) components was a very important asset in defense equipment. SPEAKeasy-II was followed by the US Navy's Digital Modular Radio (DMR), becoming a four-channel full duplex SDR, with many waveforms and many modes, able to be remotely controlled over an Ethernet interface using Simple Network Management Protocol (SNMP).

These SPEAKeasy-II and DMR products evolved, not only to define these radio waveform features in software, but also to develop an appropriate software architecture to enable porting the software to an arbitrary hardware platform and thus to achieve hardware independence of the waveform software specification. This critical step allows the hardware to evolve its architecture independently from the software, and thus frees the hardware to continue to evolve and improve after delivery of the initial product.

The basic hardware architecture of a modern SDR (Figure 1.3) provides sufficient resources to define the carrier frequency, bandwidth, modulation, any necessary cryptography, and source coding in software. The hardware resources may include mixtures of GPPs, DSPs, field-programmable gate arrays (FPGAs), and other computational resources, sufficient to include a wide range of modulation types (see Section 1.2.1). In the basic software architecture of a modern SDR (Figure 1.4), the application programming interfaces (APIs) are defined for the major interfaces to ensure software portability across many very different hardware platform implementations, as well as to ensure that the basic software supports a wide diversity of waveform applications without having to be rewritten for each waveform or application. The software has the ability to allocate computational resources to specific waveforms (see Section 1.2.3). It is normal for an SDR to support many waveforms interfaced to many networks, and thus to have a library of waveforms and protocols.

The SDR Forum was founded in 1996 by Wayne Bonser of AFRL to develop industry standards for SDR hardware and software that could ensure that the software not only ports across various hardware platforms, but also defines standardized interfaces to facilitate porting software across multiple hardware vendors, and to facilitate integration of software components from multiple vendors. The SDR Forum is now a major influence in the software defined radio industry, dealing not only with standardization of software interfaces, but many other important enabling technology issues in the industry from tools, to chips, to applications, to CR and spectrum efficiency. The SDR Forum currently has many working groups, preparing papers to advance both spectrum efficiency and CR applications. In addition, special-interest groups within the Forum have interests in these topics.

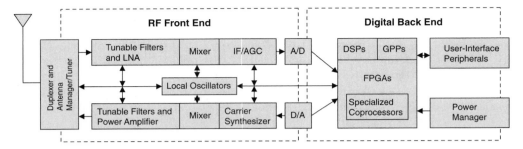

FIGURE 1.3

Basic hardware architecture of an SDR modem. The hardware provides sufficient resources to define the carrier frequency, bandwidth, modulation, any necessary cryptography, and source coding in software. The hardware resources may include mixtures of GPPs, DSPs, FPGAs, and other computational resources, sufficient to include a wide range of modulation types. *Note:* A/D = analog to digital; AGC = automatic gain control; D/A = digital to analog; DSP = digital signal processor; FPGA = field-programmable gate array; GPP = general-purpose processor; IF = intermediate frequency; LNA = low-noise amplifier; RF = radio frequency.

The SDR Forum working group is treating CR and spectrum efficiency as applications that can be added to a software defined radio. This means that we can begin to assume an SDR as the basic platform on which to build most new CR applications.

1.4 BASIC SDR

In this section, we endeavor to provide the reader with background material to provide a basis for understanding subsequent chapters.

The following definition of a Software Defined Radio is from the SDR Forum; it has been harmonized with IEEE SCC 41–P1900.1 as: "*Radio in which some or all of the physical layer functions are software defined.*" Because much of the functionality is accomplished with software, the radio platform can easily be adapted to serve a wide variety of products and applications from essentially a common hardware design. Because the hardware, and much of the software, can be reused across many products, the development cost per product can be lowered, and the cycle time to bring new products to market can be reduced.

Several manufacturers have also found it convenient to be able to revise the software in fielded equipment without having to perform a recall, thus saving huge costs of maintenance and logistics. Finally, new features and services can be added to the radio, thus future-proofing the products to have longer product life and value to the customer, and expanding the market for the product.

Within the last year, the SDR architecture has become so popular that it is now the dominant design approach. In some cases, the software is hard coded into a custom

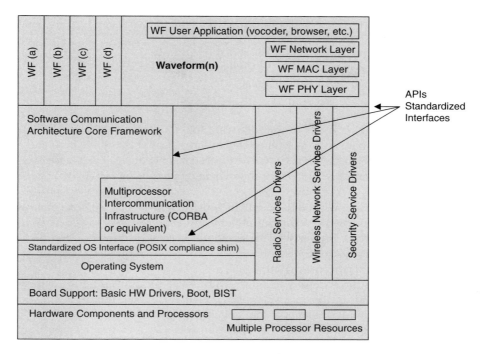

FIGURE 1.4

Basic software architecture of a modern SDR. Standardized APIs are defined for the major interfaces to ensure software portability across many very different hardware platform implementations. The software has the ability to allocate computational resources to specific waveforms. It is normal for an SDR to support many waveforms to interface to many networks, and thus to have a library of waveforms and protocols. *Note:* API = application programming interface; BIST = built-in self-test; CORBA = Common Object Request Broker Architecture; HW = hardware; MAC = medium access control; OS = operating system; PHY = physical (layer); POSIX = Portable Operating System Interface; WF = waveform.

ULSI chip, thus hiding the fact that the functionality is actually defined by software. A new industry term is also arising—*multimode* or *convergence radio*. These descriptions are intended to highlight the fact that the radio can implement a variety of waveforms and protocols.

1.4.1 Hardware Architecture of an SDR

The basic SDR must include the radio front end, the modem, the cryptographic security function, and the application function. In addition, some radios will also include support for network devices connected to either the plain text side or the modem side of the radio, allowing the radio to provide network services and to be remotely controlled over the local Ethernet.

Some radios will also provide for control of external radio frequency (RF) analog functions such as antenna management, coax switches, power amplifiers, or special-purpose filters. The hardware and software architectures should allow RF external features to be added if or when required for a particular installation or customer requirement.

The RF front end (RFFE) consists of the following functions to support the receive mode: antenna matching unit, low-noise amplifier, filters, local oscillators, and analog-to-digital (A/D) converters (ADCs) to capture the desired signal and suppress undesired signals to a practical extent. This maximizes the dynamic range of the ADC available to capture the desired signal.

To support the transmit mode, the RFFE will include digital-to-analog (D/A) converters (DACs), local oscillators, filters, power amplifiers, and antenna-matching circuits. In transmit mode, the important property of these circuits is to synthesize the RF signal without introducing noise and spurious emissions at any other frequencies that might interfere with other users in the spectrum.

The modem processes the received signal or synthesizes the transmitted signal, or both for a full duplex radio. In the receive process (Figure 1.5), the modem will shift the carrier frequency of the desired signal to a specific frequency nearly equivalent to heterodyne shifting the carrier frequency to direct current (DC), as perceived by the digital signal processor, to allow it to be digitally filtered. The digital filter provides a high level of suppression of interfering signals not within the bandwidth of the desired signal. The modem then time-aligns and despreads the signal as required, and refilters the signal to the information bandwidth. Next, the modem time-aligns the signal to the

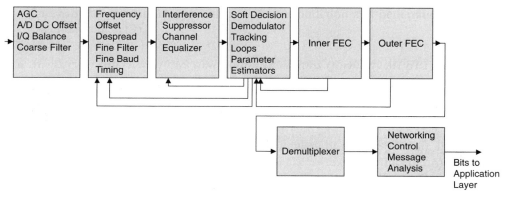

FIGURE 1.5

Traditional digital receiver signal-processing block diagram. *Note:* I/Q, meaning "inphase and quadrature," is the real part and the imaginary part of the complex valued signal after being sampled by the ADC(s) in the receiver, or as synthesized by the modem and presented to the DAC in the transmitter.

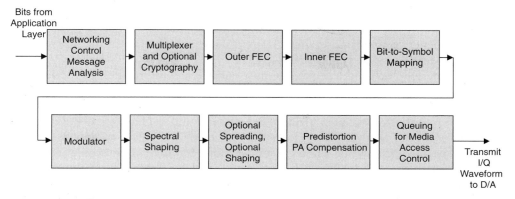

FIGURE 1.6

Traditional transmit signal-processing block diagram.

symbol[3] or baud time so that it can optimally align the demodulated signal with expected models of the demodulated signal. The modem may include an equalizer to correct for channel multipath artifacts, and filter delay distortions. It may also optionally include rake filtering to optimally cohere multipath components for demodulation. The modem will compare the received symbols with the alphabet of all possible received symbols and make a best possible estimate of which symbols were transmitted. Of course, if there is a weak signal or strong interference, some symbols may be received in error. If the waveform includes forward error correction (FEC) coding, the modem will decode the received sequence of encoded symbols by using the structured redundancy introduced in the coding process to detect and correct the encoded symbols that were received in error.

The process the modem performs for transmit (Figure 1.6) is the inverse of that for receive. The modem takes bits of information to be transmitted, groups the information into packets, adds a structured redundancy to provide for error correction at the receiver, groups bits to be formed into symbols or bauds, selects a wave shape to represent each symbol, synthesizes each wave shape, and filters each wave shape to keep it within its desired bandwidth. It may spread the signal to a much wider bandwidth by multiplying the symbol by a wideband waveform that is also generated by similar methods. The final waveform is filtered to match the desired transmit signal bandwidth. If the waveform includes a time-slotted structure, such as time division multiple access (TDMA) waveforms, the radio will wait for the appropriate time while placing samples that represent the waveform into an output first in, first out (FIFO) buffer ready to be applied to the DAC. The modem must also control the power amplifier and the local oscillators to produce the desired carrier frequency, and must control

---

[3]A symbol or baud is a set of information bits typically ranging from 1 bit per symbol to 10 bits per symbol. Since there can be many possible symbols, just as with an alphabet, each is assigned a unique waveform so that the receiver can detect which of the many possible symbols were sent, and can then decode that back to the corresponding information bits corresponding to the symbol.

the antenna-matching unit to minimize voltage standing wave radio (VSWR). The modem may also control the external RF elements including transmit versus receive mode, carrier frequency, and smart antenna control. Considerable detail on the architecture of software defined radios is given by Reed [14].

The Cryptographic Security function must encrypt any information to be transmitted. Because the encryption processes are unique to each application, these cannot be generalized. The Digital Encryption Standard (DES) and the Advanced Encryption Standard (AES) from the US National Institute of Standards and Technology (NIST) provide examples of robust, well vetted cryptographic processes [15, 16]. In addition to providing the user with privacy for voice communication, cryptography also plays a major role in ensuring that the billing is to an authenticated user terminal. In the future, it will also be used to authenticate financial transactions of delivering software and purchasing products and services. In future CRs, the policy functions that define the radios' allowed behaviors will also be cryptographically protected to prevent tampering with regulatory policy as well as network operator policy.

The application processor will typically implement a vocoder, a video coder, and/or a data coder, as well as selected Web browser functions. In each case, the objective is to use knowledge of the properties of the digitized representation of the information to compress the data rate to an acceptable level for transmission. Voice, video, and data coding typically use knowledge of the redundancy in the source signal (speech or image) to compress the data rate. Compression factors typically in excess of $10:1$ are achieved in voice coding, and up to $100:1$ in video coding. Data coding has a variety of redundancies within the message, or between the message and common messages sent in that radio system. Data compression ranges from 10 percent to 50 percent, depending on how much redundancy can be identified in the original information data stream.

Typically, speech and video applications run on a DSP processor. Text and Web browsing typically run on a GPP. As speech-recognition technology continues to improve its accuracy, we can expect that the keyboard and display will be augmented by speech input and output functionality. On CRs with adequate processors, it may be possible to run speech recognition and synthesis on the CR, but early units may find it preferable to vocode the voice, transmit the voice to the basestation, and have recognition and synthesis performed at an infrastructure component. This will keep the complexity of the portable units smaller, and keep the battery power dissipation lower.

1.4.2 Computational Processing Resources in an SDR

The design of an SDR must anticipate the computational resources needed to implement its most complex application. The computational resources may consist of GPPs, DSPs, FPGAs, and occasionally will include other chips that extend the computational capacity. Generally, the SDR vendor will avoid inclusion of dedicated-purpose nonprogrammable chips because the flexibility to support waveforms and applications is limited, if not rigidly fixed, by nonprogrammable chips.

The GPP processor is the process that will usually perform the user applications, and will process the high-level communications protocols. This class of processor is readily programmed in standard C or C++ language, supports a very wide variety of addressing modes, floating point and integer computation, and a large memory space, usually including multiple levels of on-chip and off-chip cache memory.[4] These processors currently perform more than 1 billion mathematical operations per second (mops).[5] GPPs in this class usually pipeline the arithmetic functions and decision logic functions several levels deep to achieve these speeds. They also frequently execute many instructions in parallel, typically performing the effective address computations in parallel with arithmetic computation, logical evaluations, and branch decisions.

Most important to the waveform modulation and demodulation processes is the speed at which these processors can perform real or complex multiply accumulates. The waveform signal processing represents more than 90 percent of the total computational load in most waveforms, although the protocols to participate in the networks frequently represent 90 percent of the lines of code. Therefore, it is of great importance to the hardware SDR design that the SDR architecture include DSP-type hardware multiply accumulate functions, so that the wireless signal processes can be performed at high speed, and GPP-type processors for the protocol stack processing.

DSPs are somewhat different than GPPs. The DSP internal architecture is optimized to be able to perform multiply accumulates very fast. This means they have one or more multipliers and one or more accumulators in hardware. Usually the implication of this specialization is that the device has a somewhat unusual memory architecture, usually partitioned so that it can fetch two operands simultaneously and also be able to fetch the next software instruction in parallel with the operand fetches. Currently, DSPs are available that can perform fractional mathematics (integer) multiply accumulate instructions at rates of 1 GHz, and floating-point multiply accumulates at 600 MHz. DSPs are also available with many parallel multiply accumulate engines, reporting rates of more than 8 Gmops. The other major feature of the DSP is that it has fewer and less sophisticated addressing modes. Finally, DSPs frequently use modifications of the C language to more efficiently express the signal-processing parallelism and fractional arithmetic, and thus maximize their speed. As a result, the DSP is much more efficient at signal processing but less capable to accommodate the software associated with the network protocols.

FPGAs have recently become capable of providing very significant computation of multiply accumulate operations on a single chip, surpassing DSPs by more than an order of magnitude in signal processing throughput. By defining the on-chip interconnect of many gates, more than 100 multiply accumulators can be arranged to perform multiply accumulate processes at frequencies of more than 200 MHz. In addition to the digital signal processing, FPGAs can also provide the timing logic to synthesize clocks, baud rate, chip rate, time slot, and frame timing, thus leading to a reasonably compact waveform implementation. By expressing all of the signal processing as a set of register transfer operations and multiply accumulate engines, very complex waveforms can be implemented in one chip. Similarly, complex signal processes that are not efficiently implemented on a DSP, such as Cordic operations, log magnitude operations, and dif-

---

[4]A few examples of common GPPs in use today in SDRs include Texas Instruments (OMAP), ARM-11, Intel, Marvel, Freescale, and IBM (Power PC).
[5]Mathematical operations per second take into account mathematical operations required to perform an algorithm, but not the operations to calculate an effective memory address index, or offset, nor the operations to perform loop counting, overflow management, or other conditional branching.

ference magnitude operations, can all have the specialized hardware implementations required for a waveform when implemented in FPGAs.

The downside of using FPGA processors is that the waveform signal processing is not defined in traditional software languages such as C, but in VHDL, a language for defining hardware architecture and functionality. The radio waveform description in very high-speed integrated circuit (VHSIC) Hardware Design Language (VHDL), although portable, is not a sequence of instructions and therefore not the usual software development paradigm. At least two companies are working on new software development tools that can produce the required VHDL from a C language representation, somewhat hiding this hardware language complexity from the waveform developer, and simplifying waveform porting to new hardware platforms. In addition, FPGA implementations tend to be higher power and more costly than DSP chips.

All three of these computational resources demand significant off-chip memory. For example, a GPP may have more than 128 Mbytes of off-chip instruction memory to support a complex suite of transaction protocols for today's telephony standards.

Current SDRs provide a reasonable mix of these computational alternatives to ensure that a wide variety of desirable applications can in fact be implemented at an acceptable resource level. In today's SDRs, dedicated-purpose application-specific integrated circuit (ASIC) chips are avoided because the signal-processing resources cannot be reprogrammed to implement new waveform functionality.

1.4.3 Software Architecture of an SDR

The objective of the software architecture in an SDR is to place waveforms and applications onto a software based radio platform in a standardized way. These waveforms and applications are installed, used, and replaced by other applications as required to achieve the user's objectives. To standardize the waveform and application interfaces, it is necessary to make the hardware platform present a set of highly standardized interfaces. This way, vendors can develop their waveforms independent of the knowledge of the underlying hardware. Similarly, hardware developers can develop a radio with standardized interfaces, which can subsequently be expected to run a wide variety of waveforms from standardized libraries. This way, the waveform development proceeds by assuming a standardized set of APIs for the radio hardware, and the radio hardware translates commands and status messages crossing those interfaces to the unique underlying hardware through a set of common drivers.

In addition, the method by which a waveform is installed into a radio, activated, deactivated, and de-installed, and the way in which radios use the standard interfaces must be standardized so that waveforms are reasonably portable to more than one hardware platform implementation.

According to Christensen et al., "The use of published interfaces and industry standards in SDR implementations will shift development paradigms away from proprietary tightly coupled hardware software solutions" [17]. To achieve this, the SDR radio is decomposed into a stack of hardware and software functions, with open standard interfaces. As was shown in Figure 1.3, the stack starts with the hardware and the one or more data buses that move information among the various processors. On top of the hardware, several standardized layers of software are installed. This includes the boot loader, the operating system (OS); the board support package (BSP, which consists of input/output drivers that know how to control each interface); and a layer called the Hardware Abstraction Layer (HAL). The HAL provides a method for GPPs to communicate with DSPs and FPGA processors using standardized software interfaces.

The US government has defined a standardized software architecture, known as the Software Communication Architecture (SCA), which has also been adopted by defense contractors of many countries worldwide. The SCA is a core framework to provide a standardized process for identifying the available computational resources of the radio, matching those resources to the required resources for an application. The SCA is built on a standard set of operating system features called POSIX,[6] which also has standardized APIs to perform operating system functions such as file management and computational thread/task scheduling.

The SCA core framework is the inheritance structure of the open application layer interfaces and services, and provides an abstraction of underlying software and hardware layers. The SCA also specifies a Common Object Request Broker Architecture (CORBA) middleware, which is used to provide a standardized method for software objects to communicate with each other, regardless of which processor they have been installed on (think of it as a software data bus). The SCA also provides a standardized method of defining the requirements for each application, performed in eXtensible Markup Language (XML). The XML is parsed and helps to determine how to distribute and install the software objects. In summary, the core framework provides a means to configure and query distributed software objects, and in the case of SDR, these will be waveforms and other applications.

These applications will have many reasons to interact with the Internet as well as many local networks; therefore, it is also common to provide a collection of standardized radio services, network services, and security services, so that each application does not need to have its own copy of Internet Protocol (IP), and other commonly used functions.

1.5 COGNITIVE RADIO

It is not essential, but there is broad agreement that it is most efficient, to build CR capabilities on top of an SDR radio platform. While the DSPs and FPGAs are used to implement the physical layer signal processing, additional reasoning software can be added to the GPP processor. These new functions are essentially additional user applications, but not necessarily visible to the user.

The SDR Forum and the IEEE recently approved this definition of a *cognitive radio*[7]:

> (a) *Radio in which communications systems are aware of their environment and internal state, and can make decisions about their radio operating behavior based*

---

[6]POSIX is the collective name of a family of related standards specified by the IEEE to define the API for software compatible with variants of the UNIX operating system. POSIX stands for portable operating system interface, with the X signifying the UNIX heritage of the API [18].
[7]See *http://www.sdrforum.org/pages/documentLibrary/documents/SDRF-06-R-0011-V1_0_0.pdf.*

on that information and predefined objectives. The environmental information may or may not include location information related to communication systems.

(b) *Cognitive radio (as defined in a) that uses SDR, adaptive radio, and other technologies to automatically adjust its behavior or operations to achieve desired objectives.*

As we said previously, the cognitive radio can adapt for:

- the spectrum regulator
- the network operator
- the user objectives

The first of these, the spectrum regulator, has generally allocated all the spectrum there is to existing users, and now finds it difficult to provide spectrum for new applications and users. With the global telecommunications market currently at 1.2T dollars per year, and continuing to grow, the ability to find and use spectrum is now a major issue. Consequently, international research in the subject is growing at a phenomenal pace. At the time of this writing, an Internet search on the topic "cognitive radio" produces 138,000 hits, which is nearly triple the number of hits, in only 3 years.

The ability of the CR to provide a means to negotiate for access to spectrum is therefore of huge economic value ($200M/MHz in the most recent US auction). Much of the industry has focused on this single topic. But a radio that can find and use available spectrum must have rules about what spectrum it is allowed to use. Those rules represent what the regulator would normally allow for a given application. Thus today, CRs usually also include a policy engine that provides means for the radio to behave within local regulatory constraint. In the following, we introduce the bare essentials of CR functionality, and provide much more detail in subsequent chapters.

1.5.1 Java Reflection in a Cognitive Radio

Cognitive radios need to be able to tell other CRs what they are observing that may affect the performance of the radio communication channel. The receiver can measure signal properties and can even estimate what the transmitter meant to send, but it also needs to be able to tell the transmitter how to change its waveform in ways that will suppress interference. In other words, the CR receiver needs to convert this information into a transmitted message to send back to the transmitter.

Figure 1.7 presents a basic diagram for understanding CRs. In this figure, the receiver (radio 2) can use Java reflection to ask questions about the internal parameters inside the receive modem, which might be useful to understand link performance. Measurements commonly calculated internally in the software design of a receiver, such as the signal-to-noise ratio (SNR), frequency offset, timing offset, or equalizer taps, are parameters that can be read by the Java reflection. By examining these radio properties, the receiver can determine what change at the transmitter (radio 1) will improve the most important performance objective(s) of the communication (such as saving battery life). From that Java reflection, the receiver formulates a message onto the reverse link, mul-

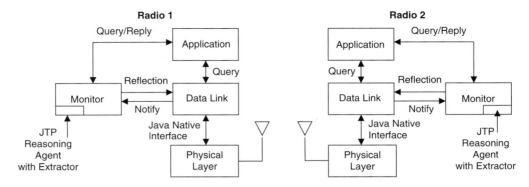

FIGURE 1.7

Java reflection, shown here, allows the receiver to examine the state variables of the transmit and receive modem, thereby allowing the CR to understand what the communications channel is doing to the transmitted signal [19].

tiplexes it into the channel, and observes whether the transmitter making that change results in an improvement in link performance.

1.5.2 Smart Antennas in a Cognitive Radio

Current radio architectures are exploring the uses of many types of advanced antenna concepts. A smart radio needs to be able to tell what type of antenna is available, and to make full use of its capabilities. Likewise, a smart antenna should be able to tell a smart radio what its capabilities are.

Smart antennas are particularly important to CR, in that certain functionalities can provide very significant amounts of measurable performance enhancement. As detailed in Chapter 5, if we can reduce transmit power, and thereby allow transmitters to be closer together on the same frequency, we can reduce the geographic area dominated by the transmitter, and thus improve the overall spectral efficiency metric of MHz * km^2.

A smart transmit antenna can form a beam to focus transmitted energy in the direction of the intended receiver. At frequencies of current telecommunication equipment in the range of 800 to 1800 MHz, practical antennas can easily provide 6 to 9 dB of gain toward the intended receiver. This same beamforming reduces the energy transmitted in other directions, thereby improving the usability of the same frequency in those other directions.

A radio receiver may also be equipped with a smart antenna for receiving. A smart receive antenna can synthesize a main lobe in the desired direction of the intended transmitter, as well as synthesize a deep null in the direction of interfering transmitters. It is not uncommon for a practical smart antenna to be able to synthesize a 20 dB null to suppress interference. This amount of interference suppression has much more impact on the users per (MHz * km^2) metric than being able to transmit 20 dB more transmit power.

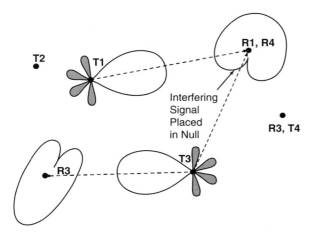

FIGURE 1.8

Utility of smart antennas. A smart antenna allows a transmitter (T) to focus its energy toward the intended recipient receiver (R), and allows a receiver to suppress interference from nearby interfering transmitters.

The utility of the smart antenna at allowing other radio transmitters to be located nearby is illustrated in Figure 1.8.

1.5.3 Policy Engine

Cognitive radios will have a policy engine that represents the voice of the local regulator about the constraints that apply to the radio. The constraints may include specifications about frequency, waveforms, transmit power level, antenna properties, applications, location, user, required licenses, roles and priorities for spectrum access, etiquettes, and even possible spectrum brokers, from whom a specific frequency, bandwidth, region of operation, and the duration of use may be negotiated. In addition to policy, the radio may have additional ability to reason and optimize for the objectives of the network operator and the objectives of the user. In the work of Chapter 17, these pieces are discussed as the Spectrum Reasoner (SR), and the System Strategy Reasoner (SSR). The reasoning function and the policy representation language are still a topic of much research, and a goal of standards activities. Generally, the reasoning process can best be understood by a review of the Prolog language, though much more sophisticated languages are now in use.

1.6 SPECTRUM MANAGEMENT

The immediate interest to regulators in fielding CRs is to provide new capabilities that support new methods and mechanisms for spectrum access and utilization now under consideration by international spectrum regulatory bodies. These new methodologies recognize that fixed assignment of a frequency to one purpose across huge geographic regions (often across entire countries) is quite inefficient. Today, this type of frequency assignment results in severe underutilization of the precious and bounded spectrum resource. The Federal Communications Commission (FCC; for commercial applications) and the National Telecommunications and Information Administration (NTIA; for federal applications) in the United States, as well as corresponding regulatory bodies of many other countries, are exploring the question of whether better spectrum utilization could be achieved given some intelligence in the radio and in the network infrastructure.

This interest also has led to developing new methods to manage spectrum access in which the regulator is not required to micromanage every application, every power level, antenna height, and waveform design. Indeed, the goal of minimizing interference with other systems with other purposes may be reasonably automated by the CR (and possible aid of associated network and system components). With a CR, the regulator could define policies at a higher level, and expect the equipment and the infrastructure to resolve the details within well-defined practical boundary conditions such as available frequency, power, waveform, geography, and equipment capabilities. In addition, the radio is expected to use whatever etiquette or protocol defines cooperative performance for network membership.

In the United States, which has several broad classes of service, the FCC has held meetings with license holders, who have various objectives. There are license holders who retain their specific spectrum for public safety and for other such public purposes such as broadcast of AM, FM, and television. There are license holders who purchased spectrum specifically for commercial telecommunications purposes. There are license holders for industrial applications, as well as those for special interests.

Many frequencies are allocated to more than one purpose. An example of this is a frequency allocated for remote-control purposes—many garage door opener companies and automobile door lock companies have developed and deployed large quantities of products using these remote control frequencies. In addition, there are broad chunks of spectrum for which NTIA has defined frequency and waveform usage and how the defense community will use spectrum in a process similar to that used by the FCC for commercial purposes.

Finally, there are spectrum commons and unlicensed blocks. In these frequencies, there is overlapping purpose among multiple users, frequencies, waveforms, and geography. An example of spectrum commons is the 2.4 GHz band, discussed in Section 1.4.1. The following sections touch on new methods for spectrum management, and how they lead to spectrum efficiency.

1.6.1 Managing Unlicensed Spectrum

The 2.4 GHz band and the 5 GHz band are popularly used for wireless computer networking. These bands, and others, are known as the industrial, scientific, and medical (ISM) bands. Energy from microwave ovens falls in the 2.4 GHz band. Consequently, it is impractical to license that band for a particular purpose because of the broadly distributed interference. However, WiFi (802.11) and Bluetooth applications are specifically designed to coexist with a variety of interference waveforms commonly found in this band as well as with each other. Various types of equipment use a protocol to determine which frequencies or timeslots to use and keep trying until they find a usable channel. They also acknowledge correct receipt of transmissions, retransmitting data packets when collisions cause uncorrectable bit errors.

Although radio communication equipment and applications defined in these bands may be unlicensed, they are restricted to specific guidelines about what frequencies are used and what effective isotropic radiated power (EIRP) is allowed. Furthermore, they must accept any existing interference (such as that from microwave ovens and diathermy machines), and they must not interfere with any applications outside this band.

Bluetooth and 802.11 both use waveforms and carrier frequencies that keep their emissions inside the 2.4 GHz band. Both use methods of hopping to frequencies that successfully communicate and to error correct bits or packets that are corrupted by interference. Details of Bluetooth and 802.11 waveform properties are shown in Table 1.1.

The 802.11 waveform can successfully avoid interference from microwave ovens because each packet is of sufficiently short duration that a packet can be delivered at a frequency or during a time period while the interference is minimal. Bluetooth waveforms are designed to hop to many different frequencies very rapidly, and consequently the probability of collision with a strong 802.11 or microwave is relatively small and correctable with error-correcting codes.

The regulation of the 2.4 GHz and 5 GHz bands consists of setting the spectrum boundaries, defining specific carrier frequencies that all equipment is to use, and limiting the EIRP. As shown in Table 1.1, the maximum EIRP is 1 watt or less for most of the wireless network products, except for the metropolitan WiMAX service, and the FCC-type acceptance is based on the manufacturer demonstrating EIRP and frequency compliance.

It is of particular interest to note that each country sets its own spectral and EIRP rules with regard to these bands. Japan and Europe each have regulatory rules for these bands that are different from those of the United States. Consequently, manufacturers may do one of the following:

- Make three models
- Make one model with a switch to select to which country the product will be sold
- Make a model that is commonly compliant to all regional requirements
- Make a model that is capable of determining its current location and then implementing the local applicable rules

The last method is an early application of cognitive techniques.

1.6.2 Noise Aggregation

Communication planners worry that the combined noise from many transmitters may add together and thereby increase the noise floor at the receiver of an important message, perhaps an emergency message. It is well understood that noise power sums together at a receiver. If a receiver antenna is able to see the emissions of many transmitters on the same desired frequency and time slot, increasing the noise floor will reduce the quality of the desired signal at the demodulator, in turn increasing the bit

Table 1.1 Properties of 802.11, Bluetooth, ZigBee, and WiMAX Waveforms

Standard	Name/ Description	Carrier Frequency	Modulation	Data Rate	Tx Pwr; EIRP
802.11a (802.11g = both 802.11a and 802.11b)	WiFi; WLAN	5 GHz, 12 channels (8 indoor, 4 point to point)	52 carriers of OFDM, 48 data, 4 pilot, BPSK, QPSK, 16-QAM, or 64-QAM, carrier separation = 0.3125 MHz, symbol duration = 4 ms, with cyclic prefix = 0.8 ms, Vitterbi R = 1/2, 2/3, 3/4	54, 48, 36, 27, 24, 18, 12, 9, 6 Mbps	12–30 dBm
802.11b	WiFi; WLAN	2.4 GHz, 3 channels	CCK, DBPSK, DQPSK with DSSS	11, 5.5, 1.0 Mbps	12–30 dBm
802.15.1	Bluetooth; WPAN	2.4–2.4835 GHz, 79 channels, each 1 MHz wide, adaptive frequency hopping at 1600 hops/sec	GFSK (deviation = 140 – 175 KHz)	57.6, 432.6, 721 Kbps, 2.1 Mbps	0–20 dBm
802.15.4	ZigBee; WPAN	868.3 MHz, 1 channel; 902–928 MHz, 10 channels with 2-MHz spacing; 2405–2483.5 MHz, 16 channels with 5 MHz spacing	32 chip symbols for 16 ary orthogonal modulation with OQPSK spreading at 2.0 Mcps (2.4 GHz); DBPSK with BPSK spreading at 300 Kcps (868 MHz) or 600 Kcps (915 MHz)	250 Kbps at 2.4 GHz; 40, 20 Kbps at 868 and 915 MHz	–3–10 dBm
802.16	WiMAX, Wibro; WMAN	2–11 GHz (802.16a), 10–66 GHz (802.16); BW = 1.25, 5, 10, 20 MHz	OFDM, SOFDM: 2048, 1024, 512, 256, and 128 FFT; carriers, each QPSK, 16-QAM, or 64-QAM; symbol rate = 102.9 ms	70 Mbps	40 dBm; EIRP = 57.3 dBm

error rate, and possibly rendering the signal useless. If the interfering transmitters are all located on the ground in an urban area, the interference power from these transmitters decays approximately as the reciprocal of $r^{3.8}$ (a detailed explanation of the exponents of range, r, is found in Chapter 5). The total noise received is the sum of the powers of all such interfering transmitters. Even transmitters the received power level of which is below the noise floor add to the noise floor. However, signals with a power level that is extremely small compared to the noise floor have little impact on the noise floor. If there are 100 signals each 20 dB below the noise, then that noise power will sum equal to the noise, and raise the total noise floor by 3 dB. Similarly, if there are 1000 transmitters, each 30 dB below the noise floor, they can raise the noise floor by 3 dB (see Chapter 5). However, the additional noise is usually dominated by the one or two interfering transmitters that are closest to the receiver.

In addition, we must consider the significant effect of personal communication devices, which are becoming ubiquitous. In fact, one person may have several devices all at close range to each other. Cognitive radios will be the solution to this spectral noise and spectral crowding, and will evolve to the point of deployed science just in time to help with the aggregated noise problems of many personal devices all attempting to communicate in proximity to each other.

1.6.3 Aggregating Spectrum Demand and Use of Subleasing Methods

Many applications for wireless service operate with their own individual licensed spectra. It is rare that each service is fully consuming its available spectrum. Studies show that spectrum occupancy seems to peak at about 14 percent, except under emergency conditions, where occupancy can reach 100 percent for brief periods of time. Each of these services does not wish to separately invest in its own unique infrastructure. Consequently, it is very practical to aggregate these spectral assignments to serve a user community with a combined system. The industry refers to a collection of services of this type as a trunked radio. Trunked radio basestations have the ability to listen to many input frequencies. When a user begins to transmit, the basestation assigns an input and an output frequency for the message and notifies all members of the community to listen on the repeater downlink frequency for the message. Trunking aggregates the available spectrum of multiple users and is therefore able to deliver a higher quality of service while reducing infrastructure costs to each set of users and reducing the total amount of spectrum required to serve the community.

Both public safety and public telephony services benefit from aggregating spectrum and experience fluctuating demands, so each could benefit from the ability to borrow spectrum from the other. This is a much more complex situation, however. Public safety system operators must be absolutely certain that they can get all the spectrum capacity they need if an emergency arises. Similarly, they might be able to appreciate the revenue stream from selling access to their spectrum to commercial users who have need of access during times when no emergency conditions exist.

1.6.4 Priority Access

If agreements can be negotiated between spectrum license holders and spectrum users who have occasional peak capacity needs, it is possible to define protocols to request access, grant access, and withdraw access. Thus, an emergency public service can temporarily grant access to its spectrum in exchange for monetary compensation. Should an emergency arise, the emergency public service can withdraw its grant to access, thereby taking over priority service.

In a similar fashion, various classes of users can each contend for spectrum access, with higher-priority[8] users being granted access before other users. This might be relevant, for example, if police, fire, or military users need to use the cellular infrastructure during an emergency. Their communications equipment can indicate their priority to the communications infrastructure, which may in turn grant access for these highest-priority users first.[9]

By extension, a wide variety of grades of service for commercial users may also prioritize sharing of commercially licensed spectrum. Users who are willing to pay the most may get high priority for higher data rates for their data packets. The users who pay the least would get service only when no other grades of service are consuming the available bandwidth.

1.7 US GOVERNMENT ROLES IN COGNITIVE RADIO

This section briefly touches on the US government role in the creation and development of CR technology. We touch on activities at Defense Advance Research Projects Agency (DARPA), Federal Communications Commission, and the National Science Foundation.[10]

1.7.1 DARPA

Paul Kolodzy was a program manager at DARPA when he issued a Broad Area Announcement (BAA) calling for an industry day on the NeXt Generation (XG) program to explore how XG communications could not only make a significant impact on spectral efficiency of defense communications, but also significantly reduce the complexity of defining the spectrum allocation for each defense user.

Shortly after proposals were sent in to DARPA, Kolodzy moved to the FCC to further explore this question and Preston Marshall became the DARPA program manager. Under Marshall's XG program, several contractors demonstrated that a CR could achieve substantial spectral efficiency in a noninterfering method, and that the spectrum allocation process could be simplified. Basic principles of spectrum efficiency are discussed by Marshall in Chapter 5. Marshall has subsequently created a second program to advance multiple cognitive pinciples called Wireless Network after Next (WNaN).

In the same time frame, Jonathan Smith worked as a DARPA project manager to develop intelligent network protocols that could learn and adapt to the properties of

---

[8]Cellular systems already support priority access; however, there is reported to be little control over the allocation of priority or the enforcement process.

[9]This technique is implemented in code division multiple access (CDMA) cellular communications.

[10]European and Asian activity in cognitive radio research, standardization, prototypes, testbeds, and demonstrations is also quite significant. We recommend the interested reader attend the yearly SDR Forum fall conference and the IEEE Dyspan conference where much of this work is published.

wireless channels to optimize performance under current conditions. Some of this work is discussed in Chapter 9.

1.7.2 FCC

On May 19, 2003, the FCC held a hearing to obtain industry comments on CR. Participants from the communications industry, radio and TV broadcasters, public safety officers, telecommunications systems operators, and public advocacy participants all discussed how this technology might interact with the existing spectrum regulatory process. Numerous public meetings were held subsequently to discuss the mechanics of such systems, and their impact on existing license holders of spectrum (see Section 1.6 and Chapter 2). The FCC has been actively engaged with industry and very interested in leveraging this technology.

1.7.3 NSF/CSTB Study

President George W. Bush established the Spectrum Policy Task Force (SPTF) to further study the economic and political considerations and impacts of spectrum policy. In addition to the FCC's public meetings, the National Science Foundation (NSF) also has held meetings on the impact of new technologies to improve spectrum efficiency. A committee chaired by Dale Hatfield and Paul Kolodzy heard testimony from numerous representatives, leading to the SPTF report, further described in Chapter 2.

The Computer Science and Telecommunications Board (CSTB) is a specific work group of the National Science Foundation. This work group produces books and workshops on important topics in telecommunications. CSTB held numerous workshops on the topic of spectrum management since its opening meeting at the FCC in May 2003. These meetings have resulted in reports to the FCC on various CR topics. A workshop was held on the topic of "Improving Spectrum Management through Economic and Other Incentives." This activity has been guided by Dale Hatfield (formerly chief of the office of Science and Technology at the FCC, now adjunct professor at University of Colorado); William Lehr (economist and research associate at MIT's Center for Technology Policy and Industrial Development); and Jon Peha (associate director of Carnegie Mellon University's Center for Wireless and Broadband Networking).

Within the few years between the first and second edition of this book, NSF has sponsored development of a number of CRs to be demonstrated as a testbed. Dr. Charles Bostian and his team of students at Virginia Tech (see Chapter 7) and Dr. Gary Minden and his team of students at Kansas University have built and demonstrated such testbeds as of the writing of this edition.

1.8 HOW SMART IS USEFUL?

The CR is able to provide a wide variety of intelligent behaviors. It can monitor the spectrum and choose frequencies that minimize interference to existing communication activity. When doing so, it will follow a set of rules that define what frequencies may be considered, what waveforms may be used, what power levels may be used for transmission, and so forth. It may also be given rules about the access protocols by which spectrum access is negotiated with spectrum license holders, if any, and the etiquettes by which it must check with other users of the spectrum to ensure that no hidden user[11] is already transmitting.

In addition to the spectrum optimization level, the CR may have the ability to optimize a waveform to one or many criteria. For example, the radio may be able to optimize for data rate, for packet success rate, for service cost, for battery power minimization, or for some mixture of several criteria. The user does not see these levels of sophisticated channel analysis and optimization except as the recipient of excellent service.

The CR may also exhibit behaviors that are more directly apparent to the user. These behaviors may include: (1) awareness of geographic location, (2) awareness of local networks and their available services, (3) awareness of the user and the user's biometric authentication to validate financial transactions, and (4) awareness of the user and his or her prioritized objectives. This book explores each of these technologies. Many of these services will be immediately valuable to the user without the need for complex menu screens, activation sequences, or preference setup processes.

The CR developer must use caution to avoid adding cognitive functionality that reduces the efficiency of the user at his or her primary tasks. If the user thinks of the radio as a cell phone and does not wish to access other networks, the CR developer must provide a design that is friendly to the user, timely and responsive, but is not continually intruding with attempts to be helpful by connecting to networks that the user does not need or want. If the radio's owner is a power user, however, the radio may be asked to watch for multiple opportunities: access to other wireless networks for data services, notification of critical turning points to aid navigation, or timely financial information, as a few simple examples.

One of the remaining issues in sophisticated software design is a method for determining whether the cognitive services the radio might offer will be useful. Will the services be accomplished in a timely fashion? Will the attempted services be undesired and disruptive? Will the services take too long to implement and arrive too late to be usable? The CR must offer functionality that is timely and useful to its owner, and yet not disruptive. Like "Radar" O'Reilly in *M*A*S*H*, we want the CR to offer support of the right type at the right time, properly prioritized to the user needs given sophisticated awareness of the local situation, and not offering frequent useless or obvious recommendations. We will explore this topic in Chapter 24.

1.9 ORGANIZATION OF THIS BOOK

In Chapter 2, Paul Kolodzy describes the regulatory policy motivations, activities, and initiatives within US and international regulatory bodies to achieve enhanced spectral efficiency.

Chapter 3, by Max Robert of Artemis Communications LLC and Bruce Fette of General Dynamics C4 Systems, describes the details of hardware and software architecture of SDRs, and explains why an SDR is the primary choice as the basis for CRs.

---

[11]Hidden in the shadow of a building or mountain and therefore not visible to the spectrum sensor.

Chapter 4, by John Polson of General Dynamics, is about the technologies required to implement basic services in a CR.

Chapter 5 was extensively revised for the second edition. In it, Preston Marshall of DARPA deals with spectrum efficiency and the demonstrated feasibility of CR principles.

Chapter 6, by Robert Wellington of the University of Minnesota, introduces the cognitive policy engine. The policy engine provides an efficient mechanism to express the rules applied to the function of the CR. This includes regulatory policy, network operator policy, radio equipment capability, and real-time checking that, at the end of all the cognitive logic, the radio's planned performance is allowed within its rules.

Chapter 7, by Tom Rondeau of Center for Communications Research and Charles Bostian of Virginia Tech, provides a detailed analysis of cognitive techniques at the physical and medium access control layers. These techniques are discussed in the context of genetic algorithms that can adapt multiple waveform properties to optimize link performance.

In Chapter 8, John Polson and Bruce Fette describe a wide variety of methods by which a radio can determine its local position, and thereby to use time and location information to assist the network or the user.

In Chapter 9, Jonathan Smith, previously of DARPA, currently on the University of Pennsylvania faculty, covers cognitive techniques in network adaptation. This technology allows the radio to be aware of local networks and their properties and services. Smith focuses on how networks apply intelligence to the selection of network protocols to optimize network performance in spite of the differences between wireless and wired systems.

Chapter 10, by Joe Campbell, Bill Campbell, Scott Lewandowski, Alan McCree, and Cliff Weinstein of Lincoln Labs, is about using speech as an input/output mechanism for the user to request and access services, as well as to authenticate the user to the radio network and services. Speech analysis tools extract basic properties of speech. These properties are further analyzed in different ways to result in word recognition, language recognition, and speaker identification.

Chapter 11, by Youping Zhao, Bin Le, and Jeffrey Reed of Virginia Tech, deals with the ways in which network infrastructure can provide cognitive functionality and support services to the user, even if the subscriber's unit is of low power or small computational capability.

Chapter 12, by Vince Kovarik of Harris Corporation, provides extensive coverage of learning technologies and techniques and how these are applied to the CR application.

In Chapter 13, Mitch Kokar, David Brady, and Kenneth Baclawski of Northeastern University give a detailed overview of how to represent the types of knowledge a radio would need to know to behave intelligently. This chapter deals with the storage and analysis of this information as additional spatial, temporal, radio, network, and application data are accumulated.

In Chapter 14, Joe Mitola of the MITRE Corporation describes how to develop a complete radio and how to make the various radio modules work with each other as an integrated cognitive system.

In Chapter 15, Jody Neel, Jeff Reed, and Allen McKenzie of Virginia Tech provide a detailed analysis of game theory and how it is used to model the performance choices and system-level behaviors of networks consisting of a mixture of cognitive and non-CRs.

In Chapter 16, Drs. Jae Moung Kim and Seungwon Choi and their student teams discuss their work to create standardized interfaces and hardware and software architectures for smart antenna systems, such as smart antenna systems for WiMAX and WiBro broadband wireless networks. They report the ability to select optimum antenna modes and MA modes given appropriate awareness of the spectral environment and activities of various spectral users.

In Chapter 17, Dr. Grit Denker and her team discuss the development of policy language, policy representation, and policy reasoning, essential to performing the selection of frequencies, waveforms, and transmit power levels within the constraints of local regulatory policy. Denker also presents policy techniques that can indicate what additional criteria must be met to enable a policy-compliant transmission when current conditions are not fully identified as sufficient.

In Chapter 18, Drs. Spooner and Nicholls develop the detailed mathematical analysis of cyclostationary analysis of communications signals. From these analyses, signals that are otherwise hidden beneath the noise floor are able to be detected, and by doing so, a CR can avoid generating interference with hidden nodes. They subsequently apply their techniques to recognition of common standard telecommunications waveforms, showing templates of performance.

In Chapter 19, Drs. DaSilva and Thomas discuss how CRs can find each other when they are not using dedicated frequencies. Performing frequency rendezvous with high probability of acquisition is important to network performance. The authors explain techniques to reduce the average and maximum time it takes for one radio to find another radio or a network, when the frequency is unknown.

In Chapter 20, Dr. Stein develops a specification of Location Based Spectrum Rights, based on the ability to model transmit power, propagation loss, antenna patterns, and signal-to-interference ratios. In addition, Dr. Stein provides considerable analytic detail on propagation performance prediction, the relationship of signal levels, interference levels, and antenna patterns.

In Chapter 21, Drs. Pursley and Royster show how to adapt the waveform and error-correcting code properties to adapt to changing channel conditions over a 30-dB range of link conditions. Their techniques lead to an efficient library of waveform and FEC choices, providing uniform steps of adjustment that minimize the need to change the transmit power level and thus the interference to other nearby receivers, and are able to do so with minimal communications channel overhead.

In Chapter 22, Drs. Thomas and DaSilva discuss cognitive networking, and in particular how to control the CR to optimize network performance rather than simply the performance of a single link.

In Chapter 23, Dr. Martinez and Ms. He discuss the IEEE standards that have begun to integrate cognitive behaviors. They discuss IEEE 802.16, 802.22, and ongoing work in SCC 41: P1900.1-P1900.5 and the CR technologies in these standardization activities. The chapter provides important guidance into the standardization of system-level architectural features of advanced cognitive network systems.

Chapter 16 was the conclusion of the book in the first edition, serving as summary and outline of the remaining hard problems. The summary is moved to Chapter 24 in this edition, and provides the overview of real implementations and solid evidence of the ability to implement the major components using standards from IEEE, SDR Forum, international research consortiums such as E2R and E3, and the defense community. From this work, real systems are being built for production and deployment.

We hope you will find this book gives you the background to start at the very beginning of radio architecture and takes you all the way to the details of effective cognitive radio and cognitive network implementation.

REFERENCES

[1] Oppenheim, A., and R. Schaefer, *Discrete Time Signal Processing*, Prentice Hall, 1989.

[2] Rabiner, L., and R. Schaefer, *Digital Processing of Speech Signals*, Prentice Hall, 1978.

[3] Rabiner, L. R., J. H. McClellan, and T. W. Parks, FIR Digital Filter Design Techniques Using Weighted Chebychev Approximations, *Proceedings IEEE*, 63(4):595–610, 1975.

[4] Parks, T. W., and J. J. McClellan, Chebyshev Approximation for Nonrecursive Digital Filters with Linear Phase, *IEEE Trans. Circuit Theory*, 19:189–194, 1972.

[5] Flanagan, J., *Speech Synthesis and Perception*, Springer-Verlag, 1972.

[6] harris, fred, *Multirate Signal Processing for Communication Systems*, Prentice Hall, 2004.

[7] *www.mathworks.com/company/aboutus/founders/jacklittle.html*.

[8] Markel, J., and A. Gray, *Linear Prediction of Speech*, Springer-Verlag, 1976.

[9] *www.intel.com/pressroom/kits/bios/moore.htm*.

[10] *www.dspguide.com/filters.htm*.

[11] Rosenblatt, F., The Perceptron: A Probabilistic Model for Information Storage and Organization in the Brain, Cornell Aeronautical Laboratory, *Psychological Review*, 65(6):386–408, 1958.

[12] Baker, J., Stochastic Modeling for Speech Recognition, Doctoral Thesis, Department of Computer Science, Carnegie Mellon University, Pittsburgh, 1976.

[13] Lee, K. F., H.-W. Hon, and R. Reddy, An Overview of the SPHINX Speech Recognition System, *IEEE Trans. Acoustic Speech and Signal Proceedings*, Jan.:34–45, 1990.

[14] Reed, J. H., *Software Radio: A Modern Approach to Radio Engineering*, Prentice Hall, 2002.

[15] *http://csrc.ncsl.nist.gov/cryptval/des/des.txt*.

[16] *http://csrc.nist.gov/CryptoToolkit/aes/*.

[17] Christensen, E., A. Miller, and E. Wing, Waveform Application Development Process for Software Defined Radios, *IEEE Milcom Conference*, Vol. 1, pp. 231–235, 2000.

[18] *http://en.wikipedia.org/wiki/POSIX*.

[19] Wang, J., The Use of Ontologies for the Self-Awareness of Communication Nodes, *Proceedings SDR Forum Technical Conference*, Orlando, November 2003.

Communications Policy and Spectrum Management

Paul Kolodzy
Kolodzy Consulting
Centreville, Virginia

2.1 INTRODUCTION

New technologies impact the worlds of commerce and policy. This is especially true of disruptive technologies that significantly alter either the realities or perceptions within these worlds. Cognitive radio (CR) technology has the potential of affecting the marketplace for radio devices and services, as well as changing the means by which wireless communications policy is developed and implemented. One of the key parameters that must be addressed to enter the radio market is access to radio spectrum. Once access is obtained, the capacity to manage interference becomes a key attribute in order to increase the number of users. Throughput is critical in order to maximize benefit (for the device) or maximize revenue (for the service). Radio frequency (RF) spectrum access and interference management are thus the primary roles of spectrum management.

CR technology has the potential of being a disruptive force within spectrum management. Spectrum management, since the dawn of radio technology, has been within the domain of management agencies, both private and government. Therefore, it has required a person-in-the-loop. The ability of a device to be aware of its environment and to adapt to enhance its performance, and the performance of the network, allows a transition from a manual, oversight process to an automated, device-oriented process. This ability has the potential to allow a much more intensive use of the spectrum by lowering the spectrum access barrier to entry for new devices and services. It also has the potential to radically change how policy should be developed in order to account for these new uses of the spectrum, and it can fundamentally change the role of the spectrum policymaker and policy regulator.

In this chapter, Section 2.2 discusses the CR technology enablers. Section 2.3 addresses spectrum access and how cognitive radio needs various types of policies, depending on the density of spectral activity and the types of usages. This section also provides examples of spectral activity measurement. Section 2.4 discusses the challenges to equipment developers associated with a policy-based approach to spectrum management. Section 2.5 presents the challenges to the regulators to manage spectrum

Fette, *Cognitive Radio Technology*

policy through radios and networks that operate based on policy. Section 2.6 discusses the global interest and activity in policy-based cognitive radio. Finally, Section 2.7 provides a summary of the chapter's major issues.

2.2 COGNITIVE RADIO TECHNOLOGY ENABLERS

The development of wideband power amplifiers, synthesizers, and analog-to-digital converters (ADCs) is providing a new class of radios: the software-defined radio (SDR) and its software and CR cousins. Although at the early stages of development, this new class of radio ushers in new possibilities, as well as potential pitfalls for technology policy. The flexibility provided by the CR class of radios allows for more dynamics within radio operations. The same flexibility poses challenges for certification and the associated liability through potential misuse.

SDRs provide software control of a variety of modulation techniques, wideband and narrowband operation, transmission security (TRANSEC) functions (such as hopping), and waveform requirements. In essence, components can be under digital control and thus defined by software. The advantage of an SDR is that a single system can operate under multiple configurations, providing interoperability, bridging, and tailoring of the waveforms to meet the localized requirements. SDR technology and systems have been developed for the military. The digital modular radio (DMR) system was one of the first SDR systems. From 1999 to 2003, the US Defense Advanced Research Projects Agency (DARPA) developed the Small Unit Operations Situational Awareness Systems (SUOSAS), which was a man-portable SDR operating from 20 MHz to 2.5 GHz. The level of success of these programs has led to the Joint Tactical Radio System (JTRS) initiative to develop and procure SDR systems throughout the US military. Further enhancement in signal-processing technology has spawned additional efforts including the DARPA NeXt Generation (XG) and Wireless Network After Next (WNAN) projects.

Software-defined radios exhibit software control over a variety of modulation techniques and waveforms. Software radios (SRs) specifically implement the waveform signal processing in software. This additional caveat essentially has the radio being constructed with a *radio frequency* front end, a downconverter to an *intermediate frequency* (IF) or baseband, an analog-to-digital converter, and then a processor. The processing capacity therefore limits the complexity of the waveforms that can be accommodated.

A CR adds both a sensing and an adaptation element to the software-defined and software radios. Four new capabilities embodied in CRs will help enable dynamic use of the spectrum: flexibility, agility, RF sensing, and networking [1].

Flexibility is the ability to change the waveform and the configuration of a device. An example is a cell tower that can operate in the cell band for telephony purposes but change its waveform to get telemetry from vending machines during low usage, or other equally useful, schedulable, off-peak activity. The same band is used for two very different roles, and the radio characteristics must reflect the different requirements, such as data rate, range, latency, and packet error rate.

Agility is the ability to change the spectral band in which a device will operate. Cell phones have rudimentary agility because they can operate in two or more bands (e.g., 900 and 1900 MHz). Combining both agility and flexibility is the ultimate in "adaptive" radios because the radio can use different waveforms in different bands. Specific technology limitations exist, however, to the agility and flexibility that can be afforded by current technology. The time scale of these adaptations is a function of the state of technology both in the components for adaptation as well as the capacity to sense the state of the system. These are classically denoted as the observable/controllable requirements of control systems.

RF Sensing is the ability to observe the state of the system, which includes the radio and, more important, the environment. It is the next logical component in enabling dynamics. Sensing allows a radio to be self-aware, and thus it can measure its environment and potentially measure its impact to its environment. Sensing is necessary if a device is to change in operation due to location, state, condition, or RF environment.

Networking is the ability to communicate between multiple nodes and thus facilitate combining the sensing and control capacity of those nodes. Networking, specifically wireless networking, enables groupwise interactions between radios. Those interactions can be useful for sensing where the combination of many measurements can provide a better understanding of the environment. They can also be useful for adaptation where the group can determine a more optimal use of the spectrum resource over an individual radio.

These new technologies and radio classes, albeit in their nascent stages of development, are providing many new tools to the system developer, while allowing for more intensive use of the spectrum. However, an important characteristic of each of these technologies is the ability to change configuration to meet new requirements.[1] This capacity to react to system dynamics will require the development of new spectrum policies in order to take advantage of these new characteristics.

The IEEE Standards Coordinating Committee on Dynamic Spectrum Access Networks (SCC-41) and the Software Defined Radio Forum—Cognitive Radio Working Group (SDR Forum—CRWG) have provided definitions for many of the critical elements of CRs. The definitions for the various radio technologies have been integrated by the International Telecommunication Union (ITU) with help from many of its member organizations. The following sidebar provides a list of radio technology definitions proposed by the Global Standards Collaboration (GSC) group within the ITU. Definitions for policy-based radios and dynamic frequency selection radios are also provided. These two new radio classes are specific implementations of CRs. Policy-based radios are discussed in Section 2.6 with an emphasis on how CR technology can impact the development and implementation of communications policy. Dynamic frequency selection radios are addressed in Section 2.3.3. These advanced radio technologies are enabling a multitude of new radio concepts, as discussed in Section 2.3.2.

---

[1]The question of how CRs would apply physical layer adaption in sensor networks that transmit and receive over very low duty cycles has not been adequately studied. Although it is assumed that they could also benefit from CR adaption, the time to reach network stability is lengthened by the low duty cycles. It seems that if the spectral dynamics exceed the time to reach stable performance, or consume more bandwidth for parametric exchange and adaption parameters than the network can effectively provide, then such sensor networks may not be able to fully benefit from CR techniques. However, if other systems operating in the same environment are CRs with stable adaption strategies, then sensor networks may still benefit.

ITU—Global Standards Collaboration (GSC)

Proposed Definitions

Software-Defined Radio:
"A radio that includes a transmitter in which the operating parameters of frequency range, modulation type or maximum output power (either radiated or conducted), or the circumstances under which the transmitter operates can be altered by making a change in software without making any changes to hardware components that affect the radio frequency emissions."

— Derived from the US FCC's Cognitive Radio Report and Order, adopted March 10, 2005

Cognitive Radio: A radio or system that senses and is aware of its operational environment and can be trained to dynamically and autonomously adjust its radio operating parameters accordingly. *Note: Cognitive* does not necessarily imply relying on software. For example, cordless telephones (no software) have long been able to select the best authorized channel based on relative channel availability.

Policy-Based Radio: A radio that is governed by a predetermined set of rules for behavior. The rules define the operating limits of such a radio. These rules can be defined and implemented:
- During manufacture
- During configuration of a device by the user
- During over-the-air provisioning and/or
- By over-the-air control

Software Reconfigurable Radio: A software defined radio that (1) incorporates software-controlled antenna filters to dynamically select receivable frequencies, and (2) is capable of downloading and installing updated software for controlling operational characteristics and antenna filters without manual intervention.

Dynamic Frequency Selection (DFS):
(1) "A general term used to describe mitigation techniques that allow, amongst others, detection and avoidance of co-channel interference with other radios in the same system or with respect to other systems."

— From current version of WP8A PDNR on SDR

(2) "The ability to sense signals from other nearby transmitters in an effort to choose an optimum operating environment."

— Derived from the US FCC's Cognitive Radio Report and Order, adopted March 10, 2005

2.3 NEW OPPORTUNITIES IN SPECTRUM ACCESS

Two general management methods allow access to the RF spectrum: spectrum access licenses and unlicensed devices. Spectrum licenses are issued by the appropriate regulatory agency within the nation. The licenses include a band, a geographic region, and the allowable operational parameters (e.g., in-band and out-of-band transmission levels).

Although the licenses have a finite duration, there is an expectation of renewal. There is also a level of protection from interference from other systems accessing the spectrum. Unlicensed devices, also called licensed-free devices and licensed-by-rule devices, are provided frequency bands and transmission characteristics (albeit at much lower transmission power levels), and are not provided regulatory protection from interference.

2.3.1 Current Spectrum Access Techniques

The RF spectrum is organized by allocations and assignments. Allocations determine the type of use and the respective transmission parameters. The allocations are specified by each individual nation determined by sovereign needs and international agreements. International agreements can be bilateral or multilateral treaties or resolutions by the International Telecommunication Union during the periodic World Radio Communication Conferences (WRCs). Figures 2.1 and 2.2 are examples of spectrum allocations for

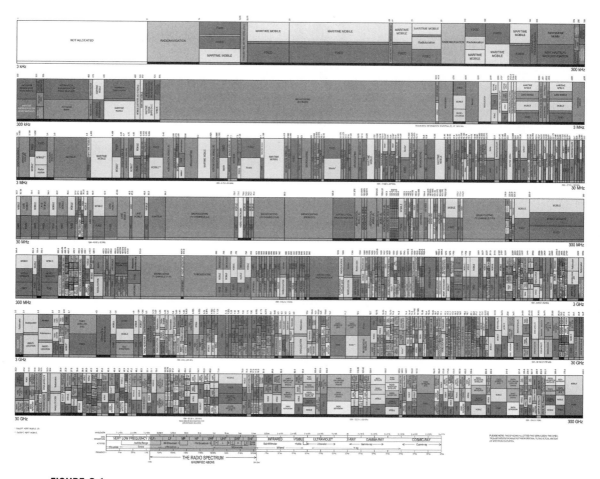

FIGURE 2.1

US spectrum allocations [2].

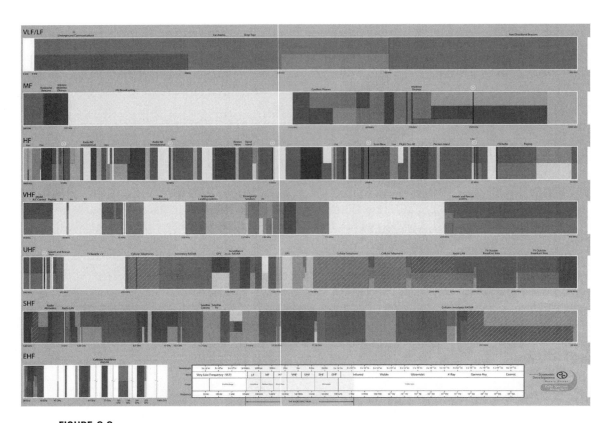

FIGURE 2.2

New Zealand spectrum allocations [3].

the United States and New Zealand, respectively. Each color[2] is an indication of a service type that is allocated to that frequency band *across the entire nation*. Many of the primary allocations—television (TV), frequency modulation (FM) radio, global positioning systems (GPS), and so on—are identical. The vertical splitting that occurs in some bands depicts the instance of multiple allocations (primary, coprimary, and secondary) within the band. The intensity of usage in the United States is much higher than in New Zealand. This is especially noticeable in the figures by the large number of multiple allocations needed to provide enough capacity for the numerous wireless services for commercial and noncommercial applications, such as defense, air traffic, and scientific exploration.

Assignments are individual licenses within an allocation. Assignments are provided by the method of choice of the sovereign state. Assignments can also be limited by geographic extent. There are many ways to obtain assignments, as depicted by

---

[2]Color in the original artwork (see References) is reproduced in the figures in the book as various shades of gray.

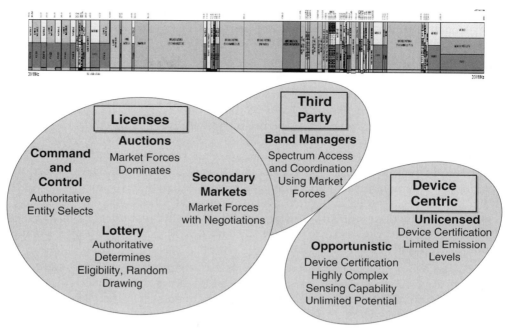

FIGURE 2.3

Spectrum access regimes.

Figure 2.3. Currently, three basic types of assignment methods are generally employed: *command and control*, *auctions*, and *protocols and etiquettes*.

Command and Control. Command and control assignments are provided by the regulatory agency by reviewing specific licensing applications and choosing the prospective licensee by criteria specific to the national goals. Detractors of this assignment technique call it a "beauty contest" because only a handful of regulators determine the relative value between potential services and license holders. Since the US Radio Act of 1934, the Federal Communications Commission (FCC) has had the regulatory authority to decide which firms should get licenses in order to bring spectrum to its highest and best use. The FCC held public hearings to gather information to make these determinations. In the early 1980s the number of applications for licenses was growing so large that the command and control system ground to a halt. The FCC, to enable more capacity for the command and control assignment mechanism, decided to start awarding licenses by lottery. The assignment process was still in complete control of the regulators; however, there were no restrictions as to who could participate in the lottery. One year, for instance, a group of dentists won a license to run cellular phones on Cape Cod, and then promptly sold it to Southwestern Bell for $41 million. This set of events disturbed regulators because the spectrum assignment process became an investment opportunity rather than providing a service (albeit for profit). This directly challenged their authority in using spectrum for the benefit of the country (i.e., its citizens).

Auctions. Spectrum assignments through auctions are fairly new. New Zealand, with the Radio Communications Act of 1989, enabled the use of market-driven allocation mechanisms for assignment of spectrum rights. Although the act did not explicitly provide the market mechanisms, by 1996 the mechanisms for selling spectrum by auction were in place. Specifically, the use of competitive bidding was authorized in granting licenses to qualified applicants.[3] In the Telecommunications Act of 1996, the United States also enabled the use of auctions for spectrum access licenses. The Wireless Telegraphy Act of 1998 enabled auctions in the United Kingdom. Auctions are now a common form for spectrum assignments throughout the industrial world. The United States, Australia, Mexico, Canada, New Zealand, and the European Union now employ auctions regularly. The current spectrum allocations in the United States and New Zealand presented in Figures 2.1 and 2.2, respectively, show the many competing applications for spectrum.

Protocols and Etiquettes. Unlicensed devices and amateur licensees do not have, per se, specific frequency assignments. The allocation allows these devices to operate within a band, and the selection of a particular frequency is accomplished through protocols and etiquettes. Protocols are explicit interactions for spectrum access. Carrier Sense Multiple Access with Collision Avoidance (CSMA/CA) is a protocol for media access and control that requires information to be communicated between devices. Etiquettes are rules that are followed without explicit interaction between devices. Simple etiquettes, such as "listen before talk," dynamic frequency selection (DFS), and power density limits are one-sided processes. The amateur radio operators have constructed a well-tested and successful manual set of etiquettes. For example, the developers of DX tuners (for long-distance transmission and reception by ham radio operators) have etiquettes such as "If there are other users logged in, *always* ask in the chat window before you tune the receiver!" Unlicensed devices are operated on a much shorter time scale. Industry groups such as the Institute of Electrical and Electronics Engineers (IEEE) form groups in order to develop standards to promote device interoperability. The 802.3 (Ethernet) and 802.11 (wireless local area network, or WLAN) are two such protocol standards. New etiquettes, such as DFS, have been developed for the new 5 to 5.8 GHz 802.11a band (IEEE 802.11h).

Spectrum access for services provided by licenses is controlled directly by the service provider. In the case of public broadcast services such as FM radio and television, the waveforms and/or protocols are defined by the rules provided by the regulators. These rules, although provided by the regulators, are developed with both the

---

[3]Section 309 [47 U.S.C. 309] Subsection J: "Use of Competitive Biding.—If mutually exclusive applications are accepted for filing for any initial license or construction permit which will involve a use of the electromagnetic spectrum described in paragraph (2), then the Commission shall have the authority, subject to paragraph (10), to grant such license or permit to a qualified applicant through the use of a system of competitive bidding that meets the requirements of this subsection."

service providers and the device manufacturers. Being a broadcast service, the broadcast station controls access simply by its own use of the frequency band. In the case of providers of two-way communications or private broadcast, the waveforms and protocols are defined independently by the license holder. However, market forces tend to drive groups of license holders to compatible technology in order to increase either the interoperability with other license holders, or to increase the geographic footprint of their service. A classic example of this independent development of industry standards has been the commercial mobile radio service (CMRS, or cellular service) and the development of the capacity to roam from service provider to service provider. The development of roaming has generally been considered to have been the threshold event in making radio telephony a viable service.

The operational envelopes for unlicensed devices are defined by rules in three general groupings: unintentional, incidental, and intentional radiators. An unintentional radiator, per FCC definition, is a device that *intentionally generates RF energy for use within the device, or that sends RF signals by conduction to associated equipment via connecting wiring, but which is not intended to emit RF energy by radiation or induction*. One example would be a television receiver with leakage from an oscillator in the receiver RF chains. Another very common example concerns a computer and display that typically exhibit harmonics up to more than 3 GHz. The first example has the emission contained in a particular band and the second example is a broadband, noise emission.

An incidental radiator, per FCC definition, is a device that *generates radio frequency energy during the course of its operation, although the device is not intentionally designed to generate or emit radio frequency energy*. Examples of incidental radiators are direct current (DC) motors, mechanical light switches, or even the radiation from a hair dryer.

In both the unintentional and incidental radiator cases, the regulatory agency has set limits on the emission levels. The allowable emission levels are generally very low, due to the general and ubiquitous deployment of these devices. The devices may be in extremely close proximity to licensed service devices. Usually, the consumer does not know the potential impact of the device and thus cannot easily resolve any interference issues. Therefore, the aggregated interference from many such devices must remain small.

Intentional radiators are devices that are specifically constructed for a communications application. A prime example is that of baby monitors, a one-way communications system for use in the home. Other examples of intentional unlicensed radiators are the data networks using the 802.11 standard. These devices are allowed to radiate at higher levels than the unintentional radiators but still at low overall power spectral density. The low-power spectral density provides some assurance that the interference potential to a primary license holder within that band is low. The regulators determine the allowable transmission parameters. There is no licensee, however, so the device manufacturer has the burden to conform to the transmission rules.

The primary difference between licensed and unlicensed device spectrum access is the afforded interference protection. The unlicensed device always operates as a secondary user within a band and is allowed to operate only in a *not-to-interfere* basis, per the transmission rules. The unlicensed device is not provided any interference assurances from either the primary use devices or other unlicensed devices. It is for this reason that device manufacturers and application designers develop standard protocols and/or etiquettes in order to reduce the potential of like-device interference. This is true with the development of the 802.11 specifications for WLANs. These specifications are widely used with the popular consumer products in the 2.4 GHz and 5.15 to 5.825 GHz bands. However, unlike devices such as 2.4 GHz cordless telephones and 802.11g devices, which operate in the industrial, scientific, and medical (ISM) band, the expanded 802.11h devices operating in the 5.15 to 5.825 GHz band have the potential to interfere with incumbent devices. The common use of the 2.4 GHz ISM band is for microwave ovens, which are not overly susceptible to radiated energy.

2.3.2 Opportunistic Spectrum Access

The current toolbox of spectrum access techniques is limited by the capacity of both the devices and the policies that are in place. As was shown in the spectrum allocations charts depicted in Figures 2.2 and 2.3, spectrum less than 6 GHz is completely allocated for various services. However, the use of the assignments in time and space is not complete. An example of a measurement of spectrum utilization is shown in Figure 2.4.

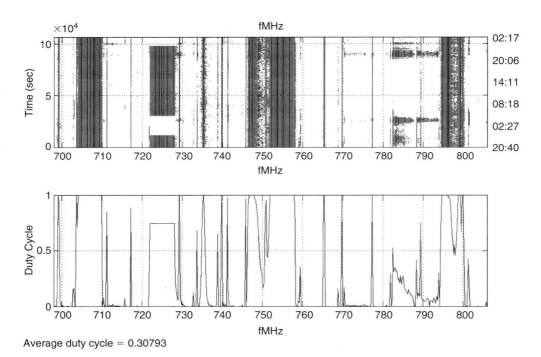

Average duty cycle = 0.30793

FIGURE 2.4

Spectrum utilization in Hoboken, New Jersey, 700–800 MHz, made by Shared Spectrum Corporation over an 18-hour period.

This time-frequency measurement of the 700 to 800 MHz band was made over an 18-hour period in Hoboken, New Jersey, which is directly across the Hudson River from New York City. It clearly indicates that there are portions of the spectrum that are continuously accessed, portions that are never accessed, and portions that are accessed for a fraction of the time. These "white spaces" in the spectrum have created interest within the US Department of Defense (DoD) to begin research into more efficiently using spectrum. The FCC also saw the potential for more intensive use of the spectrum as well as revisiting the age-old claim of scarcity of the RF spectrum.

The onset of software defined and cognitive radio technology enables *opportunistic spectrum access* (OSA). OSA looks for "holes" in the spectrum and then adjusts the link parameters to conform to the hole. That is, the radio transmits over sections of the spectrum that are not in use. However, it has the additional complexity of listening for other transmitters in order to vacate a hole when other, nonopportunistic spectrum devices are accessing it. This technology combines flexible waveform capacity with sensing and adaptive frequency technology. The potential gain is the higher utilization of infrequently used spectrum. It has been estimated that on average, less than 5 percent, and possibly as little as 1 percent, of the spectrum less than 3 GHz as measured in frequency-space-time is used.[4] Opportunistic spectrum technology is under development by the US DoD with the goal of increasing accessible spectrum by a factor of 20. The European Union Information Society Technologies (IST) as well as the US National Science Foundation (NSF) and other research organizations, as of 2009, are actively pursuing this technology. The following sidebar provides a chronology of events in OSA activities.

CHRONOLOGY OF OPPORTUNISTIC SPECTRUM

1999–2000 (United States): Localized set of measurements conducted by the Defense Advanced Research Projects Agency (DARPA) indicated that spectrum use was not very high.

2002 (United States): DARPA initiates the XG (NeXt Generation) project to investigate the potential for the military to share spectrum spatially and temporally with multiple devices.

2002 (United States): The Federal Communications Commission (FCC) Spectrum Policy Task Force (SPTF) concludes that spectrum access is a more significant problem than spectrum scarcity. The SPTF recommends that new rules be developed to allow more intensive access to spectrum, including opportunistic spectrum.

2003 (European Union): Information Society Technologies include dynamic spectrum access technologies as part of the Sixth Framework Programme of R&D.

---

[4]This percentage is a function of where the measurement is made and what sensitivity thresholds are used. Rural areas have an extremely low utilization. In the worst-case condition of downtown New York City during the Republican Convention, the utilization rose to approximately 20 percent. The peak utilization in San Diego on one sample day was 7 percent. Because the numbers are location and emergency dependent, precise numbers cannot be stated universally, but utilization is definitely low.

2003 (United States): The National Science Foundation (NSF) initiates research projects in spectrum measurements and dynamic spectrum access.

2004 (United States): The FCC issues a Notice of Proposed Rulemaking on Facilitating Opportunities for Flexible, Efficient, and Reliable Spectrum Use Employing Cognitive Radio Technologies.

2004 (European Union): End-to-End Reconfigurability (E2R) Project initiated in Information Society Technologies.

2005 (United States): DARPA XG and NSF projects complete a series of spectrum occupancy measurements indicating less than 10 percent occupancy in time–space under 3 GHz.

2006 (United States): FCC issues Rule and Order on permitting low-power devices in unused portions of the TV broadcasting spectrum (a.k.a. TV Whitespaces).

2007 (United States): FCC begins testing of prototype TV Whitespace devices.

2007 (United Kingdom): An Ofcom Consultation creates new opportunities for cognitive radio to use interleaved spectrum without causing interference.

2008 (United States): DARPA XG successfully demonstrates Opportunistic Spectrum.

The availability of SDR also allows high-priority users, such as in a public service context, to access the radio spectrum on an "as-needed" basis. This is called *interruptible spectrum access* because the normal user's spectrum access is interrupted in order to provide the spectral resource to a higher-priority user (see Figure 2.5). For example, in a major regional disaster there will be a significant increase in public safety users due

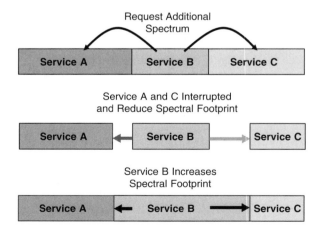

FIGURE 2.5

Interruptible spectrum: A periodic increase in spectral requirements (i.e., public safety) may be addressed by interrupting part of the band of other services. This can be accomplished through central coordination or beaconing.

to the influx of responders from outside the immediate area. Such an influx of radio users would require additional spectrum to cope with the increased load. Additional spectrum could be temporarily obtained from adjacent spectral bands in the immediate area of the disaster. After the need diminished, perhaps in minutes or hours, spectrum would be released back to the primary licensee. Mechanisms may be developed to compensate primary licensees for their inconvenience or loss of revenue. Interruptible spectrum access was proposed in the United States in 2007 for the D-Block in the upper 700 MHz band. In this case, the D-Block licensee will build a nationwide broadband network for the 5 MHz commercial band (758–763 MHz) and the 5 MHz public safety broadband band (763–768 MHz). The D-Block licensee would have access to the public safety broadband band in locations and times of low usage.

If policies can be malleable to specific conditions in order to allow greater access to the RF spectrum, then the *cost barriers for new entrants* should be reduced. Thus, new consumer products can be developed using lower-cost spectrum. The lower barriers will allow developers to use more resources in developing techniques to access spectrum instead of investing in spectrum.

2.3.3 Dynamic Frequency Selection

Opportunistic spectrum access represents the general case for accessing the spectrum using "listen before talk" etiquette. A subset of OSA is dynamic frequency selection, which also is called dynamic channel selection (DFS). DFS is used to prevent a device from accessing a specific band if it is in use and to prevent co-channel interference of the primary user from a secondary user. It differs from OSA in that it is not seeking spectrum access—it is determining whether access should be allowed. DFS has been adopted in the 802.11h standard. When the ITU allowed unlicensed WLAN in 2003, 802.11h was developed in order to satisfy the primary users within the 5.25 to 5.725 GHz band. DFS detects other devices using the same radio channel, and it switches WLAN operation to another channel if necessary. DFS is responsible for avoiding interference with other devices, such as radar systems and other WLAN devices. Specifically, for the 5.25 to 5.725 GHz service, the DFS system must be able to detect a primary user above a detection threshold of −62 dBm.

2.4 POLICY CHALLENGES FOR COGNITIVE RADIOS

The capacity to sense, learn, and adapt to the radio environment provides new opportunities for spectrum users. However, the same sensing and adaptation also creates challenges for policymakers. The primary concern is with the potential to have nondeterministic behaviors. Nondeterministic behaviors can be created by a variety of conditions:

- The allowance of self-learning mechanisms will create a condition in which the response to a set of inputs will be changing and thus unknown.
- The allowance of software changes will create conditions either from errors within the software or from rogue software, which can cause the device to not conform to the transmission rules.

■ The allowance of frequency and waveform agility will create conditions in which devices that conform to transmission rules may cause interference due to mismatch between out-of-band receivers and the in-band transmitter waveforms.

In addition to nondeterministic behaviors, another primary concern is the impact of horizontal versus vertical service structure. Vertically integrated services, such as cellular telephony, clearly delineate responsibility for spectrum management to the service provider. The service provider has the sole responsibility for problems, interference, and all other technical and service issues. However, for horizontally integrated service, which includes device-centric systems that may be the initial focus for CR technology, there isn't a single point of responsibility for interference and other problems. One example has been the issue with secondary spectrum markets. The formal responsibility of a device creating interference is the primary licensee. The rules had to be modified to allow that responsibility to follow the usage to the secondary licensee when appropriate. The extrapolation of this approach is problematic when applied to CRs, as each device is, in essence, a licensee. This is a serious problem for the policymakers that can be addressed by rules, technology, or a combination of both.

This section addresses these concerns with the deployment of CRs. In particular, the context of CR for DSA and security concerns are described. The highlighted questions are areas for research and development.

2.4.1 Dynamic Spectrum Access

The development of the OSA and DFS technology is straightforward in terms of radio technology. The discussion of policy impact is much more complex. In both cases, the dynamic system must interface seamlessly, without interference, to existing spectrum licensees. Two basic questions need to be addressed:

Question 1: What are the rights of the license holder to prevent unauthorized use by an opportunistic device? To answer this question, a clear definition of interference is needed. The interference definition must include some aspect of duration and persistence of the interferer. One draconian response would be that it is better to use less than 1 percent of the spectrum to ensure that a single packet is never dropped due to interference.

Question 2: What kind of assurance can be provided that the interference will be for only a finite and precise duration? Answering this question is complicated by the fact that as these systems become more flexible and agile, the combination of possible configurations grows exponentially.

The ability to move the operations of a radio in frequency is quite enticing. The operational rules, however, are different in each of the separate frequency bands. Therefore, a single device must: (1) know of all of the transmission rules in the bands in which it can operate; (2) have the ability to adjust its transmission parameters accordingly; and (3) ensure that device will adhere to the transmission rules.

Defining the Rules for Dynamic Spectrum Access

The policy goal of managing interference is generally defined as to the parameters for the transmitter (e.g., center frequency, bandwidth, power spectral density, antenna gain). Interference management is determining the threshold for producing harmful interference. Dynamic spectrum access (DSA) systems can move in frequency, so fixed parameters are insufficient. The metric for interference must be defined to ensure that the DSA device does not cause harmful interference. The primary question is:

1. What is an appropriate interference metric?

   With an explicit interference metric, the transmission rules could allow operation under a much broader set of transmitter parameters with the explicit goal of preventing interference above a particular threshold within a radius of the transmitter. The interference metric could also be contextually sensitive with respect to location, time, or condition of the RF environment. As the FCC has said, "Quantitative standards reflecting real-time spectrum use would provide users with more certainty and, at the same time, would facilitate enforcement."[5]

   The interference metric must be indicative of the impact of the transmitter on the surrounding region. In a static and well-defined geometry, a measurement at the transmitter can provide sufficient information from which to extrapolate the value of the interference metric across a wide area. However, the highest density of use of the RF spectrum is in areas with complex geometry and a large number of users, most of which are mobile. Although the technology for propagation models, inclusive of complex environments, is improving and can provide qualitative results for extrapolation, it is insufficient for quantitative interference analysis. Thus, another important question arises:

2. What kind of in situ measurement system is needed to provide the basis for interference analysis?

   Multiple measurements distributed across the operational area are needed to accurately measure the interference metric. One possible mechanism to obtain multiple measurements would be the development of monitoring stations such as pollution monitoring devices. Many policy and technical questions arise:

3. Who would appropriate the funds to develop and deploy such a system? If it is a federally funded system, should that system be managed by a government or independent regulatory entity?

4. Who would ensure the accuracy of the measurements? Who could challenge the accuracy of the devices?

5. What are the mechanisms to disseminate the results of the measurements? Would there be an interface to all service providers and users who could obtain the data on a near real-time basis? Would the data be broadcast to all devices or should they be on a request basis?

---

[5]See [3], Section VI: Interference Avoidance, found at *http://www.fcc.gov/sptf/reports.html*.

Liability when the policy is not followed must also be determined. Public safety communications' use of interruptible spectrum is one example in which liability would need to be explicitly addressed. The need for assured communications, free from avoidable interference, is a paramount requirement for public safety communications. So the mechanisms for obtaining and releasing spectrum need to be highly predictable. Beaconing is one such mechanism; that is, either a service would send a beacon indicating it is using the spectrum or send a beacon when the band is available for use by another entity. The question then arises as to how beaconing should be used. Propagation challenges may create shadow zones that would prevent a non–public safety user from hearing the beacon. So the basic technical challenge is to create the proper signaling technique. Thus, the following question arises:

6. Which policy will ensure the public that the system will work, determine who is liable if it does not work, and fix what type of compensation may be allowed between the public safety user and the primary licensee?

Safeguards and Incentives for Incumbent Users

Incumbent license holders that are currently using all of their rights to access the spectrum would initially see no value to having DSA [4]. The desired impact of DSA is to improve spectrum utilization and thus provide more competition and products to the market. The challenge is to determine how such impacts can be obtained while providing both assurances to current license holders and incentives for increased utilization through DSA.

7. How can assurances be made to current license holders while developing incentives for DSA?

Technology provides a potential solution to this particular challenge. The recent development of policy-enabled devices (see Section 2.4.3) can provide assurance that the device can have "fail-safe" mechanisms to prevent operation in unauthorized bands.

The problem set that has been put forward essentially requires the policymakers to provide the infrastructure to enable a technology. Generally, this is a very difficult task because regulatory agencies are not focused on providing an infrastructure or service to the commercial world. Technologists must actively address these concerns.

2.4.2 Security

The challenges of employing CRs for the policy community also include that of ensuring secure device operations. Security in this context includes enforcement of DSA rules. Enforcement for static systems is already a challenge due to the amount of resources necessary to authorize equipment, the requirement of obtaining proof that violations have occurred, and the determination of the violators' identities. As the systems become more dynamic, there is an increase in the number of potential interactions that can lead to a violation. Additionally, this leads to a decrease of the time and spatial scales of these interactions. Both of these changes will amplify the enforcement challenges.

Equipment Authorization

Initial equipment authorizations have two components that will increase significantly in complexity with the onset of dynamic policies: evaluation criteria and security certification. The capacities to modify waveforms, change operating conditions, and change transmission frequencies all contribute to an exponential growth in potentially adverse interactions between systems. Exhaustive testing becomes unrealizable due to the sheer number of combinations. Formal method techniques have been a focus to solve this problem. Formal methods provide provable atoms of code that mitigate the requirement for exhaustive testing. However, the maturity of the technology is inadequate to address the complexity of software embodied in a CR. Therefore, the challenge will be to answer a new set of questions:

1. How can realizable test plans be developed for certification that provide sufficient certainty as to the impact of the system on existing, certified equipment?
2. How can a device be certified if the software and hardware come from different manufacturers?

If self-learning mechanisms are to be employed, equipment authorization becomes even more problematic. Software and hardware certification will not provide sufficient assurances that the device conforms to the operational envelopes. New mechanisms will need to be developed, such as policy-enabled devices (see Section 2.4.3).

Software Certification

Software certification and the security of the software are also challenging areas. CR algorithms are written in software that provides the control of dynamic systems. Software can be modified to allow policies to change on either a periodic or an aperiodic basis. The security of that software is critical to ensure that rogue behavior is not programmed into the device. If the consumer can access the device's software, then the consumer can instruct the radio to perform outside of the permitted operational parameters. Thus, basic issues arise such as:

3. How is software protected to ensure that this abhorrent behavior does not occur?
4. Security is not an absolute but a level of acceptable risk. How much protection is necessary? How is it tested to ensure that it is sufficiently secure?
5. Who is liable if there is a security failure?

Monitoring Mechanisms

The number of combinations of interactions is high, and the mobility and the agility of future systems are great, so the basic issue remains:

6. How should enforcement systems be developed to observe all of these new cognitive capabilities?

Three possible mechanisms are suggested: *authority-based*, *network-based*, and *infrastructure-based* systems. The greatest challenges for development of such monitoring mechanisms are the equipment and analysis costs and the civil liberty concerns.

The *authority-based* system is for a regulatory agency[6] to deploy a national monitoring system. A secondary advantage of such a national system would be additional uses. Such a system would have capability of measuring and reporting spectrum usage, interference environment, and propagation characteristics.

The *network-based* system is to use the plethora of user devices already present to monitor the activity of the RF spectrum. The challenge would be in the methods to: (1) provide sufficient confidence in the accuracy of the measurements; (2) obtain sufficient geolocation information to make the information valuable; (3) collect and disseminate the information to enforcement organizations within the regulatory community; and (4) do these tasks with minimal overhead on network resources.

The *infrastructure-based* system is a combination of authority-based and network-based systems. The goal would be to use preexisting infrastructure such as cell towers, Federal Aviation Agency (FAA) towers, and the like. The challenge would be to obtain the authority to equip each site and to have priority access to the network from each site.

Again, as with the policy challenges for DSA systems, the challenges for CRs in the security arena are also quite great. The solution basis is also a combination of technical and policy initiatives. The technical initiatives just outlined need to be conducted and vetted through the policy communities.

2.4.3 Communications Policy before Cognitive Radio

This section addresses the role of the communications regulatory bodies on the development and deployment of CRs as well as the impact of them on regulation. Communication regulation is separate and distinct within every nation. The FCC is the regulatory agency for all wireless communications systems within the United States, except those used by the federal government (e.g., DoD, NASA, and FAA). Those systems are regulated by NTIA within the Department of Commerce (DoC). The primary goal of the regulatory agencies is to promote technology for the public interest while preventing interference between the multitudes of systems already deployed.

2.4.4 Cognitive Radio Impact on Communications Policy

The challenges that face current telecommunication regulators are great. The rapid development of new techniques to access the spectrum and applications is quickly outpacing the ability for the policymaker to *react. This* is apparent by how rapidly society is changing its views on spectrum use and management. The consumer has an expectation of untethered connectivity with more devices and applications. Technology is also lowering the barriers for new commercial entrants, therefore increasing both the

---

[6]Or an entity authorized by a regulatory agency.

amount and diversity of the uses for the RF spectrum. In a seemingly contradictory statement, given the consumer demands for using the RF spectrum, technology is also challenging the long-held views that spectrum is a scarce resource. Policymakers need a clear, nonreactive vision for future management of the RF spectrum.

The mirror image is also true. The challenges that face current technologists and entrepreneurs are also great. The life span of an individual market has shortened from decades to years and possibly months. The regulatory approval cycle can be longer than the market life span, creating risks of missing a market altogether. Although the technologists can create new capabilities and products, the regulators can have a marked impact on their ability to bring those products to the marketplace. Additionally, technology has become sophisticated and the application market is intertwined with that technology. Regulators must now be technology savvy in order to comprehend all aspects of their decisions. Technologists and entrepreneurs must now be cognizant of the role regulators have on what they can develop.

2.4.5 US Telecommunications Policy, Beginning with the *Titanic*

It is critical to understand the underpinnings of current spectrum management. The United States provides a good example of the historical context of the development of telecommunications policy. This section briefly reviews the chronology of events that have shaped US policy.

Frequently, the impetus for new policies is a catastrophe that leads to the perception of a policy failure. The sinking of the *Titanic* was the pivotal point in initiating the development of wireless communications policy. In 1910, the US Congress had mandated that passenger ships carry wireless telegraphs. Investigations into the 1912 *Titanic* disaster indicated that amateur radio operators caused interference after initial reports of the disaster. This interference, it was thought, hampered rescue efforts.

Federal Communications Commission Formation

The Radio Act of 1927 established the Federal Radio Commission and set forth as its intent to "maintain the control of the United States over all the channels of interstate and foreign radio transmission; and to provide for the use of such channels, but not the ownership thereof." The 1927 Act provided that the new Commission shall, "as public convenience, interest, or necessity requires" classify radio stations, prescribe the nature of the service, assign bands of frequencies or wavelengths, and determine the power, time, and location of stations and regulate the kind of apparatus to be used. Licenses were to be granted by the Commission for a limited duration (three years for broadcast licenses and five years for all others), but all federal government stations were to be assigned by the president.

Seven years later, the Communications Act of 1934 abolished the Federal Radio Commission and transferred the authority for spectrum management to the newly created Federal Communications Commission. The 1934 Act brought together the regulation of telephone, telegraph, and radio services within a single independent federal agency. The 1927 Radio Act was absorbed largely intact into Title III of the 1934 Act.

From 1934 to the early 1990s, the US Congress enacted many amendments to Title III, but there were no fundamental changes to the core provisions that can be traced back to the 1912 and 1927 acts. However, two noteworthy additions to the 1934 Act inserted in 1983 by Congress were:

1. that it is the policy of the United States "to encourage the provision of new technologies and services to the public" and that anyone who opposes a new technology or service will have the burden of demonstrating that the proposal is inconsistent with the public interest; and
2. "notwithstanding any licensing requirement established in this Act, [the FCC may] by rule authorize the operation of radio stations without individual licenses [in certain services]."

Therefore, new technologies will be promoted over existing services, and license by rule, or what is generally called unlicensed devices, are permissible.

National Telecommunications and Information Administration

The US Department of Commerce also has an important function in telecommunications regulation. Primarily through the NTIA, the DoC serves as the president's expert advisor on telecommunications matters and policy. NTIA is charged with reviewing policy options on behalf of the Executive Branch and communicating proposed policy decisions to Congress. NTIA also manages and administers the portion of the RF spectrum that has been set aside for exclusive use by the federal government. NTIA is also responsible for coordinating the federal government's participation in the ITU WRCs and related national and international meetings.

State Department

The US State Department is the primary representative on foreign policy matters. Through its Economics Bureau Office of International Communications and Information Policy, the State Department represents the United States in international telecommunications forums, including bilateral and multilateral negotiations, and before international organizations, such as the ITU. The WRC delegation is led by the WRC ambassador, whose role is to help negotiate a unified US position at the conference. The US president typically confers the personal rank of ambassador in connection with this special mission for a period not exceeding six months.

2.4.6 US Telecommunications Policy: Keeping Pace with Technology

The regulatory methods have gone through several changes throughout the years. During the initial era of amplitude modulation (AM) radio, the numbers of systems were small, and interference avoidance was accomplished through frequency and spatial separation. Although the frequency range was limited to low-frequency (LF) through high-frequency (HF) bands, it was sufficient to accommodate the hundreds of stations. These systems included stationary transmitters and mobile receivers. Characterizing a typical receiver and using propagation models easily allowed for the computation of potential interference. These calculations worked very well because the transmitters did not move. The benefit to the system developers was that most of the regulatory requirements were placed on the transmitters, which were few in comparison to the millions of receivers. As long as the transmitter complied with the in-band and out-of-band emission limits, interference between systems was prevented. Out-of-band emissions and antenna placement were specifically controlled to ensure that the spectral and spatial distances between two potential interferers were sufficient.

The development of television and FM radio saw the numbers of systems grow, as did the number of frequency assignments needed for these systems. If still limited to the HF bands, then a spectrum crisis would have precipitated. But fortunately, development of new RF component technology allowed the operational frequency bands to also grow into the very high frequency (VHF) and eventually the ultra high frequency (UHF) bands. Because the same topology was still being used—a single transmitter with many receivers—the same type of computations could be made to ensure a relatively interference-free operation.

Then, in the 1970s and 1980s, transistor and integrated circuit technology reduced the cost of transmitters. This provided cost-effective transmitters for the mass market in the form of Citizens' Band (CB) radios, unlicensed devices, cellular mobile radios, and numerous other devices. Protocols were developed for the devices and the users in order to provide orderly sharing of the RF channels. The channels were not all the same bandwidth. Some were wideband to accommodate high information data rate systems. Most were quite narrow. However, the main complexity of transmitter mobility had to be addressed. Therefore, the calculations for potential interference were no longer static. Out of this new complexity came ideas for minimum operational distances between potential interferers. These placed bounds on the interference. Devices that might be close in spatial proximity had to rely on spectral separation, and devices close in spectral proximity had to rely on spatial separation.

Although technology continued to march forward, the radio portion of the US communication policy has remained mostly unchanged for 70 years. Therefore, current policy is based on the technology that existed for broadcasting: one transmitter to many receivers. The evolution and revolution of radio technology over the past 15 years has significantly challenged that basis. Although the exact values can be argued, the United States had an average of more than two receivers for every person in 1980. The advent of cellular telephony, data networking, and two-way paging has now added one transceiver or more for each and every person. But the trends indicate that it will not stop there. It can be argued that a model similar to what happened to computing is taking hold in telecommunications. The evolution of technology also incurred an evolution of computer use from a corporate site, to an office group, to an individual desk, to every home, to every individual, and now to embedded computing in many consumer household products. The largest growth occurred when computers changed from devices that required interaction with the primary user to devices that work in the background, such as a refrigerator, a TV, a phone, or an automobile. The same trend is occurring within the wireless communications environment. There are now multiple wireless devices per person within the first adopters, as more consumer products are being developed that use primarily unlicensed devices.

2.5 TELECOMMUNICATIONS POLICY AND TECHNOLOGY IMPACT ON REGULATION

Regulations based on static broadcast geometries cannot address the spatial, numeric, and spectral dynamics of future radio technology. Technologists must begin to address not only how to construct such new technologies, but also how to bring dynamics into the regulatory framework.

2.5.1 Basic Geometries

Four basic geometries affect the type of technical and social/economic issues that are addressed in wireless communications policy: fixed or mobile transmitters combined with fixed or mobile receivers:

- Fixed transmitter, mobile receiver(s)
- Fixed transmitter, fixed receiver(s)
- Mobile transmitter, fixed receiver(s)
- Mobile transmitter, mobile receiver(s)

Fixed Transmitter, Mobile Receiver(s)

Fixed transmitter, mobile receiver systems include broadcasting, radio position determination, and standard time and frequency signal services. Broadcasting comprises a large fraction of the consumer devices such as radio (AM, FM, TV, etc.). Radio position determination includes radio navigation and radio beaconing services such as a global positioning system (GPS). Standard time and frequency signal services includes WWVB, the National Institute of Standards and Technology (NIST) long-wave standard time signal, which continuously broadcasts time and frequency signals at 60 kHz. The carrier frequency provides a stable frequency reference traceable to the national standard. A time code is synchronized with the 60 kHz carrier and is broadcast continuously at a rate of 1 bit per second (bps). Emission-only devices, such as those for industrial, scientific, and medical (ISM) purposes, including microwave ovens, magnetic resonance equipment, and industrial heaters, are included in this category. The important feature of these systems is that there are small numbers of high-power transmitters at fixed and potentially known locations. The economic challenge is to put most of the complexity (i.e., cost) in the transmitter since the ratio of receivers to transmitters is more than one million to one.

The policy challenge with fixed transmitter, mobile receiver systems is to determine the allowable transmission parameters (power, location) that prevent interference at the receivers and provide for the potential of an economically viable business. The trade-offs for broadcasting services include the number of stations in a given region and the viability of the service (number of customers). The trades are exceptionally complex. The station density could be increased by reducing transmission power and greater frequency reuse. However, that would decrease the coverage area and thus the number of potential customers. The station density could be increased by using closer band spacing between stations. However, that would increase the out-of-band rejection by the receivers.

Fixed Transmitter, Fixed Receiver(s)

Fixed transmitter, fixed receiver systems include point-to-point, point-to-multipoint, and radio astronomy services. Both endpoints are in fixed locations and could either be a one-way (transmitter to receiver) or a two-way (transceiver to transceiver) configuration. A point-to-point communication system is defined as having two fixed transceivers. Private operational-fixed microwave may use an operational-fixed station, and only for two-way communications related to the licensee's commercial, industrial, or safety operations. Point-to-multipoint includes multipoint distribution systems (MDSs) and multichannel, multipoint distribution systems (MMDSs) that are generally used for one-way data broadcasting. Originally, the primary MMDS application was "wireless cable" to deliver TV programs. Advances in antenna development allowed for two-way digital subscriber link (DSL) applications to be implemented with MMDS.

Radio astronomy is the scientific study of celestial phenomena through measurement of the characteristics of radio waves emitted by physical processes occurring in space. The radio telescopes that are used for astronomical work are extremely large because the signal strength coming from the distant stellar objects is low and many of the frequencies that are observed are below 3 GHz. The challenge is in addressing the location of fixed receiver, radio astronomy systems that take decades to plan and construct. Originally located in places away from population centers to minimize the potential for interference from commercial systems, these systems now find themselves surrounded by population centers. The policymakers are essentially facing an issue of whether to keep "radio-free zones" around the telescopes or to find other means to provide interference-free operation.

Fixed transmitter, fixed receiver systems are the most straightforward to determine the transmission parameters to prevent interference with other systems. However, the complexity occurs with mobile transmitters interfering with these systems. Because the location of the fixed transceivers is generally unknown to mobile users, mobile transmitters can potentially interfere with a receiver operating close to the noise floor due to out-of-band emissions or lack of out-of-band rejection by the receiver. However, the policy trades are quite straightforward because the RF environment and the geometry are fixed.

Mobile Transmitter, Fixed Receiver(s)

Mobile transmitter, fixed receiver systems include monostatic active, as well as passive, meteorological and Earth exploration systems. These systems are mobile (either airborne or space based). In the passive sensing configuration, the operational area is unknown, as it has similar characteristics to radio astronomy. In the active sensing configuration, the transmission location is unknown but the receiver is colocated with the transmitter.

Mobile transmitter, fixed receiver systems generally employ extremely sensitive receivers. Due to the mobility of the transmitter, the geographic region impacted is fixed in size and moves with the transmitter. The policy challenges are that the amount of frequency needed for these systems is large, but the specified frequency use is very small. Also, the operational frequencies are specific to the physical attributes of the chemicals that are to be sensed. The sensitivity of the receivers also requires all adjacent channel systems to have an extremely low out-of-band emission. This is usually accomplished through guard bands. The challenge to the policymaker is to determine the relative values of consumer services compared with scientific investigation and Earth exploration. Those values provide input as to whether to find mechanisms to access the unused spectrum in one location while the sensor is operating in another location.

Mobile Transmitter, Mobile Receiver(s)

Mobile transmitter, mobile receiver systems include a wide range of mobile services, as well as portable unlicensed devices. These systems include radio telephony (e.g., cellular, personal communication system (PCS), wireless communication, and specialized mobile radio (SMR)) and private land mobile radio (PLMR) services. PLMR services are for state and local governments, and for commercial and nonprofit organizations to use for mobile and ancillary fixed communications to ensure the safety of life and property and to improve productivity and efficiency. Personal radio services include CB radios, Family Radio Service (FRS), and remote control. Unlicensed devices, also known as licensed-by-rule, license-free, or "Part 15" devices as denoted in the FCC rules, are included. The important feature of these systems is that there are potentially large numbers of moderate (100 W) and extremely large numbers of low-power (1 mW–1 W) transceivers that are mobile. Both the mixture of powers and the unknown geometries between receivers and transmitters make it impossible to provide absolute assurance of interference-free operation without specifying a minimum separation distance.

Mobile transmitter, mobile receiver systems involve by far the most complex geometries for policymakers to address. Currently, all computations for interference assume a minimum separation distance between devices (and it differs from device to device). The assumption is that these distances represent the space in which the user has full control and thus can directly impact the presence of interference. As with the mobile transmitter, fixed receiver systems, the mobility of the transmitters creates the uncertainty of the geometry between the transmitters and receivers. The mobility also creates spectral regions in time, space, and frequency that are not used. The policymaking challenge is to maximize the use of the spectrum, encourage the development of new technologies and services, and provide certainty (spectrum access, interference, etc.) for the service providers to encourage investment.

2.5.2 Introduction of Dynamic Policies

Currently, the government spectrum management rules in the United States are dynamic with respect to frequency. That is, the rules for particular spectrum-based services tend to differ based on where a device is authorized to operate in the RF spectrum. For example, licensed transmitters operating in the radio broadcasting bands from 88 to 108 MHz must conform to the FM broadcasting rules of Part 73 [5]. The frequencies of such transmitters can be changed only after lengthy regulatory review to ensure such changes will not potentially cause harmful interference. Cognitive radios can change frequency readily, in seconds or milliseconds. These devices must incorporate aspects of the governing rules or policies from each of the different spectral areas in which they might operate. New devices that incorporate wireless fidelity (WiFi) with mobile telephones and "roam" between wide area networks (WANs) and local area networks

(LANs) are examples of how multiple spectrum policies can be merged within a single device. The dynamics are quite limited, yet possible under government and industry policies. The capacity to adjust spectrum policies dynamically opens the new possibility of dynamic spectrum policies. These policies can be at the device level, by which operational envelopes can be downloaded and modified by either the regulatory agency or the primary license holder. The dynamic policies can also be at the system level, whereas the network policies that are used to optimize performance can now add the parameter of spectrum access and the associated policies for using a specific band, at a location, with the particular system load. Three operational dimensions of spectrum policy avail themselves to dynamics: *time*, *space*, and *interference*.

Time—An example of using the time dynamic in spectrum policy was exhibited in the early days of radio. Particular AM stations would cease transmission late at night and resume early the next morning. Time-based dynamics can be extended significantly from this example. One extension is to include scheduled/expected interactions that are quite predictable. These may include secondary market transactions, in which a secondary provider accesses the spectrum using a separate network. These also may include the flexible access of a band by the primary user for a different application, such as reusing a cell site to provide data telemetry to/from vending machines. A further extension of this concept, which would be more opportunistic in character and less predictable, yet reliable for both primary and secondary users, may include using the spectrum for a short time or within a very limited area. One example is microtransactions within the secondary market for such "spot" use. Another could be a noncooperative use of spectrum that is currently not in use. The opportunistic use would exhibit quick transactions that could be impractical for human intervention. Automated schemes will be used similar to those used in financial transactions on the New York Stock Exchange.

Space—Spatial dynamics are depicted in cases where the location of a device would determine its operational characteristics. One proposal for spatial dynamics includes the allowance of higher-power transmission of unlicensed devices in rural environs. Another proposal is the use of unlicensed devices in bands where the device is sufficiently far away from a UHF TV transmitter. Location sensing would be necessary for the first proposal. Signal strength sensing would be necessary for the second proposal. In either case, because the transmitters are stationary, the location information is static. Therefore, once the boundaries are determined through calculation or measurement, then these boundaries could be programmed into a device. However, extending the concept to avoid mobile transmitters creates additional complexities. The distance to mobile transmitters would be constantly changing, and thus more automated sensing and interference avoidance techniques would be required.

Interference—In contrast to the spatial and temporal dynamics, interference dynamics would need to understand not only its environment but also the impact of its own transmission on the surrounding environment. The capacity to accurately measure and model the environment would be needed. A significant amount of research and development has occurred over the past decade to improve the fidelity of simulation and modeling of RF propagation. Companies such as Remcom have products that are examples of those developments. Additionally, device technology has significantly reduced the cost of RF sensing while also improving in fidelity.

There are many specific applications of dynamic spectrum policies. In the case where dynamic policies overlay current static policies, the choice for the device designer is whether to provide those new capabilities at additional cost for each device. An example of this is the case of whether to use licensed spectrum, secondary market spectrum, or unlicensed devices. Licensed spectrum has an assured quality of spectrum access and interference but is associated with higher spectrum costs. Each of the other choices has less assurance of quality but at lower spectrum costs.

It is easy to expect that with dynamic policies an explosion of new sensing devices and cooperative networks can be developed. These will be aimed at providing cost-effective solutions for both licensed and unlicensed uses. The incorporation of more processing capacity within licensed and unlicensed devices will present system developers with a large number of choices to provide new services with variable quality to the consumer.

2.5.3 Introduction of Policy-Enabled Devices

All radio devices are policy-enabled devices at the base level—the policies (or constraints) are the physical capabilities of the radio (e.g., power output, frequency range, modes of operation). CRs offer the ability to create dynamic policy-enabled devices, or in other words, CRs can realize the dynamic usage of frequency bands on an opportunistic basis by identifying and using the underutilized spectrum. Policies that determine when spectrum is to be considered as an available opportunity and that define the possibilities of using these spectrum opportunities must be specified by the regulators but implemented into the devices. The ability to encode policy into a device, thus creating a *policy-enabled device*, can potentially have a profound impact at the policy level [6], provided the device's functionality as a radio remains trustable [7].

The technology for policy-enabled devices is derived from the development of the Semantic Web (SeW). SeW is a machine for creating syllogisms. Thus the promise of SeW is to allow third parties to combine assertions to discover things that are true but not specified directly. The research has focused on the development of the Resource Description Framework (RDF) and the Web Ontology Language (OWL). The RDF is a structured environment for representing information about resources in the World Wide Web (WWW). It is particularly intended for metadata about Web resources, such as the title, author, and modification date of a Web page, and so on. OWL can be used to explicitly represent the meaning of terms in vocabularies and the ontological relationships between those terms.[7] Challenges remain in the development of policy-enabled devices. The functionality of the policy-enabled device must be "trustable" in its own operation. There has been a great deal of work in the mathematics of trustable systems and in demonstrating the impact of a flawed (e.g., not trustable) system (see, e.g., Mitola [7]).

---

[7]Note that the eXtensible Markup Language (XML) can also be used to define policies, but devices require the same parser for correct interpretation. OWL is better for distributing policy information among different types of CR devices with possibly different interpretation engines.

If the ontology for the regulations for wireless communications could be developed, then devices would be able to directly instantiate a policy into software and check that policy: (1) for self-consistency, and (2) for consistency with other policies already in use.

Two profound changes for telecommunication policymaking can result from this technology: (1) policymakers can now include policies that are distinct in both space and context, and (2) policymakers can now look at policies that change with time. The latter is much more significant. The time dimension allows for policies to expire.

These new characteristics allow for more aggressive policy concepts. Current policymakers and policy regulators are extremely cautious because the rules they develop will be long-lived and affect millions of users. Thus, the rules are very conservative and address not only well-understood interference possibilities, but also rare and statistically remote interference possibilities. In fact, with the increased density and dynamic behavior of devices, it may be practically impossible (exponentially complex) to ensure interference-free operation. With policy-enabled devices, new policies could be tried for shorter periods of time to determine the impact of the new policies. This would be equivalent to a test of the spectrum access system.

The ability to have policies instantiated in devices that can change with respect to location and context is very desirable. Power limits that change with respect to signal density or proximity to high-usage areas could enable more cost-effective deployments of communications to rural and underserved areas. In the same way, devices in high-usage areas would have to be capable of addressing interference more robustly.

Policy-enabled devices allow for the implementation of dynamic policies. Such new capabilities for the radios afford new possibilities for the regulator.

2.5.4 Interference Avoidance

The central role of spectrum management is interference management. The design and operation of RF equipment, including communications and emitting noncommunications devices, are predicated on preventing and/or mitigating electromagnetic interference. In today's RF environment, interference generally limits the usable range of communications signals. Interference protection has always been a core responsibility of communication regulatory bodies such as the FCC. Section 303(f) of the US Communications Act of 1934 as Amended directs the FCC to promulgate regulations it deems necessary to prevent interference between stations, as the public interest shall require [8]. This is still a critical aspect of the FCC operation. The FCC's strategic plan for the years 2003 to 2008 includes as a spectrum-related objective the "vigorous protection against harmful interference" [9].

In 2002, the FCC created the Spectrum Policy Task Force (SPTF) to provide recommendations for future spectrum policy [8]. The SPTF determined that there are rising concerns that current interference management paradigms will not adequately meet future spectrum demands. Four basic challenges to spectrum management were outlined:

1. Many radio communications services have grown substantially in recent years.
2. Consumer demand for RF devices has exploded.

3. The technology of waveform flexibility in radios moved from a relatively small number of waveforms to widely varying signal architectures and modulation types for voice, video, data, and interactive services.
4. The use of rapidly advancing technology, such as SDR and CR, will continue to change the interference landscape.

For example, due to advances in digital signal processing and antenna technology, communications systems and devices are becoming more tolerant of interference through their ability to sense and adapt to the RF environment.

Additionally, the increased ability of new technologies to monitor their local RF environment and operate more dynamically than traditional technologies can provide new mechanisms for interference management. The predictive models used by government regulators and network operators can be updated, and perhaps eventually replaced, by techniques that take into account and assess actual, rather than predicted, interference.

2.5.5 Overarching Impact

Regulation of spectrum will undergo revolutionary changes in the near future, allowing less restricted, more flexible access to spectrum. Such flexible spectrum usage requires regulation to realize a more open spectrum. Policies that determine when spectrum is considered as opportunity and that define the possibilities of using these spectrum opportunities are to be specified. The advent of policies that change with time and space will allow greater access to the fallow spectrum as well as lower infrastructure costs. Lower infrastructure costs are obtained through the ability to have policies change when conditions are warranted. The example of rural areas being allowed to use a higher transmit power for unlicensed devices would reduce the number of access points that are needed to service an area. The optimal density of access points is related to the number of users within the footprint of the access point. Essentially a provider wants a specific number of users per access point. Therefore, current rules have a highly suboptimal number of users per access point in low-usage areas and a highly suboptimal number of access points per user in high-usage areas. The current set of rules allows the area covered by the access point to be reduced, but it does not allow the increase of power. Dynamic policies associated with the state of the RF environment could allow a more optimal design, and thus a lower infrastructure cost, to be used.

2.6 GLOBAL POLICY INTEREST IN COGNITIVE RADIOS

Since 2000, the interest in how technology has changed both in RF spectrum needs by the user community and RF spectrum regulatory methods has been heightened within the radio community. The push to data services within the context of wireless Internet access has been seen as a fundamental shift from the wireless services to wireless access. The spectrum needs manifested themselves with the move toward third-generation (3G) services by wireless service providers as well as the plethora of unlicensed wireless devices (e.g., WiFi, cordless phones) in consumer electronics.

These trends actually followed a significant increase of wireless communications development and usage within the federal users of the spectrum, such as the military. The inherent limitation of omnidirectional RF propagation limited the new uses of the spectrum to below 6 GHz.

The development of CR technology and application concepts, especially in DSA, has received significant interest worldwide. The greatest interest has been in the United States from both the DoD and the FCC.

2.6.1 Global Interest

The following sidebar provides a list of global regulatory activities with respect to CRs DSA. Most of the interest outside of the United States and United Kingdom has been investigatory. A great deal of activity has been ongoing within the ITU and the European Telecommunications Standards Institute (ETSI) in developing definitions, standards, and regulatory regimes for using this new technology.

ETSI is officially responsible for standardization of Information and Communication Technologies (ICT) within Europe, including telecommunications. ETSI has 688 members from 55 countries inside and outside Europe, including manufacturers, network operators, administrations, service providers, research bodies, and users. ETSI plays a major role in developing a wide range of standards and other technical documentation as Europe's contribution to worldwide ICT standardization.

2008 INTERNATIONAL REGULATORY COGNITIVE RADIO ACTIVITIES

Australia
- *Australian Communications and Media Authority (formerly ACA and ABA):* Vision 20/20 report on future communications—realizing the future of ubiquitous communications with wireless expected to have an increasingly central role with increased spectrum sharing and CR technologies.

Canada
- *Canadian Radio-Television and Telecommunications Commission*
- *Industry Canada: Spectrum Policy Framework for Canada*—Implementation of new technologies and new spectrum management concepts using recent innovations in wireless technology (e.g., CR and SDR). The department solicited comments in determining to what extent these technologies might increase the use and access of the RF spectrum in the future. Industry Canada, CTRC, and CRC targeting research in CR technologies.

European Union
- *Project Team 8 (PT8) (Postal and Telecommunications Administration, Electronic Communications Committee Working Group):* Tasked with developing a report on the regulatory structure needed to enable the introduction of new radio technologies. A particular focus will be increased opportunities to share spectrum.

- *EU Commission:* Published *A Forward-Looking Radio Spectrum Policy for the European Union: Second Annual Report* with key initiatives, including the implementation of flexible spectrum usage through the development of "smart" or cognitive radios. Such efforts will be funded under the EU Research and Technology Development (RTD) Framework Programme. Now developing CR technology under the E2R programme.

India

- *Telecom Regulatory Authority of India:* Consultation paper on *Issues Relating to Private Terrestrial TV Broadcasting Service* addresses alternative technologies. Comments received state that rural communications could use CR technologies to find, in situ, better bands for foliage penetration.

International Telecommunication Union (ITU)

- *Global Standards Collaboration (GSC):* GSC-10 (radio communication items) in 2005 issued a resolution (GSC-10/6) on Global Radio Standards Collaboration on Wireless Access Systems to encourage collaboration on measurement techniques and certification requirements for cognitive capabilities including DFS; GSC-10 also called for accelerated standards development for SDR and CR; GSC-12 recognized the importance of CR technology.

- *ETSI:* Considering the impact on SDR of the Radio & Telecommunications Terminal Equipment (R&TTE) Directive with respect to electromagnetic compatibility (EMC), radio characteristics, nonconforming software, security, and integrity issues due to potential failures of the software download process.

- *ITU-R WP1B, WP8A, and WP8F:* Complete study of SDR and CR systems and their applications in 2006. In 2007, recommended approval of dynamic frequency selection in 5 GHz band and acknowledged R&D efforts within Japan and the EU. In 2008, completed a draft proposal for "Cognitive Radio Systems in the Land Mobile Services."

- *World Radio Conference (WRC):* Topics for WRC-11 included investigation of international regulations pertaining to the introduction and penetration of CRs (Agenda Item 1.19).

Japan

- *Ministry of Internal Affairs and Communications (formerly MPHPT):* Promoting R&D of technologies for efficient spectrum inclusive of CR systems to search for unused spectrum. By 2008 has multiple R&D programs for using CR technology for commercial systems.

New Zealand

- *Ministry of Economic Development:* 2005 Review of Radio Spectrum Policy in New Zealand—the role of spectrum band managers may be reduced through the use of SDRs, New Zealand's physical isolation provides for an ideal position to be a testbed for new wireless technologies, "cognitive or smart" radios will eventually have the capacity to locate and use any unoccupied spectrum, and none of New Zealand's current licensing types is well suited for managing SDRs and CRs.

United Kingdom
- *Office of Communications (OFCOM):* Technology R&D program, initiated in 2005, studies flexible, multiprotocol, multiband CR systems. Near-term investigations into band-sharing technologies. A 2007 Consultation created new opportunities for CRs to use interleaved spectrum without causing interference.
- *Commission for Communications Regulation (COMREG), Ireland:* Issued the first CR/dynamic spectrum assessment test license.

United States
- *FCC:* Report and Order, ET Docket No. 00-47—*Authorization and Use of Software Defined Radios*; Vanu Corporation gets first approved SDR in 2004; Notice of Proposed Rulemaking, ET Docket No. 03-108—*Facilitating Opportunities for Flexible, Efficient, and Reliable Spectrum Use Employing Cognitive Radio Technologies*; Notice of Proposed Rulemaking, ET Docket No. 04-186—*Unlicensed Operation in the TV Broadcast Bands and Additional Spectrum for Unlicensed Devices Below 900 MHz and in the 3 GHz Band*; Strategic Plan 2006-11—Encouraging the development of new technologies (e.g., CR and dynamic frequency selection, and so on).

2.6.2 US Reviews of CRs for Dynamic Spectrum Access

In the United States, a number of federal institutions involved with spectrum regulation, usages, or oversight undertook a wide range of studies. These studies were broken into two categories: (1) analysis of the use of needs of the RF spectrum (GAO, DSB, NTIA, and Toffler Associates; and (2) investigations of changes needed in the regulation, policy, and management of the RF spectrum (FCC, CSIS, DoC).

Government Accountability Office

The US Government Accountability Office (GAO) exists to support the US legislature through evaluation and investigations of federal programs and policies. This provides additional information for the US Congress to make oversight, policy, and funding decisions. The GAO has conducted a series of studies into wireless communications policy. During the time from 2000 to 2005, the GAO conducted a series of studies on the uses and expected needs of the spectrum.

Due to the increasing demands for RF spectrum, the GAO was asked to examine whether future spectrum needs can be met given the current regulatory framework. The ensuing report [10] focused mostly on the spectrum management structure within the United States and the decision-making impediments caused by having two regulatory agencies focusing on separate constituents: federal users versus nonfederal users. The significant increase of shared bands between federal and nonfederal users requires a consensus between the two groups. However, the report indicated that SDRs and other advanced technologies can potentially alleviate many of the conflicts by making spectrum more plentiful through more efficient access.

In 2004, the GAO performed a study [11] into how agencies within the US government that access RF spectrum can use advanced technologies to improve their "spectrum efficiency." The study clearly acknowledges that technologies such as "software-defined cognitive radios can be adapted to operate in virtually any segment of spectrum and, in the future, may be able to adapt to real-time conditions and make use of underutilized spectrum in a given location and time." The report, however, also indicates that the current allocation system does not allow these technologies to operate. The study also indicates that the lack of knowledge of the operating environment is an impediment. The report concludes that new technologies, such as CR, exhibit exceptional promise for efficient spectrum utilization but that there are few regulatory requirements or incentives that will be necessary to employ these technologies.

Defense Science Board

The Defense Science Board (DSB) is a Federal Advisory Committee for the US Department of Defense staffed by technology and business experts, who conduct studies pertinent to the DoD. The military is the primary source of research and development (R&D) funding for advanced communications due to its changing requirements and the willingness to deploy expensive communications assets.

The DSB conducted two spectrum-related studies during 2000 to 2005 on the current and expected DoD needs for accessing the RF spectrum. The DSB reported in November 2000 [12] that the management of time, space, and modulation dimensions of the RF spectrum increases the use of the scarce RF spectrum resource. This control should share the spectrum, in real time, and thus be more efficient than fixed allocations. These real-time systems dynamically control the assigned frequencies to ensure communications. The report further finds that dynamic frequency assignment by radios that can sense the spectrum could present unknown problems in spectrum management. It further questions the technical issues concerning which technologies to develop and how to apply these new capabilities for spectrum sharing and dynamic frequency allocation and/or assignment.

In July 2003, the DSB addressed dynamic access to mobile networks in a report on wideband RF modulation [13]. In that report, the DSB viewed software radios as both a risk and an opportunity. The risk lies in the potential of developing and implementing protocols and waveforms that could not be supported by existing systems and command and control techniques. There is also an interference danger with radios that can change their configuration without knowledge of how the new transmission waveforms and frequencies would impact other deployed systems. Even with these risks, the report has a strong recommendation that the US DoD "increase and focus investment in flexible and adaptive agile wideband communications technologies to achieve necessary mission capabilities in a highly dynamic radio frequency communications environment." That is, invest in CRs.

National Telecommunications and Information Administration

Although the FCC is the official US regulatory body for RF spectrum, there are cases in which it does not have jurisdiction. By law, national security use of the spectrum is under the jurisdiction of the US president. The use of spectrum for national security has been applied broadly to include federal agencies such as the DoD, FAA, and Federal

Bureau of Investigation (FBI). The jurisdiction of the federal use of the spectrum has been delegated to the Assistant Secretary of Commerce for Communications and Information, who is also the director of NTIA. NTIA performs numerous studies to investigate the impact of new technology and the needs of the RF spectrum users within federal agencies.

Toffler Associates

In 2001, a study by Toffler Associates provided recommendations as to the next steps for spectrum policy in the United States [14]. The study focused primarily on management aspects of spectrum allocations and thus did not address specific technology impacts such as those from cognitive radio. However, the recommendations from this independent group do address overall issues such as developing a long-term strategy and how to conduct spectrum reallocations. A primary focus of the proposed strategy was to develop a framework that anticipated change and remained versatile. This includes the predication of the needs for future spectrum allocations as well as the impact of future technologies such as software defined and cognitive radios.

Federal Communications Commission

The FCC is the US regulatory body for the RF spectrum as per the US Communications Act. The FCC performs numerous studies on both RF spectrum needs and technical issues such as interference. In 2002, the FCC chartered a forward-looking study to investigate the changes occurring in technology and to recommend how the FCC should develop and implement spectrum policy; more specifically, the intent of the study was to "identify and evaluate changes in spectrum policy that will increase the public benefits derived from the use of radio spectrum. The creation of the Spectrum Policy Task Force initiated the first-ever comprehensive and systematic review of spectrum policy at the FCC [15]. A primary goal was to move from a reactive spectrum management model to one more in line with a proactive spectrum policy. The SPTF noted that technological advances, specifically in software defined and cognitive radio, are enabling both the need to change spectrum policy and the capacity to look at different paradigms to implement spectrum policy.

The SPTF concluded that technology improvements have shown that the capacity is limited by the regulatory means used to access the RF spectrum. That is, access, not technical efficiency, is the limiting factor for using the spectrum. SDR and CR technology enables accessing the spectrum in multiple dimensions, such as time, frequency, bandwidth, and power.

Traditional spectrum management techniques use models and measurements to predict the interaction between different radio transmitters and receivers. Therefore, operational parameters such as transmission power, out-of-band emission, and the size of guard bands are selected to ensure interference-free interoperation of devices. The technological advances in interference rejection and digital coding will require changes to these parameters if spectrum management is to keep pace with technology. Without those changes, the operational parameters will be too conservative and thus limit the capacity to use the spectrum efficiently. Additionally, the ability to monitor the RF environment and dynamically alter radio operation make the traditional predictive interaction model of management obsolete.

Because operational values can change with technology as well as be modified *in situ*, spectrum management will need to change. The current method of explicitly regulating transmission characteristics to prevent interference will need to move to a rights and responsibilities model in which interference is explicitly defined and the operational parameters must be set by the device accessing the spectrum to be within the interference limits.

The SPTF recommended 39 changes to promote more efficient spectrum policies and more intensive use of the RF spectrum. Recommendations in four broad areas have a direct impact on CRs:

- Allow for maximum feasible flexibility of spectrum use by both licensed and unlicensed users
- Adopt a more quantitative approach to interference management based on the concept of interference temperature
- Promote access to the spectrum through secondary market policies that encourage access for "opportunistic" devices and, where appropriate, permit easements for spectral overlays and underlays
- Promote spectrum access and flexibility in rural areas such as varying power levels, leasing mechanisms, and geographic licensing

The overarching impact that the recommendations provided was that the FCC move from the single-use model of spectrum use to more of a flexible-use model. In that model, the FCC would clearly define the rights and responsibilities for access to the RF spectrum and then let the users optimize and interact and/or trade. It was recommended that policymakers address the difficult challenge of defining the appropriate engineering metrics in which to enable such a policy. Such metrics (e.g., the interference metric, which is discussed in Section 2.4.1), could then be implemented directly within a CR.

Center for Strategic and International Studies

The Center for Strategic and International Studies (CSIS) is a private, bipartisan public policy research organization that provides insights and possible solutions to current and emerging issues. In 2002 to 2003, CSIS organized a commission to address spectrum management for the twenty-first century. The commission focused on four problems for US spectrum management: (1) lack of long-range plans, (2) lack of mechanisms to resolve disputes between spectrum management organizations, (3) increased challenges in negotiating international spectrum agreements, and (4) the risk to security and economic growth due to "lag in the development and use of new technologies" [16]. Although the commission addressed concerns for US policy, the fourth problem is relevant to all spectrum policy organizations.

The CSIS commission noted that the development of spectrum-based technologies is increasing rapidly. In general, these technologies are initially developed within the military communications communities and then migrate to the commercial and consumer markets. Thus, technology is providing new ways to allow more intensive use of the spectrum and could alleviate the perception of a spectrum shortage. CR technology enables the exploitation of gaps in transmission frequencies and usage times to allow such an increase of spectrum use. Additionally, such technology can be used to provide more robust behavior to interference. The commission was concerned that the license-centric spectrum management policies cannot accommodate such new capabilities.

The CSIS commission had multiple recommendations for improving spectrum management. Two recommendations addressed organizational structure through more oversight at the White House, with new National Security Council and National Economic Council positions and a spectrum advisory board. To address the impact of the pace of technological innovation, the commission also recommended the establishment of a research consortium to support the government and private sectors. This consortium would establish goals for research in spectrum innovations as well as provide a platform for resolving technical disputes that arise as technology changes.

US Department of Commerce

In June 2003, the US president issued a memorandum recognizing how the RF spectrum contributes to "significant innovation, job creation and economic growth." He then created a Spectrum Policy Initiative, chaired by the DoC, to develop recommendations for improving spectrum management policies and procedures [17, 18]. The subsequent report from the initiative indicates: "Given the increase in new and innovative radio communication systems seeking access to the spectrum, the most challenging issue is interference problems inherent in using the latest technologies." The report continues to address the challenges from technology by stating, "The unpredictable nature of … ingenuity is not to be solved—it is a reality to be embraced."

Many of the recommendations put forth by the initiative are for modifications to the structure of the spectrum management community. However, there are two recommendation areas that could significantly impact the development of CRs: a Spectrum Sharing Innovation Testbed and Spectrum Management Tools. The report acknowledges the need for more sharing between spectrum users is inevitable due to the increased need for spectrum and the available technology to enable such sharing. In fact, 54 percent of the spectrum allocations below 3 GHz are shared and 94 percent below 300 GHz are shared. The initiative report recommends the development of a Spectrum Sharing Innovation Testbed, which consists of spectrum that will be set aside exclusively to test and evaluate new methods for spectrum sharing.

The initiative also recognized that there will be a continued explosion of spectrum uses, and that spectrum managers need the proper tools to be effective and efficient. The employment of CRs was specifically mentioned as a technology the promise and limitations of which should be better understood by both the FCC and NTIA.

In November 2006 the DoC formed the Commerce Spectrum Advisory Committee (CSAC) to help the US administration in developing wise RF policies that will promote new technologies, expand consumer choices, and enhance first-responder capabilities. In 2007, the CSAC helped initiate the Spectrum Sharing Innovation Testbed and identified four specific technology and services to be developed: DSA, multi-antenna signal processing, airborne video, and mobile satellite services.

2.7 SUMMARY

By 2005, the policy community had embraced the utility of CRs from the vantage points of new applications as well as new policy regimes. Due to the lack of operational prototypes, the community has been in a preparatory phase in developing definitions and potential operation envelopes in which CRs may function. The greatest amount of interest is in the DSA applications because these are seen as opportunities to provide more services and new technologies without having to allocate new spectrum.

This chapter has presented a review of relevant technical and policy definitions. Key among them are the following:

- CRs consist of four new capabilities: (1) *flexibility*, to change waveform and configuration; (2) *agility*, to change the spectral band in which to operate; (3) *sensing*, to observe the state of the system; and (4) *networking*, to communicate between multiple nodes for aggregating capacity.
- The Global Standards Collaboration within the ITU is developing a definition of CRs as well as extensions to also cover policy-based and dynamic frequency selection radios.
- Currently, there are three basic frequency assignment methods: command and control, auctions, and protocols and etiquettes. Command and control has been employed the most, but the use of auctions has been growing in popularity since its inception in 1994. Protocols and etiquettes are generally applied for unlicensed or license-free devices.
- OSA provides a new mechanism to dynamically obtain frequency assignments through sensing open spectral regions and adapting frequency selection in a CR. This ability has been under development within the technical community since 1999 and within the regulatory community since 2002.
- Interruptible spectrum is a special subset of frequency assignment when aperiodic increase in spectral assignment is obtained by interruption of part of the band of another service.

This chapter has also addressed the policy-relevant issues that need to be addressed if CRs are to be used to their fullest capacity:

- Policy challenges for CRs include addressing nondeterminism of self-learning algorithms, verification/validation of software, and the impact of waveform flexibility on out-of-band receivers.
- DSA policy must address the question of the rights of the license holder to prevent unauthorized use by an opportunistic device.
- DSA uses waveform flexibility and frequency agility to optimize performance. Interference metrics to determine the impact of waveforms on receivers is needed to provide a quantitative method to limit and/or eliminate interference.
- Multiple mechanisms are employed to ensure security operation of a CR: equipment authorization, software certification, and in situ monitoring of a transmitter. Equipment authorization and software certification become more impractical as the number of operations and/or software states grows exponentially. In situ monitoring is a developing technology.

Finally, this chapter provided a short tutorial as to the relevant regulatory roles with CRs:

- US spectrum policy agencies are the National Telecommunications and Information Administration and the Federal Communications Commission.

- Four basic geometries determine the type of technical and economic issues that are addressed in communications policy: (1) fixed transmitter, mobile receiver(s); (2) fixed transmitter, fixed receiver(s); (3) mobile transmitter, mobile receiver(s); and (4) mobile transmitter, fixed receiver(s).

- Spectrum management policies can be selected to optimize system and network performance given current spectrum activity and interference properties in space and time.

- Significant global interest in CR technology and policy exists, including Japan, New Zealand, Australia, Canada, the United States, the International Telecommunications Union, India, the United Kingdom, and the European Union.

2.8 EXERCISES

2.1. Which three primary spectrum access mechanisms do regulators employ?

2.2. Explain the differences between opportunistic spectrum access and interruptible spectrum access.

2.3. What are the four capabilities embodied in CRs?

2.4. What are the three primary agencies involved in spectrum regulation in the United States?

REFERENCES

[1] Kolodzy, P., Dynamic Spectrum Policies: Promises and Challenges, *CommLaw Conspectus*, 2004.

[2] *www.ntia.doc.gov/osmhome/allochrt.pdf.*

[3] *www.eps.auckland.ac.nz/poster%20comp2007/chia001.pdf.*

[4] Lynch, R., Exploring Technology Frontiers—Is There a Network in Your Future? Keynote address, Dyspan Conference, Baltimore, November 9, 2005.

[5] 47 C.F.R. § 73.200-73.600, SubPart B FM Broadcast Stations.

[6] Berlemann, L., S. Mangold, and B. Walke, Policy-based Reasoning for Spectrum Sharing in Cognitive Radio Networks, *Proceedings of First IEEE International Symposium on New Frontiers in Dynamic Spectrum Access Networks (DySPAN)*, pp. 1–10, November 2005.

[7] Mitola III, J., Software Radio Architecture: A Mathematical Perspective," *IEEE JSAC*, 17(4):514-538, 1999.

[8] FCC Spectrum Policy Task Force Report; available at *www.fcc.gov/sptf*, 2002.

[9] See Federal Communications Commission Strategic Plan FY 2003–FY 2008; available at *www.fcc.gov/omd/strategicplan2003-2008.pdf.*

[10] US General Accountability Office, Comprehensive Review of US Spectrum Management with Broad Stakeholder Involvement Is Needed, GAO-03-277, January 2003.

[11] US General Accountability Office, Better Knowledge Needed to Take Advantage of Technologies That May Improve Spectrum Efficiency, GAO-04-666, May 2004.

[12] US Department of Defense, Report of the Defense Science Board Task Force on DoD Frequency Spectrum Issues, Washington, DC, 2000.

[13] US Department of Defense, Report of the Defense Science Board Task Force on Wideband Radio Frequency Modulation, Washington, DC, 2000.

[14] Kenney, S., J. O'Connor, and R. Szafranski, Creating the Future of Spectrum Allocation (contact through tofflerassociates@toffler.com), 2001.

[15] *www.fcc.gov/sptf/reports.html*.

[16] Galvin, R., and J. Schlesinger, *Spectrum Management for the 21st Century*, CSIS Press, 2003.

[17] US Department of Commerce, Spectrum Policy for the 21st Century—The President's Spectrum Policy Initiative: Report 1, DoC, Washington, DC, 2004.

[18] US Department of Commerce, Spectrum Policy for the 21st Century—The President's Spectrum Policy Initiative: Report 2, DoC, Washington, DC, 2004.

The Software-Defined Radio as a Platform for Cognitive Radio

3

Max Robert
Artemis Communications LLC
Washington, DC

Bruce A. Fette
General Dynamics C4 Systems
Scottsdale, Arizona

3.1 INTRODUCTION

This chapter explores both the hardware and software domains of software-defined radio (SDR). The span of information that this chapter covers is necessarily broad; therefore, it focuses on some of the aspects of hardware and software that are especially relevant to SDR design. Beyond their obvious differences, hardware and software analyses have some subtle differences. In general, hardware is analyzed in terms of its capabilities. For example, a particular radio frequency (RF) front end (RFFE) can transmit up to a certain frequency, a data converter can sample a maximum bandwidth, and a processor can provide a maximum number of million instructions per second (MIPS). Software, in contrast, is generally treated as an enabler. For example: (1) a signal-processing library can support types of modulation, (2) an operating system can support multithreading, or (3) a particular middleware implementation can support naming structures. Given this general form of viewing hardware and software, this chapter presents hardware choices as an upper bound on performance, and software as a minimum set of supported features and capabilities.

Cognitive radio (CR) assumes that there is an underlying system hardware and software infrastructure that is capable of supporting the flexibility demanded by the cognitive algorithms. In general, it is possible to provide significant flexibility with a series of tunable hardware components that are under the direct control of the cognitive software. In the case of a cognitive system that can support a large number of protocols and air interfaces, it is desirable to have a generic underlying hardware structure.

The addition of a series of generalized computing structures underlying the cognitive engine (CE) implies that the CE must contain hardware-specific knowledge. With this hardware-specific knowledge, the CE can then navigate the different optimization strategies that it is programmed to traverse. The problem with such knowledge is that a change in the underlying hardware would require a change in the cognitive engine's knowledge base. This problem becomes exacerbated when one considers porting the engine to other radio platforms. For example, there could be a research and development platform that is used to test a variety of cognitive algorithms. As these algorithms mature, it is desirable to begin using these algorithms in deployed systems. Ideally, one would just need to place the CE in the deployed system's management structure. However, if no abstraction was available to isolate the CE from the underlying hardware, the cognitive engine would need to be modified to support each new hardware platform. It is clear that an abstraction is desirable to isolate the CE from the underlying hardware. The abstraction of hardware capabilities for radio software architecture is a primary design issue.

SDR is more than just an abstraction of the underlying hardware from the application. SDR is a methodology for the development of applications, or *waveforms* in SDR parlance, in a consistent and modular fashion such that both software and hardware components can be readily reused from implementation to implementation. SDR also provides the management structure for the description, creation, and teardown of waveforms. In several respects, SDR offers the same capabilities supported by operating systems; SDR is actually a superset of the capabilities provided by an operating system. SDR must support a variety of cores, some of which may be deployed simultaneously in the same system. This capability is like a distributed operating system designed to run over a heterogeneous hardware environment, where heterogeneous in this context means not only general-purpose processors (GPPs), but also digital signal processors (DSPs), field-programmable gate arrays (FPGAs), and custom computing machines (CCMs). Furthermore, SDR must support the RF and intermediate frequency (IF) hardware that is necessary to interface the computing hardware with radio signals. This support is largely a tuning structure coupled with a standardized interface. Finally, SDR is not a generic information technology (IT) solution in the way that database management is. SDR deals explicitly with the radio domain. This means that context is important. This context is most readily visible in the application programming interface (API), but is also apparent in the strict timing requirements inherent to radio systems and the development and debugging complexities associated with radio design.

This chapter is organized as follows: Section 3.2 introduces the basic radio hardware architecture and the processing engines that will support the cognitive function. Section 3.3 discusses the software architecture of an SDR. Section 3.4 discusses SDR software design and development. At present, many SDRs utilize a Software Communications Architecture (SCA) as a middleware to establish a common framework for waveforms, and the SCA is covered in some detail in this section. Section 3.5 discusses applications as well as the cognitive functionality and languages that support cognitive software as an application. Section 3.6 discusses the development process for SDR software components. Section 3.7 then discusses cognitive waveform development. Finally, Section 3.8 presents a summary of the chapter.

3.2 HARDWARE ARCHITECTURE

The underlying hardware structure for a system provides the maximum bounds for performance. The goal of this section is to explore hardware for SDR from a radio standpoint. Figure 3.1(a) shows a basic radio receiver. As an example of the basic radio receiver architecture, Figure 3.1(b) shows a design choice made possible by digital signal-processing techniques, in which the sampling process for digital signal processing can be placed in any of several locations and still provide equivalent performance.

3.2.1 The Block Diagram

The generic architecture tour presented here traces from the antenna through the radio and up the protocol stack to the application.

RF Externals

Many radios may achieve satisfactory performance with an antenna consisting of a passive conductor of resonant length, or an array of conductors that yield a beam pattern. Such antennas range from the simple quarter-wavelength vertical to the multi-element Yagi and its wide bandwidth cousin, the log periodic antenna. Antennas used over a wide frequency range will require an antenna tuner to optimize the voltage standing wave radio (VSWR) and corresponding radiation efficiency. Each time the transceiver changes frequency, the antenna tuner will need to be informed of the new frequency. Either it will have a prestored table derived from a calibration process, and will adjust passive components to match the table's tuning recommendations, or it will sense the VSWR and adapt the tuning elements until a minimum VSWR is attained.

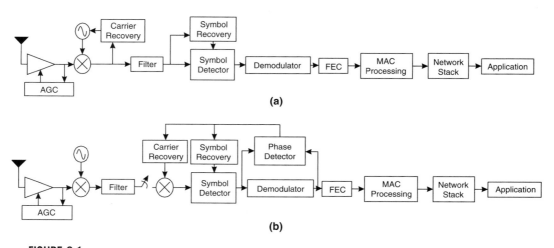

FIGURE 3.1

(a) Data flow and component structure of a generalized coherent radio receiver. (b) Data flow and component structure of a generalized coherent radio receiver designed for digital systems, with sampling as a discrete step; see Design Choices section (pp. 76–77) for an explanation.

Some modern antennas include a number of passive components spread over the length of the radiating elements that are able to present reasonable VSWR performance without active tuning. The best such units today span a frequency range of nearly 10:1. However, for radios that are expected to span 2 MHz to 2 GHz, it is reasonable to expect that the radio will need to be able to control a switch to select an appropriate antenna for the frequency band in which the transceiver is currently operating. Where beam antennas are used, it may also be necessary for the radio to be able to manage beam pointing. In some cases, this is accomplished by an antenna rotator, or by a dish gimbal. The logic of how to control the pointing depends greatly on the application; examples include:

- Exchanging the global positioning system (GPS) position of each transceiver in the network, as well as tracking the three-dimensional (3D) orientation of the platform on which the antenna is mounted so that the antenna pointing vector to any network member can be calculated.
- Scanning the antenna full circle to find the maximum signal strength.
- Dithering the antenna while tracking peak performance.
- Using multiple receive feed elements and comparing their relative strength.

Another common antenna is the electronically steered antenna. Control interfaces to these antennas are quite similar in function; however, due to their ability to rapidly steer electronically, many of these antennas update their steering angle as rapidly as once every millisecond. Thus, response time of the control interfaces is critical to the expected function.

The most sophisticated antenna is the multiple input, multiple output (MIMO) antenna. In these antennas, the interface boundary between the radio and the antenna is blurred by the complex beam-steering signal processing, and the large number of parallel RFFE receivers. For these complex antennas, the SDR Forum has developed interface recommendations that anticipate the wide variety of multi-antenna techniques currently used in modern transceivers.

Another common external component is the RF power amplifier (PA). Typically, the external PA needs to be told to transmit when the transceiver is in the transmit mode and to stop transmitting when the transceiver is in the receive mode. A PA will also need to be able to sense its VSWR, delivered transmit power level, and its temperature, so that the operator can be aware of any abnormal behavior or conditions.

It is also common to have a low noise amplifier (LNA) in conjunction with an external PA. The LNA will normally have a tunable filter with it. Therefore, it is necessary to be able to provide digital interfaces to the external RF components to provide control of tuning frequency, transmit/receive mode, VSWR and transmit power-level sensing, and receive gain control.

In all of these cases, a general-purpose SDR architecture must anticipate the possibility that it may be called on to provide one of these control strategies for an externally connected antenna, and so it must provide interfaces to control external RF devices. Experience has shown Ethernet to be the preferable standard interface, so that remote control devices for RF external adapters, switches, PAs, LNAs, and tuners can readily be controlled.

RF Front End

The RF front end consists of the receiver and the transmitter analog functions. The receiver and transmitter generally consist of frequency upconverters and downconverters, filters, and amplifiers. Sophisticated radios will choose filters and frequency conversions that minimize spurious signals, images, and interference within the frequency range over which the radio must work. The front-end design will also maximize the dynamic range of signals that the receiver can process, through automatic gain control (AGC). For a tactical defense application radio, it is common to be able to continuously tune from 2 MHz to 2 GHz, and to support analog signal bandwidths ranging from 25 kHz to 30 MHz. Commercial applications need a much smaller tuning range. For example, a cell phone subscriber unit might tune only 824 to 894 MHz and might need only one signal bandwidth. With a simplified design, the designer can eliminate many filters, frequency conversions, and IF bandwidth filters, with practical assumptions.

The RF analog front end amplifies and then converts the radio carrier frequency of a signal of interest down to a low intermediate frequency so that the receive signal can be digitized by an analog-to-digital (A/D) converter (ADC), and then processed by a digital signal processor to perform the modem function. Similarly, the transmitter consists of the modem producing a digital representation of the signal to be transmitted, and then a digital-to-analog converter (DAC) process produces a baseband or IF representation of the signal. That signal is then frequency shifted to the intended carrier frequency, amplified to a power level appropriate to close the communication link over the intended range, and delivered to the antenna. If the radio must transmit and receive simultaneously, as in full-duplex telephony, there will also be some filtering to keep the high-power transmit signal from interfering with the receiver's ability to detect and demodulate the low-power receive signal. This is accomplished by complex filters called diplexers, usually using bulk acoustic wave or saw filters at frequencies below 2 GHz, or yttrium-iron-garnet (YIG) circulators at frequencies above 2 GHz.

The typical SDR RF front end begins notionally with receiving a signal and filtering the signal to reflect the range of frequency covered by the intended signals. For a spread spectrum wideband code division multiple access (WCDMA) signal, this could be up to 6 MHz of bandwidth. In order to assure that the full 6 MHz is presented to the modem without distortion, it is not unusual for the ADC to digitize 12 MHz or so of signal bandwidth. In order to capture 12 MHz of analog signal bandwidth set by the IF filters without aliasing artifacts, the ADC will probably sample the signal at rates above 24 million samples per second (Msps). After sampling the signal, the digital circuits will shift the frequency of the RF carrier to be centered as closely as possible to 0-Hz direct current (DC) so that the signal can again be filtered digitally to match the exact signal bandwidth. Usually this filtering will be done with a cascade of several finite impulse response (FIR) filters, designed to introduce no phase distortion over the exact signal bandwidth. If necessary, the signal is despread, and then refiltered to the information bandwidth, typically with an FIR filter.

Analog-to-Digital Converters

The rate of technology improvement versus time has not been as profound for ADCs as for digital logic. The digital receiver industry is always looking for wider bandwidth and greater dynamic range. Successive approximation ADCs were replaced by flash converters in the early 1990s, and now are generally replaced with sigma–delta ADCs. Today's ADC can provide up to 105 Msps at 14-bit resolution. Special-purpose ADCs are able to provide sample rates over 5 Gsps at 8-bit resolution.

State-of-the-art research continues to push the boundaries of A/D performance with a wide variety of clever techniques that shift the boundaries between DSP and ADC.

Modem

After downconversion, filtering, and equalization, the symbols are converted to bits by a symbol detector/demodulator combination, which may include a matched filter or some other detection mechanism as well as a structure for mapping symbols to bits. A symbol is selected that most closely matches the received signal. At this stage, timing recovery is also necessary, but for symbols rather than the carrier. Then the output from the demodulator is in bits.

The bits that are represented by that symbol are then passed to the forward error-correcting function to correct occasional bit errors. Finally, the received and error-corrected bits are parsed into the various fields of message, header, address, traffic, etc. The message fields are then examined by the protocol layers eventually delivering messages to an application (e.g., Web browser or voice coder—vocoder), thus delivering the function expected by the user.

Software-defined radios must provide a wide variety of computational resources to be able to gracefully anticipate the wide variety of waveforms they may ultimately be expected to demodulate. Today, we would summarize that a typical SDR should be able to provide at least 266 MIPS of GPP, 32 Mbytes of random access memory (RAM), 100 MIPS of DSP, and 500 K equivalent gates of FPGA-configurable logic. More performance and resources are required for sophisticated waveforms or complex networking applications. Typically, the GPP will be called on to perform the protocol stack and networking functions, the DSP will perform the physical layer modulation and demodulation, and the FPGA will provide timing and control as well as any special-purpose hardware accelerators that are particularly unique to the waveform. It appears that SDR architectures will continue to evolve as component manufacturers bring forward new components that shift the boundaries of lowest cost, highest performance, and least power dissipation.

Forward Error Correction

In some instances, the demodulated bits are passed on to a forward error correction (FEC) stage for a reduction in the number of bit errors received. One of the interesting aspects of FEC is that it can be integrated into the demodulation process, such as in trellis-coded modulation; or it can be closely linked to demodulation, as in soft decoding for convolutional codes; or it can be an integral part of the next stage, medium access control (MAC) processing.

Medium Access Control

MAC processing generally includes framing information, with its associated frame synchronization structures, MAC addressing, error detection, link management structures, and payload encapsulation with possible fragmentation/defragmentation structures.

From this stage, the output is bits, which are input to the network processing layer. The network layer is designed for end-to-end connectivity support. The output of the network layer is passed to the application layer, which performs some sort of user functions and interface (speaker/microphone, graphical user interface, keypad, or some other sort of human–computer interface).

User Application

The user's application may range from voice telephony, to data networking, to text messaging, to graphic display, to live video. Each application has its own unique set of requirements, which, in turn, translate into different implications on the performance requirements of the SDR.

For voice telephony today, the dominant mode is to code the voice to a moderate data rate. Data rates from 4800 bps to 13,000 bps are popular in that they provide excellent voice quality and low distortion to the untrained listener. The G.729 standard is becoming particularly popular as an 8000-bps vocoder for Voice over Internet Protocol (VoIP) applications. The digital modem, in turn, is generally more robust to degraded link conditions than analog voice would be under identical link conditions.

Another criterion for voice communications is low latency. Much real experience with voice communications makes it clear that if the one-way delay for each link exceeds 50 ms,[1] then users have difficulty in that they expect a response from the far speaker and, hearing none, they begin to talk just as the response arrives, creating frequent speech collisions. In radio networks involving ad hoc networking, due to the delay introduced by each node as it receives and retransmits the voice signaling, it can be quite difficult to achieve uniformly low delay. Since the ad hoc network introduces jitter in packet delivery, the receiver must add a jitter buffer to accommodate a practical upper bound in latency of late packets. All of this conspires to add considerable voice latency. In response, voice networks have established packet protocols that allocate traffic time slots to the voice channels, in order to guarantee stable and minimal latency. In much the same way, video has both low error rate and fixed latency channel requirements, and thus networking protocols have been established to manage the quality requirements of video over wireless networks. Many wireless video applications are designed to accept the bit errors but maintain the fixed latency.

In contrast, for data applications, the requirement is that the data must arrive with very few or absolutely no bit errors; however, latency is tolerated in the application.

Voice coding applications are typically implemented on a digital signal processor. The common voice coding algorithms require between 20 and 60 MIPS and about 32 Kbytes of RAM on a DSP. Voice coding can also be successfully implemented on GPPs, and will typically require more than six times the instructions per second (100–600 MIPS) in order to perform both the multiply–accumulate signal-processing arithmetic and the address operand fetch calculations.

Transmitting video is nearly 100 times more demanding than voice, and is rarely implemented in GPP or DSP. Rather, video encoding is usually implemented on special-purpose processors due to the extensive cross-correlation required to calculate the motion vectors of the video image objects. Motion vectors substantially reduce the

---

[1] Some systems specify a recommended maximum latency limit, such as 150 ms for ITU-T G114.

number of bits required to faithfully encode the images. In turn, a flexible architecture for implementing these special-purpose engines is the use of FPGAs to implement the cross-correlation motion-detection engines.

Web browsing places a different type of restriction on an SDR. The typical browser needs to be able to store the images associated with each Web page in order to make the browsing process more efficient, by eliminating the redundant transmission of pages recently seen. This implies some large data cache function, normally implemented by a local hard drive. Recently, such memories are implemented by high-speed flash memory as a substitute for rotating electromechanical components, and flash memories up to 32 GB are now available.

Design Choices

Several aspects of the receiver shown in Figure 3.1(a) are of interest. One of the salient features is that the effect of all processing between the LNA and the FEC stage can be largely modeled linearly. This means that the signal-processing chain does not have to be implemented in the way shown in Figure 3.1(a). The carrier recovery loop does not have to be implemented at the mixer stage. It can just as easily be implemented immediately before demodulation. Another point to note is that no sampling is shown in Figure 3.1(a). It is mathematically possible to place the sampling process anywhere between the LNA and the FEC, giving the designer significant flexibility. An example of such design choice selection is shown in Figure 3.1(b), where the sampling process is shown as a discrete step.

The differences seen between Figures 3.1(a) and 3.1(b) are not at a functional level; they are implementation decisions that are likely to lead to a dramatically different hardware structure for equivalent systems. The following discussion on hardware concentrates on processing hardware selections because a discussion of RF and IF design considerations are beyond the scope of this book.

Several key concepts must be taken into consideration for the front end of the system, and it is worthwhile to briefly mention them here. From a design standpoint, signals other than the signal of interest can inject more noise in the system than was originally planned, and the effective noise floor may be significantly larger than the noise floor due to the front-end amplifier.

One example of unpredicted noise injection is the ADC conversion process, which can inject noise into the signal through a variety of sources. The ADC quantization process injects noise into a signal. This effect becomes especially noticeable when a strong signal that is not the signal of interest is present in an adjacent channel. Even though it will be removed by digital filtering, the stronger signal sets the dynamic range of the receiver by effecting the AGC, in an effort to keep the ADC from being driven into saturation. Thus, the effective signal-to-interference and noise ratio (SINR) of the received signal is lower than might otherwise be expected from the receiver front end. To overcome this problem and others like it, software solutions will not suffice, and flexible front ends that are able to reject signals before A/D sampling become necessary. Tunable RF components, such as tunable filters and amplifiers, are becoming available, and the SDR design that does not take full advantage of these flexible front ends will handicap the final system performance.

3.2.2 Baseband Processor Engines

The dividing line between baseband processing and other types of processing, such as network stack processing, is arbitrary, but it can be constrained to be between the sampling process and the application. The application can be included in this portion of processing, such as a vocoder for a voice system or image processing in a video system. In such instances, the level of signal processing is such that it may be suitable for specialized signal-processing hardware, especially in demanding applications such as video processing.

Four basic classes of programmable processors are available today: GPPs, DSPs, FPGAs, and CCMs.

General-Purpose Processors

General-purpose processors are the target processors that probably first come to mind to anyone writing a computer program. GPPs are the processors that power desktop computers and are at the center of the computer revolution that began in the 1970s. The landscape of microprocessor design is dotted with a large number of devices from a variety of manufacturers. These different processors, while unique in their own right, do share some similarities, namely, a generic instruction set, an instruction sequencer, and a memory management unit (MMU).

There are two general types of instruction sets: (1) machines with fairly broad instruction sets, known as complex instruction set computers (CISCs); and (2) machines with a narrow instruction set, known as reduced instruction set computers (RISCs). Generally, the CISC instructions give the assembly programmer powerful instructions that address efficient implementation of certain common software functions. RISC instruction sets, while narrower, are designed to produce efficient code from compilers. The differences between the CISC and RISC are arbitrary, and both styles of processors are converging into a single type of instruction set. Regardless of whether the machine is CISC or RISC, they both share a generic nature to their instructions. These include instructions that perform multiplication, addition, or storage, but these instruction sets are not tailored to a particular type of application. In the context of CR, the application in which we are most interested is signal processing.

The other key aspect of the GPP is the use of an MMU. Because GPPs are designed for generic applications, they are usually coupled with an operating system. This operating system creates a level of abstraction over the hardware, allowing the development of applications with little or no knowledge of the underlying hardware. Management of memory is a tedious and error-prone process, and in a system running multiple applications, memory management includes paging memory, distributed programming, and data storage throughout different blocks of memory. An MMU allows the developer to "see" a contiguous set of memory, even though the underlying memory structure may be fragmented or too difficult to control in some other fashion (especially in a multitasking system that has been running continuously for an extended period of time). Given the generic nature of the applications that run on a GPP, an MMU is critical because it allows the easy blending of different applications with no special care needed on the developer's part.

Digital Signal Processors

Digital signal processors are specialized processors that have become a staple of modern signal-processing systems. In large part, DSPs are similar to GPPs. They can be programmed with a high-level language such as C or C++ and they can run an operating system. The key difference between DSPs and GPPs comes in the instruction set and memory management. The instruction set of a DSP is customized to particular applications.

For example, a common signal-processing function is a filter, an example of which is the Direct Form II infinite impulse response (IIR) filter. Such a filter is seen in Figure 3.2.

As seen in Figure 3.2, the signal path is composed of delays of the internal state variable, z^{-1}, and the multiplication of each delayed sample by a coefficient of the polynomial describing either the poles (a) or zeros (b) of the filter. If each delayed sample is considered to be a different memory location, then to quickly implement this filter, it is desirable to perform a sample shift in the circular buffer, perform a multiply and an add that multiplies each delayed sample times the corresponding polynomial coefficients, and store that result either in the output register, in this case $y[n]$, or into the register that is added to the input, $x[n]$.

The algorithm in Figure 3.2 has several characteristics. Assuming that the filter is of length N (N-order polynomials), then the total computation cost for this algorithm can be computed. To optimize radio modem performance, filters are frequently designed to be FIR filters with only the b (all zeros) polynomial. In order to implement extremely efficient DSP architectures, most DSP chips support performing many operations in parallel to make an FIR filter nearly a single instruction that is executed in a one clock-cycle instruction loop. First, there is a loop control mechanism. This loop control has a counter that has to be initialized and then incremented in each operation, providing

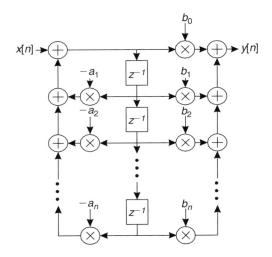

FIGURE 3.2

Structure and data flow for Direct Form II IIR filter as a basis for an estimate on computational load.

a set of $N + 1$ operations. Within the loop, there is also an evaluation of the loop control state; this evaluation is performed N times. Within the loop, a series of memory fetch operations have to be performed. In this case, there are $N + 1$ accesses to get the coefficient and move it to the multiplier.

There are also an additional $N + 1$ circular accesses of the sample data, yielding a total of $2N + 2$ memory accesses. Finally, there are the arithmetic operations: the coefficient multiplied by the signal sample, and the accumulation operation (including the initial zero operation) resulting in $2N + 1$ operations. Therefore, the DSP typically performs on the order of $6N$ operations per instruction clock cycle.

Assuming that a GPP performs a memory fetch, memory store, index update, comparison, multiplication, or addition in each computer clock cycle, then the GPP would require $6N + 3$ clock cycles per signal sample. Using these assumptions, then an FIR filter with 32 filter taps and a signal sampled at 100 ksps would yield $(6 \times 32 + 3) \cdot 100 \times 10^3 = 19.5$ MIPS. Therefore, just for the filter mentioned, almost 20 MIPS are required for the GPP to successfully filter this narrowband signal. In reality, a GPP is likely to expend more than one cycle for each of these operations. For example, an Intel Pentium 4 floating-point multiply occupies six clock cycles, so the given performance is a lower bound on the GPP MIPS load.

A DSP, in contrast, has the ability to reduce some of the cycles necessary to perform the given operations. For example, some DSPs have single-cycle MAC (multiple and accumulate). The Motorola 56001 is an example of a DSP that performs single instruction multiply and accumulation with zero overhead looping. These reductions result in a computational total of $N + 3$. Given these reductions, the computation load for the DSP is now $(32 + 3) \cdot 100 \times 10^3 = 3.5$ MIPS.

Given its customized instruction set, a DSP can implement specialized signal-processing functionality with significantly fewer clock cycles than the more generic GPP processors. However, GPPs are attempting to erode this difference using multiple parallel execution arithmetic logic units so that they can perform effective address calculations in parallel with arithmetic operations. These are called superscalar architectural extensions. They are also attempting to raise the performance of GPP multipliers through use of many pipeline stages, so that the multiplication steps can be clocked at higher speed. This technique is called a highly pipelined architecture.

Field-Programmable Gate Arrays

Field-programmable gate arrays are programmable devices that are different in nature from GPPs and DSPs. An FPGA comprises some discrete set of units, sometimes referred to as logic elements (LE), logic modules (LM), slices, or some other reference to a self-contained Boolean logical operation. Each of these logical devices has at least one logic unit; this logic unit could be one or more multipliers and one or more accumulators, or a combination of such units in some FPGA chip selections. Logical devices are also likely to contain some memory, usually a few bits. The developer then has some freedom to configure each of these logical devices, where the level of reconfigurability is arbitrarily determined by the FPGA manufacturer. The logical devices are set in a logic fabric, a reconfigurable connection matrix that allows the developer to describe connections between different logical devices. The logic fabric usually also has access to some additional memory that logical devices can share. Timing of the different parts of the FPGA can be controlled by establishing clock domains. This allows the implementation of multirate systems on a single FPGA.

To program an FPGA, the developer describes the connections between logical devices as well as the Boolean functionality of each of these logical devices. The final design that the developer generates is a circuit rather than a program in the traditional sense, even though the FPGA is ostensibly a firmware programmable device. Development for the FPGA is done by using languages such as VHSIC Hardware Design Language (VHDL), which can also be used to describe application-specific integrated circuits (ASICs), essentially nonprogrammable chips. Variants of C exist, such as System-C, that allow the developer to use C-like constructs to develop FPGA code, but the resulting program still describes a logic circuit running on the FPGA.

The most appealing aspect of FPGAs is their computational power. For example, the typical signal-processing FPGA can have anywhere from 1000 to 44,000 slices, where each slice is composed of two lookup tables, two flip-flops, some math, some logic, and some memory. To be able to implement 802.11a, a communications standard that is beyond the abilities of any traditional DSP in 2005, would require approximately 3000 slices, or less than 50 percent of the FPGA's capabilities, showing the importance of a high degree of parallelism in the use of many multiply accumulators to implement many of the complex waveform signal processes in parallel.

From a performance standpoint, the most significant drawback of an FPGA is that it consumes a significant amount of power, making it less practical for battery-powered handheld subscriber solutions. For example, an FPGA with about 9000 slices mentioned before is rated at slightly over 2 W of power expenditure, whereas a low-power DSP for handheld use is rated at 65 to 160 mW, depending on clock speed and version.

3.2.3 Baseband Processing Deployment

One of the problems that the CR developer will encounter when designing a system is attempting to determine what hardware should be included in the design, and what baseband processing algorithm to deploy into which processors. The initial selection of hardware will impose limits on maximum waveform capabilities, while the processing deployment algorithm will present significant runtime challenges if left entirely as an optimization process for the cognitive algorithm. If the processing deployment algorithm is not designed correctly, it may lead to a suboptimal solution that, while capable of supporting the user's required quality of service (QoS), may not be power efficient, quickly draining the system's batteries, or creating a heat dissipation problem.

The key problem in establishing a deployment methodology is one of scope. Once a set of devices and algorithm performance for each of these devices has been established, there is a finite set of possibilities that can be optimized. The issue with such optimization is not the optimization algorithm itself; several algorithms exist today for optimizing for specific values, such as the minimum mean square error (MMSE), maximum likelihood estimation (MLE), genetic algorithms, neural nets, or any of a large set of algorithms. Instead, the issue is in determining the sample set over which to perform the optimization.

There is no openly available methodology for establishing the set of possible combinations over which optimization can occur. However, at least one approach has been suggested by Neel et al. [1] that may lead to significant improvements in deployment optimization. The proposed methodology is partitioned into platform-specific and waveform-specific analyses. The platform-specific analysis is further partitioned into two types, DSP/GPP and FPGA. The platform-specific analysis is as follows:

1. Create an operations audit of the target algorithms (the number and the type of operations).
2. For DSP:
   a. Create a set of target devices.
   b. Establish cycle-saving capabilities of each target device.
3. For FPGA:
   a. Create a set of devices.
   b. Establish mapping between different FPGA families. This mapping can be done on the basis of logical devices, available multiplies per element, or another appropriate metric.
   c. Find logical device count for each target algorithm. FPGA manufacturers usually maintain thorough libraries with benchmarks.
   d. Use mapping between devices to find approximate target load on devices when a benchmark is not available.

Once the platform-specific analysis is complete, the developer now has the tools necessary to map specific algorithms onto a set of devices. Note that at this stage, there is no information as to the suitability of each of these algorithms to different platforms, since a base clock rate (or data rate) is required to glean that type of information.

Given the platform-specific information assembled earlier, it is now possible to create performance estimates for the different waveforms that the platform is intended to support:

1. Create a block-based breakdown of the target waveform (using target algorithms from step 1 of the platform-specific analysis as building blocks).
2. Break down target waveform into clock domains.
3. Estimate time necessary to complete each algorithm.
   a. In the case of packet-based systems, this value is fairly straightforward.[2]
   b. In the case of stream-based systems, this value is the allowable latency.
4. Compute number of operations per second (OPS) needed for each algorithm.
5. Create a set of devices from the platform-specific phase that meet area and cost parameters (or whatever other parameters are appropriate for a first cut). This set of devices can be very large. At this stage, the goal is to create a set of devices or combination of devices that meet some broad criteria.

---

[2]If the received signal is blocked into a block of many signal samples, and the receiver then operates on that block of signal samples through all of the receive signal processes, the process can be imagined to be a signal packet passing through a sequence of transforms. In contrast, if each new signal sample is applied to the entire receive process, it is referred to here as a stream-based process.

6. Cycle through the device set.
   a. Attempt to map algorithms onto given devices in sets.
      i. For DSP:
         (1) Make sure that OPS calculated in step 4 of the waveform-specific analysis are reduced by cycle-saving capabilities outlined in step 2b of the platform-specific analysis.
         (2) The result of the algorithm map is an MIPS count for each device.
      ii. For FPGA:
         (1) Mapping of the algorithm is a question of the number of occupied logical devices.
         (2) Make sure that clock domains needed for algorithms can be supported by the FPGA.
   b. If a solution set of MIPS and/or logical elements exists for the combination of devices in the set, then save the resulting solution set for this device set; if a solution does not exist, discard the device set.
7. Apply appropriate optimization algorithm over resulting solution set/device set from step 6. Additional optimization algorithms include power budgets and performance metrics.

The process described yields a coarse solution set for the hardware necessary to support a particular set of baseband processing solutions. From this coarse set, traditional tools can be used to establish a more accurate match of resources to functionality. Furthermore, the traditional tools can be used to find a solution set that is based on optimization criteria that are more specific to the given needs.

3.2.4 Multicore Systems and System-on-Chip

Even though several computing technologies have some promise in the future, such as quantum computing, it is an undeniable fact that silicon-based computing such as multicore systems and system-on-chip (SoC) will continue to be the bedrock of computing technology. Unfortunately, as technology reaches transistors under 100 nm, the key problems become the inability to continue the incremental pace of clock acceleration as well as significant problems in power dissipation. Even though the number of gates per unit area has roughly doubled every 18 months since the 1970s, the amount of power consumed per unit area has remained unchanged. Furthermore, the clocks driving processors have reached a plateau, so increases in clock speed have slowed significantly.

To overcome the technology problems in fabrication, a design shift has begun in the semiconductor industry. Processors are moving away from single-core solutions to multicore solutions, in which a chip is composed of more than one processing core. Several advantages are evident from such solutions. First, even though the chip area is increasing, it is now populated by multiple processors that can run at lower clock speeds. In order to understand the ramifications of such change, it is first important to recall the power consumption of an active circuit as:

$$P = \alpha \cdot C \cdot f \cdot V^2 \qquad (3.1)$$

As shown in Eq. (3.1), the power dissipated, P, by an active circuit is the product of the switching activity, α, the capacitance of the circuit, C, the clock speed, f, and the operating voltage, V, squared. It is then clear from Eq. (3.1) that the reduced clock speed results in a proportional reduction in power consumption. Furthermore, since a lower operating frequency means that a lower voltage is needed to operate the device, the reduction in the operating voltage produces a reduction in power consumption that follows a quadratic curve.

One of the principal bottlenecks in processor design is the input/output interface from data inside the chip to circuits outside the chip. This interface tends to be significantly slower than the data buses inside the chip. By reducing this interface capacitance and voltage swing for intercore communications, system efficiency grows. Furthermore, communication through shared memory is now possible within the chip. This capability can greatly increase the efficiency of the design.

3.3 SOFTWARE ARCHITECTURE

Software is subject to differences in structure that are similar to those differences seen in the hardware domain. Software designed to support baseband signal processing generally does not follow the same philosophy or architecture that is used for developing application-level software. Underlying these differences is the need to accomplish a variety of quite different goals. This section outlines some of the key development concepts and tools that are used in modern software-defined radio design.

3.3.1 Design Philosophies and Patterns

Software design has been largely formalized into a variety of design philosophies, such as object-oriented programming, component-based programming, or aspect-oriented programming. Beyond these differences is the specific way in which the different pieces of the applications are assembled, which is generally referred to as a design pattern. This section describes design philosophies first, providing a rationale for the development of different approaches. From these philosophies, the one commonly used in SDR will be expanded into different design patterns, showing a subset of approaches that are possible for SDR design.

Design Philosophies

Four basic design philosophies are used for programming today: linear programming (LP), object-oriented programming (OOP), component-based programming (CBP), and aspect-oriented programming (AOP).

Linear Programming

LP is a methodology in which the developer follows a linear thought process for the development of the code. The process follows a logical flow, so this type of programming is dominated by conditional flow control (such as "if-then" constructs) and loops. Compartmentalized functionality is maintained in functions, where execution of a function involves swapping out the stack, essentially changing the context of operation, performing the function's work, and returning results to the calling function, which requires an additional stack swap. An analogy of LP is creating a big box for all items on your desktop, such as the phone, keyboard, mouse, screen, headphone, can of soda, and picture of your attractive spouse, with no separation between these items. Accessing any one item's functionality, such as drinking a sip of soda, requires a process to identify the soda can, isolate the soda can from the other interfering items, remove it from the box, sip it, and then place it back into the box and put the other items back where they were. C is the most popular LP language today, with assembly development reserved for a few brave souls who require truly high speed without the overhead incurred by a compiler.

Object-Oriented Programming

OOP is a striking shift from LP. Whereas LP has data structures—essentially variables that contain an arbitrary composition of native types such as float or integer—OOP extends the data structure concept to describe a whole object. An object is a collection of member variables (such as in a data structure) and functions that can operate on those member variables. From a terminology standpoint, a class is an object's type, and an object is a specific instance of a particular class. There are several rules governing the semantics of classes, but they generally allow the developer to create arbitrary levels of openness (or visibility), different scopes, different contexts, and different implementations for function calls that have the same name. OOP has several complex dimensions; additional information can be found elsewhere (e.g., Budd [2] and Weisfeld [3]).

The differences inherent in OOP have dramatic implications for the development of software. Extending the analogy from the previous example, it is now possible to break up every item on your desktop into a separate object. Each object has some properties, such as the temperature of your soda, and each object also has some functions that you can access to perform a task on that particular object, such as drinking some of your soda. Several languages today can be classified as OOP languages. The two most popular ones are Java and C++, although there are several others.

Component-Based Programming

CBP is a subtle extension of the OOP concept. In CBP, the concept of an object is constrained; instead of allowing any arbitrary structure for the object, under CBP the basic unit is now a component. This component comprises one or more classes, and is completely defined by its interfaces and its functionality. Again extending the previous example, the contents on the desktop can now be organized into components. A component could be a computer, where the computer component is defined as the collection of the keyboard, mouse, display, and the actual computer case. This particular computer component has two input interfaces, the keyboard and the mouse, and one output interface, the display. In future generations of this component, there could be additional interfaces, such as a set of headphones as an output interface, but the component's legacy interfaces are not affected by this new capability.

Using CBP, the nature of the computer is irrelevant to the user as long as the interfaces and functionality remain the same. It is now possible to change individual objects within the component, such as the keyboard, or the whole component altogether, but the user is still able to use the computer component in the same manner as before. The primary goal of CBP is to create stand-alone components that can be easily interchanged between implementations. Note that CBP is a coding style, and there are no mainstream languages that are designed explicitly for CBP.

Even though CBP relies on a well-defined set of interfaces and functionality, these aspects are insufficient to guarantee that the code is reusable or portable from platform to platform. The problem arises not from the concept, but from the implementation of the code. To see the problem, it is important to now consider writing the code describing the different aspects of the desktop components that we discussed before, in this case a computer. Conceptually, we have a component that includes a display, keyboard, mouse, headphone, and computer. If one were to write software emulating each of these items, not only would the interfaces and actual functional specifications need to be written, but a wide variety of housekeeping functions would need to be included; for example, notification of failure. If any one piece of the component fails, it needs to inform the other pieces that it failed, and the other pieces need to take appropriate action to prevent further malfunctions. Such a notification is an inherent part of the whole component, and implementing changes in the messaging structure for this notification on any one piece requires all other pieces are informed of changes in state. These types of somewhat hidden relationships create a significant problem for code reuse and portability because relationships this complex need to be verified every time the code is changed. Aspect-oriented programming was designed to help resolve this problem.

Aspect-Oriented Programming

AOP allows for the creation of relationships between different classes. These relationships are arbitrary, but can be used to encapsulate the housekeeping code needed to create compatibility between two classes. In the messaging example, this class can include all messaging information needed for updating the state of the system. Given this encapsulation, a class such as the headphone in the ongoing example can be used not only in the computer example, but also in other systems, such as a personal music player, an interface to an airplane sound system, or any other appropriate type of system. The relationship class encompasses an aspect of the class; thus, context can be provided through the use of aspects. Unlike CBP, AOP requires the creation of new language constructs that can associate an aspect to a particular class; to this end, there are several languages (AspectJ, AspectC++, and Aspect#, among others).

Design Philosophy and SDR

The dominant philosophy in SDR design is CBP because it closely mimics the structure of a radio system, namely the use of separate components for the different functional blocks of a radio system, such as link control or the network stack. SDR is a relatively new discipline with few open implementation examples, so as the code base increases and issues in radio design with code portability and code reuse become more apparent, other design philosophies may be found to make SDR software development more efficient.

Design Patterns

Design patterns are programming methodologies used within the bounds of the language the developer happens to be using. In general, patterns provide two principal benefits: they help in code reuse and they create a common terminology. This common terminology is of importance when working on teams because it simplifies communications between team members. As will be shown in the next section, some architectures, such as the Software Communications Architecture (SCA), use patterns. For example, the SCA uses the factory pattern for the creation of applications. In the context of this discussion, patterns for the development of waveforms and the deployment of CEs will be shown. The reader is encouraged to explore formal patterns using available sources (e.g., Gamma et al. [4], Shalloway and Trott [5], or Kerievsky [6]).

3.4 SDR DEVELOPMENT AND DESIGN

From the previous discussion, it is clear that a software structure following a collection of patterns is needed for efficient large-scale development. In the case of SDR, the underlying philosophy, coupled with a collection of patterns, is called an architecture or operating environment. There are two open SDR architectures, GNURadio and SCA.

3.4.1 GNURadio

GNURadio [7] is a Python-based architecture that is designed to run on general-purpose computers running the Linux operating system. GNURadio is a collection of signal-processing components that supports primarily one RF interface, the universal software radio peripheral (USRP), a four-channel up- and downconverter board coupled with ADC and DAC capabilities. This board also allows the use of daughter RF boards. GNURadio, in general, is a good starting point for entry-level SDR and should prove successful in the market, especially in the amateur radio and hobbyist market. GNURadio does suffer some limitations—namely, (1) that it is reliant on GPP for baseband processing, thus limiting its signal-processing capabilities on any one processor, and (2) it lacks distributed computing support, limiting solutions to single-processor systems, and hence limiting its ability to support high-bandwidth protocols.

3.4.2 Software Communications Architecture

The other open architecture is the SCA, sponsored by the Joint Program Office (JPO) of the US Department of Defense (DoD) under the Joint Tactical Radio System (JTRS) program. The SCA is a relatively complex architecture designed to provide support for secure signal-processing applications running on heterogeneous, distributed hardware. Furthermore, several solutions are available today that provide support for systems using this architecture, some of which are available openly, such as Virginia Tech's OSSIE [8] or Communications Research Center's SCARI [9], providing the developer with a broad and growing support base.

The SCA is a component management architecture; it provides the infrastructure to create, install, manage, and de-install waveforms, as well as the ability to control and manage hardware and interact with external services through a set of consistent interfaces and structures. There are some clear limits to what the SCA provides. For example, the SCA does not provide such real-time support as maximum latency guarantees or process and thread management. Furthermore, the SCA does not specify how particular components must be implemented, what hardware should support what type of functionality, or any other deployment strategy that the user or developer may follow. The SCA provides a basic set of rules for the management of software on a system, leaving many of the design decisions up to the developer. Such an approach provides a greater likelihood that the developer will be able to address the system's inherent needs.

The SCA is based on some underlying technology to be able to fulfill two basic goals, namely, code portability and reuse. In order to maintain a consistent interface, the SCA uses Common Object Request Broker Architecture (CORBA) as part of its middleware. CORBA is software that allows a developer to perform remote procedure calls (RPCs) on objects as if they resided in the local memory space, even if they reside in some remote computer. The specifics of CORBA are beyond the scope of this book, but a comprehensive guide on the use of CORBA is available [10].

In recent years, implementations of CORBA have appeared on DSP and FPGA, but traditionally CORBA has been written for GPPs. Furthermore, system calls are performed through an operating system (OS), requiring an OS on the platform implementation. In the case of the SCA, the OS of choice is a POSIX PSE-52–compliant OS, but CORBA need not limit itself to such an OS. Its GPP-centric focus leads to a flow for the SCA, as seen in Figure 3.3.

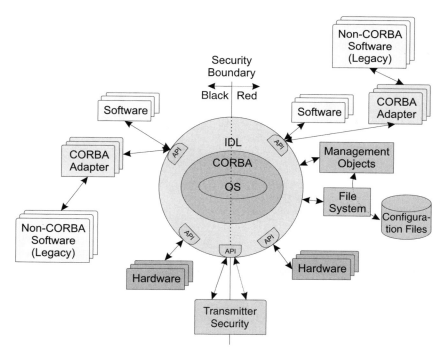

FIGURE 3.3

CORBA-centric structure for the SCA. The DoD implementation of the SCA requires that the RF modem (black) side of an SDR be isolated from the side where the plain text voice or data are being processed (red), and that this isolation be provided by the cryptographic device.

As seen in Figure 3.3, at the center of the SCA implementation is an OS, implying, but not requiring, the use of a GPP. Different pieces of the SCA are linked to this structure through CORBA and the interface definition language (IDL). IDL is the language used by CORBA to describe component interfaces, and is an integral part of CORBA. The different pieces of the system are attached together by using IDL. An aspect of the SCA not obvious in Figure 3.3 is that, since CORBA provides location independence to the implemention, there can be more than one processor at the core. Beyond this architecture constraint are the actual pieces that make up the functioning SCA system, namely SCA and legacy software, non-CORBA processing hardware, security, management software, and an integrated file system.

The SCA is partitioned into four parts: the framework, profiles, API, and waveforms. The framework is further partitioned into three parts: base components, framework control, and services. Figure 3.4 is a diagram of the different classes and their corresponding parts.

The SCA follows a component-based design, so the whole infrastructure revolves around the goal of creating, installing, managing, and de-installing the components making up a particular waveform.

Base Components

The base class of any SCA-compatible component is the Resource class. The Resource class is used as the parent class for any one component. Since components are described in terms of interfaces and functionality, not the individual makeup of the component,

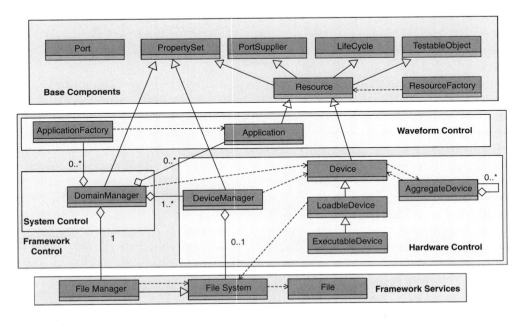

FIGURE 3.4

Classes making up the SCA core framework.

it follows that any one component comprises one or more classes that inherit from the Resource. For example, the developer may create a general filter category for components. In this example, the filter may be implemented as an FIR or IIR filter. These filters can then be further partitioned into specific filter implementations up to the developer's discretion, but for the purposes of this example, assume the developer chooses to partition the filters only into finite or infinite response times, rather than some other category such as structure (e.g., Butterworth, Chebyshev), or spectral response (e.g., low-pass, high-pass, notch). If the developer chooses to partition the filter implementation into FIR and IIR, then a possible description of the filter family of classes can be that shown in Figure 3.5.

The Resource base class has only two member methods, `start()` and `stop()`, and one member attribute, identifier. A component must do more than just start and stop, it must also be possible to set the component's operational parameters, initialize the component into the framework and release the component from the framework, do some sort of diagnostic, and connect this component to other components. In order to provide this functionality, Resource inherits from the following classes: PropertySet, LifeCycle, TestableObject, and PortSupplier, as seen in Figure 3.6.

PropertySet

The PropertySet class provides a set of interfaces that allow other classes to both configure and query the values associated with this Resource. As part of the creation process of the component, the framework reads a configuration file and sets the different values associated with the component through the PropertySet interface. Later on, during the

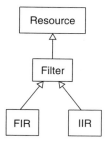

FIGURE 3.5

Sample filter component class hierarchy.

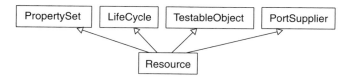

FIGURE 3.6

Resource class parent structure. The resource inherits a specialized framework functionality.

runtime behavior of the component, the Resource's properties can be queried through the `query()` interface and reconfigured with the `configure()` interface.

LifeCycle

The LifeCycle parent class provides the Resource with the ability to both initialize and release the component from the framework through the interfaces `initialize()` and `releaseObject()`. Initialization in this context is not the same as configuring the different values associated with the component. In this context, initialization sets the component to a known state. For example, in the case of a filter, the initialization call may allocate the memory necessary to perform the underlying convolution and it may populate the filter polynomial from the component's set of properties. The `releaseObject()` performs the complimentary function to initialize. In the case of a filter, it would deallocate the memory needed by the component. One aspect of `releaseObject()` that is common to all components is that it unbinds the component from CORBA. In short, `releaseObject()` performs all the work necessary to prepare the object for destruction.

TestableObject

Component test and verification can take various forms, so it is beyond the scope of the SCA to outline all tests that are possible within a component. However, the SCA does provide a simple test interface through the use of the TestableObject parent class. TestableObject contains a single interface, `runTest()`. `runTest()` takes as an input parameter a number signifying the test to be run and a property structure to provide some testing parameters, allowing the inclusion of multiple tests in the component. Even though the interface provides the ability to support multiple tests, it is fundamentally a black-box test structure.

PortSupplier

The final capability that parent classes provide to a Resource is the ability to connect to other components. Connections between components are performed through Ports (discussed later), not through the Resource itself. The use of PortSupplier allows the Resource to return one of any number of Ports that are defined within the component. To provide this functionality, the only interface provided by PortSupplier is `getPort()`. As its name implies, `getPort()` returns the port specified in the method's arguments.

ResourceFactory

The SCA provides significant latitude concerning how an instance of a Resource can be created. In its broadest context, a Resource can be created at any time by any entity before it is needed by the framework for a specific waveform. This wide latitude is not necessarily useful because sometimes the framework needs the Resource to be explicitly created. In these cases, the specifications provide the framework with the Resource-Factory class. The ResourceFactory class has only three methods, `createResource()`, `releaseResource()`, and `shutdown()`. The `createResource()` function creates an instance of the desired Resource and returns the reference to the caller, and the function `releaseResource()` calls the `releaseObject()` interface in the specified Resource. The shutdown function terminates the ResourceFactory. How the ResourceFactory goes about actually creating the Resource is not described in the specifications. Instead, the specifications provide the developer with the latitude necessary to create the Resource in whichever way seems best for that particular Resource.

Port

The Port class is the entry point and, if desired, the exit point of any component. As such, the only function calls explicitly listed in the Port definition are `connectPort()` and `disconnectPort()`. All other implementation-specific member functions are the actual connections, which, in the most general sense, are guided by the waveform's API. A component can have as many Ports as it requires. The implementation of the Port and the structure that is used to transfer information between the Resource (or the Resource's child) to the Port and back is not described in the specifications and is left up to the developer's discretion. Section 3.3.1 describes the development process and describes a few patterns that can be used to create specific components.

Framework Control

The base component classes are the basic building blocks of the waveform, which are relatively simple; the complexity in component development arrives in the implementation of the actual component functionality. When developing or acquiring a framework, the bulk of the complexity is in the framework control classes. These classes are involved in the management of system hardware, systemwide and hardware-specific storage, and deployed waveforms. From a high level, the framework control classes provide all the functionality that one would expect from an operating system other than thread and process scheduling.

From Figure 3.4, the framework control classes can be partitioned into three basic pieces: hardware control and system control. The following sections describe each of these pieces in more detail.

Hardware Control

As discussed in Section 3.2, a radio generally comprises a variety of different pieces of hardware. As such, the framework must include the infrastructure necessary to support the variety of hardware that may be used. From a software placement point of view, not all hardware associated with a radio can run software. Thus, the classes that are described to handle hardware can run on the target hardware itself, or they can run in a proxy fashion, as seen in Figure 3.7. When used as a proxy, the hardware control software allows associated specialized hardware to function as if it were any other type of hardware.

The needs of the hardware controller in the SCA are similar to the needs of components supporting a waveform. Furthermore, the component model closely fits the concept of hardware. Thus, the device controllers all inherit from the Resource base class. There are four device controllers: Device, LoadableDevice, ExecutableDevice, and AggregateDevice. Each of these classes is intended to support hardware with increasingly complex needs.

The most fundamental hardware is the hardware that performs some hardwired functionality and that may or may not be configurable. Such hardware has some inherent capacities that may be allocated for specific use. An example of such a piece of

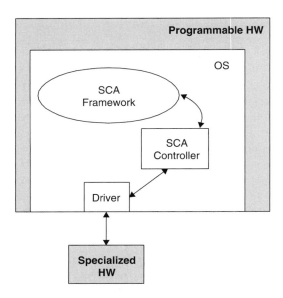

FIGURE 3.7

Specialized HW controlled by a programmable interface through a proxy structure.

hardware is an ADC. The ADC is not programmable, but it is conceivable for it to be tunable, meaning the developer may set the sampling rate or number of quantization bits. In such a case, the available capacity of this tunable hardware depends on whether it is being used for data acquisition of a particular signal or over a particular band. To this end, the only two functions that the Device class include are `allocateCapacity` and `deallocateCapacity`.

Beyond the simple ability to allocate and deallocate capacities, a particular piece of hardware may have the ability to load and unload binary images. These images are not necessarily executable code, but they are images that configure a particular piece of hardware. For example, an FPGA, once loaded with a bit image, is configured as a circuit. The FPGA is never "run," it just acts as a circuit. The LoadableDevice class was created to deal with such hardware, where LoadableDevice inherits from Device. As would be expected, the only two functions that the LoadableDevice class contains are `load()` and `unload()`.

Finally, there is the more complex hardware that not only has capacities that can be allocated and deallocated as well as memory that can be loaded and unloaded with binary images, but it also can execute a program from the loaded binary image. Such hardware is a GPP or a DSP. For these types of processors, the SCA uses the ExecutableDevice class. Much like an FPGA, a GPP or DSP has capacities that can be allocated or deallocated (like ports), and memory that can hold binary images, so the ExecutableDevice class inherits from the LoadableDevice class. As would be expected, the two functions that the ExecutableDevice class supports are `execute()` and `terminate()`.

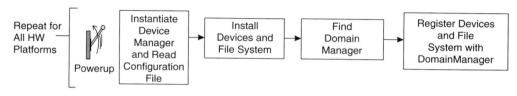

FIGURE 3.8

DeviceManager bootup sequence.

DeviceManager   The different hardware controllers behave as stand-alone components, so they need to be created and controlled by another entity. In the case of the SCA, this controller is the DeviceManager. The DeviceManager is the hardware booter; its job is to install all appropriate hardware for a particular box and to maintain the file structure for that particular set of hardware. The bootup sequence for the Device-Manager is fairly simple, as shown in Figure 3.8.

As seen in the figure, when an instance of the DeviceManager is created, it installs all Devices described in the DeviceManager's appropriate profile and installs whatever file system is appropriate. After installing the hardware and file system, it finds the central controller, in this case the DomainManager, and installs all the devices and file system(s).

In a distributed system with multiple racks of equipment, the DeviceManager can be considered to be the system booter for each separate rack, or each separate board. As a rule, a different DeviceManager is used for each machine that has a different IP address, but that general rule does not have to apply to every possible implementation.

Application Control

Two classes in the framework control section of the SCA provide application (wave-form) control: Application and ApplicationFactory. ApplicationFactory is the entity that creates waveforms; as such, ApplicationFactory contains a single function, `create()`.

ApplicationFactory   The `create()` function in the ApplicationFactory is called directly by the user, or the cognition engine in the case of a cognitive radio system, to create a specific waveform. The waveform's individual components and their relative connec-tions are described in an eXtensible Markup Language (XML) file, and the component's specific implementation details are described in another XML file. The different XML files that are used to describe a waveform in the SCA are described in the Profiles section.

Figure 3.9 shows an outline of the behavior of the ApplicationFactory. As seen in the figure, the ApplicationFactory receives the request from an external source. Upon receiving this request, it creates an instance of the Application class (discussed next), essentially a handle for the waveform. After the creation of the Application object, the ApplicationFactory allocates the hardware capacities necessary in all the relevant hard-ware, checks to see if the needed Resources already exist and creates them (through an entity such as the ResourceFactory) when they do not already exist, connects all the

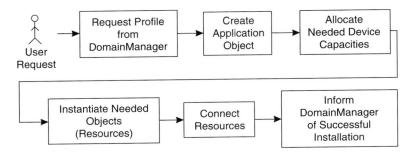

FIGURE 3.9

Simplified ApplicationFactory `create()` call for the creation of a waveform.

Resources, and informs the DomainManager (a master controller discussed in the following System Control section) that the waveform was successfully created. The other steps in this creation process, such as initializing the Resources, are not included in this description for the sake of brevity. Once the ApplicationFactory has completed the creation process, a waveform comprising a variety of connected components now exists and can be used by the system.

Application Class The ApplicationFactory returns an instance of the application class to the caller. The application class is the handle that is used by the environment to keep track of the application. The application class inherits from the Resource class, and its definition does not add any more function calls, just application identifiers and descriptors. The instance of the Application class that is returned by the ApplicationFactory is the object that the user would call start, stop, and releaseObject on to start, stop, and terminate the application, respectively.

System Control

In order for the radio system to behave as a single system, a unifying point is necessary. The specific nature of the unifying point can be as simple as a central registry, or as sophisticated as an intelligent controller. In the SCA, this unifying point is neither. The DomainManager, which is the focal point for the radio, is such an entity, and its task is as a central registry of applications, hardware, and capabilities. Beyond this registry operation, the DomainManager also serves as a mount point for all the different pieces of the hardware's system file. The DomainManager also maintains the central update channels, keeping track of changes in status for hardware and software and also informing the different system components of changes in system status.

The point at which the DomainManager is created can be considered to be the system bootup. Figure 3.10 shows this sequence. As seen in the figure, the Domain-Manager first reads its own configuration file. After determining the different operating parameters, the DomainManager creates the central file system, called a FileManager in the context of the SCA, reestablishes the previous configuration before shutdown, and waits for incoming requests. These requests can be DeviceManagers associating with the DomainManager, or new Applications launched by the ApplicationFactory.

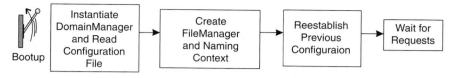

FIGURE 3.10

DomainManager simplified boot-up sequence.

In general, the bootup sequence of an SCA radio is partitioned into two different sets of steps, Domain bootup and one or more Device bootups. Once the platform has been installed (i.e., file system(s) and device(s)), the system is ready to accept requests for new waveforms. These requests can arrive either from the user or from some other entity, even a CE. Figure 3.11 shows a simplified bootup sequence for the different parts of the SCA radio.

The SCA is composed of a variety of profiles that describe the different aspects of the system. The collection of all files describes "the identity, capabilities, properties, interdependencies, and location of the hardware devices and software components that make up the system" [11] and is referred to as the Domain Profile. The Domain Profile is in XML, a language similar to the HyperText Markup Language (HTML), which is used to associate values and other related information to tags. XML is relatively easy for humans to follow and, at the same time, easy for machines to process.

The relationships among the different profiles are shown in Figure 3.12. As seen in the figure, there are seven file types: Device Configuration Descriptor (DCD), Domain-Manager Configuration Descriptor (DMD), Software Assembly Descriptor (SAD), Software Package Descriptor (SPD), Device Package Descriptor (DPD), Software Component Descriptor (SCD), and Properties Descriptor (PRF). In addition, there is the Profile Descriptor, a file containing a reference to a DCD, DMD, SAD, or SPD.

The profiles have a hierarchical structure split into three main tracks, all of which follow a similar pattern. In this pattern, there is an initial file for the track. This initial file contains information about that particular track as well as the names of other files describing other aspects of the track. These other files will describe their aspect of the system and, when appropriate, will contain a reference to another file, and so on. Three files point to the beginning of a track: DCD, DMD, and SAD. The DCD is the first file that is read by the DeviceManager on bootup, the DMD is the first file that is read by the DomainManager on bootup, and the SAD is the first file that is read by the ApplicationFactory when asked to create an application. Each of these files will contain not only information about that specific component, but one or more (in the case of the DMD, only one) references to an SPD. One SPD in this list of references contains information about that specific component. Any other SPD reference is for an additional component in that system: in the case of the DCD, a proxy for hardware; in the case of the SAD, another component in the waveform.

The SPD contains information about that specific component, such as implementation choices, and a link to both a PRF, which contains a list of properties for that component, and the SCD, which contains a list of the interfaces supported by that component. In the case of the DCD, an additional track exists to describe the actual

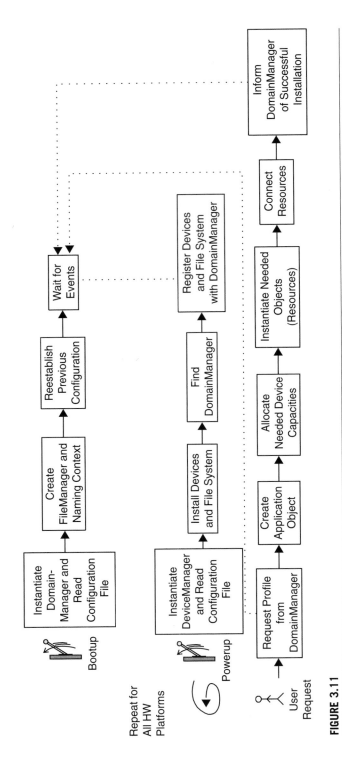

FIGURE 3.11

Simplified bootup sequence and waveform creation for SCA radio.

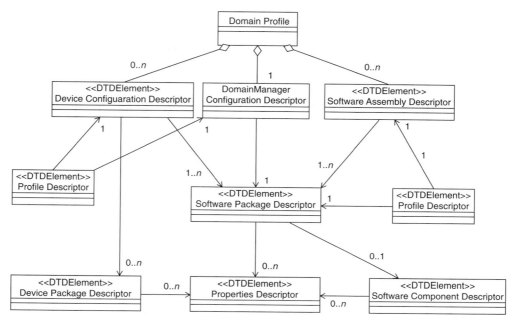

FIGURE 3.12

Files describing different aspects of an SCA radio.

hardware, the DPD. The DPD, much like the SPD, contains information about that specific component, in this case, the hardware side. The DPD, also mirroring the SPD, contains a reference to a PRF, which contains properties information concerning capacities for that particular piece of hardware.

Application Programming Interface

To increase the compatibility between components, it is important to standardize the interfaces. This requires standardizing the function names, variable names, and variable types, as well as the functionality associated with each component. The SCA contains specifications for an API to achieve this compatibility goal. The structure of the SCA's API is based on building blocks, where the interfaces are designed in a sufficiently generic fashion such that they can be reused or combined to make more sophisticated interfaces.

There are two types of APIs in the SCA: real time, shown as A in Figure 3.13, and non-real time, shown as B in Figure 3.13. Real-time information is both data- and time-sensitive control, whereas non-real time is control information, such as setup and configuration, which does not have the sensitivity of real-time functionality. An interesting aspect of the SCA API is that it describes interaction between different layers as would be described in the Open Systems Interconnection (OSI) protocol stack.

The API does not include interfacing information for intralayer communications. This means that in an implementation, the communications between two baseband process-

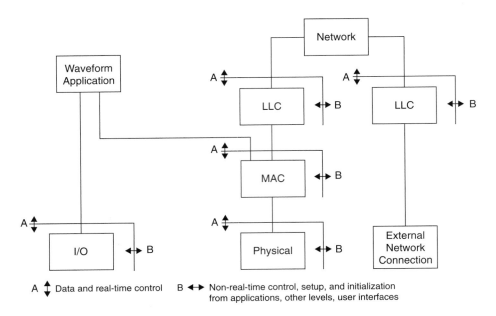

FIGURE 3.13

SCA application programming interface for interlayer communications. *Note:* I/O = input/output; MAC = medium access control; LLC = logical link control.

ing blocks, such as a filter and an energy detector, would be outside the scope of the specifications. This limitation provides a limit to the level of granularity that the SCA supports, at least at the level of the API. To increase the granularity to a level comparable to that shown in the example in Section 3.2, an additional API is required.

3.5 APPLICATIONS

In the following sections we deal with application software and the runtime environment on which the applications and the core framework will run.

3.5.1 Application Software

The application level of the system is where the CE is likely to operate. At this level, the CE is just another application in the system. The biggest difference between a cognitive application and a traditional application such as a Web browser is that the CE has some deep connections with the underlying baseband software. Unlike baseband systems, which are generally light and mainly comprise highly optimized software, application-level design has more leeway in design overhead, allowing the use of a rich and varied set of solutions that sometimes add significant performance overhead. Beyond the typical operating systems, such as products by Microsoft or products of the Linux family, and the typical toolkits, such as wxWindows for graphing, or OS-specific solutions such as Microsoft's DCOM, there exist three technologies that can have a significant impact on the application environment for CR: Java, BREW, and Python.

Java

Java is an object-oriented language developed by Sun Microsystems that allows the development of platform-independent software while also providing some significant housekeeping capabilities to the developer.

Fundamentally, Java is a compiled language that is interpreted in real time by a virtual machine (VM). The developer creates the Java code and compiles it into bytecodes by the Java compiler, and the bytecodes are then interpreted by the Java VM (JVM) running on the host processor. The JVM has been developed in some language and compiled to run (as a regular application) on the native processor, and hence should be reasonably efficient when it runs on the target processor. The benefit from such an approach is that it provides the developer the ability to "write once, run anywhere."

Java also provides some housekeeping benefits that can best be defined as dreamlike for an embedded programmer, the most prominent of which is *garbage collection*. Garbage collection is the JVM's ability to detect when memory is no longer needed by its application and to deallocate that memory. This way, Java guarantees that the program will have no memory leaks. In C or C++ development, especially using distributed software such as CORBA, memory management is a challenge, and memory leaks are sometimes difficult to find and resolve. Beyond memory management, Java also has an extensive set of security features that provide limits beyond those created by the developer. For example, Java can limit the program's ability to access native calls in the system, reducing the likelihood that malicious code will cause system problems.

Java has several editions, each tailored to a particular type of environment:

Java 2 Standard Edition (J2SE) is designed to run in a desktop environment, on servers, and on embedded systems. The embedded design of J2SE relates to reduced overhead inherent to Java, such as reduced memory or reduced computing power.

Java 2 Enterprise Edition (J2EE) is designed for large, multitier applications, as well as for the development of and interactions with Web services.

Java 2 Micro Edition (J2ME) is designed specifically for embedded and mobile applications. J2ME includes features that are useful in a wireless environment, such as built-in protocols and more robust security features.

Beyond these versions of Java, other features are available that can be useful for the radio developer. For example, under J2ME, there is a JavaPhone API. This API provides application-level support for telephony control, datagram messages (unreliable service over IP), power management, application installation, user profile, and address book and calendar. The JavaPhone API coupled with existing software components can reduce the development time associated with an application suite.

Java has been largely successful, especially in the infrastructure market. In such an environment, any overhead that may be incurred by the use of a real-time interpreter is overwhelmed by the benefit of using Java. For the embedded environment, the success of Java has been limited. There are three key problems with Java: memory, processing overhead, and predictability.

Memory: Most efforts to date in Java for the embedded world revolve around the reduction of the overall memory footprint. One such example is the Connected Limited Device Configuration (CLDC). The CLDC provides a set of API and a set of VM features for such constrained environments.

Processing Overhead: Common sense states that processing overhead is a problem with relevance that is likely to decrease as semiconductor manufacturing technology improves. One of the interesting aspects of processing overhead is that if the minimum supported system data rates for communications systems grow faster than processing power per milliwatt, processing overhead will become a more pressing problem. The advent of truly flexible software-defined systems will bear out this issue and expose whether it is in fact a problem.

Predictability: Predictability is a key real-time concern if the application software is expected to support such functionality as link control. Given that the executed code is interpreted by the JVM, asynchronous events such as memory management can mean that the execution time between arbitrary points of execution in the program can vary. A delay in execution can be acceptable because the design may be able to tolerate such changes. However, large variations in this execution time can cause such problems as dropped calls. A Real-Time Extension Specification for Java exists that is meant to address this problem.

Java is a promising language that has met success in several arenas. Efforts within the Java community to bring Java into the wireless world are ongoing and are likely to provide a constantly improving product.

Binary Runtime Environment for Wireless

Qualcomm created the Binary Runtime Environment for Wireless (BREW) for its CDMA phones. However, BREW is not constrained to just Qualcomm phones. Instead, BREW is an environment designed to run on any OS that simplifies development of wireless applications. BREW supports C/C++ and Java, so it is also largely language independent.

BREW provides a functionality set that allows the development of graphics and other application-level features. BREW is designed for the embedded environment, so the overall footprint and system requirements are necessarily small. Furthermore, it is designed for the management of binaries, so it can download and maintain a series of programs. This set of features is tailored for the handset environment, where a user may download games and other applications to execute on the handset but lacks the resources for more traditional storage and management structures.

Python

Python is an "interpreted, interactive, object-oriented programming language" [12]. Python is not an embedded language and is not designed for the mobile environment. It can, however, play a significant role in wireless applications, as evidenced by its use by GNURadio. What makes Python powerful is that it combines the benefits of object-oriented design with the ease of an interpreted language.

With an interpreted language, it is relatively easy to create simple programs that will support some basic functionality. Python goes beyond most interpreted languages by adding the ability to interact with other system libraries. For example, by using Python one can easily write a windowed application through the use of wxWidgets. Just as Python can interact with wxWidgets, it can interact with modules written in C/C++, making it readily extendable.

Python also provides memory management structures, especially for strings, that make application development simpler. Finally, because Python is interpreted, it is OS-independent. The system requires the addition of an interpreter to function; of course, if external dependencies exist in the Python software for features such as graphics, the appropriate library also has to exist in the new system.

3.6 DEVELOPMENT

Waveform development for sophisticated systems is a complex, multidesigner effort and well beyond the scope of this book. However, what can be approached is the design of a simple waveform. In this case, the waveform is designed using the SCA, bringing several of the concepts described thus far into a more concrete example. Before constructing a waveform, it is important to build a component.

3.6.1 Component Development

Several structures are possible for a component because it is defined only in terms of its interfaces and functionality. At the most basic level, it is possible to create a component class that inherits from both Resource and Port. In such a case, the component would function as a single thread and would be able to respond to only one event at a time. A diagram of this simple component is seen in Figure 3.14.

A more sophisticated approach is to separate the output ports into separate threads, each interfacing with the primary Resource through some queue. This approach allows the creation of fan-out structures while at the same time maintaining a relatively simple request response structure. A diagram of this fan-out structure is seen in Figure 3.15.

One of the problems with the component structure shown in Figure 3.15 is that it does not allow for input interfaces that have the same interface name; this limitation

FIGURE 3.14

Simple Resource and Port component. The component is both a Resource and a Port.

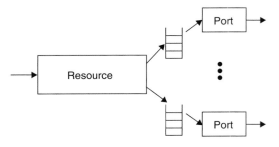

FIGURE 3.15

Fan-out structure for a component, the Resource uses output Ports but is the only entry point into the component.

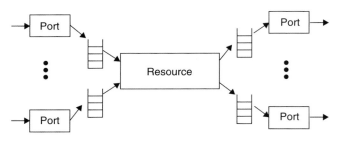

FIGURE 3.16

Fan-in, fan-out structure for a component. The Resource uses both input and output Ports.

increases the difficulty in using the API described in the specifications. To resolve this problem, a fan-in structure needs to be added to the system, even though this creates another level of complexity to the implementation. A way to implement this fan-in structure is to mimic the fan-out structure, and to place each input Port as a separate thread with a data queue separating each of these threads with the functional Resource. An example of this implementation is shown in Figure 3.16.

The approaches shown in Figures 3.14, 3.15, and 3.16 each present a different structure for the same concept. The specific implementation decision for a component is up to the developer, and can be tailored to the specific implementation. Because the components are described in terms of interfaces and functionality, it is possible to mix and match the different structures, allowing the developer even more flexibility.

3.6.2 Waveform Development

As an example of the waveform development, assume that a waveform is to be created that splits processing into three parts: baseband processing (assigned to a DSP), link processing (assigned to a GPP), and a user interface (assigned to the same GPP). A diagram of this waveform is shown in Figure 3.17.

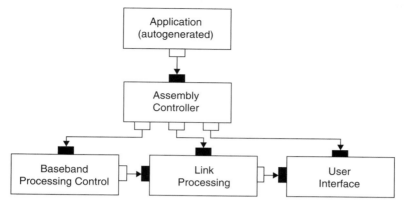

FIGURE 3.17

Simple SCA application. This example is made up of three functional components and one control component.

The application shown in the figure has several pieces that should be readily recognized; the application was generated by the framework as a result of the ApplicationFactory. Furthermore, the three proxies representing the relevant processing for the radio are also shown. Since the baseband processing is performed on a DSP, the component shown is a proxy for data transfer to and from the DSP. Link processing and the user interface, however, are actually implemented on the GPP. The assembly controller shown is part of the SCA, but not as a separate class. The assembly controller provides control information to all the different deployed components. Because the framework is generic, application-specific information, such as which components to tell to start and stop, must somehow be included. In the case of the SCA, this information is included in the assembly controller, a custom component the sole job of which is to interface the application object to the rest of the waveform.

The waveform described in Figure 3.17 does not describe real-time constraints; it assumes that the real-time requirements of the system are met through some other means. This missing piece is an aspect of the SCA that is incomplete.

3.7 COGNITIVE WAVEFORM DEVELOPMENT

In the context of SDR, a CE is just another component of a waveform, so the issue in CE deployment becomes one of component complexity. In the simplest case, a developer can choose to create components that are very flexible, an example of which is seen in Figure 3.18. The flexible baseband processor seen in this figure can respond to requests from the CE and alter its functionality. A similar level of functionality is also available in the link-processing component. The system shown in Figure 3.18 is a slight modification of the system shown in Figure 3.17. The principal difference in this system

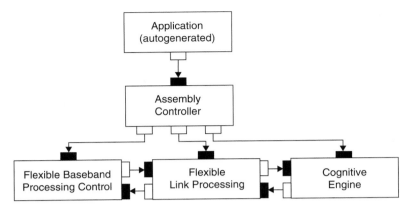

FIGURE 3.18

Simple cognitive waveform. The waveform performs both communications and cognitive functionality.

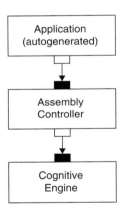

FIGURE 3.19

Cognitive engine waveform. The waveform performs only cognitive functionality. It assumes communications functionality is performed by other waveforms.

is the addition of a reverse-direction set of ports and changed functionality for each component.

The principal problem with the structure shown in Figure 3.18 is that it places all the complexity of the system onto each separate component. Furthermore, the tie-in between the CE and the rest of the waveform risks the engine implementation be limited to this specific system. An alternate structure is to create a whole waveform for which the only functionality is a CE, as seen in Figure 3.19.

The CE waveform shown in Figure 3.19 has no link to a waveform. To create this link, a link is created between the ApplicationFactory and the CE waveform. The cognitive engine can then request that the ApplicationFactory launch new waveforms. These new waveforms perform the actual communications work, which is evaluated by the functioning CE. If performance on the new waveform's communications link falls within some determined parameter, the CE can terminate the existing communications link waveform and request a new waveform (with different operating parameters or a different structure altogether) from the ApplicationFactory. This structure is seen in Figure 3.20.

One aspect of Figure 3.20 that is apparent is that the Port structure evident at the waveform level, as seen in Figure 3.19, scales up to interwaveform communications. Another aspect of this approach is that the cognitive waveform does not have to be colocated with the communications waveform. As long as timing constraints are met, the cognitive waveform can be placed anywhere within the network that has access to the system. This aspect of the deployment of the waveforms allows the concept of a CR within the SCA to easily extend to a cognitive network, where a single CE can control multiple flexible radios, as seen in Figure 3.21.

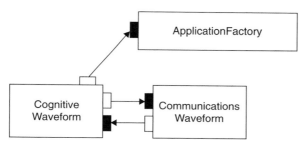

FIGURE 3.20

Multiwaveform cognitive support. The stand-alone cognitive waveform requests new waveforms from the SCA ApplicationFactory.

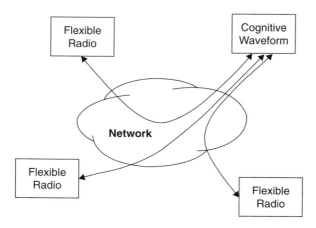

FIGURE 3.21

SCA-enabled cognitive network composed of multiple cognitive nodes.

With SDR, and an astute selection of processing and RF hardware on the part of a developer, it is possible to create highly sophisticated systems that can operate well at a variety of scales, from simple single-chip mobile devices all the way up to multitiered cognitive networks.

3.8 SUMMARY

Although a software-defined radio is not a necessary building block of a cognitive radio, the use of SDR in CR can provide significant capabilities to the final system. An SDR implementation is a system decision, in which the selection of both the underlying hardware composition and the software architecture are critical design aspects.

The selection of hardware composition for an SDR implementation requires an evaluation of a variety of aspects, from the hardware's ability to support the required signals to other performance aspects, such as power consumption and silicon area. Traditional approaches can be used to estimate the needs at the RF and data acquisition levels. At the processing stage, it is possible to create an estimate of a processing platform's ability to be able to support a particular set of signal-processing functions. With such an analysis, it is possible to establish the appropriate mix of GPPs, DSPs, FPGAs, and CCMs for a particular set of signal-processing needs.

In order to mimic the nature of a hardware-based radio, with components such as mixers and amplifiers, component-based programming is a natural way to consider software for SDR. In CBP, components are defined in terms of their interfaces and functionality. This definition provides the developer with significant freedom on the specific structure of that particular component.

Even though a developer may choose to use CBP for the design of an SDR system, a substantial infrastructure is still needed to support SDR implementations. This infrastructure must provide basic services, such as the creation and destruction of waveforms, as well as general system integration and maintenance. The goal of a software architecture is to provide this underlying infrastructure. The Software Communications Architecture is one such architecture. The SCA provides the means to create and destroy waveforms, manage hardware and distributed file systems, and manage the configuration of specific components.

Finally, beyond programming methodologies and architectures are the actual languages that one can use for development of waveforms and the specific patterns that are chosen for the developed software. These various languages have different strengths and weaknesses. C++ and Java are the dominant languages in SDR today. Python, a scripting language, has become increasingly popular in SDR applications, and is likely to be an integral part of future SDR development.

Much like the language selection, a design pattern for a particular component can have a dramatic effect on the capabilities of the final product. Design patterns that focus on flexibility can be more readily applied to cognitive designs, from the most basic node development all the way up to full cognitive networks.

REFERENCES

[1] Neel, J., J. H. Reed, and M. Robert, A Formal Methodology for Estimating the Feasible Processor Solution Space for a Software Radio, *Proceedings of Software Defined Radio Technical Conference and Product Exposition*, November 2005.

[2] Budd, T., *An Introduction to Object-Oriented Programming*, Third Edition, Addison-Wesley, 2001.

[3] Weisfeld, M., *The Object-Oriented Thought Process*, Second Edition, Sams, 2003.

[4] Gamma, E., R. Helm, R. Johnson, and J. Vlissides, *Design Patterns: Elements of Reusable Object-Oriented Software*, Addison-Wesley, 1995.

[5] Shalloway, A., and J. R. Trott, *Design Patterns Explained: A New Perspective on Object-Oriented Design*, Second Edition, Addison-Wesley, 2005.

[6] Kerievsky, J., *Refactoring to Patterns*, Addison-Wesley, 2005.

[7] *www.gnu.org/software/gnuradio/*.

[8] *http://ossie.mprg.org/*.

[9] *www.crc.ca/en/html/crc/home/research/satcom/rars/sdr/sdr*.

[10] Henning, M., and S. Vinoski, *Advanced CORBA Programming with C++*, Addison-Wesley, 1999.

[11] JTRS Joint Program Office, JTRS-5000, Software Communications Architecture Specification, SCA V3.0, August 2004.

[12] *www.python.org/*.

Cognitive Radio: The Technologies Required

John Polson
General Dynamics C4 Systems
Rhome, Texas

4.1 INTRODUCTION

Technology is never adopted for technology's sake. For example, only hobbyists used personal computers (PCs) until a spreadsheet program, a "killer" application, was developed. Then business needs and the benefits of small computers became apparent and drove PC technology into ubiquitous use. This led to the development of more applications, such as word processors, email, and more recently the World Wide Web (WWW). Similar development is under way for significant wireless communication devices.

Reliable cellular telephony technology is now in widespread use, and new applications are driving the industry. Where these applications go next is of paramount importance for product developers. Cognitive radio (CR) is the name adopted to refer to technologies believed to enable some of the next major wireless applications. Processing resources and other critical enabling technologies for wireless "killer" applications are now available.

This chapter presents a CR roadmap, including a discussion of cognitive radio technologies and applications. Section 4.2 presents a taxonomy of radio maturity, and Sections 4.3 and 4.4 present more detailed discussions. Sections 4.5 and 4.6 are about enabling and required technologies for CRs. They present three classes of cognitive applications, one of which may be the next "killer" application for wireless devices. Conjectures regarding the development of CR are included in Section 4.7 with arguments for their validity. Highlights of the chapter are discussed in the summary in Section 4.8, which emphasizes that the technologies required for CR are presently available.

4.2 RADIO FLEXIBILITY AND CAPABILITY

More than 40 different types of military radios (not counting variants) are currently in operation. These radios have diverse characteristics; therefore, a large number of examples can be drawn from the pool of military radios. This section presents the continuum of radio technology leading to the software-defined radio (SDR). Section 4.3 continues the continuum through to CR.

The first radios deployed in large numbers were "single-purpose" solutions. They were capable of one type of communication (analog voice). Analog voice communication is not particularly efficient for communicating information, so data radios became desirable, and a generation of data-only radios was developed. At this point, our discussion of software and radio systems begins. The fixed-point solutions have been replaced with higher data rate and voice-capable radios with varying degrees of software integration. This design change has enabled interoperability, upgradeability, and portability.

It will be clear from the long description of radios that follows that there have been many additional capabilities enabled in radios over their long history. SDRs and even more advanced systems have the most capabilities, and flexibility to add additional functions over time.

4.2.1 Continuum of Radio Flexibility and Capability

Basing CR on an SDR platform is not a requirement, but it is a practical approach at this time because SDR flexibility allows developers to modify existing systems with little or no new hardware development, as well as to add cognitive capabilities. The distinction of being a cognitive radio is bestowed when the level of software sophistication has risen sufficiently to warrant this distinction. Certain behaviors, discussed in this chapter, are needed for a radio to be considered a CR.

Historically, radios have been fixed-point designs. As upgrades were desired to increase capability, reduce life cycle costs, and so forth, software was added to the system designs for increased flexibility. In 2000, the Federal Communications Commission (FCC) adopted the following definition for software radios: "A communications device whose attributes and capabilities are developed and/or implemented in software" [1]. The culmination of this additional flexibility is an SDR system, as software-capable radios transitioned into software-programmable radios and finally became SDRs. The next step along this path will yield aware radios, adaptive radios, and finally cognitive radios (see Figures 4.1 and 4.2).

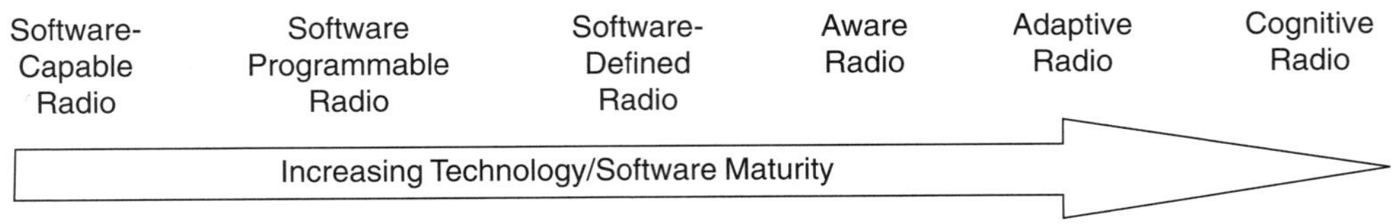

FIGURE 4.1

Software-Defined Radio Technology Continuum. As software sophistication increases, the radio system capabilities can evolve to accommodate a much broader range of awareness, adaptivity, and even the ability to learn.

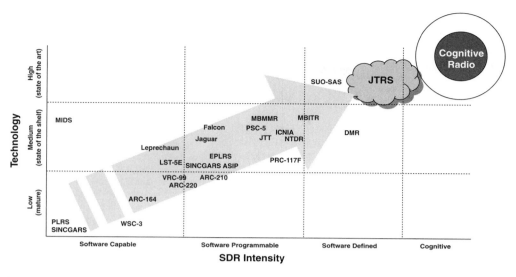

FIGURE 4.2

Examples of software radios. The most sophisticated software control and applications are reaching cognitive levels. (See Tables 4.1 through 4.3 for select descriptions.)[1]

4.2.2 Examples of Software-Capable Radios

Several examples of software-capable radios are shown in Figure 4.2 and detailed in Table 4.1. The common characteristics of these radios are fixed modulation capabilities, relatively small number of frequencies, limited data and data rate capabilities, and the ability to handle data under software control.

4.2.3 Examples of Software-Programmable Radios

Several examples of software-programmable radios are shown in Figure 4.2 and detailed in Table 4.2. The common characteristics of these radios are their ability to add new functionality through software changes and their advanced networking capability.

4.2.4 Examples of Software-Defined Radios

Only a few fully software-defined radio systems are available, as shown in Figure 4.2 and detailed in Table 4.3. The common characteristic of SDR systems is complete adjustability through software of all radio operating parameters.

Other products related to SDR include GNURadio and the Vanu Anywave Base Station. GNURadio is a free software toolkit that is available on the Internet. It allows

---

[1] Note that the radios that are part of Figure 4.2 and Tables 4.1 through 4.3 are based on the best available public information at the time this document was prepared, and are intended only to notionally indicate a distribution of some of the well-known radios. For additional, accurate information about capabilities, contact the respective manufacturers.

Table 4.1 Selected Software-Capable Radios

SINCGARS (Single-Channel Ground and Airborne Radio System)	A family of VHF-FM radios that provides a primary means of command and control. SINCGARS has frequency-hopping capability, and certain US Air Force versions operate in other bands using AM waveforms. The SINCGARS family of radios has the capability to transmit and receive voice and tactical data, and record traffic on any of 2320 channels (25 kHz) between 30 and 88 MHz. A SINCGARS with an Internet controller is software capable [2].
PLRS (Position Location Reporting System)	A command and control aide that provides real-time, accurate, 3D positioning, location, and reporting information for tactical commanders. The jam-resistant-UHF-radio transceiver network automatically exchanges canned messages that are used to geolocate unit positions to 15 meters accuracy. Commanders use PLRS for situational awareness. PLRS employs a computer-controlled, crypto-secured master station and an alternate master station to ensure system survivability and continuity of operations. The network, under master station management, automatically uses user units as relays to achieve OTH transmission and to overcome close-in terrain obstructions to LOS communications. When a rugged PC is used with the user unit, it becomes a mini–command and control station with a local area map superimposed with position and ID information. The computer interface and network control make PLRS a software-capable radio system [3].
AN/WSC-3	A DAMA SATCOM terminal. It meets tight size and weight integration requirements. This single-waveform radio has a computer (Ethernet) interface and is software capable [4].
AN/ARC-164 HaveQuick II	A LOS UHF-AM radio used for air-to-air, air-to-ground, and ground-to-air communications. ARC-164 radios are deployed on all US Army rotary wing aircraft and provide anti-jam, secure communications links for joint task force missions in the tactical air operations band. This radio operates as a single-channel (25 kHz) or a frequency-hopping radio in the 225–399.975 MHz band. The aircraft radio transmits at 10 W output power and can receive secure voice or data. The ARC-164 data-handling capability makes it software capable. It has an embedded ECCM antijam capability. The 243 MHz guard channel can be monitored. One model of the AN/ARC-164 is interoperable with SINCGARS [5].
AN/ARC-220	The standard HF (2.0–29.999 MHz) radio for US Army aviation. It includes secure voice and data communications on any of 280,000 frequencies. The system uses software-realized digital signal processing. Using MIL-STD-2217, software updates can be made over an MIL-STD-1553B bus. ARC-220 data-processing capabilities optionally include email applications. The ARC-220 is software capable. MIL-STD-148-141A ALE protocols improve connectivity on behalf of ARC-220 users. ALE executes in microprocessors and is particularly interesting as a cognitive application because of its use of a database and its sounding of the RF environment for sensing. GPS units may be interfaced with the radio, providing geolocation information [6].
AN/VRC-99	A secure digital network radio used for high data rate applications. This 1.2–2.0 GHz broadband direct-sequence spread-spectrum radio has NSA-certified high-assurance capabilities and provides users with 31 RF channels. It supports TDMA and FDMA. The networking capabilities are software programmable. AN/VRC-99 provides digital battlespace users with the bandwidth to support multimedia DTE of either an army C2V, army BCV, or a marine corps A3V with a single wideband secure waveform that supports voice, workstation, data, and imagery with growth for video requirements [7].
LST-5E	A software-capable UHF tactical SATCOM/LOS transceiver with embedded COMSEC. The LST-5E provides the user with a single unit for high-grade half-duplex secure voice and data over both wideband (25 kHz) AM/FM and narrowband (5 kHz) 1200 bps BPSK and 2400 bps BPSK. Current applications for the LST-5E include manpack, vehicular, airborne, shipborne, remote, and fixed stations. The terminal is compatible with other AM or FM radios that operate in the 225–399.995 MHz frequency band [8].

Table 4.1 Cont'd	
AN/PRC-6725, or Leprechaun	A handheld or wearable tactical radio. It can be programmed from a frequency fill device, laptop or PC, or system BS, or it can be cloned from another radio [9].
MBITR (Multiband Intra-/Inter-Team Radio)	Provides voice communications between infantry soldiers. The Land Warrior Squad Radio is a SINCGARS-compatible, eight-channel radio. MBITR is a software-capable design based on the commercially available PRC-6745 Leprechaun radio [10].
CSEL (Combat Survivor/Evader Locator)	Provides UHF communications and location for the purposes of Joint Search and Rescue Center operations. CSEL uses GPS receivers for geolocation. Two-way, OTH, secure beaconing through communication satellites allows rescue forces to locate, authenticate, and communicate with survivors/evaders from anywhere in the world. The handheld receivers are one segment of the CSEL command and control system. The satellite-relay base station, Joint Search and Rescue Center software suite, and radio set adapter units interface with the UHF satellite communications network to provide CSEL capability. Upgrades through software loads make CSEL radios software capable [11].
MIDS (Multifunction Information Distribution System)	A direct-sequence spread-spectrum, frequency-hopping, anti-jam radio that supports the Link 16 protocol for communication between aircraft. Operates in the band around 1 GHz. Originally called JTIDS, this radio waveform has been redeveloped on a new hardware platform, and is now being converted to a JTRS-compliant radio. When this conversion is completed it will be an SDR. It is currently in various stages of development and production as an interoperable waveform for Europe and the United States [12].

Table 4.2 Examples of Military Software-Programmable Radios	
AN/ARC-210	Provides fully digital secure communications in tactical and ATC environments (30–400 MHz). Additionally, the radio provides 8.33 kHz channel spacing capability to increase the number of available ATC frequencies, and provides for growth to new VHF data link modes. Currently, provided functions are realized through software and provide integrated communications systems with adaptability for future requirements with little or no hardware changes. In other words, the ARC-210 is software programmable. The ARC-210 is integrated into aircraft and operated via MIL-STD-1553B data bus interfaces. Remote control is also available for manual operation. This radio is interoperable with SINCGARS and HaveQuick radios [13].
Racal 25	A compliant Project 25 public safety radio operating in the 136–174 MHz band. Its 5 W peak transmit power, rugged and submersible housing, and digital voice with DES make it an advanced radio. The radio uses DSP and flash memory architectures, and supports 12 and 16 kbps digital voice and data modes. Racal posts software upgrades on a protected Internet site. Software programmability enables field upgrades and is possible due to its DSP and flash memory-based architecture [14].
SINCGARS ASIP (Advanced System Improvement Program)	Interoperates with SINCGARS radios and enhances operational capability in the TI environment. The ASIP models reduce size and weight and provide further enhancements to operational capability in the TI environment. SINCGARS ASIP radios are software programmable and provide improved data capability, improved FEC for low-speed data modes, a GPS interface, and an Internet controller that allows them to interface with EPLRS and BFA host computers. The introduction of IP routing enables numerous software capabilities for radio systems [2].

Table 4.2 Examples of Military Software-Programmable Radios—*Cont'd*	
EPLRS (Enhanced Position Location Reporting System)	An AN/TSQ-158 that transmits digital information in support of tactical operations over a computer-controlled, TDMA communications network. EPLRS provides two major functions: data distribution and position location and reporting. EPLRS uses a frequency-hopping, spread-spectrum waveform in the UHF band. The network architecture is robust and self-healing. Radio firmware may be programmed from external devices. The peak data rate of 525 kbps supports application layer radio-to-radio interaction [15].
AN/PRC-117F	A software-programmable, multiband, multimission radio. The PRC-117F operates in the 30–512 MHz band. Embedded COMSEC, SATCOM, SINCGARS, HaveQuick, and ECCM capabilities are standard. Various software applications (e.g., file transfer, TCP/IP, and digital voice) are included in this software-programmable radio [7].
Jaguar PRC-116	In service in more than 30 nations, including the UK (Army) and US (Navy). It is a frequency-hopping, software-programmable radio with considerable ECCM capability, resisting jamming by constantly shifting hopsets to unjammed frequencies. Security is further heightened by use of a scrambler. The Jaguar-V can also be used for data transmission at a rate of 16 kbps, and it may tolerate up to 50 radio nets, each with dozens of radios, at once; if each net is frequency-hopping in a different sequence, it will still transmit to all of them [16].
JTT (Joint Tactical Terminal)	A high-performance, software-programmable radio. Its modular functionality is backward- and forward-compatible with the IBS. Using a software download, JTT can accept changes in format and protocol as IBS networks migrate to a common format. Subsequent intelligence terminals require total programmability of frequency, waveform, number of channels, communication security, and transmission security, making subsequent terminals SDRs [17].
NTDR (Near-Term Digital Radio) System	A mobile packet data radio network that links TOCs in a brigade area (up to 400 radios). The NTDR system provides self-organizing, self-healing network capability. Network management terminals provide radio network management. The radios interface with emerging ABCS automated systems and support large-scale networks in mobile operations with efficient routing software that supports multicast operations. OTA programmability eliminates the need to send maintenance personnel to make frequency changes [18].
AN/PRC-138 Falcon	A human-pack or vehicular-mounted HF and VHF radio set. Capabilities include frequency coverage of 1.6–60 MHz in SSB/CW/AME and FM in the VHF band and 100 preset channels. In the data mode, the AN/PRC-138 offers a variety of compatible modem waveforms that allows it to be integrated into existing system architectures. Specific features include embedded encryption, ALE, embedded high-performance HF data modems, improved power consumption management, and variable power output [19].
MBMMR (Multiband, Multimission Radio)	An AN/PSC-5D(C) that enhances interoperability among SOF units. The MBMMR supports LOS and SCs voice and data in six basic modes: LOS, Maritime, HQ I/II, SINCGARS, SATCOM, and DAMA. Additional features include embedded Tactical Internet Range Extension and MELP voice coding. OTAR, extended 30 to 420 MHz band, MIL-STD-188 to 181B high data rate in LOS communications and SATCOM, and MIL-STD-188-184 embedded advanced data controller are supported [20].
MBITR (Multiband Intra-/Inter-Team Radio)	Designated AN/PRC-148, the MBITR provides AM/FM voice and data communications in the 30 to 512 MHz band. Development of the MBITR is an outgrowth of Racal's work on DSP and flash memory. MBITR is JTRS SCA 2.2-compliant and does not require a JTRS waiver. The INFOSEC capabilities are software programmable [21].

Table 4.3 Selected Examples of Software-Defined Radios	
SUO-SAS (Small Unit Operations—Situation-Awareness System)	Developed by DARPA to establish the operational benefits of an integrated suite of advanced communication, navigation, and situation awareness technologies. It serves as a mobile communications system for small squads of soldiers operating in restrictive terrain. SUO-SAS provides a navigation function utilizing RF ranging techniques and other sensors to provide very high accuracy [22].
DMR (Digital Modular Radio)	A full SDR capable of interoperability with tactical systems such as HF, DAMA, HaveQuick, and SINCGARS, as well as data link coverage for Link-4A and Link 11. These systems are programmable and include software-defined cryptographic functions. The US Navy is committed to migrating DMR to SCA compliance to allow the use of JTRS JPO-provided waveforms. The DMR may be reconfigured completely via on-site or remote programming over a dedicated LAN or WAN. The four full-duplex programmable RF channels with coverage from 2.0 MHz to 2.0 GHz require no change in hardware to change waveforms or security. The system is controlled, either locally or across the network, by a Windows-based HMI [23].
JTRS (Joint Tactical Radio System)	A set of current radio procurements for fully SDRs. These radio systems are characterized by SCA compliance that specifies an operating environment that promotes waveform portability. The JTRS JPO is procuring more than 9 increment one waveforms and more than 16 increment two waveforms that will ultimately be executable on JTRS radio sets. System packaging ranges from embeddable single-channel form factors to vehicle-mounted multichannel systems. JTRS radios are the current state-of-the-art technology and have the highest level of software sophistication ever embedded into a radio [24].

anyone to build a narrowband SDR. Using a Linux-based computer, an RF front end, and an analog-to-digital converter (ADC), one can build a software-defined receiver. By adding a digital-to-analog converter (DAC) and possibly a power amplifier, one can build a software-defined transmitter [25].

The Vanu Anywave Base Station is a software-defined system that uses commercial off-the-shelf (COTS) hardware and proprietary software to build a wireless cellular infrastructure. The goal is simultaneous support for multiple standards, reduced operating expenses, scalability, and future proofing (cost-effective migration) [26].

4.3 AWARE, ADAPTIVE, AND COGNITIVE RADIOS

Radios that sense all or part of their environment are considered aware systems. Awareness may drive only a simple protocol decision or may provide network information to maintain a radio's status as aware. A radio must additionally autonomously modify its operating parameters to be considered adaptive. This may be accomplished via a protocol or programmed response. When a radio is aware, adaptive, and learns, it is a cognitive radio [27]. Only cutting-edge research and demonstration examples of aware, adaptive, or CRs are available currently. Research and demonstration examples of aware, adaptive, and CRs are currently available and a few companies have products in various levels of production.

4.3.1 Aware Radios

A voice radio inherently has sensing capabilities in both audio (microphone) and RF (receiver) frequencies. When these sensors are used to gather environmental information, it becomes an aware radio. The local RF spectrum may be sensed in pursuit of channel estimates, interference, or signals of interest. Audio inputs may be used for authentication or context estimates or even natural language understanding or aural HMI interactions. Added sensors enable an aware radio to gather other information, such as chemical surroundings, geolocation, time of day, biometric data, or even network quality-of-service (QoS) measures. The key characteristic that raises a radio to the level of aware is the consolidation of environmental information not required to perform simple communications. Utilization of this information is not required for the radio to be considered aware. There is no communication performance motivation for developing this class of aware radios, and it is expected that this set will be sparse.

One motivation for an aware radio is providing information to the user. As an example, an aware radio may provide a pulldown menu of restaurants within a user-defined radius. The radio may gather this information, in the future, from low-power advertisement transmissions sent by businesses to attract customers. A military application may be a situational awareness body of information that includes a predefined set of radios and their relative positions. As an example, the radios exchange GPS coordinates in the background, and the aware radios gather the information for the user and provide it on request. The radio is not utilizing the information but is aware of the situation.

One example of an aware radio is the code division multiple access (CDMA)-based cellular system proposed by Chen et al. [28]. This system is aware of QoS metrics and makes reservations of bandwidth to improve overall QoS. Another example of an aware radio is the orthogonal frequency-division multiplexing (OFDM)-based Energy Aware Radio Link Control discussed by Bougard et al. [29].

4.3.2 Adaptive Radios

Frequency, instantaneous bandwidth, modulation scheme, error-correction coding, channel mitigation strategies (e.g., equalizers or RAKE filters), system timing (e.g., a TDMA structure), data rate (baud timing), transmit power, and even filtering characteristics are operating parameters that may be adapted. A frequency-hopped spread spectrum radio is not considered adaptive because once programmed for a hop sequence, it is not changed. A frequency-hopping radio that changes hop pattern to reduce collisions may be considered adaptive. A radio that supports multiple channel bandwidths is not adaptive, but a radio that changes instantaneous bandwidth and/or system timing parameters in response to offered network load may be considered adaptive. If a radio modifies intermediate frequency (IF) filter characteristics in response to channel characteristics, it may be considered adaptive. In other words, if a radio makes changes to its operating parameters, such as power level, modulation, frequency, and so on, it may be considered an adaptive radio.

At this time, two wireless products exhibit some degree of adaptation: the digital European cordless telephone (DECT) and 802.11a. DECT can sense the local noise floor and interference of all the channels from which it may choose. Based on this sensing capability, it chooses to use the carrier frequencies that minimize its total interference. This feature is built into hardware, however, and not learned or software adaptive; thus, DECT is not normally considered an adaptive radio.

802.11a has the ability to sense the bit error rate of its link, and to adapt the modulation to a data rate and a corresponding forward error correction (FEC) that set the bit error rate (BER) to an acceptably low error rate for data applications. Although this is adaptive modulation, 802.11 implementations generally are dedicated purpose fixed, application-specific integrated circuit (ASIC) chips, not software defined, and thus 802.11 is not normally considered to be an adaptive radio.

4.3.3 Cognitive Radios

A cognitive radio has the following characteristics: sensors creating awareness of the environment, actuators to interact with the environment, a model of the environment that includes state or memory of observed events, a learning capability that helps to select specific actions or adaptations to reach a performance goal, and some degree of autonomy in action.

Since this level of sophisticated behavior may be a little unpredictable in early deployments, and the consequences of "misbehavior" are high, regulators will want to constrain a CR. The most popular suggestion to date for this constraint is a regulatory policy engine that has machine-readable and interpretable policies.

Machine-readable policy-controlled radios are attractive for several reasons. One feature is the ability to "try out" a policy and assess its system performance. The deployment may be controlled to a few radios on an experimental basis, so it is possible to assess the observation and measurement of the behaviors. If the result is undesirable, the policies may be removed quickly. This encourages rapid decisions by regulatory organizations. The policy-driven approach is also attractive because spatially variant, or even temporally variant, regulations may be deployed. As an example, when a radio is used in one country, it is subject to that country's regulations, and when the user carries it to a new country, the policy may be reloaded to comply in the new jurisdiction. Also, if a band is available for use during a certain period but not during another, a machine-readable policy can realize that behavior.

The language being used to describe a CR is based on the assumption of a smart agent model, with the following capabilities:[2]

- Sensors creating awareness in the environment
- Actuators enabling interaction with the environment
- Memory and a model of the environment
- Learning and modeling of specific beneficial adaptations
- Specific performance goals
- Autonomy
- Constraint by policy and use of inference engine to make policy-constrained decisions

---

[2]A smart agent also has the ability to *not* use some or all of the listed capabilities.

The first examples of CR were modeled in the Defense Advanced Research Projects Agency (DARPA) NeXt Generation (XG) radio development program. These radios sense the spectrum environment, identify an unoccupied portion, rendezvous multiple radios in the unoccupied band, communicate in that band, and vacate the band if a legacy signal reenters that band. These behaviors may be modified as the radio system learns more about the environment; the radios are constrained by regulatory policies that are machine interpretable. The first demonstrations of these systems took place late in 2004 and in 2005 [30].

4.4 COMPARISON OF RADIO CAPABILITIES AND PROPERTIES

Table 4.4 summarizes the properties for the classes of advanced radios described in the previous sections. Classes of radios have "fuzzy boundaries," and the comparison shown in the table is broad. There are certainly examples of radios that fall outside the suggestions in the table. A cognitive radio may demonstrate most of the properties shown, but is not required to be absolutely reparameterizable. Note that the industry

Table 4.4   Properties of Advanced Radio Classes

Radio Property	Software-Capable Radio	Software-Programmable Radio	Software-Defined Radio	Aware Radio	Adaptive Radio	Cognitive Radio
Frequency hopping	X	X	X	X	X	X
Automatic link establishment (i.e., channel selection)	X	X	X	X	X	X
Programmable crypto	X	X	X	X	X	X
Networking capabilities		X	X	X	X	X
Multiple waveform interoperability		X	X	X	X	X
In-the-field upgradable		X	X	X	X	X
Full SW control of all signal-processing, crypto, and networking functionality			X	*	*	*
QoS measuring/channel state information gathering				X	X	X
Modification of radio parameters as function of sensor inputs					X	X
Learning about environment						X
Optimizing with different settings						X

The industry standards organizations are in the process of determining the details of what properties should be expected of aware, adaptive, and cognitive radios.

consensus is that a cognitive radio is not required to be a software-defined radio, even though it may demonstrate most of the properties of an SDR. However, there is also consensus that the most likely path for development of CRs is through enabling SDR technology.

4.5 AVAILABLE TECHNOLOGIES FOR COGNITIVE RADIOS

The increased availability of SDR platforms is spurring developments in cognitive radio. The necessary characteristics of an SDR required to implement a practical CR are excess computing resources, controllability of the system operating parameters, affordability, and usable software development environments including standardized application programming interfaces (APIs). This section discusses some additional technologies that are driving CR. Even though this is not a comprehensive list of driving technologies, it includes the most important ones.

4.5.1 Geolocation

Geolocation is an important CR-enabling technology due to the wide range of applications that may result from a radio being aware of its current location and possibly being aware of its planned path and destination.

The global positioning system (GPS) is a satellite-based system that uses the time difference of arrival (TDoA) to geolocate a receiver. An overview of this system is presented in Chapter 8. The resolution of GPS is approximately 100 m. GPS receivers typically include a one-pulse-per-second signal that is Kalman filtered as it arrives at each radio from each satellite, resulting in a high-resolution estimate of propagation delay from each satellite regardless of position. By compensating each pulse for the predicted propagation delay, the GPS receivers estimate time to approximately 340 nanoseconds (ns, or 10^{-9} sec) of jitter [31].

In the absence of GPS signals, triangulation approaches may be used to geolocate a radio from cooperative or even noncooperative emitters. Chapter 8 discusses the classical approaches of TDoA, time of arrival (ToA), and, if the hardware supports it, angle of arrival (AoA). Multiple observations from multiple positions are required to create an accurate location estimate. The circular error probability (CEP) characterizes the estimate accuracy.

4.5.2 Spectrum Awareness/Frequency Occupancy

A radio that is aware of spectrum occupancy may exploit this information for its own purposes, such as utilization of open channels on a noninterference basis. Methods for measuring spectrum occupancy are discussed in Chapters 5 and 19.

A simple sensor resembles a spectrum analyzer. The differences are in quality and speed. The CR application must consider the quality of the sensor in setting parameters, such as maximum time to vacate a channel upon use by an incumbent signal. It ingests a band of interest and processes it to detect the presence of signals above the noise floor. The threshold of energy at which occupancy is declared is a critical parameter. The detected energy is a function of the instantaneous power, instantaneous bandwidth, and duty cycle. Spectrum occupancy is spatially variant, time variant, and subject to observational blockage (e.g., deep fading may yield a poor observation). Therefore, a distributed approach to spectrum sensing is recommended.

The primary problem associated with spectrum awareness is the hidden node problem. A receive-only node (e.g., a television set) may be subjected to interference and may not be able to inform the CR that its receiver is experiencing interference. Regulators, spectrum owners, and developers of CR are working to find robust solutions to the hidden node problem including techniques discussed in detail in Chapters 18 and 23. Again, a cooperative approach may help to mitigate some of the hidden node problems.

In addition to knowing the frequency and transmit activity properties of a radio transmitter, it may also be desirable for the radio to be able to recognize the waveform properties and determine the type of modulation, thereby allowing a radio to request entry into a local network. Many books and papers have been published on this topic [32], and Chapter 18 of this book addresses this as well. Once the modulation is recognized, then the CR can choose the proper waveform and protocol stack to use to request entry into the local network.

4.5.3 Biometrics

A cognitive radio can learn the identity of its user(s), enabled by one or more biometric sensors. This knowledge, coupled with authentication goals, can prevent unauthorized users from using the CR. Most radios have sensors (e.g., microphones) that may be used in a biometric application. Voice print correlation is an extension to an SDR that is achievable today. Requirements for quality of voice capture and signal-processing capacity are, of course, levied on the radio system. The source radio can authenticate the user and add the known identity to the data stream. At the destination end, decoded voice can be analyzed for the purposes of authentication.

Other biometric sensors can be used for CR authentication and access control applications. Traditional handsets may be modified to capture the necessary inputs for redundant biometric authentication. For example, cell phones recently have been equipped with digital cameras. This sensor, coupled with facial-recognition software, may be used to authenticate a user. An iris scan, or retina scan, may also be feasible for some applications. Figure 4.3 shows some of the potential sensors and their relative strengths and weaknesses in terms of reliability and acceptability for biometric authentication [33].

4.5.4 Time

Included in many contracts is the phrase "time is of the essence," testament to the criticality of prompt performance in most aspects of human interaction. Even a desktop computer has some idea about what time it is, what day it is, and even knows how to utilize this information in a useful manner (date and time stamping information). A radio that is ignorant of time has a serious handicap in terms of learning how to interact and behave. Therefore, it is important for the CR to know about time, dates, schedules, and deadlines. Reasoning about time and schedules is discussed in Chapter 12.

Biometrics in Order of Effectiveness	Biometrics in Order of Social Acceptability
1. Palm scan	1. Iris scan[a]
2. Hand geometry	2. Keyboard dynamics
3. Iris scan	3. Signature dynamics
4. Retina scan	4. Voiceprint[b]
5. Fingerprint	5. Facial scan[a]
6. Voiceprint	6. Fingerprint[c]
7. Facial scan	7. Palm scan[c]
8. Signature dynamics	8. Hand geometry[c]
9. Keyboard dynamics	9. Retina scan[a]

[a] Requires a camera scanner.
[b] Uses a copy of the voice input (low impact).
[c] Requires a sensor in the push-to-talk (PTT) hardware.

FIGURE 4.3

Biometric sensors for CR authentication applications. Several biometric measures are low impact in terms of user resistance for authentication applications.

Time-of-day information enables time division multiplexing on a coarse-grained basis, or even a fine-grained basis if the quality of the time measurement is sufficiently accurate. Time-of-day information may gate policies in and out. Additionally, very fine knowledge of time may be used in geolocation applications.

GPS devices report time of day and provide a one-pulse-per-second signal. The one-pulse-per-second signal is transmitted from satellites, but does not arrive at every GPS receiver at the same time due to differences in path lengths. A properly designed receiver will assess the propagation delay from each satellite and compensate each of these delays so that the one-pulse-per-second output is synchronous at all receivers with only a 340-ns jitter. This level of accuracy is adequate for many applications, such as policy gating and change of cryptographic keys. Increased accuracy and lowered jitter may be accomplished through more sophisticated circuitry.

The local oscillator in a radio system may be used to keep track of time of day. The stability of these clocks is measured at approximately 10^{-6}. These clocks tend to drift over time, and in the course of a single day may accumulate up to 90 ns of error. Atomic clocks have much greater stability (10^{-11}), but have traditionally been large and power-hungry. Chip-scale atomic clocks have been demonstrated and are expected to make precision timing practical. This will enable geolocation applications with lower CEPs.

4.5.5 Spatial Awareness or Situational Awareness

A very significant role for a CR may be viewed as a personal assistant. One of its key missions is facilitating communication over wireless links. The opposite mission is impeding communications when appropriate. As an example, most people do not want to be disturbed while in church, in an important meeting, or in a classroom. A CR could learn to classify its situation into "user interruptible" and "user noninterruptible." The radio accepting aural inputs can classify a long-running exchange in which only one person is speaking at a time as a meeting or classroom and autonomously put itself into vibration-only mode. If the radio senses its primary user is speaking continuously or even 50 percent of the time, it may autonomously set the radio to a no-interference mode.

4.5.6 Software Technology

Software technology is a key component for CR development. This section discusses key SW technologies that are enabling CR. These topics include policy engines, artificial intelligence (AI) techniques, advanced signal processing, networking protocols, and the JTRS Software Communications Architecture.

Policy Engines

Radio transmitters are a regulated technology. A major intent of radio regulatory rules is to reduce or avoid interference among users. Currently, rules regarding transmission and reception are enumerated in spectrum policy as produced by various spectrum authorities (usually in high-level, natural language). Regulators insist that even a CR adhere to spectrum policies. To further complicate matters, a CR may be expected to operate within different geopolitical regions and under different regulatory authorities with different rules. Therefore, cognitive radios must be able to dynamically update policy and select appropriate policy as a function of situation.

Spectrum policies relevant to a given radio may vary in several ways:

1. Policies may vary in time (e.g., time of day, date, and even regulations changing from time to time).
2. Policies may vary in space (e.g., radio and user traveling from one policy regulatory domain to another).
3. A spectrum owner/leaser may impose policies that are more stringent than those imposed by a regulatory authority.
4. The spectrum access privileges of the radio may change in response to a change in radio user.
5. Additional policies may be imposed by hardware and software developers.

As a result, the number of different policy sets that apply to various modes and environments grows in a combinatorial fashion. It is impractical to hard-code discrete policy sets into radios to cover every case of interest. The accreditation of each discrete policy set is a major challenge. SDRs, for example, would require the maintenance of downloadable copies of software implementations of each policy set for every radio platform of interest. This is a configuration management problem.

A scalable expression and enforcement of policy is required. The complexity of policy conformance accreditation for cognitive radios and the desire for dynamic policy lead to the conclusion that CRs must be able to read and interpret policy. Therefore, an accredited (endorsed by an international standards body) language framework is needed to express policy. For example, if an established policy rule is constructed in the presence of other rules, the union of all policies is applicable. This enables hierarchal policies and policy by exception. As an example, suppose the emission level in band A is X dBm, except for a subband A′ for which the emission level constraint is Y dBm if a Z KHz guard band is allowed around legacy signals. Even this simple structure is multidimensional. Layers of exceptions are complex. The policy engine must be able to constrain behavior according to the intent of the machine-readable policy. An inference capability is needed to interpret multiple rules simultaneously.

In the case of spectrum subleasing, policies must be delegated from the lessor to the lessee, and a machine-readable policy may be delegated. When a CR crosses a regulatory boundary, the appropriate policy must be enabled. Policies may also be used by the system in a control function.

The policy should use standard tools and languages because the policy engine must be able to perform automatic interpretation to achieve the goals of cognitive radio applications. Policies may be written by regulatory agencies or by third parties and approved by regulators, but in all cases policy is a legal or contractual operating requirement and provability in the policy interpretation engine is needed for certification [34]. Chapters 6, 17, and 20 each deal with various aspects of policy.

Artificial Intelligence Techniques

The field of artificial intelligence has received a great deal of attention for decades. In 1950, Alan Turing proposed the Turing Test, regarding interacting with an entity to determine if that entity is human or machine. The AI techniques that work are plentiful, but most are not widely applicable to a wide range of problems. The powerful techniques may even require customization to work on a particular problem.

An agent is an entity that perceives and acts. A smart agent model is appropriate for cognitive radio. Figure 4.4 explains the four models of smart agents: (1) simple reflex agents, (2) model-based reflex agents, (3) goal-based agents, and (4) utility-based agents, defined as follows.

A *simple reflex* agent is a simple mapping from current sensor inputs to actuator settings. This is a stateless agent model that neither learns nor adapts to the environment.

A *model-based reflex agent* is still a simple mapping, but now includes memory of past inputs. The actions are a function of the current sensor inputs and the recent past inputs, making it a finite-memory agent model. There is still no learning. Adaptation is limited, but this is the minimum level of sophistication for an adaptable radio.

A *goal-based agent* adds to the memory of past inputs; it is a "realistic" model of the environment. Now a sequence of actions may be "tested" against a goal and an appropriate next action may be selected. The level of sophistication for the model of the environment is not well defined. These agents have increased capability of adapting because a prediction about the consequences of an action is available. There is no feedback, and learning is therefore limited. This is the minimal level of sophistication for a cognitive radio.

A *utility-based agent* maps the state sequence (memory state) to a "happiness" value and therefore includes feedback. The more sophisticated environment model may experiment with sequences of actions for selection of the best next action. This model of a CR has the ability to learn and adapt [35].

A smart agent model for CR is appropriate. The agent framework supports the continuum of radio maturity, and it allows the modular introduction of various AI tech-

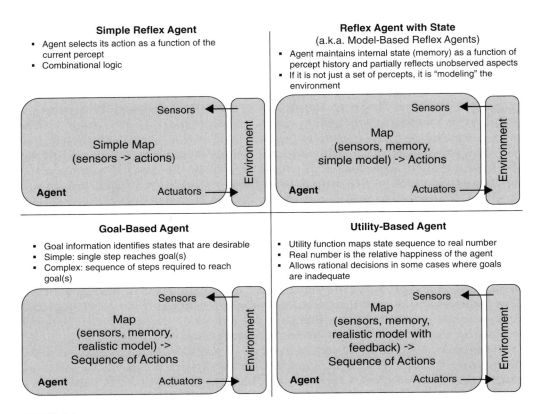

FIGURE 4.4

Four smart agent models. Smart agents provide a framework that is consistent with the continuum of software maturity in software radios.

niques such as fuzzy control, genetic algorithms (GAs), or case-based reasoning (covered in detail in Chapters 7 and 12). Agents may be tailored to the application's environment. In this sense, the environment may be characterized in the following dimensions: fully observable versus partially observable, deterministic versus stochastic, episodic versus sequential, static versus dynamic, discrete versus continuous, and single agent versus multiagent.

The following is an incomplete list of AI techniques likely to find applicability to cognitive radio; they are described further in subsequent chapters:

- State–space models and searching (Chapter 17)
- Ontological engineering/declarative reasoning (Chapters 4–6, 12–14, 17, 24)
- Neural networks in CR (Chapter 7)
- Fuzzy control in CR (Chapter 7)
- Genetic algorithms in CR (Chapter 7)
- Game theory in CR (Chapter 15)

- Knowledge-based reasoning in CR (Chapter 12)
- Case-based reasoning (Chapter 7)

Advanced Signal Processing

Digital signal-processing technology enables rapid advances in CRs. Intellectual property resources are widely available for signal-processing functions. In GPP or DSP resources, libraries of routines realize functions in efficient assembly language. In field-programmable gate array (FPGA) or ASIC resources, licensable cores of signal-processing engines are available in VHSIC Hardware Design Language (VHDL) or Verilog. Signal-processing routines are available for communication signal processing (i.e., modulation/demodulation, forward error correction, equalization, filtering, and others); audio signal processing (i.e., voice coding, voice generation, natural language processing); and sensor signal processing (i.e., video, seismic, chemical, biometric, and others).

Synthesizing signal-processing functions together to form a system is a complex task. A process for algorithm development, test case generation, realization, verification, and validation eases the process of building a waveform or a cognitive system. Integrated tools for system development are available. Many of the tool sets include automatic generation of high-level language source code or hardware definition language code. A bit-level accurate simulation environment is used to develop the system algorithms and to generate test cases for postintegration verification and validation. This environment may be used for cognitive radios to synthesize communications or multimission waveforms that enable a CR to achieve specific goals.

Networking Protocols

Cooperative groups (a multiagent model) have the potential to increase capabilities in a variety of ways. For example, a lone cognitive radio is limited in its ability to access spectrum, but a pair of CRs can sense the spectrum, identify an unused band, rendezvous there, and communicate. A network of CRs enables other significant increases in capabilities. Software for Mobile Ad Hoc Networking (MANET), although maturing slowly, is a key enabling technology.

The medium access control (MAC) layer is critical in CR networks. If the CR is employing advanced spectrum access techniques, a robust MAC that mitigates the hidden node problem is needed. In a "static spectrum access" environment, a more traditional MAC is possible. A carrier sense collision detection (802.11 MAC) mode is not possible because a radio cannot receive in the exact same band in which it is transmitting, so a carrier sense collision avoidance approach is frequently used. Request-to-send (RTS) and clear-to-send (CTS) messaging are popular wireless MACs amenable to MANETs. Other approaches include TDMA or CDMA MACs.

The architecture for routing packets is important for performance in MANETs. The approaches are generally divided into proactive and reactive algorithms. In a proactive routing environment, routing data, often in the form of a routing table, are maintained so that a node has a good idea of where to send a packet to advance it toward its final destination, and a node may know with great confidence how to route a packet even if one is not ready to go. Maintaining this knowledge across a MANET requires resources. If the connection links are very dynamic or the mobility of the nodes causes rapid handoff from one "network" to another, then the overhead to maintain the routing state may be high. In contrast, reactive routing approaches broadcast a short search packet that locates one or more routes to the destination and returns that path to the source node. Then the information packet is sent to the destination on that discovered route. This causes overhead in the form of search packets. Proactive and reactive routing both have pros and cons associated with their performance measures (e.g., reliability, latency, overhead required, and so on). A hybrid approach is often best to provide scalability with offered network load. Cognitive networking is covered in greater detail in Chapter 22.

An interesting application of CR is the ability to learn how to network with other CRs and adapt behavior to achieve some QoS goal such as data rate below some bit error rate bound, maximum latency, minimum jitter, and so forth. Various cognitive-level control algorithms may be employed to achieve these results. As an example, a fixed-length control word may be used to parameterize a communications waveform with frequency, FEC, modulation, and other measures. The deployment of a parameterized waveform may be controlled and adapted by using a generic algorithm and various QoS measures to retain or discard a generated waveform.

JTRS Software Communications Architecture

The primary motivations for SDR technology are lower life cycle costs and increased interoperability. The basic hardware for SDR may be more expensive to develop than for a single-purpose radio system, but the resulting radio platform performs many radio functions and can be reused for many applications. The radio hardware platform requires only one logistics tail for service, training, replacement parts, and so on, and is produced in higher volume, often resulting in lower production cost. One of the driving costs in SDR development is that of software development. The US Department of Defense (DoD) JTRS acquisitions are controlling these costs by ensuring software reuse. The approach for reuse is based on the Software Communications Architecture (SCA), which is a set of standards that describes the software environment (see Chapter 3). It is currently in release 2.2.2. Software written to be SCA compliant is more easily ported from one JTRS radio to another. The waveforms are maintained in a JTRS Joint Program Office (JPO) library.

A CR can be implemented under the SCA standards. Applications that raise the radio to the level of a CR can be integrated in a standard way. It is expected that DARPA's XG program will provide a CR application for policy-driven, dynamic spectrum access on a noninterference basis.

4.5.7 Spectrum Awareness and Potential for Sublease or Borrow

The Spectrum Policy Task Force (SPTF) recommends that license holders in exclusive management policy bands be allowed to sublease their spectrum. Figure 4.5 shows a sequence diagram for spectrum subleasing from a public safety spectrum owner. During the initial contact between the service provider and the public safety spectrum owner, authentication is required. This ensures that spectrum use will be accomplished according to acceptable behaviors and that the bill will be paid [36].

For a subleasing capability to exist in a public safety band a shut-down-on-command function must be supported with a bounded response time. There are three approaches to this: continuous spectrum granting beacon, time-based granting of spectrum, and

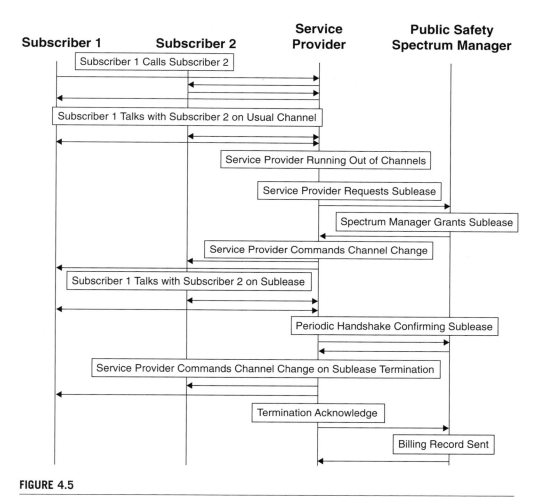

FIGURE 4.5

Spectrum subleasing sequence. One motivation for CR is the potential income stream derived from subleasing idle spectrum on a noninterference basis.

request to send–clear to send–inhibit send. Figure 4.5 shows the time-based granting of spectrum, described as a periodic handshake confirming sublease.

Even though public safety organizations may not sublease spectrum, other organizations may choose to do so. Subleasing has the benefit of producing an income stream from only managing the resource. Given proper behavior by lessees and lessors, the system may become popular and open up the spectrum for greater utilization.

4.6 FUNDING AND RESEARCH IN COGNITIVE RADIOS

DARPA is funding a number of cognitive science applications, including: the XG program, Adaptive Cognition-Enhanced Radio Teams (ACERT), Disruption Tolerant

Networking (DTN), Architectures for Cognitive Information Processing (ACIP), Real World Reasoning (REAL), and Wireless Network After Next (WNAN). DARPA research dollars under contract are easily in the tens of millions. Good results have been achieved in many of these efforts.

The National Science Foundation (NSF) is also funding cognitive science applications including grants to the Virginia Polytechnic Institute and State University (i.e., Virginia Tech, or VT). Additionally, NSF has sponsored Information Theory and Computer Science Interface workshops that communicate CR research results.

The Software Defined Radio Forum (SDR Forum) has a Cognitive Radio Working Group that is investigating various CR technologies such as spectrum access techniques, and a Cognitive Applications Special Interest Group that is working with regulatory bodies, spectrum owners, and users to communicate the potential of CR technologies. Numerous organizations participate at SDR Forum meetings.

Both the FCC and the National Telecommunications and Information Administration (NTIA) have interest in cognitive radio. The FCC has solicited various comments on rule changes and an SPTF report. NTIA has been involved in the discussions as they relate to government use of spectrum (see Chapter 2).

4.6.1 Cognitive Geolocation Applications

If a CR knows where it is in the world, myriad applications become possible. The following is a short set of examples. Figure 4.6 shows a use case-level context diagram for a cognitive radio with geolocation knowledge.

Establishing the location of a CR enables many new functions. A cognitive engine for learning and adapting can utilize some of the new functions. There are multiple

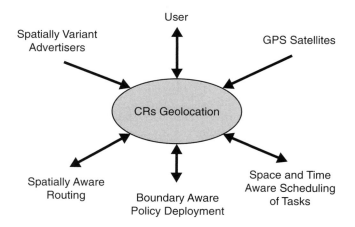

FIGURE 4.6

Cognitive radio use case for geolocation. Several interactions between the CR and the world are enabled or required by geolocation information.

methods for determining location. For example, GPS receiver technology is small and inexpensive. Given an appropriate API, numerous applications including network localization (discussed next) and boundary awareness are available to the cognitive engine in the CR. Other methods for determining location are discussed in Chapter 8.

Network localization is a term to describe position-aware networking enhancement. For example, if a radio knows it is in a car, that it is morning rush hour, and that the path being taken is the same as that of the last 200 days of commuting, it can predict being in the vicinity of the owner's office wireless LAN within a certain time period. The radio may wait until it is in the office to download email and upload the pictures of the accident taken a few minutes ago. This is an example of the highest level of management of radio functions (assuming the radio is used for email, photos, etc.).

Spatial awareness may be used for energy savings in a multiple short hop-routing algorithm with power management because just closing the wireless link with multiple short hops is usually more energy efficient than one long hop. Knowing the position of each node in a potential route allows the CR to take energy consumption into consideration when routing a packet.

Spatially variant searching is a powerful concept for increasing a user's operating efficiency. If it is time for supper, the CR may begin a search for the types of restaurants the user frequents and sort them by distance and popularity. Other spatially variant searches are possible.

A radio aware of boundaries may be able to invoke policy as a function of geopolitical region. When passing from one regulatory jurisdiction to another, the rules change. The radio can adopt a conservative operational mode near the boundaries and change when they are crossed. The radio must have the ability to distinguish one region from another. This may require a database (which must be kept up-to-date) or some network connectivity with a boundary server (see Chapter 11). Figure 4.7 shows a simplified sequence diagram in which a cognitive radio accesses spectrum as a function of a spatially variant regulatory policy.

A sequence of position estimates may be used to estimate velocity. Take a scenario in which teenagers' cell phones would report to their parents when their speed exceeds a certain speed. For example, suppose tattle mode is set. The radio is moving at 45 mph at 7:30 a.m. The radio calls the parent and asks a couple of questions such as: "Is 45 mph okay at this time? Is time relevant?" The questions will be in the same order each time and the parent won't have to wait for the whole question, just the velocity being reported. Then the parent keys in a response. After a few reports, the radio can develop a threshold as a function of time. For example, during the time the radio is heading for school, 45 mph is okay. During the lunch break (assuming a closed campus), 15 mph might be the threshold. An initial profile may be programmed, or the profile may be learned through tattling and feedback to the reports. Vehicle position and velocity might also be useful after curfew.

The CR application uses special hardware or customized waveforms that return geolocation information. This information is used to access databases of policies or resources to make better decisions. Dynamic exchange of information may be used for other networking actions. The set of CR applications that are enabled by geolocation capability is large and has many attractive benefits.

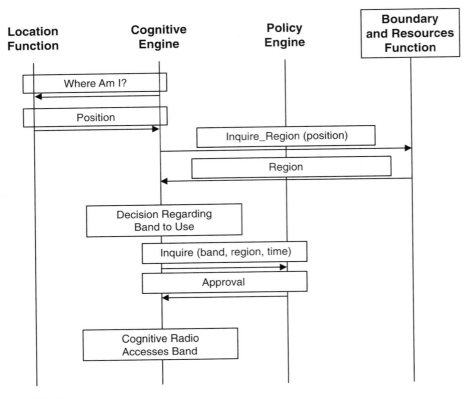

FIGURE 4.7

Spectrum access sequence diagram. A policy engine uses regional inputs to select spatially variant policies to approve or disapprove a requested spectrum access.

4.6.2 Dynamic Spectrum Access and Spectrum Awareness

One of the most common capabilities of a CR is the ability to intelligently utilize available spectrum based on awareness of actual activity. Current conservative spectrum management methods (static spectrum assignments) are limited because they reduce spatial reuse, preclude opportunistic utilization, and delay wireless communication network deployment. Without the need to statically allocate spectrum for each use, however, networks can be deployed more rapidly. A CR with spectrum-sensing capability, and cooperative opportunistic frequency selection, is an enabling technology for faster deployment and increased spatial reuse.

Spectrum access is primarily limited by regulatory constraints. Recent measurements show that spectrum occupancy is low when examined as a function of frequency, time, and space [36]. Cognitive radios may sense the local spectrum utilization either through a dedicated sensor or by using a configured SDR receiver channel. Uses of this information may create increased spectrum access opportunities.

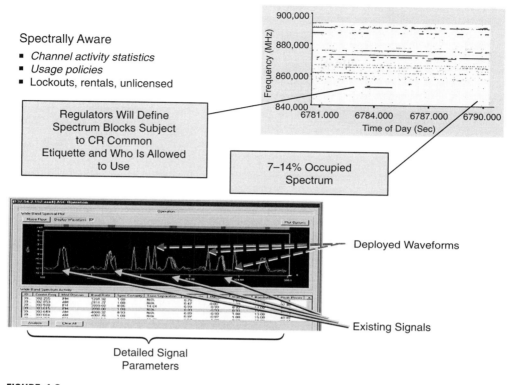

FIGURE 4.8

Spectrum awareness. A CR, or a set of CRs, may be aware of the spectrum and may exploit unoccupied spectrum for its own purposes.

One of the primary considerations for such a cognitive application is noninterference with other spectral uses. Figure 4.8 shows local spectrum awareness and utilization. If the regulatory body is allowing cognitive radios to utilize the unoccupied white space, increased spectral access can be achieved. The CR can examine the signals and may extract detailed information regarding use. By estimating the other uses and monitoring for interference, two CRs may rendezvous at an unoccupied "channel" and be able to communicate.

Sophisticated waveforms that have the ability to periodically stop transmitting and listen for legacy users are called for in this application, as well as waveforms that can adapt their spectral shape over time. Dynamic selection of channels to utilize or vacate is important and these principles have been demonstrated by example equipment. Another advantageous waveform characteristic is discontinuous spectrum occupancy. This allows a wideband communication system to aggregate available spectrum using frequencies between other existing signals. Careful analysis is needed to ensure that sufficient guard bands are utilized.

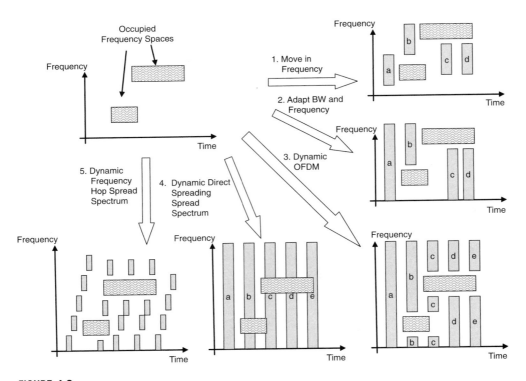

FIGURE 4.9

Noninterference methods for dynamic spectrun access. Different strategies for deploying noninterfering waveforms have been proposed.

Figure 4.9 shows five suggested alternatives for utilizing spectrum in and around legacy signals. The characteristics of the legacy signals may be provided to the cooperating CRs by federated sensors. An alternative method for characterizing the legacy signals is time-division sharing of a channel as a sensor and providing a look-through capability by duty-cycling transmit and monitor functions. The five methods shown in the figure avoid the legacy signal in various ways. The first method assumes a fixed bandwidth waveform, where the center frequency may be adapted. The second method assumes a variable bandwidth signal, such as a direct-sequence spread-spectrum waveform, where the chip rate is adapted and the center frequency is adapted. The third method uses a water-filling method to populate a subset of the carriers in an OFDM waveform. These three methods impact legacy signals very little if appropriate guard bands are observed. The fourth method is a direct-sequence spread-spectrum waveform that underlies the legacy signals. The interference from this underlay must be very small so that legacy systems do not experience noise and consequently reduction of communications range; thus the processing gain of the spread-spectrum underlay must be very high. The last method shown also avoids the legacy signals by frequency hopping into only unoccupied channels.

The spread-spectrum method deserves some elaboration at this point. Legacy receivers will perceive the spread-spectrum signal as an increase in the noise floor, which may result in a decrease in the link margin of the legacy signal. The spread-spectrum receiver will perceive the legacy waveform(s) as a narrowband interference. The despreading of the desired signal by the spread-spectrum receiver will cause the narrowband signals to spread. This spreading will cause the narrowband signals to appear at a signal level reduced in power by the spreading gain, and thus appear as noise to the CR, resulting in reduced link margin. The ability of each of these communication systems to tolerate this reduced link margin is link specific and therefore a subject of great concern to the legacy system operators.

An OFDM waveform, the third method in Figure 4.9, has several benefits including flat fading subchannels, simplification of equalizer complexity due to long symbol time and guard intervals, the ability to occupy a variety of bandwidths that "fit to the available opportunity," and the ability to null subcarriers to mitigate interference. Variable bit loading enables prenulling and dynamic nulling. Table 4.5 compares several methods for dynamic bit loading of an OFDM waveform. Not loading subcarriers occupied by legacy signals with guard bands around them minimizes interference between CR and non-CR systems [37–39].

Because spectrum utilization is a spatially and temporally variant phenomenon, it requires repeated monitoring and needs cooperative, distributed coordination. The familiar hidden node and exposed node problems have to be considered. Figure 4.10 shows a context diagram in which external sensor reports are made available to the CR and may be considered when selecting unoccupied bands.

Figure 4.11 shows a sequence diagram in which a set of cognitive radios is exchanging sensor reports and is learning about local spectrum occupancy. At some time, a pair of CRs wishes to communicate and rendezvous at a band for that purpose. When a legacy signal is detected, the pair of CRs must vacate that band and relocate to another.

Table 4.5   Comparison of Variable Bit-Loading Algorithms

Method	Characteristic	Complexity
Water-filling	Original approach, optimal, frequently used for comparison	$O(N^2)$
Hughes-Hartogs [37]	Optimal, loads bits serially based on subcarrier energy level, slow to converge, repeated sorts	$O(SN^2)$
Chow et al. [38]	Suboptimal, rounds to integer rates using signal-to-noise gap approximations, some sorting required	$O(N\text{Log}N + N\text{Log}S)$
Lagrange (Unconstrained) Krongold et al. [39]	Optimal, computationally efficient, efficient table lookup, Lagrange multiplier with bisection search, integer bit loading, power allocation, fast convergence	$O(N\text{Log}N)$, Revised $O(N)$

Note: N = number of subcarriers; S = number of bits per subcarrier.

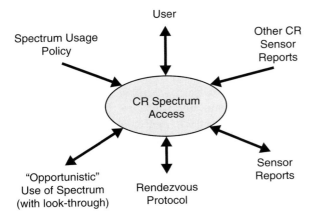

FIGURE 4.10

Spectrum access context diagram. A spatially diverse sensing protocol is required to mitigate such problems as hidden node or deeply fading RF channels for CR access to spectrum on a noninterference basis.

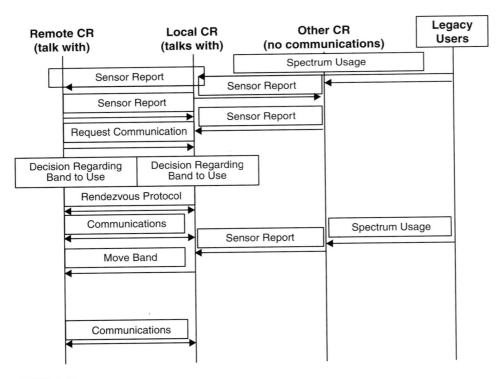

FIGURE 4.11

Dynamic spectrum access sequence diagram (assumes a control channel). A pair of CR communicating on an unoccupied channel must vacate when interference potential is detected.

The sensor technology utilized for spectrum awareness should be of high quality to mitigate the hidden node problem. For example, if a CR wants to use a television (TV) band, its sensor should be significantly more sensitive than a TV set so that if it detects a TV signal, it will not interfere with local TV set reception, and if it does not detect a signal, there is a high probability that no TV set is near enough to demodulate a TV signal on that channel.

Dynamic spectrum access as a result of learning about the spectrum occupancy is a strong candidate for a CR application. Numerous discussions in regulatory organizations have involved whether to allow this behavior. If implemented correctly, it is a win-win situation in which more spectrum is utilized and very little additional interference is suffered.

4.6.3 The Rendezvous Problem

The difficulty for the cognitive radio network is the radios locating each other and getting the network started. Before any transmission can occur, all radios must survey the spectrum to determine where the available "holes" (frequency reuse opportunities) are located. However, each receiver–sensor will perceive the spectrum slightly differently because each will see different objects shadowing different transmitters, and will either see each transmitter with a different signal strength, or will not see some transmitters that other nodes are able to see. So while node A may see an open frequency, node B may consider that frequency to be in use by a distant transmitter node or network. To get the network started, the cognitive radios must agree on a protocol to find each other.

There are several possible methods by which to do this. These methods depend strongly on whether there is an infrastructure in place to help start up a CR network, or whether no infrastructure can be assumed.

Infrastructure-Aided Rendezvous

Infrastructure can improve the rendezvous process. First, a more complete picture of the RF environment can be gathered to an infrastructure node or nodes, and thus, less interference is created by the CR systems. Second, the infrastructure node(s) acts as a broker between CRs wishing to communicate. Each node within the infrastructure node's or nodes' area of responsibility registers with the infrastructure. The infrastructure's antenna location is published/known and this removes the spatial component from the search. The infrastructure broadcasts "probe" signals that include information about how to register. Once the "user" nodes are registered, the system brokers communication parameter selections (including frequency) between users (see Chapter 11).

If we assume that there is an infrastructure component, we must also assume it behaves just like the CR and does not interfere with legacy systems. We assume, however, that it periodically transmits a beacon signal, and we assume that this beacon includes reference time, next frequency hop(s), and a description of frequencies in use within the local region. Furthermore, we assume that the infrastructure beacon is followed by an interval in which CRs request an available time-frequency slot and associate a net-name to that slot, as well as to their location and transmit power, followed by a response from the infrastructure recommending which frequency to use and when to check back. Subsequent requests by other net members can be directed to the proper time-frequency slot, and the geographic distribution of net members can be tracked, allowing the infrastructure to assess interference profiles.

Unaided Rendezvous

Defense systems are rarely able to assume support infrastructure. Similarly, early deployments of commercial CR equipment will not be able to assume infrastructure. Consequently, it is important to have an unaided method for rendezvous. Several methods exist for systems to find each other.

The problem is somewhat like two men searching for each other when they are a mile apart on a moonless dark night in the desert. Each has a flashlight the other may see, but only when it is pointed in the right direction. They can finally signal each other, but only when each has noticed the other's flashlight, so each must look in the proper direction at the proper time.

In the case of CRs trying to find each other, one must transmit and the other must receive in the proper frequency at the proper time. Several methods are feasible. All involve one node transmitting "probe" signals (a uniquely distinguishing waveform that can be readily correlated) until receiving a response from a net member. In all cases, the problem is that the frequencies the transmitter sees as being "usable" are different from those the receiver sees. Thus, neither can be stationary and just sit on one frequency hoping the other will find it.

Case 1: Node A transmits probes in randomly selected frequencies, while node B listens to random frequencies considered to be unoccupied. The search time can be dramatically reduced if node B is able to listen to many simultaneous frequencies (as it might with an OFDM receiver). Node B responds with a probe response when node A is detected.

Case 2: Node A selects the five largest unoccupied frequency blocks, and rotates probe pulses in each block. Similarly, node B scans all of the unoccupied frequency blocks that it perceives, prioritized by the size of the unoccupied frequency block, and with a longer dwell time.

Both of these cases are similar, but minor differences in first acquisition and robustness may be significant.

After node B hears a probe and responds with a probe acknowledge, the nodes will exchange each node's perception of locally available frequencies. Each node will "logical OR" the list of active frequencies, leaving a list of frequencies that are not in use according to all sensing reports of all nodes. Then node A will propose which frequency block to use for traffic for the next time slot, at what data rate, and in what waveform. From this point forward, the nodes will have sufficient connectivity to track changes in spectral activity and to stay synchronized for which frequencies to use next. A detailed discussion of rendezvous is provided in Chapter 19.

4.6.4 Cognitive Radio Authentication Applications

A cognitive radio can learn the identity of its user(s). Authentication applications can prevent unauthorized people from using the CR or the network functions available to

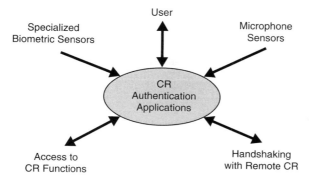

FIGURE 4.12

Authentication context diagram. A CR application that learns the identity of its user(s) and has an Access List Security Protocol can minimize fraud and maximize secure communications.

the CR. This enhanced security may be exploited by the military for classified communications or by commercial vendors for fraud prevention.

Because many radios are usually used for voice communications, a microphone often exists in the system. The captured signal is encoded with a vocoder and transmitted. The source radio can authenticate the user (from a copy of the data) and add the known identity to the data stream. At the destination end, decoded voice can be analyzed for the purposes of authentication, and the result may be correlated with the sent identity.

Other sensors may be added to a CR for the purposes of user authentication. A fingerprint scanner in a push-to-talk (PTT) button is not intrusive. Automatic fingerprint correlation software techniques are available and scalable in terms of reliability versus processing load required. Additionally, cell phones have been equipped with digital cameras. This fingerprint sensor coupled with facial-recognition software may be used to authenticate a user. Again, the reliability is scalable with processor demands.

Figure 4.12 shows a context diagram for a cognitive radio's authentication application. The detailed learning associated with adding a user to an access list through a "third-party" introduction is not shown. The certificate authority could be the third party, or another authorized user could add a new user with some set of authority. The CR will learn and adapt to the new set of users and to the changing biometric measures of a user. For example, if a user gets a cold, his voice may change, but the correlation between the "new voice" and the fingerprint scanner is still strong and the CR may choose to temporarily update the voice print template. There is more on this topic in Chapter 10.

4.7 TIMELINE FOR COGNITIVE RADIOS

Cognitive radio development will be a spiral effort. The when, where, who, and how of this development are as follows:

When: Currently, several CR initiatives are under way. Progress is evident every day, and more sophisticated demonstrations have been performed in 2008. Some of these demonstrations include policy conformance characteristics.

Where: The FCC and NTIA are currently discussing a test band for CR. They have just made some blocks of spectrum available to cognitive techniques. This promising development must be exploited. As policy is written for this band, it can be deployed and policy violations may be assessed.

Who: Vendors, regulators, service providers, and users are highly interested in CR systems. A great deal of discussion on exactly what that means has already taken place and continues. Academic researchers using COTS SDR demonstrations and government-sponsored demonstrations using custom-developed SDRs are now reducing cognitive radio to practice. Where the technology will progress is difficult to predict. As an example, the following organizations are some of the better known organizations working with CR: General Dynamics; Shared Spectrum; Raytheon; Lockheed Martin; Bolt, Beranek, & Newman; Rockwell Collins; Harris; Virginia Tech; and Northeastern University (and the list has doubled during the preparation of this edition).

How: The easiest experiments utilize COTS SDR hardware and execute cognitive applications as demonstrations. Custom-developed hardware is more expensive, but is better tailored to show the benefits of cognitive radio. Spectrum awareness using a spectrum sensor has the best information to exploit unoccupied spectrum. This custom capability also has computational resources sized to execute a policy-constraint engine.

4.7.1 Decisions, Directions, and Standards

Numerous organizations and standards bodies are working in the area of CR. The SDR Forum, the Institute of Electrical and Electronics Engineers (IEEE), the Federal Communications Commission (FCC), the National Telecommunications and Information Administration (NTIA), and the International Telecommunication Union (ITU) all have interests in this area.

4.7.2 Manufacture of New Products

Many products have new sensors and actuators in them. Cellular telephone handsets are a high-volume, highly competitive product area that are driving innovation. In fact, a large fraction of new cell phones have an integrated digital camera. These are manually operated today, but CR applications may take advantage of this sensor for more "cognitive" operation modes.

Chemical sensors have been integrated to some cell phone models. The purpose of the chemical sensors is to report on important "problems" such as "bad breath" and "blood alcohol level." This is a manually operated device, but future applications may be more autonomous.

Among the new applications in cellular telephones is a Bluetooth-like waveform to introduce single people. "Flirting radios" may subsequently need to have AI technology added to filter out the "undesirable" introductions.

Much to the dismay of teenagers and employees everywhere, phone-tracking applications are now available. Although these capabilities are used "manually" now, learning algorithms can be applied to the interface to create a new "filtered" report stream. There is serious interest in tracking other sets of people (e.g., first responders, delivery people, service people, or doctors). However, this function is accepted differently by different cultures, and may not be universally attractive.

4.8 UPDATE OF CR-SPECIFIC TECHNOLOGIES

There has been a proliferation of work in the area of CR and the supporting technologies. The Software Defined Radio Forum, Dyspan, Globecomm, and Milcom conferences have all supported CR technical tracks in the last couple of years, and many CR patent applications have been filed. The work can be categorized any number of ways, but the following areas seems to be emerging from the work:

- Sensors and infrastructure
- PHY, MAC, and network layers
- Reasoners
- Policy
- Hardware and demonstrations

4.8.1 Sensors and Infrastructure

The Mobile and Portable Radio Research Group (MPRG) at Virginia Tech has studied the spectrum access problem discussed in Section 4.6.2. Group members have developed a Radio Environment Map (REM) that is spatially variant and temporally variant. Additionally, they have studied distributed and centralized approaches (global works better) [40–42]. Their simulations and REM development are clear progress in the CR field of study (see Chapter 11).

IEEE 802.22 for Wireless Regional Network (WRAN) is an emerging CR standard for spectrum access on a noninterference basis. The WRAN premise is the reuse of television bands (54–862 MHz in roughly 6 MHz channels) without interfering with primary users (PUs). The IEEE standard committee and FCC are pursuing a centralized spectrum discovery (base station, BS) approach and fixed, GPS-enabled access points (APs), or customer premises equipment (CPE). The CPEs send remote spectrum sensing information to the BS for evaluation, control, and assignment of channels. The physical (PHY) layer uses OFDMA and may occupy multiple TV channels. The MAC approach is a TDMA structure using frames and superframes [43].

4.8.2 PHY, MAC, and Network Layers

Multicarrier waveforms for the PHY layers are a hot topic in CR. The ability to deploy equally spaced carriers in a band and to selectively include or exclude carriers creates the opportunity for filling in around detected legacy signals. Orthogonal frequency-division multiplexing is a popular choice for multicarrier waveform structure, and OFDM is frequently realized using Fast Fourier Transform (FFT) channelization techniques. Alternative architectures that are tailored for CR and are lower in computational complexity have been recently studied [44]. While the idea of discontinuous multicarrier waveforms sounds attractive on the surface, there are issues that must be addressed. For example, the spectrum shaping requirements for noncontiguous multicarrier waveforms have been studied in [45]. Also, synchronization at the transmit and receive nodes to the exact subset of carriers used must be addressed to create a valid link.

The MAC protocol for CR has also garnered a lot of attention. Time division multiple access, busy signal in adjacent control channels, multiple channel protocols, request to send/clear to send, centralized master/slave, and other approaches have been published. Given the many approaches under consideration, there is no single obvious choice for MAC solutions in CR.

The rendezvous and re-rendezvous problems are inherently coupled to the MAC protocol. A pair of nodes wishing to communicate must converge or rendezvous on the same channel, at the same time, with the same waveform parameters (e.g., error-correction coding, packet structure, modulation type, etc.). Converging in an open channel is just a generalization of the MAC problem in higher dimensions.

The protocol that specifies where the communication should shift, in frequency for example, when a primary user enters the channel in use is the re-rendezvous problem. One approach that has been discussed is negotiating a backup channel that can be utilized very quickly when a PU reenters a channel. After the "hop" to the backup channel, the protocol finds and renegotiates a new backup channel.

There are many routing algorithms in use by mobile ad hoc networks. These algorithms can usually be divided into proactive and reactive. The proactive algorithms maintain routing tables all of the time, and when a packet is ready to be sent to a particular node, it is sent out and the intermediate nodes already have direction about where to forward the packet. The reactive algorithms flood discovery packets prior to sending out the data packet and must wait on a route to be returned. Depending on the environment, load offered, and topology dynamics, reactive may be better than proactive and vice versa. All of them are valid CR choices and should be a configurable characteristic of a cognitive network. If the routing algorithm is considered part of the waveform, then the routing algorithm choice becomes part of the rendezvous/re-rendezvous problem.

4.8.3 Reasoners

A number of reasoners and requirements for them have been identified. Without a clear partitioning of CR workload this remains an open research area. Specifically, how smart should an environment sensor be as compared to the reasoner and which decisions are made by which reasoners are open questions. There are a number of suggestions being vetted in the literature. Additionally, the learning algorithm used by a cognitive engine (CE) is also in the trade space.

Two of the important reasoners under consideration are the spectrum availability/interference avoidance reasoner and the waveform configuration reasoner. These two CR capabilities are recognized for their importance. If a CR has these two working reasoners, it will be recognized widely as a cognitive radio because it would be able to adapt its waveform to the available channel.

All reasoners must be constrained by a radio policy and the interface between reasoners and the rest of the CR is critical. The interface to the policy engine must be low latency to support real-time reconfiguration and robust to gain regulatory approval and support the required functionality.

If a CR follows the level 2, 3, or 4 smart agent model discussed in Section 4.5.6, then performance prediction models are needed. Which models, at what level of sophistication, and at what accuracy level, are open questions. The response time for a performance prediction model is important for making real-time adaptations to the CR. There has been some work toward placing a communications channel model into the radio that supports the waveform configuration reasoner. Since there are many mature channel models that consume a variety of computational resources and require a variety of parameters, this is an excellent place to start exploring the sophistication level necessary to achieve CR goals.

One of the clever applications garnering attention is cognitive establishment of an RF channel. In certain geometries and physical environments, some frequencies/channels behave and work well while others near them (in frequency) do not. A CR has the potential to sound the channel and select a good one. There are several examples of this. A first example is the Automatic Line Establishment (ALE) algorithm used in HF bands, which is a well-known version of channel sounding and channel selection. Another example is sounding in tunnels, such as subways or mines, which is frequently required for establishing communications. A CR reasoner could perform this task for its users better than the users themselves.

4.8.4 Policy

Several papers argue for policy-based dynamic spectrum reuse CR applications. This approach is well accepted because of the regulatory certification issues and the ability of a policy engine to constrain a radio's transmissions. Secure distribution of policies and certification of the policy engine are recognized as necessary for CR success.

Multiple policy languages are being considered. If multiple languages are adopted, the exchange of policy between different regulatory bodies and different CR families must be accommodated. Rule sharing among CR families is discussed in Ginsberg et al. [46]. Interchange of rules is not limited to CR applications. The Semantic Web is working toward standards for machine-usable formal languages, knowledge representations, and methods. These standards are a supporting technology for CR.

4.8.5 Hardware and Demonstrations

It is clear that SDR technology is penetrating the radio market, especially the military radio market. The Joint Tactical Radio System (JTRS) procurement programs are proceeding with significant successes. Several high-speed radio links are now SDR realizations, and the CR community is realizing cognitive applications using SDR platforms in several organizations. Many of these SDRs have the excess processing capability needed to support CR applications.

Two of the interesting CR hardware discussions in the literature are: smart antennas for CR base transceivers and the international collaboration for a CR testbed [47]. It has been noted that spatial diversity can be better utilized with a smart antenna, and that coupling this subsystem to a CR enables new optimizations for the CR [47]. Current size and processing requirements for smart antennas limit their application to fixed sites and vehicular systems. The collaboration for a CR testbed is a worldwide effort to enable flexible experimentation. Two bands have been allocated in Ireland, and multiple radios have been used to experiment there. Implementing Radio in Software (IRIS), Kansas University Agile Radio (KUAR), Winlab at Rutgers, and the GNURadio have all been used in these experiments. Sensing, learning, optimization, and adaptation have all been exercised in this effort. It is clear that CR development is under way and technological breakthroughs are being discovered regularly.

4.9 SUMMARY

Radio evolution has taken the path toward more digital realizations and more software capabilities. The original introduction of software in embedded microprocessors made software-capable radios that communicate and process signals digitally an economical reality. In the pursuit of flexibility, software-programmable radios, and the even more flexible SDRs, have become the standard in the military arena and are starting to gain favor in the commercial world, as explained in Section 4.2. We are now seeing the emergence of aware radios, adaptive radios, and finally cognitive radios, and we have traced the continuum among these various degrees of capability, as well as providing a few examples of each. Sections 4.3 and 4.4 explored the properties and capabilities of each of these classes of radios.

Section 4.5 outlined the enabling technologies for cognitive radios. Numerous technologies have matured to the point where CR applications are possible, and even attractive. The ability to geolocate a system, sense the spectrum, know the time and date, sense biometric characteristics of people, access new software capabilities, and determine new regulatory environments are all working together to enable cognitive radio.

Geolocation through the use of GPS or other methods is now available. Chapter 8 discusses how a radio can know where it is located. This enabling technology allows a radio to make spatially variant decisions, which may be applied to the selection of policy or networking functions.

Sensing of the local RF environment is available. This information may be used to mitigate a deeply fading channel or may be used to access locally unoccupied spectrum. Noninterference is particularly important, and protocols for sensing, deciding, and accessing spectrum are being designed, developed, and demonstrated today.

Increased robustness in biometric sensor technology provides a whole new dimension to CR applications. The most likely initial use of this technology is in user authentication applications, such as the purchasing of services and products.

Knowledge of time has been available in many forms, but integration into a broader range of full-function capabilities will enable all new applications. Stable time knowledge enables a CR to plan and execute with more utility and precision. Using this capability for non-infrastructure-based geolocation, dynamic spectrum access, or AI planning is envisioned for near-term CR functions.

A smart agent model of CRs is attractive. An agent is an entity that perceives and acts on behalf of another. This is where CRs are going. Smart agent models of the world enable a radio to provide services for its user. Improved performance or new capabilities may be provided. As the CR's smart agent model of the world becomes more sophisticated and realistic, situational awareness will increase. As the models improve, the ability of the CR to act effectively over a broader range of user services will improve.

Maybe the most important software technology is a policy engine that enables a CR to know how to behave where it is right now, given the priorities of current circumstances. Artificial intelligence applications at a very high level, networking services in the middle levels, and signal-processing primitives at a very low level are all available for a CR developer to utilize in creating new capabilities. Finally, middleware technology enables greater software reuse, which makes CR development economical.

Modern regulatory philosophy is starting to allow CR to deploy new services and capabilities. As the trend continues, there are economic motivations for deploying CR systems. Section 4.6 covered research in cognitive radio technologies, and presented three significant classes of CR applications. Geolocation-enabled applications and authentication applications were discussed in some detail. The most promising CR application is dynamic spectrum access. Suggestions for using OFDM waveforms along with dynamic bit loading are included in this chapter. Solutions to the rendezvous problem are suggested, and the hidden node problem is described.

Section 4.7 covered the timeline in which these technologies will roll out and be integrated into radio equipment and products. Many of the technologies required to provide some of the useful and economically important CR functions already exist, so some of these features should begin to appear within the timeline of the next development cycle.

The bottom line is that the enabling technology for cognitive radio applications is available. There is interest in integrating the technologies to build cognitive applications. Finally, the emergence of cognitive radios and their cognitive applications is imminent.

EXERCISES

4.1 Compare and contrast the use of a "local" or distributed radio environment map and a "global" or centralized radio environment map.

4.2 Sketch a Venn diagram of CR functionality with the following sets: sensors, reasoners, policy engine, and configuration parameters. Connect elements from each set to elements of the other sets with a dotted line to show possible interfaces. Write a description and discussion of your figure in less than one typed page. Be careful to consider the partitioning of workload.

4.3 Discuss the need for a predicate calculus engine in a policy engine. Limit the discussion to one page or less of text and one page or less of figures.

REFERENCES

[1] Federal Communications Commission, Notice of Proposed Rule Making, August 12, 2000.
[2] *www.fas.org/man/dod-101/sys/land/sincgars.htm.*
[3] *www.fas.org/man/dod-101/sys/land/.*
[4] *www.fas.org/spp/military/program/com/an-wsc-3.htm.*
[5] *www.fas.org/man/dod-101/sys/ac/equip/an-arc-164.htm.*
[6] *www.fas.org/man/dod-101/sys/ac/equip/an-arc-220.htm.*
[7] *www.harris.com.*
[8] *www.columbiaelectronics.com/motorola_lst_5b_lst_5c.htm*
[9] *army-technology.com.*
[10] *www.thalescomminc.com.*
[11] *www.fas.org/man/dod-101/sys/ac/equip/csel.htm.*
[12] *www.jcs.mil/j6/cceb/jtidsmidswgnotebookjune2005.pdf.*
[13] *www.fas.org/man/dod-101/sys/ac/equip/an-arc-210.htm.*
[14] *www.afcea.org/signal.*
[15] *www.fas.org/man/dod-101/sys/land/eplrs.htm.*
[16] *www.nj7p.org/history/portable/html.*
[17] *www.raytheon.com/capabilities/products/cibs/.*
[18] *www.acd.itt.com.*
[19] *enterprise.spawar.navy.mil/body.cfm?type=c&category=27&subcat=60.*
[20] *enterprise.spawar.navy.mil/body.cfm?type=c&category=27&subcat=60.*
[21] *www.thalescomminc.com.*
[22] *www.comsoc.org.*
[23] *www.gdc4s.com/.*
[24] *enterprise.spawar.navy.mil/body.cfm?type=c&category=27&subcat=60.*
[25] *www.gnu.org/software/gnuradio.*
[26] *www.vanu.com/?page_id=14.*
[27] Polson, J., Cognitive Radio Applications in Software Defined Radio, Software Defined Radio Forum Technical Conference and Product Exposition, 2004.
[28] Chen, H., S. Kumar, and C.-C. Jay Kuo, QoS-Aware Radio Resource Management Scheme for CDMA Cellular Networks Based on Dynamic Interference Guard Margin (IGM), *Computer Networks*, 46:867–879, 2004.
[29] Bougard, B., S. Pollin, G. Lenoir, L. Van der Perre, F. Catthoor, and W. Dehaene, Energy-Aware Radio Link Control for OFDM-Based WLAN; available at *www.homes.esat.kuleuven. be/~bbougard/Papers/sips04-1.pdf.*
[30] *www.darpa.mil.*
[31] Dana, P. H., The Geographer's Craft Project, Department of Geography, The University of Colorado at Boulder; available at *www.colorado.edu/geography/gcraft/contents.html.*
[32] Azzouz, E., and A. Nandi, *Automatic Modulation Recognition of Communication Signals*, Springer, 1996.
[33] Harris, S., *CISSP All-in-One Exam Guide*, Second Edition. McGraw-Hill, 2002.
[34] *www.bbn.com.*
[35] Russell, S. J., and P. Norvig, *Artificial Intelligence: A Modern Approach*, Second Edition, Pearson Education, 2003.
[36] FCC, Spectrum Policy Task Force Report, ET Docket No. 02-135, November 2002.
[37] Hughes–Hartogs, Ensemble Modem Structure for Imperfect Transmission Media, US Patent 4,679,227, July 7, 1987.
[38] Chow, P., J. Cioffi, and J. Bingham, A Practical Discrete Multitone Transceiver Loading Algorithm for Data Transmission over Spectrally Shaped Channels, *IEEE Transactions on Communications* 43(2/3/4), 1995.

[39] Krongold, B., K. Ramchandran, and D. Jones, Computationally Efficient Optimal Power Allocation Algorithms for Multicarrier Communications Systems. *IEEE Transactions on Communication,* 48:23–27, 2000.

[40] Zhao, Y., J. Gaeddert, K. K. Bae, and J. H. Reed, Radio Environment Map-Enabled Situation-Aware Cognitive Radio Learning Algorithms, SDR Forum Technical Conference, November 2006.

[41] Zhao, Y., L. Morales, J. Gaeddert, K. K. Bae, J. S. Um, and J. H. Reed, Applying Radio Environment Maps to Cognitive Wireless Regional Area Networks, *Proceedings of DYSPAN Conference,* April 2007.

[42] Zhao, Y., D. Raymond, C. daSilva, J. H. Reed, and S. F. Midkiff, Performance Evaluation of Radio Environment Map-Enabled Cognitive Spectrum-Sharing Networks, *Conference Record of IEEE Milcom,* November 2007.

[43] *www.ieee802.org/22/.*

[44] Jaeger, D., M. Vondal, P. Ring, and F. Harris, A Low-Complexity Architecture for Multicarrier Cognitive Radio, SDR Forum Technical Conference, November 2007.

[45] Clancy, T. C., and B. D. Walker, Spectrum Shaping for Interference Management in Cognitive Radio Networks, SDR Forum Technical Conference, November 2006.

[46] Ginsberg, A., W. D. Horne, and J. D. Poston, Cognitive Radio, Spectrum Policy Specification, and the Semantic Web, SDR Forum Technical Conference, November 2006.

[47] Briasco, M., A. Cattoni, G. Oliveri, M. Raffetto, and C. S. Regazzoni, Sensorial Antennas for Radio-Features Extraction in Vehicular Cognitive Applications, SDR Forum Technical Conference, November 2006.

Spectrum Awareness and Access Considerations

Mr. Preston Marshall
*STO, Defense Advanced Research Projects
Agency, Arlingtom, Virginia*

"Spectrum is the 'life blood' of RF communications."

5.1 DYNAMIC SPECTRUM AWARENESS AND ACCESS OBJECTIVES

The wireless designer's paraphrase of the classic New England weather observation could be *"Everyone complains about spectrum availability (or at least the lack of it), but no one does anything about it!"* Cognitive radio (CR) technology offers the opportunity to do something about it (spectrum shortage). Spectrum-aware radios offer the opportunity to fundamentally change how we manage interference, and thus transition the allocation and utilization of spectrum from a command and control structure, dominated by decade-long planning cycles, assumptions of exclusive use, conservative "worst-case" analysis, and litigious regulatory proceedings, to one that is embedded within the individual radios, each of which individually and collectively, implicitly or explicitly, cooperate to optimize the ability of the spectrum to meet the needs of all the RF devices. As we look at this opportunity, we will investigate solutions that range from local brokers that "deal out" spectrum, to totally autonomous systems that operate completely independently of any other structures. In its ultimate incarnation, it is possible to actually use spectrum awareness and adaptation to relax the hardware physical layer performance requirements by avoiding particularly stressing spectrum situations. As such, a CR could ultimately be of lower cost than a less intelligent but higher-performance radio design.

Work on Dynamic Spectrum Access (DSA) was initiated to provide new approaches to managing the scarcity of spectrum and the difficulty of adapting manual and static spectrum planning to increasingly dynamic spectrum dependent systems, and increasing demand for the resource. These objectives are still important drivers for research in the field, but additional rationales for dynamic spectrum behaviors have emerged. The author has argued that the greater benefits of DSA may actually arise from its ability to enable the network to dynamically provide spectrum to enable dynamic network topologies and to address shortfalls in the performance of the analog components [1]. For purposes of this chapter, the following are the three objectives.

1. Provide access to spectrum on an as-needed basis without interference to other users and relinquish this spectrum when needed or used by a protected user.
2. Adapt the selection of spectrum to ensure that the radio only operates within spectrum choices that are within its operating capability and limitations.
3. Provision alternative spectrum choices to a network layer that can adapt its network strategy based on dynamic availability of spectrum.

5.2 PRIOR WORK IN SPECTRUM AWARENESS AND ACCESS

Since the inception of the term *cognitive radio* [2], a number of different approaches to, and definitions of, CR have appeared. For this chapter, the working definition of a CR is a radio that is aware of its environment, and constantly uses this awareness to adjust its operating characteristics and behaviors to provide effective performance in a wide range of environments. In contrast, a conventional or non-CR has these same decisions made during the design process, or at least in advance of its operation, based on assumptions about the likely future environment.

The recent upsurge in interest in DSA and CR technology attests to the rapid growth in understanding and appreciation of its potential to revolutionize the interaction between a radio, its environment, and its user. Although it has been known that the environment has strong (and usually deleterious) effects on the operation of a radio; only recently have we become able to exploit awareness of this environment. Traditionally, we have considered how a fixed mode of operation is degraded in a variety of less than ideal situations. In cognitive radio, there is now the opportunity to investigate how the use of adaptive modes maintains and optimizes performance across a range of actual environments.

Previously, design studies stated an assumed environment for which performance was optimized; a CR design instead can argue for a range of environments, each of which will have individually and continually optimized performance. Radio operation thus transitions from being designed by the radio implementer based on a projected environment, to being dynamically determined based on the radio's continually updated perception of its environment. We can approach this as an opportunity to achieve incremental gains in radio performance, or as an opportunity to fundamentally change the relationship of radios, their environment, and the networks and applications that run over them.

The transition of cognitive radio from concept to reality has made considerable progress in the last several years, driven down two fundamental paths.

The first is the drive to mature, demonstrate, and ultimately deploy DSA systems. A number of commercial and governmental organizations have completed prototype implementations and submitted them for varying degrees of technical or operational scrutiny. This effort broadly clusters into two applications, with the majority of commercial and industry focused on television white space, such as contemplated by the IEEE 802.22 Wireless Regional Area Networks (WRAN) [3], and research into more broad-based spectrum sharing, generally focused on military or public safety as the initial applications.

Progress specific to 802.22 applications has been driven by the opportunities provided by technical achievements and potential regulatory acceptance of television white space utilization [4]. Proposed approaches vary from access to central databases to fully distributed and fused sensing. Of particular interest is the reliance on multiple resolutions of spectrum sensing to control the operation of the device, with both fast (1 ms) and slow sensing cycles (25 ms) driving the adaptation of the access point and customer equipment. Implementations of receivers supporting this architecture have been reported [5], supporting the technical feasibility of this approach. One approach relies on analog spectral analysis and temporal correlation, with only the correlation difference requiring digitization at a very low rate, a departure from the assumption of all digital sensing [6]. Increased processing in the analog domain, and processing of temporal signatures, promises much lower energy consumption and potentially lower cost than high-speed digital approaches. Scheduling of CR sensing within and across communities has been explored and shown to be a feasible approach to the implementation of multidomain CR operation [7].

While much research and discussion has focused on the white space arguments, important and relatively neglected effects of adjacent channel energy are beginning to be discussed within the framework of CR. A simple "occupied/not-occupied" categorization has been the basis of much of the discussion of spectrum availability. Consideration of adjacent band effects [8] showed that secondary use of television frequencies in adjacent bands has minimal effect on television reception, and that reasonable performance is achievable at relatively close distance to an adjacent channel television transmitter.

Perhaps in recognition of both the likely deployment of sensing-based systems, and the history of malicious applications of technology, methods of detection and mitigation of falsified incumbent signals and trust of distributed CR-sensing data have now been proposed. These certainly need to be on the technology agenda if the sensing scope of CRs is to transcend individual trust domains [9]. Without these features, it is difficult to imagine how any user or operator would be willing to trust the operation of a network to a technology so vulnerable to denial of service.

The second major research topic is the increased understanding and initial prototyping of learning-based CR algorithms. Reported work by a number of researchers [10, 11] shows levels of performance that should be sufficient to make the argument for incorporation into operational radios in the near future.

Early work on the representation of knowledge within a CR has also been extended. Yarkin and Arslan [12] described a Radio Knowledge Representation Language that provides a mechanism to organize multiple sources of environmental and location awareness into an integrated representation. Of particular importance, this work provided a canonical method to organize the channel awareness into a structure on which a CR can make operating decisions. In a similar vein, other researchers have extended knowledge representation and encapsulation to describe the state of knowledge of the radio components of the transceiver [13].

Not only has research in DSA advanced the technology over the last several years, but also the regulatory and spectrum community has become increasingly involved in the process, evidenced by the success of the policy program at the First and Second IEEE Conference on New Frontiers in Dynamic Spectrum Access Networks (DYSPAN).

Corresponding interest in industry and regulatory standards is reflected in transition of the IEEE P1900 to a Standards Coordinating Committee (SCC41) for Dynamic Spectrum Access Networks [14]. The breadth and international scope of membership offer the opportunity to advance DSA as an integrated package of technologies, rather than as individual and stand-alone efforts addressed on a nation-by-nation basis.

5.3 SOME END-TO-END DSA EXAMPLE IMPLEMENTATIONS

Several complete purpose-built CR experimental platforms have been developed based on DSA approaches. The Defense Advanced Research Projects Agency (DARPA) has initiated a "second-generation" purpose-built CR program [1]. This program, Wireless Network after Next (WNaN), uses DSA as its fundamental operating principle, and trades high-performance individual transceivers for replicated, but lower-performance transceivers. An example of this philosophy would be adapting around situations that would otherwise cause front-end overload, as will be discussed later in this chapter. One of the stated objectives of this program is to demonstrate that CR can produce at least the performance of a conventional radio, but accomplish this with reduced performance components through use of adaptation to mitigate the performance-stressing environments that would otherwise drive energy consumption and cost. The WNaN radio has four independent transceivers, covers 900 MHz to 6 GHz, has high-quality front-end filters to enable the DSA functions to identify and utilize low-energy preselector bands, and uses commercial-quality Radio Frequency Integrated Circuits (RFICs) for the transceiver functionality. This platform is intended to provide not only the DSA functionality, but cognitive topology management and content management as well.

One of the objectives of the development of the NeXT Generation (XG) program was to provide a clear partitioning of the CR reasoning infrastructure. On one side, there are functional behaviors (e.g., sharing and etiquette) that focus on the device's impacts on other users in the environment. Presumably this behavior is externally mandated, as the radio does not benefit from the constraints, and might have its performance reduced. On the other side, one or more adaptation mechanisms can optimize performance, while constrained by external policies. Not surprisingly, these two research areas have unique approaches to exploiting environmental awareness.

Another of the program's objectives was developing a declarative language and processing tool set for external policy control, such as required by spectrum sharing. The program has developed two radio-borne reasoning engines that provide inferential processing of predicate calculus policies based on a spectrum management ontology and rule set [15, 16].

It is appropriate to partition the intelligence model of a CR into endogenous and exogenous components. Previously, the author had proposed the partitioning of DSA aspects of CR into two essential elements: a System Strategy Reasoner (SSR) that optimizes the spectrum decisions and performance of the device, and a Policy Conformance Reasoner (PCR) that executes and enforces externally provided policies, as shown in Figure 5.1 [17].

The PCR provides the external (exogenous) policy component(s) that address the device's impact on the external environment; primarily avoidance of spectrum

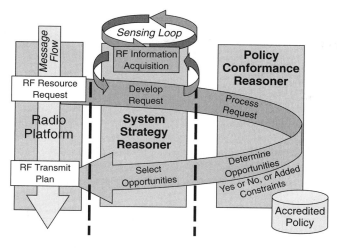

FIGURE 5.1

PCR and SSR model of a DSA cognitive radio.

interference. The SSR provides the internal optimizing behaviors (endogenous) to maximize performance through selection of operating mode and other parameters. This partition reflects the fact that the external effects of the device are difficult, and often impossible to detect from the reference frame of the individual radio. Therefore, it is difficult to argue for an exclusively learning model of a CR, or one based solely on performance-optimizing algorithms.

An example of this partitioning is apparent in the three principles of cognitive radio, which the author proposed as the standard for evaluating the performance of the XG DSA technology [18]. The following are the principles:

1. Do No Harm (to other users of the spectrum)
2. Add Value (to the user, operator, or owner who invested in the technology)
3. Perform (robustly and reliably in a range of environments and user mission needs)

5.4 DYNAMIC SPECTRUM AWARENESS

In the following sections, we study the properties of the spectral environment including understanding how bandwidth interacts with sensitivity, availability, energy levels, and how all these factors interact with the nonlinearities of the radio frequency (RF) front end.

5.4.1 Spectrum Environment Characterization Summary

This section provides an analysis and characterization of the spectrum environment in which a CR will operate. The different spectrum environments characterize the performance-constraining regions that impact wireless systems operation, and must be addressed or mitigated by CR techniques and technology. Environments will be char-

Table 5.1 Spectrum Collections

Sample	Location	Date(s)
Chicago	Illinois Institute of Technology, Chicago [19]	November 16–18, 2005
Riverbend	Riverbend Park, Great Falls, VA [20]	April 7, 2004
Tysons	Tysons Square Center, Vienna, VA [21]	April 9, 2004
New York	Republican National Convention, New York (Days 1 and 2) [22]	August 30, 2004–September 2, 2004
NRAO	National Radio Astronomy Observatory (NRAO), Green Bank, WVA [23]	October 10–11, 2004
Vienna	Shared Spectrum Building Roof, Vienna, VA [24]	December 15–16, 2004

acterized by both empirical and experimental measures, and closed-form approximations that can be applied to generalize the results of the following chapter to any arbitrary environment.

The National Science Foundation (NSF) sponsored a set of spectrum measurements in a number of city and rural environments. The data from these collections were reported by McHenry. The locations are shown in Table 5.1.[1] These data are the basis for the empirical examination of the spectrum due to the comprehensiveness of the surveys, the technical and methodological consistency across collections at multiple sites, and their consistency with reports of other researchers. The technical details regarding the collection of each data set are provided in the references. In general, they consist of fixed antenna locations, a collection duration of 12 to 24 hours, and scan rates of 30–120 seconds.

The sampled environments span the range of potential CR environments. Two of the locations are located in or near the centers of major US cities (Chicago and New York). In contrast, one location, the National Radio Astronomy Observatory (NRAO) in Green Bank, West Virginia, is one of the quietest RF locations on Earth, due to direct actions to eliminate potential RF sources from the environment. The Riverbend location had high foliage absorption, so the upper frequencies were highly attenuated, and therefore quite vacant. The other sites were located within more urban environments, and correspondingly higher signal densities. Note that none of the sites were in proximity to emitters, so that the high collocation energy often encountered by radios was not present in any of the samples. They therefore understate the peak energy that a device immersed in a dense user environment will be subjected to. In examining energy-related performance (such as front-end overload), the measurements reported here should be considered to be a floor on the likely energy density.

Although the measurement sets are technically quite consistent, there are some differences. In some cases, no FM broadcast was recorded. FM broadcast is both a

[1] Note that one site (NSF roof in Arlington, VA) reported in the NSF study was not used in this analysis because the fixed antenna spectrum data above 1 GHz were reported to be unreliable, and therefore no comparable set of values could be obtained. Data from Chicago were substituted for the Arlington data to provide a similar high-energy environment.

high-power and very narrow signal, so its elimination from several of the measurement sets does reduce the high-energy portion of the distribution significantly, although it has minimal impact on spectrum occupancy and availability.

5.4.2 Signal Bandwidth Spectrum Environment Characterization

Dynamic spectrum access has generally been the first fundamental benefit of CR identified, and has been the first that the wireless community has approached. Early work by Mitola [2] pointed out that spectrum occupancy sensing was one of the obvious and potentially useful benefits of placing cognitive features within a radio.

An important metric to determine is the probability that a given frequency, channel, or range of frequencies will be available for use by a CR, consistent with whatever policy controls are imposed on the devices operation. For the purposes of this chapter, this measure is referred to as *SpectrumOpportunity*, which is the probability that a given frequency or channel will be usable by a CR at any given time. *Spectrum Opportunity* is one of the basic characteristics of the environment of a CR, and is simply the ratio of frequencies (at a given fixed bandwidth, b_0) that have power below the threshold compared to the total extent of frequencies. If b_0 is equal to the smallest resolution bandwidth, then this function approaches the uncorrelated mean value of the distribution. As b_0 increases, then the probability of *SpectrumOpportunity* becomes influenced by the degree of correlation of the spectral openings, and is significantly effected by the minimum size of openings (b_0). Both terms will be used interchangeably to describe spectrum characteristics, as convenient.

Although this chapter develops arguments for the performance relationships that are closed-form and independent of any specific measurement set or assumptions, the spectral characteristics shown in the following sections are important to ground the consideration of CR operation, and to instantiate specific modes of performance within these typical environments.

In assessing the occupancy of spectrum, there are three driving parameters: the threshold level at which spectrum is declared occupied, the extent of the spectrum that must be contiguously available, and the contiguous time duration during which the spectrum must be available. A summary of the relative energy distributions of the spectrum samples examined is provided in Figure 5.2. The process used to analyze these will be described in the upcoming section using the Chicago sample set, but the process was performed for all samples, and they differ only in parametrics, not in fundamental character. The sharp rise of the NRAO curve is due to the very low density of signals.

This process begins with characterizing individual spectrum time/frequency/energy profiles. Such a profile is shown in Figure 5.3, in this case Chicago from 30 MHz to 3 GHz approximately every 30 minutes for 24 hours.

Figure 5.4 illustrates a depiction of the distribution of spectral energy as a function of bandwidth and energy. This figure depicts the energy in contiguous bands from 25 kHz through 10 MHz and demonstrates the expected dependence of energy to the contiguous bandwidth.

These fundamental relationships of frequency, energy, bandwidth, and time form the relationships needed to assess the operation of a CR within a given spectrum environment. The rest of this section will establish empirical measures of these relation-

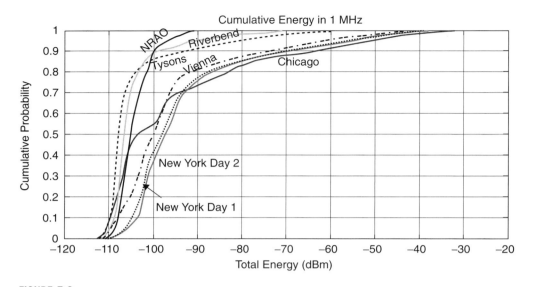

FIGURE 5.2

Summary of spectrum samples for 1-MHz signaling bandwidth.

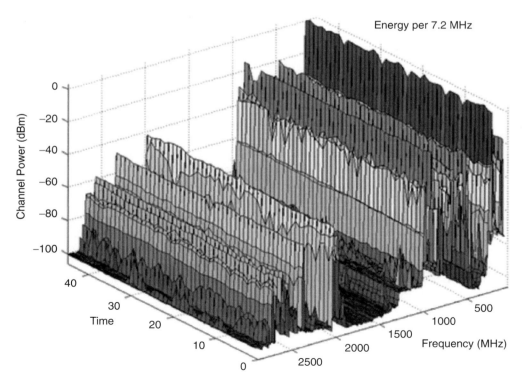

FIGURE 5.3

Twenty-four hours of Chicago spectrum dynamics.

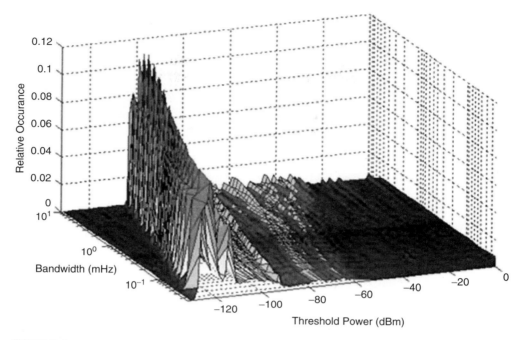

FIGURE 5.4

Spectral distribution of instantaneous energy in the Chicago spectrum sample.

ships, and derive closed-form approximations for the performance estimation of a CR. The dependency of the *SpectrumOccupancy* function on contiguous bandwidth (b_0) is also important. Figure 5.5 illustrates the value of *SpectrumOpportunity* for various ranges of b_0 and threshold power in this same sample. A *SpectrumOccupancy* of zero indicates no opportunities, while one indicates that all spectrum is available.

The *SpectrumOpportunity* function (for small values of occupied bandwidth) is essentially slices of information from Figure 5.5, and forms the basis of a closed-form estimate of spectrum availability for a range of threshold and bandwidth parameters that are spaced at 3 dB intervals. Correspondingly, the resulting energy thresholds themselves are spaced at approximately 3 dB through the usable range of unoccupied spectrum.

The following analysis is performed as a function of signal and front-end bandwidth rather than the more generalized energy per bandwidth, which would be appropriate for independent signals, but would fail to reflect the correlated nature of the signals that will be encountered by a CR.

These data show that a threshold just at the noise floor shows the expected 100 percent occupancy (0 percent or no opportunities), and that the probability of occupancy for higher threshold values decreases quite rapidly once outside of the Gaussian noise region. A 5 dB above the noise level threshold shows quite low occupancy, and would appear to be a reasonable threshold to determine occupancy. The basic shape

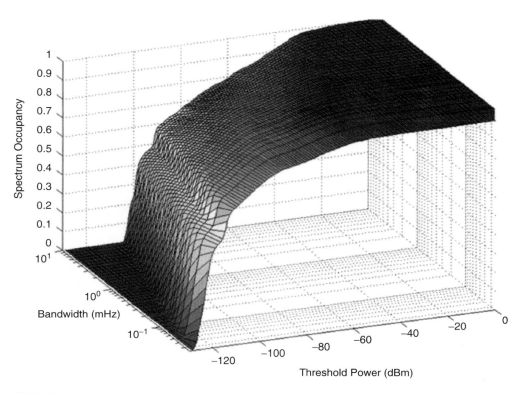

FIGURE 5.5

SpectrumOccupancy for Chicago collection over a range of threshold and occupied bandwidth values.

and character of the spectral density distribution is common to the set of spectrum surveys investigated.

The distributions clearly demonstrate that the shape of spectrum occupancy is essentially indifferent to bandwidth (when sensed as an aggregate, or integrated across the bandwidth), and the integrated bandwidth value (b_0) simply displaces the intercept with the noise floor. The spectrum available at any given point is therefore driven by the level of the threshold above the noise floor associated with the noise bandwidth.

5.4.3 Front-End Energy Distributions and the Importance of Front-End Nonlinearity

In the previous sections we have discussed spectrum occupancy from a perspective that each frequency operated independently of every other frequency, and that the process of examining spectrum was separable, in that each individual frequency was an independent decision. In this section we address the practical effects and constraints of realistic device performance. One of the most important is the effect of total energy on the performance of the receiver front end. Advocates of DSA have often assumed that any available white space would be usable for communications. In fact, white space next to a strong emitter may be unexploitable, regardless of policy. Previously, we examined the energy environment for small signaling bandwidths; in this section, we examine the typically larger front-end preselector bandwidths that drive the intermodulation response of the radio. Another important distinction is that preselector operation is generally proportional to the carrier frequency, whereas in spectrum occupancy, we considered the signaling bandwidth as a constant.

In fact, the drive for addressing these concerns is not specific to DSA systems. In the United States, a long regulatory proceeding was initiated to resolve interference between the towers of a mobile service provider, NEXTEL, and numerous local public safety systems. In the end, the US regulator (the Federal Communications Commission, FCC) elected to relocate the cellular systems through a mix of spectrum offerings and cellular provider contributions to public safety frequency relocation [25]. These systems did not overlap in spectrum, so this was not a frequency management issue as typically defined. But the placement of high-power cellular basestations did have a significant impact on the performance of the public safety radio systems due to the very high energy level in adjacent frequency bands. Similar anecdotal experience is often referred to as co-site interference, as it is often the product of placement of a receiver in close proximity to a relatively strong emitter within the frequency response range of the receiver's front end. It is the contention of this chapter that this front-end overload because of adjacent channel energy[2] is a more common and generally less recognized experience, and that with denser spectrum assignments and as technologies such as DSA become more prevalent, this phenomenon will become a fundamental constraint on wireless networking.

A representative quantitative example of this situation is shown in Figure 5.6. In Figure 5.6(a) the input spectral distribution of eight (unequally spaced) signals is shown as a typical input to a low-noise amplifier (LNA). In this case, the total energy is just below the IIP3 of the amplifier. Figure 5.6(b) illustrates the spectral distribution of the same signal set after amplification by the receiver front-end LNA. In this case, the input signals have separations ranging from 40 to 100 MHz. This generates intermodulation products that intermix to produce artifacts essentially every 10 MHz throughout the octave depicted and beyond. A denser and more complex signal mix would create correspondingly denser and even more complex sets of intermodulation products.

The analysis of spectrum environments that impact front-end linearity is an analog of the spectrum occupancy considerations discussed in the previous section with two important distinctions. First, in addressing linearity concerns, the performance is driven by the high-energy region of the environmental distribution. Therefore, the criteria for the characterizations to be developed in this section are driven by fidelity in the high-energy region of distribution. A second distinction is the differences in how bandwidth selection is performed for demodulation and for front-end energy control.

---

[2]As compared to the intended signal, this is in-band and can be addressed through gain control or attenuation without significant impact on the reception quality.

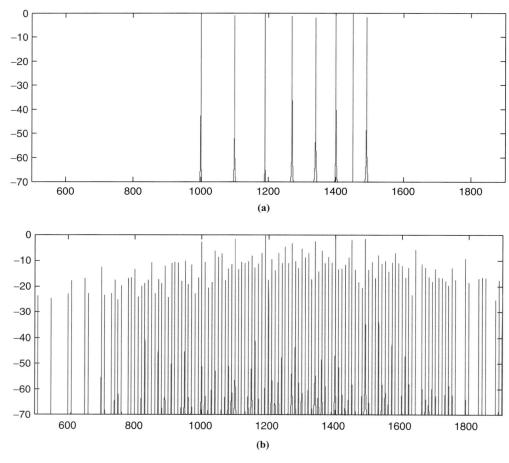

FIGURE 5.6

Effect of low-noise amplifier distortion on spectral distribution: (a) input signal spectral distribution, and (b) LNA output signal distribution.

In a demodulator, it is reasonable to treat the response bandwidth as independent of frequency.[3] In a front-end filter, the bandwidth is both significantly larger (generally), and it is typically proportional to the tuned frequency, assuming a fixed filter Q (or pole count). A filter with a given complexity might have a bandwidth of 100 MHz at 1 GHz, and one of similar complexity would be expected to have a bandwidth in the range of 200 MHz if tuned to 2 GHz. For that reason, front-end filter bandwidth is treated as a constant proportion of the center frequency (BW), rather than an absolute bandwidth.

---

[3] For example, a 10-MHz intermediate frequency–stage response range is constant whether the receiver is tuned to 300 MHz or 3 GHz.

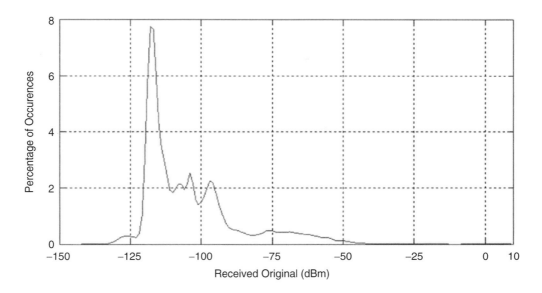

FIGURE 5.7

Histogram of spectral power levels for Chicago Collection (linear).

A summary of spectral power density is shown in Figure 5.7 for a representative collection of spectrum measurements of 25-kHz channels. Although this chart would appear to indicate that low energy dominates the spectrum, the same sample in a log expression demonstrates that a large amount of energy is received in a few sparse, but highly energetic frequencies. More than 1 in 10^5 individual frequencies (25-kHz resolution channels) have power in excess of 1 dBm! The same data are depicted in a log format in Figure 5.8. Additionally, any practical filtering system would be very likely to allow multiple signals to enter the front end, so the energy could be additive in driving the receiver front end into nonlinear behavior. However, the CR is free to select the front-end band of interest, therefore providing the flexibility to avoid these relatively rare, but highly significant situations.

Equally interesting is the aggregation of energy in closely spaced frequencies. Even with reasonable front-end filter performance levels, energy from multiple channels would be aggregated when presented to the LNA and first mixer stage. The correlation of the energy in individual frequencies is thus critically important to understanding the LNA environment.

There are several intermodulation products to consider. The second-order intermodulation (IIP2) creates intermodulation artifacts that are at the sum and difference frequencies. For a receiver with preselector selectivity of less than an octave, these artifacts must fall outside the band of interest ($f_1 \pm f_2$, when f_1 and f_2 are close in frequency) and any signals that would have artifacts within the band of interest are sufficiently attenuated by the filter. Less amenable to filtering strategies, and of particular concern to a less than octave filtered radio, is the third-order intercept product (IIP3), the intermodulation products of which fold back into the original tuning range of the radio ($f_1 \pm f_2 \pm f_1$),

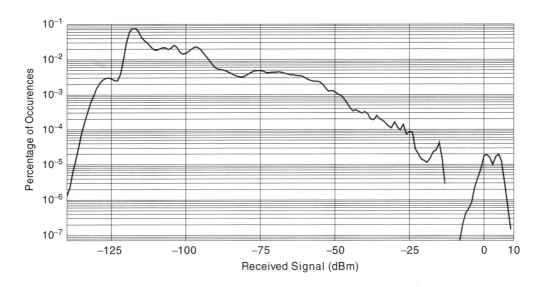

FIGURE 5.8

Histogram of spectral power levels for Chicago collection (log).

creating products mixing the original frequencies with the differences between the original frequencies. Typically, IIP3 values range from –12 dBm for lower cost, low-power consumption devices, to many dBm for high-cost, higher-performance devices.

The cumulative distribution of the total power provided from the antenna through the front-end filter to the front-end LNA stage is shown in Figure 5.9 for a range of filter bandwidths. Overload occurs if the total energy exceeds the IIP3 value of the front end, which for this figure is annotated for a –5 dBm front end.

In the chart in the figure, it is worth noting the intercept of the power surface with the –5 dBm contour, typically the higher end of the IIP3 for medium-cost consumer devices. It thus reflects a value far in excess to the amount of energy an LNA can be exposed to. In a manually assigned spectrum environment, any situation in which this energy level was exceeded, or even approached, would have significant link margin degradation for operation on any frequency within the same filter band. This operating point is thus a critical measure of RF component performance.

An important note is that receiver dynamic range is not just a product of good (or in contrast, bad) design. There is an inherent relationship between the linearity achieved in an amplifier stage and the energy required to power it, for any given value of gain. Additionally, the energy required is not limited to the front-end amplifier alone, but also has an almost linear effect on energy required in other stages, such as the local oscillator and mixer stages.

Historically, a number of strategies have been employed to mitigate the receiver desensitization and nonlinear environment. One approach that is commonly applied is automatic gain control (AGC) circuitry. Such circuitry adjusts one or more stages of receiver gain to remain within the dynamic range of the circuit. AGC to address a stron-

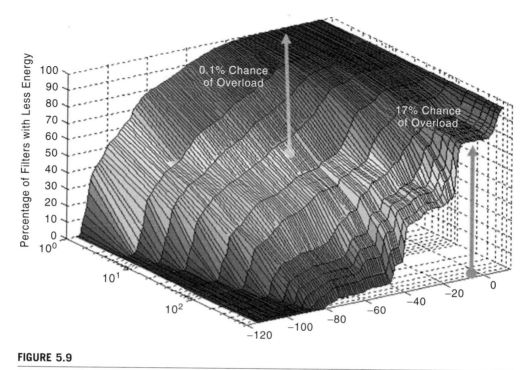

FIGURE 5.9

Total spectral power as a function of filter bandwidth for Chicago collection.

ger than necessary desired signal has negligible effect on link operation, as it is only required in the relatively rare situation in which excess signal is present. The receiver's dynamic range is generally far in excess of the flat bit error rate (BER) portion of the Shannon curve (E_b/N_o), so reducing the gain is acceptable. However, if the AGC operation is required due to an in-band, but adjacent, interferer that is beyond the linear range, then the loss of gain raises the effective receiver noise floor, and thus reduces link performance. AGC is an effective technique for in-channel energy, but compromises link performance if required to reduce the energy of adjacent channel interfering signals.

When AGC response causes the demodulator to loose performance in demodulating the signal (that in the absence of an in-band interferer it could acceptably demodulate), this effect is referred to as desensitization. Desensitization from the demodulated signal itself is not a concern because the receiver operates in a high SNR.

The effect of linearity performance is seen in two areas. The first and most commonly addressed area is in the receive performance, and the second is apparent in the impact on spectrum sensing performance. Nonlinearity effects, or intermodulation on receive performance, generally can be considered to be an elevation of the noise floor as long as the input signals are sufficiently random and decorrelated so that their mixing products are randomized. Assuming that reasonable filtering is provided, the major impacts are due to the third-order products (IIP3), which fall into the original received band regardless of filter width. The second-order effects fall at harmonic distances from the

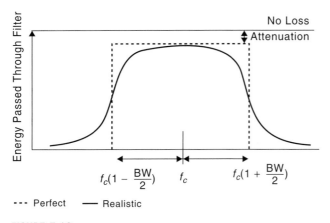

FIGURE 5.10

Perfect filter equivalent compared to a realistic filter.

signal of interest, so only are an issue when over an octave of signal is presented to the LNA. This is not useful configuration for a CR, and is not further considered.

We can now assess the opportunities for effective dynamic range management by a CR. The key elements of this technique are selection of filter frequency, and sometimes, filter width along with associated insertion loss. Filter width is expressed generally as a filter Q (or quality factor), which describes the equivalent Resistor–Capacitor–Inductance (RCL) circuit. One important characteristic of filters is the shape of the filter's skirts, or how much attenuation is provided for signals not within the passband of the filter. To be able to characterize all of these effects from the perspective of a CR, we model the filter to be idealized, as shown in Figure 5.10.

Since our major concern is energy passing through the filter, we reflect the nonideal performance of the filter by characterizing it in terms of the equivalent energy passed through an "ideal" and realistic filter, given the same central frequency attenuation. We examine this in an evenly distributed spectral noise distribution environment, to reflect the mean case. This characterization is an appropriate one, since the CR will make decisions based on environments as they are measured by the device, and our overload function is driven by voltage, not spectral distribution. This characterization is considered to be the "noise bandwidth" of the filter, and closely approximates the 3-dB bandwidth points of a realistic filter with more than two poles. The filter loss is characterized by a_{f0}, and the bandwidth is f_{BW}. By convention, we will characterize the filter bandwidth ratio as BW, which is a fixed proportion of the center frequency (f_c). This representation is relatively proportional to filter complexity, as measured in filter poles.[4]

Considerable design evolution has gone into controlling the relative bandwidth of the passband and the shape of the filter skirts, but for this chapter, we will consider a

---

[4]By contrast, measuring it in absolute bandwidth would mean that the filter's complexity (as measured by poles) would have to vary with center frequency, as a filter at one octave higher in frequency would have to be twice as selective to maintain a constant absolute bandwidth. The selected measure is appropriate to examine the relative trades between performance and complexity in a frequency-independent framework, without constraints of specific filter implementation technique.

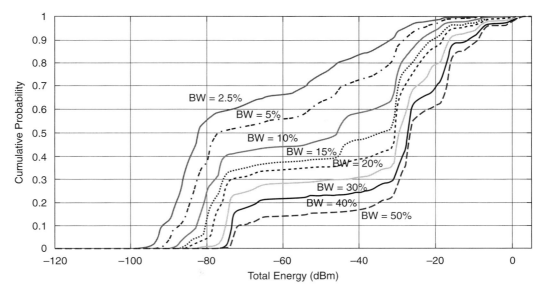

FIGURE 5.11

Illustrative cumulative distribution for bandwidth-constrained spectral energy.

simple equation of filter response as a percentage of center frequency. This function can be applied to any set of spectral occupancy data.

Filter bandwidth is important to us for two reasons. First, as will be shown later, the total energy into the front end is clearly impacted by the width of the filter. If the signals in a segment of spectrum are typically spaced less than the bandwidth of the filter, it is reasonable to assume that the total energy is proportional to the width of the filter. If the energy is dominated by a single or small set of signals, the introduction of narrow filters provides the CR more choices (statistically independent trials) that can enable it to select a frequency in which strong signals are not present.

The cumulative distribution function (CDF) of the Chicago spectrum for the underlying distribution is shown in Figure 5.11 for a range of filter bandwidth ratios. Figure 5.12 illustrates the 10 percent bandwidth energy for a number of the spectrum collections. Similar to the analysis of spectrum occupancy, this distribution is clearly skewed to the low-energy end of the distribution at small bandwidths, and to the high-energy end for larger bandwidths.

An instructive overview of the data presented is to examine the peak and median values (in dB between the minimum and maximum power in the band, similar to the x in the Beta distribution) of each sample and bandwidth. The indices for a range from 0.025 to 10 MHz bandwidth are shown in Figure 5.13. Note that each spectrum sample occupies a unique region that is not highly influenced by the choice of signal bandwidth.[5] The two days of the New York City collection are tightly clustered.

The number of spectrum sample sets currently available is quite limited, so the approach demonstrated here will significantly benefit from additional spectrum samples,

---

[5] The samples are grouped from low bandwidth at the lower median levels, up to the larger bandwidths at the higher median levels.

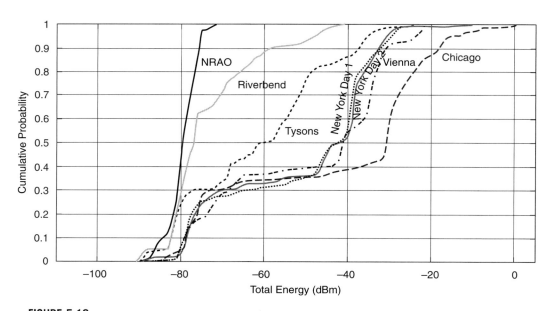

FIGURE 5.12

Energy distributions for a number of locations with 20 percent preselector bandwidth.

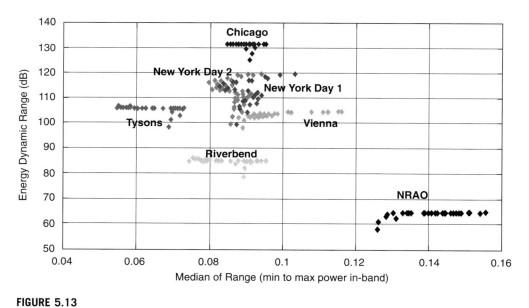

FIGURE 5.13

Spectrum characterization indices.

primarily to remove the inherent correlation present in the small sample base. Additionally, the inclusion of additional environments will enable extension of this approach to deal with the nonindependent, correlated nature of the spectrum environment, which could not be performed with the limited experimental base currently available. Of particular importance is additional characterization of "outlier" environments. It is likely that the NRAO sample is as sparse an environment as is present in normally populated regions, but the other end of the scale, the effects of emitter density in urban and highly colocated environments, is necessary to complete understanding of CR performance in all environments.

5.5 FRONT-END LINEARITY MANAGEMENT

The effects of receiver front-end energy loading is currently a significant factor in the performance of conventional radio devices that inhabit densely populated (with RF devices) environments with either discrete high-power emitters or aggregations of emitters collectively achieving high-energy levels. The future deployment of DSA and CRs will only tend to make this situation become more stressing, as spectrum density is increased by technologies such as DSA. It has been suggested [1] that a CR can adapt the use of spectrum to avoid situations in which the front end will be overloaded or desensitized (e.g., by AGC behavior) by the energy of other than the desired signal.[6] As a minimum, this will reduce performance by raising the effective noise floor, and often may preclude operation of the device at any capacity level.

If the device is operating with static frequency assignments, there is little effective mitigation of this front-end overload condition other than high-performance receiver front-end performance. On the other hand, if the device is permitted to select its own frequencies, then implementing the following basic principle would appear to offer mitigation:

> Frequency should be selected such that the preselector tuning can ensure that the total energy of all signals (intended and adjacent channels) passing through the preselector is constrained to no more than a certain ratio of the overload input energy (e.g., 20 dB below IIP3).

The actual spectrum distribution is outside of the control of the receiver, and is given by the real-world occupancy. The primary control that the receiver has over this is the operation of the preselector parameters. The implementation of CR techniques in this area is specific to the hardware capability and organization, but the following analysis is generally applicable to typical configurations and filter technology. It is not the intention of this chapter to provide an in-depth analysis of filter technology; instead, it is intended to develop the fundamental relationships between the capabilities of the filters, the energy behavior of the LNA, the cognitive radio's algorithms, and the performance impact on the device in a range of environments.

The input preselector bandwidth (typically a significant fraction of the carrier frequency) is the constraint on energy entering the front end, and drives the degree to

---

[6]This effect is commonly referred to as adjacent channel, cosite interference, or receiver desensitization.

which the front-end LNA and mixing stages are overloaded by total energy. The CR must first be able to determine the relationship between the total signal environment it can sense and the amount of noise generated through the intermodulation of signal products prior to the intermediate frequency (IF), or the digitization stages of the receiver, where bandwidth is much more limited, and the signal levels can be tailored through automatic gain control without impact on signal-to-noise ratio.

The standard engineering measurement of intermodulation typically inserts two pure tones, and measures the energy in the intermodulation products, which are therefore also tonal. This situation is of reduced concern to a CR, since the resulting tonal intermodulation products can be avoided through the DSA algorithms by the same mechanisms that avoid any other occupied frequency. However, in complex environments, it is likely that the intermodulation products will be present in large numbers, and also contain significant bandwidth due to the underlying modulation products. When these factors are present, the effect is much less correlated, and approaches an additive white Gaussion noise (AWGN) noise source, composed of many individual intermodulation distortion (IMD) products, with energy falling throughout the tuning range.

Since the nth intermodulation product has a bandwidth n times the original (for equal bandwidths), the effect of the intermodulation is to broaden the intermodulation product significantly, as well as to create multiple occurrences of intermodulation products. For example, if a given piece of spectrum was occupied by 10 signals that collectively occupied 6 percent of the bandwidth, the intermodulation would create 100 signals each of which occupied 1.8 percent individually! Collectively, they would appear as noise to any wideband receiver.

Intermodulation products can fall in a wide range of frequencies, but the ones of most concern to a CR are those generated by LNA intermodulation that fall within an octave range of the inputs—the effect of the third-order intermodulation. From Rodhe and Newkirk [26], we obtain the time domain output of an LNA as:

$$V_{out} = Gain * V_{in} + \frac{-2 * Gain^2}{IIP2_{volts}} V_{in}^2 + \frac{-4 * Gain^3}{3 * IIP3_{volts}} V_{in}^3, \qquad (5.1)$$

where IIP3 and IIP2 are in volts and Gain is absolute.

When examined in the frequency domain, the definition of the IIP3 point allows us to write the two intermodulation products of the third-order intermodulation products of two equal-strength signals ($Energy_{Signal}$) as:

$$Energy_{IMD3} \approx Gain_{LNA} \frac{Energy_{Signal}^3}{IIP3^2} \qquad (5.2)$$

This relationship would provide a simple and straightforward method to compute intermodulation effects, except that radios are generally subject to more complex situations, with a large number of unequal amplitude signals present in the spectrum provided to the LNA and subsequent stages. However, it is still instructive to understand the general effect of the spectral distribution of front-end energy.

If we distribute a fixed amount of front-end energy evenly among n equal-amplitude signals, the resulting total intermodulation distortion (IMD3) induced noise is given by the individual distortion products of the signals times the number of permutations of the two unique input signals, as shown:

$$Energy_{IMD3} = n(n-1)Gain_{LNA}\left(\frac{\left(\frac{Energy_{Totalin}}{n}\right)^3}{IIP3^2}\right) \qquad (5.3)$$

It is clear that for a given amount of input energy, the IMD3 produced is not constant, but is a factor of the distribution of the energy. The IMD3 is different in the situation in which all the energy is concentrated into two signals from the case in which all the energy is distributed among many signals. This is readily apparent by examining the derivative of total IMD3 (Eq. 5.3) taken against the number of signals, for a constant total input energy, as shown:

$$\frac{\partial Energy_{IMD3}}{\partial n} = \left(\frac{(n-2)Gain_{LNA}Energy_{Totalin}^3}{IIP3^2 n^3}\right) \qquad (5.4)$$

Clearly, total IMD3 is maximized when only two signals are present, and minimized when the energy is most evenly $\left(\lim_{n\to\infty}\right)$ distributed. The real-world operating points between these extremes is a function of the characteristics of the individual environments. For a CR to assess the impact of the IMD, it must be able to assess the noise impact of an environment in a straightforward manner. Adaptation mechanisms are dependent on the radio's ability to predict the effect of energy on front-end performance, and thus the effectiveness of the CR adaptations in mitigating these effects.

Assuming a constellation of n signals of amplitudes $\{a_1, a_2, a_3, ..., a_n\}$ and corresponding frequencies $\{f_1, f_2, f_3, ..., f_n\}$, the third-order IMD product of the mixing of any two of them (a_i, a_j) has amplitude $a_{i,j}$ and frequency $f_{i,j}$:

$$a_{i,j} = Gain_{LNA}\frac{a_i^2 a_j}{IIP3^2} \text{ band } a_{j,i} = Gain_{LNA}\frac{a_j^2 a_i}{IIP3^2} \qquad (5.5)$$

$$f_{i,j} = 2f_i - f_j \quad \text{and} \quad f_{j,i} = 2f_j - f_i. \qquad (5.6)$$

The total intermodulation energy is therefore given by:

$$Energy_{IMD3} = \frac{Gain_{LNA}}{IIP3^2}\sum_{i=1}^{n}\sum_{j=1}^{n}a_i^2 a_j \qquad (5.7)$$

5.5.1 **Representative Front-End Linearity of Experimental Collections**

Applying Eq. (5.7) to the spectrum environments described previously yields the IMD3 energy distribution. The BW parameter interpretation is as defined as the fractional bandwidth. For convenience, the energy distributions in this chapter are shown for the equivalent of unity gain, in order to relate the IMD energy products to the equivalent input energy from the antenna (OIP3 = IIP3), and to reflect the equivalent impact as energy at the antenna.

The bulk of the third-order intermodulation energy is distributed over the frequencies between and around the intermodulating frequencies, which are the ones passed by the filters. Evenly distributed, this energy appears as additional noise in the demodulator, as if it had arrived through the antenna.

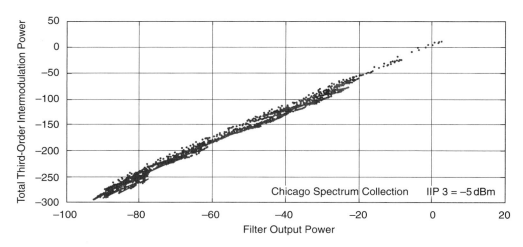

FIGURE 5.14

IMD3 energy as a function of LNA input energy.

We can extend Eq. (5.7) to estimate the energy for any given signaling bandwidth b_0 within the range of: $f_c\left(1\pm\dfrac{BW}{2}\right)$, where $b_0 < f_c\, BW$.

$$Energy_{b_0} = \frac{b_0}{f_c BW}\,\frac{Gain_{LNA}}{IIP3^2}\left(\sum_{i=1}^{n}\sum_{j=1}^{n}(a_i^2 a_j)\right) \qquad (5.8)$$

The predictability of the mapping of input energy to the mean of the distribution of IMD3 product is important, as it is potentially constraining if we are to apply total input energy (after the filter) as a criteria for preselector selection. Figure 5.14 illustrates the relationship between the input energy and IMD3 energy for the same Chicago spectrum measurements and a specific value of IIP3. These results directly scale with IIP3 as shown previously.

The clusters associated with the different IIP3 values are separated by the square of the IIP3 value difference (two times the difference in dB) as would be expected. Applying least squares fit to these collections yields a first-order polynomial fit to estimate the IMD3 noise from the input energy:

$$IMD3 = k_1 Power_{in} - 2\,IIP3 - k_2$$
$$k_1 = 3.25,\ k_2 = 11.8 \qquad (5.9)$$

where IMD3, IIP3, and $Power_{in}$ are in dBm by convention.

These coefficients are consistent with the closed-form (which has a first-order coefficient (k_1) of 3.0) and provide an average of approximately 12-dB less intermodulation energy (k_2) than would be expected if all of the input energy was concentrated into the minimum of just two tones. The increased slope (3.25 vs. 3.0) reflects a slight shift in energy distribution of higher-energy environments toward more energy concentrated (or correlation) in stronger signals, which have a disproportionate impact on IMD3.

To assess the quality of this estimate, we need to verify that the range of intermodulation energies that result from the inputs are within the uncertainty range of likely inputs. For a typical collection, such as Chicago, the standard deviation of the IMD3 estimate is 5.8 dB. This corresponds to an input energy standard deviation of 1.78 dB. The energy detected on any single frequency has a standard deviation of 1.49 dB, so the error in estimating the intermodulation energy is only slightly more than the inherent uncertainty in knowing the exact energy in the channel. The estimator provides us with a tool that can closely estimate intermodulation noise over the usable operating range from where it will be below the front-end noise temperature to approaching the point where the amplifier is clipping as it approaches P_{sat}.

The remaining step is to determine what portion of the energy falls within the signal bandwidth (b_0). This energy is distributed across the range of the filter bandwidth BW f_c. The simplified determination of total energy in the channel is therefore given by Eqs. (5.8) and (5.9). This calculation is practical for even low-performance processors and energy-limited platforms.

In summary, the effect of front-end overload is significant for many of the environments in which a cognitive (or noncognitive) radio will operate. It is possible to establish a straightforward and readily computable polynomial relationship between a course measurement of total energy in each of the front-end filter pass bands and a high confidence estimate of the total energy that will be distributed across the signal bandwidth of interest.

Assessment of the performance of the algorithms will be in terms of the environmentally induced probability of front end overload ($P_{frontendoverload}$) function and intermodulation induced mean noise floor (IIMNF). The first function quantifies the likelihood that a given algorithm and design will experience overload in any specific environment (as described in Chapter 3) and the second determines the likely noise floor induced by intermodulation levels even when below the IIP3 value.

To assess the effectiveness of these techniques, our assessment criteria compare methods that create equivalent reliability and performance operating in the identical environments, but through both cognitive and noncognitive means. There are a large number of elements that enter into consideration of front-end overload; however, we need concern ourselves only with the most driving ones, as many of the secondary ones are monotonic with these first-order effects.

The first metric is the overload probability, which combines the effect of the actual spectral power distribution, the effect of the filtering process on this distribution, and the overload characteristics of the front end. The distribution of this probability for a range of IIP3 and bandwidth ratio values are shown in Figure 5.15. This function is clearly consistent with experience, and with closed-form expectations of the relative participation of IIP3 and filters in overload performance.

In this model, only the third-order distortion is considered, however, if the bandwidth of the signals provided to the LNA was sufficiently wide (more than one octave), then the same function would be replicated for the IIP2 performance. However, an octave filter width in a CR would be a poor performer in most any environment investigated. Although these results appear severe, they match anecdotal experience with wideband receivers in the presence of strong broadcast signals, such as TV or FM radio.

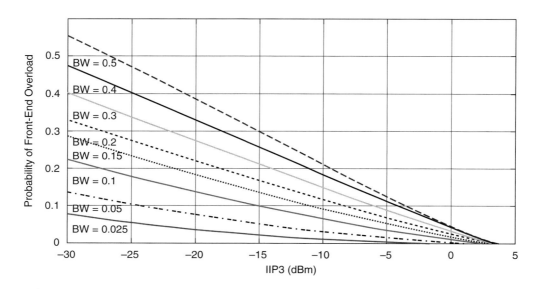

FIGURE 5.15

Non-CR $P_{overload}$ for a range of IIP3 and filter bandwidth ratio.

In the non-CR baseline, we consider the case in which the radio is assigned to any frequency within an operating range, and the filter is adjusted accordingly (or alternatively, is fixed, tuned, and discretely selected). The percentage of time the power in the filter distribution exceeds the limit of the LNA's linear limits (without desensitizing it by AGC, which would reduce adjacent channel sensitivity) is $P_{frontendoverload}$. Since overload conditions are generally long compared to symbol time, this measure should be considered in the context of the packet error rate (PER) of the link or the probability of even closing the link.

It is clear that even high levels of LNA performance (specifically IIP3) are not adequate to ensure high reliability and performance communications in bands that do not have homogenous usage, such as cellular up- and downlinks, satellite links, or other spectrum that has been segregated for "likes with likes" [27]. Section 4.3 compares CR algorithms against these performance benchmarks and determines the required IIP3 to support identical $P_{overload}$ values.

Even if the front end is not driven into a totally distorting region, the intermodulation noise can be a significant contributor to the noise floor encountered by the receiver. From Eq. (5.9), for a given bandwidth and IIP3, the total overload energy is the composite function of the energy distortion function operating on the total energy distribution function. The intermodulation noise induced by various performance front ends is shown as:

$$IMD3(dBm) = \frac{b_0}{BWf_c} k_1 \ FEpower_{in} - 2IIP3 + k_2 \qquad (5.10)$$

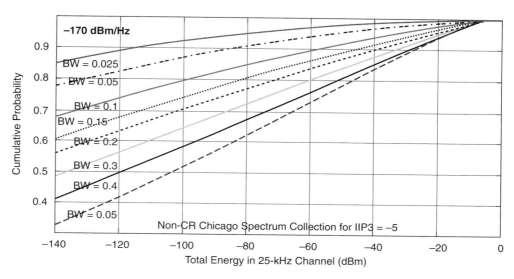

FIGURE 5.16

Non-CR front-end intermodulation-induced noise floor elevation for IIP3 = −5 dBm in Chicago spectrum.

Solving for $FEpower_{in}$ provides:

$$FEpower_{in} = \left(\frac{BWf_c}{k_1 b_0}\right)(IMD_3 - 2IIP3 + k_2) \tag{5.11}$$

The third-order intermodulation energy for the specific case of a −5 dBm front end in the Chicago environment is shown in Figure 5.16. The dotted line represents the point at which the noise energy generated in the front end exceeds the level of the typical receiver background noise value ($\approx$ 170 dBm/Hz), and the link margin would begin to be impacted. For purposes of demonstration, the figure depicts operation centered at 1.5 GHz, and with an IIP3 value typical of a quality consumer-level device. These values can be adjusted to other IIP3s by translating the curves to account for the difference in IIP3.

Figure 5.16 can be read as providing the probability of noise floor elevation for a given front-end performance level. Traditionally, link planners allocate margin for potential in-band interference, and other attenuative phenomena, but they rarely allocate margin to address the impact of adjacent (or within the filter band) channel effects that are internally generated within the radios front end. Yet, even reasonable performing filters have a 20 percent chance of 10 dB increase in noise floor if assigned throughout the spectrum.

A reasonable reaction to this figure might be that radios do not typically fail as often as this metric would imply. In fact, the effect of front-end noise is often not recognized, or addressed by relocation of nodes to resolve cosite. (Note that a CR cannot request other radios to physically move!) Another reason that it is less common in daily use is that in the non-CR environment, spectrum allocations and service architectures inherently avoid this situation. For example, cellular handsets only have to typically receive the closest basestation, and do not have to listen on the same frequencies that the surrounding handsets are transmitting on. The up- and downlink frequencies are sufficiently separated so that fixed filters can reject the energy that would otherwise overload the relatively low IIP3 front ends. However, when contemplating ad hoc, peer-to-peer, and CRs, these architectural protections are not present. A node needs to be able to hear the nodes distant from it, and can reasonably expect to be immersed in other networks doing the same thing. The transition to a more ad hoc and opportunistic vision of wireless networking will force us to abandon the architectural protections of today's services and rely on either greatly increased component performance or adaptation technologies.

5.5.2 Front-End Linearity Management Algorithms and Methods

The algorithms for front-end linearity are an extension of the ones for dynamic spectrum. Even though the dynamic spectrum algorithms look for available bandwidth within the bandwidth of the preselector, the front-end linearity management function must look for preselector bands that have the minimum energy that would cause intermodulation, and thus raise the effective front-end noise floor. In the case of front-end linearity, the bands are defined by the performance characteristics of the preselector filters. This section is based on the assumption that the linearity of the receiver is driven by the wideband RF stages prior to the first mixer stage, or the A-to-D conversion stage.

The operation of the radio in a noncognitive mode is assumed to be assigned randomly over the operating range of the device. Its "average" operating situation is thus at the median value of the spectral energy distribution. The CR is assumed to be able to select the most apparently optimal operating point, and therefore its choices are driven by the filter resolution (essentially its bandwidth) and the tuning range of the device. For example, if the filter bandwidth was 10 percent, and the radio tuned over two octaves, then there would be approximately 15 discrete and statistically independent choices.[7]

Our baseline spectrum density mitigation algorithm is Pick Quietest Band First or PQBF. Intuitively, the algorithm consists of applying all possible filter selection points to the incoming (unfiltered) spectrum and measuring the total power. We imagine that the CR has a set of filter-tuning parameters that each has discrete center frequencies and effective Q factor. This is certainly a reasonable assumption for typical varactor or micro electro-mechanical (MEM) devices, tuned devices that have a digital-to-analog converter to drive the varactor bias or the MEM device's state. Sole reliance on high- and low-pass filters is not considered, as their performance is generally unacceptable in any sophisticated conventional radio or CR.

---

[7]The independence of this selection is somewhat constrained by the statistical characteristics of the spectrum assignment process. For example, if one preselector band is tuned to a TV broadcast signal, it certainly raises the probability that the adjacent one is also a TV band.

PQBF simply locates the lowest energy filter option and thus the minimum front-end linearity challenge. This structure supports both fixed and variable Q filter structures, and cascaded combinations of top hat and tunable filters in series. By implication, this decision is reached on the instantaneous value of the spectrum that was sensed. The number of filter settings (in an idealized, full access, and decorrelated environment) is given by:

$$\text{Preselector Settings} = \text{PSS} = \frac{\log\left(\dfrac{f_{\text{high}}}{f_{\text{low}}}\right)}{\log\left(\dfrac{1 + \dfrac{BW}{2}}{1 - \dfrac{BW}{2}}\right)} \tag{5.12}$$

In fact, the number of preselector settings that are likely to be available are typically less than this because of limitations on the use of spectrum, and correlation of usage (some bands may be less likely to be usable if their neighbor is unusable). This can be reflected in reducing the value of PSS accordingly.

The performance of the PQBF algorithm is shown in Figure 5.17 for the same spectrum data sets and filter bandwidth values as in Figure 5.15 for one octave of tunable range. Note that this allows significant reduction in the level of IIP3 required for equivalent levels of intermodulation events. The effect of filter bandwidth is compounded, since the filter bandwidth has a significant effect on the amount of energy admitted to the front end, and as the filter becomes narrower, it also provides more

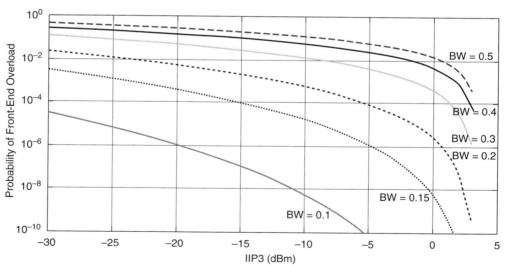

Note: Bandwidths below 10% (2.5% and 5% cases) are too low to depict on scale.

FIGURE 5.17

Probability of overload for PQBF algorithm over one octave for various bandwidth ratios.

trials in which to select a band without excess energy. Note that several of the band-widths do not appear within the range of the graph because the values of $P_{overload}$ for all IIP3 levels are below the lowest axis (10^{-10})!

Inspection of these results leads directly to the conclusion that the most important resource to avoid front-end overload in a CR is the filter; increases in IIP3 are much less significant in reducing the probability that a given node will be overloaded. There is no obvious comparison between a cognitive and non-CR; the exponentially better perfor-mance of the CR cannot be achieved by any reasonable noncognitive alternative. For example, for an IIP3 of –5 dBm and a filter bandwidth of 20 percent, the non-CR has a 0.04 probability chance of overload, while the equivalent CR would have one of 10^{-7}.

A similar probability distribution can be developed for the distribution of inter-modulation induced noise floor elevation. In this case, the probability is the possibility that for a given amount of noise, none of the filter selections had lower energy. This is inherently a binomial probability of determining that the number of settings drawn would have less than a given threshold of noise. Considering it as the probability that all choices were above a given threshold enables us to adapt the single band probabil-ity of overload by raising it to the power of the number of preselector settings.

Figure 5.18 illustrates the distribution of noise floor elevations for a set of represen-tative component performance levels. The CR is able, with high probability, to locate

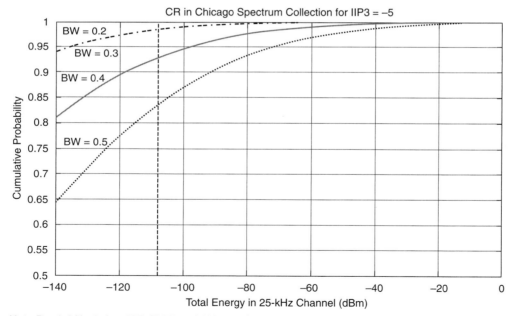

Note: Bandwidths below 20% (2.5% and 5% cases) are too close to 1.0 to depict on scale.

FIGURE 5.18

Probability distribution of intermodulation-induced noise floor elevation when using PQBF algorithm.

spectrum that has minimal intermodulation noise, given that it has both a reasonable filter selectivity and a frequency range to provide selection candidates from. This curve is essentially unchanged for levels of IIP3 as low as –20 dBm, with typically a 3 dB increase in IMD3 through the range. Filter bandwidth values better than 30 percent show low IIP3 product noise even with low IIP3 performance in the densest environment investigated.

Radio designers have generally been reluctant to include high-selectivity filters in front of the LNA stages in order to avoid the noise figure degradation that the insertion loss of the filters would cause. It is clear that in dense and stressing environments the use of high-selectivity filters has significant benefits for both reliability of operation and for noise floor reduction, despite the signal attenuation that this design choice implies. In dense radio environments, intermodulation effects dominate over the effects of a fixed attenuation in the signal path.

A more complex algorithm recognizes that while there is certainly a unique lowest-energy band, there are probably a large number of selections that are essentially indistinguishable from each other in terms of their likely total intermodulation energy; however, they may actually be beneficial choices for other reasons such as propagation characteristics. The Consider Marginal Noise Impacts (CMNI) algorithm defers selection of specific bands, and provides the topology determination layer with a set of choices, along with the likely noise floor impact of each choice.

Figure 5.19 shows the probability of various levels of noise for the Chicago sample, for an IIP3 of –5 dBm. Even the worst filter has a range of choices that are within a single dB of the noise floor, and a further range of 5 percent or so with slightly elevated noise levels, but which still could be attractive for other reasons (antenna performance, propagation, etc.).

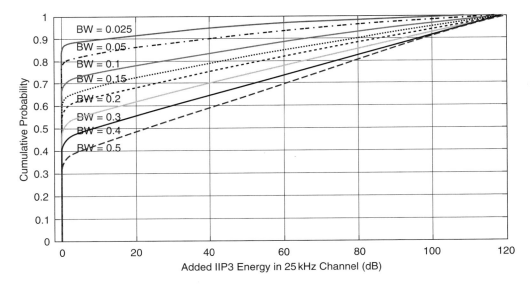

FIGURE 5.19

Probability of additional intermodulation-induced noise.

The overload and noise impact performance of this algorithm is essentially unchanged from the PQBF algorithm. It has the advantage that we can apply the noise distributions developed previously and characterize a set of choices that a network could make without significant regret.

Substituting an acceptable noise floor impact for $IMD3_{CR}$ yields the CDF of the spectrum energy distribution from no noise impact to the noise floor, and thus the number of preselector choices that are available. For example, setting this equation equal to the noise floor would result in the ratio of choices yielding a maximum 3 dB increase in the noise predicted for the CR. In contrast to PQBF, this algorithm fully implements the philosophy and structure of Section 5.4.2, in that it defers the choice of frequency until the considerations from other layers are concatenated, and an overall performance metric can be determined. As such, it serves as an extensible building block for CRs that can address upper-layer considerations of range, power and energy, topology, and reliability in a globally optimized network structure.

5.5.3 Front-End Linearity Management Benefits

There are two measures of benefit for front-end linearity that we will examine. The first is the reduction in noise floor that can be achieved by selecting filter settings that minimize the noise generated in the receiver's front end; the second is the corresponding ability to reduce the performance of the receiver's components and still maintain equivalent performance. These are two sides of the same coin. However, a high-performance application may look to CR to resolve issues of reliability that are induced by cosite interference, while more cost-sensitive applications may look to CR to enable lower cost or energy-consumption devices to perform equivalently.

The performance effects of linearity management are highly dependent on the signal environment. Underneath a cell tower, the downlink may have a very uniform signal density. On the other hand, in a public safety band, there may be a great mix of signal strengths. Therefore, each technique must be considered in a variety of environments, as characterized previously.

The effect of front-end overload is so significant that it is highly likely that an appropriately sized communications link will either fail to acquire or fail to achieve operation within the error-correcting regime for which it was designed. For that reason, we treat the extreme overload case as a link-reliability issue rather than an incremental contributor to BER, or as an adjustment to the link throughput. This reflects the inherently bistatic nature of the link closure process; either it forms a link above a certain threshold or it completely fails to do so.

To understand the performance improvement offered by CR, we can examine the improvement offered by the adaptations, which is

$$\text{Benefit}_{Poverload} = \frac{\text{Probability of Overload of non-CR}}{\text{Probability of Overload of CR}} \quad (5.13)$$

which simply reduces to:

$$\text{Benefit}_{Poverload} = (P_{overloadNCR})^{-PSS+1} \quad (5.14)$$

where PSS is the number of preselector settings.

Note that for PSS = 1, the performance is identical (= 1). The benefits of the adaptation mechanism are directly related to the performance of the nonadaptive radio. If the chance of overload in a given environment is 5 percent, then with PSS = 10, the chance of overload in this environment is $(0.05)^9$. These performance relationships are quite extreme, and essentially unbounded, since this benefit equation is the improvement in reliability that is achieved through adaptation of the preselector setting. Some illustrative curves of the benefits of adaptation are shown in Figure 5.20. Note that the gains for low-filter bandwidths are irrelevant because the underlying performance is already generally acceptable, and there are orders of magnitude increases in reliability even for one octave of coverage with filter bandwidth ratios in the 20 percent range.

Another way to exploit the adaptation is to examine the levels of IIP3 that create equivalent performance. This is an equally important consideration by providing affordability benefits achieved through reductions in performance of the underlying components, through use of CR technology, while still maintaining equivalent performance levels. To analyze the component performance alternatives, the performance of each approach is set equal. These relationships are shown in Figure 5.21.

A cognitive radio with only a 30 percent filter and IIP3 of –27 dB has the equivalent probability of overload as a noncognitive one at –5 dBm IIP3! The performance of a CR with a 20 percent filter shows more than 30 dB of benefit at $P = 2$ percent.

Eq. (5.10) established the performance of a cognitive and non-CR within given spectrum environments. The process for computing this performance metric is similar in approach to the one described for front-end overload, except focused on the linear effects rather than the discrete overload versus not overloaded conditions examined

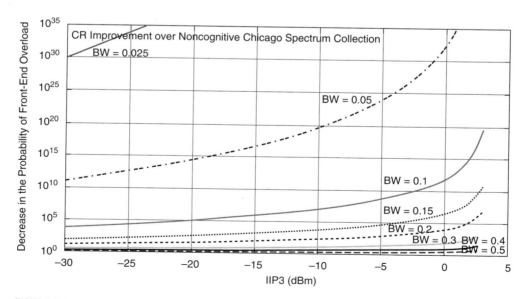

FIGURE 5.20

Illustrative improvement in the probability of front-end overload.

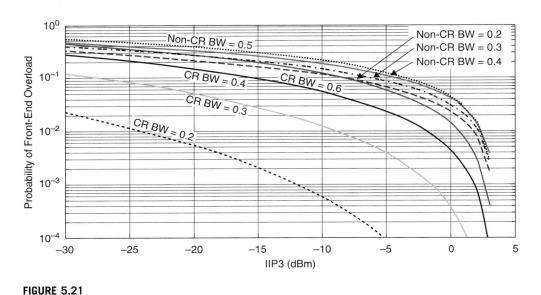

FIGURE 5.21

Values of IIP3 and PSS for equivalent values of $P_{overload}$ in conventional radios and CRs.

previously. Since we consider that significantly elevated noise levels are not typical in any radio, the mean or median is not particularly insightful to understanding the benefits of adaptation. The crucial issue is to achieve equivalent confidence in performance. Therefore, we wish to measure the decrease in noise for a fixed probability. A dB ratio of non-CR to cognitive radio IIP3-induced noise is shown in Figure 5.22.

Much of this benefit is reduction below the effective noise floor, and therefore yields no performance advantage despite the very noise floor differences. Figure 5.23 illustrates the 90 percent confidence level (10 percent of the environments will be worse than depicted) and capping the noise floor at –170 dBm/Hz.

The noise energy is sufficiently high so that all of the IIP3 values generate IMD3 above the 170 dB/Hz level for the 40 and 50 percent bandwidth filter. The 30 percent filter has constant benefit until the IIP3 is sufficiently high to reduce the IMD below the noise, and the extremely narrow filters (2.5 to 20 percent) have diminished improvement as the IIP3 is improved. The benefits of the CR adaptation become even more extreme in high-performance environments. Throughout the typical operating region of –10 to –5 dBm IIP3 and 25 percent filters, the benefits in the 90 percent case exceed 40 dB.

The introduction of CR technology enables significant reductions in the component performance needed to meet identical performance levels. The primary interest is reduction of required LNA linearity, as reflected in IIP3 levels and in filter resolution, or BW factor. Solving for identical noise ratios (benefit = 1) and varying IIP3 shows the anticipated reduction in component performance for LNAs as a function of the difference in the energy distribution. A numerical analysis is provided in Figure 5.24.

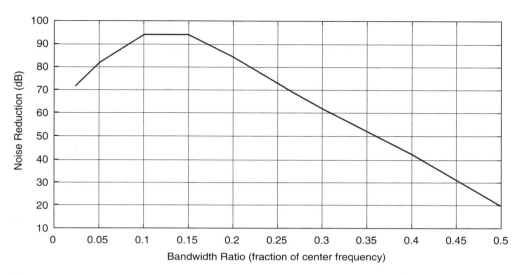

FIGURE 5.22

Median IMD3 noise reduction through cognitive radio.

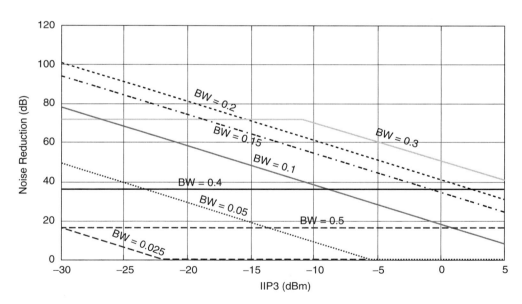

FIGURE 5.23

Beneficial region of CR noise reduction.

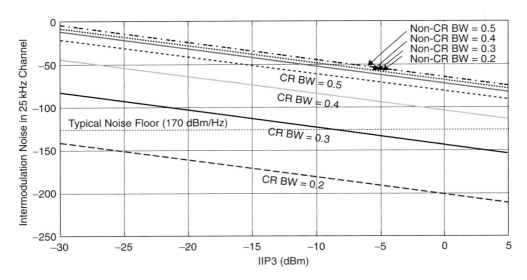

FIGURE 5.24

The 90 percent high-energy equivalently performing combinations of IIP3 and PSS for cognitive and non-CRs.

The benefits of adaption increase rapidly with even poor-performing preselector filters, or more specifically with values of PSS (preselector option that can be obtained through filter performance or octave coverage). Use of adaptation with even 30 percent filters yields equivalent performance at 30 dB less IIP3 performance. The sensitivity to PSS is clear, since the 50 percent filter only has a 5 dB advantage!

5.6 DYNAMIC SPECTRUM ACCESS OBJECTIVES

This section establishes the process and anticipated performance of dynamic spectrum access within CR systems. This is important for three reasons:

1. The DSA mechanism is an enabling requirement for many of the techniques and technologies discussed in other chapters, and it is an inherent element in any adaptive wireless structure. The DSA-enabled techniques include front-end linearity management, dynamic topology, and waveform selection. Even if the spectrum management benefits were nonexistent, the flexibility it affords to manage other aspects of the wireless environment justifies its inclusion.

2. DSA offers capability benefits through pooling of spectrum resources and managing spectrum conflict resolution dynamically.

3. DSA radios are inherently interference tolerant, and therefore radios sharing spectrum with CRs can be more aggressive in their spectrum reuse policies since, if they occasionally cause interference, the interference can be resolved directly by the CR.

In examining spectrum utilization for CRs, the following four spectrum usage density (SD) situations should be considered.

1. SD_{NC-NC} : The spectrum density of non-CRs that must be guaranteed a high level of interference protection, since they cannot be adaptive and mitigate in-channel interference for stationary nodes the location of which is known.
2. SDM_{NC-MNC}: As before, but for nodes that are also mobile within an operating region, need protection throughout the region, and could cause interference to other nodes throughout the region.
3. SD_{CR-NC}: The spectrum density of CRs sharing spectrum with non-CRs that must be guaranteed a high level of interference protection, since the radios cannot be adaptive and mitigate in-channel interference. No guarantees are assumed for the CRs.
4. SD_{CR-CR}: The spectrum density of CRs sharing spectrum with other CRs that are not guaranteed interference protection, since CRs can adaptively mitigate in-channel interference. The ceiling on capacity is given by maximizing the aggregate effective throughput of the radios in the band, rather than the performance of any given radio.

Before discussing a cognitive spectrum process, consider the classical spectrum management and assignment case. Once radios moved beyond spark-gap techniques (the original impulsive ultra-wideband radio), use of the spectrum has been deconflicted to avoid interference. Spectrum and frequency managers assign individual radios or networks discrete frequencies and attempt to ensure that the emissions from one do not adversely impact others. A not insignificant legal (and seemingly smaller technical) community has grown up around this simple principle.

A key measure of a DSA system is its ability to provision more spectrum access in order to create a corresponding increase in network capability. In examining the operation of DSA performance, we will consider three cases:

1. A manually de-conflicted spectrum that is statically assigned without specific (real-time) knowledge of user location, actual usage, or bandwidth needs.
2. Cognitive radios that share spectrum with, and avoid, interference with non-cooperative incumbent systems.
3. Cognitive radios sharing spectrum with other CRs, minimizing, but not necessarily avoiding, interference with other users, on the assumption that they can resolve interference autonomously.

It is important to recognize that this conservative manual planning is not inherent in the operation of the radio links; statistically it is significant only because it represents a set of cases in which the radio system would have no ability to operate since, without adaptation even if it recognizes that an interference condition exists, it cannot unilaterally implement and coordinate a strategy for migrating to a clear channel. In this chapter, we mostly consider spectrum strategies that use awareness to locate spectrum holes that are themselves often the result of the essential conservative nature of the planning process. However, an equally important rationale for their inclusion in real systems is the ability to locally resolve interference by using the same behaviors as used to locate new and unblocked spectrum. This feature of interference adaptive radios offers all users of the spectrum the ability to back off of the currently conservative assumptions that underlay spectrum planning.

Spectrum and frequency planners are inherently disadvantaged by a number of factors. For one, they have to assume that: interfering signals will propagate to the maximum possible range; and desired signals will need to be received without unacceptable link-margin degradation in the worst-case propagation conditions.

In practice, this means that interference analysis is often driven by two unlikely conditions, maximal propagation of interfering signals, and minimal propagation of the desired signal. Although individual situations vary widely, the range of conditions has been measured and its distribution characterized in a number of environments [28]. A summary of one set of measurements of the propagation exponent (n) and fade loss (σ) random variables is shown in Figure 5.25 (from [28]).

Figure 5.26 illustrates the relationship between the various communications and interference ranges involved in DSA. Case (a) is the desired high-assurance communications range, which must assume worst-case propagation (α_{wc}) and fade, and still ensure a signal level above $E_{receive}$. This is a conservative range, but typical for the assumptions required for high-assurance link planning. Case (b) is the range by which radios must be separated for manual frequency de-confliction, reflecting that the victim radio may be in

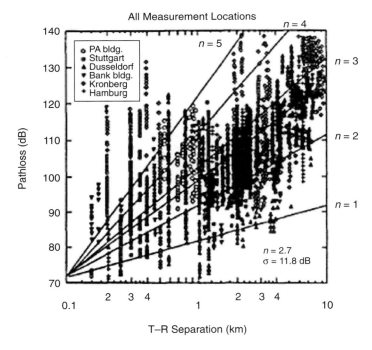

FIGURE 5.25

Illustrative spectrum measurements of propagation exponent and fade random variable.

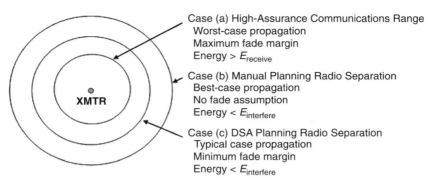

FIGURE 5.26

Practical interference margins.

an advantaged position to the transmitter (α_{bc}) with no fade condition present, and that the signal level in that situation must be below the interference threshold ($E_{interfere}$). The DSA separation, shown in case (c), is between the best and worst-case propagation (α_{tc}); significant fading is not present, and the interference is also limited to $E_{interfere}$.

Most of this pessimism is not inherent in the operation of radios; it is strictly a consequence of having to plan for "edge cases" in advance of knowledge of the actual conditions. Static spectrum planning must assume that links operate at the maximally stressing conditions, while interference will occur when links are maximally configured to cause interference.

Dynamic spectrum access enables CRs to reduce both of these margins to reflect the actual conditions present for both intended and victim links. This produced capability through two mechanisms:

1. The CR can exploit spectrum by reflecting other spectrum users actual usage (time, frequency, power, etc.), and thus create more assignments in the same spectrum.
2. CRs can tolerate more aggressive spectrum reuse because they can move their spectrum assignments in response to any interference they do receive, allowing other radios to be less conservative in sharing spectrum with them.

The deployment of CR will thus yield two increments of benefit: the first when CRs more effectively share spectrum with non-CRs, and the second when CRs can assume that other radios are cognitive and thus can mitigate any (hopefully rare) situations of interference that they may have caused. This latter assumption allows the radio to reduce its protection margin since the network as a whole can tolerate occasional interference without disruptive consequences.

5.6.1 Interference-Intolerant Operation

As stated, the mobility metric is driven by the requirement that spectrum be deconflicted over the entire range the device might be located in, and out to the interference range, extending the interference range of the device, as depicted in Figure 5.27.

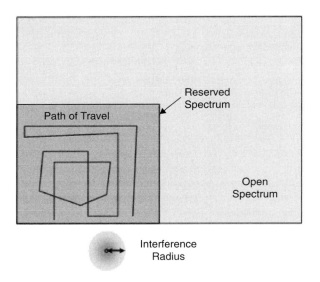

FIGURE 5.27

Determination of mobility factor in spectrum management.

In this figure, the spectrum manager is aware of all of the paths (or locations of operation) that the device might operate, plus the interference radius by which it must be de-conflicted. This creates a reserved area (actually a volume, but the third dimension is rarely exploitable, so is ignored for this discussion) for which spectrum cannot be used for any other purpose without risk of mutual interference. If there is no pre-knowledge of mobility, the spectrum must be reserved over the entire extent for which operation is permitted to the level of operational availability required (A_0). Note that the interference radius must be bilateral; it must include the effect of the power spectral density of both devices. In considering the determination of interference radius, it must reflect the larger of the two directionalities; that is, the greater of the range at which A might interfere with B, or B might interfere with A.

Power management enables the link to operate with power appropriate to the actual conditions of the link, rather than those of the worst-case conditions and fade margin. The conditions described here are rare enough not to influence the mean or median value of the operating network, but are significant enough to substantially influence its reliability. In many networks it is impractical to provide power management. The hardware used in the network is often not capable of significant ranges of output power; the feedback mechanisms cannot be provided in a simplex device; there are a significant number of receivers that must receive the broadcast; or some of the receivers have no feedback mechanism (e.g., a receive-only terminal). Although power management may be present, the spectrum planning process must assume that it does not reduce the maximal radiated power.

The *dutycycle* is the amount of time for which spectrum must be reserved compared to the actual time during which it is actually used. The definition of the term *used* is not necessarily just the time for which energy is being transmitted. Its evaluation is somewhat more subtle because of two effects. First, the time during which the receiver is sensitive to interference is certainly use of the spectrum and should be considered as "used." Also, time periods that are too short to be exploited by other users are essentially "used." Interleaving independent (noncooperative) users within medium access control (MAC) layer intervals is generally not practical, and thus the entire operating time could be considered as being used.

We postulate that CR offers the ability to manage this situation more effectively by using the ability to sense the actual propagation conditions that occur and to adjust the radio dynamically to best fit these conditions. To do this, we distinguish its operation with two objectives. In the first, it attempts to minimize its own spectral "footprint," consistent with the environment and needs of the networks it supports. In the second, it adapts itself to fit within whatever spectrum is available based on local surveillance. When put together, we can conceive of a radio that can find holes and morph its emissions to fit within one or more of the holes. Such radios could offer radio services without any explicit assignment of spectrum, and still be capable of providing high-confidence services.

This author proposed a structure for segregating these two operating policies in 2004, and proposed it to the ITU as a starting point for regulatory consideration of CRs, shown in Figure 5.1. This partitioning was later adopted as the basis for the DARPA XG demonstrations. In the model the DSA spectrum reasoner is free to locate solutions that maximize the performance of the wireless device.

One fundamental difference in performance between cognitive and non-CRs is in how they obtain and access spectrum. Non-CRs generally obtain spectrum in one of two ways:

Assigned. Assigned spectrum is typically assigned to a given user or usage from a regulatory authority (or by delegation from one) and is typically assumed to be exclusive or preemptive use. Typically there is an assurance of noninterference with this class of spectrum. Broadcast services, cellular, satellite, and public safety are examples of this class.

Commons. Commons spectrum is provided for use by a number of users, generally with some technical or operational constraints. There is no assurance of availability or noninterference. Examples of this class include the industrial, scientific, and manufacturing (ISM) bands commonly referred to as unlicensed.

The effect of duty cycle is somewhat subtle. If the duty cycle is 25 percent, then does that mean that there is an opportunity to load four times as many radios? This is possible on a mean value, assuming that the system can use spectrum whenever available. This is certainly an appropriate model for a system that is intrinsically tolerant of access delays, such as a delay-tolerant network (DTN) [29], but for most applications would not be acceptable. The extent to which duty cycle can be exploited is a function of the size of the pool of spectrum. For example, 10 channels shared among 40 users is quite different statistically from 100 channels shared among 400 users in terms of the reliability it can deliver. Therefore, we introduce two additional parametrics to fully specify an environment: the required availability (A_0) and the pool size (pool).

The availability of a given number of channels (needed) at a given duty cycle and pool size is the binomial distribution, where a success is that the channel is accessible, the probability of success is $1 - duty$.

$$A_0 = \sum_{k=needed}^{Npool} \binom{Npool}{k} (1-duty)^k \, duty^{Npool-k} \text{ for needed} \leq N_{pool} \quad \textbf{(5.15)}$$

For large pool sizes, this becomes a normal distribution, and a more convenient description of the CDF uses the regularized incomplete beta function (I_x):

$$A_0 = 1 - P_x (X \leq needed - 1) = 1 - I_{1-duty} (\text{pool-needed-1 needed}). \quad \textbf{(5.16)}$$

The benefits of a statistically large pool of spectrum are clear in the values of A_0 for various values of relative spectrum availability, and are shown in Figure 5.28. The horizontal axis is the degree of "excess" spectrum provided. *Excess* is spectrum beyond the expected value of the product of the number of nodes and the duty cycle.

Without a sufficiently large number of radios sharing spectrum, the pool size is too small statistically to ensure access to spectrum at high enough confidence to support reliable operations. For example, in the above case of a 25 percent radio, high-confidence operation ($A_0 > 98$ percent) requires essentially spectrum for each radio in a pool of 10 radios, at least twice the mean value for a pool of 20 radios, but only 25 percent margin above the mean for a set of 160 radios sharing the spectrum. All users benefit from large-scale spectrum pooling.

The metric by which these will be assessed is *SpectralDensity*. This metric is highly sensitive to individual designs, environments, and usage patterns, but its derivative

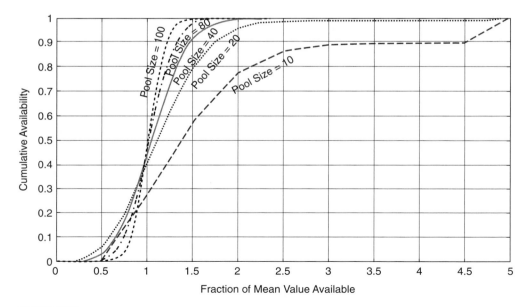

FIGURE 5.28

Representative values of spectrum probability of availability (A_0) for 20 percent duty cycle.

(holding these assumptions constant) will be shown to be insightful into the ability of CRs to achieve higher usage densities from fixed portions of the spectrum.

$$SpectralDensity = \frac{\sum_{i=1}^{n} dutycycle_i \cdot bandusage_i}{Bandwidth_0 \cdot Area_0} \qquad (5.17)$$

where:

n = number of users in the spectrum
$dutycycle_i$ = the duty cycle of user i
$bandusage_i$ = the instantaneous spectral bandwidth used by user i
$bandwidth_0$ = the total bandwidth made available to the set of users
$Area_0$ = the geographic area over which the spectrum is used

Clearly, the optimum situation is to have spectrum available for use and allocation in all locations not within the interference radius of the potential emitter. This density is thus one device per the interference area. The density of a de-conflicted mobile device is the operating area, plus the region surrounding the perimeter of the operating area out to the interference radius. Our measure is the ratio of the maximal possible device density over the achievable device density.

$$Area\ Effectiveness = \frac{InterferenceArea}{DeconflictedArea}$$

$$= 1 + \frac{2\pi r_{interference}^2}{DeconflictedArea + r_{interference}Deconflicted_{perimeter}} \qquad (5.18)$$

where

$r_{interference}$ = interference radius of the worst-case link pairing
$Deconflict_{area}$ = the de-conflicted mobility area
$Deconflict_{perimeter}$ = the de-conflicted mobility perimeter

The first case is a simple two-way communications between two vehicles that will navigate around the continental United States. The communications range of the device is 7 kilometers, so a reasonable estimate of its interference radius is 22 kilometers. (estimated by assuming diffractive propagation in an r^4 environment and a 20-dB SNR at the edge of the operating range). Since the United States has an approximate area of 9.6 million square kilometers, the spectrum effectiveness is 1.5×10^{-4}. It is even lower when considering the effect of duty cycle. If we assume that the vehicles operate for 8 hours a day, and talk at most 10 percent of the time, the actual utilization drops to 5.2×10^{-6}! This is a very good reason to believe that the CR technology has a target-rich environment to improve these practices.

A more reasonable approach might be to only dedicate a piece of spectrum for a single city. Los Angeles has an area of 12,000 square kilometers. Adding the perimeter that must also be protected from use, the total reservation increases to 22,000 square kilometers. Using the same operating assumptions as before, the effectiveness is approximately 2.3×10^{-3}.

The benefits of increased spectrum utilization can be expressed as either additional wireless capability, or the reduction in constellation order, or through introduction of spreading. These can be equated to capacity linearly (assuming spectrum usage is reduced), or to reduction in energy (using the same spectrum, but at lower modulation order) through Shannon limit analysis.

It should be noted that many systems of wireless devices already implement more advanced techniques than the baseline. For example, wireless hubs may search for open channels, cellular devices may have open slots or frequencies assigned, and the 802.11a DFS. These examples do not argue against this baseline; they represent the first (albeit simple) implementations of dynamic spectrum systems, and thus cognitive radio.

5.6.2 Interference-Tolerant DSA Operation

In the previous section, we considered the initial case of DSA radios sharing spectrum with devices that were not tolerant of any interference to their operation, as they were presumed not to have DSA capability, and therefore any energy in their communications channel was considered to be noise that would degrade their link margin. In this section, we instead consider the case where the incumbent radios have DSA capability, and use that capability to not only locate open spectrum, but also to mitigate the effects of interference with their own network. The effect of this change is profound: instead of having to create essentially near-zero interference operation, they are allowed to create a possibility of interference, with that interference level constrained to be low enough so that the aggregate network costs of frequency relocation do not exceed the additional capability created by these more aggressive spectrum usage practices.

The approach of spectrum outage probability (SOP) provides a model for determining the probability that the aggregate signal strength from a set of homogenous emitters exceeds a set emissions mask at a fixed location within a network [30, 31]. We consider the difference in node density that is permitted in spectrum environments that must provide incumbent users confidence of noninterference, such as in shared spectrum, and the maximum density for nodes in the spectrum in which nodes must accept a probability of having some level of interfering signals. In this latter case, it is assumed that the device can relocate itself in the spectrum after determining that it is interfered with. For purposes of analysis, we consider this as a set of discrete choices, but in practice, a device could select multiple contiguous or noncontiguous opportunities.

Pinto and Win [30] show that the SOP of the total environment of an infinite plane of a Poissonly distributed field of nodes is given by an alpha-stable distribution in the general case. Although the SOP formulation considers a range of frequencies and interference masks, we need consider only the primary frequency in this analysis and, for both simplicity and generality, assume a flat energy distribution within the transmission band. Our interest is the relative density of CRs and non-CRs, and the results scale with differing values of interference threshold. We also look at the situation from the perspective of the decisions of the interferer, so we will examine only the effects of a single interferer.

The distinction of this analysis from the similar intent of Gupta-Kumar is that in Gupta-Kumar, the assumption was that an interference event caused a loss of throughput due to the failure of the information transfer. In this analysis, we will assume that the consequence of a interference event is a forced transition to a new frequency, and the cost of this transition is not a failure but a temporary loss of capability while the transition is performed. The mean effect on throughput in a statistically independent environment is thus:

$$Capacity = \frac{T_{\text{sensing}}}{T_{\text{sensing}} + SOP \cdot T_{\text{rendezvous}}} \tag{5.19}$$

where:

SOP = spectrum outage probability
T_{sensing} = the interval between sensing intervals
$T_{\text{rendezvous}}$ = the time to re-rendezvous the physical layer

This capacity measure is 1 if there are no occurrences of a spectrum outage. In fact, for short intervals of time (T_{sensing}) the assumption of independence is very conservative, as the movement of nodes is typically much slower than the sensing rate. This value is thus an upper bound on the rate of disruption and thus is a lower bound on the capacity. A unitless generalization of this relationship can be constructed by substituting the relationship of the rendezvous time to the sensing interval, referred to as the DSA index.

$$Capacity = \frac{1}{1 + SOP \times I_{\text{DSA}}}, \quad \text{where } I_{\text{DSA}} = \frac{T_{\text{rendezvous}}}{T_{\text{sensing}}} \tag{5.20}$$

Figure 5.29 illustrates the relationship of capacity, SOP, and the DSA index for some representative values of each (a sensing interval of 100 ms and a re-rendezvous time of 186 ms, or an I_{DSA} of 1.86). This range of I_{DSA} values can operate at even a SOP of 10^{-1} with only a 15 percent degradation of channel access performance.

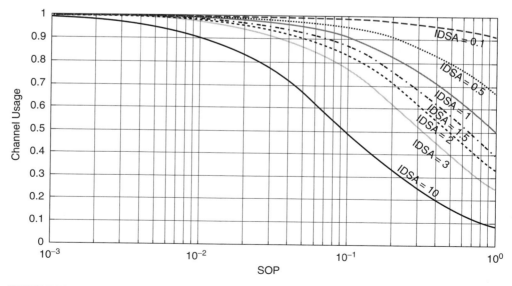

FIGURE 5.29

Capacity for a typical set of sensing rate and rendezvous times.

While a 10^{-1} chance of causing interference would be unacceptable to a non-DSA radio sharing a band, it is not a significant performance impediment to a DSA radio sharing the same band. The opportunity provided by DSA-to-DSA spectrum sharing is to greatly increase the density of nodes, maximizing the aggregate throughput. The product of node density times channel access provides the total traffic density per unit area and spectrum allotment.

5.7 SPECTRAL FOOTPRINT MANAGEMENT OBJECTIVES

There are two sides to the CR problem. The well-known one is fitting into the spectral footprint of other radios. The second, and more subtle one, is to minimize the radio's own footprint. For years, modulation designers have defined spectrum efficiency as the number of bits/hertz of bandwidth and have often used this metric as a scalar measure of the best modulation. The assumption has been that the design that used the least spectrum was intrinsically less consumptive of shared spectrum resources. This proposition is worth examining. The Shannon bound argues that essentially infinite bits/hertz can be achieved, but it can only be accomplished by increasing the energy per bit exponentially; thus the radiated spectral power increases nearly at the third power of the spectral information density. We broaden the view of spectrum impact to include not only the amount of spectrum used, but also the area over which it propagates until it is no longer significant in relationship to the noise floor. It is worth noting that volume would be a more generalized measure; however, since most spectrum usage is on the surface of the Earth, we will limit our consideration to two rather than three dimensions.

One problem is how to define spectral efficiency metrics. Classically, digital radio engineers have attempted to minimize the spectrum used by signals through maximization of bits/hertz. This is a simple and readily measured metric. Shannon [32] shows a basic relationship between energy and the maximum possible bits/hertz.

We can see that the proportionate cost in energy of going from 1 to 2 bits/hertz is essentially the same as that required to go from 6 to 7. Essentially, arbitrary bits/hertz are possible if the channel is sufficiently stable and power is available. But it is tough on battery-powered devices!

It is equally tough on the other users of the spectrum. To increase the bits/hertz, the radio must now transmit both slightly more bits, and vastly more energy per bit to meet the Eb/N_o requirements of the receiver. If we look at the spectral energy density distribution, we see the increase joules/hertz as we increase the constellation depth or order. This shows spectral energy increase as we increase the constellation order (bits/hertz). The progression of spectrum energy is even more progressive than the Shannon bound itself because of the multiplicative effect of the number of bits being transmitted.

Using Shannon, we can compute the increase in spectral energy required to increase spectrum efficiency from 0.5 bits/hertz up to 10. If simple spectrum were the measure, such a radio strategy would be effective. However, for most frequencies, spectrum is a valuable asset over a given region. Therefore, *bits per unit area* is an appropriate measure for how effectively we use spectrum. Extending this bound to consider propagation is important if we want to understand how this denies spectrum usage to the radios. Ignoring multipath and absorption, propagation between terrestrial antennas can be simply modeled as r^α, with α varying from 2 to 4 depending on distance, antenna height, and frequency [33].

In this case, we compute the change in effective interference region for each point in the Shannon curve and for cardinal values of the propagation constant. Clearly, simplistic spectrum strategies that only look at bits/hertz are not effective and are quite counterproductive for the all spectrum users as a whole. The better the propagation, the more ineffective increasing energy is a strategy for spectrum usage. Note that even small increases in spectral efficiency have major and disproportionate impacts on the interference footprint of the radio transmissions.

For most propagation conditions it is clear that by increasing the spectral efficiency of one radio, we have disproportionately reduced the ability of the spectrum to support other users. Since α generally increases with frequency (for propagation in VHF and above), this implies that more sophisticated strategies are needed. Increasing modulation constellation depth is a poor solution for this radio because it greatly increases the energy it needs; and it is a poor solution for the other radios sharing the spectrum because it essentially raises the noise floor throughout a greatly increased region or precludes operation at rapidly increasing radiuses from the transmitter. The effect of modulation order on the system's carrying capacity of a number of nodes is quite severe.

This leads to a measure that recognizes this trade. In this case, bits/hertz/area reflects that the critical issue in spectrum optimization is not spectrum use: It is spectrum reuse, and we want to measure and optimize not just how the radios perform themselves, but also how the radios allow other radios to share the spectrum in a close to globally optimal manner. The relative value of this measure is shown in Figure 5.30. When one

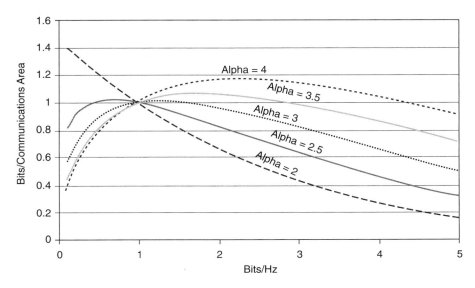

FIGURE 5.30

Impact of constellation order on bits/hertz/area.

includes path attenuation in the determination of effectiveness, the conclusion can change dramatically.

Whereas a wireless node in an r^2 environment is only 40 percent as efficient in going from 2 to 4 bits/hertz, while in a r^4 environment, the same change is efficiency neutral. This adds another dimension to the algorithms for a CR. If it, or its spectrum access protocol or etiquette is attempting to optimize overall spectrum utilization, then awareness of the relative antenna height and range is required to determine between alternatives that are either significantly right or dramatically wrong. Measuring propagation loss is not enough because the critical issue is how fast the signal is attenuated on the far side of the receiver. Logically, if we have 12-db SNR, then the signal is above the noise for another two times the link range in the r^2 environment, and only one in the case of r^4. The preceding discussion assumes that the power was perfectly matched to the channel; in practice, such an alignment is impossible to achieve or maintain over any degree of time-varying channel, so real results will fall above these values.

When discussing footprint management, we must recognize that the application of this optimization is appropriate for those situations in which the radio has some specific benefit from being cooperative to the other users of the spectrum. Although many authors have assumed that a shared spectrum would resemble the historic "tragedy of the commons," there are fundamental reasons to believe that even uncooperative users would not intentionally destroy the utility of the spectrum for all. A dramatic difference is that radios stop "polluting" once the transmission stops. There is no permanent degradation (once the multipath ringing time is over) that was the derivation of the original term. In fact, there are two situations where we can conceive that the footprint would be something a radio would desire to control. The first is that the radio was part of a network, and there is a high probability that the network would benefit from the decreased footprint. The other is when there is an established or mandated protocol or etiquette that requires or causes the devices to minimize their impact on other users. Work by Neel is also supportive of the concept that nondestructive operation is inherent in the operation of self-interested devices in the spectrum [34].

5.8 IMPLICATIONS ON NETWORK-LEVEL DECISION MAKING

The preceding sections have made a quantitative argument that lower-layer CR adaptations have a significant positive and profound effect on network performance and/or affordability. Lower-layer adaptations provide a corresponding opportunity for profound changes in the upper layer [35–38]. These adaptations also are enabling for a corresponding set of network and upper-layer technology changes that can further enhance capability. This section assesses additional optimizing methods that can be applied above the unilateral physical and media access decision making as discussed previously. Because the benefits of these decisions cannot be determined for specific environments and network application traffic characteristics, they are not integrated into the quantitative assessment. Despite their inherently situational nature, they provide a strong argument that likely CR performance can exceed the benefits developed in the previous sections by extending the framework and principles to upper layers. The methods described here are transitional between the physical layer–focused CR and the network and upper layer–focused cognitive wireless network device that has not yet been described but is the logical next step in this technology path.

In this chapter, we investigated the benefits to the device and to the network of violating the principles of layer abstraction. We assumed that a CR has insight into at least the type of upper-layer traffic that is being offered to the device in the simplest case. A more complex case is possible when the node also has knowledge of some aspects of the topology by which packets are being delivered and can consider these facts in determining the device's operating point.

5.8.1 DSA-Enabled Dynamic Bandwidth Topology

Previously in this chapter, we discussed the opportunities for dynamic spectrum assignment in a CR. We now extend that concept to dynamically vary the bandwidth associated with the communications link, presumably in response to instantaneous demands on the network. The benefits of this is that within a given spectrum usage, the dynamics of the user demand can be reflected in the bandwidth allocated (and deallocated) to each link. Although the core networks of the Internet are planned with the principle that network demand is Gaussian or Poisson, as one moves closer to the edge of these networks, the traffic demand becomes highly time correlated. A cell phone is quite correlated; if it is in use at one second, it is very likely that it will be in use the next, and if not in use, remain not in use. Even packet services have this characteristic, with users going through alternating page retrieval and reading phases, and in most cases, completely inactive modes. With thousands of users sharing portions of the Internet, the sum of these variations are Gaussian, and average into the classic Gaussian stream, but at the wireless edge they are far from even.

Leveraging this characteristic has been difficult because most wireless systems have a fixed spectrum environment, forcing the network to be constructed of fixed and invariant pipes. Much of the wireless networking infrastructure is derived from the technologies first developed for, and employed in, the fixed Internet. Therefore, the ability to adapt bandwidth dynamically is not present in these wireless systems. Even the wireless-specific technology developed for Mobile Ad Hoc Networking (MANET) focuses on routing decisions and awareness, not topology optimization.

In this section we consider how CR wireless network nodes can balance spectrum and throughput bandwidth to respond to the dynamics of the traffic. This is one capability that wireless networks uniquely enjoy and can exploit. The simplest version of this technology would be an access point that varied the spectrum in use as the traffic queue in the router or a client terminal increased and relinquished that spectrum to other access points when lower bandwidth footprints could meet the quality of service thresholds and minimize any queuing.

The key question in examining node bandwidth adaptation is the time constant of the network traffic. Fixed networks typically replan their bandwidth on intervals in the range of six months. The planning, execution, and routing information exchange has both time delay and resource cost implications, so the CR must constantly balance the time correlation of the traffic load, the cost of adaptation, and the benefits of adaptation. As in dynamic spectrum, the benefits of tighter bandwidth allocation can be used to either improve capability or to lower constellation sizes due to spectrum availability.

5.8.2 **DSA-Enabled Dynamic Topology and Network Organization**

Although much of the published literature for cognitive radios has focused on the characteristics of physical layer adaptation, it can be argued that the network layer can leverage the flexibility enabled by the DSA and CR model providing significant benefits. DSA, by eliminating the specific assignment of individual links to specific frequencies and bandwidths, enables adaptation not only of the physical layer but also in the organization of the network itself. Some of the adaptations that WNaN will perform to address and mitigate the effects of a range of environments, device limitations, and network delivery requirements are shown in Figure 5.31. These adaptations have a prerequisite that the network be able to unilaterally select frequency and bandwidth for each link.

If bandwidth in one region of the network is inadequate, the network can "move" spectrum through locating or reassigning spectrum to supplement the throughput. If a link is interfered with by other signals or overloaded by adjacent channel energy, the CR locates spectrum more suitable to its operation and is free to implement that decision locally and unilaterally. If there is no way to locate enough spectrum in a congested region of the network, the network changes its topology to route out of and around the congested region.

The DSA community has advocated the acceptance of DSA technology to address many of the shortfalls in manual frequency planning and management. The DARPA

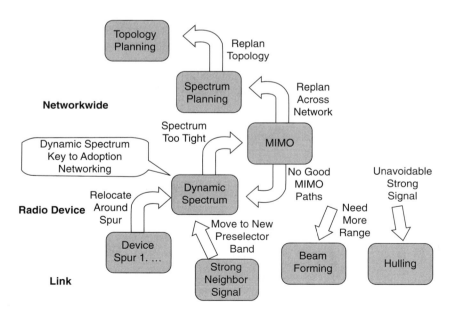

FIGURE 5.31

Upper-layer exploitation of DSA flexibility. *Note:* Each technology can throw "tough" situations to other more suitable technologies without impact on user QoS.

WNaN program approach additionally argues for the adoption of DSA because of the flexibility it affords:

- To the device for self-management of the environment and the resulting reduction in component performance requirements, and thus cost.
- To the spectrum manager to not have to separate receivers from strong sources of adjacent band energy.
- To the network and applications to reconfigure and reprovision wireless bandwidth dynamically.

Simply put, it argues that a DSA cognitive radio has advantages, even if access to spectrum is neutral.

5.9 SUMMARY

Spectrum awareness and DSA are important tools in the CR toolkit for ensuring access to spectrum and are logical substitutions for much of the manual planning that is now the basis for spectrum management. However, the spectrum access benefits are only a subset of the potential benefits of CR and the new opportunities it provides to both improve performance and to reduce the implementation cost of wireless devices. There are essentially two strategies for the application of DSA. The first is as an extension of noninterfering operation; this offers significant increases in spectrum access. The second is when one radio can assume that another is cognitive and resolve interference locally. The allowable interference rate then is not driven by reliability but by overall effectiveness, much the same as in MAC-layer design where occasional collisions are tolerated in the pursuit of high channel utilization.

EXERCISES

5.1 Using available equipment at your facility, perform a survey of local spectrum usage there.
  (a) Compare these to the cities shown in this chapter.
  (b) Identify strong local signals.
  (c) Determine if the noise floor was within the actual noise level, or was the result of intermodulation distortion.
5.2 Relocate the spectrum sensor and obtain the same data.
  (a) Compare these with the data from Exercise 10.1, and determine how correlated the two locations are.
  (b) Plot the degree of correlation as a function of frequency.
5.3 Set up multiple signal generators and a spectrum analyzer.
  (a) Inject two low-level signals that do not raise the noise floor.
  (b) Increase the injection energy until the noise floor is impacted. Compute the effective two-tone IIP3. Add additional signals and observe the resulting noise level. Recompute the effective IIP3.
5.4 There are a number of different models of spectrum energy distributions. Describe the phenomenological rationale for selection.

5.5 Take a spectrum analyzer into the proximity of strong broadcast transmitters (television or FM broadcast) and plot the strength of the strongest broadcast signal, and also a number of weaker ones. Insert a 5 to 10 dB attenuator in the signal path and replot the measurements. Identify which signals were "true" and which were second- and third-order intermodulation artifacts. (*Hint:* The true ones will be attenuated by the value of the attenuator, the second-order ones by twice (in dB) and third-order ones by three times.) Repeat adding attenuation until intermodulation products disappear into the noise. Identify the attenuation levels needed to reduce intermodulation below the noise floor. Measure the effective spectrum occupancy for each overload state and plot how perceived spectrum occupancy changes as a function of overload condition.

REFERENCES

[1] DARPA, Wireless Network after Next, Proposer Information Pamphlet; available at *www.darpa.mil/sto/solicitations/WNaN/pip.htm*.

[2] Mitola, J., Cognitive Radio: An Integrated Agent Architecture for Software Defined Radio, Dissertation, Doctor of Technology, Royal Institute of Technology (KTH), Sweden, May 2000.

[3] IEEE 802.22 Working Group on Wireless Regional Area Networks, Functional Requirements for the 802.22 WRAN Standard; available at *www.ieee802.org/22*.

[4] FCC, Notice for Proposed Rulemaking (NPRM 03 322): Facilitating Opportunities for Flexible, Efficient and Reliable Spectrum Use Employing Spectrum Agile Radios Technologies, ET Docket No. 03 108, December 2003.

[5] Park, J., T. Song, J. Hur, S. M. Lee, J. Choi, K. Kim, J. Lee, K. Lim, C. H. Lee, H. Kim, and J. Lasker, A Fully Integrated UHF Receiver with Multi-Resolution Spectrum Sensing (MRSS) Functionality for IEEE 802.22 Cognitive Radio Applications, IEEE International Solid State Circuits Conference, February 2008.

[6] Hur, Y., J. Park, W. Woo, J. S. Lee, K. Lim, C. H. Lee, H. S. Kim, and J. Laskar, A Cognitive Radio System Employing a Dual-Stage Spectrum Sensing Technique: A Multi-Resolution Spectrum Sensing and a Temporal Signature Detection Technique, Global Telecommunications Conference, pp. 1–5, November 2006.

[7] Hu, W., D. Willkomm, M. Abusubaih, J. Gross, G. Vlantis, M. Gerla, and A. Wolisz, Dynamic Frequency Hopping for Efficient IEEE 802.22 Operation, *IEEE Communications Magazine*, May 2007.

[8] Petty, V., et al., Feasibility of Dynamic Spectrum Access in Underutilized Television Bands, Second IEEE International Symposium on New Frontiers in Dynamic Spectrum Access Networks, pp. 331–339, April 2007.

[9] Chen, R., J. Park, Y. Hou, and J. H. Reed, Toward Secure Distributed Spectrum Sensing in Cognitive Radio Networks, *IEEE Communications Magazine*, April:50–55, 2008.

[10] Rondeau, T., Application of Artificial Intelligence to Wireless Communications, Dissertation. Virginia Polytechnic Institute and State University, September 2007.

[11] Mody, A., S. Blatt, N. Thammakhoune, T. McElwain, J. Niedzwiecki, D. Mills, M. Sherman, and C. Myers, Machine Learning Based Cognitive Communications in White As Well As Gray Space, *IEEE Military Communications Conference*, 26(1):2321–2327, October 2007.

[12] Yarkin, S., and H. Arslan, Exploiting Location Awareness Toward Improved Wireless Systems Design in Cognitive Radio, *IEEE Communications Magazine*, 46(1):128–135, 2008.

[13] Nolan, K., P. Sutton, and L. Doyle, An Encapsulation for Reasoning, Learning, Knowledge Representation, and Reconfiguration Cognitive Radio Elements, First International Conference on Cognitive Radio Oriented Wireless Networks and Communications, pp. 1–5, June 2006.

[14] IEEE Standards Coordinating Committee 41 (Dynamic Spectrum Access Networks); available at *www.scc41.org/*.

[15] Elenius, D., G. Denker, M. Stehr, R. Senanayake, C. Talcott, and D. Wilkins, CoRaL—Policy Language and Reasoning Techniques for Spectrum Policies, IEEE Workshop on Policies for Distributed Systems and Networks, June 2007.

[16] Perich, F., Policy-based Network Management for NeXt Generation Spectrum Access Control, IEEE International Symposium on New Frontiers in Dynamic Spectrum Access Networks, April 2007.

[17] Marshall, P., XG Communications Program. Information Briefing, Semantic Web Applications for National Security Conference, April 2005.

[18] McHenry, M., E. Livsics, T. Nguyen, and N. Majumdar, XG Dynamic Spectrum Access Field Test Results, *IEEE Communications Magazine*, 45(6):51–57, 2007.

[19] McHenry, M., P. Tenhula, D. McCloskey, D. Roberson, and C. Hood, Chicago Spectrum Occupancy Measurements and Analysis and a Long-Term Studies Proposal, *Proceedings First International Workshop on Technology and Policy for Accessing Spectrum*, Boston, 2006.

[20] McHenry, M., and K. Steadman, Spectrum Occupancy Measurements, Location 1 of 6: Riverbend Park, Great Falls, VA, NSF Shared Spectrum Company Report, August, 2005.

[21] McHenry, M., and K. Steadman, Spectrum Occupancy Measurements, Location 2 of 6: Tysons Square Center, Vienna, VA, April 9, 2004, NSF Shared Spectrum Company Report, August 2005.

[22] McHenry, M., and K. Steadman, Spectrum Occupancy Measurements, Location 4 of 6: Republican National Convention, New York, August 30, 2004–September 3, 2004, Revision 2, NSF Shared Spectrum Company Report, August 2005.

[23] McHenry, M., and K. Steadman, Spectrum Occupancy Measurements, Location 5 of 6: National Radio Astronomy Observatory, Green Bank, WVA, October 10–11, 2004, Revision 3, NSF Shared Spectrum Company Report, August 2005.

[24] McHenry, M., and K. Steadman, Spectrum Occupancy Measurements, Location 6 of 6: Shared Spectrum Building Roof, Vienna, VA, December 15–16, 2004, NSF Shared Spectrum Company Report, August 2005.

[25] Luna, L., Nextel Interference Debate Rages On, *Mobile Radio Technology*, 1 August 2003.

[26] Rhode, U. L., and D. P. Newkirk, *RF/Microwave Circuit Design for Wireless Applications*, Wiley, 2000.

[27] Federal Communications Commission, Spectrum Policy Task Force Report, ET Docket No. 02-135, p. 22, November 2002.

[28] Andersen, J. B., T. S. Rappaport, and S. Yoshida, Propagation Measurements and Models for Wireless Communications Channels, *IEEE Communications Magazine*, 33(1):42–49, 1995.

[29] Cerf, et al., Delay-Tolerant Network Architecture, IETF RFC 4838, April 2007.

[30] Pinto, P. C., Communication in a Poisson Field of Interferers, Master's thesis, Department of Electrical Engineering and Computer Science, MIT, 2006.

[31] Pinto, P. C., and M. Z. Win, Spectral Outage Due to Aggregate Interference in a Poisson Field of Nodes, *GLOBECOM*, 2006.

[32] Shannon, C. E., A Mathematical Theory of Communication, *Bell System Technical Journal*, 27(July):379–423, (October):623–656, 1948.

[33] Marshall, P., Spectrum Awareness, *Cognitive Radio Technology*, Chapter 5, B. Fette (ed.), Elsevier, 2007.

[34] Neel, J., Analysis and Design of Cognitive Radio Networks and Distributed Resource Management Algorithms, Dissertation at Virginia Polytechnic Institute, 2006.

[35] Marshall, P., Unmet Challenges in the Technology to Exploit Dynamic Spectrum and Networking, First International Workshop on Technology and Policy for Accessing Spectrum (TAPAS), at the IEEE Wireless Internet Conference, Boston, August 2006.

[36] Marshall, P., Spectrum Management and Other Ongoing Cross-Layer DARPA Initiatives, NSF NETS Focus Area Symposium, Arlington, VA, February 2004.

[37] Marshall, P., Adaptation and Integration Across the Layers of Self-Organizing Wireless Networks, MILCOM Conference, Armed Forces Communications and Electronics Association (AFCEA), Washington, DC, October 2006.

[38] Marshall, P., Connectionless Networks: Radios and Networking Technology for Sensor Applications, Conference on Unattended Ground Sensor Technologies and Applications VII, International Society for Optical Engineering Symposium on Defense and Security, Orlando, March 2005.

Cognitive Policy Engines

Robert J. Wellington
University of Minnesota
St. Paul, Minnesota

6.1 THE PROMISE OF POLICY MANAGEMENT FOR RADIOS

In familiar usage, policies are procedural statements expressing administrative conventions that are adopted by various organizational entities. The concept of automatic policy management of resources has its commercial roots in the administration of information systems and networks. Policy management refers to a particular approach for automating network management activities by specifying organizational objectives that can be interpreted and enforced by the network itself. The automatic application of management policies provides flexibility to change the configuration of network devices at runtime to satisfy administrative goals and constraints regarding security, resource allocation, application priorities, or quality of service. A *policy engine* is a program or process that is able to ingest machine-readable policies and apply them to a particular problem domain to constrain the behavior of network resources.

This chapter concerns the application of policy management to cognitive radio (CR) technology in general, and to spectrum management for frequency-agile radios in particular. It focuses on what lessons can be learned from prior applications of policy management to network resource problems. It reviews and leverages previous standards, research, and commercial implementations for policy engines, and applies them to the architecture and design of policy engines for CRs.

6.2 BACKGROUND AND DEFINITIONS

The policy engine is the main inference component that triggers responses to events that require changing the resource configuration. Often the output of the policy engine amounts to configuration commands or authorizations that are tailored to specific kinds of network devices. In this sense, the policy engine bridges the gap between domain-specific objectives and device-specific capabilities. Despite the intrinsic need for interfacing with particular vendor devices, a popular research trend has been to postulate the policy engine as a general-purpose tool capable of deductive reasoning based on rules. Seen in this manner, policy engines are descendents of the rule-based programming frameworks that were popular in the 1980s. Bemmel et al. [1] describe an expert system as simulating human reasoning using heuristic deduction rules, where knowledge is stored as facts and new facts are derived by using a set of deduction rules.

Much of the research in the area of policy-based networking has focused on the specification of formal languages for expressing complex policies for various domains and network management problems. Policies are expressed as sets of rules about how to change the behavior of the network. Chadha et al. [2] define a policy to be "a persistent specification of an objective to be achieved or a set of actions to be performed in the future or as an on-going regular activity." Carey et al. [3, p. 58] explain that "policies are expressed in terms of an event that triggers the evaluation of a policy rule, a set of conditions that must be met prior to changing the behavior, and a set of actions that are performed to change the behavior." Two trends are evident in the literature: (1) the deconstruction of policies into sets of conditional rules of varying degrees of complexity, and (2) the use of object-oriented representations to support machine readability.

Figure 6.1 calls out commonly recognized functions and relations in a conceptual architecture for a network policy management system. The interpretation and application of these functions for cognitive radio networks requires revisions to this conceptual model, which are explored in this chapter. Evidently, the network resource is the CR, and the policy decision point (PDP) and policy enforcement point (PEP)

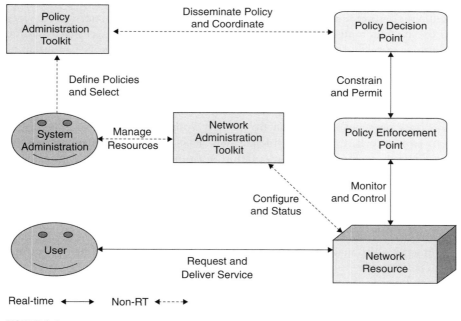

FIGURE 6.1

Policy management system concept.

represent new functions that enable policy management of cognitive radio networks. The purpose of the PDP and PEP functions are to interpret policies to control the behavior of the network devices to satisfy both the users and administrators of network resources.

To implement the policy management architecture shown in Figure 6.1, the system must monitor real-time network events and trigger the policy engine to decide how current device states (policy conditions) should be mapped into desired policy actions that can be quickly enforced by controlling specific device operations. Carey et al. [3] note that policies allow less centralized and more flexible management architectures by enabling administrative decisions to be made closer to where the event and conditions are actually detected. We must examine how this approach applies to the particular problem of spectrum management for CRs.

6.3 SPECTRUM POLICY

The usable frequency range of radio spectrum is divided into frequency bands called allocations for particular types of use. The US frequency allocations are available online [4]. Within a particular allocation, an allotment is a frequency channel designated for a particular user group, or service, in some country. An assignment is a license that grants authority to a specific party to operate a transmitter on a specific channel under specific conditions. The allotments and assignments are associated with a particular geographic area.

Historically, the allotments for broadcast services have been de-conflicted to ensure that the signal strengths in one area will not create interference for signals in another area. However, it is not easy to define interference unless certain assumptions are made about the capability of radio receivers to reject interference in adjacent bands. Other technical considerations will include uncertain propagation characteristics, locations of nearby receivers or transmitters, and limitations of waveforms.

6.3.1 Management of Spectrum Policy

The International Telecommunication Union (ITU) radio communication sector [5] plays a global role in the management of the radio frequency (RF) spectrum and satellite orbits. Around the world, RF spectrum is considered a finite natural resource that is increasingly in demand from a large number of services, such as fixed, mobile, broadcasting, amateur, space research, meteorology, global positioning systems (GPS), environmental monitoring, and, last but not least, those communication services that ensure safety of life at sea and in the skies. World Radio Communication Conferences (WRCs) are held every two to three years to review, and, if necessary, revise the Radio Regulations, the international treaty governing the use of the RF spectrum.

Within the United States, the Federal Communications Commission (FCC) decides spectrum policy for commercial radio communications, and the National Telecommunications and Information Administration (NTIA) plays a complementary role for the federal sector. Due to the perceived spectrum shortage resulting from prior policies and the burgeoning demand, there has been great interest in rethinking the leasing and allocation of spectrum. The FCC's Spectrum Policy Task Force (SPTF) [6] reported in 2002 that "spectrum policy is not keeping pace with the relentless spectrum demands of the market." Two recurring recommendations have been to "migrate from the current command and control model to [a] market-oriented exclusive rights model and unlicensed device/commons model" and to "implement a new paradigm for interference protection." The FCC intends to facilitate cognitive radio technologies to this end. In particular, the discussion has often focused on spectrum-enhancing technologies, including software-defined radios (SDRs), leasing certain spectrum bands or white space, and sensory or adaptive devices that could find unused spectrum. For example, to identify unused RF channels, a CR might be able to measure an interference temperature for the ambient spectral environment.

Deciding which policies apply to a particular CR will require understanding the role the radio is playing for the user of some service at a particular location. The policy language must be rich enough to express the semantics of the possible spectrum management policies, and the cognitive policy engine must be able to automatically interpret and enforce the applicable policies. The FCC and NTIA are contemplating pilot projects to explore options for encoding selected spectrum policies in a machine-readable language. Section 6.4 looks more closely into these languages.

6.3.2 System Requirements for Spectrum Policy Management

A comprehensive policy management system that could autonomously control interference created by frequency-agile radios would have to include an extensive syntactic capability to specify policies and policy engines that can interpret policy semantics expressing technical concepts such as authorization, frequency bands, channels, propagation conditions, signals and noise, waveforms, geopolitical boundaries, geographic locations, dates, times, types of services, and possible roles for the various radios in the environment.

The policy engine must be able to automatically modify the runtime configuration of the CR to sense the spectrum environment; to detect other local radio networks; and to control its own transmissions by adjusting frequency, power, modulation, signal timing, data rate, coding rates, and antenna.[1] In summary, the spectrum policy management system for CRs shall:

- Be distributed across multiple radio platforms with autonomy at the radio level.
- Permit network administrators to determine applicable spectrum policies.
- Represent spectrum policy rules in a machine-readable format.
- Resolve conflicts and inconsistencies in sets of policy rules.
- Designate specific roles and services for each radio.
- Require authorization for policy changes or updates for radio.
- Monitor spectrum utilization and detect RF interference.
- Control runtime configuration of CRs to satisfy spectrum policies.

---

[1] Multiple policy engines—one for the physical layer, one for the network, and one for the user—are possible; this chapter mainly covers the physical layer policy engine. In addition, there may be engines for equipment-specific implementation, behavior-specific policies, or other policies.

- Support heterogeneity and diversity of legacy vendor radios.
- Support continual growth and increasing complexity of CRs.

6.4 ANTECEDENTS FOR COGNITIVE POLICY MANAGEMENT

The last decade has seen significant research and applications for policy management technology in the area of network management. Tacit assumptions about the design of the Internet and related information technology (IT) infrastructures have greatly influenced the development of policy management approaches. It is not surprising that much of the research has focused on the limitations of descriptive ontologies for Web-enabled applications and packet network resources. However, Kagal et al. [7] rightly caution that reasoning about policies generally requires application-specific information, forcing researchers to create policy languages that are bound to the domains for which they were developed.

Chadha et al. [2] have surveyed policy-based network and distributed systems management approaches that have been the subject of extensive research over the last decade (see also [8]). The Internet Engineering Task Force (IETF) [9] has sponsored standardization efforts for object-oriented models for representing policies as well as a framework and protocols for managing Internet Protocol (IP) networks.

This section examines how well the existing state of the art can be adapted to support the performance characteristics of radio networks and the specialized requirements for spectrum resource management. It draws from government projects, commercial applications, academic research, and standardization efforts that provide a context for designing a cognitive policy engine specialized for spectrum management.

6.4.1 DARPA Policy Management Projects

The Defense Advanced Research Projects Agency (DARPA) has funded research and development (R&D) efforts that push forward the technology frontiers in the area of network policy management in general and even the particular application to spectrum management. This section reviews some of that work that is in the public domain.

Funding for contractors involved in the DARPA NeXt Generation (XG) radio communications program [10] has been a very significant, if not the principal, driving force behind the development of policy management techniques for radios. The FCC now has complementary projects to define policies for frequency-agile radios. The XG program followed on the heels of the DARPA Policy-Based Survivable (PolySurv) communications program, which demonstrated that increased military survivability and real communication performance gains could be brought about by downloading dynamic mission policies to automatically manage radio networks. XG is now focused on system concepts and enabling technology to dynamically redistribute allocated spectrum in operating radio networks in order to address rapidly growing requirements for communications bandwidth. The program goals are to enable radios to automatically select spectrum and operating modes in a manner that increases the survivability of communication networks and minimizes disruption to existing users.

The DARPA Dynamic Coalitions program [11] has strongly influenced the technical approach for XG by offering policy representations and policy engines that support other network management techniques. For example, Phillips et al. [12] describe constraint-based models and the application of role-based access control (RBAC) for implementing security policies in the context of dynamic coalitions. Uszok et al. [13] describe a Semantic Web (SeW) language [14] called KAoS that has proliferated with support from DARPA. In fact, KAoS was based on the DARPA Agent Markup Language (DAML) [15], and KAoS has capabilities that support both the expression and enforcement of policies in a software agent context. The policy language has always been based on the eXtensible Markup Language (XML) to support common Web services, but due to shortcomings of the DAML description logic, KAoS now relies on the Web Ontology Language (OWL) [16] to represent knowledge about domains and rule-based policies. In the KAoS environment [17], domain managers act as PDPs and are responsible for administering policy for entire domains.

DARPA has also been active in funding more traditional network policy management approaches for the Next Generation Internet (NGI), and Stone et al. [18] provide background information that is relevant for policy management of CR networks.

6.4.2 Academic Research in Policy Management

This section looks at what a spectrum management implementation might leverage from the research community concerning existing policy languages and frameworks for network policy management. Carey et al. [3] include an overview of the state of the art in policy languages that addresses access control and resource management. Rules governing spectrum access can be enforced by a radio that subjects itself to access control policies. In computer network management systems, only users associated with accounts included in an access control list (ACL) can access the resource. The next enhancement is association of users with groups and roles, leading to RBAC systems. Spectrum resources are already assigned to particular radio services, so the cognitive policy engine can process attributes associated with roles for the radio to provide a context for evaluating spectrum access control rules. Ideally, we would want users to authenticate themselves to the radio and associate different types of authentication with different roles for the user, the radio, the network, and network resources.

Strauss [19] describes the requirements and architecture of a policy management system based on the IETF script management information base (MIB) infrastructure. An MIB is a device-specific database for remotely managing a network resource using the Simple Network Management Protocol (SNMP) based on IP communications. Most IP-capable devices support an MIB that provides a standard interface to monitor and configure the device. For the CR, we could envision a "spectrum MIB" with a standard interface supported by an underlying device-specific mechanism to actually configure the network elements. Strauss [19] used this approach to implement network quality-of-service (QoS) control with the Jasmin Script MIB agent [20]. In this case, the policy engine is just the Java runtime engine executing policy scripts supported by policy-class libraries for different device capabilities.

Ponder is a different policy management framework that includes a relatively mature policy language with a suite of tools and source code that has been freely available to download from the Internet [21]. Ponder tools support administration of domain hierarchies, positive and negative authorization policies, delegation policies, and event-triggered condition–action rules. Dulay et al. [22] describe how to use the Ponder framework to encode, disseminate, and process security and management policies for distributed applications. In fact, Ponder would be an initial starting point for developing an administrative toolkit that could be used to specify, compile, maintain, and disseminate spectrum policies for cognitive radios. It integrates with a domain server and supports role abstractions that could be used to manage spectrum policies for multiple communications services that might eventually be supported by cognitive radios.

Ponder has been well tested in various applications [23], and "back ends" (i.e., application-specific PDP and PEP functions) have been implemented to generate firewall rules, Windows access control templates, Java security policies, and obligation policies for a policy agent. The Ponder language for representing policies has been described as a declarative, object-oriented language that can express both "obligation" and "authorization" policies [24]. To be specific, Damianou et al. [23] explain that "policies define choices in behavior in terms of the conditions under which predefined operations or actions can be invoked rather than changing the functionality of the actual operation themselves." Obligation policies are defined as "event-triggered condition–action rules that can be used to define adaptable management actions," and authorization policies are "used to define what services or resources a subject (user or role) can access."

The policy research community in general is particularly concerned about the difficult task of analyzing the meaning of groups of policies to determine the implications for particular agents and resolving possible conflicts between policies. Even if spectrum management objectives are clearly stated in a policy, the implications for device configurations or required actions are not always obvious in practice. For example, consider a long-standing policy that authorizes a particular frequency band for some type of messaging service, and specifies service-specific protocols for users to share air time. Suppose a newer policy for cognitive radios specifically authorizes a class of CR users to share an overlapping band of spectrum subject to different limitations on availability of channels for legacy users. Is it clear what the air-time restrictions are for a particular cognitive device that performs a similar type of messaging using the legacy channels and protocols? Which rule takes priority, or must both usage restrictions be observed by the cognitive radio? Do permissions take precedence over prohibitions? Policy refinement is the process of deriving lower-level, more specific policies that the device can enforce in order to completely meet the requirements of a group of management policies.

Damianou et al. [24] describe other problems with policy refinement, and Ponder provides tools for policy analysis and refinement to assist administrators in detecting and resolving policy conflicts. In particular, Ponder supports the introduction of priorities and preferences. A simple method to resolve policy conflicts for a device is to assign explicit priorities to every policy so it is clear which policies overrule others. Locally, rules can be prioritized for the device in order to reflect local management priorities, such as ensuring that efficiency is more important than reliability, or vice versa. For a specific device, sets of rules are also ordered by update times, particularly if partial updates of the rule base are accepted practice. DAML relies on update times as well as numeric priorities to determine priority [24].

Stone et al. [18] suggest the idea of differentiating policies "by their granularity, such as the application level, user level, class level, or service level," and letting spectrum managers designate certain mission applications for priority. Hierarchical policy management and domain groupings provide another degree of flexibility, permitting a PDP to branch beyond the linear ordering of priorities. The device may give priorities to policies originating within a more local domain, given an inheritance hierarchy. Similar to the manner in which federal policies overrule state policies, Uszok et al. [17] anticipate a "policy harmonization" process that invalidates portions of the lower-priority policy to resolve the conflict. Decisions about how a PDP should handle inheritance must be made at the time that the policy hierarchy is established. In addition, these decisions should belong to the human realm of policy administration. Experience tells us that it takes a human judge to decide how to invalidate portions of state policies to eliminate conflicts with federal rules.

Another way that a PDP can handle nonlinear priorities and resolve conflicts and ambiguities between overlapping rules is to permit the specification of "metapolicies" constraining the interpretation of groups of policies. For example, Kagal et al. [7] describe another policy language, Rei, which supports metapolicies for conflict resolution. Rei was designed for general application and permits domain-specific information to be added without modification. Tonti et al. [25] provide a comparison of the capabilities and shortcomings of KAoS, Ponder, and Rei, rendering a valuable perspective of the various approaches.

Clearly, the techniques for resolving policy conflicts could be a fruitful area of research for a long time to come. As far as spectrum management is concerned, it appears that the research community has already done enough work to get started expressing FCC policies in a machine-readable format. A number of abstract policy languages already exist, and the task ahead is to introduce terminology that is directly applicable to spectrum management and the cognitive radio domain. The challenge will be to come up with a reasonably useful and self-consistent rule base for CRs that does not present an opportunity for litigation about the meaning, implications, or applications of the rules.

6.4.3 Commercial Applications of Policy Management

This section examines what can be learned from commercial products that could be useful for designing cognitive policy engines. No one will be surprised to discover that major network vendors (e.g., Cisco, Nortel, and Lucent Technologies) have already developed policy management products to support administration of local and wide area networks (LANs and WANs).

Damianou et al. [23] indicate that Lucent has used a Policy Definition Language (PDL), similar to Ponder, to program Lucent switching products. PDL "uses the event–condition–action (ECA) rule paradigm of active databases to define a policy as a function that maps a series of events into a set of actions." This approach is interesting because in addition to policy rules, there are policy-defined event propositions that allow groups of simple events to trigger more complex events.

Nortel [26] advertises its Optivity Policy Services with system components consisting of a policy server, management console, directory server, and policy-enabled network components. This distributed IP network architecture interoperates using

Lightweight Directory Access Protocol (LDAP), Common Open Policy Service (COPS), and command-line interface (CLI). The administrator is required to select from a number of predefined policies (i.e., "if condition, then action" rules) that relate to the roles of various network devices identified in the directory server. The policy server then issues COPS or CLI commands to configure devices such as routers, IP telephone gateways, and firewalls.

Damianou et al. [23] observe that a common component of commercial tools is a graphical user interface (GUI), which typically allows the administrator to visually select a network device or other managed element from a hierarchically arranged tree-view of policy targets, and specify the policies in the form of "if conditions, then action rules for the selected targets."

What comes across in all the commercial examples is that the network devices must be designed to support policy rules that can configure their behavior by some mechanism. The sophistication of the devices determines the nature of the interface with a policy server that either directly issues device configuration commands, supports a COPS dialog to make policy decisions for the device, or disseminates defined policies that are recognized by the device.

Ultimately, the nuances of the policy language seem to be relatively unimportant compared to the sophistication of the network devices.

6.4.4 Standardization Efforts for Policy Management

The proliferation of vendor architectures for policy management of telecommunications networks has motivated the IETF to address standards for interoperability. Using the Policy Core Information Model (PCIM), Moore et al. [27] begin the process of standardizing policy management terminology and representations of network management policies to provide an accepted framework for vendor-specific implementations.

To what extent can these standardization efforts be applied to policy engines for spectrum resource management for CR operations? Initial cognitive radio implementations will necessarily focus on radio-specific features, and as the technology proves successful, the focus will extend to larger radio networks rather than individual radios. The IETF policy domain is already oriented toward management of large-scale telecommunications networks, with the goal to assist human network administrators to define machine-readable policies and architectures for the automatic control of network resources. The CR is analogous to a particular network device.

Snir et al. [28] envision a physical network architecture in which the PDP translates abstract policy constructs into configuration commands for multiple devices (e.g., a router, switch, or hub) where policy decisions are actually enforced (i.e., PEP). Although there may be compelling arguments for the architectural assumption that one PDP services multiple PEPs in the case of high-speed, high-reliability network architectures, this is not so clear for the CR application. Stone et al. [18] point out that an underlying assumption of PCIM is that policies are stored in a centralized repository, and the PDP is the entity in the network where policy decisions are made using information retrieved from policy repositories. The PEP requires a policy decision about a new flow of traffic or authentication—for example, "the PEP will send a request to a PDP that may reside on a remote server."

Moore et al. [27] indicate the PCIM standard fits into an overall framework for representing, deploying, and managing policies that are being developed by the IETF Policy Framework Working Group. In Figure 6.1, the link between the PDP and PEP has two characteristics: (1) it needs to operate in near real time for timely enforcement decisions, and (2) it is conceived to be a query-response dialog. For the CR application, due to concerns about link reliability and bandwidth, the first assumption is tenable only if the PDP and PEP functions are colocated on the radio platform. Furthermore, the query–response design is also very natural, given the prevalence of client–server and three-tier database transaction architectures of the Internet. In fact, two important Internet applications for policy management involve access control (i.e., security) and admission control (i.e., QoS), and both involve permission. However, there is no reason to assume this Internet design is optimal for a CR application in which the PDP and PEP functions are colocated.

In PCIM, the policy-controlled network is modeled as a state machine in which policy rules control which device states are allowed at any given time. Each policy rule consists of a set of conditions and a set of actions. Policy conditions are constructs that can select states according to complex Boolean expressions. Policy actions are device behaviors, such as selecting or prohibiting certain frequency bands, bandwidths, protocols, coding, or data rates. When events lead to certain conditions, then certain actions can or must be performed by the device (depending on whether the actions are obligated or simply permitted for the device).

Bemmel et al. [1] describe these policy rules as examples of event–condition–action rules, by which changes within a system or the environment trigger adaptation of the system's behavior. Such systems are called event-driven, and many programming languages and environments support compatible software development techniques. For example, in the Microsoft Windows architecture, events are basically messages that are routed between processes called event handlers. Unfortunately, if the policy rules were coded in this familiar manner, the component's behavior would basically be hardcoded when the program is compiled. We need a more flexible way, however, to bind actions to the event handlers at runtime—for example, Java Remote Method Invocation (Java RMI) or Common Object Request Broker Architecture (CORBA).

The PCIM defines class representations and abstract attributes for policies, but not the algorithms or design of the policy engine. Figure 6.2 depicts the ontology for the object-oriented design of policy classes in the PCIM. Policy rules are aggregated into policy group classes. These groups may be nested to represent a hierarchy of policies. Although retaining all the same attributes of the policy classes is not particularly important, a similar class diagram will be suitable for representing policies for the CRs.

One important attribute of the policy rule is the ability to associate a "role" for the radio. The policies (e.g., assigned frequency bands) for land mobile radios (LMRs) are different from those for air traffic control (ATC). Moore et al. [27] stress that rather than configuring—and then later having to update the configuration of—hundreds or thousands (or more) of resources in a network, a policy administrator assigns each resource to one or more roles, and then specifies the policies for each of these roles. The policy framework is then responsible for configuring each of the resources associated with a role in such a way that it behaves according to the policies specified for that role. When network behavior must be changed, the policy administrator can perform a single

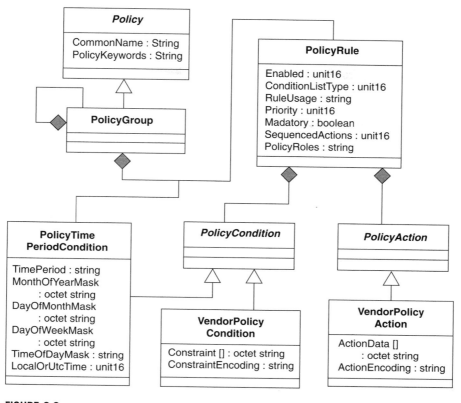

FIGURE 6.2

UML diagram for PCIM policy classes.

update to the policy for a role, and the policy framework will ensure that the necessary configuration updates are performed on all the resources playing that role.

6.5 POLICY ENGINE ARCHITECTURES FOR RADIO

In this section we will explore the main issues of designing the policy engine, beginning with the concept, then exploring technical approaches, and finally discussing the enabling technologies.

6.5.1 Concept for Policy Engine Operations

This section begins to synthesize a concept for how the cognitive policy engine will operate. Section 6.3 shows how network policy management systems operate in a distributed fashion with administrators using policy languages to express network objectives, policy servers that act as PDPs deciding how to configure devices to achieve these goals, and policy-enabled devices that are configured to enforce the policy (i.e., PEP). The spectrum management objectives for the CR are slightly different than network access and resource control, for which the primary goals are security and network application performance. The primary interest here is ensuring that the frequency-agile radio is utilizing available spectrum resources in accordance with regulatory licenses and prohibitions while acting as a good neighbor to avoid interference with other users. So there is a need to distinguish between a PDP that can optimize global network performance, and a PEP that configures the CR to obey locally enforceable policy constraints. In other words, the cognitive policy engine is responsible for device configuration.

The relationship between the PDP and PEP functions is also different for the CR. Carey et al. [3] describe a typical network policy management dialog in which events detected by the device cause the PEP to formulate a request for a policy decision and send it to the PDP (usually a policy server), such as when Resource ReSerVation Protocol (RSVP) is employed by a user application to advertise its QoS requirements. COPS may be used as a policy transaction protocol between the PDP and PEP for transporting the policy requests and decisions. The PDP returns the policy decision and the PEP then enforces the policy decision and responds to the RSVP from the user application accordingly.

In our concept, the cognitive policy engine will perform both PDP and PEP functions for the radio using local platform resources. The operation will generally accord with Figure 6.1, and involve the steps shown in Table 6.1. There is no necessity for radio device functions to request a policy decision from the policy engine because the policy engine is already responsible for enforcing the policy on the platform. As shown in Figure 6.1, the policy functions (PDP and PEP) act to monitor and control the radio platform to satisfy any policy constraints or obligations.

If we assume the CR is administered by a central policy authority in a domain that recognizes the capabilities of the particular cognitive radio and disseminates all enforceable policies directly to the CR, then the complexity of the policy engine can be reduced. For example, the policy engine does not have to determine whether the policies can be enforced on the platform. Any ECA rule sent to the platform could (but not should!) even be compiled and executed by a Java engine on the radio.

Table 6.1 Concept of Cognitive Policy Engine's Operation

Step 1	The policy engine receives a download of policies to be observed by the radio.
Step 2	The policy engine determines what information to monitor on the radio platform (e.g., location, time).
Step 3	The CR provides the requested situation data (e.g., location, time) to the policy engine as required.
Step 4	The arrival of new information is an event triggering some processing of policy rules.
Step 5	The policy engine formulates applicable constraints on spectrum usage for the radio.
Step 6	The policy engine configures the radio to observe spectrum constraints and perform obligatory activities (e.g., communication protocols).
Step 7	The CR operates within defined constraints and performs obligatory activities.
Step 8	The policy engine performs any obligatory hierarchical reporting activities for network management.

In the case of radio networks, such as an Internet community, no centralized authority will own all of the resources. It will not even be possible to guarantee efficient connectivity among the worldwide users of cognitive radio technology. Feeney et al. [29] argue against a centralized policy administrative authority in cases of great organizational diversity, and propose a concept of operations with hierarchies of policy authorities. This approach reflects a real-world community of users, but it leads to the possibility of policy conflicts that must be resolved among organizations. These conflicts must be resolved at the network level, and the cognitive radio should not be forced to handle this difficult problem.

6.5.2 Technical Approaches for Policy Management

This section proposes a technical approach for designing the cognitive policy engine to satisfy the concept of operations presented so far in this chapter. We look first at what must be done to design the policy language and then examine alternatives for the technology behind the policy engine.

Multiple approaches for policy specification have been proposed that range from formal policy languages that can be processed and interpreted easily and directly by a computer, to rule-based policy notation using an if-then–else format, to the representation of policies as entries in a table consisting of multiple attributes [30]. This chapter has already looked at alternatives, including the compiled policies of Ponder, Java scripts supported by interpreted classes, KAoS with its OWL semantics, RBAC, Rei, and PDL.

Whatever language is selected, the syntax will have to be extended to support the technical jargon of spectrum policy with attributes such as frequency bands, channels, propagation conditions, signals and noise, waveforms, geopolitical boundaries, geographic locations, dates, times, and types of services. The language should support authorization and obligation policies, and roles for the various radios in the environment. In addition, the language should make a clear distinction between management policies and the resources and activities being managed [31]. The plan from the beginning should consider eventual "use of domains as a means for grouping resources with dependencies reflecting both hierarchical interactions (e.g., control of resources, authority delegation to subordinate managers) as well as peer-to-peer interactions (e.g., negotiations between peer managers to prevent/resolve management conflicts)" [31].

Looking forward, the ontology should support standardization efforts for multiple vendors of CRs, even though it will likely be redesigned in the future. For this reason, we should eschew compiled and interpreted languages, and start with the flexibility provided by Semantic Web languages such as OWL. Then the ontology will be built up as capabilities and understanding will increase over time.

The choice to start with a Semantic Web language does not exclude the use of compilers and interpreters; it just focuses this technology on the implementation of actions and behaviors with processes named in the policies. For example, useful radio protocols should be named in the ontology, and policies relating to standard protocols should be defined. As more device-specific actions are represented with new terminology, the generality of the policy architecture will decrease because the language must support increasingly complex and specialized actions for proprietary device behaviors [32]. Another shortcoming of this approach is that the policy engine and ontology have difficulty reasoning about complicated, composite behaviors. Selecting the right balance between generality and specificity in referencing CR behavior is an area for further research that need not obstruct initial efforts to create a spectrum policy language for the policy engine.

For policy engine development, there are also several approaches to consider. Note that the engine itself need not internally use the policy language as its input, because the policies can be interpreted when they are downloaded to the cognitive radio. Named behaviors can be bound in the radio to specific procedures and algorithms. For example, a compiler or just-in-time interpreter can create "tokens" from the input and link together whatever processes or logical structures are used internally to the cognitive radio.

Turning to the technical approach for the cognitive policy engine, it is important to recognize that the policy engine will be performing both PDP and PEP functions on the same platform. The concept of operation proposed here is that the policy engine will be responsible for both interpreting and enforcing spectrum policies, as well as monitoring platform events to trigger changes in the configuration of the radio functions. Again, according to Figure 6.1, the functional interface between the policy engine and the radio platform is defined as a relationship of monitoring and controlling the radio. This is the situation shown in Figure 6.3(a).

An alternative approach characterizes the purpose of the interface as validating that the operation of the radio complies with all relevant policies. This situation is shown in Figure 6.3(b). Uszok et al. [17] argue that the interface between a PEP and a native environment can be standardized to answer the question, "Is a given action authorized or not?" This approach posits a query-response dialog between the policy engine and the native device. Thus, the policy engine explicitly polices actions by the radio, requir-

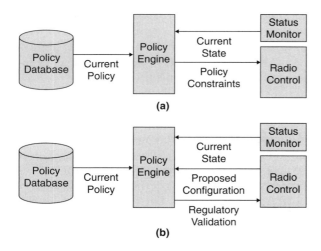

(a)

(b)

FIGURE 6.3

Alternative policy engine architectures.

ing the radio to create objects that describe proposed actions so that the policy engine can pass judgment on whether the action is permissible.

Some involved parties (but not members of the FCC) have publicly voiced the opinion that validation is the only technical approach that will support certification of a frequency-agile radio by the FCC. Were this opinion correct, the policy engine architecture would have to combine both functions in Figure 6.3. Function (a) performs vendor-specific configuration and control of a proprietary device and function (b) certifies policy compliance based on names for generic behaviors. In the latter case, the policy engine can have only a partial understanding of the device's proprietary behavior. This leads to requirements for default authorization policies in which actions that are not defined in the policy language should generally be denied, or are assumed to have been deemed permissible when the radio platform was certified by the FCC.

The technical approach presented here for the cognitive policy engine starts with configuration rather than validation. The emphasis is on supporting obligation policies and ECA rules in addition to authorization policies typical of security platforms. As Stone et al. [18] state, "a policy specifies what action(s) must be taken when a set of associated conditions are met." Bemmel et al. [1] consider this policy-based approach with ECA rules to be "useful in cases when a controlled part has many choices/options and a controller intervenes, enforcing a choice in accordance with a particular policy goal." This seems to be the likely situation as CR technology will increase in agility over time.

An important part of this approach at both the device level and the network level is status monitoring. According to Lee et al. [33], "feedback is a critical part of the process where monitoring activities by the management system lead to changes in the low-level policies." This permits the policy management system to adjust the radio according to changes in demand for bandwidth or interference conditions.

6.5.3 Enabling Technologies

This section completes the examination of what particular technologies will be required to implement this technical approach. Carey et al. [3] summarize two challenges for cognitive policy management: "Currently policy-based management suffers from fragmentation of approach and there is no commonly accepted policy language and no common approach to the engineering of policy based systems."

Regarding the technology behind the policy engine, Khurana and Gligor [34] approach the problem by modeling the network as a state machine and using policy to place limitations on the state transitions that are allowable at any given time. Any model for the CR that is internal to the policy engine will tend to become more complex and specialized over time. The same issue arose with the increasing specialization of the policy language over time. More R&D will determine whether this requires a custom design for the policy engine that is matched to the radio's capabilities.

The state transitions in the model can occur only when the constraints are satisfied, and this may require logical deductions at some point. Damianou et al. [23] note that logic-based policy languages have a well-understood formalism that is amenable to analysis. For example, a policy engine supporting RBAC can be implemented in the Prolog language because the policy constraints are equivalent to restricted first-order predicate logic (RFOPL) statements [35]. Kagal et al. [7] say that the Rei policy engine was developed in Java and used Prolog as a reasoning engine. Stone et al. [18] summarize the situation: "One clear method is to use formal logic to represent network policies. Although this method would make conflict detection much easier with the use of existing theorem provers, most network policy implementers are not as comfortable with this representation."

The policy engine behind KAoS is an online theorem prover that permits logical reasoning about domains and policies [17]. The enforcement mechanisms are Java classes that are specific to a resource platform, but capable of interpreting and enforcing policies from the PDPs. The Java Theorem Prover (JTP) [36] supports queries consisting of properties defined in a given namespace and selection of all possible values. The JTP will also provide a response regarding whether a given assertion is true. For conditional policies, JTP can store a state consisting of certain assertions that are true. These assertions can be removed when the condition is determined to be false by the radio or event monitor.

Another area for research and development is motivated by Carey et al. [3]: "Problems with policy decomposition and hierarchical translation still present difficulties." Daminaou et al. [23, p. 31] characterize the situation:

> Despite significant efforts in developing different policy specification techniques, the ability to refine such goals into concrete policy specification would be useful. Policies can be used to support adaptability at multiple levels in a network.... Research is needed on defining interfaces for the exchange of policies between these levels. It is not easy to map the semantics of the policies between the different levels.

The basic problem is that the higher-level management applications "will not be aware of the network components and cannot specify policies to be interpreted by them" [23].

At this time, the need for extensive logical reasoning capabilities in the policy engine has been oversold, particularly at the level of the device. Whenever a problem requires complex reasoning, it is generally solved with a special-purpose algorithm or application. Two techniques are now available for use to help resolve ambiguities in the rule base. First, policy rules can be associated with a priority value to resolve conflicts between rules [24]. Second, metapolicies are policies about policies, and may be used to detect semantic conflicts between policies [3]. Kagal et al. [7] describe this capability in Rei.

6.6 INTEGRATION OF POLICY ENGINES INTO COGNITIVE RADIO

In the end, the integration of the policy engine into the CR is primarily an issue of software system engineering. The process will involve deriving a complete set of software requirements from the system concepts; defining a software architecture that is compatible with CR platform software architecture; defining the software interfaces; and designing, coding, and testing the software components. Some assumptions must be made about the software architecture and application programming interface (API) provided by the services on the host platform. This section, for example, makes the assumption that the first cognitive radio implementation will likely evolve from the efforts of the Software Defined Radio Forum (SDR Forum [37]). It begins with platform integration issues and then moves on to issues related to integration into a policy-managed network.

6.6.1 Software Communications Architecture Integration

The JTRS JPO defines the Software Communications Architecture (SCA) based on CORBA middleware [38]. The SCA is a layered architecture with well-defined APIs for integration with new applications (called components). The applications are supported by a core framework that provides interfaces for exchanging information, controlling software processes, configuring the platform, and accessing files. The policy engine will be another component in this architecture supported by the core framework with interfaces to other components. One of the most important architectural considerations is the definition of the interface between the policy engine and software environment of the radio.

The policy engine can be divided into three main policy management functions: policy service, policy decisions, and policy enforcement.

Policy service maintains the policies that are downloaded to the radio frame by a policy authority, and it requires an external network interface communication. The policy service also performs any necessary policy life cycle management operations, such as parsing the policy language, keeping track of whether policies are enabled or disabled, and removing policies if they become obsolete.

The *policy decisions* function fulfills the requirement that the device have a local PDP. To make these decisions, it needs information about the communication conditions and spectrum events in the radio environment. Thus, it must interface to some kind of status monitor that can report conditions and real-time events that require a policy decision.

The *policy enforcement* function acts as the PEP for the device, and it needs a control interface to constrain the radio's behavior and perform any obligated actions demanded by policy.

Figure 6.4 shows how the policy engine functions are integrated into the SCA environment by creating the necessary functional interfaces.

The coupling between the policy engine and the radio will be rather complex because there is likely to be a big gap between abstract policy syntax and names, and because of the detailed design of the radio software functions and interfaces. At least one of the interfaces should be relatively simple to accomplish. The radio is already capable of network communications, presumably through an API for data input/output (I/O), and a policy network interface can be implemented by software that will interact with the SCA to securely download policies to the policy server.

The policy enforcement interface will be more complicated because the policy actions will not generally correspond to particular controls for the attributes of the waveforms that are being transmitted by the radio. The waveform control interface must

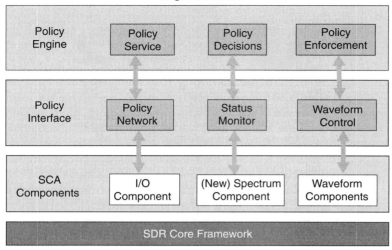

FIGURE 6.4

Software design for policy management.

interact with the SCA waveform components to bridge this gap between behavior specifications and control parameters. Depending on the type of policies that will be enforced, it is likely that the waveform component may have to be adapted so that it is "policy enabled." This adaptation would likely involve exposing new API methods that were designed to support the actions required by the policies. For example, the waveform control interface will have to communicate any limitations on transmission frequencies and power levels, limitations on protocols, or times when transmission is permitted.

What is entirely new is the spectrum component, which is the focus of the status monitor. The spectrum component will require signal-processing capabilities to determine channel occupancy and any other information about radio conditions and events that will be necessary to evaluate the SCA policy rules. The status monitor interface must be able to "observe" the radio operation and its spectrum environment and maintain information about the state of the device. It will have to notify the policy decision function about any relevant changes to the state that may trigger policy rules. This functional interface must aggregate digital signal-processing results into higher-level event objects and provide event notifications to the policy engine. Because the events and conditions of concern depend on the policy specification, the policy engine must inform the status monitor what radio and spectrum attributes and conditions to observe, and what kind of notifications to return for consideration by the policy engine. Examples of the types of data that may flow across this interface are current frequency band, current geographic location, current time or data, and the communication roles of the radio in the network.

6.6.2 Policy Engine Design

Having discussed the external software interfaces for the policy engine, this section now looks deeper into the internal functional design of the policy engine by looking at what others have done, proposing a top-level design, trying to define software requirements, and building up a detailed design introducing new functionality as necessary.

Montanari et al. [39] describe a useful programming environment called Policy-Enabled Mobile Applications (POEMA), which gives one example of designing policy-based "middleware." Boutaba and Znaty [31] insist that a policy engine design must have three logical components: execution engine, situation matcher, and observer. The observer is responsible for defining the situation in terms of "combinations of the values of observables (an attribute of some object which has a value which can be measured) and times of observations." The situation matcher makes policy decisions by "matching the observed state of the system with stored situation specifications" that correspond to the policy conditions and policy events discussed previously.[2] "The language used to express situations must be able to represent observables and provide operators for obtaining the value associated with an observation, and the time an observation was made" [31]. The execution engine operates on an "algorithmic block" (i.e., policy action) and produces a "sequence of instructions" that the policy enforcement function discussed previously will use to control the radio.

Figure 6.5 offers a tentative functional design for the cognitive policy engine. Because of the importance of the status monitor interface for the policy engine, it is included in the figure because Boutaba and Znaty [31] consider it to be an integral part of the policy engine. Note that two types of policy actions are produced by the policy decisions function. According to Stone et al. [18]: "Policies can be triggered in two ways, either statically or dynamically.... Static policies apply a fixed set of actions in a predetermined way according to a set of pre-defined parameters that determine how the policy is used." In other words, these are ECA rules with no policy events to trigger a change in policy. "Dynamic policies are only enforced when needed, and are based on changing conditions; actions can be triggered when an event causes a policy condition to be met" [18].

Some lessons about software design that have been learned from prior efforts to prototype a policy engine are informative. Strauss [19] lists some useful design requirements for a policy management engine, which are adapted to the CR application and are listed here without further comment.[3]

- A policy condition must allow read access to variable attributes in a way that the policy action can reference those attributes that matched the condition and the attributes of the event that triggered the rule. Thus, the run-time system must bind variables to instances when passed to the condition and action.

- There must be a construct to specify the value space in which the free variables of conditions are evaluated that span all instances within a certain table or all instances of a class.

---

[2]The situation matcher may be implemented with a variety of technologies, ranging from complex Boolean specification, to fuzzy logic, to neural nets, or even to a Markov model.
[3]Any errors in adaptation are the responsibility of this author.

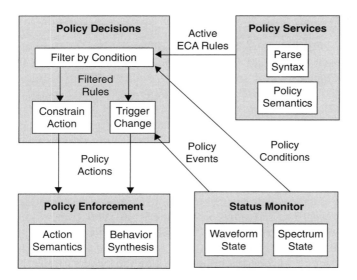

FIGURE 6.5

Functional design for policy engine.

- A class that models a certain object must support accessor methods that allow a policy to retrieve and manipulate an element and support computations. This is not a requirement for the policy engine architecture itself but for the design of the interface.

- Three types of time events may be required: periodic events that trigger continuously for a given period, calendar events that trigger periodically at points in time specified by calendar-type attributes, and one-shot events that trigger exactly at a point in time specified by calendar-type attributes.

- Another type of event is based on the reception of external notification, such as state reports, and these must be mapped to specific events. Details of the initiating notifications should be accessible through accessor methods of the events.

- The policy run-time engine must support a mechanism to report errors and optional tracing and debugging information so that users can monitor the policy engine and the authors of policies can test and debug their policy codes.

- The access to variables in conditions and actions may fail, and the policy runtime engine must be able to handle these situations in a way that policy code can catch the error conditions and bring the variable to a determined state.

- It must be possible to store and execute multiple policies independently. Their codes must not share any namespace.

- A security mechanism is required to differentiate which users have access to which roles on which policies (build on the existing SCA security component).

- The notation of policies should remain declarative as much as possible, but the programmatic policy system has to implement complicated actions with code instead of just the goal.

- It should be possible to avoid redundancy in a way that policies, or policy groups sharing rules and rules-sharing conditions or actions, can be built by referring common code instead of copying code fragments. This will increase reusability and avoid some errors.

- The policy engine should be able to pass arguments to policies when they are activated by conditions based on messages or shared memory to reduce redundancy.

6.6.3 Integration of the Radio into a Network Policy Management Architecture

Once PDP and PEP functionalities are accommodated in the policy engine, the first question concerns what can be gained by embedding the CR in networked management architecture. The discussion of the policy service functionality makes it clear that, at a minimum, the network policy architecture needs to be concerned with policy dissemination. As Stone et al. [18] put it: "Communication is needed to and from the policy repository.... In many proposals the policy repository is a directory, and therefore the appropriate access protocol would be the Lightweight Directory Access Protocol."

Of course, the actual "minimum" must do a little bit more. There must be some kind of secure authorization to download policies to the radio. One way to handle this is to architect the CR in such a manner that the only way to update the policy database is to physically wire a connection to a special interface, thus relying on physical security to ensure the integrity of the policy database. This makes it very tedious and inconvenient to update the policy database, however, particularly if a large number of radios are involved.

Since the purpose of the radio is to communicate, it makes sense to rely on network communications for policy updates. Sheridan-Smith [40] points out that peer-to-peer management communications also eliminate problems with synchronization of policy repositories because the radios can be informed of changes. Whenever radios enter a new network, part of the session initiation can include setting up a policy management interface to synchronize policy databases as required.

Naturally, no one would permit an unknown radio to remotely update the policy database without some kind of exchange of credentials to resolve trust issues. Li et al. [41] introduce a family of role-based trust management languages for representing policies and credentials. They define a "decentralized collaborative system" being formed by several autonomous organizations desiring to cooperate and share resources for their mutual benefit. Sheridan-Smith [40] also studies synchronization techniques for distributing the responsibility for policy-based network management (PBNM) across a set of independent autonomous PDPs: "The use of the pull model is more efficient than a push model. Alternatively, the network management system is required to poll each device in turn to ensure that they are operating correctly." The push model is particularly problematic for the cognitive radio because it will often be turned off and restarted or reinitialized.

Damianou et al. [23] introduce the next level of complexity: "It should be possible to dynamically update the policy rules interpreted by distributed entities to modify their behavior." When the concept of roles is introduced into the policy language, the following premise is tacitly accepted: "It is not practical to specify policies relating to individual entities—instead it must be possible to specify policies relating to groups of entities and also to nested groups such as sections within departments, sites within organizations and within different countries" [23].

A simple view of policy in regard to cognitive radio networks is that network policies constrain network communications. Specifically, network policy defines the relationship between clients (users, applications, or services) using spectrum resources and the radios that provide access to those resources. (See Chapter 9 in this book for additional details on how the policy engine may manage network protocols.) Lee et al. [33] take the argument to the next step: "Ideally, the management system will use feedback from monitoring and network measurement to influence its decisions about how to control the network configurations to improve efficiency, manage service quality, or to deal with changes in the network environment."

Thus, a network policy management should incorporate a policy engine operating on network policies, receiving status updates from the policy service on individual CRs, and downloading policy instructions to the members of the network. Now the policy management architecture is self-similar, and as Boutaba and Znaty [31] state: "The whole network policy management system is then logically constructed in a hierarchical domain structure where low level domains provide their services to those of the upper layers. Domains are managed according to a set of harmonized policies."

At the network level, the policy manager must be focused on managing services by creating policies for individual devices based on the roles they will play in the network architecture. This is the root of the distinction between network-level policies and device-level policies. Sheridan-Smith [40] comments:

In general the network policies can apply to more entities and are easier to read, write and comprehend, precisely because they are not specific about how the goal should be achieved. Device-level policies can be specific about what needs to be done, but they are difficult to read and complex to write and can apply only to a subset of entities in the system.

This, then, is the state of the art in automated policy management of complex networks. The difficulty involves how to automatically refine higher-level policies into more specific management policies for domain members. This process involves planning for network operations and deriving specific policies for all the network elements to satisfy network goals. Stone et al. [18, p. 18] define the research challenge as follows:

Refining the goals, partitioning the targets the policies affect, or delegating responsibility to another manager who can perform this derivation. The main motivation for understanding hierarchical relationships between policies is to determine what is required for the satisfaction of policies. If a high-level policy is defined or changed, it should be possible to decide which lower-level policies must be created or changed.

6.7 THE FUTURE OF COGNITIVE POLICY MANAGEMENT

Having examined functionality and designs for cognitive policy management, this section returns to the fundamental question: "Why bother?" Damianou et al. [23] reply that the "motivation for policy-based services is to support dynamic adaptability of behavior by changing policy without recoding or stopping the system." This chapter has already indicated how this is accomplished. This section considers promising military and commercial applications of policy management technology and examines the challenges.

6.7.1 Military Opportunities for Cognitive Policy Management

Military network communications are managed by a hierarchy of network operations centers that fit naturally into a recursive management model. In fact, the military is unique in that it can hierarchically manage both the network resources and the demand by users for resources, unlike commercial market applications in which the financial goals always involve increasing user demand for services.

Figure 6.6 depicts the policy management hierarchy for communications resources. This hierarchy also handles application performance, but a parallel command structure exists for military personnel (users). Higher-level network domains delegate communication management tasks to their subordinates to be planned and executed on lower-level network resources. In this type of architecture, coordination is necessary, as Boutaba and Znaty [31] note: "Peer-to-peer interactions take place in case of overlapping between domains to optimize plans and de-conflict policies."

To handle the complexities of policy refinement that occur in this hierarchical architecture, "the concept of management policy is introduced as an intermediate step between goals and plans" [31]. In this case, "plans" refer to complex coordinated actions, so the refinement process becomes: (1) lower-level policies are derived from

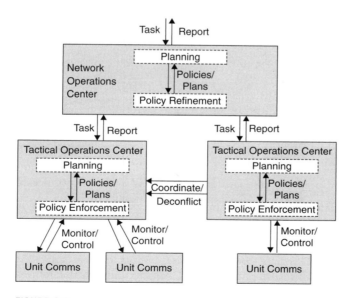

FIGURE 6.6

Military policy management hierarchy.

goals and missions passed down the hierarchy as tasks, (2) officers define action plans consistent with the policies, and (3) subordinates are tasked to accomplish the plans.

The fact that DARPA has invested significant funds to support the development of various policy management applications for the military suggests that this technology will filter down to the operational level. As network communications systems such as the tactical Warfighter's Internet (WIN-T) and the JTRS cluster procurements come online, policy management should play an increasing role.

6.7.2 Commercial Opportunities for Spectrum Management

It is widely recognized that the FCC's spectrum licensing and leasing practices have become highly inefficient and have led to artificial shortages due to the dedication of frequency bands for underused services. As a result of the significant financial investments in this area, progress will probably depend primarily on the perception of financial incentives for the current spectrum lessees.

In particular, there has been some discussion of the possibility of spectrum "microleases" that would permit occasional opportunistic use of certain spectral bands in specific locations. Chiang et al. [42] describe protocol primitives to support a "Vickery auction" as a rational allocation of resources in a distributed system [43]. The concept requires creation of a "broker space" and an "auction space" by which buyers and sellers could automatically purchase portions of the time–space–frequency spectrum. In this protocol, the following are the necessary steps:

1. A selling agent informs the market of its intent to sell spectrum by issuing an *offer* message.
2. Brokers *link* buyers to market opportunities.
3. Buyers submit *bid* messages to the seller.
4. The selling agent *chooses* its preferred buyer.
5. The transaction is *settled* (with a message) and the lease and moneys are conveyed between buyer and seller.

In some manner, it would be necessary to tie the policy management system into this market to enable and disable the use of spectrum resources.

6.7.3 Obstacles to Adoption of Policy Management Architectures

This section summarizes a few of the challenges to widespread adoption of policy management techniques for CRs. Concerns about potential difficulty in getting FCC certification for a radio managed by a cognitive policy engine have already been addressed and will not be further discussed here.

It is one thing to design a policy management architecture and another thing to demonstrate that its functionality is worth the effort to implement it. Lee et al. [33] propose the design of experiments to test hypotheses and evaluate whether the actions specified actually improve or worsen system performance; "the evaluation should proceed by examining a particular approach and making quantitative measurements according to a well-defined methodology." The DARPA PolySurv program made some efforts in this direction, but the published literature consists only of marketing brochures and vendors' advertisements.

Verma [44] recalls the need for policy-transformation logic that translates high-level network administration policies to device-specific policies that can be automatically disseminated. The approach also ensures that high-level policies are mutually consistent, correct, and feasible. "The heart of policy management lies in the policy translation logic as to how the policies will be represented and how they will be managed" [44].

As Boutaba and Znaty [31] note: "Today's challenge is to provide automated support not only for detecting and responding to trivial network and system events, but also for the process of planning and policy making to handle more complex situations." They enlarge the scope of the automation problem as follows: "Peer negotiations for conflict avoidance should be handled as part of the proactive management process, whereas negotiations for conflict resolution should be handled as part of the reactive management process. Both approaches should involve all derivation of goals, policies, plans, and actions." Sheridan-Smith [40] also expresses concern:

> If multiple autonomous management entities are making independent decisions, then each participant in the network might behave in a manner that is not aligned with other parts of the network, or other functions of the network [that] are independently managed. Coordination will ensure that the behavior that is stipulated in the policies is correctly met by all of the individual management entities.

The perspective put forth in this chapter is that the adoption of a policy-based spectrum management approach for CRs will require standardization of policy languages and dissemination techniques designed specifically for cognitive radios and driven by market concerns. It seems likely that a cognitive policy engine can be designed to function usefully in the context of CRs. The real question is probably not the technical feasibility of automatic policy management architectures; instead, it is whether market forces will be favorable to adoption of this technology and lead to integration across multiple platforms and vendors.

6.8 SUMMARY

The focus of this chapter has been on the design of a practical cognitive policy engine that would enable the cognitive radio to operate reasonably within a suitable policy space. The approach has been to explore the engineering requirements for the policy engine in the context of a radio policy management domain that might reasonably reflect the operational environment for a CR. The scholarly literature shows what enabling technology is available and necessary to realize this objective in the near term.

The basic syntactic and semantic mechanisms for representing spectrum and network management policies were found to already be in existence in several forms. Although there are limits on the semantic capabilities of extant policy engines and languages, it is difficult to argue convincingly that there exists a significant technology gap in this area that would prevent CRs from being deployed in a fruitful manner.

If anything, the opposite conclusion is much more compelling. There is no demonstrated requirement that calls for the representation of highly complex logical constraints on the operation of the cognitive radio. In fact, there is every reason to believe that implementations for policy engines in the first few generations of CRs will be more than capable of enforcing restrictions on policy to constrain and optimize the behavior of the radio. It seems highly unlikely that communication regulations will call on the radio to dynamically resolve complex and sometimes inconsistent policies.

This chapter has reviewed the application of policy engines in commercial networking environments and has found that the most difficult challenge lies not in processing or enforcing the policies, but rather in coherently, unambiguously specifying which behavior is desired. For this reason, existing policy-enabled network systems are generally focused on well-defined access control or network configuration and performance optimization problems for which the desired behaviors can be defined and agreed on by users of the system.

The critical engineering problems that remain to be resolved for implementing cognitive policy engines are primarily software design questions. Almost all of these questions will be answered in terms that are specific to the particular kinds of real-time software platforms and policy instantiations that have already been addressed in this chapter. The interesting academic questions addressed primarily relate to theoretical limitations of policy languages and abstract processing architectures that are much more complex than what will be initially embodied in early deployments of CRs. They are worthy of note, but should not be construed as obstacles for implementing a cognitive policy engine. The real challenge lies in unambiguously specifying what behaviors are required of the cognitive radio, and achieving consensus that such behaviors are sufficient for licensing the operation of the radio.

REFERENCES

[1] Bemmel, J., P. Costa, and I. Widya, Paradigm: Event-Driven Computing, Freeband Awareness deliverable D2.7a, October 2004; available at *http://awareness.freeband.nl*.

[2] Chadha, R., G. Lapiotis, and S. Wright, Policy-Based Networking, *IEEE Network*, 16(2):8–9, 2002.

[3] Carey, K., K. Feeney, D. Lewis, State of the Art: Policy Techniques for Adaptive Management of Smart Spaces, *State of the Art Surveys*, Release 2, pp. 58–66, Trinity College Dublin, May 2003.

[4] *www.ntia.doc.gov/osmhome/allochrt.html*.

[5] *www.itu.int/ITU-R/*.

[6] *www.fcc.gov/sptf/reports.html*.

[7] Kagal, L., T. Finin, and A. Joshi, A Policy Language for a Pervasive Computing Environment, *Proceedings IEEE Fourth International Workshop Policies for Distributed Systems and Networks*, pp. 63–77, June 2003.

[8] *www.dse.doc.ic.ac.uk/Research/policies/index.shtml*.

[9] *www.ietf.org/html.charters/policy-charter.html*.

[10] *www.darpa.mil/sto/smallunitops/xg.html*.

[11] *www.darpa.mil/sto/strategic/dc.html*.

[12] Phillips, C., S. Demurjian, and T. Tang, Towards Information Assurance for Dynamic Coalitions, *Proceedings IEEE Workshop Information Assurance*, June 2002.

[13] Uszok, A., et al., KAoS Policy Management for Semantic Web Services, *IEEE Intelligent Systems*, 19(4):32–41, 2004.

[14] *www.swsi.org*.

[15] *www.daml.org*.

[16] *www.w3.org*.

[17] Uszok, A., et al., KAoS Policy and Domain Services: Toward a Description-Logic Approach to Policy Representation, Deconfliction, and Enforcement, *Proceedings IEEE Fourth International Workshop Policies for Distributed Systems and Networks*, p. 93, June 2003.

[18] Stone, G., B. Lundy, and G. Xie, Network Policy Languages: A Survey and a New Approach, *IEEE Network*, (January):10–21, 2001.

[19] Strauss, F., Java Policy Management System: Design and Implementation Report, Computer Science Department, Technical University, Braunshweig, September 2001.

[20] *www.ibr.cs.tu-bs.de/projects/jasmin/*.

[21] *www.dse.doc.ic.ac.uk/research/policies*.

[22] Dulay, N., et al., A Policy Deployment Model for the Ponder Language, *Proceedings IEEE/IFIP International Symposium on Integrated Network Management*, May 2001.

[23] Damianou, N., et al., A Survey of Policy Specification Approaches, Imperial College of Science Technology and Medicine, April 2002.

[24] Damianou, N., et al., The Ponder Policy Specification Language, *Proceedings Policy 2001: Workshop Policies for Distributed Systems and Networks*, vol. 1995, p. 18, Springer-Verlag, 2001.

[25] Tonti, G., et al., Semantic Web Languages for Policy Representation and Reasoning: A Comparison of KAoS, Rei, and Ponder, *Proceedings Second International Semantic Web Conference*, Springer-Verlag, October 2003.

[26] *www.nortel.com*.

[27] Moore, B., et al., Policy Core Information Model—Version 1 Specification, *IETF RFC 3060*, February 2001; available at *www.rfc-editor.org/rfc/rfc3060.txt*.

[28] Snir, Y., et al., Policy Quality of Service (QoS) Information Model, *IETF RFC 3644*, November 2003; available at *www.rfc-editor.org/rfc/rfc3644.txt*.

[29] Feeney, K., D. Lewis, and V. P. Wade, Policy-Based Management for Internet Communities, *Proceedings Fifth IEEE International Workshop Policies for Distributed Systems and Networks*, p. 23, June 2004.

[30] Uszok, A., et al., Policy and Contract Management for Semantic Web Services, *Proceedings AAAI Spring Symposium on Semantic Web Services*, 2004.

[31] Boutaba, R., and S. Znaty, An Architectural Approach for Integrated Network and Systems Management, *ACM-SIGCOM Comp. Communications Review*, 25(5):13–39, 1995.

[32] Uszok, A., et al., Applying KAoS Services to Ensure Policy Compliance for Semantic Web Services Workflow Composition and Enactment, *Proceedings Third International Semantic Web Conference*, vol. 3298, pp. 425–440, Springer-Verlag, LNCS, November 2004.

[33] Lee, S., et al., Managing the Enriched Experience Network—Learning-Outcome Approach to the Experimental Design Life-Cycle, *Proceedings Australian Telecommunications, Networks and Applications Conference*, December 2003.

[34] Khurana, H., and V. D. Gligor, A Model for Access Negotiations in Dynamic Coalitions, Enterprise Security Workshop, *Proceedings 13th IEEE International Workshop Enabling Technologies: Infrastructures for Collaborative Enterprises*, June 2004.

[35] Lloyd, J. W., *Foundations of Logic Programming*, Second Edition, Springer-Verlag, 1987.

[36] *www.ksl.stanford.edu/software/JTP*.

[37] *www.sdrforum.org*.

[38] *Software Communications Architecture*, prepared by Joint Tactical Radio System Joint Program Office, version 2.2.1, April 2004.

[39] Montanari, R., G. Tonti, and C. Stefanelli, Policy-Based Separation of Concerns for Dynamic Code Mobility Management, *Proceedings 27th Annual International Computer Software and Applications Conference*, IEEE CS, 2003.

[40] Sheridan-Smith, N., A Distributed Policy-Based Network Management (PBNM) System for Enriched Experience Networks, Assessment of Proposed Doctoral Research, University of Technology, Sydney, November 2003.

[41] Li, N., J. C. Mitchell, and W. H. Winsborough, Design of a Role-Based Trust Management Framework, *Proceedings IEEE Symposium on Security and Privacy*, IEEE Computer Society Press, May 2002.

[42] Chiang F., et al., Autonomic Service Configuration for Heterogeneous Telecommunication MASs with Extended Role-Based GAIA and JADEx, *Proceedings International Conference on Service Systems and Service Management*, June 2005.

[43] *http://teleholonics.eng.uts.edu.au*.

[44] Verma, D., Simplifying Network Administration Using Policy Based Management, *IEEE Network*, 16(2):20–26, 2002.

Cognitive Techniques: Physical and Link Layers

Thomas W. Rondeau
Center for Communications Research
Princeton, New Jersey

Charles W. Bostian
Bradley Department of Electrical and Computer
Engineering, Virginia Tech, Blacksburg, Virginia

7.1 INTRODUCTION

This chapter discusses the expectation of a fully functional cognitive radio (CR), including the cognitive decision-making process using case-based theory and genetic algorithms (GAs), to solve the multiobjective optimization problem posed by such a radio. The presentation has as its basis the cognitive engine (CE) developed at the Virginia Polytechnic Institute and State University (Virginia Tech)–Center for Wireless Telecommunications (VT-CWT).

The focus of this chapter is intelligent cross-layer optimization of physical (PHY) and link (or medium access control, MAC) layers. The reader is encouraged to think beyond these two layers and consider how other layers, particularly network and transport, can be adapted and optimized by using the same techniques.

Section 7.2 defines optimization for a cognitive radio. A discussion of the cognitive radio as a mix of artificial intelligence (AI) and wireless communications follows in Section 7.3. Section 7.4 addresses the PHY and MAC layers, and considers which measurable radio settings and specifications ("knobs") and which radio and channel performance measures ("meters") fall into which layer. Section 7.5 introduces multiobjective decision-making (MODM) theory to analyze the radio's performance, and presents the analogy of GA to represent the methodology. In Section 7.6, the tiered algorithm structure of the cognition loop, based on modeling, action, feedback, and knowledge representation, is explored in detail.

Tailoring GA techniques for a cognitive radio, based on the simple genetic algorithm discussed in these earlier sections, is addressed in Section 7.7 by looking at case-based reasoning (CBR) and case-based decision theory (CBDT). The need for a higher level of intelligence is the topic of Section 7.8, and Section 7.9 shows how the collection of techniques addressed in this chapter creates a CE capable of controlling and adapting a CR. Section 7.10 then summarizes the processes and ideas brought forth in this chapter.

7.2 OPTIMIZING PHYSICAL AND LINK LAYERS FOR MULTIPLE OBJECTIVES UNDER CURRENT CHANNEL CONDITIONS

The goal of a cognitive radio is to optimize its own performance and support its user's needs. But what does "optimize" mean? It is not a purely selfish adaptation where the radio seeks to maximize its own consumption of resources.

Consider the "tragedy of the commons" metaphor as it is often applied to wireless communications [1].[1] If two pairs of nodes are communicating on different networks using transmissions that overlap in time and frequency, they will interfere. The nodes observe the interference as a low signal to interference and noise ratio (SINR), and the classic response is then to increase transmitter power to obtain a corresponding increase in SINR. As the transmitter on one link increases its power, the other link will experience a lower SINR and respond by increasing its power. Each radio will respond in turn to maximize its SINR at its intended receiver by increasing its own transmit power. Each transmitter will ultimately increase its power to the limitations of the hardware. At this point, either both links will have low SINR and therefore poor performance, or the link with the more powerful transmitter will completely drown out the other. This is obviously a poor solution. Even in the latter scenario, a second glance shows this to be bad for all concerned because now each radio is transmitting much more power than is required, raising power consumption, reducing battery life, and increasing potential interference to other users. This scenario is regularly reenacted in the 2.4 GHz industrial, scientific, and medical (ISM) band in which Bluetooth and IEEE 802.11 devices are constantly creating cross interference. Here 802.11 has a higher transmit power, but Bluetooth has a protocol to continually repeat packet transmissions until a successful transmission occurs.

In contrast, a radio capable of understanding its environment and making intelligent adaptations will recognize the problem it encounters with a competing link trying to use the same band. While observing the other link, the CR will not be limited by a simplistic understanding that "low SINR means I should increase my transmitter power." Instead, it will try other solutions, such as altering the modulation or channel coding in ways that will improve frame error rate (FER) performance in the channel. Or it will seek a channel free of interference and change its center frequency, thus relieving both radios of the burden of fighting for the spectrum.

In a heavily congested spectral environment, changing frequency might not be an acceptable solution. This is why it is important to look at all the possible adjustments to the PHY and MAC layers to improve performance. A situation might arise when all possible frequencies are in use, or the bandwidth required is not available free from interference. At such times, a mix of cooperative techniques could be used, perhaps involving quadrature amplitude (and phase) modulation (QAM), spread-spectrum techniques, orthogonal frequency division multiplexing (OFDM), clever timing mechanisms, smart antenna beamforming or null forming, or other operations that will allow sharing.

---

[1]Hazlett's paper [1] does not give a highly technical overview of this concept, but rather a regulatory analysis of the spectrum issues in general.

This chapter addresses methods of how to find a local or global optimum for the current channel environment. A generalized solution is not usually the answer; for example, spreading techniques might better share spectrum than narrowband techniques, but a spread-spectrum system has its limits and might unnecessarily waste system resources. A dynamic combination of techniques is really required to best adapt the radio in real time for the specific local problems at hand. A well-designed CR understands situations and analyzes how to best adapt all available radio communications parameters to present conditions for optimum performance.

First, a definition of optimization is in order. Within this chapter, a radio is optimized when it achieves a level of performance that satisfies its user's needs while minimizing its consumption of resources such as occupied bandwidth and battery power. To apply this, we need to understand what needs the user has and how the radio performance can be adapted to meet these needs. The bottom line, in general, is that the radio not overoptimize its external performance (its performance as observed by other radios) because this will have a negative impact on its internal performance (computational power, complexity, and available internal resources).[2] The next section clearly defines the cognitive radio. The succeeding sections then discuss what parameters can be altered and how a cognitive radio can use them to optimize its performance.

7.3 DEFINING THE COGNITIVE RADIO

Cognitive radios merge artificial intelligence and wireless communications. The field is highly multidisciplinary, mixing traditional communications and radio work from electrical engineering while applying concepts from computer science. Interestingly, Claude Shannon, one of the early giants of communication theory, spent some of his time thinking about and discussing intelligent machines, specifically, chess-playing machines. In a paper published in 1950, he discusses how computers could be made to intelligently play chess, but he also lays out reasons why this has implications in other areas, such as designing filters, equalizers, relays, and switching circuits, routing, translating languages, organizing military operations, and making logical deductions [2].

The cognitive radio architecture envisioned in the discussion in this chapter is shown in Figure 7.1. Here, the intelligent core of the cognitive radio exists in the CE. The CE performs the modeling, learning, and optimization processes necessary to reconfigure the communication system, which appears as the simplified open systems interconnection (OSI) stack [3]. The CE takes in information from the user domain, the radio domain, the policy domain, and the radio itself. The user domain passes information relevant to the user's application and networking needs to help direct the CE's optimization. The radio domain information consists of radio frequency (RF) and environmental data that could affect system performance such as propagation or interference sources. The policy engine receives policy-related information from the policy domain. This information helps the cognitive radio decide on allowable (and legal) solutions and blocks any solutions that break local regulations.

---

[2] Other, more sophisticated considerations are beyond the scope of this chapter.

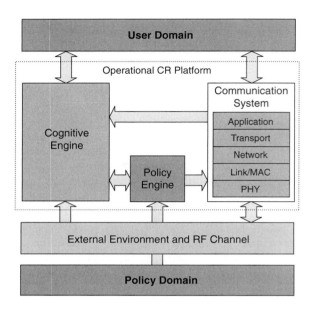

FIGURE 7.1

Generic CR architecture. This architecture has a CE to observe behaviors of the OSI protocol stack and propose optimizations based on the current environment. The policy engine determines whether the hardware can support those optimizations as well as whether it is allowed to by regulatory and network control.

Most of the topics shown in Figure 7.1 are covered in more detail in this chapter as their relationships to the CE are developed. The policy engine and policy domain will be left to experts of that field, but are included here for completeness.

7.4 DEVELOPING RADIO CONTROLS (KNOBS) AND PERFORMANCE MEASURES (METERS)

The first problem in dealing with cognition in a system is to understand (1) what information the intelligent core must have and (2) how it can adapt. In radio, we can think of the classical transmitters and receivers as having adjustable control parameters (knobs) that control the radio's operating parameters. Think of a frequency modulation (FM) broadcast receiver with a tuning knob to select which station you are listening to as well as equalizer knobs to adjust the sound quality to your liking. Radio performance metrics are referred to as meters. What follows is an analysis of the knobs and meters important to a cognitive radio on the physical and link layers.

Huseyin Arslan has developed a useful layered classification of knobs and meters (*writable parameters* and *observable parameters*, respectively, in his notation), which is summarized and expanded in Table 7.1.[3]

---

[3] Huseyin Arslan, personal communication.

Table 7.1 Example Tabulation of Knobs and Meters by Layer

Layer	Meters (Observable Parameters)	Knobs (Writable Parameters)
NET	Packet delay Packet jitter	Packet size Packet rate
MAC	Cyclic redundancy check (CRC) Automatic repeat request (ARQ) Frame error rate Data rate	Source coding Channel coding rate and type Frame size and type Interleaving details Channel/slot/code allocation Duplexing Multiple access Encryption
PHY	Bit error rate (BER) Signal-to-noise ratio (SNR) Signal-to-interference and noise ratio (SINR) Received signal strength indicator (RSSI) Pathloss Fading statistics Doppler spread Delay spread Multipath profile Angle of arrival (AOA) Noise power Interference power Peak-to-average power ratio Error vector magnitude Spectral efficiency	Transmitter power Spreading type Spreading code Modulation type Modulation index Bandwidth Pulse shaping Symbol rate Carrier frequency Dynamic range Equalization Antenna beam shape

7.4.1 Physical and Link Layer Parameters

In the following we explain the fundamental sensing and control mechanism through which the reasoner understands what is going on in the spectrum and in the communication channel, and by which it performs changes to the control mechanisms to improve communication performance. We refer to these as knobs and meters.

Knobs

The knobs of a radio are any of the parameters that affect link performance and radio operation. Some of these are normally assumed to be design parameters, and others are usually assumed to be under real-time control of either the operator or the radio's real-time control processes. Figure 7.2 shows a simple system diagram of the physical and link layer portions of a transmitter. In the physical layer, center frequency, symbol rate, transmit power, modulation type and order, pulse shape filter (PSF) type and order, spread-spectrum type, and spreading factor can all be adjusted. On the link layer are variables that will improve network performance, including the type and rate of the channel coding and interleaving, as well as access control methods such as flow control, frame size, and the multiple access technique.

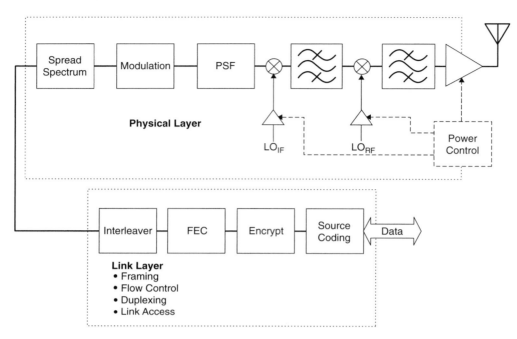

FIGURE 7.2

Generic transmitter PHY and MAC layers. Many of the radio control parameters (knobs) apply to these elements of the block diagram, resulting in profound impact on radio performance (meters).

Meters

Once we understand what knobs are available to optimize the radio system performance, we must understand how these changes affect the radio channel and system performance to allow an autonomous, intelligent decision maker to adapt the radios.

Performance is a measure of the system's operation based on the meter readings. In optimization theory, the meters represent utility or cost functions that must be maximized or minimized for optimum radio operation. All of these performance analysis functions constitute objective functions. In an ideal case, we can find a single objective function the maximization or minimization of which corresponds to the best settings. However, communication systems have complex requirements that cannot be subsumed into a single objective function, especially if the user or network requirements change. Metrics of performance are as different for voice communications as they are for data, email, Web browsing, or video conferencing.

The types of meters represent performance on different levels. On the physical layer, important performance measurements deal with bit fidelity. The most obvious meters are the signal-to-noise ratio, or a more complex signal-to-interference and noise ratio. The SINR has a direct consequence on the bit error rate (BER), which has different meanings for different modulations and coding techniques, usually nominally determined by the SINR ratio, $E_b/(N_0 + I_0)$, where E_b is energy per bit, N_0 is noise power per bit, and I_0 is interference power per bit. On the link layer, the packet fidelity is an important metric, specifically the packet error rate.

There are more external metrics to consider as well, such as the occupied bandwidth and spectrum efficiency (number of bits/hertz) and data rate. The growth of complexity to optimize multiple metrics quickly becomes apparent. Metrics have independent relationships, and knobs affect certain metrics in different ways. For example, altering the modulation type to a higher order will increase the data rate but worsen the BER.

Internal metrics also are involved in decision making. To decrease the FER, we could use a stronger code, but this increases the computational complexity of the system, increasing both latency as well as the power required to perform the more complex forward error correction (FEC) operation. Decreasing the symbol rate or modulation order will decrease the FER as well without increasing the demands of the system, but at the expense of the data rate. Figure 7.3 begins to expand on these relationships,

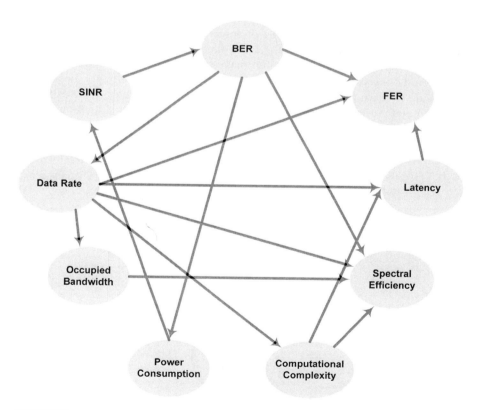

FIGURE 7.3

Directed graph indicating how one objective (source) affects another objective (target).

where the direction of the arrow indicates that optimization in the source objective affects the target objective. Ongoing work fully defining these relationships should lead to more knowledge for the adaptation and learning system to use.

7.4.2 Modeling Outcome as a Primary Objective

The basic process followed by a CR is that it adjusts its knobs to achieve some desired (optimum) combination of meter readings. Rather than randomly trying all possible combinations of knob settings and observing what happens, it makes intelligent decisions about which settings to try, and observes the results of these trials. Based on what it has learned from experience, and on its own internal models of channel behavior, it analyzes possible knob settings, predicts some optimum combination for trial, conducts the trial, observes the results, and compares the observed results with its predictions, as summarized in the adaptation loop of Figure 7.4. If results match predictions, the radio understands the situation correctly. If results do not match predictions, the radio learns from its experience and tries something else.

This operational concept employed for the CR closely resembles some of the current thinking about how the human brain works [4]. The argument holds that human intelligence is derived from predictive abilities of future actions based on the currently observed environment. In other words, the brain first models the current situation as perceived from the sensor inputs, and it then makes a prediction of the next possible states that it should observe. When the predictions do not match reality, the brain does further processing to learn the deviation and incorporates that learning with its future modeling techniques. Although knowledge of how the human brain actually works is still uncertain, this predictive model is a good one to work from because it brings together the necessary behavior required from the cognitive radio.

As an example of the mathematics involved in this process, consider observations of BER and SINR. BER formulas are generally represented by the complementary error function or the Q function, Eq. (7.1) as a function of the SINR.

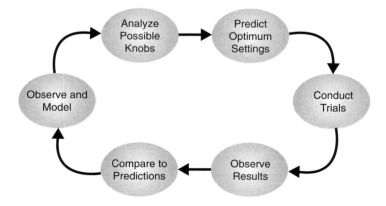

FIGURE 7.4

Adaptation loop.

$$erfc(x) = \frac{2}{\sqrt{\pi}} \int_x^\infty e^{-t^2} dt$$

$$Q(x) = \frac{1}{\sqrt{2\pi}} \int_x^\infty e^{-\frac{t^2}{2}} dt \quad x \geq 0 \qquad \text{(7.1)}$$

$$Q(x) = \frac{1}{2} erfc\left(\frac{x}{\sqrt{2}}\right),$$

where x is the SINR. A computationally efficient calculation for BER formulas uses an approximation to the complimentary error function. Eq. (7.2) shows two approximations, one for small values of x (less than 3) and one for large values of x (greater than 3) [5]. Figure 7.5 compares the results of the formulas to the actual analytical function.

erfc *Approximation Eq. (7.2) Compared to Analytical Formula Eq. (7.3)*

The following formulas are useful because the normal approximation for the *erfc* function is valid only for large $x(x > 3)$, and the Q function is too computationally intensive to calculate.

$$erfc(x) = \begin{cases} 1 - \frac{1}{\sqrt{\pi}}\left(x - \frac{x^3}{3} + \frac{x^5}{10}\cdots\right) & \text{for } x < 3 \text{ with 40 items in the series} \\ \frac{e^{-x^2}}{x\sqrt{\pi}}\left(1 - \frac{1}{2x^2} + \frac{1\cdot3}{2^2x^4} - \frac{1\cdot3\cdot5}{2^3x^6} + \cdots\right) & \text{for } x \geq 3 \text{ with 10 items in the series} \end{cases}$$

$$\text{(7.2)}$$

Because a CR needs to perform a lot of these calculations, we need efficient equations; these equations trade accuracy for computational time based on the number of terms

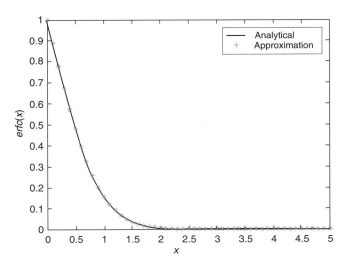

FIGURE 7.5

erfc approximation Eq. (7.2) compared to analytical formula Eq. (7.3).

included in the series expansion. Similar lines of thought must go into developing each objective calculation.

The exact representation of the BER formula depends on the channel conditions and modulation being used. A standard BER formula for binary phase shift keying (BPSK) signals in an additive white Gaussian noise (AWGN) channel is:

$$P_e = \frac{1}{2} erfc\left(\sqrt{T_0 B \frac{C}{N}}\right), \tag{7.3}$$

where T_0 is the symbol period, B is the bandwidth, C is the signal carrier energy, and N is the noise power. In a fading channel with a probability density function of $p(x)$, the BER of a signal is approximated as:

$$P_e = \int_0^\infty P_{AWGN}(x)\, p(x)\, dx, \tag{7.4}$$

where $P_{AWGN}(x)$ is the BER formula in an AWGN channel.

The radio observes the BER and SINR value. If these are consistent according to the preceding formulas, the radio can assume that the channel is behaving predictably. It can then turn knobs that directly affect SINR, for example starting with the easiest, transmitter power.[4] If the transmitter power is already at the allowable limit, the radio may lower the data rate to change the occupied bandwidth and therefore increase the average energy per bit. If the BER and SINR are not consistent with the known formulas, the radio might assume, for example, that the channel is dispersive and opt to change the carrier frequency rather than the transmitter power.

This analysis has dealt with only a single objective. The radio can, in fact, read a number of meters, and each of these can be some objective function we may wish to optimize. Standard communications theory can lead us to the methods of mathematically modeling each objective [7]. The communications analysis tools are fairly standard, and another aspect is the consideration for how to efficiently realize each objective function. For each, we must carefully choose the proper analytical expression that is not too computationally complex.

7.5 MULTIOBJECTIVE DECISION-MAKING THEORY AND ITS APPLICATION TO COGNITIVE RADIO

The wireless optimization concept has already been described through an analysis of the many objective functions (dimensions) of inputs (knobs) and outputs (meters). In this scenario, the interdependence of the objectives to each other and to various knobs makes it difficult to analyze the system in terms of any one single objective. Furthermore, the needs of the user and of the network cannot all be met simultaneously, and

---

[4] The difference between predicted link performance and actual link performance includes both errors in the estimate of propagation channel losses, and nonlinear effects arising from interference and multipath. Performance on an AWGN channel in the absence of multipath is predictable. When this performance difference is significantly large, it may become clear that transmit power alone is inadequate to achieve the necessary performance. Thus, the performance difference is a good indicator of the need to invoke these CR techniques to optimize performance in the presence of unusual channel behavior.

these needs can change dramatically with time or between applications. For different users and applications, radio performance and optimum service have different meanings. As a simple example, email has a much different performance requirement than voice communications, and a single-objective function would not adequately represent these differing needs.

Without a single-objective function measurement, we cannot look to classic optimization theory for a method to adapt the radio knobs. Instead, we can analyze the performance using MODM criteria. MODM theory allows us to optimize in as many dimensions as we have objective functions to model.

MODM work originated about 40 years ago and has application in numerous decision problems from public policy to everyday decisions (e.g., people often decide where to eat based on criteria of cost, time, value, customer experience, and quality). An excellent introduction to MODM theory is given in a lecture from a workshop held on the subject in 1984 [8]. Schaffer then applied MODM theory to create a genetic algorithm capable of multiobjective analysis in his doctoral dissertation [9]. Since then, GAs have been widely used for MODM problem solving. Genetic algorithms are addressed in detail in Section 7.5.5.

7.5.1 Definition of MODM and Its Basic Formulation

At their core, MODMs are a mathematical method for choosing the set of parameters that best optimizes the set of objective functions. Eq. (7.5) is a basic representation of a MODM method [10].

$$\min/\max\{y\} = f(\overline{x}) = [f_1(\overline{x}), f_2(\overline{x}), \ldots, f_n(\overline{x})]$$
$$\text{subject to: } \overline{x} = (x_1, x_2, \ldots x_m) \in X \tag{7.5}$$
$$\overline{y} = (y_1, y_2, \ldots, y_n) \in Y$$

Here, all objective functions are defined to either minimize or maximize y, depending on the application. The x values (i.e., x_1, x_2, etc.) represent inputs, and the y values represent outputs. The equation provides the basic formulation without prescribing any method for optimizing the system. Some set of objective functions combined in some way will produce the optimized output. There are many ways of performing the optimization. Section 7.6 discusses one of the more complex, but useful, methods of solving MODM problems for cognitive radios.

7.5.2 Constraint Modeling

An added benefit of MODM theory implicit in its definition is the concept of constraints. The inputs, x, are constrained to belong to the allowed set of input conditions X, and all output must belong to the allowed set Y. This is important for building in limitations for hardware as well as setting regulatory bounds.

7.5.3 The Pareto-Optimal Front: Finding the Nondominated Solutions

In an MODM problem space, a set of solutions optimizes the overall system if there is no one solution that exhibits a best performance in all dimensions. This set, the set of

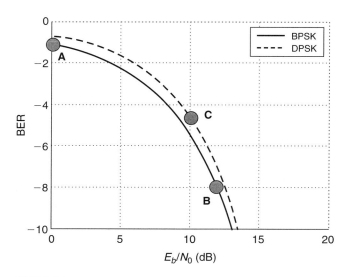

FIGURE 7.6

Pareto front of a BPSK BER curve compared to a dominated solution of DPSK. Condition A is least power, condition B is lowest BER, and condition C is least complexity.

nondominated solutions, lies on the Pareto-optimal front (hereafter called the *Pareto front*). All other solutions not on the Pareto front are considered dominated, suboptimal, or locally optimal. Solutions are nondominated when improvement in any objective comes only at the expense of at least one other objective [11].

The most important concept in understanding the Pareto front is that almost all solutions will be compromises. There are few real multiobjective problems for which a solution can fully optimize all objectives at the same time. This concept has been referred to as the utopian point [12]; this point is not considered further here, because in radio modeling problems, only very rarely do situations have a utopian point. One only has to consider the most basic radio optimization problem to see this: simultaneously minimize BER and power. Figure 7.6 shows the ideal BER curve of a BPSK signal. Here, point A is a nondominated point that minimizes power at the expense of the BER, and moving down the curve to point B, which is also nondominated, optimizes for BER at the expense of greater power consumption. Point C is a dominated point that represents a suboptimal solution of using differential phase-shift keying (DPSK) to minimize the complexity of carrier phase tracking in high-multipath mobile applications. The MODM problem then reduces to a trade-off decision between low BER, low power consumption, and complexity due to other system constraints.

7.5.4 Why the Radio Environment Is a MODM Problem

The primary objectives developed thus far have different meanings and importance, depending on the user's needs and channel conditions. To optimize the radio behavior for suitable communications, we must optimize over many or all of the possible radio objectives. Take, as an example, the BER curve. In the two-dimensional plot, the result of using a differential receiver was suboptimal because we were concerned with BER and power. But if we add complexity as an objective, or the need for a solution that does not require phase synchronization, such as might be necessary in highly complex and dynamic multipath, then using a differential receiver for lower complexity becomes an important decision along a third dimension. The resulting search space is N-dimensional and, due to the complex interactions between objectives and knobs, the space is difficult to define and certainly not linear, or even simply convex as is desirable in optimization theory [13]. These interactions are often difficult to characterize and predict, and so we must analyze each objective independently and use MODM theory to find an optimal aggregate set of parameters.

What further enhances the complexity of the search space is that it will change depending on the user and the application. For certain users or applications, different objectives will mean different levels of quality. As the overall optimization is to provide the best quality of service (QoS) to the user, there is no single search space that can account for all the variations in needs and wants from a given radio.

From this analysis, a few important points emerge about how to analyze the multiple objectives used in optimizing a radio:

- Many objectives exist, creating a large N-dimensional search space.
- Different objectives may be relevant for only certain applications/needs.
- The needs and subjective performances for users and applications vary.
- The external environmental conditions determine what objectives are valid and how they are analyzed.
- We may search for regions where multiple performance metrics meet acceptable performance, rather than searching for optimal performance.

This leads to a need for a MODM algorithm capable of robust, flexible, and online adaptation and analysis of the radio behavior. The clearest method of realizing all the needs of the problem statement is the genetic algorithm, which is widely considered the best approach to MODM problem solving [9, 10, 14–16]. Section 7.5.5 discusses the approach to genetic algorithms, and Section 7.9.1 shows how to add user and application flexibility into the algorithm.

7.5.5 Genetic Algorithm Approach to the MODM

Analyzing the radio by using a GA is inspired by evolutionary biological techniques. If we treat the radio like a biological system, we can define it by using an analogy to a chromosome, in which each gene of the chromosome corresponds to some trait (knob) of the radio. We can then perform evolutionary-type techniques to create populations of possible radio designs (e.g., waveform, protocols, and even hardware) that produce offspring that are genetic combinations of the parents. In this analogy, we evolve the radio parameters much like biological evolution to improve the radio "species" through successive generations, with selection based on performance guiding the evolution. The traits represented in the chromosome's genes are the radio knobs, and evolution leads toward improvements in the radio meters' readings.

GAs are a class of search algorithms that rely on both directed searches (exploitation) and random searches (exploration). The algorithms exploit the current generation of chromosomes by preserving good sets of genes through the combination of parent chromosomes, so there is a similarity between the current search space and the previous search space. If the genetic combination is from two highly fit parents, it is likely that the offspring is also highly fit. The algorithms also allow exploration of the search space by mutating certain members of the population that will form random chromosomes, giving them the ability to break the boundaries of the parents' traits and discover new methods and solutions. While providing the iterative solution through genetic combination, the randomness helps the population escape a possible local optimum or find new solutions never before seen or tried, even by a human operator. In effect, this last quality provides the algorithm with creativity.

Introduction to Genetic Algorithms

Genetic algorithms are often useful in large search spaces, which can enable their use in many situations. A GA is a search technique inspired by biological and evolutionary behavior. The GAs use a population of chromosomes that represent the search space and determine their fitness by a certain criterion (fitness function). In each generation (iteration of the algorithm), the most fit parents are chosen to create offspring, which are created by crossing over portions of the parent chromosomes and then possibly adding mutation to the offspring. The crossover of two parent chromosomes tries to exploit the best practices of the previous generation to create a better offspring. The mutation allows the search algorithm to be "creative"—that is, it can prevent the GA from getting stuck in a local maximum by randomly introducing a mutation that may result in improved performance metrics possibly closer to the global maximum, according to the optimization criteria.

To realize the GA, we follow the practices described by Goldberg [17]:

1. Initialize the population of chromosomes (radio/modem design choices)
2. Repeat until the stopping criterion
   a. *Choose* parent chromosomes
   b. *Crossover* parent chromosomes to create offspring
   c. *Mutate* offspring chromosomes
   d. *Evaluate* the fitness of the parent chromosomes
   e. *Replace* less fit parent chromosomes
3. Choose the best chromosome from the final generation

This process is illustrated in detail in the next section.

Knapsack Example

To explain the operation of a simple GA, we examine the knapsack problem [18], which is a classic NP-complete[5] problem [19], also called the subset-sum problem (SSP). The

---

[5]Nondeterministic polynomial (NP) time is a class of decision problems the positive solutions of which can be verified in polynomial time given the right information, but finding a computational solution is extremely difficult.

x_1	x_2	x_3	$\cdots$	x_{N_s}

FIGURE 7.7

Chromosome representation of knapsack item vector.

knapsack problem is defined by the task of taking a set of items, each with a weight, and fitting as many of them into the knapsack while coming as close to, but not exceeding, the maximum weight the knapsack can hold. Mathematically, the knapsack problem is shown by Eq. (7.6), where K is the maximum weight the knapsack can hold, and N_s is the number of items in the set, S. The problem is represented by a weight vector, w, and a vector x that is a vector of 1's and 0's that indicates whether an item is present in the knapsack.

$$\max \sum_{i=1}^{N_s} x_i w_i$$
$$\text{subject to: } \sum_{i=1}^{N_s} x_i w_i \leq K \tag{7.6}$$

The following practices were enumerated previously.

Step 1. Initialize Chromosomes

We introduce radio chromosome selection as a problem similar to knapsack selection. In this case, the problem consists of choosing the right set of items to place in the knapsack, so the chromosome will represent the vector x and consist of 1's and 0's, as shown in Figure 7.7. Each gene is 1-bit wide and thus the chromosome is very compact and mathematically easy to manipulate.

Step 2a. Choose

The choice of the parent chromosomes determines how random the population will remain and how much memory the biological system retains of its fitness for its environment. Like biological evolution, the most fit parents are more likely to be chosen to produce offspring; however, it is still possible to choose an unfit parent. The more random the selection process, the more random the population will be in each generation. Conversely, the more the decision is weighted toward the most fit parents, the faster the convergence to local optima will occur. This trade-off is referred to as selection pressure.

As in all GA properties, the choice of how parents are chosen comes down to how random the population will remain and how fast convergence will occur. Although convergence time is highly important, fast convergence may not always be beneficial, depending on the shape of the fitness landscape or the solution space. If the fitness landscape has many local maxima and the GA is searching for the global maximum, fast convergence is more likely to lock on to a local maximum and take many generations and mutations to get beyond the local maximum to find the global maximum. A more diverse population will generate many more solutions that are more likely to find the global maximum. Five different methods of selection were analyzed by De Jong in his doctoral dissertation [20].

The method used in this GA is called tournament selection. During reproduction, this selection scheme chooses as many parents as there are members in the population with replacement (i.e., the same parent may be chosen multiple times). In the selection, two parents are randomly chosen and the parent more fit wins the tournament and is selected for reproduction.

Step 2b. Crossover

Crossover is performed on two parents to form two new offspring. The GA has a crossover probability that determines if crossover will happen. A randomly generated floating-point value is compared to the crossover probability, and if it is less than the probability, crossover is performed; otherwise, the offspring are identical to the parents. If crossover is to occur, one or more crossover points are generated, which determines the position in the chromosomes where parents exchange genes. In this GA, once two parents are selected, two crossover points are randomly generated.

Figure 7.8 illustrates the crossover operation, but, for the simplicity of the figure, the genes represent a knapsack problem with only eight items. The genes after the first crossover point and before the second crossover point are interchanged in the parents to form the new offspring.

Step 2c. Mutate

After the offspring are generated from the selection and crossover, the offspring chromosomes may be mutated. Like crossover, there is a mutation probability. If a randomly selected floating-point value is less than the mutation probability, mutation is performed on the offspring; otherwise, no mutation occurs. Mutation is performed by randomly

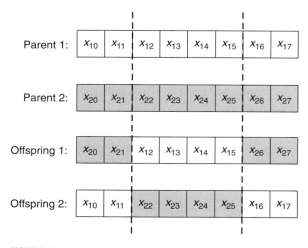

FIGURE 7.8

Parent chromosomes crossover. Crossover occurs at points 2 and 6 to create offspring chromosome sets.

selecting a gene in the offspring's chromosome and generating a new (uncorrelated) value based on a prescribed probability density function (often a uniform or Gaussian distribution). In the knapsack problem, a gene may be reset to either a 1 or 0 at random (other techniques invert the gene).

Steps 2d and 2e. Evaluate and Replace

Evaluation (step 2d) is done on the multiobjective performance of the offspring from the most recent genetic cycle to determine whether to continue the genetic process (step 2e). Evaluation is probably the most important piece of the GA aside from the initial chromosome definition. Choice of the fitness evaluation is vital to convergence and proper application.[6] The best set of offspring from among all those evaluated are ordered by performance and added to the rank-ordered list of chomosomes, for example, the top 10 or the top 100.

Evaluation determines whether the gene pool is improving with the previous cycles, such as the last 5 cycles N generations. If the performance is sufficiently above the goal behavior, the iterative process may terminate and the procedure goes to step 3. If, however, the performance over the previous cycles continues to improve but the goal performance is not yet met, the algorithm iterates back to step 2a. If the performance is unable to improve in the previous cycles, the mutation rate may be increased to a higher mutation rate. The evaluation function is very problem-specific, and is one of the main sources of research in the multiobjective GA to optimize cognitive radios.

Step 3. Results: Choose Best Chromosomes

To analyze the genetic algorithm performance, we often graphically represent the performance over many generations, such as the knapsack example shown in Figure 7.9. The figure shows the general trends of a genetic algorithm as a maximization problem with steady increase in the fitness over generations. In this example, we set the crossover probability at 0.95 and mutation probability at 0.05. There are 20 members of the population, and in each generation 17 members of the population are replaced by offspring. The problem contains 200 items, and each item could take on any floating-point weight value between 0 and 20. The maximum weight the knapsack can hold is a randomly generated integer of 810. The trends displayed in Figure 7.9 show steady improvement toward the goal of 810, coming very close by the 155th generation with a value of 809.87.

As shown in Figure 7.9, a GA performs by making great initial progress, but then slows down as it approaches the optimal value, and even then, it does not perfectly achieve the optimum. Section 7.6 explores the use of GAs in radio optimization problems. Radio optimization must be real time, but it does not have to completely optimize under real-time constraints. Instead, we want to provide *better* responses as opposed to a *best* response for the immediate future. GAs quickly improve their performances in the first few generations; by the time the knapsack problem reaches its first plateau around generation 30, we have a useful solution. Looking at what we *need* out of the performance of a radio instead of just what we *want*, the GA can give us very usable

---

[6]For the knapsack problem, we try to find the chromosome with fitness defined by Eq. (7.6). If the constraint condition is not met, the fitness is set to 0.

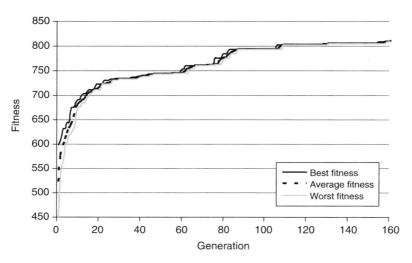

FIGURE 7.9

Performance graph of knapsack GA. The fitness metric improves with each generation, nearly reaching the maximum value after 155 generations.

solutions very quickly. As time and resources permit, the final iterations of the GA can find increasingly better solutions; other techniques, introduced in Section 7.6, such as case-based decision theory, adaptive genetic algorithms, and distributed computing, can improve the solutions even faster.

Another point to consider involves the temporal features of optimum solutions. In the future, situations envisioned for the cognitive radio, the environments, both propagation and interference, as well as the user's and application's needs, will change, and with them the optimum solution. We therefore want a system that will continue to change and adapt with the needs of the situation for continuous improvement in performance.

7.6 THE MULTIOBJECTIVE GENETIC ALGORITHM FOR COGNITIVE RADIOS

The primary goal of the CE is to optimize the radio, and the secondary functions are to observe and learn in order to provide the knowledge required to perform the adaptation. A CR becomes a learning machine through a tiered algorithm structure based on modeling, action, feedback, and knowledge representation, as can be seen in the cognition loop shown in Figure 7.10.

7.6.1 Cognition Loop

This section presents the high-level structure of the Virginia Tech CR solution as well as the background of the basic theories and procedures that were followed in its development. The following sections discuss the system in detail.

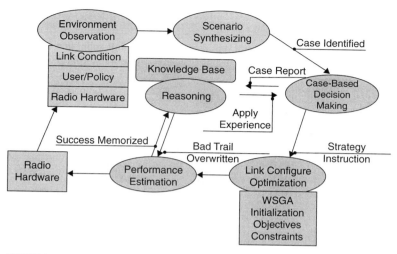

FIGURE 7.10

Cognition loop. Note the outer loop to observe and adapt a radio and the inner loop for learning. (*Note:* WSGA = wireless system genetic algorithm.)

Modeling

Modeling means that the machine must have some representation of the outside world to which it can respond. Models represent external influences, such as the user's network needs and actions, the radio spectrum and propagation environment, and the governing regulatory policy. As each of these influences changes, the radio must adjust itself to satisfy the new operating environment.

A primary function of the modeling system is to provide environmental context to the CR, such as propagation effects and the presence of other radios, which might be either cooperative or potential interferers.

Another goal of the modeling system is to monitor the data sent by the user and use them to determine the QoS parameters the CR must provide. This is a learning domain concept that a neural network could be employed to help solve, and the research group at Virginia Tech is currently investigating the possibilities. The regulatory policy modeling and interpretation will come from other efforts, such as the Defense Advanced Research Projects Agency (DARPA) NeXt Generation (XG) project [21].

Figure 7.11 provides background about how Virginia Tech's work fits into the literature of machine learning and artificial intelligence. It shows the VT-CWT cognitive engine in more detail, emphasizing its three major subsystems.

The cognitive system module (CSM) is responsible for learning, and the WSGA handles the behavioral adaptation of the radio, based on what it is told to do by the CSM. The modeling system observes the environment from many different angles to develop a complete picture. The CSM holds two main learning blocks: the evolver and the decision maker, which takes feedback from the radio that allows the evolver to properly update the knowledge base to respond to and direct system behavior.

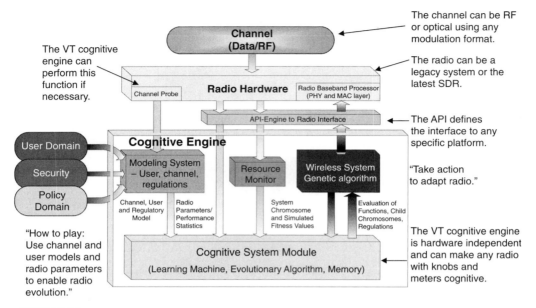

The channel can be RF or optical using any modulation format.

The VT cognitive engine can perform this function if necessary.

The radio can be a legacy system or the latest SDR.

The API defines the interface to any specific platform.

"Take action to adapt radio."

"How to play: Use channel and user models and radio parameters to enable radio evolution."

The VT cognitive engine is hardware independent and can make any radio with knobs and meters cognitive.

FIGURE 7.11

The Virginia Tech Cognitive Engine. This hardware-independent modeling, learning, and adaptation system interfaces to radio hardware via a radio-specific application programmable interface (API).

Action

Actions are taken by the WSGA in creating a new radio configuration. This system is instructed by the CSM when a new model is observed and change is required in the radio operation. The WSGA creates a set of parameters that best optimize the radio for the new settings.

The WSGA treats the radios like biological creatures, where genes represent radio traits such as power, frequency, modulation, FEC coding, filtering, spreading, and so on. Groups of genes constitute chromosomes. The WSGA applies a multiobjective genetic algorithm to a population of chromosomes and evaluates the results by using the objective functions and weights provided by the knowledge base. The functions determine which outputs are important, such as the BER, packet error rate (PER), power consumption, interference potential, and so forth, and the weights determine the relative importance of each function. After running the WSGA through some number of generations (e.g., terminating after convergence has been observed or following a predetermined number of generations, whichever occurs first), the chromosome that performs best in the largest number of objectives is chosen as the new set of radio parameters. Throughout this process, the chromosomes will be analyzed continually with respect to validation routines to ensure that no combination is chosen for the final radio setting that would violate regulatory authority rules or the network operator's operational procedures. Detailed information about the WSGA and experimental results have been published [22], and a summary is provided in an example in Section 7.6.5.

Feedback

Unsupervised learning is accomplished through feedback and a series of rewards and punishments. Alan Turing proposed this idea in his paper on computational intelligence [23], and it is a major theme in expert system and neural network learning algorithms [24]. In the Virginia Tech learning machine, feedback information containing the radio's actual observations of its operation is provided. These actual observations of the performance parameters are then compared to simulated parameters from the WSGA, which uses the performance parameters within its GA to hypothesize the best new radio settings. When the simulated parameters are compared to the actual meters, the evolver in the CSM can determine how well the simulated performance matched the actual performance.

This comparison helps the CE to understand how well its action analysis and selection block are working. Deviations between the actual and simulated performance measurements are penalized. The feedback and comparison mechanism allows the evolver to update the knowledge base to better represent the environment for improved future performance. This process requires an analysis of the model, action, and feedback system to know what is not modeled accurately, and then it specifically corrects this part of the knowledge base—probably by taking a new action from another part of the knowledge base that has performed well in these areas. This type of adaptation has been studied and used previously in genetic algorithm work through techniques such as chromosome tagging and templates or in knowledge-based genetic techniques [17]. The evolver is a highly directed adaptation mechanism to correct for specific mistakes in judgment.

Knowledge Representation

Knowledge is represented as a database of past models and actions to take, along with past actions taken, and any estimates of the level of success associated with these actions. The knowledge base is a set of these models and actions that allows the WSGA to perform better as improved knowledge is gathered or generated by the evolver. As models are observed, actions are taken that best respond to the model's requirements for the radio. As these correlations are learned, the CE can more easily respond to situations as they arise. Likewise, past actions are kept in short-term memory to provide the WSGA with a set of recent actions; when a change is required, it may not have deviated too much from the immediate past, and the recently used actions may already work to guide the WSGA. The knowledge base is designed to provide the WSGA with as much relevant information as possible to reduce the computational complexity of the genetic algorithm. If the population is primed with good solutions, the algorithm can terminate early and adapt the radio faster.

Case-Based Decision Theory

The core of the decision-making process is rooted in case-based decision theory [25], a technique that grew out of economic decision-making research and is similar to the case-based reasoning used in some AI implementations. CBR recognizes similarity between the current event and past events in its memory, and actions taken by the machine for that past event are used to determine the actions the machine will take in the new event (these actions may not be the same). CBDT, in contrast, not only calculates a similarity of events in memory but also assesses the utility of the past actions. The added knowledge of CBDT allows the decision maker to act autonomously, more intelligently, and with better results than a CBR system or other memory-based approaches. CBDT is discussed in more detail in Section 7.7.2.

The key responsibility of the CSM is to maintain the knowledge base of model-action pairs, which ties the abstract models developed by the modeling system with the set of actions taken by the implementation system, which is the WSGA shown Figure 7.11. The actions consist of genetic parameters such as crossover and mutation rates; previously used chromosomes that are related to both the specific channel and network model (exploiting benefits from long-term memory) and the most recently used chromosomes (exploiting benefits from short-term memory); and a set of objective (or fitness) functions and function weights. The actions provided by the CSM define the fitness landscape [17] and prime the genetic algorithm.

Learning

Learning is the aggregation of the preceding processes. This learning machine mechanism is similar to processes known to occur in human learning: sensing, acting, reasoning, feedback, and accumulating knowledge and experience. It is a loop that continuously improves the CR's ability to perform. An even more powerful technique arises when the knowledge and learning capabilities are shared across the network. As any one radio gains knowledge about the models and actions to take, all radios benefit by the distribution of this knowledge. Likewise, the computational processes involved in algorithms such as the WSGA can be shared among many radios in the network, reducing the computations any one radio must perform as well as increasing the speed of the algorithm. Another benefit to distributing the CE is to share radio capabilities. For instance, if one radio in the network has a sounder, such as that described by Rondeau et al. [26], then the modeling capabilities could be shared among all other radios on the network without such equipment.

7.6.2 Representing Radio Parameters as Genes in a Chromosome

The previous sections of this chapter introduced the possible knobs of the physical and link layers that a cognitive radio can adapt, and we argued that we can treat the radios as biological systems with chromosomes that use genes to represent the different knobs. Here, both of these thoughts are expanded to show the representation.

The greatest difficulty in the representation is the relatively large difference in the range of possible values (settings) for each knob. With a frequency-agile radio spanning frequencies from, say, 1 MHz to 6 GHz and a step size of 1 Hz, we must represent approximately 6×10^9 discrete frequencies in the gene. In contrast, the number of different modulation types is limited to a few dozen. Within each of these are only a handful of modulation indices or orders. Thus, modulation offers far fewer genetic possibilities than does operating frequency.

For the frequency representation, we could use clever programming techniques to realize different length genes for different traits; however, this quickly becomes problematic when a 37-bit gene represents the frequency (37 bits allows 1 Hz resolution from direct current (DC) to 100 GHz)—a gene this size is very difficult to explore in a genetic algorithm because a random search through this space is unrealistic (2^{37} possibilities). One practical approach is to segment these into multiple genes, where one gene provides a certain bandwidth of spectrum in which to search and the other provides the resolution within that bandwidth. Because both of these are genes, the search responsibility is distributed between the genes, and a usable band and center frequency can be found faster.

Knobs with smaller search spaces can be realized with smaller genes, and some traits can be combined. Modulation is segmented into the type of modulation—such as amplitude shift keying (ASK), frequency shift keying (FSK), phase shift keying (PSK), quadrature amplitude (and phase) modulations, and so forth—and the order (2-, 4-, 8-, 16-point constellations, etc.). A 16-bit gene representation (65,536 possibilities) is far too large for either type or order, but the gene could be split into two 8-bit pieces. A system could then adapt these parameters together (choosing BPSK or 16-QAM) or as two separate genes (PSK with order 2 or QAM with order 16).

The challenge lies in scaling the GA to the full-scale factors of real-world applications, dealing with the difference between generality and domain-specific knowledge. Segmenting the genes into different sizes depending on traits leaves us with the problem of deciding how to best do this. If we overshoot our domain range, we can end up with large genes and overly complex search spaces. For example, if we decide to have a 37-bit gene for frequency, what do we do if we are applying the algorithm to a radio that can only realize 1 to 2 GHz center frequencies? We could easily represent the frequency range with a 30-bit gene (for just over 1×10^9 possibilities), but the remaining search space would be detrimental to the algorithm. However, setting a static field of 30, or the easier-to-realize 32-bit values, we could lose resolution for more complex or flexible radios. The chromosome can therefore give a large number of bits to the frequency gene and a small number to the modulation gene and transmit power gene as shown in Figure 7.12. A key result of this structure is that it makes the GA independent of which radio it is optimizing.

The variable bit representation is a result of the SDR platform definition file [48]. The platform definition file includes an eXtensible Markup Language (XML) file to represent the waveform bounds as well as a document type definition (DTD) file to represent the basic waveform structure. Instead of providing an explicit value for each knob, though, the definition XML file provides the range of values each knob may have. The

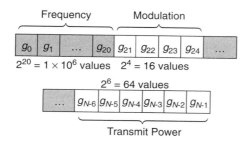

FIGURE 7.12

Sketch of the WSGA's chromosome with variable bit representations of genes.

definition file can be thought of as representing the possible genes in the chromosome while the waveform XML file represents the specific gene. As an example, the following listing is the representation of the frequency range from a definition file for a radio capable of transmitting from 400 to 500 MHz in steps of 1 kHz and from 2.3 to 2.5 GHz in steps of 100 kHz. It also says that the radio can transmit in the 400 MHz range from 0 to 100 dBm in steps of 0.1 dBm and at 2.3 GHz from 0 to 20 dBm in steps of 0.1 dBm. This XML example shows how both continuous values as well as discrete jumps in values can be represented.

The XML file provides the bounds and step size, and therefore the number of bits required to represent any possible genetic value for the knob in the chromosome. The DTD file provides the minimum representation of the waveform to structure the chromosome and understand how to parse the XML file. A brief slice of the DTD file is shown next while the full listing is in Appendix D of Rondeau [49]. The WSGA uses the DTD representation to know what genes are available and builds each gene from the XML file above. The logic to accomplish this first parses the XML and DTD files into trees that can be walked. The algorithm removes any elements that contain #PCDATA, which indicates that the element contains data (and is therefore a min, max, or step element). The algorithm then walks the DTD tree looking for leaf nodes on the tree, which are now the elements that describe the trees after the min, max, and step elements are removed. Each of the leaf nodes are names of genes. Each gene has the minimum and maximum value it can contain as well as a step value between the two endpoints. The range represented is easily calculated as (max−min)/step and a `log2()` of this value provides the minimum number of bits required to represent the possible values of the gene. Likewise, this information is used in reverse to decrypt the genetic code; the bits representing the gene are an index of step number of steps above the min value. As long as the result is less than max, the gene representation is valid.

Any time more than one node of the same name exists, an index is used to address the gene. Multiple nodes of this sort are used when continuous ranges do not exist in a radio, such as support for different frequency bands. The number of nodes is indexed into a set of bits and the rest of the gene is made up of the minimum number of bits required to represent the maximum range of any of the nodes. The index position then identifies which range should be used to decrypt the gene.

This method to build chromosomes allows easy representation of a radio's capabilities in XML. The GA behaves exactly the same regardless of the radio attached as long as the description is valid in the XML file. The XML and DTD are powerful tools to represent the radio that effectively allows the genetic algorithm to design itself around the radio system, making it truly platform independent.

7.6.3 Multidimensional Analysis of the Chromosomes

Section 7.5 showed how different applications and users will have different performance criteria from the radio. This section discusses ways to capture these effects in the genetic algorithm-based CE. When evaluating each chromosome in all of the objectives, we must find some way to analyze the performance of all radio chromosomes in such a way as to reflect the user needs and radio performance under current channel conditions.

Again, by using dynamic programming techniques, the CSM of the CE can learn the user and application needs and instruct the WSGA about which objectives matter and how much they matter. This information is passed from the CSM to the WSGA as a set of objective functions to use along with relative weighting values for each objective. For a video link, BER will receive a smaller weight than data rate and latency, and a standard data application, such as File Transfer Protocol (FTP) or HyperText Transfer Protocol (HTTP), will have a higher weighting for BER and PER. It is the job of the modeling and learning machines to make the necessary distinctions and learn the appropriate responses.

When evaluating the chromosomes based on the relevant objectives, the most obvious way to compare chromosomes is based on a simple summation of the weighted objectives. The next subsection discusses and shows how, once again, the trade-off between generalization and domain-specific knowledge causes us to rethink this standard method of evaluating MODM problems.

Objective Function Definition

XML and DTD files provide the mechanism for generic representation of radio platform capabilities. The GA is fed the objective functions in the form of an XML file that describes what functions the library holds. The library is compiled as a shared object library that allows dynamic, run-time linking. The GA can then link to the library, pull out the required objectives, and close the library as required. Then, when new objectives are introduced, or better mathematical representations are found for the existing objectives, the library can be recompiled separately from the rest of the CE and transmitted to the cognitive radios for the next optimization process.

The XML file that the GA receives contains the list of functions available, so during evaluation, the objective functions are referenced by name in the library. The library processes the objective and returns a solution. The important aspect of this is in the function prototype. Each function representation is in the form:

$$\text{float} < functionname > (\text{radio hw def} * \text{knobs, radio meters} * \text{meters})$$

The radio `hw def` data structure is a class that contains the information to map the chromosome representation to the radio platform capabilities. The structure is built from the XML and DTD files that define the chromosome as discussed previously. The radio meters are simple data structures that hold the results of the objective function calculations. To allow the dynamic representation of fitness values for use as maximization metrics in the WSGA problem, each objective function returns a real number assessing the performance of this system with respect to its objective. Simply put, when an objective function should be maximized (e.g., throughput, SINR), the returned fitness is the objective itself. When the objective should be minimized (e.g., BER, power), the returned fitness is the inverse of the objective. These fitness values are then stored as credits to represent an individual's fitness. The ranking and selection is based on the values of the credits, which, again, are designed to be compared in a maximization problem. Meanwhile, each member holds a copy of its meters data structure used in the postanalysis of the algorithm's performance, for use in the cognitive engine's learning routine.

Relative Tournament Evaluation

We use a relative tournament selection method similar to that proposed by Lu and Yen [15], except that the fitness of the winner from a single comparison is scaled by the weight associated with that objective. After all of the single comparisons in all dimensions, the winning member, and the one that becomes a parent to the next generation, is the one with the highest fitness. Although this method does not guarantee that all winners are the best, or nondominated, members of the population (only better relative to their ancestors), it maintains species diversity within the population while still pushing toward a Pareto front (see Section 7.5.3). Diversity in the population allows different solutions to be tried and helps to prevent the algorithm from getting stuck in a local optimum.

Unlike the linear combinations of objectives, this method compares like objectives only. There are no issues, then, of improperly comparing dimensions and ranges. If we are trying to minimize BER, 10^{-6} is obviously better than 10^{-4} without any obfuscation by additively or multiplicatively combining these values with data rates of 10^4 and 10^6. We are now comparing objectives of similar dimensions and ranges. There will be no need to normalize the BER values to compare against transmit power values of 10 dBm or 1 dBW.

If we define $F(C_i)$ as the overall fitness of chromosome i, and w_k as the weight associated with an objective $k \in K$, then we compare the chromosomes in each dimension k with the algorithm shown in Eq. (7.7). The value I is 1 if the fitness of a chromosome of interest is greater than another chromosome, otherwise, it is 0 (i.e., the chromosome is awarded points for winning the tournament).

$$F(C_i) = \sum_k I_i \cdot w_k \qquad I_i = \begin{cases} 1, & f_{i,k} > f_{j,k} \\ 0, & f_{i,k} < f_{j,k} \end{cases}$$

$$\text{where} \qquad (7.7)$$

$$F(C_j) = \sum_k I_j \cdot w_k \qquad I_j = \begin{cases} 1, & f_{j,k} > f_{i,k} \\ 0, & f_{j,k} < f_{i,k} \end{cases}$$

So the winning chromosome has its weight adjusted by the weight associated with the objective. The weightings then help the decision maker adjust its choices by putting more or less emphasis on objectives as their importance to the solution changes.

7.6.4 Relative Pooling Tournament Evaluation

In our tournament selection process, we randomly choose between two chromosomes to select one as a parent. Another method is to randomly choose two parents and a random pool of chromosomes from the remaining population [27]. Each parent will compete independently with each member of the population pool. Whichever parent wins the most number of tournaments wins the selection and ability to reproduce. This method leads to parents who are more fit for the user's needs and the communication channel within the population as a whole, and puts a lot of selection pressure on the algorithm.

Here we did not apply Horn et al.'s [27] method of fighting two individuals against a subset of the population because their method, although they claim it produces better results, calls for a larger population and more fitness comparisons. This becomes computationally intensive and we have yet to test whether the improvement is worth the computational cost.

7.6.5 Example of the WSGA

This section presents the results described by Rondeau et al. [22]. The first test conducted employed a real hardware test bed. However, these radios were Proxim Tsunami radios, which are hardware-based platforms with limited knobs and tuning range; all adjustable physical layer properties are listed in Table 7.2.

Even with this limited radio platform, the genetic algorithm optimization technique was found to be viable. The scenario design consisted of a point-to-point radio link controlled by the WSGA and a third radio of the same type that acted as an interferer. The scenario is shown in Figure 7.13, and results are shown in Figure 7.14.

The test consisted of setting up an initial video streaming link with high-quality throughput (Figure 7.14 (a)). When the interference started, the signal quality quickly

Table 7.2 Proxim Tsunami Adaptable Parameters

Parameter	Range
Frequency	5730–5820 MHz
Power	6–17 dBm
Modulation	QPSK, QAM8, QAM16
Coding	Rate 1/2, 2/3, 3/4
Time division duplex (TDD)	29.2–91% (basestation unit to subscriber unit)

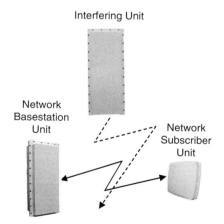

FIGURE 7.13

Initial test of WSGA hardware platform on Proxim Tsunami radios. The test used two radios as friendly net members and another radio as an interference source.

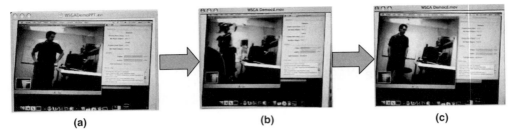

FIGURE 7.14

Results of the initial test of the WSGA hardware platform. The radio experienced degraded performance with onset of interference and then found acceptable waveform modifications to avoid interference.

dropped and became barely distinguishable (Figure 7.14 (b)). The WSGA was then run with the objectives set to minimize BER, minimize transmit power, and maximize data rate, yet the radio was prevented from switching frequencies because that was the easy solution and the test was for how the radio would handle its other parameters. The result was that it automatically reestablished the video link (Figure 7.14 (c)).

Although this test showed good performance in a live test, a more flexible platform was desired, so a PHY layer simulation of an SDR was developed with more knobs to tune and larger tuning ranges, as listed in Table 7.3. The resulting radio simulation's objective functions were BER, bandwidth occupancy, spectral efficiency, power, data rate, and amount of interference. This test had three goals: (1) minimize spectrum occupancy, (2) maximize throughput, and (3) avoid interference. The weighting values for the available dimensions are shown in Table 7.4.

The results are graphically illustrated in Figure 7.15, which shows that each objective was met, and each test resulted in a BER of 0. The first test (Figure 7.15 (a)) minimized the spectrum occupancy to 1 MHz; the second test (Figure 7.15 (b)) maximized the throughput to 72 Mbps; and the third test (Figure 7.15 (c)) found a solution that fit into the white space left by the interferers.

Table 7.3 Simulation Adaptable Parameters	
Parameter	**Range**
Power (dBm)	0–30 dBm
Frequency (MHz)	2400–2480 MHz
Modulation	M-PSK, M-QAM
Modulation, M	2–64 M
PSF roll-off factor	0.01–1
PSF order	5–50
Symbol rate	1–20 Msps

Table 7.4 Simulation Test Conditions

	Weights		
Functions	**Minimize Spectral Occupancy**	**Maximize Throughput**	**Interference Avoidance**
BER	255	100	200
BW	255	10	255
Spectral efficiency	100	200	200
Power	225	10	200
Data rate	100	255	100
Interference	0	0	255

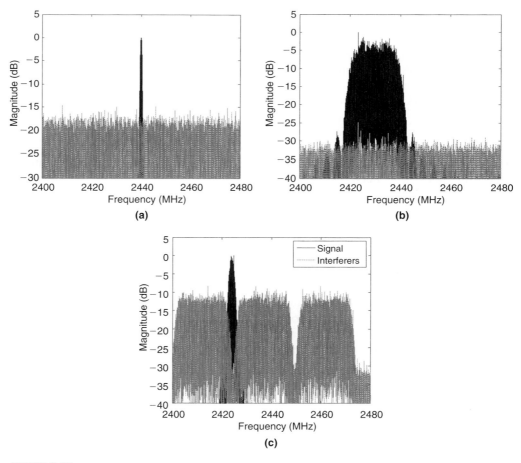

FIGURE 7.15

Results from WSGA simulation tests. The tests showed that the algorithm could find acceptable solutions under a variety of different interference conditions and objective criteria.

7.7 ADVANCED GENETIC ALGORITHM TECHNIQUES

The previous sections have introduced the genetic algorithm mostly through the example of a simple GA. This GA is only the starting point for real genetic algorithms, and adjustments have already been made in the simple GA to make the multiobjective genetic algorithm. However, far more techniques are available to improve the performance or tailor the GA performance to a specific problem. Rather than presenting here an entire treatise on the subject of GA techniques, this section is restricted to some of the more popular and useful techniques. The reader is referred to the literature for details; two excellent general sources of GA techniques are Chambers [28] and Goldberg [17].

Niching is a popular technique to ensure population diversity throughout the GA. In a population, we wish to spread our chromosomes around the solution space, especially if the space is multimodal (has many local optima). A niche is a particular range within the entire solution space, and we want the GA to maintain some presence in all niches to create many localized search spaces in hopes of finding the global optimum. Niching has a long history in GAs, going back to De Jong's crowding [20] in 1975. A multiniche crowding algorithm is described in detail with many useful examples in Vemuri and Cedeno [29]. Fonseca and Fleming [30] and Horn et al. [27] use niching in multiobjective GA searches. Niching is one way of maintaining diversity along the Pareto front.

Dividing a total population into some smaller subset of populations is a technique used to reduce the computational complexity of each algorithm. These techniques have been tested to run different groups of solutions simultaneously to find the same optimum. These methods have been performed, both on the same platform, or in a parallel method on multiple parallel platforms. Not only do these populations allow more searches to occur, they also result in populations that migrate between the "islands" to share highly fit members and maintain diversity. Many researchers have studied this issue [17, 31, 32].

Genetic algorithms are sensitive to parameters such as crossover and mutation rate, population size, and replacement size, but nothing stops a GA from altering its parameters throughout its lifetime [33, 34]. One method is to monitor the performance improvement over the generations. As the improvement slows down, we might have converged to a local optimum or a global optimum. Not being sure, it would be wise to keep looking, for a while at least, but an increased mutation rate would enhance the exploration features of a GA to let it search in new areas of the search space. If the GA was caught in a local optimum, a high mutation rate will help it break out of that cycle to find the path toward a new optimum, hopefully the global optimum this time.

7.7.1 Population Initialization

Although these different methods have been proposed to enhance the performance of genetic algorithms, overall only a rather small subset focuses on the population's initialization strategy. A few good representatives of the initialization methods include biasing the initial population using domain knowledge [35], creating unbiased or diverse populations [36], and using case-based initialization techniques [37]. The case-based systems, in particular, have been shown to lend themselves to successful operation in changing situations, or "anytime learning," as Ramsey and Grefenstette [37] define it. These systems are very similar to the online learning we wish to accomplish. The technique in this chapter is similar in that it presents an advanced understanding of the operation and benefits that case-based, or memory-feedback, GAs offer.

7.7.2 Priming the GA with Previously Observed Solutions

The method of priming the GA with previously observed solutions falls under the category of case-based intelligence in the AI literature. Here we can look into the two methods discussed in the Case-Based Decision Theory subsection: case-based reasoning [38] and case-based decision theory [25]. The application of CBDT to the initialization problem, as presented here, adds to other case-based methods [37]. Those systems are usually inspired by the field of case-based reasoning [38], which uses case lookup based on similarity, but the application in this chapter is a decision-making theory based in both similarity and utility [22]. The added dimension aids the decision-making process such that cases in memory are not only similar to the current situation, but they have also performed very well in past situations. This section presents the basics behind the CBDT application to genetic algorithms and its improvement to online GA optimization systems. In doing so, it provides a more complete analysis of how the memory system works.

CBDT was introduced by Gilboa and Schmeidler in the mid-1990s [25]. The method accesses past knowledge to help make decisions about present situations. The theory defines the set of cases to include all possible problems, actions, and results in which a decision maker contains a memory M, which is a subset of the total number of possible cases, C. Current decisions are based on the knowledge of past cases by defining the amount of similarity of the current situation to the past cases and the utility of each case in memory. The case that maximizes the desirability based on the utility and similarity is chosen as the case that best applies to the current situation.

Formally, CBDT is defined to have $q \in P$ set of problems, $a \in A$ set of actions, and $r \in R$ set of outcomes. A case, c, is a tuple of a problem, an action, and a result such that $c \in C$, where $C \in P \times A \times R$. Furthermore, memory, M, is formally defined as a set of cases c currently known such that $M \subset C$.

Similarity is defined in Eq. (7.8) as how similar two decision problems are to each other:

$$s : P \times P \to [0, 1] \qquad (7.8)$$

Utility is defined in Eq. (7.9) as the level of desirability the outcome represents:

$$u : R \to \Re \qquad (7.9)$$

We concentrate on the simplest formulation, where the most desirable act is chosen from the set of known acts in memory by using the product of similarity and utility [25], as defined in Eq. (7.10):

$$U(a) = s(p, q)u(r) \quad \text{where } (q, a, r) \in M \qquad (7.10)$$

For a given problem, p, this equation determines the desirability of action a. The action that maximizes this function is the chosen action to respond to problem p. Section 7.7.3 introduces how the GA is aided by CBDT.

7.7.3 CBDT Initialization of GAs

If the CR system builds a memory database that represents the outcomes of continually running the GA for real-time optimization, then as the system learns and optimizes, each new optimization would take significantly less time to run; or, conversely, it would find a more optimal solution in the same amount of time.

The Theory

For the genetic algorithm, we represent a new input as the problem p. We have previously observed problems, q, in a database associated with a set of chromosomes, a, used to model the problem along with their fitness values, r. When the system receives this new problem, it finds the action within its memory that maximizes the product of the utility and similarity functions. The action is a set of chromosomes that partially initializes the GA population while the rest of the population is randomly generated.

Chou and Chen [36] discuss the use of uniformly generating a population to have an equal representation of all points in space, which should lead to at least one chromosome close to the global optimum. As such, by initializing with both the case-based solutions and randomly generated solutions, we maintain some diversity for proper searching while focusing the search in a region of presumably high fitness, the concept illustrated in Figure 7.16.

The initial population is of size N_p, where M_a members are initialized from the case base and $N_p - M_a$ members are randomly initialized. To maintain the diversified population, we ensure that M_a is always less than N_p. We define the action associated with

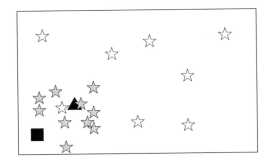

FIGURE 7.16

Search space. The square is the new target optimum, the triangle is the previous target optimum, and the stars are the initialized population. The initialized solutions from the CBDT system (*gray*) concentrate around the previous optimum while other initial solutions (*white*) are spread out randomly to cover the whole search space.

each case in memory to consist of a number of chromosomes, M, and the results are therefore the fitness of each chromosome as fed back from the GA along with the respective chromosome. The utility of the case k, shown in Eq. (7.11), is then defined as 1 minus the sum of all of the solutions' fitnesses stored in each case.

$$u(r_k) = 1 - \sum_{i=1}^{M} f_{k_i} \qquad (7.11)$$

The winning case, k', is the case that maximizes the product of the similarity and utility functions as defined by Eq. (7.10).

From here, we define two initialization methods. First, some number of solutions, $M > 1$, are associated with each case, and all M solutions are used to initialize the GA. The other method sets $M = 1$ and the solution of the winning case is used to initialize the GA, along with some number of surrounding cases. Other ways to mix these two techniques exist, but are not discussed here.

The Implementation

The design of the information flow is shown in Figure 7.17. Here, the input is compared to the cases in memory through both similarity and utility calculations. The chromosomes (action space) of the winning case are sent to the GA as well as the input. The GA then runs to optimize the result within a set period of time. The GA outputs a solution and its fitness value, which are both fed back to memory. The initialization can happen in two ways:

1. The case base reasoner stores multiple solutions to one case in memory and sends all these solutions to the GA.
2. The case base reasoner stores one solution per case and sends solutions from multiple cases to the GA.

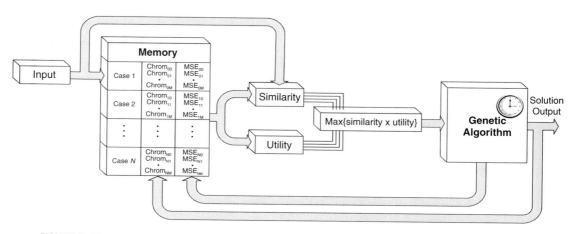

FIGURE 7.17

System diagram of CBDT initialization of GA.

The number of chromosomes associated with any case in memory is limited to some number, M, and when a new chromosome is inserted because it is locally optimal for current conditions, the oldest chromosome (or least suitable) is replaced. When inserting a new case to the finite memory space of size N, the oldest case in the memory is forgotten and replaced by the most recently observed case. The new memory case is then associated with both its most recent solution and the solutions of the case originally selected from memory if $M > 1$. Through this update procedure, the memory is kept current and positive actions used in the past are preserved.

Other possible methods for replacing cases and solutions include replacing the case with the worst overall utility or a case in memory that has been used less frequently than the others. Both long-term and short-term memories with different forgetting methods to exploit both time (short-term) and utility (long-term) benefits have been considered.

Two important issues to consider are the size of the memory and the initialization method. The size of the memory has a direct effect on the computational complexity of the decision maker. The choice of the memory size should reflect the trade-off between a large memory that will add to the time required to both learn and process the input signal, versus the increased efficiency to find solutions similar to current conditions. If this added time is too great, it begins to break down the benefits achieved through the system. The initialization method is the choice between associating multiple solutions with a single case and sending only those solutions to the GA, or associating only a single solution per case and sending multiple cases to the GA.

7.8 NEED FOR A HIGHER-LAYER INTELLIGENCE

An important goal of the cognitive engine's higher-layer intelligence is to autonomously set the weights for the GA to both reduce limitations in design and remove operator interaction with the radio. The preference is that the CE have the ability to set weights on its own.

7.8.1 Adjusting Parameters Autonomously to Achieve Goals

By observing the user and the needs of the system, the radio software should be able to adjust the weights accordingly. The system designer could establish a preset table of weights for different applications, but these presets are limited by the imagination of the designer and current understanding of the uses of the CR. If new situations and needs arise in the field, creative solutions should be developed as they are needed.

The cognitive engine should reflect the user's needs without burdening the user with developing the weights themselves or constant communication between the radio and the user. Purpose, intent, and lower-layer needs should be inferred from the higher layers and not from the user answering questions from the radio such as, "Would you like to increase your data rate?"

The last requirement, to infer user and application needs, is possibly the most challenging concept introduced so far and is a rich subject for research. The current design uses a simplified model, as discussed in the next section.

7.8.2 **Rewards for Good Behavior and Punishments for Poor Performance**

In his landmark paper, "Computing Machinery and Intelligence," Alan Turing [23] recalls his British public school days, where discipline was taught through physical punishment, which he himself both gave and received, but we can adapt this and think of it as positive and negative reinforcement. Provided the input from the user and the radio, and having produced the WSGA output, the engine now requires a means to tie the information together and learn from its past behavior, both its mistakes and successes.

A reinforcement loop must exist between the WSGA solution based on its simulated meters when solving for the objectives and the actual meters observed by the radio while operating under the parameters set by the WSGA. If the WSGA's optimization was incorrect, the system should be adjusted (punished) for improved performance next time. If the WSGA's optimization was correct, the system should be rewarded to make sure it does the same in future situations. Two ways of distributing these rewards and punishments are discussed next.

Rewards and Punishments Can Be Inflicted by Algorithms

The CR must have some sense of its performance and adjust itself accordingly to improve its performance. The cognitive radio must know its current internal performance and capabilities in its approach to intelligent optimization, an approach that has been called "self-aware" in current CR circles. This does not mean that the radio is self-aware in any sense that humans are self-aware; instead, it refers to the radio's ability to keep track of its performance and available resources (e.g., battery life, geography, computational power, use-case scenario) and alter its behavior to account for needs and changes required by this understanding. On this level, the radio will use the knowledge of itself in the optimization problem, where battery life and computational power of waveform generation can play major roles in the radio performance.

On another level, the radio must be conscious of its external performance measures. When the GA adjusts the radio, the adjustments are expected to produce some known result, such as a target BER or data rate. The CR then monitors the actual observed performance and will either reward or punish itself for differences between the actual and simulated performances.

For this to work, the radio must be able to read the same meters used in the GA optimization process. The farther a meter deviates from the simulated performance, the more the CR process is punished. Again, punishment results in changes to the radio decision-making process to help it change itself for improved performance later.

Because the radio will be able to measure some meters itself, such as SINR and PER, the amount of deviation between these observed meters and the WSGA's simulated meters should tell the CE how best to adapt. It would do this by asking questions such as, "Is this PER right for this SINR?" If the answer is "yes," then the PER can possibly be improved by increasing transmitter power. If the answer is "no" (the PER is too high for the given SINR), then the problem is interference or multipath, and the PER can possibly be improved by changing frequency or modulation parameters. This method suggests the use of an expert system AI technique [24]. By programming just a few basic rules of radio operation, we can infer a lot and adjust the system appropriately. Expert systems are difficult to apply, especially to a general problem setting, because they use very specific, preprogrammed knowledge. Mixing expert systems with fuzzy logic [39, 40] allows much more flexible reasoning. Also, an implementation of an expert system in this case should be very general, using only the most basic wireless communication rules to make inferences. Other techniques, such as evolutionary strategies [24], could then be tiered with the expert system to make more creative assertions about the problem and how to fix it.

Rewards Can Be Inflicted by Users

Although we advocate minimizing interaction between the user and the radio, this can be helpful in teaching the CR. The self-reflexive method just described is unsupervised learning, but sometimes a supervised learning process is necessary, especially when trying to tailor behavior to a particular user. Using the same punishment metric of credit deduction as used before, credits will be altered based on user satisfaction. This is another area of great interest and research in the CR community. Rather than promoting a particular method here, Section 7.9 looks to the field of human–computer interface (HCI) for how to best facilitate the interaction and collect the data required to perform the credit adjustments.

7.9 HOW THE INTELLIGENT COMPUTERS OPERATE

This chapter has discussed many aspects we expect to see from a fully functional CR capable of altering the PHY and MAC layers with respect to the user and external environment. It has discussed the cognitive decision-making process using case-based theory and GAs to solve the multiobjective optimization problem. It has also referred to concepts of user and channel modeling, which require some advanced computer techniques, and mentioned the need for self-awareness in the radio. In the end, though, there is no single method of AI capable of performing all possible tasks required by the CR. Therefore, we look to a tiered approach of computational intelligence and machine learning algorithms to perform their tasks to create a fully cognitive radio. This section revisits these concepts to show how the collection of techniques creates a CE capable of controlling and adapting a CR.

Ultimately, all work involving cognitive radios really comes down to the multiobjective optimization problem, which is that the MODM search space:

- Has no *utopian point*; that is, it has no point that fully optimizes all objectives.
- Is nonlinear and nonconvex; indeed it is very complicated to model with complex relationships between all inputs and outputs.

All other aspects of the CR are inputs to the MODM problem in order to either set up the optimization problem or speed up convergence. (Refer back to Figure 7.1 for the graphical representation of the items discussed here.)

7.9.1 Sensing and Environmental Awareness

The radio must first understand what is required of it and what it must be capable of, thereby defining the optimization space. This requires sensing or awareness on four levels: (1) recognizing the needs of the user, (2) understanding the limitations imposed on the radio operation by the channel and external environment, (3) realizing its own limitations in flexibility and power, and (4) conforming to local regulations and policy. Although this chapter discusses a few implementation methods, this is not the central focus, and they are only briefly covered here.

User Awareness

The user domain provides subjective information to the CR, by which the cognitive radio is given instructions on how to behave and what to optimize. User awareness could be either passive or active. Passive awareness requires learning user needs and behavior by inferring concepts from the applications being used and data being sent through the radio. Active learning is done through direct communication with the user and HCI techniques to collect information.

Inference can be accomplished by sampling the packet headers as well as the packet and bit rate required by the application sending the data. Such learning could be accomplished using case-based techniques or even online, unsupervised neural network learning through bidirectional associative memories [41]. These techniques could learn that a header containing the protocol values for File Transfer Protocol (FTP), Transmission Control Protocol (TCP), Internet Protocol (IP), and so forth is a file transfer application requiring full packet fidelity, retransmission for failed packets, and high data rate with acceptable tolerances in latency. A similar packet that contains a protocol value for ReSerVation Protocol (RSVP) instead of FTP would be remembered as dealing with real-time operation, which would require different services. Some of these protocols and applications can be preprogrammed into the user modeling system, but a true unsupervised learning machine is preferred that can learn new and unknown protocols required by the user.

Environmental Awareness

The external, RF, and geographical domains provide the environmental context. Sensing and modeling the channel can be done in a number of ways. Sending packets of training data through a channel and observing the bit errors and burst error rates is one method. Another uses channel sounding to capture a snapshot of the channel's propagation information [26]. Other methods exist for calculating noise floor and interference power to understand and recognize the presence of other users and radios [42].

Neural networks are well known for their ability to classify information and for pattern recognition. Neural networks have been used to classify modulation types in signals with great accuracy. Understanding not only how many other radios are nearby but also the types of communication the radios employ can be of great use in determining the optimization. If a frequency is in use by another radio network, finding an orthogonal modulation, antenna polarization, and so on could lead to highly efficient spectrum sharing [43–45].

External radio and propagation information is also useful when looking at the current internal radio resources. If the radio is operating in a clean channel with few interferers, then a high-order, narrowband modulation technique could be chosen over advanced spread-spectrum signals to reduce computational complexity (and therefore prolong the battery life) while providing high data rates.

7.9.2 Decision Making and Optimization

Using the knowledge gathered in the sensing mechanisms, the CR must then make a decision as to how to optimize radio performance. The optimization problem is multi-objective, and solved by using the well-known GA. Genetic algorithms are not, of course, the only method to do this, but they are well understood and easy to implement as well as flexible and robust.

7.9.3 Case-Based Learning

We now have a modeling and adaptation system, but we need some way to tie them together. A knowledge-based system to store information about the modeled environment can be tied to a set of objectives to be solved by the GA. Another entry to the knowledge base is memory of past actions taken by the GA. When similar environmental situations are observed by the modeling systems, similar actions should produce comparable results. If these past solutions gave positive results, we can aid the GA optimization process by starting with this knowledge. Genetic algorithms are notorious for their length of time to find an optimal solution, but this chapter has shown how this knowledge-based approach reduces the problem quite nicely to a localized optimization search.

7.9.4 Weight Values and Objective Functions

The CR bases its decisions on a set of metrics to best represent the requirements of the whole radio system, the network, and the user. To do this properly, the decision maker must value objectives differently. In the optimization process, different needs of the radio should be handled by using different objectives. The objectives used for a given optimization problem can be learned through successes and failures, and the weights associated with each objective can likewise be altered to best represent the situation.

7.9.5 Distributed Learning

Learning situations arise through modeling of the environment, inferring needs of the user, or simply asking the user for input. However, another way of learning that could increase the power of the CR greatly is peer learning. Children learn through their own play [46] and through teacher instruction; they also learn from each other [47]. Similarly, what one radio has already experienced and learned, it can share with newer, inexperienced radios, which removes their need for trial and error, possibly successive mistakes, and greatly accelerates optimal network performance behaviors.

7.10 SUMMARY

The goal of a CR is to optimize its own performance and support its user's needs. A radio is optimized when it achieves a level of performance that satisfies its user's needs while minimizing its consumption of resources such as occupied bandwidth and battery power. The intelligent core of a CR exists in the CE, which performs the modeling, learning, and optimization processes necessary to reconfigure the communication system in which the radio operates.

The first problem in dealing with cognition in a system is to understand (1) what information the intelligent core must have and (2) how it can adapt. In radio, we can think of the classical transmitters and receivers as having adjustable control parameters (knobs) that control the radio's operating parameters, and observable metrics (meters) that measure its performance. The knobs are any of the parameters that affect link performance and radio operation. Meters are indicators of performance on a particular level; thus, at the link layer, packet error rate is an important metric.

The basic process followed by a CR is this: It adjusts its knobs to achieve some desired (optimum) combination of meter readings. Rather than randomly trying all possible combinations of knob settings and observing what happens, it makes intelligent decisions about which settings to try and observes the results of these trials. Based on what it has learned from experience and on its own internal models of channel behavior, it analyzes possible knob settings, predicts some optimum combination for trial, conducts the trial, observes the results, and compares the observed results with its predictions as summarized in an adaptation loop.

Without a single-objective function measurement, we cannot look to classic optimization theory for a method to adapt the radio knobs. Instead, we can analyze the performance using multiobjective decision-making criteria. MODM theory allows us to optimize in as many dimensions as we have objective functions to model. CR operation requires a MODM algorithm capable of robust, flexible, and online adaptation and analysis of the radio behavior. The clearest method of realizing all of these needs is the genetic algorithm, which is widely considered the best approach to MODM problem solving. In it, we represent the radio parameters as genes in a chromosome and select the fittest chromosomes through a process called relative tournament evaluation.

The primary goal of the CE is to optimize the radio, and the secondary functions are to observe and learn in order to provide the knowledge required to perform the adaptation. A CR becomes a learning machine through a tiered algorithm structure based on modeling, action, feedback, and knowledge representation, as shown in the cognition loop of Figure 7.10. In the Virginia Tech cognitive engine, these functions are realized through the algorithmic structure of Figure 7.11. Its main parts are the cognitive system module, responsible for learning, and the wireless system genetic algorithm, which handles the behavioral adaptation of the radio based on what it is told to do by the CSM. The modeling system observes the environment from many different angles to develop a complete picture. The CSM holds two main learning blocks: the evolver and the decision maker, which takes feedback from the radio that allows the evolver to properly update the knowledge base to respond to and direct system behavior.

The algorithm structure discussed in this chapter provides the groundwork for solving the problems of the PHY and MAC layer adaptation. Many areas are left to explore in the machine learning for these layers, and many potential improvements and research directions have been outlined and discussed.

Bringing all of these ideas together is one of the great challenges of CR. By its nature, it is a multidisciplinary problem, just as the subjects of the other chapters in this book suggest. We need solutions that can mix these disciplines into a coherent system solution that learns and adapts to the user to satisfy all of the problems encountered when attempting to optimize the point-to-point link, as the PHY and MAC layer algorithms presented here do, as well as the end-to-end solution of a CR network—all the while both enhancing and protecting the regulatory structure of spectrum allocation. This chapter has provided an analysis of part of this solution and presented one possible approach to it.

REFERENCES

[1] Hazlett, T. W., Spectrum Tragedies, *Yale Journal on Regulation*, 22:2–5, 2005.

[2] Shannon, C. E., Programming a Computer for Playing Chess, *Philosophical Magazine*, 41:256–275, 1950.

[3] Open Systems Interconnection—Reference Model (OSI-RM). FED-STD-1037C: Telecommunications: Glossary of Telecommunication Terms, pp. vii–xiv, National Communications Systems, 1996.

[4] Hawkins, J., *On Intelligence*, Times Books, 2004.

[5] Decker, D. L., Computer Evaluation of the Complimentary Error Function, *American Journal of Physics*, 43:833–834, 1975.

[6] Abramowitz, M., and I. A. Stegun, *Handbook of Mathematical Functions with Formulas, Graphs, and Mathematical Tables*, Dover, 1972.

[7] Proakis, J. G., *Digital Communications*, Fourth Edition, McGraw-Hill, 2000.

[8] Zionts, S., Multiple Criteria Mathematical Programming: An Overview and Several Approaches, *Mathematics of Multi-Objective Optimization*, P. Serafini (ed.), pp. 227–273, Springer-Verlag, 1985.

[9] Schaffer, J. D., Multiple Objective Optimization with Vector Evaluated Genetic Algorithms, *Proceedings International Conference Genetic Algorithms*, pp. 93–100, 1985.

[10] Zitzler, E., and L. Thiele, Multiobjective Evolutionary Algorithms—A Comparative Case Study and the Strength Pareto Approach, *IEEE Trans. Evolutionary Computation*, 3:257–271, 1999.

[11] Fleming, P., Designing Control Systems with Multiple Objectives, *IEE Colloquium Advances in Control Technology*, pp. 4/1–4/4, 1999.

[12] Lim, Y., P. Floquet, and X. Joulia, Multiobjective Optimization Considering Economics and Environmental Impact, *ECCE2*, Montpellier, pp. 1–10, 1999.

[13] Pedregal, P., *Introduction to Optimization*, Springer, 2004.

[14] Fonseca, C. M., and P. J. Fleming, Genetic Algorithms for Multiobjective Optimization: Formulation, Discussion, and Generalization, *Proceedings International Conference Genetic Algorithms*, pp. 416–423, 1993.

[15] Lu, H., and G. G. Yen, Multiobjective Optimization Design via Genetic Algorithm, *IEEE Proceedings International Conference on Control Applications*, pp. 1190–1195, 2001.

[16] Hiroyasu, T., M. Miki, and S. Watanabe, Distributed Genetic Algorithms with a New Sharing Approach in Multiobjective Optimization Problems, *IEEE Proceedings Congress on Evolutionary Computation*, pp. 69–76, 1999.

[17] Goldberg, D. E., *Genetic Algorithms in Search, Optimization, and Machine Learning*, Addison-Wesley, 1989.

[18] Spillman, R., Solving Large Knapsack Problems with a Genetic Algorithm, *IEEE Proceedings Systems, Man and Cybernetics*, pp. 632–637, 1995.

[19] Garey, M. R., and D. S. Johnson, *Computers and Intractability: A Guide to the Theory of NP-Completeness*, W. H. Freeman & Company, 1979.

[20] De Jong, K. A., An Analysis of the Behavior of a Class of Genetic Adaptive Systems, Ph.D. dissertation, The University of Michigan, 1975.

[21] *www.darpa.mil/sto/smallunitops/xg.html.*

[22] Rondeau, T. W., B. Le, C. J. Rieser, and C. W. Bostian, Cognitive Radios with Genetic Algorithms: Intelligent Control of Software Defined Radios, *Software Defined Radio Forum Technical Conference*, pp. C-3–C-8, Phoenix, 2004.

[23] Turing, A. M., Computing Machinery and Intelligence, *Mind*, 59:433–460, 1950.

[24] Negnavitsky, M., *Artificial Intelligence: A Guide to Intelligent Systems*, Addison-Wesley, 2002.

[25] Gilboa, I., and D. Schmeidler, *A Theory of Case-Based Decisions*, Cambridge University Press, 2001.

[26] Rondeau, T. W., C. J. Rieser, T. M. Gallagher, and C. W. Bostian, Online Modeling of Wireless Channels with Hidden Markov Models and Channel Impulse Responses for Cognitive Radios, *International Microwave Symposium*, pp. 739–742, 2004.

[27] Horn, J., N. Nafpliotis, and D. E. Goldberg, A Niched Pareto Genetic Algorithm for Multiobjective Optimization, *IEEE Proceedings World Congress on Computational Intelligence*, pp. 82–87, 1994.

[28] Chambers, L., *Practical Handbook of Genetic Algorithms: New Frontiers*, CRC Press, 1995.

[29] Vemuri, V. R., and W. Cedeno, Multi-Niche Crowding for Multi-Modal Search, *Practical Handbook of Genetic Algorithms: New Frontiers*, vol. 2, L. Chambers (ed.), pp. 5–29, CRC Press, 1995.

[30] Fonseca, C. M., and P. J. Fleming, Multiobjective Optimization and Multiple Constraint Handling with Evolutionary Algorithms. I. A Unified Formulation, *IEEE Transactions Systems, Man, and Cybernetics*, 28:26–37, 1998.

[31] Cohoon, J. P., W. N. Martin, and D. S. Richards, Punctuated Equilibria: A Parallel Genetic Algorithm, *Proceedings Second International Conference on Genetic Algorithms*, 2:148–154, 1987.

[32] Lin, W.-Y., T.-P. Hong, and S.-M. Liu, On Adapting Migration Parameters for Multi-population Genetic Algorithms, *IEEE Proceedings Systems, Man and Cybernetics*, pp. 5731–5735, 2004.

[33] Srinivas, M., and L. M. Patnaik, Adaptive Probabilities of Crossover and Mutation in Genetic Algorithms, *IEEE Transactions on Systems, Man and Cybernetics*, 24:656–666, 1994.

[34] Cao, Y. J., and Q. H. Wu, Convergence Analysis of Adaptive Genetic Algorithms, *IEE Proceedings Genetic Algorithms in Engineering Systems: Innovations and Applications*, pp. 85–89, 1997.

[35] Arabas, J., and S. Kozdrowski, Population Initialization in the Context of a Biased Problem-Specific Mutation, *IEEE Proceedings Evolutionary Computation World Congress on Computational Intelligence*, pp. 769–774, 1998.

[36] Chou, C., and J. Chen, Genetic Algorithms: Initialization Schemes and Gene Extraction, *IEEE Proceedings Fuzzy Systems*, pp. 965–968, 2000.

[37] Ramsey, C. L., and J. J. Grefenstette, Case-Based Initialization of Genetic Algorithms, *Proceedings Fifth International Conference on Genetic Algorithms*, pp. 84–91, 1993.

[38] Kolodner, J., *Case-Based Reasoning*, Morgan Kaufmann, 1993.

[39] Black, M., Vagueness: An Exercise in Logical Analysis, *Philosophy of Science*, 4:427–455, 1937.

[40] Zadeh, L. A., Fuzzy Sets, *Information and Control*, 8:338–353, 1965.

[41] Kosko, B., Bidirectional Associative Memories, *IEEE Transactions Systems, Man and Cybernetics*, 18:49–60, 1988.

[42] Mo, T., and C. W. Bostian, A Throughput Optimization and Transmitter Power Saving Algorithm for IEEE 802.11b Links, *IEEE Proceedings WCNC*, pp. 57–62, 2005.

[43] Le, B., Modulation Identification Using Neural Network for Cognitive Radios, *Software Defined Radio Forum Technical Conference*, Anaheim, CA, 2005.

[44] Nandi, A. K., and E. E. Azzouz, Algorithms for Automatic Modulation Recognition of Communication Signals, *IEEE Transactions Communications*, 46:431–436, 1998.

[45] Cheung, V., K. Cannons, W. Kinsner, and J. Pear, Signal Classification through Multifractal Analysis and Complex Domain Neural Networks, *IEEE Proceedings CCECE*, pp. 2067–2070, 2003.

[46] Piaget, J., *Biology and Knowledge*, University of Chicago Press, 1971.

[47] Vygotsky, L. S., *Mind in Society: The Development of Higher Psychological Processes*, Harvard University Press (original works published 1930, 1933, 1935), 1978.

[48] Scaperoth, D., B. Le, T. W. Rondeau, D. Maldonado, C. W. Bostian, and S. Harrison, Cognitive Radio Platform Development for Interoperability, *IEEE Milcom*, pp. 1–6, Washington, DC, October 2006.

[49] Rondeau, T., Application of Artificial Intelligence to Wireless Communications, Ph.D. Dissertation, Virginia Tech, 2007.

Cognitive Techniques: Position Awareness

8

John Polson, Bruce A. Fette
General Dynamics C4 Systems
Scottsdale, Arizona, and Rhome, Texas

8.1 INTRODUCTION

For cognitive radio (CR) to reach its full potential as an efficient member of a network or as an aid in users' daily tasks, and even to conserve the precious spectrum resource, a radio must primarily know its position and what time it is. From position and time, a radio can (1) calculate the antenna-pointing angle that best connects to another member of the network; (2) place a transmit packet on the air so that it arrives at the receiver of another network member at precisely the proper timeslot to minimize interference with other users; or (3) guide its user in his or her daily tasks to help achieve the user's objectives, whether it be to get travel directions, accomplish tasks on schedule, or any of a myriad of other purposes. Position and time are essential elements to a smart radio. Furthermore, from position and time, velocity and acceleration can be inferred, giving the radio some idea about its environment.

Geolocation applications are also a key enabling technology for such applications as spatially variant advertisement, spatially aware routing, boundary-aware policy deployment, and space- and time-aware scheduling of tasks. These capabilities enable a CR to assist its user to conveniently acquire goods and services as well as to communicate with other systems using minimal energy (short hops) and low latency (efficient directional propagation of packets through a network). Geolocation applications in a CR enable the radio to be carried throughout the world and used without any manual adjustment or modification to maintain compliance with local regulations. Finally, space- and time-aware scheduling of tasks improves the efficiency of CR operations by managing vital resources and accomplishing goals "at the right place" and "just in time."

Section 8.1 covers various techniques for a CR to geolocate itself and other systems, and Section 8.2 addresses network-aided position awareness. Section 8.3 presents the mathematics of time of arrival, time difference of arrival, and frequency difference of arrival types of systems. Section 8.4 discusses the method of converting global positioning system (GPS) *x*-, *y*-, *z*-coordinate locations into latitude and longitude, or geopolitical region localization. Section 8.5 provides an example of boundary decisions, and an example for 911 geolocation for first responders is given in Section 8.6. Section 8.7 discusses interfaces with other cognitive subsystems. Finally, Section 8.8 summarizes the chapter.

8.2 RADIO GEOLOCATION AND TIME SERVICES

Several radio transmission services are located throughout the world to aid in geolocation and accurate time tracking. The best-known system is GPS[1] [1], which is based on a constellation of satellites constantly broadcasting time and position (see the next section for details on GPS). The National Institute of Standards and Technology (NIST) radio station WWV, which continuously broadcasts time with high accuracy, is somewhat well known in the Western Hemisphere. (WWV consists of stations WWVB and WWVH [2–4].) However, without knowing position, it is difficult to determine how long the transmission took to propagate to the receiver, and thus it renders only a coarse time.

Less well known are the very high frequency (VHF) omnidirectional ranging (VOR) transmitters used by aircraft to locate current position [5], LOng RAnge Navigation (LORAN) used by ships at sea to calculate position, and the geopositional services of cellular telephone systems [6]. Also quite widely deployed is geolocation by wireless local area network (WLAN) Internet Protocol (IP) address [6], in which the IP address of a WiFi access point is directly translated into a geolocation [7]. The most recently published technique is the use of television (TV) broadcasts for time, frequency, and position [8]. The attempt to address these location techniques in this section will be presented only to the degree that it is clear that a software-defined radio (SDR) can participate in these time and location processes, and can thereby know geolocation and time sufficiently to aid its network, to aid its use of spectrum, and to aid its user.

8.2.1 Global Positioning System

GPS is without a doubt the best-known location system in the world. GPS is a satellite navigation system funded and controlled by the US Department of Defense (DoD) and the US Department of Commerce (DoC). The system comprises a constellation of satellites, ground control stations, and GPS receivers. At most points on Earth (other than in the deep urban canyons between skyscrapers), there is a high probability of line-of-sight (LOS, meaning unobstructed) contact with multiple GPS satellites. Given LOS with four or more satellites, three-dimensional (3D) position and time can be determined.

Satellite System Architecture
The GPS system is readily divided into three segments: space, control, and user.

Space segment. Not counting orbiting spares, there are normally 24 active GPS satellites. The orbital period is nominally 12 hours. The 24 satellites are distributed evenly in 6 orbital planes with 60-degree separation between each of the four satellites in each

---

[1] Similar systems have also been launched by Europe (Galileo) and by Russia (Glonass).

plane. The inclination is about 55 degrees off the equator. This geometric distribution provides between 5 and 8 satellites in view from any point on Earth. An LOS view of four or more satellites is needed to process the signals and calculate location.

Control segment. Ground-tracking stations are positioned worldwide to monitor and operate the constellation of GPS satellites. The master control station is located at Schriever Air Force Base in Colorado. The stations monitor the satellites' signals, incorporate them into orbital models, and calculate ephemeris data that are transmitted back to the space vehicles. Ephemeris data are, in turn, transmitted to GPS receivers.

User segment. GPS receivers and their operators form the user segment. The receivers process the signals from four or more satellites into 3D position and time. In the differential mode, a reference GPS receiver communicating with another GPS receiver, where the position of one node is known to high accuracy, can then improve on the inherent accuracy of another stand-alone GPS receiver. GPS receivers also produce a precise one-pulse-per-second signal. The one-pulse-per-second timing pulse as seen by the receiver experiences a propagation time delay from each satellite.

Because each satellite is at a different distance from the satellite, these time pulses are seen at different times for different satellites. Once the receiver solves for its location and the corresponding propagation time delay, it can then attribute an additional delay to all of these pulses that will exactly align all of them to when the next pulse should occur. Thus, a sophisticated receiver can reproduce one-pulse-per-second time-aligned pulses on all ground units over a large regional area by adjusting the pulses to compensate for propagation delay, and then identify the time that corresponds to those pulses, just as if the atomic clock in the satellite were connected by a wire to the receiver. As a result, all receiver units will be able to have precision time, regardless of their position. Given the internal logic that performs this alignment, the precision of the output pulse is reduced to around 340 nanoseconds (ns, equal to 10^{-9} sec).

The operators of GPS receivers include military users as well as civilian users. The military operators, including those of the United States and its allies, use a different signal-processing architecture that includes cryptographic decoding and increased accuracy.

Accuracy-Obtained and Coordinate System

The two classes of GPS geolocation capabilities are the precise positioning service (PPS) and the standard positioning service (SPS). The PPS capability requires cryptographic technology and achieves 22 m horizontal accuracy, 27.7 m vertical accuracy, and 200 ns time accuracy. The SPS capability is available to any user and achieves 100 m horizontal accuracy, 156 m vertical accuracy, and 340 ns time accuracy. These specifications are defined by the 1999 Federal Radionavigation Plan and are 95 percent accuracy figures (two standard deviations of radial error).

GPS Satellite Signals

GPS satellites transmit two spread-spectrum signals, one at 1575.42 MHz (L1 for SPS) and the other at 1227.60 MHz (L2 for PPS). There is a unique 1 MHz wide, 1023-chip–

long coarse acquisition pseudorandom (Gold code) spreading code for each satellite [9]. A Gold code is a spreading code synthesized by exclusive ORing of the output of two linear feedback shift register (LFSR) pseudorandom (PN) generators together. Each LFSR uses a carefully selected tap and initialization to ensure that the autocorrelations of the Gold code are small at all delays except perfect time alignment, and are not confused with cross correlation from other spreading codes.

The coarse acquisition code on L1 is used for civilian GPS. There is also a precision code (P-code), 10 MHz wide and repeating every seven days, that is superimposed on L1 and L2. The P-code is used in precise mode. Encryption converts the P-code into the Y-code. The P- and Y-codes are used for military GPS. Additionally, there is a 50 Hz signal on the L1 coarse acquisition code that transmits such ancillary data as orbit parameters and clock data.

GPS Navigation Message

A data frame (1500 bits) is transmitted every 30 seconds and consists of five 300-bit subframes.

- Subframe 1 is Telemetry Word | Handover Word | Space Vehicle Clock Correction Data
- Subframe 2 is Telemetry Word | Handover Word | Space Vehicle Ephemeris Data, part 1
- Subframe 3 is Telemetry Word | Handover Word | Space Vehicle Ephemeris Data, part 2
- Subframe 4 is Telemetry Word | Handover Word | Other Data
- Subframe 5 is Telemetry Word | Handover Word | Almanac Data for All Space Vehicles

Other data and almanac data are spread over 25 data frames and take 12.5 minutes to transmit. The 12.5-minute period is the complete navigation message.

Signal Processing of GPS Signals

The GPS receiver correlates known coarse acquisition spreading codes (with a 1-millisecond period of 1023 chips) from each of the GPS satellites with the processed signal from the GPS satellites. The known spreading codes are very short and may be generated or stored in memory. Because each satellite uses a different Gold-word spreading code, when the receiver has a peak correlation it knows which satellite sent the signal. This despreading produces a full-power signal. This signal is tracked using a phase locked loop (PLL), and the 50 Hz navigation message is demodulated from each satellite. Time of arrival (ToA) information is extracted when a correlation peak is measured.

Given the ToA information measured from the correlation peak and the GPS time embedded in the signal, the GPS receiver can measure range to each satellite in view. An intersection of multiple range spheres determines where the GPS receiver is located. Four satellites must be in view to estimate x, y, and z coordinates along with a time estimate. A precise estimate of the position of each space vehicle in view is determined from the broadcast ephemeris data.

Table 8.1 GPS Errors	
Error Source	**Approximate Error (m)**
Noise (PN code and receiver)	2
Selective availability[a]	100
Uncorrected clock errors (in space vehicle clocks)	1
Ephemeris data errors	1
Tropospheric delays	1
Unmodeled ionosphere delays	10
Multipath	0.5

[a] Selective availability is now off.

Note: The intentional degradation from selective availability is a significant source of position error in GPS systems.

Reference axes. The x, y, and z estimates are computed in Earth-centered fixed (ECF) coordinates. ECF is a right-hand orthogonal Cartesian coordinate system with the origin at the center of Earth, the z-axis increasing through the rotational North Pole of Earth, the x-axis increasing through the prime meridian (Greenwich, England) at latitude zero and longitude zero, and the y-axis increasing through 90 degrees longitude and 0 degrees latitude.[2]

Differential GPS. Position accuracy may be improved through the use of differential GPS processing. Correcting bias errors using a known location accomplishes this. A known location receiver measures its position and calculates a correction for each satellite that is passed to other GPS receivers in the local area. This is a sophisticated solution that requires more capability at both the reference and mobile GPS receiver and a data link between the reference receiver and the mobile receiver.

Another form of differential GPS is the measurement of carrier phase. This capability is used in surveying and can generate subfoot accuracies over short distances. Again, at least two receivers, a reference and a mobile, are required. Due to ionospheric effects, the set of receivers doing carrier phase geolocation must be relatively close together (approximately a 30 km limit).

Potential GPS position error sources are summarized in Table 8.1.

8.2.2 Coordinate System Transformations

Satellite positions and GPS receiver position are reported in ECF coordinates (x, y, z). Navigators, however, are frequently interested in latitude, longitude, and height. Eq. (8.1) is the conversion from ECF to latitude, longitude, and height (ϕ, λ, and h, respectively).

---

[2] Located in the Atlantic Ocean, 0 latitude and 0 longitude is 380 miles south of Ghana and 670 miles west of Gabon. Located in the Indian Ocean, 0 latitude and 90 longitude is between Singapore and Mogadishu.

$$\text{Latitude} \quad \phi = \tan^{-1}\left(\frac{z + e^2 b \sin^3 \theta}{P - e^2 a \cos^3 \theta}\right)$$

$$N(\phi) = \frac{a}{\sqrt{1 - e^2 \sin^2 \phi}}$$

$$\text{Longitude} \quad \lambda = \tan^{-1}\left(\frac{y}{x}\right)$$

$$\text{Height} \quad b = \frac{P}{\cos \phi} - N(\phi)$$

(8.1)

where

$$P = \sqrt{x^2 + y^2}$$

a = semimajor Earth axis (ellipsoid equatorial radius)

b = semiminor Earth axis (ellipsoid polar radius)

$$\theta = \tan^{-1}\left(\frac{za}{Pb}\right)$$

$$e^2 = \frac{a^2 - b^2}{b^2}$$

A GPS receiver in a CR is one way to let a cognitive radio know where it is. Adding this information to an interradio data stream enables other CRs to know where a particular radio is located. Even though GPS is not tremendously precise without differential mode or precision mode modules, it is good enough for policy enabling and relatively long range (similar to cell phone to basestation) communication optimization.

8.3 NETWORK LOCALIZATION

Geolocation of the radio (the subscriber or the user) units enables a number of useful services.

8.3.1 Spatially Variant Network Service Availability

There may be, for example, services that the user wishes to find in his immediate area. Consider the international traveler who has just landed after a long flight from a foreign country, and who is unfamiliar with the airport. He may have several immediate needs including changing his currency, finding a restroom, finding food, finding a power plug adapter, and finding a train to his final destination. If he is not fluent in the local language, his radio's geolocation may be able to give him the relative coordinates to find such services nearby, and perhaps even help him negotiate the correct elevators.

Such processes require the ability to geolocate the user and then the requested services. The requested services may come with caveats such as "the closest," "the least expensive," or "near the route or path to be followed to some other service or objective." Having located the user, the system needs to be able to access local networks and perform inquiries about available services and their respective locations, and then sort to the user's preference criteria for presentation.

The preceding examples do not involve unsolicited advertising. However, there may be a way for a cognitive radio to accept unsolicited advertising, filter it against a list of topics of interest to the user, and present only those advertisements that pass the interest filter. Those advertised services that do pass the interest filter test can be offered to the user, prioritized with the user's other objectives and any route planning, so as to make them convenient.

Mechanization of service offerings becomes the major issue to enable this degree of cognitive service based on geolocation. Assuming the CR has geolocated itself with one or more of the techniques previously described, it needs to find a gateway to local networks. If the CR has several personal, local, regional, and cellular network access points available to it, it may then query among these networks to find those that offer cognitive support to the types of queries that support the user.

Assume, for the moment, that the CR has found an access point on a local network that offers such query services. Then the two radios may refine the location of the user to sufficient accuracy to provide appropriate directions to find the requested services. Those directions should be scaled by a scale factor appropriate to the distance as well as the local knowledge and experience of the user. Once the radio finds a service access point with the proper query service, the user may in turn be directed to a different access point where that specific type of knowledge is served. If necessary, it may include a change to whichever wireless access point is used to access the required service, as directed by the access process defined from the first server.

Finally, notice that this database may be built in part by the success of other users who have found useful services and captured the location of those services into the database. Thus, the query engine may be able to build a significant library of services without a significant system initialization effort.

EXAMPLE 1

A student is connected through her wireless local area network (WLAN) to a campuswide server that answers a query about a restaurant with a special musical offering this evening, and the time of the date. When seeking a new restaurant in an otherwise familiar city, the student need only know to go to a nearby street location, and then be given the final guidance of where that restaurant is to be found (e.g., "The stairs to the second floor restaurant are around the back of the clothing store on the first floor, and you must go around the left side of the store to get to them."). However, when the student is off campus, but near the restaurant, the CR switches to regional service, by which the restaurant has a local query responder that can provide an applet to help find the restaurant. The CR requires 3D positional knowledge, as well as 3D navigation driving or walking, and, of course, the ability to understand the positional uncertainty of the user, the user's current orientation and velocity, and the relative position between the user and the objective. It is likely that local navigation may consist of applets offered by the endpoint server, in order to provide for an endless supply of unique specialized access requirements (e.g., "Take the last elevator because it's the only one that leads to the top floor").

EXAMPLE 2

Another example of geolocation-based services is emergency alert messaging. Suppose that all users in a geographic region must be alerted to imminent danger. Radios that know their position can be alerted, and the priority of all tasks in that region can be completely changed. It may be necessary to provide such messages over multiple networks, so there is a high probability that all CR radio users will receive the message. It is also feasible to determine how many users remain in an area and what their locations are, to expedite any remaining evacuation.

8.3.2 Geolocation-Enabled Routing

In addition to the user's geolocation support, the radio network functionality may benefit from geolocation knowledge. Chapter 11 provides a detailed analysis of the radio environment map (REM), which is an infrastructure to enable cognitive network functionality.

However, in addition to the REM, the routing functionality of an ad hoc network, and the cellular handover of cellular networks, may be improved by explicit geolocation knowledge, velocity vector knowledge, and planned route path knowledge.

One can envision that an ad hoc network could use destination location addressing rather than medium access control (MAC) and IP addressing. In such a network, messages propagate only to nodes that know they are on a path to that location. It is expected that by reducing the number of nodes that are off course for a destination, the end power drain from all nodes and the radio interference level can be reduced. Cellular networks can improve their handover process to anticipate handovers and reduce dropouts and dead zones by being aware of geolocation, velocity vector, and planned path.

8.3.3 Miscellaneous Functions

A number of niche applications have been proposed that would be enabled by CRs with geolocation knowledge, including radios that allow users to: (1) flirt or reject flirtation; (2) recognize when someone nearby is either desirable to meet or should be avoided; (3) recognize individuals with common or special interests who are nearby; (4) identify individuals with limited time budgets who cannot support any extraneous interactions; (5) determine the location of criminals who have been released after serving time, but who may still pose a threat under certain conditions. This list is likely to grow as the economics of CR applications begin to benefit users.

8.4 ADDITIONAL GEOLOCATION APPROACHES

Terrestrial radio geolocation is accomplished through one or more of the following techniques: time of arrival (ToA), time difference of arrival (TDoA), angle of arrival (AoA), or received signal strength (RSS). Normally, a strong LOS signal is needed for accurate measurements.

8.4.1 **Time-Based Approaches**

Time-based approaches for geolocation may be divided into ToA and TDoA approaches. Both approaches require a high-resolution system clock. ToA and TDoA approaches to geolocation are based on the propagation speed of light, which is defined to be 3×10^8 m/sec, and "straight-line" LOS propagation paths so that the signal time delay relates directly to the LOS distance. Propagation is analyzed by multiple receivers or by multiple antennas converting time delay to phase difference, time difference, or frequency shift. These are then translated into equations of range, range ratio, angle of arrival, and/or other parameters. Most systems also track not only the estimated value but the error bounds, so the location can also be represented as a circular error probability (CEP) ellipse. At the propagation speed of light, 1 ns of time measurement error creates 0.3 m of ranging error.

Time of Arrival Approach

The time of arrival approach is centered on the ability to time-tag a transmitted signal and measure the exact ToA of that signal at a receiver. The propagation time, at the speed of light assuming LOS propagation, is a direct measure of the propagation distance. This provides a receiver with an iso-range sphere for a given transmitted and received signal. If multiple receivers at known locations receive the same signal, generally at different times, the multiple iso-range spheres intersect at the transmitter's location. It requires four receivers to geolocate one transmitter in three dimensions. A reverse problem may be constructed in which four transmitters provide their location in a time-tagged transmitted signal and the receiver can geolocate itself. More complex situations for which only relative positions can be determined can be constructed. This problem is useful in sensor fields and also for large numbers of cognitive radios. A two-dimensional (2D) depiction of this process is shown in Figure 8.1.

Round-Trip Timing and Distance Measuring Equipment

A valuable capability is being able to locate a transmitter, for example, to rescue a sinking ship or a wrecked aircraft. Some aircraft are equipped with distance-measuring equipment (DME) transponders. These systems respond to a transmission by transmitting a response pulse the timing of which is precise relative to detecting an arriving signal. By measuring the round-trip time of the interrogation pulse and its arriving response, and then subtracting the response time of the transceiver's receiver–detector–transmitter, the total distance between the two radios can be estimated. If the search transponder also uses three or more antennas, it can estimate both the angle of arrival and the distance to the distressed ship or aircraft relative to its current location and can immediately fly directly to it or otherwise dispatch aid directly to it. This has resulted in such products as the General Dynamics PRC-112 transceiver and its interrogator.

Certain cellular radio systems have developed a similar DME approach to subscriber localization. In these systems, the cellular subscriber responds to transmissions from each of several cellular base stations of known position. Each cellular base station can then estimate the range to that subscriber. By combining the range estimates from each base station, the subscriber can be accurately located. This method eliminates the

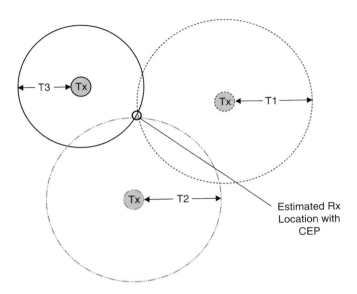

FIGURE 8.1

Two-dimensional ToA. The intersection of ISO-range spheres (3D) or ISO-range circles (2D) may not be a point, introducing a circular error probability. *Note:* Rx = receiver, Tx = transmitter.

requirement to install a separate GPS receiver function in each subscriber unit and shifts the location complexity to the base station, thus minimizing subscriber unit cost.

Time Difference of Arrival Approach

The time difference of arrival approach[3] is centered on the ability to measure the time difference between the reception of a signal at one location, and the time of reception of the same signal at another location. Again, propagation time, at the speed of light, is assumed to provide a direct measure of the LOS propagation distance. The constant difference in arrival time produces a hyperboloid surface with the foci at the two receivers. If multiple pairs of receivers at known locations receive the same signal, the multiple iso-range hyperboloids intersect at the transmitter's location. It requires three surfaces to geolocate one transmitter in a 2D plane, and it requires four surfaces to geolocate one transmitter in three dimensions.

Fitting a TDoA Curve with Two Receivers

Suppose two radio transmitter–receivers (named 1 and 2), separated by a known distance $D_{1,2}$ in meters, each record the time of arrival of a signal from a source (named *S*) in an unknown location, indicated by time stamps t_1 and t_2 in nanoseconds. Assuming the signal propagated uniformly at the speed of light, $c = 0.300$ m ns^{-1}, then the speed of propagation multiplied by the difference between the values of these time stamps is the difference of the distances from the source to the two radio transmitter–receivers:

---

[3] This section, Time Difference of Arrival Approach, was contributed by Nicholas W. Fette, July 2005, personal communication.

$$c(t_1 - t_2) = c\Delta t_{1,2} = D_{1,S} - D_{2,S}. \tag{8.2}$$

This is one form of an equation for a hyperbola, the graph of which shows the possible locations of S based on the two radio receivers. If both receivers and the source are taken to be in the same plane, the hyperbola is roughly V-shaped (Figure 8.2), with an axis of symmetry through the two receivers.

The hyperbola can be written in terms of a coordinate system (x', y') with radio 1 (R_1) at $(0, -\tfrac{1}{2}D_{1,2})$ and radio 2 (R_2) at $(0, \tfrac{1}{2}D_{1,2})$. A familiar form for a hyperbola with these foci is

$$y'^2 = a^2 x'^2 + k^2 \tag{8.3}$$

with the constraints that y', a, and k carry the same sign as $\Delta t_{1,2}$. The constants a and k can be determined by inspection and are dependent only on $D_{1,2}$ and $c\Delta t_{1,2}$.

Consider the end behavior of the hyperbola. The signal from a hypothetical distant transmitter (named HT) placed on the hyperbola at (x'_{HT}, y'_{HT}) for some infinitely high x'_{HT} value travels in parallel paths through radios 1 and 2 and the origin. A right triangle can then be formed (Figure 8.3) with the hypotenuse as the segment from radio 1 (R_1) to radio 2 (R_2), one leg as a segment of length $c\Delta t_{1,2}$ along the path to R_1, and the second leg

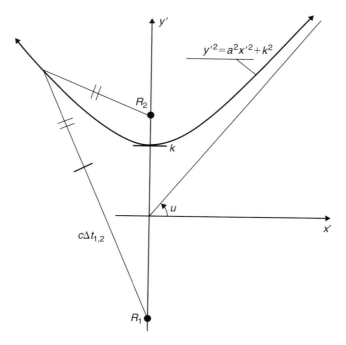

FIGURE 8.2

Hyperbola with two receivers and a source in the same plane.

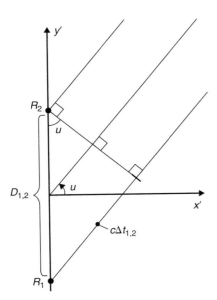

FIGURE 8.3

Right triangle in the coordinate system (x', y').

closing the right triangle at R_2. Let θ be the angle between the hypotenuse and the second leg. By trigonometry, $\sin\theta = c\Delta t_{1,2}/D_{1,2}$, and defining $f = c\Delta t_{1,2}/D_{1,2}$, the value of θ is:

$$\theta = \arcsin(f) \tag{8.4}$$

By geometry, θ is also the angle from the x'-axis to the infinite-distance transmitter, so $\tan\theta = y'_{HT}/x'_{HT}$. Thus, $\lim_{x\to\infty} y'/x' = a$, with a constant such that:

$$a = \tan\theta = \tan(\arcsin f) = f/\sqrt{1-f^2}. \tag{8.5}$$

Whereas constant a is evaluated by considering infinite x', k is evaluated by moving the hypothetical transmitter to $(0, k)$. Since HT is in line with radios 1 and 2, $D_{1,2} = D_{1,HT} + D_{2,HT}$. Thus, $D_{1,HT} = \dfrac{1}{2}D_{1,2} + k$ and $D_{2,HT} = \dfrac{1}{2}D_{1,2} - k$. By Eq. (8.2),

$$c\Delta t_{1,2} = D_{1,S} - D_{2,S} = 2k$$

or

$$k = \frac{1}{2}c\Delta t_{1,2} \tag{8.6}$$

In summary, the hyperbola in this system is

$$y'^2 = \frac{c^2\Delta t_{1,2}^2}{1 - c^2\Delta t_{1,2}^2}x'^2 + \frac{1}{4}c^2\Delta t_{1,2}^2 \tag{8.7}$$

where y' carries the sign of $\Delta t_{1,2}$.

Transforming to a Common Coordinate System

The usefulness of the coordinate system used so far in Section 8.3 is limited to communication between two radios only. To locate the signal source, at least three radios must be employed. Therefore, all pairwise hyperbolas must be transformed into a common coordinate system to solve for the source location. That system may be chosen arbitrarily such that it enables conversion to and from GPS coordinates recognized by each radio set; however, the latitude–longitude system is warped and adds difficulty to the algebra in solving for the source location. For example, the common coordinate system (x, y) could be oriented with $+y$ directed north, $+x$ directed east, and the origin $(0, 0)$ at the location of R_1, and it would form a valid system if the curvature of Earth is negligible, given fairly even terrain. To transform Eq. (8.7) to this form, a rotation followed by a translation would suffice. The counterclockwise angle of rotation ψ is determined by the angle formed by the ray from R_1 through R_2 and the ray from R_1 pointing northward (Figure 8.4).

In this example, $\tan\psi = \Delta x_{1,2}/\Delta y_{1,2}$, and the old temporary axes convert by

$$x' = x\cos\psi - y\sin\psi,$$
$$y' = y\cos\psi + x\sin\psi,$$

(8.8)

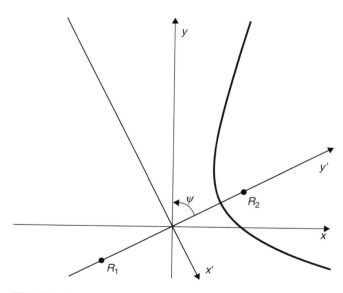

FIGURE 8.4

Geometry for determining counterclockwise angle of rotation ψ. This process would be used to convert the hyperbolic equations in a radio-centered coordinate system to a north-centric coordinate system.

where these substitutions are made into Eq. (8.3), in the manner

$$(-a^2 \sin^2 \psi + \cos^2 \psi) y^2 + (a^2 \sin 2\psi + \sin 2\psi) xy + (\sin^2 \psi - a^2 \cos^2 \psi) x^2 - k^2 = 0 \quad \textbf{(8.9)}$$

The solution of a system of at least two distinct equations in the form of Eq. (8.9) can determine the possible position(s) (x, y) of S, the source of the transmission.

Solving for Position of Source Transmitter

The number of solutions that can result from systems of equations such as Eq. (8.9) depends on the arrangement of the receivers and which pairs of receivers were fitted with hyperbolas. As a rule, if all the receivers are in a line, there will be two solutions, one on either side of the line. If all the receivers are on a single plane, then there may be two height solutions, one of which may be underground.

LORAN

LOng RAnge Navigation systems transmit a known burst signal from multiple transmitters with a known and published periodicity. Furthermore, the exact location of each transmitter is known. Three such transmitters cooperate to enable TDoA measurements. Ships at sea receive these transmissions and measure the time difference between each received signal. From these time differences, ships are able to calculate the TDoA hyperbolas. To simplify the process, the TDoA hyperbolas have been converted into published charts so that a navigator can directly look up the time differences for each transmitter pair and find an intersection of two time difference pairs to perform a location at sea.

Television Broadcast [4]

Recently, Rosum Corporation (Mountain View, CA) announced they had developed a technique to recognize the ghost canceling reference (GCR) signal chirp that is included on the 19th line of the vertical retrace [8, 9]. By measuring the exact time that this chirp is received for several TV stations, and comparing the time delay between the measured arrival and the precalculated transmit time for that pulse, they can estimate a ToA for each detectable TV signal. TV transmitter locations are known with high accuracy, and they transmit with high power, so this TDoA system can work in urban areas where satellite signals fail. Since the bandwidth is 4.5 MHz versus the 1 MHz bandwidth of the standard GPS signal, it may also improve spatial resolution by a factor of approximately 4. Finally, because networks include precision time information as a data component during vertical retrace, all the information necessary for a sophisticated TDoA system is present in urban areas. The GCR chirp described above can be used to recognize and suppress multipath components, and therefore can be used in urban canyons.

Timing Estimates

Time can be derived from a number of sources, including atomic clocks, standard clocks, GPS time, disciplined GPS, phase estimation techniques, and correlation tech-

---

[4] The move to digital TV, scheduled for June 2009 in the United States, will render this technique obsolete, though similar techniques in other broadcast formats may be useful.

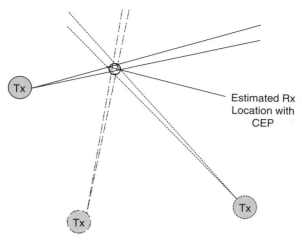

FIGURE 8.5

Angle of arrival. The geometry affects the accuracy of AoA-based geolocation estimates.

niques in wideband transmission environments. The accuracy of the time estimate directly influences the ranging estimate accuracy at approximately 1 ns per 30 cm.

A popular way to obtain TDoA information is through cross correlation. At a fixed point in time (maybe on the GPS one pulse per second boundary), two stations digitize their received signals and pass the measurements to a common processing location. The processor calculates the cross correlation of the two signals. The peak of the cross correlation reveals the TDoA.

8.4.2 Angle of Arrival Approach

The angle of arrival approach requires an antenna array at the receivers. Multiple receivers estimate the AoA of a signal. Combining the bearing to the signal with the known location of multiple receivers yields an intersection point of the transmitter. This is simple triangulation.

Geometry of AOA Approach

Geometry, of course, affects how well any time or angle measurements work, including the AoA approach (Figure 8.5).[5] If the "baseline" of receivers does not have a sufficiently large angle of observation, the result is poor. If the accuracy of the AoA sensors is poor and the range to the receiver is long, the "angular dispersion" impacts the measurement.

---

[5] Geometric dilution of precision (GDOP) is one way of relating approximately the ranging errors to position errors. Even though the distance measurement to each transmitter and its corresponding time may be measured quite accurately, the geometry of the position of the transmitter as observed by the receiver may find that the lines of possible position intersect at shallow angles. This will cause the intersection to have much larger positional error than the error from the range estimate alone.

However, the computations are simple if the AoA is known to high accuracy and the interprocessor communications data volume is low.

VHF Omnidirectional Ranging

Aircraft navigate using a number of transmitters located around the globe, frequently near airports. VOR transmitters transmit an amplitude modulated (AM) signal with a constant frequency. In addition, they transmit this signal through a series of many antennas, each selected sequentially around a circle of radius approximately 5 m. The energy of the transmission is thus electronically swept around in a circle such that it introduces frequency modulated (FM) doppler onto the received signal. Each receiver will perceive the relative phase angle between AM and FM as a different phase, reflecting the relative location of the receiver relative to the transmitter. VOR receivers convert this modulation phase angle into the angle bearing to the VOR. By locating two such VOR transmitters, the receiver is able to estimate its location.

8.4.3 Received Signal Strength Approach

If the transmit power on a signal is accurately known, the patterns of the transmit and receive antenna gains are known accurately, and the receiver is able to measure receive signal strength accurately (all generally difficult assumptions to ensure), then a propagation model can be used to estimate the distance to the transmitter as a function of RSS. But propagation channels are very dynamic, so this approach is problematic. The greatest source of error in this approach is multipath fading and shadowing, and the effect can easily be a 40 dB impairment or a 6 dB increment over the direct path LOS in as little as half of a wavelength. Mobility allows a receiver to average out these effects to some degree for FM music, but multipath represents a significant impairment to data services.

This location approach is similar to the ToA approach. Four estimates of range (upper and lower bounds of range) to four known transmitter locations are used to calculate the intersection of iso-range spheres. This yields an estimate of position and position error in three dimensions.

Simple propagation channel models predict pathloss attenuation as an exponent ranging from 2 to 4.5, depending on terrain, foliage, and building shadowing and blockage losses. Since the physical environment is such a key component to other more realistic channel models, some averaging over position is required and some setting of the attenuation exponent must be assumed. The setting may be a table lookup as a function of rural, suburban, or urban environments and frequency.

If a correlation process based on a licensed transmitter's database is undertaken, a RSS-based receiver application could determine in which regulatory region it is located. For example, if a cognitive radio is receiving certain TV channels and certain AM and FM stations all at the same time, it can infer its city location. If the location of the transmitters is included in the database along with transmission levels, the RSS process could improve this estimate due to the fairly large number of measurements.

The quality of RSS-based geolocation estimates is fairly low. It is useful to CRs for some applications but not for others. For example, it may be close enough to regionalize an estimate to a city or a part of a city, and thus to a regional accuracy of 30 km.

8.5 NETWORK-BASED APPROACHES

Wired networks have already developed a database that allows a table lookup translation from IP addresses to geolocation. This database provides to the subscriber component the immediate ability to determine a geolocation to the accuracy of a service region. For WLAN devices, this allows the device to know its location to within a radius of 100 m or so. Services offered from longer range connectivity, such as fiber-optic or wired service, may be able to infer regional positional accuracy from such a table lookup. At the present time, the database does not track its geographic uncertainty, nor does the database guarantee that it is currently up to date. Chapter 11 provides more information on how infrastructure-supported data services can provide geopositional support and timing to wireless subscribers.

8.6 BOUNDARY DECISIONS

The ability to determine the location of a cognitive radio to a geographical region enables at least one beneficial capability. The benefit of automatic compliance with spatially variant regulations has been discussed extensively in the cognitive radio community. Other applications may develop from this capability as well.

8.6.1 Regulatory Region Selection

The capability of a CR to determine which geopolitical region it is operating in enables spatially variant policy selection, and therefore worldwide mobility, and radio compliance with local regulatory rules. One approach for this capability is a map database that contains a boundary set for each of the regulatory region. A hierarchal search starting with continent, focusing on country (i.e., regulatory authority), and finally selecting a specific policy region would minimize the time and the computational load required to determine in which region a radio is located. The regulatory region returned is then used to select a set of policies. If the radio does not have the correct policy, it may be able to acquire it from a locally available infrastructure, such as a policy database server, or to apply a default set applicable to the largest number of regions in the last known continent.

Border Database Representation Analysis

This section explores several methods to determine the applicable geographic region, and therefore the associated geopolitical region. Various methods are feasible for defining a geopolitical region based on current GPS coordinates of a radio node. The objective is to find methods that minimize memory resources to represent the regions a radio will experience during its product lifetime or mission lifetime, so that it can choose corresponding policies for the associated region.

We assume in this analysis that the usual preference is for the mechanism that consumes the least memory resource to represent each country's borders. (There are currently 192 countries in the world [11].) Specifically, we explore three methods to determine whether the current GPS location is inside a given complex polygon defining the borders of the region:

    A. Successive tiling using latitude and longitude (east–west and north–south) boundaries, in which the aggregated tiles define the geopolitical region
    B. A list of endpoints of successive line segments defining the borders
    C. A set of K nearest neighbor (KNN) position points

Method A

In method A, we define a large rectangular region in the center of a country, and then smaller rectangular tiles to fill in the shape of the irregular border regions. If we assume one central region and an average of 100 smaller tiles, and four coordinates per tile, then each country would require, on average, $A = 101 \cdot 4 = 404$ points.

    Searching this database can be quite efficient, particularly if we assume an outer constraint rectangle as well. So, for example, if Figure 8.6 defines the maximum northern and southern extent, as well as the maximum eastern and western extent of each country, then we can perform a subset analysis for any given GPS point that can determine within which countries it may possibly lie. This is likely to include a maximum of six countries. We then search the interior subtiles for each of those possible countries, performing east, west, north, and south boundary comparisons, and stop as soon as we determine that a point is within a specific subtile. The resulting subtile is then associated to a country and the country to a policy set.

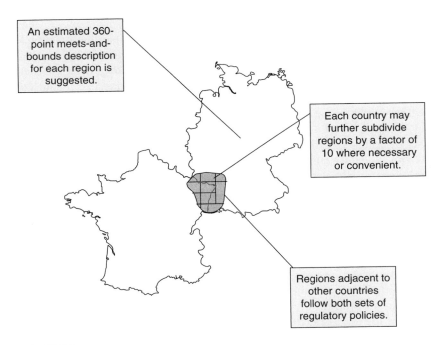

An estimated 360-point meets-and-bounds description for each region is suggested.

Each country may further subdivide regions by a factor of 10 where necessary or convenient.

Regions adjacent to other countries follow both sets of regulatory policies.

FIGURE 8.6

Defining geopolitical regions.

Method B

In method B, countries may define their borders as a linked list of straight line segments using GPS coordinates. We assume that sufficient resolution requires a maximum of 360 2D GPS locations, plus a country geographic centroid. Therefore, $B = 2 \cdot 361 = 722$ points.

Although it may appear that this method is more frugal than method A in data that must be stored, the search problem is more complex because the search algorithm must determine whether a point is enclosed inside or outside the region defined by many line segments. To perform this calculation, we can begin, as in method A, by determining which countries must be searched by comparing the putative GPS coordinate against the extrema of all countries and only searching those that are relevant. Again, we assume a maximum search size of six countries. Next we calculate the perpendicular vector to each line segment. Along that vector, we calculate the minimum distance from the point to the line, and we save the identity of the two closest line segments.

For the sake of a simple explanation, we generalize that there are very few instances of spiral borders. We can then determine whether most of the country lies to the east, west, north, or south of the two line segments, and whether the putative point is likewise to the east, west, north, or south of the two line segments. Failing that, we use the borders of each country until we find a case in which the radio node location lies in the same direction as the centroid of the country relative to the two nearest line segments. Note that this is equivalent to dividing each country into a set of 100 triangular (pie-shaped) regions and determining whether a point is within the triangle.

Method C

In method C, we define a K nearest neighbor approach to determine whether a point is in a country. In this case, all countries must define a set of points approximately equidistant from the border that result in a KNN border that smoothly approximates the true border.

We assume that the countries agree to place their KNN points at 10 km distance from the border and at approximately 10 km spacings. Furthermore, to ensure fairness, these points are placed on each side of the border along the same perpendicular bisectors of border boundary line segments. This has the result of smoothing fine-grain border detail to approximately 3 km feature sizes. The database size is now proportional to the length of the border. If we assume 100 sample points for the average border, we must store $C = 100 \cdot 2 = 200$ points.

Furthermore, let us assume that $K = 2$ in our KNN search. Now, for each country, we calculate the Euclidean distance to all KNN locations, and we save the final distance as the average of the two smallest distances. We compute this calculation for all countries, and select the correct country as the one with the smallest Euclidean distance. As with the other search processes, the search computation can be reduced to a subset of local neighbors by selecting to search only those countries the maximal extents of which cover the range where our point of interest is currently located.

Anomalies

There may be countries that have anomalous shapes that may not behave correctly when using the algorithms, such as spiral arms brought about by the circuitous path

Table 8.2 Database Sizes for Three Representation Methods

Representation Method	Database Size (bytes)
A	323,200
B	577,600
C	160,000

Note: These methods are used to define whether a position is inside of a country.

of rivers through mountain ranges (see previous section). In addition, islands are anomalies, but an island can be handled as a second country with the same name and same policy.

More likely sources of trouble, however, are disagreements about what the actual border really is, such as overlapped regions claimed by more than one country. Methods A and B can be extended to report that two countries both satisfy the criterion, thereby defining the location of a radio. Under these conditions, it may be possible to define a common subset policy or to report to the user that there are no applicable rules in disputed zones.

The format for the latitude and longitude is negotiable. One suggestion is a 32-bit floating-point number (4 bytes) representing degrees, minutes, and seconds of latitude or longitude. This would result in a worst-case resolution along the equator of about 5 m.

For each of the representation databases, and the conservative assumption of 200 countries, we can estimate the size of the database to be as shown in Table 8.2.

Note that at any one time only a subset of such a database is needed and the meets-and-bounds database can be loaded into a CR similar to loading a working set into a cache memory.

8.6.2 Policy Servers and Regions

Possible sources of policy are: (1) a periodic broadcast (e.g., a protocol that is not defined at this time[6]); (2) a load from a certified source (e.g., this could be realized in the form of kiosks); or (3) a wired database server that the device must proactively acquire, interrogate, and capture.

A policy region is not a defined entity at this time. An example of a policy region might be the greater Washington, DC, area. In a specific regulatory region, legacy transmitters are licensed, thus defining other broadcast bands that could be harvested (only under specific technical circumstances). Another example of something defined in a regulatory region is a licensed spectrum holder and a registry of frequency sublease server(s) (frequency and protocol) that may be contacted for leasing a channel for a short period of time. Both of these examples are "reach-out" capabilities and have not

---

[6] At the time of this writing, there is ongoing serious discussion in the ITU and in the European research community (E3) about a Cognitive Pilot Channel (CPC) to provide this function.

been approved in the United States. These types of technologies have been discussed only in limited circles (see, e.g., Chapter 2).

In the short term, a small set of policy regions could be defined as boxes in space and a subset of a small experimental band (i.e., suggestions for an experimental CR band) where demonstrations are conducted. Field demonstrations of these capabilities could show the degree of spatial and frequency band reuse achievable. Demonstrations also will show the ability of CRs to share the space on a noninterference basis and the ability to comply with machine-readable policies.

8.6.3 Other Uses of Boundary Decisions

Radio policy selection is the frontrunner of CR applications that are dependent on knowing in which geopolitical region a radio is located. Other potential applications also exist. Given natural language processing capability, a CR that knows it is in Germany might offer to translate an English-speaking user's word from English into German when transmitting and could translate German words into English when receiving. Although this could be accomplished through language recognition more efficiently than through political boundary decisions, a boundary decision could be used to improve the language recognition search by prioritizing by local languages.

If a tourist is using the CR, the boundary decision engine could be used to display local points of interest such as historical markers. On a more personal note, the boundary decision engine could be correlated with the user's address book to locate friends and family in the vicinity, and the user can apply this information appropriately.

8.7 EXAMPLE OF CELLULAR PHONE 911 GEOLOCATION FOR FIRST RESPONDERS

The productivity of ubiquitous cellular phone and handheld communications devices such as pagers or BlackBerry-type devices have dramatically increased. Additional problems, however, have been introduced. For example, with wired telephones emergency calls were traditionally easily located to the phone where the call was placed and the location was very reliable. At the time of the initial deployment of cellular telephones, a call placed to an emergency response center did not come with a location attached for the emergency operator. The problem prompted the Federal Communications Commission (FCC), in 1996, to mandate geolocation services in the cellular network infrastructure. The FCC mandate was for 125 m accuracy in 67 percent of all measurements by October 31, 2001.

Multiple cellular telephone interfaces exist including Advanced Mobile Phone System (AMPS), code division multiple access (CDMA) EIA/TIA IS-95, time division multiple access (TDMA) EIA/TIA IS-136, the older TDMA IS-54, and the Global System for Mobile Communications (GSM). The common characteristic of these systems is a set of basestations that communicate directly with mobile stations or handsets. The frequencies, bandwidths, modulations, and protocols vary from standard to standard.

The two broad categories of geolocation techniques for cellular telephones are (1) network- or infrastructure-based approaches, and (2) handset-based approaches. The advantage of the network-based approach is there are no requirements placed on the owners of the cell phones. The advantage of the handset-based approach is the precision available.

Infrastructure-based approaches include ToA, TDoA, and AoA. The characteristics and techniques for these approaches have been discussed in Section 8.3.1. The geolocation application executes on cooperating basestations, measures one or more of the essential physical properties, exchanges information, and processes the signals to produce a location estimate.

The best example of a handset-based approach is putting a GPS receiver into the handset and interrogating the handset for its location when an emergency call is placed.

8.8 INTERFACES TO OTHER COGNITIVE TECHNOLOGIES

This section discusses interfaces between the geolocation engine and other entities in a CR. The policy engine, networking functions, planning engines, and user, at least, will interface with the geolocation engine.

8.8.1 Interface to Policy Engines

One possible interface of the geolocation engine (server) and the policy engine (client) is a client-server model. For example, the client requests from the server a location in terms of latitude, longitude, and altitude. Another request would be for a coded geopolitical regulatory region. These two requests would spawn different signal processing steps to answer the inquiry.

Another interface to the policy engine is an "interference analysis" request. This requires the relative location of other users to process the "message." The policy engine may request relative position directly to estimate the possibility of interference. However, this does not address hidden node issues that are inherent in the harvesting of spectrum.

8.8.2 Interface to Networking Functions

The networking functions may use a similar client server model to request relative position of other cognitive radio and non-CR devices. The resulting information may be used to direct steerable antenna beams and nulls or to select next-hop destinations in energy-efficient routing algorithms.

Networking functions may also request absolute position and use this in a search for local services, such as access points. The database of services is contained in the networking engine in this case. If the database of services is contained in the geolocation engine, the networking function could request location of the closest service provider, and the engine would return that. The partitioning of these functions to the two engines has not yet been discussed in the literature.

8.8.3 Interface to Planning Engines

A new capability for the geolocation engine is a distance, "as the crow flies," from the CR's current location to a designated position. An address to coordinate system function could be included. A variation of this request is from position A to position B. The positions would need to be provided in a variety of coordinate systems including ECF, latitude–longitude, or addresses. A planning engine executing a traveling sales person's algorithm uses the geolocation engine for metrics. The distances returned could be straight line or driving distances and paths, which may be obtained from the Internet.

8.8.4 Interface to User

A user interface to the geolocation engine may be a simple "Where am I?" that is used to select a digital map segment. The map is overlaid with a "You are here" marker. If relative positions of other emitters are requested, their positions may be overlaid as well. This capability is very useful for navigation applications and may be integrated with time and space management functions.

8.9 SUMMARY

A CR that is aware of its position is able to demonstrate spatially variant behavior. This is a critical capability for new functions, such as spatially variant regulatory policy or spatially variant networking functions. If the radio knows where it is located, it can self-report for such important functions as 911 responder interfaces.

Section 8.2 and 8.4 covered a number of ways to determine where a system is located. An inertial navigation system can be used to integrate a current position relative to a known starting point, but this approach is fraught with unreliability and excessive expense. A better alternative is a GPS receiver. This inexpensive subsystem provides 3D position and current time. The system is based on a constellation of satellites and it has the capability for two resolutions, precise positioning system and standard positioning system. Very precise GPS location can be obtained using differential approaches.

Noninfrastructure-based approaches include time of arrival, time difference of arrival, angle of arrival, and even received signal strength. The first two approaches use spatial diversity and interradio communications to estimate a position for an emitter. A circular error probability may be calculated a priori and is accurate with the exceptions introduced by high multipath channels. The AoA approach uses spatial diversity and interradio communications to make a geometric interpretation of the location of an emitter. The last approach, RSS, yields a regional estimate of position, but also uses spatial diversity and interradio communication to complete its estimate.

Section 8.3 touched on the value of geolocation knowledge to enable spatially aware networking functions. Routing may take position for energy savings or other purposes into account. Services may be accessed as a function of location. New novelty functions, such as spatially variant advertising, are possible when a CR knows its location.

Section 8.5 briefly looked at how networking can provide a coarse degree of localization, and Section 8.6 explored how geolocation is extended to become boundary analysis. One of the key interfaces to a geolocation engine is a boundary decision. A meets-and-bounds database may be employed for a CR to obtain its current geopolitical region. This approach may be used for a variety of other regions of interest decisions as well.

Section 8.7 provided examples of geolocation in the context of cellular emergency location, and Section 8.8 reviewed the many interfaces to other supporting subsystems a CR will need to build well-integrated, systems-level functionality. These interfaces will be the areas for significant standards development as CR technology evolves.

EXERCISE

Given: GPS position in Earth-centered fixed (ECF) coordinates ($-566,000$ m, $5,384,000$ m, and $3,360,000$ m), find the position in Geodetic (latitude, longitude, height) coordinates. *Hint:* This could be the location of a natural gas deposit in or around the Barnett Shale.

REFERENCES

[1] *www.colorado.edu/geography/gcraft/notes/gps/gps_f.html.*
[2] *http://tf.nist.gov/timefreq/general/pdf/1383.pdf.*
[3] *http://tf.nist.gov/timefreq/stations/wwvb.htm.*
[4] *http://tf.nist.gov/timefreq/stations/wwvb.htm.*
[5] *www.navfltsm.addr.com/vor-nav.htm.*
[6] *www.lucent.com/press/0699/990630.bla.html.*
[7] *www.linuxjournal.com/article/7856.*
[8] *www.catchoday.com/archives/39006.html.*
[9] *www.uspto.gov* (search on patent #6,879,286).
[10] Gold, R., Optimal Binary Sequences for Spread Spectrum Multiplexing, *IEEE Trans. Information Theory IT-13*, pp. 619–621, 1967.
[11] *www.google.com.*

Cognitive Techniques: Three Types of Network Awareness

Jonathan M. Smith
University of Pennsylvania, CIS Department
Philadelphia, Pennsylvania

9.1 INTRODUCTION

Users see the network through the window of distributed applications, which carry out some combination of communication and computation to meet user needs. Familiar examples include Web applications for shopping and social networking, interactive applications such as video-teleconferencing, Voice over Internet Protocol (VoIP) and chat, as well as music sharing and massively multiplayer games. Each of these applications has an application-specific model of the interactions among the user, the system, and other users.

This diversity of applications suggests that these communication models might differ substantially. These applications share the need for a set of communications protocols that deliver network services required by the applications. As our understanding of networks and the variety of services they can provide has improved with experience, the common elements among many application requirements have been exploited to build general-purpose network protocols. The general-purpose nature of such "one size fits all" protocols is extremely attractive but may sacrifice the ability to support specific applications effectively [1].

9.2 APPLICATIONS AND THEIR REQUIREMENTS

There is an increasing trend toward mobile applications. Familiar mobile telecommunications systems, such as cellular telephones, are supported by a substantial infrastructure of base stations supported in turn by a telephone system. Emergency search and rescue communications challenges may include lack of predeployed infrastructure, long latencies, and higher bit error rates than commercial software systems assume. Consider, for example, an emergency scenario (e.g., rescue of people imperiled by natural disasters such as typhoons or earthquakes), where a mobile satellite terminal is used to request data over a satellite communications channel that may have limited bandwidth and substantial propagation delay. If conventional wired Internet Protocol (IP) transport protocols, such as the Transmission Control Protocol (TCP) [2], are deployed end-to-end (i.e., between the mobile satellite terminal and an intercontinental data source), data throughput can be unacceptable to the application.

In this scenario, the application makes a relatively compact request for needed data, and the request is forwarded over the satellite channel. The data source responds with the data, often imagery such as a contour map. Such data can be substantial in size and will certainly require multiple packets to transport. The application requirements are reliable delivery *and* minimal latency. When TCP is used to obtain a reliable stream of bytes to support the reliable delivery requirement, it also provides a congestion control feature desired by the network operator, but not, in general, by an individual user. The TCP congestion control scheme interprets packet loss as indicative of congestion, and paces the number of packets in flight during a round-trip time (the "window size") according to its perception of whether the network is congested or not. Thus, even when capacity is available, the conventional protocol architecture cannot exploit it. While new protocol solutions, such as TIA-1039 [3], have been devised to address weaknesses of implicit congestion control architectures, there is considerable opportunity to improve the performance of applications if more is known about application requirements, *and* this knowledge can be brought to bear in the network [1].

9.2.1 Layering and Information Hiding

Important engineering design choices are embedded in conventional protocol architecture implementations. Conceptually, these protocol architectures can be represented as a vertical "stack" of layers. At the interface between each layer, each layer provides services to the next higher protocol layer in the stack. For example, a reliable transport system will require access to more fundamental services such as addressing and packet delivery.

The IP stack is often represented conceptually as an "hourglass" of layers, as in Figure 9.1, where the hourglass shape indicates that multiple logical "stacks" can exist, sharing a common "waist," the interoperability layer. For the Internet, the interoperability layer (here labeled IP) is used to coerce multiple incompatible subnets into a "virtual" network, known as the internetwork [2], the properties of which can then be assumed

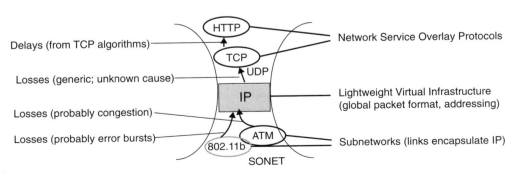

FIGURE 9.1

The IP "hourglass"—an interoperability solution.

to be present by end-to-end protocols. At a minimum, these properties are a common packet formatting and addressing convention, thus the need for new versions of the IP, such as IPv6, when these change.

The internetwork is overlaid on multiple subnetworks, and routes through the internetwork are used to provide end-to-end connectivity to protocols such as TCP, which provide an ordered reliable stream of bytes between applications. HyperText Transfer Protocol (HTTP) employs the TCP protocol.

This architectural model has consequences for network awareness. The layering model employed reflects the software engineering mindset of the time, "modular programming" [4] and "information hiding" [5]. The idea of modular programming is that more understandable, and therefore more robust, software can be written when the software structure is decomposed into modules of small size, with localized concerns (e.g., a certain class of operation) or a particular data object. Information hiding is a discipline that limits the flow of information between modules by limiting data sharing to a small number of carefully type-checked parameters.

In the case we are examining here, the concerns are localized in the *layers*, and the information-hiding discipline is enforced by the implementation, which passes information either in the data structure for representing the packets (e.g., the `mbuf` in UNIX implementations) or the limited additional information used in procedure calls between layers. It is clear that such a discipline enhances interoperability, a major goal achieved by the IP architecture. However, like most architectural choices, it represents a particular choice in a rich space of trade-offs, and as a consequence may introduce some undesirable properties.

As one example, we can consider the issue of packet loss. In many cases, if all the information available to the end host was employed, an intelligent statistical model for causes of packet loss could be developed. For example, the failure of link-layer checksums computed in the device driver for the particular line card employed to connect the host to the network would be indicative of error bursts on the link, while a recent trend of increasing delays in communication to a particular host, or set of hosts, might be indicative of congestion.

However, since the IP does not indicate why packets are lost (it could be burst errors, either causing whole packet loss or damage to a checksum, or it could be congestion), TCP must make an estimate. On commercial networks, the reasonable assumption as to the cause of loss is congestion. Thus, TCP slows down ("sends less," in the decision matrix of Figure 9.2) using a sophisticated algorithm that is clocked in units of round-trip times (RTTs). Since TCP's control algorithms are clocked in RTTs, TCP's "discovery" cycle is much slower in long-latency networks such as those that incorporate satellite communications links. If in fact the cause of loss is *not* congestion but some other cause, as is more common in mobile communications systems found in emergency and rescue scenarios (as discussed in Section 9.1), then communications havoc may ensue.

9.3 NETWORK AWARENESS: PROTOCOLS

The sections that follow discuss dynamic protocol composition, interaction features in dynamic protocol composition, and cognitive control.

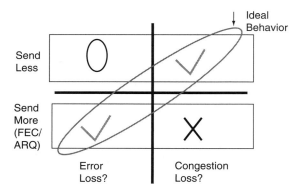

FIGURE 9.2

Packet loss causality versus flow control responses. The horizontal axis depicts hypothesized reasons for bit errors, while the vertical axis represents possible responses to the hypothesized reasons for the errors.

9.3.1 Dynamic Protocol Composition

One possible approach to overcoming certain limitations of information hiding as reflected in protocol architectures is *dynamic protocol composition*. Protocol architectures need not be limited to a "stack" [6], but this representation of protocol structure is convenient and it is clear how to generalize it in terms of more elaborate data and control flows. The idea is that a protocol stack can be constructed dynamically [7], in response to the application needs in terms of properties such as reliability, throughput, security, and so on. The construction of a stack could be done starting with no stack at all, or might be used to augment an existing stack. In such circumstances, a protocol architecture could be adjusted [8] by inserting a thin "shim" module into the protocol stack [9], which is compliant with the interfaces defined above and below it in the layered architecture. For example, an additional protocol function, such as automatic repeat reQuest (ARQ) or forward error correction (FEC), can be inserted beneath the IP layer to reduce the observed losses due to errors. The addition of such protocol functions reduces the probability of burst errors being observed by the IP layer. Thus, the assumption of the TCP layer that packet loss is because of congestion is made more probable, and the congestion control strategy is therefore much more likely to respond correctly to the packet loss events that it observes post-FEC.

Since the ARQ protocol seems useful, why not always employ it at the link layer? This is in fact done for certain types of links. For example, the 802.11 medium access control (MAC) layer employs an ARQ-like strategy to emulate the reliability of a wired local area network (LAN). Some satellite communications links employ performance-enhancing proxies (PEPs) at the link endpoints to provide a reliable substrate to end-to-end protocols. Such link protocols are not universally employed, however, due to performance consequences such as increased latency, because the 802.11 MAC can account for much more latency than the propagation delay. It should be clear, then, that an *adaptive* protocol architecture [8], which employs the error compensation mechanism "as-needed," is desirable. For example, we might learn that there has recently been an increase in the checksum failures on link-layer frames, and deploy ARQ or an FEC scheme. Over time, the wisdom of this decision can be balanced against the cost of the deployment, likely duration of error bursts, hysteresis models for the module removal decision, and so forth.

Structural adaptation of protocols [8, 10] is a powerful technique for improving application performance in spite of an extremely wide range of application requirements and network conditions, by adding and removing protocol elements in response to changes in the situation. While this problem might be solved in specific circumstances by event-driven adaptations such as the insertions of ARQ or FEC as we have discussed here, there is a wide variety of additional information that must be available to fully exploit structural adaptation in network protocols.

9.3.2 Feature Interaction in Dynamic Protocol Composition

Consider an application that requires data privacy and integrity. For such an application, it may be desirable to employ an encryption algorithm, which protects data privacy through a key-controlled data transformation, and as a consequence of this transformation, detects data tampering as a failure to successfully reverse the transformation at the receiver. Redundancy present in the application data and limited network throughput might suggest that a data-compression protocol be employed.

Consider the two possibilities for ordering compression and encryption protocol elements illustrated in Figure 9.3. In the compress–encrypt composition of two protocol elements, as might occur in the course of a structural adaptation process like the one discussed earlier, if compression is performed first, then data size is reduced by removing redundancy, and then an encryption transformation is performed to hide the bits of the compressed data. If, on the other hand, the encryption is performed first, then the structure exploited by a compression algorithm is obscured, meaning that no data size reduction will be accomplished in spite of the considerable computation typically required for protocol elements of this type. So the first composition is desirable and meets the goals of "compressed + encrypted," while the second composition meets only the "encrypted" goal.

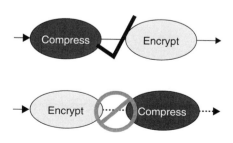

FIGURE 9.3

Protocol composition must be aware of protocol element interactions. Encryption of a data source renders the redundancy that enables compression invisible to the compression algorithm.

This example illustrates the point that several kinds of knowledge must be available in a general structural adaptation scheme. The first is knowledge of what external conditions the protocol element is responsive to, such as network conditions and application requirements. The second is knowledge of the assumptions inherent in particular protocol elements responsive to external conditions. This knowledge affects its suitability and role in a complete protocol architecture that results from a multiplicity of structural adaptations. In the past, the daunting complexity of managing the many possible adaptations and their possible interactions has inhibited adaptations for which simple robust models, such as the control laws employed for TCP/IP's adaptation to network congestion or the definition of a "spanning tree" [10], have not been devised.

9.3.3 Cognitive Control

Automating the management of complexity now seems possible. Automation does not seek to eliminate or obscure the complexity inherent in a system (as the information-hiding approach might be seen to). Rather, automation seeks to remove the detailed management of complexity from the purview of the application programmer or protocol designer. Since many protocol design issues are quite subtle, as has been exposed in this discussion, any such protocol automation strategy must have the property that many kinds of knowledge can be represented, and that empirical knowledge, such as observed performance, can be "fed back" into the knowledge employed by the automation strategy. This feedback seems to create protocol structures suitable for particular applications encountered in particular network conditions.

The opportunity for automation is considerable [1]. Figure 9.4 illustrates the possibilities for structural adaptations that closely couple application requirements to the encountered network conditions (which, of course, as in the packet loss example, may vary considerably with time). When the network conditions are very attractive, as might be the case on the interconnection network of a parallel computer or a set of hosts resident on a common LAN, very little protocol support may be necessary for typical desiderata such as high "goodput" with low latency. Where conditions degrade, more

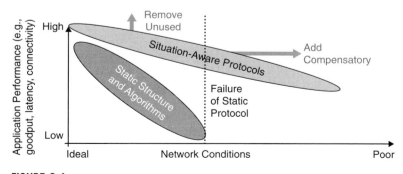

FIGURE 9.4

Situation-aware protocols in a conceptual trade space.

and more protocol mechanisms may be employed to adapt the application to the degraded network, with these compensatory mechanisms permitting the application to operate in domains unreachable by traditional architectures (the limitations of which are illustrated by the demarcation of the vertical dushed line in Figure 9.4).

A detailed example of the possibilities can be given using the TCP/IP congestion control algorithm's response to packet loss. Both simulations and laboratory experiments [11] have shown that this protocol's response to error conditions above a certain threshold (e.g., a bit error rate greater than 10^{-6}) causes the TCP/IP algorithm to close its congestion window, and consequently to reduce its performance to the level of a "stop-and-wait" protocol, the performance of which is extremely sensitive to the delay inherent in the link. If error compensation is inserted, the windowing strategy is made effective, and relatively high throughput can be achieved in operating regimes with high bit error rates. Hadzic and Smith [11], for example, have shown experimentally that a protocol stack incorporating TCP/IP and adaptive compensation can operate with bit error rates as large as 10^{-4}.

Cognitive systems are systems that apply human-style reasoning and capabilities. In the case of the structural adaptations we have discussed, the cognitive system would be required to first identify the combination of application requirements and network conditions, then select or devise an appropriate composition of protocol elements, and finally deploy that composition. Notions, such as similarity, trial and error, and improvement by learning appropriate responses over time, help make a cognitive system both less brittle and more evolvable than a conventional protocol design. It provides a clean *global* separation of policy and mechanism, which is otherwise only done locally in modules, if at all.

9.4 SITUATION-AWARE PROTOCOLS IN EDGE NETWORK TECHNOLOGIES

The clear utility of such an approach [1] in extreme networks (e.g., for disaster relief), particularly those at the logical "edge," suggested a research program [12] to investigate the effectiveness of cognitive approaches to rapid adaptive composition and adaptation of protocol structures.

The approach, situation-aware protocols in edge network technologies (SAPIENT), identified three core challenges that must be addressed in a working system:

1. Knowledge representation
2. Learning
3. Selection and composition

Presuming that these challenges are addressed, a conceptual architecture for a SAPIENT system might be structured as in Figure 9.5. In the figure, an application that employs the network to carry out its intended task is shown in the upper left.

The requirements of the application might be characterized explicitly by a designer or user; see, for example, Nahrstedt and Smith's OMEGA [13] system that manages quality-of-service (QoS) trade-offs. Although it is more attractive, of course, if the

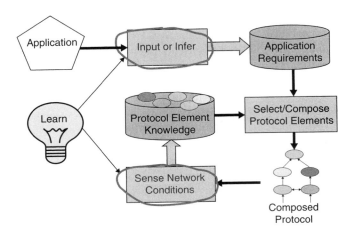

FIGURE 9.5

One possible structure for a situation-aware protocol system.

SAPIENT system deduces or infers such requirements automatically. Automatic inference might be done by technologies, such as statistical machine learning, operating on information derived from the packet stream emitted by the application.

As a simple example of automatic inference, consider observing packet size and a timestamp. A relatively constant packet size and interarrival delay might indicate the packets are part of a media stream and the protocol support might be adjusted to support this inference. On the other hand, great variability in timing, small size of transmitted packets, and highly variable sizes in received packets might indicate a transactional application.

The goal is to produce a knowledge base of application requirements, combined with other knowledge about network conditions and available protocol elements, to produce an effective adaptive system. This protocol element knowledge, illustrated at the center of the diagram in Figure 9.5, is used to react to network conditions. It is brought together with the application information in the process of selecting and composing protocol elements, which results in the composed protocol illustrated at the lower right of the figure.

There are many opportunities for learning. Learning to automatically identify applications and their requirements has already been discussed, but the sensing of network conditions (losses, timing, load, etc.) is central to understanding whether a new protocol is needed and, if so, which network conditions it must address.

Sensing of network conditions can involve a variety of sensors, including those that monitor packet checksums, packet ordering, packet sequencing, packet loss, congestion indications, latencies, and relative application usage of link capacity, as well as alerts such as might occur on the loss of a link or a dramatic reduction in signal strength on a wireless link. Sophisticated sensors might include hysteresis or damping algorithms to compensate for stochastic "noise" that might otherwise induce unwanted responses. Input from multiple low-level sensors might be fused to gain a better overall estimate of the state of the network; such sensory information could include information obtained through active probes as well as from remote systems, whether or not they employ a SAPIENT architecture.

The sensing of network conditions is a central element of a SAPIENT system's "situation awareness," since it provides the measurement basis for maintaining the current state or pursuing an adaptation. When the system is initialized, it might start using a composition equivalent to TCP/IP as a baseline, and then modifying it while a positive performance gradient is realized. If a protocol composition is not working, then the network-sensing process should deduce this and stimulate production of a more appropriate protocol. When a performance "plateau" is achieved, the SAPIENT system would preserve the status quo until a significant change in network conditions occurred.

Over time, the protocol element knowledge base incorporates this sensed information in its learned responses to detected network conditions. Here, statistical machine learning can prove powerful, for example, in classifying situations in which the operating protocol should be left untouched and when it should be adjusted. One property possible in such a system are the ability to recover from protocol elements with errors, presuming that the sensed data are correct and properly interpreted.

The use of these varieties of knowledge in creating situation awareness, resulting in application-specialized protocols, offers a new way to support networked applications.

9.5 NETWORK AWARENESS: NODE CAPABILITIES AND COOPERATION

A radio device interacts with the physical world by transmitting and receiving energy in radio frequency (RF) bands. The free-space propagation characteristics of radio waves have made RF the basis of most wireless communication. Information is encoded in a radio channel using the many properties associated with a wave, such as amplitude, frequency, and phase. Digital transmission systems [14, 15] seek to modulate and demodulate discrete quantities using these properties and sophisticated coding schemes, and a variety of protocols (e.g., packet radio and TCP/IP) can then be overlaid on the digital channel.

The performance of the channel can be effectively characterized [16] by a signal-to-noise ratio (SNR), which indicates how much information capacity the channel has available to exploit. The coding scheme is the means by which the SNR is exploited; it is affected by a variety of factors such as transmit power, receiver sensitivity, distance, and the propagation environment [17, 18]. If sufficient SNR is available, then a link can be established between a transmitter and a receiver.

Figure 9.6 gives a map of grid squares indicating the received power levels at various points in a two-dimensional (2D) physical space. Red (hot, *black*) indicates very good received power (making it easy to close a communications link), and blue/violet (cold, *grays*) indicate very poor power levels (making it difficult to close a communications link). At whatever scale these grid squares represent, motion (e.g., of a group of users) will affect the sample of grid squares experienced by the group of users.

Generally, digital communications systems welcome a high SNR, can adapt within limits to a degradation in SNR (e.g., by increasing transmit power, increasing the protec-

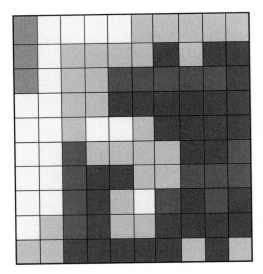

FIGURE 9.6

RF received signal strength–power grid illustration.

tion of error correction codes), and have a lower limit beneath which they cannot function. When the SNR drops below this "floor," the transmitter and receiver are unable to communicate. When the intervals of low SNR are very short, they can be viewed as error bursts. For somewhat longer intervals, rapid retransmission may prove effective. The upper limit on acceptable intervals is usually defined by user annoyance.

The propagation environment is increasingly complex. This is a function of the increasingly many devices exploiting the spectrum. For example, there are many devices operating in the 2.4 GHz industrial scientific and medical (ISM) band, as well as complex physical surroundings (e.g., inside buildings [19, 20], non-line of sight, urban [21, 22]) in which the devices are used. For example, in an urban environment, the many reflective surfaces can lead to multipath phenomena, where many copies of the transmitted signal, taking different paths through the propagation environment, can arrive at a receiver. Sophisticated techniques, such as the multiple input multiple output (MIMO) [22] space-time adaptive processing concept, can be used to interpret the combined energy or even create more diversity that can be exploited. As conventional rake fingers have a fixed spacing, the implementation of MIMO tends to be within a constrained range of frequencies to which this spacing is well suited. The idea of exploiting diversity, however, is a key motivating factor for the radio team.

9.6 A DISTRIBUTED SYSTEM OF RADIOS—THE RADIO TEAM

The increasing capability of mobile devices is accompanied by, and often enabled by, an increase in *programmability*. The most aggressive vision for programmability, the software radio, was proposed by Mitola [23], and has now been realized in research efforts, military radios, commercial radios, and even open-source software radios [24]. The software radio can be viewed as a polymorphic input/output (I/O) device. This has many advantages in a radio-rich environment, where the user's needs vary with time and circumstances, and the behavior of the radio system can vary to meet these needs. If the radio's adaptation is automated and spectrum aware, a cognitive radio (CR) [12, 25] may result; while CR does not require software radio, the polymorphic property is a powerful attribute for cognitive radios. For users, the ability to carry a Swiss Army Knife-like device that meets many needs in a single package has many obvious advantages in total size, weight, and power (SWaP) in spite of the extra complexity of a software radio. However, software provides another advantage because implementing the functionality in software makes the system more adaptive, like clay, with its late-binding functionality.

While many radio functions historically implemented in hardware have been realized in software radios, the computational capability of these devices has rarely been viewed as a resource to be exploited apart from its role in the realization of RF waveforms. Viewing software radios as computers with an extremely powerful capability to control their I/O (waveforms) can lead to new radio capabilities. The *distributed computing* model, where computers are interconnected to achieve resource sharing, has proven to be an extremely powerful (if still not fully realized) vision. Examples include networked storage, cluster computing, and application-specific server farms providing Web services. Illustrated in part by these examples, some distributed computing advantages include:

- The use of many simple elements to build a powerful aggregate (clusters)
- Resource sharing, including sharing of specialized resources (storage, I/O)
- Decentralized locations, able to exploit advantages of locality (caches)

The following are the three main technical challenges in realizing the radio team's distributed radio architecture:

1. Maintaining a collaborative channel with uncertainty in communication
2. Distributing work within the team
3. Distributed learning of the environment and effective adaptations

9.6.1 Maintaining a Collaborative Channel

The fundamental goal of a communications system is to move information. In the radio team's model there are two key forms of communication. First, there is communication with elements outside the team, which without the team structure would be attempted by individual devices acting on their own, rather than in concert. Then, there are inter-node communications that support communication of team members with elements outside the team. These elements, as examples, might include communications satellites, GPS satellites, 802.11 access points, or cellular base stations.

This model is unabashedly adapted from that of the Ethernet and its uses in larger networks such as those constructed with transparent bridges and IP routers. There is some irony in the fact that Ethernet carrier sense multiple access with collision-detection (CSMA/CD) MAC is modeled on Abramson's ALOHA protocol, and radio teams model their local radio communication on a broadcast model. The broadcast channel, even if its availability is only probabilistic, is an extremely powerful mechanism for collaboration, since most participants in a global exchange are listeners.

The members of a radio team can use the collaboration channel to improve their ability to collaborate. For example, they can choose frequencies that are more effective in a given setting at overcoming phenomena such as fast fades, choosing codes to suppress the effects of interference, and choosing timers for MAC events to reflect actual distances rather than worst-case distances. Aggressive implementations might even dynamically adapt the protocol in use, as in the SAPIENT concept discussed in Section 9.2.

9.6.2 Distributing Work within the Team

One of the most powerful conceptual advantages of teamwork is that the strengths of one team member can be used to compensate for the weaknesses of another team member [26]. This same advantage is an important goal for radio teams and requires three things: (1) knowledge of each team member's capabilities, (2) knowledge of the assigned tasks for the team, and (3) knowledge of the radio environment.

Mitola's CR [25] is environmentally aware, and in the most ambitious version of his vision may automatically determine many of its tasks, assigned or unassigned. More conventional tasking can still be quite powerful, as it might include concrete objectives (e.g., applying cognition to maintain communications with a specific server, or in a military radio setting, communication with higher echelons).

9.6.3 Distributed Learning of the Environment

Learning is best measured by its effects, with improvements in performance over time the most pertinent to engineers. To build a learning system, it must first be determined what is to be learned, and then how it is to be learned. Then the system must be designed so that the learning can be applied to one or more tasks.

With a CR, the sensor $\rightarrow$ computer $\rightarrow$ actuator system is particularly interesting in that the major sensor is a radio receiver, and the major actuator is a radio transmitter. The system can also be fruitfully viewed as input $\rightarrow$ decision $\rightarrow$ output, where the decision making performed by the computer system can be extremely complex.

Several system architectures have been explored to test these ideas [24, 27] and have shown significant improvements in continuous connectivity.

9.7 NETWORK AWARENESS: NODE LOCATION AND COGNITION FOR SELF-PLACEMENT

The combination of many uses for RF, many people, and many objects that affect RF signals [17, 19, 20] make cities [21, 22] extremely challenging environments for communications systems. Building materials create attenuation losses and, at the same time, may introduce geometric phenomena such as multipath through their surface reflectivity. These losses are affected by both position and frequency and can have a severe effect on closing the link in a communications system. We can quantify the gain associated with a particular motion, known as *mobility gain* [28].

Urban cell phone users are familiar with this complexity and the process of trying to find a good location to carry out their communications. A typical search method is to move and observe signal strength indications such as the number of "bars" in the cell phone display. Once an acceptable location is found so that a call can be placed or received, the user stays in this position and the "search" terminates.

Automation of this process (e.g., by a robot) would consist of a search strategy, a search goal, and a termination criterion. Example strategies might include random walks, bounded linear motions, spirals, and so on. A strategy can be evaluated by the amount of motion required to locate the "goal" of improved mobility gain. The search will be bounded by the travel required to apply the search strategy in a constrained area. The distance traveled in the strategy can also serve as a bound, in which case the efficiency with which the search of the volume can be achieved is important.

Integration of the radio system and the robot into a single package provides an engineering trade space at least comprising size, weight, cost, battery lifetime, RF power, and antenna capability. For example, some robots may be attractive from a robustness and packaging standpoint but may be severely limited in their ability to achieve acceptable antenna height (see, for example, the RHex [29, 30] bio-inspired robot).

Robot capabilities can dictate the motion strategies affecting mobility gain-based searches and optimizations. Low antenna heights will demand some combination of high power, novel antenna designs (e.g., self-raising antennae), and/or high node-deployment density. Another possible approach is robots, such as RISE [30, 31], that can climb surfaces. Climbing gives extremely attractive antenna heights and an additional dimension in which to search for desirable locations. Climbers may demand considerable power to climb the surface, have limited weight-carrying capacity, and limit mobility gains achievable through horizontal motions to those that can be attained along a single planar surface such as a wall (which may also induce RF interactions).

Communication among the robots provides the opportunity for optimizing the search strategy using the communications channel both as a means of sharing knowledge and as a sensor. That is, the value of the received signal strength informs the search strategy's optimization and termination. The shared knowledge among cooperating nodes is used to prevent a solitary node from improving its mobility gain at the expense of other nodes that comprise an urban mesh network. The distributed control algorithms necessary to build a scalable robotic mesh of nodes with a meshwide improvement in mobility gain are a major research challenge. To illustrate this, consider two nodes concurrently seeking the highest mobility gain. If they are completely independent, then they can proceed independently. However, as soon as they interact (i.e., where they are near elements in a potential urban mesh), then each move they make has an impact on the neighboring node [32]. Thus, the global optimization of the mesh reliability and performance as an ad hoc infrastructure may demand at least local cooperation, where a node uses both its own RF-sensing capability and communication achieved as a result of its success in searching for good locations to optimize the RF mesh for the common good.

On a large scale, this optimization may require economic algorithms or other schemes for large-scale distributed control that communicate via an agreed-on mechanism such as resource pricing. A key question is the pricing mechanism. In a robotic repeater-based ad hoc mesh infrastructure, battery power would be a fine candidate.

9.8 SUMMARY

This chapter has presented three forms of network awareness: protocols running at the network edge, other nodes to form teams, and location for node placement decisions. Cognition used in such network-aware manners will create new classes of evolvable communications systems. As new protocol elements and knowledge about them expand, productive group behaviors, goals, and applications for robotic nodes can be inserted directly into the system. This would permit data from "experienced" cognitive networking systems to be loaded into a newly created system, in essence sharing the learned knowledge. The basic learning structure illustrated in Figure 9.5 permits the cognitive networking system, after an initial load, to continue learning as it encounters novel combinations of network situations local to its operating environment.

Network awareness and cognitive approaches to managing complexity represent a pathfinder to new methods of building robust systems that can operate successfully in extreme environments and unforeseen conditions, such as those faced by military tactical and strategic communications systems on a daily basis.

9.9 EXERCISES

9.1 The cognitive approach to networking outlined in the first part of this chapter has at least implicitly presumed that the cognitive system always converges successfully. What factors (referring to Figure 9.5) impact convergence? Which strategies could be employed to make this approach more fault-tolerant?

9.2 Cognitive approaches may address complexity in network systems with simple general strategies and a fact base. However, as sensed data and learned responses proliferate, the system may itself become harder to manage. Which strategies could address scaling issues in cognitive systems design, or challenges in the scale of systems in which cognitive approaches might be applied?

9.3 The radio team's concept appears effective at small scales (e.g., for 5–10 mobile nodes). Propose a methodology where, for a set of communications problems, the effectiveness of the technique could be evaluated. Is there a performance/scalability trade-off as a broadcast channel becomes harder to maintain, due to distance and the number of nodes involved? Could an autonomous cognitive process determine when the threshold in performance/scalability is reached and "split" a radio team into two teams?

9.4 What level of application awareness is necessary for application-specific optimizations? Can these optimizations be composed into a system that works well overall?

REFERENCES

[1] Smith, J. M., Application-Private Networks, *Computer Systems: Theory, Technology and Applications: A Tribute to Roger Needham*, A. Herbert and K. Sparck Jones (eds.), pp. 273–278, Springer-Verlag, 2004.

[2] Cerf, V. G., and R. E. Kahn, A Protocol for Packet Network Intercommunication, *IEEE Transactions on Communications*, pp. 637–648, May 1974.

[3] QoS Signaling for IP QoS Support, Telecommunications Industry Association Standard TIA-1039, August 28, 2006.

[4] Brooks, F. P., Jr., *The Mythical Man–Month: Essays on Software Engineering (Anniversary Edition)*, Addison-Wesley, 1995.

[5] Parnas, D. L., On the Criteria to Be Used in Decomposing Systems into Modules, *Communications of the Association for Computing Machinery*, December 1972.

[6] Hutchinson, N. C., and L. L. Peterson, The x-Kernel: An Architecture for Implementing Network Protocols, *IEEE Transactions on Software Engineering*, 1991.

[7] Ritchie, D. M., A Stream Input/Output System, *AT&T Bell Laboratories Technical Journal*, 1984.

[8] Feldmeier, D. C., A. J. McAuley, J. M. Smith, D. S. Bakin, W. S. Marcus, and T. M. Raleigh, Protocol Boosters, *IEEE Journal on Selected Areas in Communications*, 16(3):437–444, 1998.

[9] Mallet, A., J. D. Chung, and J. M. Smith, Operating Systems Support for Protocol Boosters, *Proceedings HIPPARCH Workshop*, June 1997.

[10] Alexander, D. S., M. S. Shaw, S. M. Nettles, and J. M. Smith, Active Bridging, *Proceedings ACM SIGCOMM*, pp. 101–111, Cannes, 1997.

[11] Hadzic, I., and J. M. Smith, Balancing Performance and Flexibility with Hardware Support for Network Architectures, *ACM Transactions on Computer Systems*, 21(4):375–411, 2003.

[12] Fette, B., *Cognitive Radio Technology*, Newnes/Elsevier, 2006.

[13] Nahrstedt, K., and J. M. Smith, Design, Implementations and Experiences of the OMEGA Endpoint Architecture, *IEEE Journal Sel. Areas Communications*, 14(7):1263–1279, 1996.

[14] Sklar, Bernard, *Digital Communications: Fundamentals and Applications*, Second Edition, Prentice Hall, 2001.

[15] Wozencraft, J. M., and I. M. Jacobs, *Principles of Communication Engineering*, Waveland Press, 1965.

[16] Shannon, C., A Mathematical Theory of Communication, *Bell System Technical Journal*, 27:379–423, 623–65, 1948.

[17] Hata, M., Empirical Formulae for Propagation Loss in Land Mobile Radio Services, *IEEE Transactions on Vehicular Technology*, Vol. VT-29(3):317–325, 1980.

[18] Jakes, W. C., *Microwave Mobile Communications*, Wiley, 1975.

[19] Walker, E. H., Penetration of Radio Signals into Buildings in the Cellular Radio Environment, *Bell System Technical Journal*, 62(9):2719–2734, 1983.

[20] Savov, S. V., and M. H. Herben, Modal Transmission-line Modeling of Propagation of Plane Radiowaves through Multilayer Periodic Building Structures, *IEEE Transactions Antennas and Propagation*, 51(9):2244–2251, 2003.

[21] Seidel, S. Y., et al., Path Loss, Scattering and Multipath Delay Statistics in Four European Cities for Digital Cellular and Microcellular Radiotelephone, *IEEE Transactions on Vehicular Technology*, 40(4):721–730, 1991.

[22] Chizhik, D., J. Ling, P. W. Wolniansky, Reinaldo, A. Valenzuela, N. Costa, and K. Huber, Multiple-Input–Multiple-Output Measurements and Modeling in Manhattan, *IEEE Journal on Selected Areas in Communications*, 21(3):321–331, 2003.

[23] Mitola, J. III, *Software Radio Architecture*, Wiley-Interscience, 2000.

[24] Troxel, G. D., E. Blossom, S. Boswell, A. Caro, et. al, Adaptive Dynamic Radio Open-source Intelligent Team (ADROIT): Cognitively Controlled Collaboration among SDR Nodes, *Proceedings First IEEE Workshop on Networking Technologies for Software Defined Radio Networks*, Reston, VA (held in conjunction with IEEE SECON 2006: Third Annual IEEE Communications Society Conference on Sensor, Mesh and Ad-Hoc Communications and Networks), Invited Paper, September 2006.

[25] Mitola, J. III, *Cognitive Radio Architecture*, Wiley, 2006.

[26] Fette, B. A., J. R. Miller, P. A. D'Antonio, M. L. Wormley, and J. Huie, Definable Radio and Method of Operating a Wireless Network of Same, US PTO, Appl. 20040264403, December 2004.

[27] Lau, R., S. Demers, Y.-L., B. Siegell, E. Vollset, K. Birman, R. van Renesse, H. Shrobe, J. Bachrach, and L. Foster. Cognitive Adaptive Radio Teams, *Proceedings Second International Workshop on Wireless Ad-hoc and Sensor Networks*, New York, June 2006.

[28] Smith, J. M., M. P. Olivieri, A. Lackpour, and N. Hinnerschitz, RF Mobility Gain: Concept, Measurement Campaign and Exploitation, *IEEE Wireless Communications Magazine* (Special Issue on Wireless Communications in Networked Robotics), February 2009.

[29] Saranli, U., M. Buehler, and D. E. Koditschek, RHex: A Simple and Highly Mobile Hexapod Robot, *International Journal of Robotics Research*, 20:616, 2001.

[30] Boston Dynamics Web site (*www.bostondynamics.com/content/sec.php?section=robotics*), accessed May 2008.

[31] Spenko, M. J. et al., Biologically Inspired Climbing with a Hexapedal Robot, *Journal of Field Robotics*, 25:223–242, 2008.

[32] Sherrington, D., and S. Kirkpatrick, Solvable Model of a Spin-Glass, *Physical Review Letters*, 35(26):1792–1796, 1975.

Cognitive Services for the User

**Joseph P. Campbell, William M. Campbell,
Alan V. McCree, Clifford J. Weinstein**
MIT Lincoln Laboratory, Lexington, Massachusetts

Scott M. Lewandowski
The Wynstone Group, Inc., San Antonio, Texas

10.1 INTRODUCTION

Software-defined cognitive radios (CRs) use voice as a primary input/output (I/O) modality and are expected to have substantial computational resources capable of supporting advanced speech- and audio-processing applications. This chapter extends previous work on speech applications (e.g., [1]) to cognitive services that enhance military mission capability by capitalizing on automatic processes, such as speech information extraction and understanding the environment. Such capabilities go beyond interaction with the intended user of the software-defined radio (SDR)—they extend to speech and audio applications that can be applied to information that has been extracted from voice and acoustic noise gathered from other users and entities in the environment. For example, in a military environment, situational awareness and understanding could be enhanced by informing users based on processing voice and noise from both friendly and hostile forces operating in a given battle space. This chapter provides a survey of a number of speech- and audio-processing technologies and their potential applications to CR, including:

- A description of the technology and its current state of practice.
- An explanation of how the technology is currently being applied, or could be applied, to CR.
- Descriptions and concepts of operations for how the technology can be applied to benefit users of CRs.
- A description of relevant future research directions for both the speech and audio technologies and their applications to CR.

Note: This work was sponsored by the Department of Defense under Air Force contract FA8721-05-C-0002. Opinions, interpretations, conclusions, and recommendations are those of the authors and are not necessarily endorsed by the US government.

A pictorial overview of many of the core technologies with some applications presented in the following sections is shown in Figure 10.1. Also shown are some overlapping components between the technologies. For example, Gaussian mixture models (GMMs) and support vector machines (SVMs) are used in both speaker and language-recognition technologies [2]. These technologies and components are described in further detail in the following sections.

Speech and concierge cognitive services and their corresponding applications are covered in the following sections. The services covered include speaker recognition, language identification (LID), text-to-speech (TTS) conversion, speech-to-text (STT) conversion, machine translation (MT), background noise suppression, speech coding, speaker characterization, noise management, noise characterization, and concierge services. These technologies and their potential applications to CR are discussed at varying levels of detail commensurate with their innovation and utility.

10.2 SPEECH AND LANGUAGE PROCESSING

The following speech- and language-processing technologies begin with the acoustic speech signal collected from a single microphone (multisensor collection using a variety of sensors is also shown). All the speech and language technologies described here can be viewed as being abstractly related to the anatomy and dynamics of the vocal apparatus and the behaviors expressed via speech and language and/or viewed as statistical modeling methods.

10.2.1 Speaker Recognition

Speaker recognition can enable a CR to authenticate users for access control, identify communicating parties, personalize the device, and adapt the device and its applications to individuals. Speaker-recognition technologies allow systems to automatically determine who is talking or, to be precise, whether the incoming voice compares favorably with an enrolled user's voice. This determination can then be used to provide user authentication for access control, identification of communicating parties, and personalization and adaptation of the radio and its applications. Figure 10.2 shows the basic operations of a speaker-recognition system and its two phases of operation: enrollment and verification.

Enrollment and Verification

In the enrollment phase, voice samples from the subject are used to create a model or template, which is sometimes improperly referred to as a voiceprint, for the specific speaker. In the verification phase, the unknown voice is compared with the model of the claimed identity. A verification decision can be made to accept or reject the identity claim.

Speaker recognition is imperfect and is characterized by two types of errors: false alarm (FA—meaning false recognition) and miss (failure to recognize the claimed enrolled speaker). These systems are characterized by whether the speech they use is text dependent (e.g., phrase prompted or pass phrases) or text independent (e.g., conversational speech). The performance of these systems is quantified by a plot of the

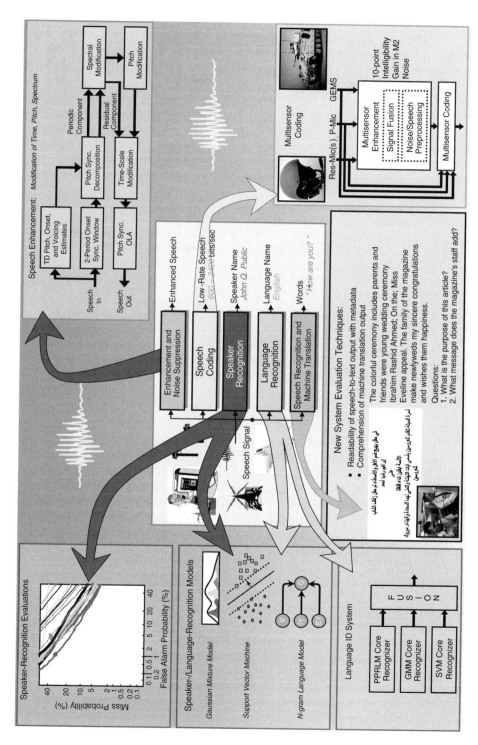

FIGURE 10.1

Speech and language technologies. A wide variety of speech and language technologies (e.g., speaker and language recognition and machine translation) can be used to enable cognitive-like services for SDRs. In addition, multisensor speech coding and enhancement technologies can aid users, especially in harsh noise environments.

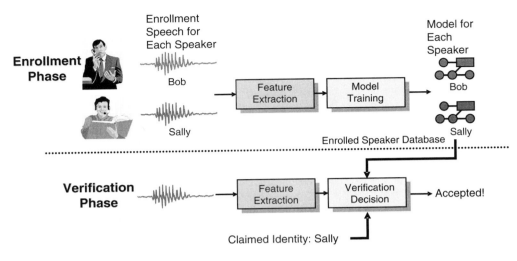

FIGURE 10.2

Speaker recognition. Speaker identification is based on extracting features from speech that best characterize the anatomical and behavioral differences of each individual when compared to the distributions of those features among the entire speaker population. Those differences are used to identify speakers or verify the claim of a speaker's identity.

miss rate versus the false alarm rate, which vary as a parametric trade-off of a control threshold selected by the system designer called the operating point. The operating point is adjusted by varying an acceptance threshold that shifts trade-offs between the two error types up or down the curve of a given system. Often, a combined measure is cited to provide an approximate representation of overall system accuracy; this measure, known as the equal-error rate (EER), indicates the operating point at which the miss and false alarm rates are equal. The state-of-the-art text-independent speaker-recognition performance for conversational telephone speech of a few minutes in duration is in the range of 7 to 12 percent EER [3]. Given an extended duration of enrollment speech, even better performance can be achieved, as shown in Figure 10.3.[1]

As introduced in Campbell et al. [4], voice biometrics[2] are well suited to radios that already incorporate microphones and speech coders[3] (necessary for encrypted voice communications). Additionally, voice biometrics can be combined with other biometrics (e.g., face recognition as shown in Figure 10.4) for increased security and/or backup

---

[1] *Explanation of terms in figure:* slope = the rate of change of a feature usually with respect to time (e.g., the rate of change of Cepstrum (Cep) coefficient *n*); Gaussian Mixture Model (GMM) = a multimodal distribution resulting from accumulating the distribution properties of sounds uttered by many speakers; Support Vector Machine (SVM) = a hyperplane providing the maximum separation between two data sets in a pattern-recognition application.

[2] *Biometrics* is the automated recognition of individuals based on behavioral and biological characteristics.

[3] Low-rate speech coding vocoders parameterize the speech signal for transmission. This parameterization can be shared with speaker-recognition engines, typically with some loss in performance, to conserve processing.

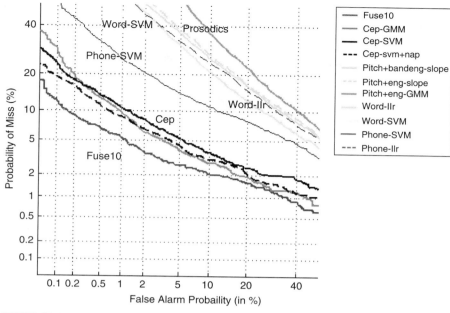

FIGURE 10.3

Performance curves of speaker-recognition systems. Shown here are the performance trade-offs between probability of miss and probability of false alarm of 10 algorithms and their fusion. The speaker models were trained on approximately 20 minutes of speech and tested on about 2 minutes of speech. *Note:* eng = energy; llr = log likelihood ratio; nap = nuisance attribable projection; phone = different speech sounds or phones.

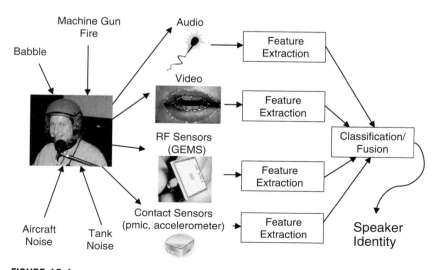

FIGURE 10.4

Speaker identification combined with facial and other biometrics. This combination enhances the accuracy, reliability, and survivability of the biometric system. Additionally, the suite of sensors shown enables robust low-rate speech coding in extremely harsh military noise environments [5].

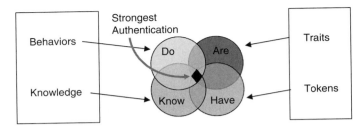

FIGURE 10.5

The four pillars of user authentication. Recognizing people based on what they do, what they know, who they are, and/or what they have. Using multiple attributes can increase reliability and survivability. Using all four attributes can provide the strongest authentication [4].

operating modes (e.g., face recognition could be more reliable than speaker recognition in high-noise environments or vice versa in adverse lighting conditions). Given the proliferation and improving quality of cameras in cellular phones, CRs are also likely to have cameras that might be suitable for face recognition.

Voice biometrics can provide access control via biometric logins and screen locks. This includes guarding against unauthorized use of lost CRs, such as disabling an idle radio that has been left behind. Voice biometrics can also enable user conveniences, such as recalling preferences and adapting to users (as do modern operating systems and application software). Conventional biometrics are generalized here to incorporate an additional authentication factor by learning and recognizing the users' distinctive behaviors. Future research directions for speaker recognition focus on making it more robust to mismatched channel conditions and applying high-level features that resemble those used by humans [6–8].

User Authentication

This section presents various means of user authentication, introduces generalized biometrics, and illustrates continuous and confidence-based user authentication. As shown in Figure 10.5, the four pillars of user authentication are *knowledge* (e.g., PIN or password), *tokens* (e.g., key or badge), *behaviors* (e.g., usage patterns or outcomes), and *traits* (e.g., voice or fingerprint). The proper combination of all four pillars provides the strongest user authentication. Biometrics are used to authenticate users,[4] as opposed to authenticating something they know (e.g., a password, which can be forgotten or compromised) or possess (e.g., an identification card, which can be lost or stolen). Unlike knowledge- and token-based authenticators, however, the inability of users to transfer biometrics can lead to difficulties; as in, for example, an emergency transfer of operation of a radio with biometric access control to an unenrolled user. To solve this difficulty, knowledge- and token-based authenticators can be used to authenticate users in these situations.

---

[4] Strictly speaking, biometric verification is a binary hypothesis test. Here, the hypothesis is that the live voice matches an enrolled and stored voice model. The biometric system decides to accept or reject this hypothesis.

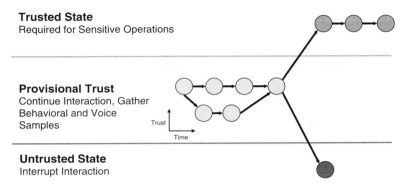

Trusted State
Required for Sensitive Operations

Provisional Trust
Continue Interaction, Gather
Behavioral and Voice
Samples

Untrusted State
Interrupt Interaction

FIGURE 10.6

Continuous user authentication and trust. Varying levels of trust assessed over time via continuous authentication of a user are represented as state transitions. With sufficiently long samples of speech, the system is generally able to make a confident decision about the speaker's identity and therefore whether to trust the user [4].

Popular biometrics include voice, face, fingerprint, and iris.[5] Voice and face biometrics (possibly in combination) are well suited to radios that already incorporate microphones and cameras. Some biometrics lend themselves to continuous user authentication (e.g., to guard against lost or stolen radios being misused) and allow a system to assess varying levels of trust. For example, voice verification can be used to continuously authenticate a user while the user is talking; this can be useful if the voice quality makes it difficult for the interlocutor to detect a change in talkers. Figure 10.6 shows an example of an authentication process over time with varying levels of trust [4]. This example begins in a state of provisional trust and, over time, proceeds in continued states of provisional trust and then to a trusted or untrusted state. While in a state of provisional trust, benign operations (e.g., adjusting radio volume) can be performed, whereas sensitive operations (e.g., downloading an SDR waveform) require a trusted state.

Behavior-based user authentication recognizes users via their actions, interests, tendencies, preferences, and other patterns. Examples of distinctive behaviors include:

1. *How* a user does something (e.g., speed and pattern of typing, stylus angle and intensity, use of menus versus keyboard shortcuts).
2. What a user *does* (e.g., patterns of applications use, program features used, patterns of collaboration).
3. What a user *causes to happen* (e.g., sequences of system calls, patterns of resource access).

These behaviors include not only a user's local actions, but also network interactions and outcomes. Behavior-based user authentication, like voice verification, has minimal

---

[5] See *www.biometrics.org* for additional biometrics.

adverse impact on a user's normal activities. The authentication is inherent and transparent; there is continuous mode operation and modest resource utilization, and user acceptance is likely to be high.

Monitoring these behaviors can be combined with situational awareness to fuse multiple factors into the authentication process. A cognitive approach allows for many interesting possibilities. First, the threshold to reach the trusted state of user authentication can be adapted based on situational, environmental, and mission awareness and the risk of the requested operation (e.g., ranging from benign volume adjustment to sensitive security operations). Second, authentication can be performed over time by combining available information—voice communication, mouse/stylus movement, dialog structure, and so on.

Some issues and questions in biometric deployments involve:

- Whether to use remote versus distributed versus network enrollment and verification.
- Where user models are created and stored.
- How models are maintained and updated.
- How enrollment is conducted.
- How models are bound to users.
- What the tolerable verification time or rate is.
- How models of new users are distributed and their integrity assured.
- Whether there are accuracy or policy requirements.
- What the architecture to support the biometrics is.

Biometric Sensors

There are many approaches to biometric-based user authentication and all require some form of hardware input device to gather the required information about the user to be authenticated. For example, fingerprint recognition requires a fingerprint scanner, user voice recognition requires a microphone, and user behavior monitoring requires a traditional user input device (e.g., a keyboard or mouse). These hardware devices must be an integral part of the CR platform and must communicate with their software counterparts over a secure channel. Many of these devices are high bandwidth, although the utilization is often in bursts. The channel connecting the hardware and software must be capable of supporting the data transfer requirements without an undue performance impact on the device's core functionality.

Security Architecture with Biometric Processing

Once data have been gathered from a biometric sensor, they must be processed to determine user identity (or other user characteristics that the sensor has been designed to assess). Such processing can occur either in software, as shown in Figure 10.7, or in specialized hardware. Although the use of specialized hardware provides the advantages of increased tamper-proofing and higher performance, the complexity of managing updates and modifications to the functionality of the hardware often outweighs these benefits. Implementing the biometric processor in software does not fundamentally diminish the overall security of the system, given that the overall security of the radios

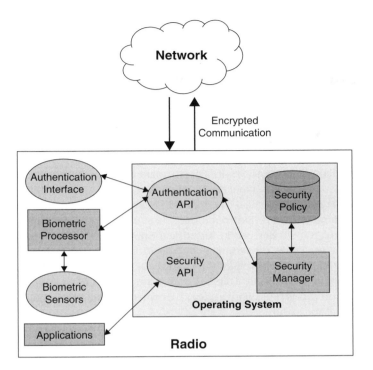

FIGURE 10.7

Notional radio security architecture supports continuous biometric processing and trust assessment over time for CR security [4].

(and the network services provided by radios) is also built on software components. Although not shown in Figure 10.7, most biometric processors require access to a database that contains information required to authenticate users, such as biometric models or templates, profiles, or logs of past behavior.

Applications

Although the platform has been designed to accommodate existing applications without modification, new applications can be designed to leverage the capabilities exposed by the security application programming interface (API) to improve the user experience and to improve overall system performance. Designing applications to leverage this architecture is critical to its success [4].

Legacy applications can automatically benefit from continuous authentication, but will be completely unaware of confidence-based authentication. Therefore, the platform will need to define a confidence level at which the user is considered authenticated, and any level of confidence below that will cause the user to be considered completely unauthenticated by legacy applications. Applications that are aware of confidence-based authentication, in contrast, can enable functionality or access to data based on the confidence of the user's identity.

10.2.2 Language Identification

Language identification (LID) can enable a CR to infer whether its user or a remote communicator is a friend, a neutral party, or a foe based on identifying the language transmitted over a radio channel. LID technologies allow systems to automatically determine the language of the user from a list of possibilities. These technologies are available for more than a dozen languages. These systems usually require about 30 seconds of speech to obtain good spoken language identification performance.

Methods for spoken language recognition have traditionally been based on phonetic transcription of different languages [9]. By discovering the relation between occurrences of phones[6] in different languages (i.e., phonotactics[7]), a statistical model can be constructed of a particular language for later use in identification.

An emerging class of recent methods for language recognition is based on novel features [10]. One of these new features—shifted-delta cepstral coefficients[8]—measures changes in the speech spectrum over multiple frames[9] of speech to model long-term language characteristics. These methods need only a speech corpus labeled by language for training in order to achieve good results.

As mentioned previously, current system performance [10] is measured in terms of false alarm rate and target miss rate for detectors of individual languages. State-of-the-art error rates for speech from telephone environments are shown in Figure 10.8. This plot shows average results for the following 12 languages: Arabic, English, Farsi, French, German, Hindi, Japanese, Korean, Mandarin, Spanish, Tamil, and Vietnamese. Results are shown for male (m), female (f), and both male and female (b) speakers for 3-, 10-, and 30-second utterances. EERs are less than 3 percent for 30-second utterances.

LID has many potential applications in CR. First, LID could be used as a defense against system overrun; that is, the system could allow only certain languages to be used for radio communications. A more experimental strategy may be to look for *shibboleths*[10] to recognize the actual dialect or accent of the speaker; for example, whether the speaker has a foreign accent. A second application of LID is in situational awareness. If speech communications can be intercepted (and decrypted), their language could be identified to aid in the recognition of friends and foes and to alert soldiers of changes in the languages spoken by nearby forces.

---

[6] Phones are subunits of the basic sound units of speech (e.g., "ah" or "t").

[7] The set of allowed arrangements or sequences of speech sounds in a given language.

[8] The cepstral coefficients are common features used in speech signal processing that are derived from a speech signal's spectral content. The cepstrum, pronounced *kepstrum*, and an anagram of spectrum, is the result of taking the Fourier transform of the decibel spectrum, as if it were a signal. The mel-frequency cepstrum coefficients are a common variation and popular features used in speech and speaker recognition.

[9] Each frame of speech is on the order of 20 ms in duration.

[10] This biblical term has come into modern usage as a linguistic test to determine members of one group versus outsiders.

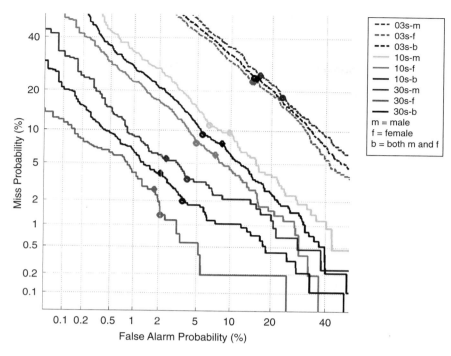

FIGURE 10.8

Typical performance of a language-recognition system. Results are taken from the 2003 NIST Language Recognition Evaluation. Performance improves (curves shift toward origin) as the test speech duration increases and performance is fairly independent of the talker's gender.

10.2.3 Text-to-Speech Conversion

Text-to-speech (TTS) conversion can provide status information to an eyes-busy CR user. TTS technology automatically speaks textual information. Textual information could originate from text-based communications or equipment display readouts. Text-based examples of communications that could be spoken via TTS include email, news, Web, rich site summary (RSS), Web logs (blogs), instant messaging (IM), Internet relay chat (IRC), and short message service (SMS). Traditional equipment display readouts could also be spoken via TTS, such as radio frequency, battery power, signal strength, network data rate, time, velocity, location, and bearing.

By providing status information to an eyes-busy user, TTS enables soldiers to focus on their mission while hearing an explanation of their battle space and status. Different synthesized voice types (e.g., male and female) are usually employed to convey various types of information. For example, routine or urgent information could be conveyed in male or female voices, respectively.

The current state of TTS technology produces mostly reasonable-sounding speech; however, it does not yet sound quite human. Future research directions in TTS are focusing on improving the quality of voice synthesis, pronunciation of named entities, conveyance of expression, and integration with machine translation and speech-to-text conversion.

10.2.4 Speech-to-Text Conversion

Speech-to-text (STT) conversion can recognize a CR user's voice commands, take dictation, and compose text messages from voice. STT technology attempts to automatically convert speech into a form that can be read by a user, produce entire transcripts of a conversation (continuous speech recognition), perform word spotting (looking for particular words), and provide command and control functions via voice. As CR keypads and displays shrink, the voice control modality becomes even more useful.

Recently, speech recognition has developed along several paths. One path is to work on large vocabulary for continuous speech recognition for conversational situations. This work has been funded through projects such as the Defense Advanced Research Projects Agency's (DARPA) Effective Affordable Reusable Speech Recognition (EARS) program; work in this area can be found in, for example, Schwartz et al. [11]. Progress in STT has brought error rates down to less than 12 percent word error rate for telephone speech. Another recent path for STT work is in noise robustness. An overview of some of these methods can be found in Junqua and Haton [12]. Noise robustness has been studied extensively for standardization by the European Telecommunications Standards Institute (ETSI) for distributed speech recognition (DSR), as exemplified by Parihar and Picone [13]. DSR's goal is to make STT a client–server application, in which the client uses the DSR front end to parameterize the speech, while recognition is done on the server.

Speech-to-text conversion has many possible applications in CR:

- First, STT can be used for gisting—rather than having a user listen to the complete conversation, a summarized version of the output can be produced.
- Second, STT can be used to route certain conversations to appropriate users (see Riccardi and Gorin [14] and related references).
- Third, STT can be used for data mining speech. If radio communication is processed by STT and stored, then text-retrieval techniques (e.g., those used to search for documents on the Internet) can be a quick and efficient way of searching for content.
- Fourth, STT can be used for command and control of a CR, as described by Broun and Campbell [15]. In this scenario, a speech interface frees up tactile and visual modalities so that the user can more effectively multitask. The speech interface can be used to control various aspects of the cognitive radio: radio modes, sensor interfaces, sensor analysis, etc.

10.2.5 Machine Translation

Machine translation (MT) automatically converts words or phrases from one language into another. This is generally done on text; however, MT can be combined with STT and/or TTS conversions to provide mixed-mode translation [16].

MT technology could help a soldier during operations in foreign-language environments. For example, foreign-language signs, news, and radio intercepts could be roughly translated to the soldier's language to aid in understanding the battle space.

Current MT technology, as typified by various Web-based systems, can be helpful for extracting some of the key words and phrases from foreign language material, but such translations are by no means transparent, as they generally contain many errors. Transcription problems are frequent and are often, but not always, easily detectable by users (it could be argued that it is more problematic when users are unable to detect transcription problems). Future MT-related research will likely be aimed at improving basic MT performance, automatically extracting meaning, gisting, and summarization.

10.2.6 Background Noise Suppression

Background noise suppression enhances communication in the presence of noise but extracts useful background noises to enhance situational awareness. Background noise suppression is primarily used in conjunction with STT conversion and voice communication. For voice communication, many new technologies have become available over the last few years.

Noise suppression can be used in voice communication to enhance the effectiveness of a speech coder. In this case, a noise-suppression system attempts to improve both the quality and the intelligibility of coded speech. These methods fall into several categories. First, methods that attempt to "subtract out" the noise spectrum have achieved considerable success (e.g., [17]). A sophisticated form of spectral subtraction [18, 19] is integrated in the enhanced mixed-excitation linear-predictive (MELPe) 1200 to 2400 bps speech coder [20].

A second class of noise-suppression algorithms is based on computational auditory scene analysis (CASA) [21]. The idea, in this case, is to use algorithms inspired by human processing—people effectively separate a sound field into multiple components such as music, voice, and noise, among others. CASA methods use techniques (e.g., independent component analysis and array processing) to achieve noise suppression. A third class of noise-suppression methods is based on multimodality. A well-known phenomenon for humans is that visual processing and audio processing of speech are fused, as evidenced by the McGurk effect [22].[11] Several systems have been developed to take advantage of the visual component (e.g., [23]). Alternate nonacoustic modalities have also been explored; these include electromagnetic, accelerometer, and electroglottogram (EGG) sensors. Significant improvement in noise suppression has been achieved with these approaches [5].

Active noise suppression is another technology that is being incorporated into radio systems. It reduces the noise that a user perceives by emitting sound to cancel the undesired noise field. Active noise suppression can be used to decrease fatigue caused by exposure to high noise levels and to reduce the Lombard effect.[12] Headphones that incorporate active (and passive) noise suppression are commercially available.

---

[11] The McGurk effect is a phenomenon that indicates an interaction between hearing and vision in speech perception.

[12] The Lombard effect is the tendency to increase one's vocal effort, which alters speech, in noise.

Noise suppression is a critical component of a CR with a speech user interface. Although not usually perceived as a cognitive capability, noise suppression is ultimately a test of a system's ability to deal with real-world conditions. Techniques, such as multimodality and CASA, show the sophistication and the challenge of matching human processing in this task.

10.2.7 Speech Coding

Speech coding is the process of compressing a speech waveform into a digital representation appropriate for efficient transmission or storage [32]. Speech coders rely on such well-established quantization techniques as differential coding for high-quality applications, while speech-specific models are typically used at lower rates. The key attributes of a speech coding algorithm are bit rate, complexity, delay, speech quality, and robustness. Since different applications have significantly different needs and priorities, there is a wide variety of speech coding algorithms currently in use.

For example, in the public telephone network and high-bandwidth Voice over Internet Prototcol (VoIP), users expect high quality and the communication network does not require aggressive compression. Speech coder attributes (e.g., low delay and robustness to talker, language, signal level, and multiple encodings) are important. Therefore, these applications use higher bit rate speech coders (e.g., 64 kbps G.711) and recent research focuses on achieving even higher quality, using both more efficient quantization and wider signal bandwidth to produce a richer speech output.

By contrast, in digital cellular systems, the priority is on efficient compression to allow as many simultaneous users as possible, while still providing quality nearly as good as landline telephones. In addition, low complexity is important to increase battery life in the handset, and robustness to bit errors and frame losses is crucial to deal with radio link fades and dropouts. These applications commonly use speech coders based on the ACELP algorithm operating in the range from 8 to 12 kbps, for example, the 12.2 kbps GSM-EFR coder.

In military communications and satellite communication systems, bandwidth is often at a premium, so that low bit rate speech coding is attractive. These coders rely on parametric models for the speech signal, encoding only the parameters of the model rather than the entire waveform, to achieve greater compression with only a modest drop in quality. Existing systems use coders like the 2.4-kbps NATO standard MELPe [20]. In many military applications, the coding emphasis is on speech intelligibility, communicability, robustness in acoustic noise, and recognizability of the talker. Special applications (i.e., the avoidance of interception, detection, or jamming) require even lower bit rates to keep the processing gain as high as possible, resulting in the development of coders as low as 300 bps [33].

A CR could dynamically adjust the speech coding algorithm used based on the current user environment. For example, the trade-off between speech and error-correction bits could change in response to current radio channel conditions, as in the 4.7 to 12.2 kbps GSM-AMR cellular algorithm, allowing higher quality over clean channels, while maintaining graceful degradation with link quality. As another example, if the CR determines that the user has a special need for clandestine communication, it could automatically switch to a low-power transmission with a very low bit rate coder.

This clandestine mode would conserve battery life and allow for high processing gain to decrease the probability of a communication being intercepted, detected, or jammed, and provide improved and safer voice communications or voice messaging for the soldier. More generally, there are many possible scenarios where a CR could adaptively select the best speech coding algorithm to optimize the user experience.

Future research will likely focus on developing and using improved models for speech coding, applying language-processing methods, widening the analysis bandwidths, and fusing multiple sensor streams.

10.2.8 Speaker Stress Characterization

Speaker characterization is the process of automatically determining the state of a user by using speech-processing techniques. Typically, this has meant trying to determine a person's emotional state, which is often related to the stress a user is experiencing.

Speaker characterization is still a developing science. One of the difficulties is elicitation of an emotional state for corpus collection—how can an experimenter truly ensure that a participant is stressed and still comply with Institutional Review Board human subject testing requirements? Another difficulty is the definition of emotional states. For example, stress can take many forms: physical stress, emotional stress, task-based stress, noise-induced stress, and so on. Should all of these be separate categories of stress? Regardless of the experimental difficulties, several practical techniques for stress recognition and compensation have been examined; for examples, see the earlier work [24, 25] and work on the speech under simulated and actual stress (SUSAS) corpus [26].

Speaker characterization is related to CR in various ways. Speaker characterization can be part of a broader strategy of affective computing [27]. Some examples include:

- Knowing the stress state of the local user as well as other users in the field to improve situational awareness.
- Knowing if a user is irritated by a particular feature by relying on the user's voice characteristics.
- Using stress level to determine appropriate modality (e.g., visual versus audio) for response to a query.
- Using verbal cues to determine if the CR made a correct decision.

Ultimately, if a CR is able to perceive and exploit a user's emotional state, it will make better decisions and be a more effective device.

10.2.9 Noise Characterization

Although noise has been considered a nuisance up to this point, it can be a useful source of information. Noise can be exploited in several ways. First, noise characterization can provide situational awareness; it can infer the location and situation of friends and foes by characterizing the acoustic environment in which the radio is operating. The CR, or a user, could catalog and track noise types in an environment to recognize anomalies that might indicate the presence of friend or foe. In this case, a noise characterization system would have to find features and provide recognition of different types of noise sources: vehicles, guns, planes, and so on. Also, the directionality of noise sources in the acoustic and radio frequency (RF) environments would be a critical property to assess. Second, noise characterization can provide diagnostics. Noise analysis can be used to detect imminent mechanical failure of common military equipment [28] or provide a quick diagnosis of mechanical problems.

10.3 CONCIERGE SERVICES

With CR, there is an opportunity to provide sophisticated human interfaces. Broadly speaking, these interfaces fall into two different categories: transparent and concierge. For a transparent interface, the goal is to provide services to the user without the user explicitly invoking the interface. A simple example of this service would be automatic detection of input modality. A more advanced interface example is an augmented reality interface where the user sees information overlaid on real objects through a heads-up display. The other interface type, concierge, requires the user to explicitly invoke and respond to cognitive services. An example of this type of service is an agent that searches for information on a topic based on a verbal request. Hybrid architectures between these two interfaces are also a possibility—we may not want services automatically invoked, but we may want the cognitive radio to transparently know how to support the user.

A unique opportunity in CR is the combination of advanced interfaces with a portable radio. A simple example that illustrates some of the ideas involved is shown in Figure 10.9. Suppose two users are communicating, and that they want to decide on a rendezvous for lunch (time, place). Both CRs have knowledge of their users' preferences and current status (money, schedule, etc.). Both radios also have a common ontology for communicating about the users' preferences. When one user asks to have lunch, the two radios can communicate and reason about a rendezvous place. This problem has many facets and can be quite sophisticated; the system will have to reason about:

Location: The location of the two users and how long it would take to go to a common restaurant.

Schedule: The available schedules of the two users, and how traveling to lunch would impact these schedules.

Resources: How much money do the two users have? Is the restaurant available?

User preferences: What type of restaurant is reasonable for both users? What prior choices have the users made?

Social network: If they invite other people, how does this impact the choices?

Dynamic rescheduling: The CR could suggest an alternative rendezvous, if new information was discovered en route to the first choice.

Network resources: The CR will access a comprehensive list of restaurants stored on the Web.

User intent: Is this meeting business or personal? Can the radio anticipate a lunch date from the user's schedule?

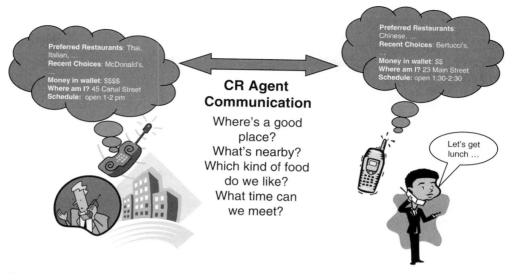

FIGURE 10.9

An example concierge service for a CR. The radios' concierge services make arrangements for a rendezvous for lunch. In a cognitive-like manner, the radio might anticipate the users' needs for lunch and sense nearby associates to invite.

Ideally, the CR would find a list of prioritized restaurant choices. The users could then verbally negotiate the remaining details.

The example highlights the need for several technologies for cognitive interfaces. First, a common method of describing the user state and preferences is needed. For many situations, a language, such as OWL, would be appropriate. A second technology of importance would be modeling user intent. Several options exist in this area: Hidden Markov Models or Dynamic Bayesian Network models, described in [29] and [30], and plan–goal recognition [31]. These technologies attempt to predict which task the user is trying to accomplish and what the current state of the task is. Third, technologies that mimic a human interface may be appropriate: avatars, speech technologies (as previously discussed), visual interfaces, haptic interfaces, and so on. Finally, sophisticated reasoning technologies must be able to handle time, geography, and uncertainty.

Reasoning and learning in a CR to support the user interface consists of many different types; three significant examples are reactive, adaptive, and network-based. Reactive reasoning is quick and does not involve explicit planning. A cognitive system would be expected to respond immediately to user voice commands, user preferences, and so forth (i.e., a reactive reasoner would be local to the CR). An adaptive cognitive radio would learn from its past experiences by adapting user preferences, speech-recognition models, user intent models, and so on. It could also pass this knowledge on to other cognitive radios. Finally, one of the large potentials for CR would be in network-based reasoning. A CR could access a large network of services to respond to user requests. For instance, in the concierge example, the agent representing the user would be expected to obtain information about location, restaurants, travel time, and more. The benefit of network-based reasoning is flexibility and minimization of local resource usage; the drawback is latency and protocol standardization (a common communication ontology).

10.4 SUMMARY

This chapter has provided an overview of several speech- and audio-processing technologies exploiting the voice and acoustic noise streams that are likely to be available to a software-defined CR.

An integrated approach to user authentication and architecture to enhance trusted radio communications networks has been presented. User authentication, via generalized biometrics, can be combined with other authenticators to provide a continuous, flexible, and strong system. This biometrically enhanced authentication system approach can be extended to become part of a CR system that learns about users, situations, and surroundings and takes appropriate proactive or reactive actions. One kind of learning presented here was generalized biometric authentication, where the users' distinctive behaviors and traits are learned and recognized. Cognitive-like applications to CRs were given using speaker recognition, language identification, TTS conversion, STT conversion, machine translation, background noise suppression, speech coding, speaker characterization, noise management, noise characterization, and concierge services technologies. An advanced CR will capitalize on these technologies to learn about and take action based on user preferences, availability of resources, and other elements of the situation and environment.

These technologies leverage the significant computational capabilities of future CRs to improve the capability, effectiveness, and efficiency of cognitive radio users. A high-impact implication of this technology is the ability to drastically reduce the complexity of the human–machine interface with the radio. As these technologies mature and become more robust, they will provide significant efficiency for the user, which will better enable the user to accomplish his or her most important tasks.

REFERENCES

[1] Weinstein, C. J., Opportunities for Advanced Speech Processing in Military Computer-Based Systems, *Proceedings of the IEEE*, 79(11):1626–1641, 1991.

[2] Campbell, W. M., J. P. Campbell, D. A. Reynolds, E. Singer, and P. A. Torres-Carrasquillo, Support Vector Machines for Speaker and Language Recognition, *Computer Speech and Language*, 20(2–3):210–229, 2006.

[3] Przybocki, M. A., and A. F. Martin, NIST Speaker Recognition Evaluation Chronicles, *Proceedings of Odyssey: The Speaker and Language Recognition Workshop*, pp. 15–22, Toledo, Spain, 2004.

[4] Campbell, J. P., W. M. Campbell, D. A. Jones, S. M. Lewandowski, D. A. Reynolds, and C. J. Weinstein, Biometrically Enhanced Software-Defined Radios, *Proceedings Software Defined Radio Technical Conference*, Orlando, 2003.

[5] Quatieri, T. F., K. Brady, D. Messing, J. P. Campbell, W. M. Campbell, M. S. Brandstein, C. J. Weinstein, J. D. Tardelli, and P. D. Gatewood, Exploiting Nonacoustic Sensors for

Speech Encoding, *IEEE Transactions Audio, Speech, and Language Processing*, 14(2):533–544, 2006.

[6] Campbell, J. P., D. A. Reynolds, and R. B. Dunn, Fusing High- and Low-Level Features for Speaker Recognition, *Proceedings of Eurospeech*, pp. 2665-2668, 2003.

[7] Campbell, W. M., J. P. Campbell, D. A. Reynolds, D. A. Jones, and T. R. Leek, High-Level Speaker Verification with Support Vector Machines, *Proceedings of International Conference on Acoustics, Speech, and Signal Processing*, pp. 73-76, 2003.

[8] Campbell, W. M., Compensating for Mismatch in High-Level Speaker Recognition, *Proceedings of the IEEE Odyssey: The Speaker and Language Recognition Workshop*, pp. 1-6, San Juan, 2006.

[9] Zissman, M. A., Comparison of Four Approaches to Automatic Language Identification of Telephone Speech, *IEEE Transactions Speech and Audio Processing*, 4(1):31-44, 1996.

[10] Singer, E. P., A. Torres-Carrasquillo, T. P. Gleason, W. M. Campbell, and D. A. Reynolds, Acoustic, Phonetic, and Discriminative Approaches to Automatic Language Recognition, *Proceedings of Eurospeech*, pp. 1345-1348, 2003.

[11] Schwartz, R., T. Colthurst, H. Gish, R. Iyer, C.-L. Kao, D. Liu, O. Kimball, J. Makhoul, S. Matsouka, L. Nguyen, M. Noamany, R. Prasad, B. Xiang, D. Xu, J.-L. Gauvain, L. Lamel, H. Schwenk, G. Adda, L. Chen, and J. Ma, Speech Recognition in Multiple Languages and Domains: The 2003 BBN/LIMSI EARS System, *Proceedings International Conference on Acoustics, Speech, and Signal Processing*, pp. 753-756, 2004.

[12] Junqua, J.-C., and J.-P. Haton, *Robustness in Automatic Speech Recognition*: Kluwer/Academic Publishers, 1996.

[13] Parihar, N., and J. Picone, Analysis of the Aurora Large Vocabulary Evaluations, *Proceedings of Eurospeech*, pp. 337-340, Geneva, 2003.

[14] Riccardi, G., and A. L. Gorin, Stochastic Language Adaptation over Time and State in Natural Spoken Dialog Systems, *IEEE Transactions Speech and Audio Processing*, 8(1):3-10, 2000.

[15] Broun, C., and W. M. Campbell, Force XXI Land Warrior: A Systems Approach to Speech Recognition, *Proceedings International Conference on Acoustics, Speech, and Signal Processing*, pp. 973-976, 2001.

[16] Shen, W., B. Delaney, and T. R. Anderson, The MIT-LL/AFRL MT System, *Proceedings of International Workshop on Spoken Language Translation*, pp. 71-76, 2005.

[17] Ephraim, Y., and D. Malah, Speech Enhancement Using a Minimum Mean-Square Error Log-Spectral Amplitude Estimator, *IEEE Transactions Acoustics, Speech, and Signal Processing*, 33(2):443-445, 1985.

[18] Martin, R., I. Wittke, and P. Jax, Optimized Estimation of Spectral Parameters for the Coding of Noisy Speech, *Proceedings of ICASSP*, pp. 1479-148, 2002.

[19] Martin, R., D. Malah, R. V. Cox, and A. J. Accardi, A Noise Reduction Preprocessor for Mobile Voice Communication, *EURASIP Journal on Applied Signal Processing*, 2004(8):1046-1058, 2004.

[20] Wang, T., K. Koishida, V. Cuperman, A. Gersho, and J. S. Collura, A 1200/2400 bps Coding Suite Based on MELP, *Proceedings IEEE Workshop on Speech Coding*, pp. 90-92, Tsukuba, 2002.

[21] Cooke, M., and D. Ellis, The Auditory Organization of Speech and Other Sources in Listeners and Computational Models, *Speech Communication*, 35(3-4):141-177, 2001.

[22] McGurk, H., and J. MacDonald, Hearing Lips and Seeing Voices, *Nature*, 264:746-748, 1976.

[23] Zhang, X., C. C. Broun, R. M. Mersereau, and M. A. Clements, Automatic Speech Reading with Applications to Human–Computer Interfaces, *EURASIP Journal on Applied Signal Processing*, pp. 1228-1247, 2002.

[24] Cummings, K., and M. A. Clements, Analysis of the Glottal Excitation of Emotionally Stressed Speech, *Journal of the Acoustical Society of America*, 98(1):88–98, 1995.

[25] Hansen, J., and M. Clements, Source Generation Equalization and Enhancement of Spectral Properties for Robust Speech Recognition in Noise and Stress, *IEEE Transactions of Speech and Audio Processing*, 3(5):407–415, 1995.

[26] Bou-Ghazale, S. E., and J. H. L. Hansen, Speech Feature Modeling for Robust Stressed Speech Recognition, *Proceedings of ICSLP*, pp. 887–890, Sydney, 1998.

[27] Picard, R. W., *Affective Computing*, MIT Press, 2000.

[28] Chen, C., and C. Mo, A Method for Intelligent Fault Diagnosis of Rotating Machinery, *Digital Signal Processing*, 14:203–217, 2004.

[29] Oliver, N., E. Horvitz, and A. Garg, Layered Representations for Human Activity Recognition, *Proceedings of Fourth IEEE International Conference on Multimodal Interfaces*, pp. 3–8, 2002.

[30] Albrecht, D. W., I. Zukerman, and A. E. Nicholson, Bayesian Models for Keyhole Plan Recognition in an Adventure Game, *User Modeling and User-Adapted Interaction*, 8(1-2):5–47, 1998.

[31] Blaylock, N., and J. Allen, Corpus-Based Statistical Goal Recognition, *Proceedings 18th International Conference on Artificial Intelligence*, pp. 1303–1308, 2003.

[32] Kondoz, A. M., *Digital Speech: Coding for Low Bit Rate Communication Systems*, Wiley, 2004.

[33] McCree, A., K. Brady, and T. Quatieri, Multisensor Very Low Bit Rate Speech Coding Using Segment Quantization, *Proceedings of ICASSP*, pp. 3997–4000, Las Vegas, 2008.

[34] *www.biometrics.org*.

Network Support: The Radio Environment Map

11

Youping Zhao
Shared Spectrum Company, Vienna, Virginia

Bin Le
Comcast NE&TO, Philadelphia, Pennsylvania

Jeffrey H. Reed
*Virginia Tech, Bradley Department of Electrical
and Computer Engineering, Blacksburg, Virginia*

11.1 INTRODUCTION

This chapter discusses the strategy of exploiting network support in cognitive radio (CR) systems architectures introducing the radio environment map (REM) as an innovative vehicle of providing network support to CRs. The REM is an approach that allows "knowledge" of the radio frequency (RF) signal environment, policy, historical performance, and network or node limitations to be shared throughout the region serviced by the wireless network. As shown in Sections 11.2 and 11.3, by leveraging local and global REMs, CRs can achieve desired cognition capabilities in a top-down and cost-efficient approach. Section 11.3 summarizes various types and different levels of situation awareness for CRs and presents the generic architecture of an REM-enabled cognitive engine (CE). Furthermore, Section 11.4 discusses the high-level REM design issues, the relevant enabling technologies and the supporting elements, and REM information dissemination schemes.

Section 11.5 illustrates that the REM can be employed as the network manager of CRs for various application scenarios. Section 11.6 offers example applications of REM in both infrastructure-based and ad hoc cognitive wireless networks. Experimental results from a CE testbed and from a network simulator are also presented, which demonstrate the significant performance improvement through REM-enabled cognitive learning and adaptation algorithms. Section 11.7 summarizes this chapter as well as the major open issues to be addressed in the future. Some review questions and discussion problems are provided in Section 11.8.

The motivations for network support for CR are threefold:

1. With powerful network support, the requirements on CR user equipment could be significantly relaxed because many computation-intensive cognition functionalities can be realized within the network, thus reducing computational loading on

handheld and mobile equipment. Distributed and collaborative information processing over the network can reduce the workload of a single user's equipment and speed up the adaptation process of each CR node and thus improve performance of the whole network. This is an important strategy to facilitate the commercialization of CR technology, considering the many constraints imposed on cost-sensitive user equipment such as limited battery power, signal-processing capability, memory footprint, and device form factor.

2. Key cognitive functionality, such as incumbent primary user detection, cannot be reliably accomplished by the user equipment itself due to the shadowing and/or fading effects of radio propagation and the practical system limitations of the sensitivity, dynamic range, and the noise floor [1, 2]. The radio has to resort to network support for many situations to solve the hidden node problem and achieve the operational goals of CR.

3. Network support is critically important to the evolution of wireless communications from legacy radios to CRs, and from the coexistence of various disparate radio networks to converged cooperative networks. As explained later in this chapter, network-enabled cognitive radio offers maximal flexibility to the government, regulator, and service provider by supporting dynamic policies on spectrum access and utilization.

The REM is proposed as a vehicle of network support for CR [3, 4]. The REM is a synthesized abstraction of real-world radio scenarios: it characterizes the environment of CRs across multiple domains such as geographical features, regulation, policy, radio equipment capability profile, and RF emissions. The REM is an integrated spatio-temporal database that can be exploited to host cognitive functionalities of the user equipment, such as situation awareness (SA), reasoning, learning, and planning, even if the subscriber unit is relatively simple. The REM can also be viewed as an evolution of the available resource map (ARM), which is proposed to be a real-time map of all radio activities in the local network for CR applications in unlicensed wide area networks (UWANs) [2, 5].

11.2 REM: THE VEHICLE FOR PROVIDING NETWORK SUPPORT TO CRS

This section first discusses the internal and external network support to CRs, and then introduces the REM as a vehicle of providing such network support to CRs. Furthermore, the role of an REM in the cognition cycle is analyzed.

11.2.1 Internal and External Network Support

From the CR user's point of view, the network support to the cognitive radio can be classified into two categories: internal network support and external network support. The internal network refers to the radio network with which the CR is associated. Along with various communication services, the internal network can provide some cognitive functionality as well. For example, the CR network can provide location information

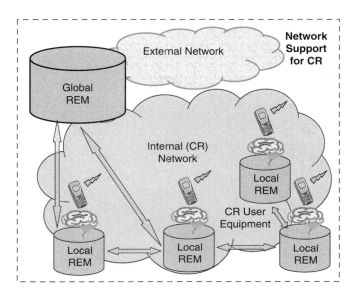

FIGURE 11.1

Radio environment map. The REM provides cognitive services to both the associated internal networks and a useful awareness of external networks such as noncognitive legacy systems.

and location-based services to the user; it can also characterize the usage pattern of other users in the neighborhood. The external network refers to any other networks that can provide meaningful knowledge to support the cognitive functionalities of the radio network. For example, a separate sensor network could be dedicated to gather information for CR networks [6]. The external network could be a legacy network or other CR networks.

Both internal and external networks can contribute to building up the REM and can be employed in a collaborative way. For instance, location information needed for a CR can be obtained either from internal network support through a network-based positioning method for indoor scenarios or from external network support through the global positioning system (GPS) for outdoor scenarios.

As depicted in Figure 11.1, network support can be realized through a global REM and local REMs. In this figure, the CR is symbolized as a "brain"-empowered radio. The global REM maintained by the network keeps an overview of the radio environment, while the local REMs stored at the network node or user equipment only present site-specific views to reduce the memory footprint and communication overhead. The local REMs and global REM may exchange information in a timely manner to keep the information stored at different entities current. The figure shows that a global REM can be aggregated from the combined experiences of several local REMs.

11.2.2 Introduction to the REM

This subsection starts with an insightful analogy about CR, and then proposes the REM as a cost-effective network manager for CRs [3, 4].

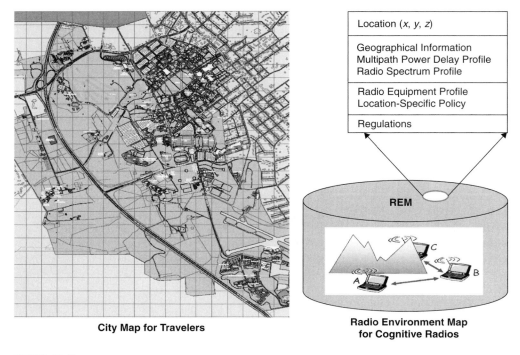

| Location (x, y, z) |
| Geographical Information
Multipath Power Delay Profile
Radio Spectrum Profile |
| Radio Equipment Profile
Location-Specific Policy |
| Regulations |

REM

City Map for Travelers

**Radio Environment Map
for Cognitive Radios**

FIGURE 11.2

City map versus REM. An REM provides services to local CRs similar to the way a city map aids a driver with local navigation. In contrast to the city map, the REM database must be current to be useful.

Similar to how a city map helps a traveler, the REM can help the CR know the radio environment by providing information on the following:

- Spectral regulatory rules and user-defined policies to which the CR should conform
- Spectrum opportunities, where the radio is now and where it is heading
- The appropriate channel model to use
- Current and expected pathloss and signal-to-noise ratio (SNR)
- Hidden nodes present in the neighborhood
- Usage patterns of primary users[1] (PUs) and secondary users (SUs)
- Interference or jamming sources

Figure 11.2 shows an insightful analogy. In fact, there are many commonalities between transportation and telecommunication such as the concept of throughput, channel

---

[1] The PU usually holds the license to use certain spectrum. Therefore, the PU has first rights to the spectrum and is typically an incumbent user of the spectrum. The SU must yield the spectrum to a PU when the PU begins transmission.

capacity, routing, signaling, traffic rules, etiquette, and utilization efficiency of resources. Enlightened by this analogy, some lessons can be drawn from the field of transportation and employed in the field of CR.

The REM can provide both current and historic radio environment information, so that most cognitive functionalities in the CR network can be realized in a cost-efficient way. For example, by querying the REM, CRs can conduct spectrum sensing with prior knowledge rather than blindly scanning over the whole spectrum continually. Thus, observation time and energy consumption in the portable radio front end can be significantly reduced. By incorporating collaborative information-processing techniques, the costs of a CR system can be further reduced by relaxing the requirements on transmission power, dynamic range, and sensitivity of the individual radio device.

By combining reasoning and learning with data mining in a REM, network intelligence can directly enable and/or support capabilities for network nodes whether they are cognitive or not. In this sense, even legacy networks can become cognitive by resorting to an REM. The REM also supports a system-level solution to the fundamental CR technical challenges, including SA, cross-layer optimization, hidden- and exposed-node problems, load balancing across the network, opportunistic spectrum access, dynamic spectrum regulation, and policy management.

11.2.3 The Role of the REM in Cognition Cycle

The REM plays an important role in the cognition cycle of CRs, as illustrated in Figure 11.3. Both direct observations (e.g., spectrum sensing) from the radio and knowledge derived from network support (e.g., querying a local or global REM) can contribute to the radio's environment awareness. Reasoning and learning help the CR to identify the specific radio scenario, learn from past experience and observation, and make decisions

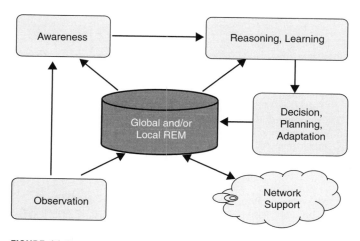

FIGURE 11.3

Role of REM for cognitive radio. The REM provides the infrastructure to support the fusion of multiple cognitive services to the local subscribers.

and plans to meet both individual subscriber and system goals. The global REM and/or the local REM should be updated once action is taken or scheduled by the radio to keep the REM's information current.

11.3 OBTAINING COGNITION WITH REM: A SYSTEMATIC TOP-DOWN APPROACH

This section first explains the meaning and importance of SA to CRs. It also shows how radios become situation aware with the help of the REM. As discussed in the previous section, the REM can present multidimensional information for CRs, such as geographical environment, location and activities of radios, regulations, and policy of the user or service provider. One of the most important features of the REM is that it is transparent to the specific application of CRs. The architecture of an REM-enabled CE is presented at the end of this section.

11.3.1 Awareness: Prerequisite for Cognitive Radios

The REM presents comprehensive system-level knowledge for a CE to exploit. The cognitive engine, which is the cognition core of the CR, is typically implemented as a software system consisting of learning and adaptation algorithms.

An interesting analogy between two intelligent agents—a taxi driver and a CE—makes this clear (Figure 11.4). The comparisons of cognition components between these two intelligent agents are listed in Tables 11.1 and 11.2. The situation awareness and performance measures for a taxi driver or a pilot have been extensively discussed [9, 10]. Situation awareness is one of the most important features that differentiate a CR from an adaptive radio.

Table 11.1 Comparison of Two Intelligent Agents: Taxi Driver and CR Engine

Agent Type	Environment	Performance Measure	Sensors	Actuators
Taxi driver [9]	Roads, other traffic, pedestrians, customers	Safe, fast, legal, comfortable trip, maximize profits, minimize collisions	Cameras, sonar, speedometer, GPS, odometer, engine sensors, accelerometer, keyboard	Steering, accelerator, brake, signal, horn, display
CR engine	Radio spectrum, other traffic by PU and SU, jammer, RF noise and interference	Spectrum utilization, reliable, fast, legal, cost-efficient, low power consumption, minimize interference	GPS, antenna, BER/PER/FER, interference temperature, QoS	Transmission power control, MAC, beamforming

Note: BER = bit error rate; FER = frame error rate; PER = packet error rate; QoS = quality of service; MAC = medium access control.

Cognitive Radio's Capabilities	Taxi Driver's Capabilities
• Senses, and is aware of, its operational radio environment and its capabilities • Can dynamically and autonomously adjust its radio-operating parameters accordingly • Learns from previous experiences • Deals with situations not planned at the time of initial design	• Senses, and is aware of, the driving environment and capabilities • Can dynamically and autonomously adjust the driving operation accordingly • Learns from previous experiences • Deals with situations not planned at the time of initially learning to drive

FIGURE 11.4

Capabilities of two intelligent agents: CR engine and taxi driver.

It is apparent that both intelligent agents observe the environment and become aware of their situations; make in situ decisions according to their observations, anticipations, and experiences; and then execute intelligent adaptations to reach their goals. They evolve by a spiral-learning process (also known as the "cognition cycle") throughout their lifetimes. Like the taxi driver, the CR may also have three levels of situation awareness (SA), as shown in Table 11.2. To appreciate the significance of three-level awareness, consider the following scenario.

Table 11.2 Three-Level Situation Awareness

Agent Type	Level 1 Awareness	Level 2 Awareness	Level 3 Awareness
Taxi driver (or pilot) [10]	*Looks* and *perceives* basic information	*Thinks* about and *understands* the information's meaning	Uses the environmental understanding to *anticipate* what will happen ahead in time and space
CR engine	*Observes* the RF spectrum, waveform, and surrounding activities with focused attention	Performs calculation, estimation, reasoning, and *understands* the radio activity and what it represents	Performs prediction and planning to *anticipate* radio performance, network requirements, and user needs

Suppose you are driving from your office to a restaurant for lunch. You may make many "cognitive" decisions and adaptations according to your current situation, observations, and previous experience:

- If you have a tight schedule for lunch, you may prefer to drive through a fast-food restaurant nearby.
- If it is a sunny Friday, and you want to enjoy the lunch with your buddies, you may carpool with your friends to a nice restaurant in another town.
- If it is a rainy day and the road is curvy, you may turn on the headlights and drive slowly to avoid an accident.
- As you drive, you still need to be aware of the traffic lights, road signs, and speed limits.
- You may look at a map for directions, or recall from your memory the most convenient way (short-cut) to the restaurant.
- When you learn from the radio news broadcast that there is an accident ahead, you may anticipate that it will result in slow traffic and choose an alternate route.

This list could be extensive, considering many other possible situations and their adaptations for a simple automotive errand. Obviously, the CE has both awareness and learning capability, but with an REM the ability to predict radio performance is facilitated as readily as predicting the path distance for the car trip.

11.3.2 Classification of Awareness

Awareness means the understanding of the situation [11]. Both geographical and RF environment-related information (e.g., radio propagation characteristics, waveform, and spectral regulations) play major parts in CR knowledge [12]. In addition, policies, goals, and contexts are also important to CRs.

Situation awareness here means that the CR knows its current radio scenario, the intent of the user, and the regulations with which it must comply. Situation awareness may include, but is not limited to, the following key types of awareness.

Location awareness: The CR knows where it is, in the form of latitude, longitude, and altitude, or relative location to some reference nodes.

Geographical environment awareness: The CR knows the terrain and geographical information related to the radio propagation and channel characteristics. This awareness is critically important for a CR to choose the appropriate spectrum, channel model, or radio access technology (RAT), antenna configuration, and networking techniques.

RF environment and waveform awareness: The cognitive radio knows the spectrum utilization, the existence of primary and/or secondary users, the topology of the user group, the interference profile, and other RF characteristics that may be of concern.

Mobility and trajectory awareness: The CR knows its moving speed and direction. For example, in conjunction with geographical awareness, the CR can know it is moving south along Main Street at a speed of 45 mph, and it can "foresee" the radio environment ahead, such as the available channel after the user passes over the next hill, or the radio standards supported along the route.

Power supply and energy efficiency awareness: The CR knows the source of its power supply, the remaining battery life, and the energy efficiency of alternative adaptation schemes.

Regulation awareness: The CR knows the spectrum allocation and emission masks at specific locations and frequency bands regulated by government authorities such as the US Federal Communications Commission (FCC).

Policy awareness: The CR knows the policy defined by the user and/or the service provider. For example, the user may prefer to use the wireless local area network (WLAN) from a specific service provider at some locations for quality-of-service (QoS) or security reasons.

Capability awareness: The CR knows its own capabilities as well as those of its team members and/or the network. Such awareness may include knowing which waveforms are supported, the maximum transmit power, and the sensitivity of the CR.

Mission, context, and background awareness: The CR understands the intent of the user, and knows which mode and volume of traffic it is going to generate and what the impact of that traffic will be to the local networks. The CR understands the QoS requirements, and how overhead activities may trigger additional network traffic and latency.

Priority awareness: The CR knows the user's priorities and habits. For example, the user may prefer to use low-cost services whenever possible (e.g., to switch to WLAN from a third-generation (3G) system when entering a WiFi zone) or may prefer reliability over cost.

Language awareness: The CR knows the signs, ontologies, and etiquette used among CRs to communicate with each other.

Past experience awareness: The CR remembers past experience and learns from it.

Note that all of these items can be interrelated and employed together in various ways. For example, the CR may adapt network topology and make dynamic spectrum access (DSA) decisions based on the user's location, mission, goal, RF environment, regulations, and service priority. Table 11.3 summarizes the various types of awareness a CR may have, the relative significance of each type of awareness, and how to obtain such awareness.

11.3.3 Obtaining SA with REM: A Top-Down, Cost-Efficient Approach

The CR can obtain situation awareness through different approaches:

- Direct observation (e.g., through field measurement)
- Inference from network support (e.g., through a network-maintained REM)

Table 11.3 Summary of Situation Awareness for CRs

Type	Significance	Approaches to Obtain Awareness	Current Status
Location	High	GPS (or assisted GPS) Network-based positioning Landmark or RF fingerprint matching Combination of inertial navigation and GPS	Many kinds of positioning techniques are commercially available.
Geographical environment	Low to medium	Query and exploit GIS database Terrain recognition Site-specific propagation prediction	GIS database is available (e.g., from U.S. Geological Survey [13]). Many site-specific propagation tools can predict pathloss, delay spread, and service coverage.
RF environment/ waveform	Medium to high	Radio transceiver database Collective observations by CRs Sensor network Field measurement	Microwave point-to-point radio, FM, and TV station databases are available and maintained by the FCC, which provides radio station information (e.g., site location and transmission power) [14].
Mobility and trajectory	Low to medium	Estimate moving speed and trajectory of the user by analyzing the change of locations over a period of time and correlating with GIS	Can be addressed with current technologies together with the REM.
Power supply and energy efficiency	Battery: high AC: low	Measure the voltage and/or current of power supply	Mature technique (e.g., the cellular phone knows the source of its power supply and the remaining power).
Policy	High	Defined by the service provider and/or the user	Can be addressed with the policy database in the REM.
Regulation	High	Defined by the government authorities	Can be addressed with the regulation database in the REM.
Capability	High	Provided by the CRs and/or networks	Can be addressed with the capability database in the REM.
Mission/context/ background environment	Low to medium	Using machine intelligence, various sensors Applying speech and/or image recognition and understanding techniques	Common industry software tools demonstrate some context awareness, and even interact with the user occasionally.
Priority	Low to medium	Defined by the service provider and/or the user	Can be addressed with the priority database in the REM.
Language	Medium to high	Standardizing CR languages and etiquettes	This is under development (e.g., by the DARPA XG program [15]).[a]
Past experience	Medium to high	Long- and short-term memory of experience for recall	Can be realized with CBDT and other technologies.

[a] *Defense Advanced Research Projects Agency NeXt Generation (XG) program.*
Note: GIS = geographical information systems.

■ Analysis of local terrain propagation models combined with existing database structures defining known communication systems

Following the previous analogy of driving a car, you cannot just rely on your own vision when you drive, especially under unfavorable weather conditions, such as a snowstorm, or at night. The map complements your limited local vision and helps you know the road conditions ahead and how far to the destination. You can make an informed decision as to whether you need to stop at a gas station. You may take extra caution because you know the road ahead will be winding. You may schedule to have a dinner at the next rest area ten miles ahead. In summary, you easily obtain helpful context information. Therefore, while traveling, you can examine the map, refresh a previous experience if you had been there before, or just take a "trial-and-error" learning approach if the map or experience is insufficient.

For the CR, shadowing, fading, and Doppler shift are the most common degradation or distortions imparted by the radio channel. REM can provide the CR with channel characteristics associated with the location and direction of a mobile user. Channel information can be obtained through observing instantaneous measurements of the environment as well as long-term measurements and learning of the general characteristics of the environment. Once these channel measurements are available, models can be created to predict the performance of the link. Usually, these models are stochastic and produce outputs that are random variables. The variance of model outputs can be incorporated into the decision process of CR. Furthermore, cognitive radios can take advantage of channel awareness for planning.

With the awareness of shadowing and fading characteristics, CR may adopt the appropriate waveform (i.e., PHY or MAC layer) to adapt to or take advantage of the propagation characteristics. For example, in a multipath-intense environment, CR may choose to apply multiple input, multiple output (MIMO) techniques to improve its performance. A CR can also anticipate a call drop due to multipath or shadowing and take preemptive measures, such as switching on the backup power amplifier, increasing the number of RAKE fingers, leveraging smart antenna resources or spreading gain, altering the power control policy, or making an intersystem handoff.

The REM can exist at the user's terminal equipment (local REM) and/or at the network level (global REM). The local REM may be unique to each user's device. Each CR can use its own local REM to memorize its past experience, particularly for the region that it usually operates within as well as its current status, and a high-quality CR could store a wider variety of experiences and effective responses than a less expensive radio could.

As shown in Figure 11.5,[2] by using a global REM, it is possible that even the legacy radio network can be upgraded to support some cognitive functionality and behave as if it were a CR network. For instance, through a software upgrade to the network-level radio resource management system, the legacy network can know the subscriber's location and the interference environment, and then instruct the radio to use the most effective physical (PHY) and medium access control (MAC) layers supported by the radio device. A simple radio with limited information-processing capability can become

---

[2] In Figure 11.5, the term *radio* refers to any type of radio device, even a cognitive RF identification (RFID) tag.

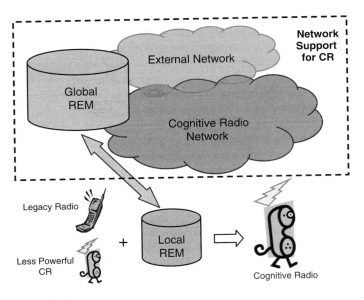

FIGURE 11.5

SA through the REM. REM helps radios become situation aware by bootstrapping them with local accumulated experience, customized to each radio's functionality requirements.

more cognitive by leveraging the REM-based network support. The local REM may exchange information with the global REM—for example, through a common control channel.

Cognitive behavior, usually taking place in a rich and complex environment, is goal oriented and a function of the dynamic environment [38]. Therefore, obtaining necessary radio environment information is imperative for CRs. The idea behind the REM is digitizing and indexing radio environment information. The more clearly the radio environment is characterized and modeled, the better the CR can learn from experience and environment. In addition, the REM can incorporate policy layer, application layer, optimization layer, topology, and network layer information, all of which are important to CR networks. Table 11.4 illustrates the information domains of an example REM and the index to each REM information element. Enabled by advanced database technologies (e.g., a Web-based one), an REM can be accessed in a centralized or distributed way.

Rather than observing the radio environment with blind and wide spectrum sensing, an REM-enabled CR may choose to have a scalable view of the radio environment with an application-specific observation range. For example, for indoor wireless access, REM information pertaining to only a few rooms could suffice; whereas for outdoor wireless ad hoc networks, a large-scale REM would be appropriate to provide more information coverage. To obtain SA, not every CR needs to conduct sophisticated spectrum sensing as long as it maintains or has access to an up-to-date REM through the network support. In other words, the REM enables a system-level, top-down approach for CR nodes to obtain SA in a cost-efficient way. For example, the REM can inform the CR which kind of radio networks could be in service at a certain location.

Table 11.4 Digitizing and Indexing REM Information Elements

Domain and Index Range	Syntax and Index
Application type => 700–799	Voice (701), packet data (702), video conference (703), etc.
Optimization layer => 600–699	Minimize interference to PU (600), maximize SU throughput (601), etc.
Topology and network type => 500–599	Infrastructure-based network—WCDMA (500), cdma2000 (501), WRAN (502), etc.; ad hoc network (510), mesh network (520), etc.
MAC and duplex => 400–499	TDMA (400), FDMA (401), CDMA (402), OFDMA (403); FDD (410), TDD (411), etc.
Geography and mobility information => 300–399	Indoor—home (300), office (301), airport (302), factory (303), etc.; outdoor—urban (310), suburban (311), open rural (312), highway (313), etc.; in-vehicle—train (320), bus (321), car (322), plane (323), etc.; etc.
Modulation type => 200–299	AM (200); FM (210); MPSK—BPSK (220), QPSK(221), etc; M-QAM—16-QAM (230), 64 QAM (231), etc.; etc
Radio device capability => 100–199	Channel coding—convolutional coding (100), Turbo Coding (110), etc.; maximum RF transmit power (120), sensitivity (130), operational bands (140); antenna type (150), etc.
Experience => 0–99	Blind zone (10), hot spot (20), hidden node (30), etc.

Based on the radio interface specifications stored in the REM database, the CR will know the possible frequency bands and modulation types used by PUs. The CR can even obtain some prior knowledge of PUs by analyzing the historical REM data and learning from past experience. Therefore, it can conduct PU detection with focused attention instead of spending excessive processing time performing complex spectrum sensing and signal classification algorithms. This top-down approach for PU detection and classification is time saving, cost effective, and energy efficient. Furthermore, REM technology shows the potential of supporting global cross-layer optimization and cognitive networking functions by enabling CRs to "look" through various layers, from policy layer, application layer, optimization layer, topology layer, down to the network, MAC, and PHY layers [27, 28].

11.3.4 Architecture of REM-Enabled Cognitive Engines

The building blocks and overall system architecture of the REM-enabled CR model are illustrated in Figure 11.6 [30, 35]. In this model, the REM consists of multidomain information. During operation, the CR observes its operational environment via sensor(s), and synthesizes necessary SA of the current radio scenario by leveraging the sensing results and REM information. The CR reasoner then determines an appropriate utility function based on the policy and the goals, by considering the specific application and radio scenario. The utility function maps the current state of the CR, usually represented by an array of chosen metrics, to a value for indicating how close the state is toward the desired (or optimal) CR state. The most pertinent performance metric(s) should be taken into account and incorporated into a utility function to meet the CR's goal for this application and radio scenario. By leveraging experience and knowledge, the CE

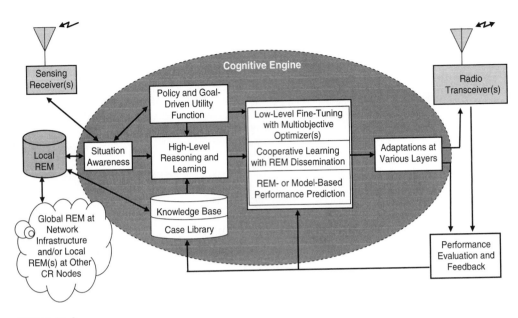

FIGURE 11.6

Architecture and building blocks of the REM-enabled cognitive engine model [35].

can choose the most efficient reasoning and learning method and make improved cross-layer adaptations (if necessary) subject to the constraints of regulation, policy, and radio equipment capability.

11.4 HIGH-LEVEL SYSTEM DESIGN OF REM

This section first discusses the various types of REMs that could be implemented in practice, and then explains three key components of REM system design; that is, database design, application programming interfaces (APIs), and supporting elements (e.g., REM information indexing and retrieving, REM memory management, and data mining over REM). Finally, REM dissemination schemes and overhead analysis are discussed.

11.4.1 Classifications of REM

In practice, we may design various types of REMs as follows:

- According to the application of concern, the REM can be classified into an application-specific REM (e.g., REM designed for a specific region or for a specific application) and a general-purpose REM (e.g., REM designed for a variety of potential applications).
- According to the host location and network topology, the REM can be classified into local (distributed) REM and global (centralized) REM.

- According to the implementation approach, REM can be classified into embedded (virtual) REM (where the REM information is embedded into the CE) and stand-alone REM (where the REM is maintained at a server that interfaces with the CE through APIs).

11.4.2 REM Database Design Guidelines

The REM system design essentially consists of three key components: database design, APIs, and supporting functionalities (e.g., memory management and data mining).

The database design can be conducted using the following six steps.

1. *Requirements analysis:* determine what information CRs need and understand what the database should provide.
2. *Conceptual design:* develop a high-level description, often done with an entity-relation (E-R) model.
3. *Logical design:* translate an E-R model into the database management data model.
4. *Schema refinement:* check consistency and normalization.
5. *Physical design:* develop indexes.
6. *Security design:* create constraints on what information can be accessed by each application and each user.

For more detailed explanations about database design methodology and the design process, readers can refer to Connolly and Begg [36].

E-R Model and REM Information Elements

This subsection provides a reference E-R model for an REM design and its information elements for cognitive wireless regional area network (WRAN) systems. As discussed in detail in Section 11.5, IEEE 802.22 WRAN is the first worldwide commercial application of CR networks to utilize the unused TV broadcast channels [40].

Tables 11.5 and 11.6 show the attributes, estimated memory size, and index of REM information elements for WRAN basestation (BS) and customer premise equipment (CPE), respectively. Note that additional information elements could be added in the future along with the further development of WRAN systems (especially the mobile ad hoc operational mode of WRAN systems) and the size of the information element is a rough estimation. Most of these information elements are incorporated in the 802.22 WRAN BS cognitive engine testbed developed by the Mobile and Portable Radio Research Group (MPRG) within Wireless@Virginia Tech [37]. Figure 11.7 shows the E-R model for the REM maintained at the WRAN BS that can be used in the REM design for the 802.22 WRAN BS cognitive engine.

REM Database Implementation Options

The REM characterizes the real-world radio scenarios for CR. Ideally, it can be a highly comprehensive and integrated database. However, when being implemented, according to the specific application and system design requirements, the REM can be realized with any appropriate data format, as long as it provides all needed radio environment

Table 11.5 Preliminary REM Information Elements for the WRAN BS

Attribute Index	Syntax	Estimated Size	Notes
1	BS_ID	2 bytes	BS_ID = WRAN Network ID + internal BS ID
2	BS_Location	12 bytes	GPS coordinates (longitude, latitude, and altitude)
3	BS_TxPower	1 byte	Transmit power of BS
4	Active channel set*	16 bytes	* indicates the information element is of "short-term memory" nature
5	Candidate channel set*	16 bytes	
6	Occupied channel set*	16 bytes	
7	Geographical environment type	2 bytes	Type of geographical environment (e.g., mountainous, open rural, suburban, or urban), corresponding to different radio propagation channel models
8	CPE_Information {CPE_ID, CPE_Type, CPE_Location, Status of Connection}	44 bytes per CPE entry	CPE ID = BS_ID + subscriber ID
9	TV_Station_Information {TV Station_ID, TV_Station_Location, TV_Station_Channel, TV_Station_Power}	32 bytes per TV station entry	
10	WM_Information {WM_Location, WM_Channels, WM_in_use_likelihood, etc.}	1 byte per waveform (WM) entry	The value of likelihood can range from 1 to 0, which indicates the different likelihood of the presence of WM (e.g., highly likely, possible, or rare)
11	Timestamp	4 bytes	The time when this observation is made

information. For example, a class-structure, a multidimensional array (vector), a data file, or a database are all candidate implementation options. Commercial databases may be appropriate for managing huge radio environment information; however, they usually require a much larger memory footprint and tend to be "heavy" for the WRAN BS cognitive engine testbed. Therefore, for the WRAN testbed developed at Virginia Tech, the REM is implemented using a hybrid approach: the current REM information is implemented with a C++ class *RadioEnvironmentMap*, as shown in Figure 11.8, while historical REM information is periodically stored in data files. This REM implementation approach is fairly general and allows easy extension or updating to incorporate new environmental information elements or functionality.

The REM contains information at multiple layers, as illustrated in Figure 11.9 (see page 344). By integrating various existing databases, the REM enables or supports cognitive functionality for radios with different levels of intelligence. The REM helps a CR to be aware of situations and make optimal adaptations according to its goals; for legacy or hardware reconfigurable radios, the REM facilitates smart network operations by providing cognitive strategies to the network radio resource management control. Just

Table 11.6 Preliminary REM Information Elements for the WRAN CPE

Attribute Index	Syntax	Estimated Size	Notes
1	CPE_ID	2 bytes	Unique ID for each radio device
2	CPE_Location	12 bytes	GPS coordinates (e.g., longitude, latitude, altitude, or relative position in the network)
3	CPE location accuracy	1 byte	Indication of the accuracy of the position estimation
4	CPE_Tx_Power	1 byte	Transmit power of CPE
5	Favorite channel set (i.e., "channel reputation"[a] graded by this CPE)	16 bytes	Observed spectrum usage at this location; actually only about 80 TV channels are allocated between 54–862 MHz; more bits are reserved for error correction or future use
6	Geographical environment type	1 byte	Type of local radio environment: indoor, mountainous, open rural, suburban, dense urban, etc., which indicates the appropriate radio propagation channel model to apply
7	Interference temperature	2 bytes	The interference temperature estimated by the wireless node at this location
8	CPE type and capability	2 bytes	CPE type and its capability (e.g., interference excision capability)
9	Timestamp	4 bytes	The time when this observation is made
10	Channel reciprocity flag	1 byte	The confidence in reciprocity of the channel
11	Channel model and statistics	4 bytes	The proper channel model that can be applied for the CPE's current scenario (e.g., Rician, Rayleigh channel model) and the corresponding parameters (e.g., Rician K factor)
12	Pathloss correlation	1 byte	The confidence in determining the pathloss based on the geographical locations of the radios
13	CPE velocity, acceleration, and orientation	6 bytes	This set of parameters provides mobility support
14	CPE carrier and timing stability and offset	2 bytes	The carrier and timing accuracy is important information for improving network synchronization and radio resource management

[a] TV "channel reputation" represents the probability of presence of PUs at a TV channel, which can be obtained by analyzing the historical radio environment information.

like the city map that is informative to every traveler, no matter whether driving a car or taking the bus, the REM is transparent to the specific RAT being employed regardless of whether the subscriber radio is cognitive or not.

11.4.3 Enabling Techniques for Implementing REM

Figure 11.10 shows the interdisciplinary nature of the REM design for CRs (see page 345). To implement, populate, and exploit the REM, various technologies must be

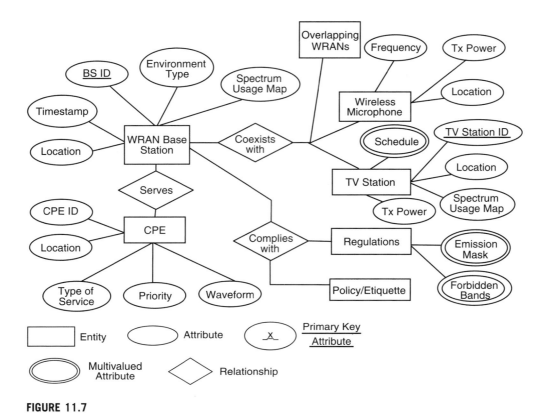

FIGURE 11.7

E-R diagrams of the WRAN basestation REM.

employed—for example, artificial intelligence (AI), detection and estimation, pattern classification, cross-layer optimization, database management and data mining, site-specific propagation prediction, and network-based ontology.

CRs and their required infrastructure are enabled by a collection of many new technologies: data storage technology; very large scale integration (VLSI); RF-integrated circuits (RFICs); geographical information systems (GIS), such as the three-dimensional (3D) digital map; terrain-based radio propagation modeling; and GPS receivers. For example, several electronics manufacturers have recently developed handheld products with gigabyte storage capacity, and many cellular phones have already been equipped with GPS receivers. Therefore, it is quite possible that future CRs will be allocated sufficient memory to contain a comprehensive database.

Radio propagation simulation tools can predict many important parameters, such as pathloss, SNR, or signal-to-interference and noise ratio (SINR), as long as the geographical environment information is available [7]. Compared to the empirical channel model-based prediction, the advanced site-specific radio propagation techniques and software tools using 3D terrain maps have been successfully employed to provide much more reliable radio propagation predictions for system planning and management [8]. This

```
class RadioEnvironmentMap
{
  public:
     RadioEnvironmentMap();                                  // Default constructor
     ~RadioEnvironmentMap();                                 // Destructor
     void AddEntry(REMDataEntry & _rde);                     // Add entry to database
     bool IsEntry(const REMDataEntry & _rde);                // Query database to see
                                                             // if entry exists
     void Merge(RadioEnvironmentMap & _rem);                 // Merge REM databases
     bool LoadDatabaseFromFile(const char * filename);       // Load database from file
     bool StoreDatabaseToFile(const char * filename);        // Store database to file
     void AnalyzeChannelStatistics(int *& tv_channels, int & num_channels);
                                                             // Rank channel reputations
     float GetChannelReputation(unsigned int tv_channel_id);
                                                             // Retrieve the particular
                                                             // TV channel
                                                             // reputation from REM
                                                             // database
     void PrintDatabase();                                   // Print database to screen
     RadioEnvironmentMap & operator = (const RadioEnvironmentMap & _rde);

  protected:
     RadioEnvironmentMap(RadioEnvironmentMap & );
     REMDataEntry * database;
     unsigned int database_length;
     unsigned int database_length_max;
};
class REMDataEntry
{
  public:
   REMDataEntry();              // Default constructor
   ~REMDataEntry();             // Destructor
   void Print();                // Print entry
   REMDataEntry & operator = (const REMDataEntry & _rde);
   bool operator == (const REMDataEntry & _rde);
   bool operator != (const REMDataEntry & _rde);
   int timestamp;
   int device_id;
   DeviceType device_type;
   ObjectPosition device_location;
   int geo_environment_type;
         // corresponding to the channel model or path loss exponent index for the
         // BS-CPE radio link
   int frequency;
   int bandwidth;
   int tx_power_dBW;
};
```

FIGURE 11.8

REM implemented as a C++ class RadioEnvironmentMap.

makes it possible to embed multiple contours into the REM (e.g., the service contour, blind zone, and interference region), which will enable CRs to make decisions and adaptations that overcome the most common user complaints.

Information to populate the REM can be obtained by:

- Integrating and/or correlating various existing databases (e.g., GIS databases or radio equipment databases).

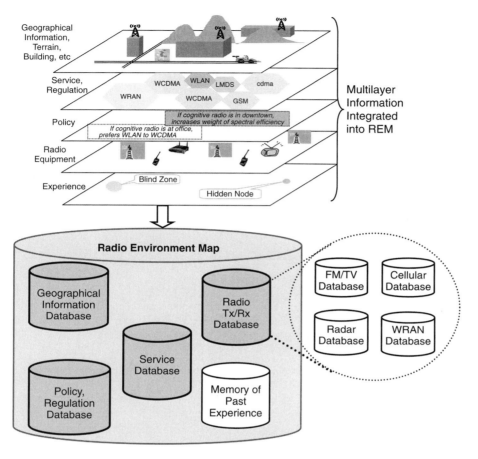

FIGURE 11.9

Integrating various databases for building up the REM.

- Sensing with collaboration among distributed nodes, which may or may not be CRs.
- Observing from a dedicated sensor network and/or other external networks.
- Probing the radio environment.
- Estimating the radio propagation characteristics with software tools.

11.4.4 Supporting Elements for Exploiting REM

The underlying techniques to support an REM include, but are not limited to, database design, information management and protocols, database transactions (e.g., query, search, and update), and data mining. Learning, reasoning, and decision mechanisms should be architected to facilitate a system that can serve a variety of radio types and networks and support a variety of applications.

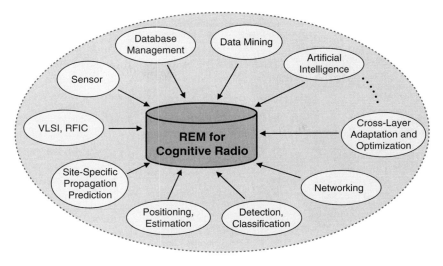

FIGURE 11.10

Interdisciplinary research for developing and exploiting REM.

Learning, Reasoning, and Decision Mechanisms

Reasoning, data mining, and querying capability to external databases can furnish information for the REM; this composite capability is called the self-informing system (SIS) and is shown in Figure 11.11. Decision making outside of the SIS is the key to integrating heterogeneous networks or even dissimilar radios within a homogenous network. Given the information provided by the SIS, a radio with its own decision-making capability can make decisions relevant to the radio and/or network capabilities and the supported application. Even for non-CRs, a cognitive network can be achieved by interfacing the SIS with the network radio resource management system. Note that the SIS does not need to reside at a central location. It can in fact be distributed among the nodes in the network.

More complicated learning and reasoning techniques can be practically implemented on the network server rather than on individual CR user nodes. Data mining is a powerful tool to extract hidden information from a huge information base and then make meaningful predictions. For example, data mining over the global REM can help to find the usage pattern of primary users, secondary users, and other interferers from large quantities of data (i.e., observations over an extended period of time), using various techniques, such as decision trees, neural networks, and neural-fuzzy systems.

As discussed in more detail in Chapter 12, the CR can employ various learning methods such as statistical learning, instance-based learning, reinforcement learning, and decision tree induction learning. Two basic optimization approaches can be employed by CRs to optimize complex parameters: (1) the classical optimization based on the properties of the objective function and (2) heuristic optimization. Common classical optimization routines include the cyclic-coordinate method, the steepest descent method, and the quasi-Newton and conjugate-gradient methods. Heuristic techniques include evolutionary algorithms (e.g., the genetic algorithms (GAs) discussed in

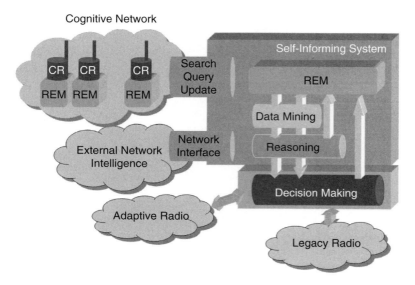

FIGURE 11.11

Cognitive radio self-informing system block diagram.

Chapter 7), simulated annealing, ant-colony optimization, Tabu search, and neural networks, among others [4, 24].

Classic optimization is more analytically tractable compared to the heuristic approach; however, neither of them can be guaranteed to yield optimal solutions. As explained in Chapter 7, the CR environment is typically characterized by multiobjective decision making (MODM). Without a single objective function measurement, it is difficult to apply classic optimization theory to adapt the radio parameters. Evolutionary algorithms are suitable for searching the optimal PHY/MAC/NET combinations and other parameters for CRs [4, 25]. Hidden Markov models (HMMs) can be used for radio scene classification, case recognition, and making meaningful predictions based on training data (past experience).

Artificial intelligence techniques are useful to fully exploit the REM to obtain SA and to enable reasoning, decision making, and self-learning. For example, a CR that learns to foresee where and when a spectrum hole or interference will appear will do better than one that does not. Just as evolution provides animals with enough built-in reflexes so that they can survive long enough to learn for themselves, it would be reasonable to provide a CR with some initial knowledge as well as an ability to learn. After sufficient experience in its environment, the behavior of the CR can become effectively independent of its prior knowledge. The incorporation of learning allows one to design a CR that is able to evolve and succeed in a vast variety of environments.

Memory Management

Memory requirements of the REM may be highly dependent on the following factors:

- Scale of the CR network (e.g., the number of nodes, the number of channels, the size of network coverage, etc)
- Granularity of REM information
- Application-specific requirements on a CR's behavior (e.g., the expected capability and functionality of CRs)

Furthermore, the REM memory footprint may grow larger and larger with continued network operation, which indicates more and more radio scenarios will be stored—therefore, more and more experience will be accumulated. A rough estimation of the REM memory footprint for a typical WRAN application is outlined next.

Suppose an REM (database) is developed for a WRAN BS supporting up to 600 subscribers and 12 simultaneous active CPEs, the radius of the cell is 30 km and the service area of the BS is about 2826 km². Each CPE entry in the REM may require 55 bytes of memory, as estimated in Tables 11.5 and 11.6. Therefore, supporting 600 CPEs requires roughly 33 KB memory for one record. Storing one record per minute and 1440 records per day will require about 48 MB memory for one day and 336 MB memory for one week's worth of records. Considering that the trend is for storage devices to become cheaper, higher capacity, and smaller in size, the memory requirements for the REM will not be a limiting factor. Furthermore, by applying memory-management techniques, as described later, the memory footprint of the REM could be further reduced.

To efficiently manage memory for storing and accessing information, the REM information can be generally classified into two broad types: (1) long-term, static or slowly changing information and (2) short-term, fast changing information. Different forgetting factors (or, equivalently, different updating rates) can be applied to these two types of information elements, such that the required memory and the access delay can be minimized. For the WRAN application, some radio environment information, such as the location of TV stations and TV receivers, WRAN BSs, and CPEs, is static or quasi-static. The locations of TV stations are available from the FCC database, whereas the locations of TV receivers can be roughly identified with the demographical information (e.g., population and location and distribution of residence). This kind of information can be stored in long-term memory, and the update period for such kinds of information tends to be longer than that for the short-term memory.

For the WRAN application, the TV channel "reputation" can be determined based on the historical radio environment observation (sensing) information and should be stored in long-term memory [32]. Some radio environment information, such as the PU activity and the location of a mobile device, may dynamically change. This kind of information is short term in the sense that the updating period could be quite short. Some types of long-term REM information that require a huge amount of memory (e.g., geographical information) could be stored in a database, whereas the short-term radio environment data could be stored in the main memory of the processor for quicker access. The partition and update rates for long-term and short-term radio environment information are implementation issues, which could depend on many systems' design considerations (e.g., the specific application of the CR and the available memory at the CR node).

Indexing and Retrieving REM Information

Indexing and retrieving are two closely related issues. A good indexing scheme enables efficient retrieving. The key idea behind REM is digitizing and indexing radio environment information. Just as we can "Google" the needed information from the Internet for various applications, future CRs can autonomously google the desired radio environment information from the global or local REMs for various cognitive functionalities. Table 11.4 presents a simple indexing scheme for radio environment information elements. Various retrieval algorithms can be employed for REM information retrieving (e.g., k-nearest neighbor, binary tree, and cover tree). Furthermore, if the REM is implemented with a commercial relational database, many existing information retrieval tools (e.g., SQL) are ready to use, and relational database theories (e.g., relational algebra and relational calculus) can be further exploited for finding the needed radio environment information.

11.4.5 APIs between REM and Cognitive Engine

Common APIs between the REM and the CE need to be defined to allow independent and flexible development of different CR building blocks (e.g., the REM module, CE module, and spectrum-sensing module). REM access could be classified into two broad categories: queries and updates.

For the WRAN BS cognitive engine testbed, the REM has been implemented with a C++ abstract class that contains both static radio environment information and dynamic information (i.e., the data part) and various APIs (i.e., the function part). APIs between the REM and the CE enable flexible deployment of the REM for various applications. Furthermore, both the REM and CE can be developed and/or updated independently of each other due to the use of APIs.

Some typical API functionalities have been defined for the WRAN BS cognitive engine testbed as follows. For example, in the WRAN BS testbed, an API function has been developed between the CE and the REM; called *AnalyzeChannelStatistics* it can retrieve the historical REM data file and rank the TV (WRAN) channel reputation based on the probability of being occupied by incumbent PUs.

```
AddEntry(REMDataEntry & _rde);              // add one entry to REM data structure
                                            //    array
LoadDatabaseFromFile(const char * filename);  // load database from file
StoreDatabaseToFile(const char * filename);   // store database to file
MergeREM(RadioEnvironmentMap & _rem);         // merge REM files
AnalyzeChannelStatistics( );                  // rank the TV channel reputation
GetChannelReputation();                       // retrieve the TV channel reputation
                                              //    from REM database
```

If the spectrum-sensing module is already integrated into the WRAN BS cognitive engine testbed, the following APIs can be added:

```
REM_Update_PU_Detection (Location, Freq_Channel, PU_ID, timestamp);
REM_Update_SU_Detection (Location, Freq_Channel, SU_ID, timestamp);
REM_Query_WhiteSpace (Location, Freq_Channel, Max_TxPwr, timestamp);
```

11.4.6 REM Dissemination Schemes and Overhead Analysis

Disseminating and sharing REM information is a vehicle for *network support* for CRs. The REM information can be disseminated in infrastructure-based or ad hoc networks by various approaches [29]. In this subsection, three schemes are discussed for REM dissemination in ad hoc CR networks. The overhead traffic is one of most important performance measure for REM dissemination and indicates the efficiency of REM dissemination algorithms [29].

REM Dissemination Schemes

A first and naive scheme is to periodically broadcast REM information to the entire network via plain flooding. This approach is straightforward although it tends to be prohibitively costly in terms of energy and spectrum consumption. The merit of this scheme is that it is simple, easy to implement, and requires no specific protocol support. It may be appropriate for small CR networks consisting of a limited number of nodes.

Motivated by the analogy between routing information dissemination in link state-based routing protocols and REM dissemination in CR networks, the Optimized Link State Routing Protocol (OLSR) can be extended for REM dissemination. More specifically, REM dissemination in ad hoc networks can be viewed as setting up a "generalized routing table" in cognitive radio networks. Considering that OLSR is a proactive protocol and uses the link-state scheme in an optimized manner to diffuse topology information, the topology control (TC) message in OLSR can be extended to support other dimensions of REM information in addition to topology; TC messages are broadcast and retransmitted by the multipoint relays (MPRs) in order to diffuse the messages in the entire network.

The OLSR protocol developed for mobile ad hoc networks is an optimization of the classical link-state algorithm tailored to the requirements of a mobile wireless LAN. The key concept used in the protocol is MPRs—selected nodes that forward broadcast messages during the flooding process. This technique substantially reduces message overhead compared to a classical flooding mechanism in which every node retransmits each message when it receives the first copy of it. In OLSR, link-state information is generated only by nodes elected as MPRs. Thus, a second optimization is achieved by minimizing the number of control messages flooded in the network [23].

The OLSR protocol provides optimal routes (in terms of number of hops). The protocol is particularly suitable for large and dense networks because the mobile and portable radio technique works well in this context. Figure 11.12 illustrates the MPR employed in OLSR and how it works. As we can see from this figure, MPRs are selected in a way such that messages from current nodes can reach all of its two-hop neighbors via retransmissions by MPRs only.

Finally, application-specific ad hoc methods can be used to further reduce the control overhead. For example, the REM dissemination rate can be adaptively adjusted according to the features of the incumbent PUs or interference; we may also disseminate only the selected REM information (rather than complete map) in an "on-demand" mode. A clustered approach may also be effective in optimizing the cooperative CR network performance while keeping the overhead to a minimum.

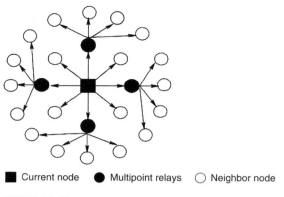

■ Current node ● Multipoint relays ○ Neighbor node

FIGURE 11.12

Illustration of MPRs employed in OLSR protocol.

The REM information can be disseminated through a common control channel, which could be implemented with one of the following four options:

1. A (narrowband) channel in licensed band
2. A channel in the license-free ISM or UNII band
3. An ultrawideband (UWB) channel
4. Sharing with the traffic channel(s)

In summary, to minimize the traffic load of REM dissemination, we may try to minimize the number of retransmissions, reduce the message size of REM, and limit the sources of REM (i.e., the number of overhead originators).

REM Dissemination Overhead Analysis

Jacquet et al. [46] present a theoretical analysis on the performance of OLSR multipoint relay-based flooding and develop two graph models for indoor and outdoor scenarios, respectively. These analytical models can also be used for REM dissemination overhead analysis for the scheme based on OLSR and are summarized next [29].

Case I—Random Graph Model for indoor ad hoc networks: The overhead of topology broadcast via MPRs in the random graph model is $O((\log N)^2)$ when N tends to infinity. Note that N is the number of wireless nodes in the network. This is a large reduction in overhead compared to $O(N^2)$ associated with the plain link-state algorithm.[3]

Case II—Random Unit Graph Model for outdoor ad hoc networks: The overhead of topology broadcast via MPRs in the random unit graph model is $O((N)^{2/3})$ when N tends to infinity.

Figure 11.13a plots the analytical upper bound of topology broadcasting overhead via plain flooding and MPR flooding, respectively. As can be seen, the broadcasting

---

[3] Classic link-state algorithms declare all links with neighboring nodes and flood the entire network with routing messages.

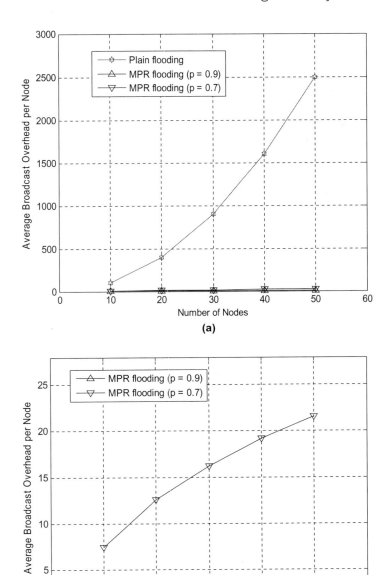

FIGURE 11.13

Broadcast overhead comparison: (a) plain flooding versus (b) MPR flooding.

overhead can be significantly reduced via MPR flooding by orders of magnitude relative to plain flooding, especially for a network consisting of a large number of nodes. Figure 11.13b shows the MPR flooding overhead for ad hoc networks with different link probabilities based on the random graph model. The link probability is p; in other words, a link exists between two nodes with probability p. Higher link probability indicates that the radio nodes are closely connected with each other and therefore result in much less MPR flooding overhead than that when the nodes are loosely connected. Figure 11.13b is a close-up of the details of the overlapping curves in Figure 11.13a.

The overhead of REM disseminations for different network size, topology, node density, and mobility is also simulated with NS-2 [29]. It shows that the overhead of REM dissemination can be significantly reduced by exploiting the MPR flooding method.

11.5 NETWORK SUPPORT SCENARIOS AND APPLICATIONS

This section illustrates the implementation of the REM for CR network support through the analyses of various application scenarios requiring different network structures and services [16]. It shows that REM-based network support has the following key features:

- It fits into the core of CR functional architecture.
- It is independent from specific network topology.
- It is compatible with hybrid node technology and intelligence.

With the help of the REM, a radio can become aware of the performance metrics and the application, topology, and networking (routing), as well as the MAC and PHY layers of the communication stack under various dynamic radio environments. For example, if the radio is used on the battlefield, reliability and security are of utmost importance. Therefore, special source coding, encryption, antijamming channel coding, frequency planning, and routing algorithms could be employed accordingly. The REM can support various network architectures: centralized, distributed, or heterogeneous networks, or even point-to-point communications. It can also support collaborative information processing among multiple nodes for obtaining comprehensive awareness. With the REM, a CR can choose an access network based on cost, data rate, spectral efficiency, and many other performance metrics. The optimal adaptation is subject to the constraints of various radio scenarios (e.g., available services, available spectrum, user policy, and capability of the radio equipment).

11.5.1 Infrastructure-Based Network and Centralized Global REM

The centralized global REM can play an important role in many infrastructure-based radio systems such as the IEEE 802.22 wireless regional area network (WRAN) and cellular radio systems. For example, the 802.22 WRAN is the first worldwide wireless standard based on CRs. It is composed of WRAN basestations, repeaters, and consumer premise equipment. The 802.22 systems are primarily targeted at rural and remote areas

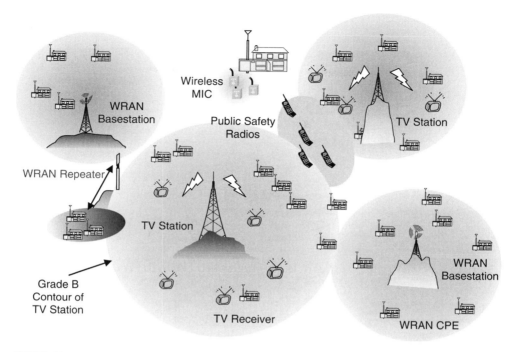

FIGURE 11.14

A typical radio environment for cognitive WRANs. It aggregates the knowledge of all local wireless activity. (Modified from slide 9 in [18]; used with permission.)

offering fixed wireless access services. Figure 11.14 shows a typical WRAN scenario in which TV stations, TV receivers, wireless microphones, and public safety systems operating on certain TV channels are primary users and 802.22 system subscribers are secondary users [17, 18]. Coexistence is the key requirement for 802.22 systems because the SUs must avoid generating harmful interference to the PUs and/or other collocated secondary users [19, 20].

With a global REM maintained at a WRAN infrastructure, the WRAN BS can know the location, antenna height, and transmit power of nearby TV stations; the local terrain; the Grade B service contours of TV stations; the forbidden spectrum used by public safety radios and satellite communications; the demographical distribution of TV receivers; and available TV channels to use. It can also know the distribution and usage patterns of other WRAN service subscribers. Such information helps the WRAN BS choose the best spectrum opportunity to use at the optimal transmission power. Smart antenna techniques can be more efficiently employed at the BS and/or the CPE with the radio environment information from the REM.

Another example application of the global REM is cognitive WLAN supporting network-based interference management. Figure 11.15 shows a typical 802.11 WLAN radio environment in an office building where various industrial, scientific, and medical (ISM) band interferences (e.g., microwave oven leakage, cordless phones, Bluetooth

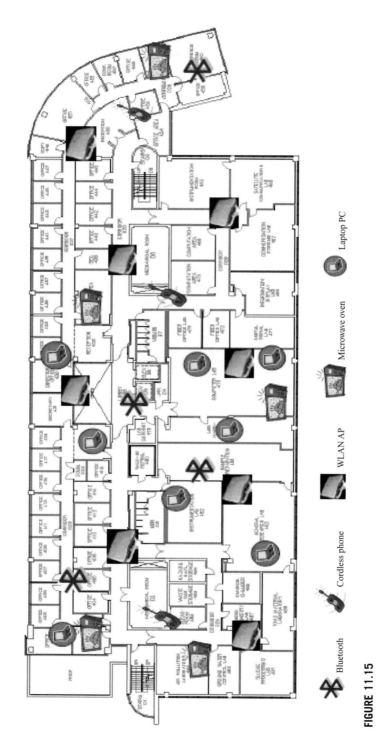

Bluetooth   Cordless phone   WLAN AP   Microwave oven   Laptop PC

FIGURE 11.15

A typical REM of 2.4-GHz WLAN activity. This map provides interference avoidance and management services.

devices, and ZigBee devices) coexist [21]. A global REM can be built up at the network infrastructure and used as a WLAN interference management tool. WLAN interference is detected, classified, and located by the cognitive subscriber nodes or access points (APs). Such information is stored and updated in the global REM. In this way, the WLANs can cooperate to effectively identify and report the sources of interferences and to make adaptations at APs and/or subscriber devices to mitigate or avoid the interference—that is, changing the channel, managing the diversity mechanisms, selecting proper MIMO schemes, or switching off some APs. Note that no changes to the current WLAN PHY–MAC standards are required; however, flexibility in the APs and/or subscriber devices may be needed.

11.5.2 Ad Hoc Networks and Distributed Local REMs

Local REMs can be used in ad hoc (mesh) networks that consist of CRs. The Wireless World Research Forum (WWRF) has established a working group (WG3) to study cooperative and ad hoc networks as an integral and evolving part of the future communication infrastructure [22]. In such networks, the topology can be self-organized and optimized; various modulation and coding schemes can be adapted; smart antenna and MIMO techniques can be selected dynamically; and information-sharing and collaboration methods (e.g., maximal throughput, improved reliability, or extended transmit range) can be employed to achieve the goal(s) defined by the network or participating radios. As pointed out in Chapter 7, for ad hoc CR networks, distributed learning is another effective way of learning that can greatly increase the power of CRs. By sharing and exchanging the local REM information, CR nodes may dynamically employ the most appropriate routing protocol and air interface. Master nodes may be elected to be responsible for collecting the distributed local REMs and combining them into a global REM for the whole network. Such a complete map can be accessed by each individual node, similar to the routing table for an ad hoc network.

11.6 EXAMPLE APPLICATIONS OF REM TO COGNITIVE WIRELESS NETWORKS

This section provides more detailed examples of applying REMs to both infrastructure-based and ad hoc cognitive spectrum-sharing wireless networks. Preliminary experimental results from a CE testbed and from a network simulator are presented, respectively, which demonstrate that significant performance gains can be achieved with REM-enabled cognitive learning and adaptation algorithms.

11.6.1 Applying REM to 802.22 WRANs

The following sections explore the principles of REM applied to 802.22 and various testbeds experiments with REM concepts.

Overview of Potential Applications of REM in 802.22 WRANs

Compared to traditional fixed wireless access systems, the most distinctive features of cognitive WRANs are the awareness and protection of PUs, as well as the autonomous management and optimization of radio resources. Potential applications of REM in cognitive WRAN systems may include, but are not limited to, the following [32]:

- Network initialization – "The WRAN BS starts by consulting the TV usage database and the regional WRAN information base to find potentially empty channels" [40]. The BS can be initialized by downloading relevant REM entries.
- Efficient spectrum sensing and optimal channel assignment – From historical information about REM, the basestation can derive the usage patterns of both PUs and SUs, which can be exploited for efficient spectrum-sensing and channel state prediction [41].
- CPE transmission power control.
- Awareness and protection of PUs.
- Fast adaptation – By leveraging prior knowledge of the radio environment and applying REM-enabled case- and knowledge-based learning algorithms (REM-CKL, to be further explained in this subsection), the BS can make significantly faster adaptations for channel evacuation when PUs reappear.
- Radio resource management and optimization.
- REM-based CPE positioning enhancement.
- REM-enabled WRAN interference management and network security – A detection scheme for malicious PU emulation attacks could be based on location verification, as proposed in [39].
- Performance testing and evaluation of CRs – Testing CRs is much more challenging, due to their flexibility, learning capabilities, and the demanding or unpredictable operating environments. REM-based radio scenario-driven testing (REM-SDT) is a viable approach to evaluating the performance of the WRAN cognitive engine [35].

Experimental Results of a Cognitive Engine Testbed

REM-Enabled Case- and Knowledge-Based Learning

As discussed in Chapter 12, case-based learning (CBL) is an inductive learning technique that recalls previous experiences to solve current problems [42, 43]. Knowledge-based learning (KBL), also known as rule-based systems, is a deductive learning technique that employs prior knowledge for similar purposes. REM-CKL is proposed to leverage both CBL and KBL techniques in conjunction with the REM.

Motivations for Applying REM-CKL to Cognitive WRANs

WRAN systems have the following unique features as compared to cellular systems:

- The demanding coexistence requirements of WRANs require that the WRAN BS and CPEs be able to make prompt complicated adaptations in multiple domains including time, frequency, power, modulation, and coding.
- The radio links of WRAN systems are quasi-static; considering that both PU and SU nodes could be assumed to be stationary, the spectrum usage patterns of PUs and SUs are usually periodic over one day and/or one week.

- The nature of TV signals is not bursty, and TV signals usually have predefined schedules that operate on fixed, preallocated TV channels and are not changing dynamically. Wireless microphone operational bands are also fixed and known.

All these features motivate us to employ the CKL in a WRAN cognitive engine. REM provides an efficient tool to characterize radio scenarios, which is a prerequisite for employing CKL.

How REM-CKL Works

A system approach that formalizes the REM-CKL is as follows. First, the CR needs to be aware of its situation by characterizing and indexing the current radio scenario, and then retrieving the case library and/or knowledge base for applicable experiences and/ or rules. For WRAN systems, the radio scenario can be identified by the combination of active CPE identities, the requested services, active/candidate/occupied channel sets, which altogether present the geographical layout of radio links and the required and available radio resources at the BS.

An important step for the REM-CKL is to determine the similarity between radio scenarios (i.e., the correlation between REMs). In other words, it is the similarity between two "cases": the current case and a previously experienced and stored case. A generic approach to accomplish this is to define the similarity function as follows:

$$f(\cdot) = w_1 * \Delta_1 + w_2 * \Delta_2 \tag{11.1}$$

where Δ_1 and Δ_2 represent the similarity of RF environments (e.g., the RF emission requirements) and the similarity of problems (e.g., responding to the service requests from CPEs subject to available radio resource at WRAN BS), respectively; w_1 and w_2 are weights assigned to Δ_1 and Δ_2, respectively. In general, $f(\cdot)$ is an application-specific function that can be heuristically determined. How to define the similarity function for complicated radio scenarios is a remaining research issue.

For 802.22 systems, a unitless metric, called radio resource units (RRU), is proposed to measure the radio resource capacity at a WRAN BS [31, 33]. The difference between emission masks at the BS and CPEs represent the similarity between radio environments:

$$\Delta_1 = \frac{1}{N \times P_0} \sum_{i=1}^{N} |\Delta P_{max}^i| \tag{11.2}$$

where N is the total number of operational channels at the WRAN BS, P_{max}^i represents the maximum allowed transmit power on the ith channel at BS, and P_0 is a normalization factor.

The difference between RRU profiles (i.e., the distribution of available RRU among all operational channels) at a WRAN BS indicates the similarity of radio resource optimization problems:

$$\Delta_2 = \frac{1}{N \times RRU_0} \sum_{i=1}^{N} |\Delta RRU_{available}^i| \tag{11.3}$$

where $RRU_{available}^i$ represents the available radio resource for the ith channel in terms of RRU, and RRU_0 is a normalization factor.

The required *RRU* for setting up a connection between the basestation and CPE is estimated by

$$RRU_{req} = (1 + \alpha) \frac{R}{\eta BW_{sc}} \qquad (11.4)$$

where α is the overhead factor (unitless) that takes the overhead of the WRAN protocol into consideration and can be determined by the WRAN system specifications; R is the data rate of the new connection (in units of bits per second, bps) and is determined by the service type; η is the spectral efficiency (in units of bps/Hz) jointly determined by the highest applicable modulation level and channel coding rate; BW_{sc} is the bandwidth of the WRAN OFDM subcarrier (in units of Hz) that is defined by

$$BW_{sc} = \frac{TV\ Channel\ Bandwidth}{FFT\ Mode} \qquad (11.5)$$

For the OFDMA/TDD-based WRAN, the RRU_{req} indicates the number of OFDM subcarriers that need to be allocated.

REM-CKL can provide a good starting point for CRs and help accelerate the optimization process. For WRAN applications, after the current radio scenario is characterized, KBL is applied with the current domain knowledge to seek an appropriate solution. If none can be found, the CE searches its previous experience through the CBL. Emulations and/or field testing can be employed to initialize the case library, which should be updated continually with feedback information from the network operations.

Preliminary Experimental Results

Preliminary experiments are conducted to provide concept proof of the REM-CKL cognitive engine. For the WRAN BS cognitive engine testbed, the preset simulation parameters and tunable parameters are summarized in Tables 11.7 and 11.8, respectively. Figure 11.16 shows the WRAN BS cognitive engine performance simulation results in terms of adaptation time under five different radio scenarios (as depicted in Table 11.9). For each scenario, a number of new connections are added to an existing WRAN network. Each scenario was run for 20 times, and the new CPEs have random locations with respect to the BS for each run. Compared to GAs, when using the REM-CKL algorithm, the WRAN cognitive engine makes adaptation significantly faster, which is critical for time-sensitive CR applications. In the meantime, the REM-CKL can achieve the near-optimal global utility by leveraging the REM-CKL and a simple local search. For a more detailed description of the WRAN cognitive engine testbed and further experimental results, readers are referred to other studies [31–33].

11.6.2 Applying REM to Ad Hoc Spectrum-Sharing Networks

This subsection presents some preliminary experimental results from a network simulator developed with Network Simulator 2 (NS-2) [45].

The simulation scenario for ad hoc spectrum-sharing networks is shown in Figure 11.17, where 20 SU nodes are moving along the streets in a dense urban area and another 20 PU nodes are stationary and clustered at a street crossing. The objective is to have the SUs share the spectrum with the PUs with minimal mutual interference.

Table 11.7 Simulation Parameters for the WRAN BS Cognitive Engine Testbed

Parameter	Value or Range
Number of BSs	1
Cell radius	33 km
Distribution of CPEs	Random uniformly distributed or clustered
Types of service (data rate) requested from CPEs and QoS (target BER)	■ Voice: 10 kbps; target BER: 10^{-2} ■ Video: 100 kbps; target BER: 10^{-3} ■ Low data rate: 250 kbps; target BER: 10^{-6} ■ High data rate: 750 kbps; target BER: 10^{-6}
Channel model	AWGN channel
Multiplexing/duplexing	OFDMA/TDD (downlink-to-uplink ratio is $3:1$)
FFT mode	2048 (each TV channel has 2048 subcarriers)
Channel (TV) bandwidth	6 MHz
Number of total TV channels supported at the BS	8
Protocol overhead (α)	0.1
$RRU_{capacity}$ per TV channel	2048

Table 11.8 Adjustable Parameters (Knobs) at WRAN BS and CPE

Parameter	Value or Range
Frequency channel	VHF/UHF (54–862 MHz)
CPE and BS transmission power	Up to 4 Watts, subject to the EIRP profile (i.e., emission mask)
Modulation schemes	QPSK, 16-QAM, and 64-QAM
Channel coding	None, 1/2, 2/3, and 3/4 (convolutional coding rate)
Number of UL/DL subcarriers allocated to the new connection	Variable from 4 to 256

Table 11.9 Testing Scenarios for WRAN BS Cognitive Engine Performance Evaluation

Scenario	Number of Existing CPEs	Number of CPEs to Add to Network	Number of Initial Active Channels	Number of Initial Candidate Channels
1	2	3	1	9
2	10	5	2	8
3	10	10	2	8
4	10	20	3	7
5	10	40	3	7

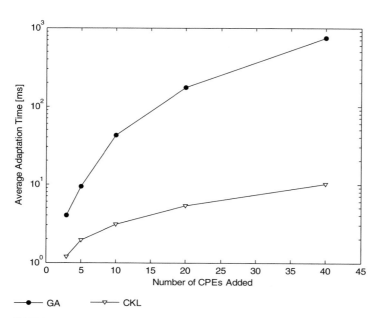

FIGURE 11.16

The required adaptation time for WRAN cognitive engine can be significantly reduced when using REM-CKL as compared to that when using traditional GA [32].

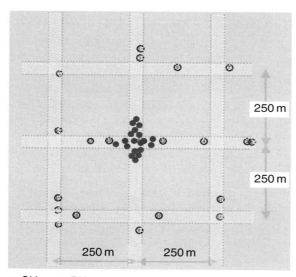

○ SU   ● PU

FIGURE 11.17

Network-level simulation scenario for ad hoc spectrum-sharing networks in a dense urban area.

Table 11.10 Network-Level Simulation Parameters for the Performance Utility Experiments

Parameter	Value
Number of PUs and mobility of PUs	20, stationary
Routing protocol for PUs	AODV
Number of SUs and mobility model in use	20, Manhattan mobility model
Routing protocol for SUs	OLSR
Transmission range of PU	250 meters
Transmission range of SU	Adaptive (0–250 meters)
Free space interference range[a] of SU	Selected from 250–400 meters
Speed of SUs	Uniformly distributed in (0, 10 m/s)
Data rate of wireless link	2 Mbps
Interface queue length	50 packets
Radio channel model	Two-ray ground reflection model and two-ray Manhattan model [44]
Simulation period	480 seconds
Warm-up period	20 seconds
Number of replications	20
Weights	$w_1 = w_2 = 1$

[a] *The interference range of a secondary node is defined as the minimum free-space distance to avoid resulting harmful interference to the PU from the secondary node.*

Constant bit rate (CBR) traffic is generated by two connections between two pairs of PU nodes. The simulation parameters are detailed in Table 11.10. Twenty simulation runs were made to estimate the confidence interval of performance measures. The following utility function is proposed to evaluate the performance of overall networks, including both PU and SU ones.

$$u = \frac{(u_1)^{w_1}}{(u_2)^{w_2}} \tag{11.6}$$

where u_1 denotes the sum throughput (in units of bps), including the throughput from both PU and SU networks; u_2 denotes the average packet delay of the PU network (in units of seconds); and w_1 and w_2 are the weights[4] assigned to u_1 and u_2, respectively. In some sense, u_1 is an indicator of the overall spectrum utilization, whereas u_2 is an indicator of the performance degradation experienced by the PU network [34, 41]. Therefore, u is the overall network utility. In practice, the weight vector ($[w_1, w_2]$) could be determined according to the spectrum "subleasing" agreement between the PU operator and the SU operator or determined by the RF spectrum regulators (e.g., the FCC).

   Figure 11.18 shows that significantly increased network utility can be obtained when CR nodes (i.e., the SU nodes) exploit the REM, in the context of spectrum sharing with

---

[4]Typically, $w_1 \geq 1$; $w_2 \geq 1$.

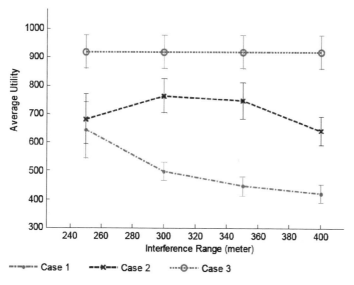

FIGURE 11.18

Network utility comparison when the CR nodes adopt three different adaptation schemes for case 1, case 2, and case 3, respectively. From this figure it is clear that network utility can be significantly increased by exploiting the REM information of spectrum-sharing networks.

incumbent PUs. As depicted in this figure, three different adaptation schemes are evaluated:

Case 1: The CRs are unaware of the topographical environment. Therefore, they take a conservative approach; when any PU node lies in an SU cognitive radio's free-space interference range, the SU CR will stop transmission.

Case 2: The CRs first estimate the pathloss to the PU nodes by using the two-ray ground model and then adaptively adjusting their transmit power if any PU is within their interference range according to the REM.

Case 3: REM-enabled CRs are fully aware of the radio environment and apply the Manhattan propagation model for pathloss prediction. Based on this estimate, the CRs adaptively adjust their transmit power if any PU is within their interference range. The Manhattan propagation model differentiates the line-of-sight (LOS) and non-line-of-sight (NLOS) conditions for appropriate pathloss prediction. The simulation results show that the high penetration loss due to the buildings in a dense urban area creates many "spectrum holes" that enable much higher spectrum reuse by the REM-enabled CRs. Therefore, the network utility for Case 3 is higher than that for Cases 1 and 2.

Readers are referred to [33, 34, 41] for a more detailed description and further results of this research.

11.7 SUMMARY AND OPEN ISSUES

This chapter has addressed the important role of network support in developing CR systems for various application scenarios, including infrastructure-based and ad hoc networks. As a systematic top-down approach to providing network support to CRs, the radio environment map is proposed as an integrated database consisting of multi-domain information such as geographical features, available services, spectral regulations, locations and activities of radios, policies of the user and/or service provider, and past experience. An REM can be exploited by a CE to enhance or achieve most of cognitive functionalities such as SA, reasoning, learning, planning, and decision support. Leveraging both internal and external network support through global and local REMs presents a sensible approach to implementing CRs in a reliable, flexible, and cost-effective way.

Network support can dramatically relax the requirements on a CR device as well as improve the performance of the whole CR network. Considering the dynamic nature of spectral regulation and operation policy, the REM-based CR is flexible and future-proof in the sense that it allows regulators or service providers to modify or change their rules or policies simply by updating REMs accordingly. Because the REM is a comprehensive database that contains multidomain information, even low-cost–low-complexity radio devices can obtain certain cognitive functionalities by making use of an REM.

As a system-level solution to cognitive networking, the REM not only makes cognitive applications feasible, but, more importantly, it facilitates the evolution and convergence of wireless communication networks with cost-efficient databases. The REM presents a smooth evolutionary path from the legacy radio to the CR and can also be viewed as a natural, but major, evolution of radio resource management used in today's commercial wireless networks.

The REM also has great potential in bridging or converging different wireless communication networks, and thus facilitating the integration of various radio network architectures and access technologies, such as WRAN, WiMAX, WLAN, WPAN, B3G, and 4G systems, to support heterogeneous seamless cognitive wireless communications [26]. In summary, employing REMs in CR networks leverages prior knowledge and collective intelligence.

Although the preliminary experimental results from a CE testbed and a network simulator show that a network-enabled approach for CR looks fairly promising, many technical issues currently remain open.

Some important open questions include:

1. How can the global REM and the local REMs keep their information current with minimal communication overhead under various network topologies?
2. How current and what level of granularity does the information contained in the global and local REMs need to be to provide desired performance?
3. What is the impact on CR network performance due to imperfect REM information, and what is the constraint on REM dissemination latency?
4. How can the integrity, security, privacy, and reliability of the REM be assured?

5. How would it be possible to standardize the REM and its API interface to a CE for widespread applications and compatibility? Supposing the REM and its APIs have been standardized, the REM will be better exploited by the CR community and further facilitate the research and development of the cognitive radio [35].

Certainly there are many challenges ahead. Although these technical challenges seem achievable, business, regulatory, and political challenges can be much more difficult to address and predict.

EXERCISES

11.1 Explain why network support is so important for developing CRs.

11.2 Design an REM for a WRAN BS. List the main attributes and estimate the memory footprint.

11.3 Design an REM for a MANET node supporting multiple channels. List the main attributes and estimate its memory footprint.

11.4 Estimate the overhead of REM dissemination in an ad hoc network and an infrastructure-based network, and list at least three methods to reduce the overhead.

11.5 Estimate the memory footprint of REM for various CR nodes sharing the spectrum with other wireless networks, such as WLAN, WiMAX, and WiMedia, and discuss efficient memory-management schemes for CR nodes.

11.6 (a) Discuss the security issue of REM information and REM dissemination. (b) Discuss the possible measures to secure REM information and REM integrity.

11.7 Discuss the remaining open issues summarized in Section 11.7. Do you have some solutions or ideas to these issues? Can you list some other open issues?

REFERENCES

[1] Reed, J. H., and C. W. Bostian, Understanding the Issues in Software Defined Cognitive Radio, *Tutorial for First IEEE International Symposium on New Frontiers in Dynamic Spectrum Access Network*, Baltimore, November 2005.

[2] Krenik, W., and A. Batra, Cognitive Radio Techniques for Wide Area Networks, *42nd Design Automation Conference*, pp. 409–412, Anaheim, CA, June 2005.

[3] Zhao, Y. and J. H. Reed, Radio Environment Map Design and Exploitation, *MPRG Technical Report*, Virginia Tech, December 2005.

[4] Reed, J. H., C. Dietrich, J. Gaeddert, K. Kim, R. Menon, L. Morales, and Y. Zhao, Development of a Cognitive Engine and Analysis of WRAN Cognitive Radio Algorithms, *MPRG Technical Report*, Virginia Tech, December 2005.

[5] Krenik, B., and C. Panasik, The Potential for Unlicensed Wide Area Networks, Wireless Advanced Architectures Group, Texas Instruments White Paper, October 2004.

[6] Sai Shankar, N., C. Corderio, and K. Challapali, Spectrum Agile Radios: Utilization and Sensing Architectures, *First IEEE International Symposium on New Frontiers in Dynamic Spectrum Access Network*, pp. 160–169, Baltimore, November 2005.

[7] Rappaport, T. S., *Wireless Communications Principles and Practice*, Prentice Hall, 2002.

[8] Bauer, G., R. Bose, and R. Jakoby, Three-Dimensional Interference Investigations for LMDS Networks Using an Urban Database, *IEEE Transactions on Antennas and Propagation*, 53(8, Part 1):2464–2470, 2005.

[9] Russell, J. S., and P. Norvig, *Artificial Intelligence: A Modern Approach*, Second Edition, Pearson Education, 2003.

[10] *www.2pass.co.uk/awareness.htm#SAdefinition.*

[11] Mitola, J. III, and G. Q. Maguire Jr., Cognitive Radio: Making Software Radios More Personal, *IEEE Personal Communications*, 6(4):13–18, 1999.

[12] Fette, B. The Promise and the Challenge of Cognitive Radio; available at *www.sdrforum.org/MTGS/mtg_40_sep04/fette_cognitive_radio.pdf*, March 2004.

[13] *www.usgs.gov/.*

[14] *www.fcc.gov/mb/databases/cdbs/.*

[15] *www.darpa.mil/ato/programs/xg/.*

[16] Polson, J., Cognitive Radio Applications in Software Defined Radio, *Software Defined Radio Technical Conference*, Phoenix, November 2004.

[17] Cordeiro, C., K. Challapali, D. Birru, and S. N. Shankar, IEEE 802.22: The First Worldwide Wireless Standard Based on Cognitive Radio, *First IEEE International Symposium on New Frontiers in Dynamic Spectrum Access Network*, pp. 328–337, Baltimore, November 2005.

[18] Kim, C.-J., et al., WRAN PHY/MAC Proposal for TDD/FDD, doc.: *IEEE 802.22-05/0109r0*, November 7, 2005.

[19] FCC, In the Matter of Facilitating Opportunities for Flexible, Efficient, and Reliable Spectrum Use Employing Cognitive Radio Technologies, Authorization and Use of Software Defined Radios, FCC NPRM 03-322, December 30, 2003.

[20] FCC, Facilitating Opportunities for Flexible, Efficient, and Reliable Spectrum Use Employing Cognitive Radio Technologies, FCC NAO 05-57, March 11, 2005.

[21] Zhao, Y., B.G. Agee, and J. H. Reed, Simulation and Measurement of Microwave Oven Leakage for 802.11 WLAN Interference Management, *IEEE International Symposium on Microwave, Antenna, Propagation and EMC Technologies for Wireless Communications*, pp. 1580–1583, Beijing, August 2005.

[22] *www.wireless-world-research.org/index.html.*

[23] Clausen, T., and P. Jacquet, Optimized Link State Routing Protocol (OLSR), IETF RFC 3626, October 2003.

[24] Wang, T., *Global Optimization for Constrained Nonlinear Programming*, Ph.D. thesis, Department of Computer Science, University of Illinois, Urbana, December 2000.

[25] Rondeau, T. W., B. Le, C. J. Rieser, and C. W. Bostian, Cognitive Radios with Genetic Algorithms: Intelligent Control of Software Defined Radios, *Software Defined Radio Technical Conference and Product Exposition*, November 2004.

[26] Wireless World Research Forum Working Group 6 (WWRF-WG6), Cognitive Radio, Spectrum and Radio Resource Management, WWRF-WG6 White Paper, 2004.

[27] Zhao, Y., and J. H. Reed, Radio Environment Map-enabled Cognitive Radios (Poster), Wireless@ Virginia Tech Wireless Personal Communications Symposium, Blacksburg, June 2006.

[28] Marshall, P., and S. Griggs, DARPA Wireless Networking Vision (DWNV), Making Network Centric Accessible for the Warfighter, WANN BAA 06-26, Proposers' day presentation, March 16, 2006.

[29] Zhao, Y., J. H. Reed, S. Mao, and K. K. Bae, Overhead Analysis for REM-enabled Cognitive Radio Networks, *Proceedings First IEEE Workshop on Networking Technologies for Software Defined Radio Networks*, Reston, VA, September 2006.

[30] Zhao, Y., L. Morales, K. K. Bae, J. Gaeddert, J. H. Reed, and J. Um, A Generic Cognitive Engine Design Based on Radio Environment Map (REM), Patent Application, VTIP:07-060; available at *www.vtip.org/availableTech/technology.php?id=192679*, 2007.

[31] Zhao, Y., J. Gaeddert, L. Morales, K. K. Bae, and J. H. Reed, Development of Radio Environment Map Enabled Case- and Knowledge-based Learning Algorithms for IEEE 802.22 WRAN Cognitive Engines, *Proceedings CROWNCOM*, pp. 44–49, Orlando, August 2007.

[32] Zhao, Y., L. Morales, J. Gaeddert, K. K. Bae, J. Um, and J. H. Reed, Applying Radio Environment Maps to Cognitive WRAN Systems, *Proceedings IEEE DySPAN*, pp. 115–118, Dublin, April 2007.

[33] Zhao, Y., Enabling Cognitive Radios through Radio Environment Maps, Ph.D. Dissertation, Virginia Tech, 2007.

[34] Zhao, Y., D. Raymond, C. da Silva, J. H. Reed, and S. Midkiff, Performance Evaluation of Radio Environment Map Enabled Cognitive Spectrum-Sharing Networks, *Proceedings IEEE MILCOM*, pp. 1–7, Orlando, October 2007.

[35] Zhao, Y., S. Mao, J. Neel, and J. H. Reed, Performance Evaluation of Cognitive Radios: Metrics, Utility Functions and Methodologies, submitted to *Proceedings IEEE,* Special issue on cognitive radio, pp. 1–16, January 2008.

[36] Connolly, T., and C. E. Begg, Database Systems: *A Practical Approach to Design, Implementation, and Management,* Fourth Edition, Addison-Wesley, 2005.

[37] Gaeddert, J., K. Kim, R. Menon, L. Morales, Y. Zhao, K. K. Bae, and J. H. Reed, Development of a Cognitive Engine and Analysis of WRAN Cognitive Radio Algorithms, Mobile and Portable Radio Research Group (MPRG) Technical Report, Virginia Tech, December 2006.

[38] SOAR—A General Cognitive Architecture for Developing Systems That Exhibit Intelligent Behavior; available at *www.sitemaker.umich.edu/soar*.

[39] Chen, R., and J. Park, Ensuring Trustworthy Spectrum Sensing in Cognitive Radio Networks, *Proceedings First IEEE Workshop on Networking Technologies for Software Defined Radio Networks*, pp. 110–119, Reston, VA, September 2006.

[40] IEEE P802.22 Draft Standard for Wireless Regional Area Networks Part 22: Cognitive Wireless RAN Medium Access Control (MAC) and Physical Layer (PHY) Specifications: Policies and Procedures for Operation in the TV Bands, 2006

[41] Zhao, Y., J. Gaeddert, K. K. Bae, and J. H. Reed, Radio Environment Map-enabled Situation-aware Cognitive Radio Learning Algorithms, *Proceedings of Software Defined Radio (SDR) Technical Conference*, Orlando, November 2006.

[42] Xu, L. D., Case-Based Reasoning: A Major Paradigm of Artificial Intelligence, *IEEE Potentials*, pp. 10–13, Dec. 1994–Jan. 1995.

[43] Gaeddert, J., K. Kim, R. Menon, L. Morales, Y. Zhao, K. K. Bae, and J. H. Reed, Applying Artificial Intelligence to the Development of a Cognitive Radio Engine, Technical Report, Mobile and Portable Radio Research Group, June 2006

[44] Raymond, D., I. Burbey, Y. Zhao, S. Midkiff, and C. P. Koelling, Impact of Mobility Models on Simulated Ad Hoc Network Performance, *Proceedings Ninth International Symposium on Wireless Personal Multimedia Communications*, pp. 398–402, San Diego, September 2006.

[45] The Network Simulator (NS-2); available at *www.isi.edu/nsnam/ns/*.

[46] Jacquet, P., A. Laouiti, P. Minet, and L. Viennot, Performance Analysis of OLSR Multipoint Relay Flooding in Two Ad Hoc Wireless Network Models, Research Report 4260, INRIA, *RSRCP Journal Special Issue on Mobility and Internet*, September 2001.

Cognitive Research: Knowledge Representation and Learning

12

Vincent J. Kovarik, Jr.
Harris Corporation
Melbourne, Florida

12.1 INTRODUCTION

Generally speaking, a cognitive radio (CR) is a radio system that can adapt its behavior or operational characteristics in response to changes in the radio's internal state or external environment based on encoded knowledge. The ability to adapt to situations and observed conditions may address a range of capabilities, from providing better operational performance to making the radio system more "personal," as suggested by Mitola and Maguire [1]. The following are examples of internal states that may alter the operational behavior of the radio.

Low battery power: A reduction in available power may result in the inability to support multiple waveforms or to provide sufficient transmit power. In response, the radio may select and disable a low-priority waveform in order to conserve power or maintain operation of a mission-critical waveform.

Component failure: The radio may reroute around a failure point; it may provide an alternative service possibly at a lower data rate, or it may terminate a lower-priority service in favor of mission-critical communication.

Some examples of environmental states that can affect operational behavior include:

Cosite interference: Use of the same radio frequency (RF) by multiple radios or by a single radio attempting to exercise multiple frequencies through a single RF transmitter will diminish or negate successful communications. The radio may select an alternate operating frequency or change other waveform characteristics.

Background noise and interference: Noise and interference from other RF sources can impede the effectiveness of communications on a given channel or frequency. Again, the radio may sense the level of background noise and change operational frequency, perform frequency hopping, or initiate other adaptive transmission actions to enable effective communications.

Although, a CR system may exhibit intelligent behavior by adapting to changes in its environment, it is, nonetheless, following a fixed set of behavioral guidelines. As long as the environmental and internal states remain within the boundaries of its established guidelines, the radio system will continue to successfully modify its operational behavior in response to those changes. However, if the internal state or the environmental conditions fall outside the range of the precoded patterns and responses, the radio system cannot identify a feasible solution path to follow. Without a prescribed set of actions, the system typically ceases to operate or vacillates between states in an attempt to find a stable operational mode.

Every CR system consists of several common functional elements, including:

- Sensors or other methods for gathering information about the external environment and internal state of the system.
- A corpus of knowledge that defines a set of behaviors to be performed in response to some set or pattern of inputs from the external environment and/or internal states of the radio.
- A reasoning engine or algorithm that applies the knowledge to the current state of the system and reaches one or more conclusions.
- A control interface that enables the cognitive system to change the radio's operational characteristics or, in some way, act on the conclusions.

The reasoning engine modifies the operation of the system based on the application of knowledge to the combined state information. Reasoning is the process by which the system has an existing set of knowledge, applies it to a current situation, and identifies a course of action.

Figure 12.1 illustrates the basic operational control of a CR. The radio is performing some communications function represented by the data plane box in the figure. The waveform implementation and general radio hardware information provide the radio system's internal state data. One or more sensors provide information regarding the environment in which the radio is operating; this state information provides input data to the reasoning engine. The reasoning engine applies the collection of knowledge to the current state of the system. Based on the current combined state and the knowledge, the reasoning system changes one or more operational parameters for the function being performed.

As shown in Figure 12.1, a CR can intelligently adapt its operational behavior in response to external and internal influences. However, such a system has a critical limiting factor: The range of adaptation the system can perform is based solely on existing conditions and actions as represented in the knowledge base. Thus, even though the radio can react intelligently to external environment and internal state changes, it can adapt its behavior only within the bounds of the previously defined knowledge; that is, it cannot adapt to new situations. To modify its behavior in response to new situations, the radio must *learn*.

Learning entails not only the ability to sense and adapt based on algorithms, or heuristics, but also the ability to analyze sensory input, recognize patterns, and modify internal behavioral specifications based on some type of comparative analysis of the new situation. The learning algorithm can improve the selection of a candidate action, and improve the results of the selected action versus the projected outcome.

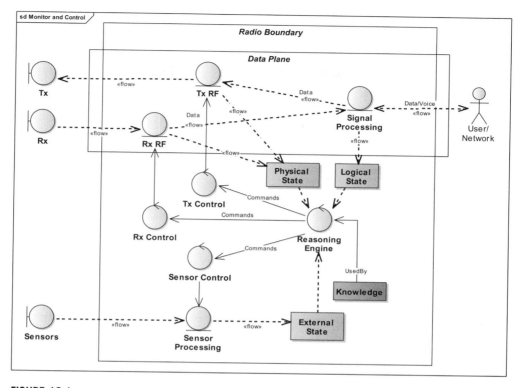

FIGURE 12.1

Cognitive radio monitoring and control.

A simple machine-learning architecture is illustrated in Figure 12.2. The architecture has a set of control parameters (e.g., rules, functions, algorithms) that provide some control output to the external environment. The essential concept is that the machine-learning system has some perception of the environment, gained through sensors, external data input, and other means, that provides the baseline truth on which the system asserts some conclusion or action. Coupled with the action is some predictive assertion regarding the anticipated impact or change to the environment as a result of the action. The action in turn alters or modifies the environment in some fashion. As the environment is modified, changes are received by the learning component and compared against the expected changes. If the resultant changes are the same as the anticipated changes, then, based on the proximity of the actual changes to the expected changes, the system reinforces the parameters that led to the decision. If the resultant changes are different than the predicted values, then the learning system will alter the parameters of the decision process to more closely fit the actual results.

Figure 12.3 shows the learning architecture illustrated in Figure 12.2 merged with the basic cognitive architecture of Figure 12.1. To learn, there must be some method for assessing the operational performance of the reasoning engine and, based on that

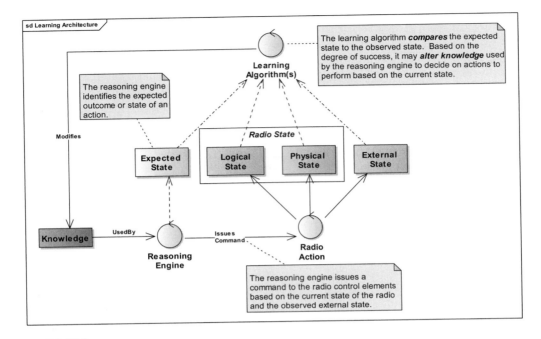

FIGURE 12.2

General machine-learning process.

assessment, a method for modifying the set of knowledge that is applied by the reasoning engine. So, as illustrated in Figure 12.3, a learning algorithm is integrated into the cognitive decision process of the radio system. The learning algorithm, although potentially complex, has a simple mission:

1. To *observe* the state of the radio system and the actions selected by the reasoning engine.
2. To *compare* the resultant state after the action selected by the reasoning engine is performed to the anticipated state.
3. To *modify* the knowledge base to reflect the success (or failure) of the selected action.

Alteration of the knowledge base can take multiple forms, from modifying existing knowledge, to changing the action-selection policy, to generating entirely new knowledge entries for the system to apply in subsequent actions.

The balance of this chapter is organized into five sections. Section 12.2 provides a brief overview of knowledge representation and reasoning paradigms. Understanding the different representation mechanisms is an important prerequisite to discussing learning algorithms because the algorithm may be inextricably tied to the underlying representation and reasoning mechanism. Section 12.3 presents several approaches to

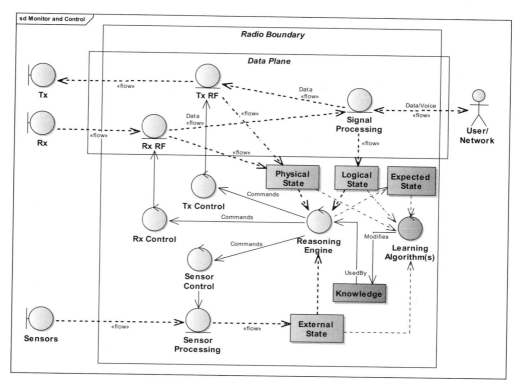

FIGURE 12.3

Learning integrated into CR architecture.

machine learning and their application within a CR system. Section 12.4 presents some considerations regarding the implementation of CR technology within a radio system. Section 12.5 summarizes the chapter followed by exercises in Section 12.6.

This chapter provides a representative cross-section of machine learning with an emphasis on application within the radio communications domain. It does not, however, provide an exhaustive overview of the multiple areas of machine learning, related research, and approaches; rather, some approaches are discussed in depth, whereas others may be discussed at a superficial level or not at all.

12.2 KNOWLEDGE REPRESENTATION AND REASONING

A common theme across the machine-learning methods discussed in this chapter—and, to a great extent, across machine-learning research—is the underlying tenet to enable a machine to react intelligently to some future situation based on knowledge it has gained through experience. Knowledge can have multiple forms based on its type and the selected reasoning mechanism. How the reasoning system represents and applies the knowledge underlying its decision process has an impact on the learning process applied to improve the system. This section looks at several common implementation methods and their potential uses within a CR system.

In humans, knowledge is represented as a complex set of neural connections that are activated in an associative manner. The set and pattern of the neural activation is experiential in nature and represents a thing or concept and the properties that describe the concept's state—a relationship or context describing how the concept is related to other things, or an action or event defining a temporally ordered set of related actions with associated pre- and postconditions that describe the changes resulting from the event or action. Thus, when we are faced with a situation that we remember encountering before, the set of neural connections activated by the current state also activates a set of connections representing the action or response performed when previously encountered.

The different sensory inputs interact to form a multidimensional perspective of a concept or object that is in the physical world. For example, waking up in the morning to the smell of bacon cooking evokes a set of associated references. The olfactory sensory input simulates other neural associations that were activated as part of previous, similar experiences. Thus, we can visualize cooking bacon, remember the taste of the bacon, and perhaps a specific memorable occasion. However, more significant than simply remembering the situation is the ability to recall actions associated with a particular situation. Continuing with our breakfast scenario, we may also remember the last time we cooked bacon and the steps involved: getting the bacon out of the refrigerator, getting the frying pan from the cabinet, getting tongs or another utensil to manipulate it, placing the bacon in the pan, flipping it over as it cooks, placing the cooked bacon on a plate, and then placing the plate on the table. The key concept here is that, in addition to remembering the *objects* or concepts, we also remember the *process* or action sequence associated with achieving a particular goal.

Supporting a reasoning mechanism requires two critical capabilities. First, there must be some form of capturing, categorizing, and storing the knowledge. Second, stored knowledge must be activated and applied to a given set of parameters or conditions. Learning provides a method for extending the set of knowledge structures without explicit programming on the part of a human developer. The types of learning that can be supported depend on the underlying structure of the knowledge. Representing knowledge can take many forms, and in some cases, it may take multiple forms within a single system. For the objectives of this chapter, the knowledge representation forms are limited to two basic categories: *declarative* and *behavioral*.

Declarative knowledge and reasoning refers to any method of representation that asserts and reasons over descriptive factual knowledge about an object or entity. For example, the statement, "The radio can transmit and receive in the VHF frequency band," provides a simple declarative fact: that the radio is operable from 30 to 300 MHz. In this example, the fact describes a property or attribute about the radio. This type of declarative knowledge is different from that provided by the statement, "The radio is operated by John." In this statement, declarative knowledge is also provided, but, whereas the first example provides information that describes a property of the radio, the second asserts a relationship between two entities—the radio and John. Thus, the second assertion of declarative knowledge provides contextual or semantic knowledge about the radio and its relationship to other entities within the knowledge space. The key distinction is that the radio's frequency range is independent of contextual relationships and is a primitive property of the radio entity. The "operated" relationship, on the other hand, describes semantic connections between two objects.

Behavioral knowledge and reasoning are concerned with actions and the effect they have on the knowledge system. The actions may be initiated by some agent or actor within the system or they may be external events that alter the system's environment. In both cases, the event has some initial state or set of assertions at the start of the action, some state at the end or completion of the action, and some temporal extent over which the action occurs.

The remainder of this section briefly discusses several different representation and reasoning approaches that intersect the preceding knowledge types. It should be noted that a particular approach may encompass reasoning, representation, or both. Furthermore, one reasoning approach may use different representation approaches, depending on the situation.

12.2.1 Symbolic Representation

In a symbolic representation and reasoning system, extensible data structures capture the salient facts, descriptions, and properties associated with a concept. This encapsulation of multiple descriptive information forms a *symbol*, or a unit of knowledge. Symbolic knowledge representation can have any of multiple forms. For example, knowledge may be represented symbolically as a conceptual entity that has a set of properties describing the entity, or it may represent a semantic association between two entities. For example, Figure 12.4 illustrates a simple assertion in symbolic form expressing a *transmits* relationship between Radio 1 and Radio 2. This concise pictorial view captures the fact that Radio 1 transmits some information or signal to Radio 2. The assertion may be represented in basic data structures within a high-level programming language, such as C++, Java, or LISP, to name a few.

Each of the radios in Figure 12.4 can be represented by a data structure that may have additional descriptive information or properties about each of the radios. This is illustrated in Figure 12.5, in which Radio 1 now has additional information associated with it that describes its operational frequency and the power level of its transmitter amplifier.

The benefits of a symbolic knowledge representation approach include that it is relatively easy for humans to understand, and it is easily captured in graphical and/or natural language notation. Symbolic knowledge representation also provides a more natural means of conveying the underlying information for human understanding and interpretation.

Knowledge can be represented symbolically by using a variety of approaches, including semantic nets, rules, frames, and objects, among others. The key underlying concept is that the representation formalism uses a symbolic form to represent declarative knowledge.

As a CR system functions, part of the process, as described in the previous section, is to observe the actual response or result of its interaction with the environment, to

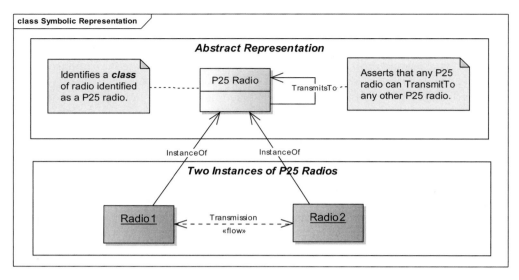

FIGURE 12.4

Symbolic knowledge assertion.

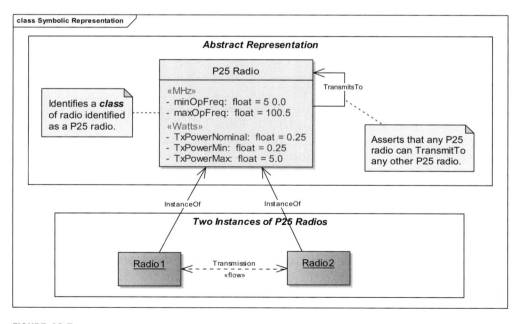

FIGURE 12.5

Properties associated with Radio 1.

analyze the result with respect to its expectations, and then to adapt it based on some learning algorithm. One of the applications of learning is to extend existing knowledge based on newly observed data or patterns.

12.2.2 Ontologies and Frame Systems

Declarative knowledge is of little value when it is simply a disjoint collection of facts and assertions. The data must be organized into a useful form in order to support reasoning and learning. This organization is typically referred to as an *ontology*. Ontology-based knowledge representation and reasoning for software-defined radios (SDRs) have been investigated by a number of individuals, including Wang et al. [2, 3], and in this text by Kokar et al. [4].

The common *is-a-kind-of* relationship organizes classes or types of entities into a type–subtype hierarchy similar to object-oriented programming languages. The differences between the type and subtype can be structural or behavioral. For example, a pager and a cell phone may be both classified as *a-kind-of* personal communications device.

Ontology-based representation can be traced back to early frame systems [5], and both have several aspects or viewpoints that can be applied to the organization of a corpus of knowledge into an ontology. The Web Ontology Language (OWL) of Bechhofer et al. [6] has been used by a number of individuals to represent declarative knowledge about CR structures and relationships. For example, Baclawski et al. [7] explore the use of ontology-based reasoning for communications protocol interoperability. Figure 12.6 illustrates a possible ontology of communications devices according to a type hierarchy. Two common types of devices—a cell phone and a pager—are described within this contextual organization.

The value of this type of organization from a learning system perspective is that, as new concepts and situations are encountered, the existing knowledge ontology can be searched to identify an appropriate classification for the new situation or concept. For example, if a Family Radio Service (FRS) walkie-talkie is introduced into the system, it may be described as being used by an individual capable of either sending or receiving at any given point in time. Using this knowledge to guide the traversal of the knowledge hierarchy, the system would follow the hierarchical links based on the known attributes of the new entity and reach the conclusion that the cell phone radio system should be classified as a *PersonalCommunicationsDevice* and a *FullDuplex* device.

Learning within an ontology-based system is performed by incorporating and integrating new concepts into the existing set of entries. New concepts are incorporated into the ontology based on their similarity to existing entities; that is, new information is processed by a classifier. The classifier analyzes the properties and the values associated with the new concept and compares it against the existing set of concepts within the ontology. Wherever there is a similarity, the similar concept or concepts are candidates for associating the new concept into the ontology.

This is nearly the same as the method by which we incorporate new concepts within the framework of our knowledge and experience. Using this approach, new entities can be linked into an ontology, enabling the system to extend its set of knowledge. Note that, although the example shown describes physical entities, the ontology can

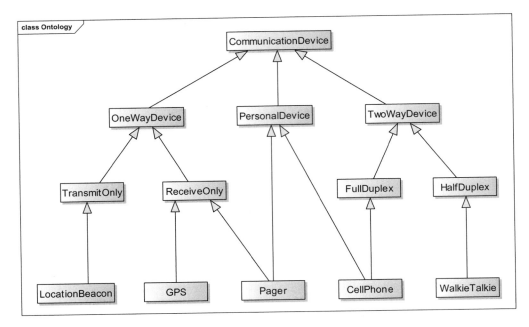

FIGURE 12.6

Communications device ontology.

also represent related actions, events, or abstract entities, such as waveforms, and the attributes associated with each of these types of entities. Thus, organizing new knowledge can be performed across a range of conceptual domains.

Also note, however, that learning based on ontology extensions is predicated on the existence of an initial set of entities already organized into a hierarchical network of concepts.

12.2.3 Behavioral Representation

Behavioral knowledge refers to the embodiment or description of actions that an object or entity can perform or has performed. This may be behavior executed in response to a simple stimulus, as in the case of object-oriented programming languages; it may be described within a neural net or genetic algorithm (GA) as a set of patterns and actions; or it may be a more complex behavior embodied within a set of rules or heuristics that allow additional options or degrees of flexibility in the entity's response to a given stimulus.

Although each type of behavioral knowledge may vary significantly in both flexibility and the type of behavioral patterns it supports, each has the same fundamental objective: describing behavioral responses that an intelligent system can exhibit in response to a stimulus it receives.

Behaviors may be realized as control functions that the CR is capable of performing —for example, setting the operational frequency, changing automatic gain control (AGC) parameters to boost weak signals, or reduce audible noise on high-interference channels. The behaviors are represented in an implementation-independent form within the knowledge representation form. These are then mapped to a control function or application programming interface (API) call. The control API interfaces to the device-specific driver or interface that effects the change to the physical device or software to initiate the behavior within the radio.

Autonomously extending behavioral knowledge requires the system to incorporate a learning algorithm implementation that monitors the results of the system's actions and adapts or modifies the decision parameters based on the specific approach implemented by the learning algorithm. Typically, this is accomplished by strengthening those decisions or actions that lead to achieving a system goal, weakening those choices that did not lead to the attainment of the goal, or both.

12.2.4 Case-Based Reasoning

When faced with a new situation or problem, people typically attempt to find some prior experience or situation that is similar and recall which actions were taken in response to the situation. In some instances, there is a high degree of similarity between the current situation and a prior experience in memory. When there is a high degree of similarity, the actions previously taken can usually be applied to the current situation. In those instances where there is some similarity to prior experience, then typically only some subset of the prior actions may be applicable or some may be applicable but require modification. This process of comparing current situations to prior experience and then applying, potentially with adaptations, prior actions to the current situation is the fundamental approach of case-based reasoning (CBR). The origin of CBR is generally attributed to Schank [8].

Figure 12.7 illustrates the general architecture of a CBR system. Performing case-based resoning consists of several steps:

1. The current situation or state must be categorized and quantified. This entails collecting the set of external and internal sensory input, whether obtained autonomously or through external input, and coercing it into a canonical form that is consistent with the set of previously stored cases.
2. The current case is compared against the set of stored cases to identify previous experience that is similar to the current state.

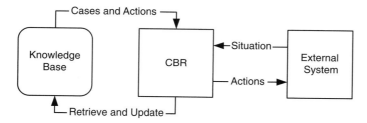

FIGURE 12.7

CBR applies prior experience to the current situation.

3. The actions taken for one or more of the most similar cases are retrieved and applied to the current situation. If there are no matching cases, then a set of actions must be adapted from a similar case or new actions must be proposed [9].

4. The case and the actions taken are updated or stored. If the case performed is new, then it must be classified and inserted into the library of cases with the appropriate semantic relationships to the other cases.

Implementations of CBR systems may vary as widely as the systems to which they are applied. The underlying representation may be frame based, organized into ontologies, or based on symbolic knowledge representations. Regardless of the underlying representation, the essential reasoning mechanism is to perform a pattern-matching process that codifies the current state of the system, both internally and externally, and to compare that against stored representations of previously encountered or codified states entered as a baseline set of knowledge.

This last point raises a key aspect of CBR: the accuracy and value of the reasoning process increases proportionately to the breadth and depth of the cases available. In other words, to be effective, a case-based system must have a significant set of prior experiences from which it can search and identify similar situations that can be applied to the current state.

The second key aspect of an effective CBR system is its ability to adapt the actions associated with a prior case that is close to but not a clear match for the current situation. The ability to identify specific action steps requiring modification, to change those actions identified in a meaningful way, and to perform the actions is a central CBR capability.

Depending on the underlying representation of the cases and the number of cases stored in the knowledge base, the memory requirements for a comprehensive CBR system can be fairly significant. Another performance consideration involves the computational requirements to (1) perform the pattern matching required to identify potential cases that are similar to the current state, and (2) perform the modifications necessary to the stored set of actions, when there is not a highly similar match, in order to apply them to the current situation.

Case-based reasoning would typically not be a candidate for low-end, resource-limited systems because of its memory and processing requirements. It is more suitable for larger systems with adequate computing resources, such as fixed systems in vehicles, larger aircraft, and ships.

12.2.5 Rule-Based Systems

Rule-based (production) systems have a long history [10] and have been applied to a variety of applications. A rule-based system has a knowledge base represented as a collection of "rules" that are typically expressed as "if-then" clauses. The set of rules forms the knowledge base that is applied to the current set of facts. Rule-based systems provide a method for representing inferential knowledge by using a simple "if-then" form, which is relatively easy to state and understand. The rule paradigm is naturally understood by humans. The basic architecture of a production rule system is shown in Figure 12.8.

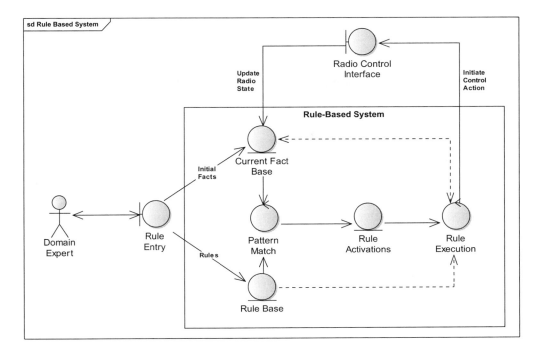

FIGURE 12.8

Basic architecture of a rule-based system. The production rule matches the current state in working memory to one or more rules in the knowledge base.

As illustrated in the figure, the current state of the system is represented as a set of facts or assertions in working memory. The corpus of knowledge is stored within a set of rules that form the rule base. The inference engine performs a pattern match of the antecedents or conditional portion of the production rule against the set of assertions in the working memory. When there is a match, the rule is tagged for possible execution, building a set of rule activations. Because more than one rule may match the current state in working memory, some mechanism is usually provided for resolving the conflict and deciding which rule should be executed from the set of possible rule activations. The selected rule is then executed, or "fired," resulting in some action being performed or one or more facts being asserted to (added) or retracted from (removed) the working memory and performing any control calls to the radio through the API. Any external data (e.g., radio state and sensory input) that are represented in working memory are updated. The updated statistical information in the current fact base is input into the pattern-matching process and the cycle continues.

Figure 12.9 illustrates a rule-based implementation. The syntax shown is a common form for production systems. The antecedents are expressed as a set of tuples in parentheses. The implied relationship between the tuples is usually AND. The use of the question mark (?) in front of a term (e.g., ?waveform) denotes that the item is a variable.

```
(defrule is-spectrum-available
      "Check current set of sensed frequencies for availability"
    (not (spectrum-sensed ?freq))
    (spectrum-requested ?freq ?waveform)
    =>
    (assert (spectrum-available ?freq)))

(defrule issue-start-waveform
        "Initialize state information to start waveform
        on the available frequency"
    ?fact1 <- (spectrum-requested ?freq ?waveform)
    ?fact2 <- (spectrum-available ?freq)
    =>
    (retract ?fact1)
    (retract ?fact2)
    ?wf-status <- (start-wf ?waveform)
    (assert (spectrum-in-use ?freq ?waveform))
    (assert (start-waveform-status ?wf-status)))
```

FIGURE 12.9

An example of rule-based reasoning.

So, any assertion in the fact base that matches the pattern results in the variable being set to the value in the fact base.

There are two rules defined in the knowledge base: *is-spectrum-available* and *issue-waveform-start*. The first rule checks to see whether there is an existing request for a waveform to be started on a particular frequency, (spectrum-requested ?freq ?waveform), and that the frequency is available, (not (spectrum-sensed ?freq)). If so, it asserts, that is, adds to the fact base, that the request is allowable: (spectrum-available ?freq).

The second rule is activated when it has been confirmed that there is a specific frequency request by a waveform, (spectrum-requested ?freq ?waveform), and that the spectrum is available, (spectrum-available ?freq), which was asserted by the first rule. Both these assertions are remembered by the rule in temporary variables: ?fact1 and ?fact2. The rule retracts ?fact1 and ?fact2, removing them from the working memory, and then issues a call to start the waveform through an external function, start-wf. The result of the start request to the radio is provided as a return value and stored in the ?wf-status variable. The rule then asserts that the spectrum is in use, (spectrum-in-use ?freq ?waveform), and also inserts the status of the call to start the waveform, (start-waveform-status ?wf-status).

One of the shortcomings of the rule-based paradigm, however, is that it typically has no means to introspect the knowledge within the system. In other words, the system's knowledge provides guidance regarding the domain of the system and responds based on the set of knowledge and the current state of the environment. However, the rule engine cannot scan through the rules to adjust them, add rules, or delete rules. To accomplish these activities, an additional set of reasoning must be implemented. Thus, the learning algorithm would be applied to monitor which rules were applied in a particular situation, assess the success of the decision process based on the actual outcome versus the predicted outcome, and then have at its disposal access to the rules in a form that the learning algorithm can understand and process.

12.2.6 Temporal Knowledge

Temporal reasoning provides the ability for a system to reason about its operational characteristics within the context of time. More than simply a discrete set of points, temporal reasoning that captures relationships between temporal intervals provides a basis for qualitative temporal reasoning that does not require assignment of specific time values. The notion of temporal interval logic proposed by Allen [11] comprises a set of 13 relationships that captures the relation between any two temporal intervals. For example, interval A may overlap interval B, implying that A starts before B and A ends sometime between the start and end of B. A number of researchers have developed additional aspects of temporal interval logic, including Kovarik and Gonzalez [12], resulting in a rich variety of temporal representation and reasoning approaches, as described by Allen [13].

Dynamic spectrum utilization can be mapped into a temporal representation paradigm by mapping the times associated with observed frequency usage. Once a map of spectrum usage over time has been built, intervals of time can be identified when a particular portion of the spectrum is underutilized. These predictive temporal "holes" can then be used to more intelligently allocate spectrum in a collaborative fashion across a number of devices. This concept was proposed by Kovarik [14] and is illustrated in Figure 12.10. Temporal reasoning also has applications in the area of cosite interference mitigation and concurrent requests for physical and processing resources.

12.2.7 Knowledge Representation Summary

Section 12.2 has thus far presented a brief overview of representation and reasoning paradigms. Each of the different implementation approaches has benefits and limitations. Although there are multiple reasoning paradigms from which to choose, several key tenets can be asserted regarding the use of a reasoning system within a CR system.

First, even though any one of the approaches described, as well as those not covered in this section, will provide a measurable degree of intelligent behavior within a CR, no single representation and reasoning system is capable of embodying all of the types of knowledge and reasoning expressiveness required for the range of operational situations that a radio system must be capable of handling.

Second, although each reasoning mechanism provides a degree of adaptability to operational situations not explicitly defined, when any of the mechanisms encounters a totally new situation or scenario, the ability of the system to reason and decide on an appropriate behavior or response drops off sharply. This brittleness at the edge conditions leads to the pattern of oscillation between stable and unstable states, or termination of operation, because the system has encountered a situation for which it does not have any precoded knowledge.

Finally, the performance of each of the reasoning systems discussed is highly dependent on the breadth and depth of the knowledge codified within the system. The more knowledge encoded within a system, the better it performs. This axiom has its conse-

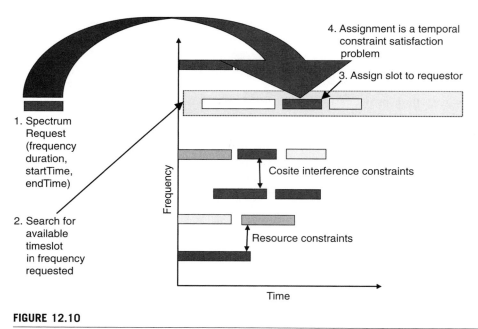

4. Assignment is a temporal constraint satisfaction problem

3. Assign slot to requestor

1. Spectrum Request (frequency duration, startTime, endTime)

Frequency

Cosite interference constraints

2. Search for available timeslot in frequency requested

Resource constraints

Time

FIGURE 12.10

Dynamic spectrum assignment based on temporal mapping of frequency usage.

quences, of course, in that the cost in terms of memory and computational resources required to support a specific reasoning system may be significantly more than is available on some resource-limited systems.

The next section discusses different learning mechanisms. Each of the methods can typically be applied to different representation and reasoning implementations, as just discussed in this section.

12.3 MACHINE LEARNING

Research in machine learning has grown dramatically during the past decade, with a significant amount of progress along several research paths. Multiple strategies exist for implementing machine learning, each with some degree of benefits as well as drawbacks both in general and in the context of a CR system. This section addresses common learning approaches and provides some analysis of their applicability to CR systems.

One of the important aspects of the learning mechanism is whether the learning performed is supervised or unsupervised. In a supervised-learning system, as the name implies, learning is performed through a set of predetermined conditions, and the system is trained to select the right choice or outcome. In effect, the learning mechanism learns to associate a particular set of input stimuli with a learned response or output. In the context of a CR system, either technique may be applied as the radio's method of learning. Supervised learning may take the form of a dialog between the operator and the radio, in which the radio may develop some new assertion or operational behavior model based on the learning mechanism within the radio and then ask for confirmation from the operator that its conclusion is correct.

Conversely, the radio system may extend its knowledge through the learning algorithm and simply add the new knowledge to its existing base of knowledge assertions and behaviors. In operational radio systems, there will likely be some level of protection or guard to limit the degree of behavioral modifications that a CR can incorporate without external verification and validation. In general, this will come from a policy system that controls the learning process to ensure accuracy.

The potential problem with autonomous or unsupervised learning is that the learning process may result in many wrong choices before an acceptable decision path is developed. This can be computationally intense and time consuming. Learning algorithms, such as reinforcement and temporal difference—discussed in Sections 12.3.5 and 12.3.6, respectively—are applicable to this type of learning.

The following sections provide an overview of several types of learning mechanisms. As previously noted, the focus is on those learning mechanisms that are applicable to CR systems. Consequently, not all learning methods are covered within this chapter.

12.3.1 Memorization

One of the most basic learning mechanisms is memorization, or rote learning. This approach captures a sequence of steps or a response to a specific set of conditions and then, when the same task or set of conditions is again encountered, the memorized responses are applied. This can be an effective method of learning if the range of situations encountered by the radio system is limited and well defined. Incomplete and partial data and situations, however, are not amenable to rote learning.

Rote learning forms a set of steps that represent the observed or described sequence of actions for a given set of conditions. Applying the sequence learned requires the system to take a set of measurements, sensor data, or descriptive input data provided by an external source, and then use that information as an associative key to look up the information within an internal database of patterns. When the matching set of descriptive data is found, then the set of associated actions is activated and applied to the current state.

An example of memorization would be the selection of a particular waveform or specific adjustments to the operating parameters of a waveform in response to measured interference. Although the memorization would allow for multiple response sequences, each associated with a specific interface value, each sequence of actions would be explicitly tied to the measured values. Memorization does not allow for generalization of responses based on similar responses to different measured values. This capability could be thought of as a computational implementation of conditioned response, as pioneered by Ivan Pavlov [15]. Pavlov's research on conditional reflexes greatly influenced not only science but also popular culture. The phrase "Pavlov's dog" is often used to describe someone who merely reacts to a situation rather than using critical thinking. When the radio senses some specific combination of stimuli, it performs some specific conditioned response based on the association of the response to the stimuli over time. The problem is, of course, that if the set of stimuli that may be sensed by the radio is insufficient to delineate two distinct conditions that may exist given the same set of stimuli, the radio will execute a learned behavior that may not be correct for the actual condition.

Due to the explicit and inflexible nature of memorization learning, it is best suited for relatively fixed sequences of actions in which there is not a great deal of variation in the set of choices. Examples would be the ordered set of steps associated with radio start-up and initialization, such as the rendezvous process described in Chapters 4 and 20.

12.3.2 Classifiers

Classifiers refer to algorithms that analyze two or more units of data or knowledge and identify similarities or patterns in the structure and content of the data. Classifiers are applied to areas (e.g., data mining) and, in the case of declarative knowledge structures, provide a learning mechanism by extending an ontology with new concepts based on the similarity of the new concepts to those already within the system.

Similarities can be discovered between two distinct pieces of declarative knowledge by a number of methods. A classifier can compare the set of properties associated with each of the concepts and, based on the similarities of the two concepts, item B may be categorized as a *subtype* of item A. This would result in the insertion of item B into the ontology as a subtype or subclass of item A.

The process of quantitatively assessing the similarity between two concepts can be attributed to Tversky in his "Features of Similarity" paper [16]. A computational implementation of a common method for assessing the similarity between two objects is to take the set of properties associated with each of the objects and create an n-dimensional space, with each dimension representing a property, and then assign a number between 0 and 1 to represent the value of each property for each object.

Once this structure has been built for each object, the center-of-gravity point can be calculated as the sum of the distances between each of the property values in the N-dimensional space. Then, the relative similarity between the two concepts is the distance (i.e., difference in magnitude) between the two center-of-gravity points. The shorter the distance between the points, the more similar the concepts are.

In addition to classifiers being applied to declarative knowledge structures in an ontology, they can be applied to identifying similarities and patterns in declarative structures representing temporal relationships. Learning as applied to temporal knowledge enables the radio system to enhance or extend its repertoire of known behavior patterns and then apply those patterns to new situations. For example, the system may observe a particular pattern of interference on a regular set of frequencies at a particular time of day. Or, a system may observe that a particular pattern of activities and events precedes the initiation of a communications activity that is of interest from a signal-analysis perspective. Repeated observation of the temporal patterns reinforces the validity of the set of observed temporal events. The stored temporal pattern can then be applied by matching observed events against the collection of known temporal patterns to predict a specific activity and initiate an appropriate action or countermeasure based on the temporal prediction.

12.3.3 **Bayesian Logic**

Statistical learning methods employ some method of probability of a given outcome for a given set of input stimuli. The system matches a set of active input stimuli to one or more sets of statistical functions having the same input parameters, and then applies the function to the input values, thus generating an expected outcome, course of action, or classification assignment. The probabilities applied to the calculation affect the result.

Learning is accomplished through the system observing the *actual* outcome as opposed to the *expected* outcome and adjusting the weights accordingly. Here the range of statistical functions that can be applied is significantly large.

One of the fundamental behaviors of a reasoning system is the tendency to alter the prediction or expectation of the outcome of an event based on prior history. This concept is the fundamental idea of Bayesian logic. The probability behind Bayes' theorem is:

$$P(\phi|x) = \frac{P(x|\phi)P(\phi)}{P(x|\phi)P(\phi) + P(x|\sim\phi)P(\sim\phi)} \tag{12.1}$$

The event or occurrence of interest is represented as ϕ, and x is the observed phenomenon or evidence of the occurrence. $P(\phi|x)$ is the probability that the event ϕ has occurred given the observation x or, in other words, a measure of the reliability of the assertion that ϕ has actually occurred given the observation x. The probability of the observation for a specified set of the occurrence of interest is $P(x|\phi)$, and $P(\phi)$ is the overall probability of the occurrence within a sample population. Conversely, $P(x|\sim\phi)$ is the probability that the observation will manifest itself in a sample population that does not contain the event or occurrence of interest, and $P(\sim\phi)$ represents the percentage of the set that does not contain the event or occurrence of interest. An example in the context of a CR illustrates how Bayes' theorem might be applied as a learning mechanism.

Suppose the problem domain is predicting whether there will be interference, ϕ, encountered for a given geographical region. Further, assume that the radio has one or more sensors that can detect the interference. When the interference is detected, a memory location is set to a specified value, flagging the detection that provides the evidence, x, of the interference phenomenon.

As a starting point, assign a value, $P(\phi)$, representing the probability that there actually is interference in a given region. The probability that the interference is not present would thus be $P(\sim\phi)$. However, because detection equipment is not perfectly accurate all the time, there is a finite probability that it will detect interference when it is present, $P(x|\phi)$, and also a finite probability that it will detect interference when it is not present, $P(x|\sim\phi)$, which is also called a false positive.

Assume that the detection equipment is calibrated in the lab and is shown to be accurate 80 percent of the time. So, in general terms, out of every 100 detection attempts when interference is present, 80 will be detected and 20 will not. Assume also, however, that for every 100 readings when there is no interference, the detector registers interference (false positive) 10 percent of the time. Now assume a total population of 10,000 samples of the environment and that the probability interference is actually present is 30 percent. So, of the 10,000 samples, 3000 are likely to be samples containing interference and 7000 are not.

Inserting these values into Eq. (12.1) yields

$$P(\phi/x) = \frac{0.8 \cdot 0.3}{0.8 \cdot 0.3 + 0.1 \cdot 0.7} \tag{12.2}$$
$$= 0.774$$

Therefore, based on an observed interference phenomenon, it can be asserted with roughly a 77 percent degree of confidence that the interference is actually present when the detector reports interference.

Bayesian logic can be adapted as part of the learning mechanism through adjustment of the weights based on observed phenomenon. The limiting factor in general use is the required verification of the prediction in order to adjust the probabilities. Thus, the learning mechanism must have either external input or some internal method of verification of the predicted value in order to adjust the weighted values. Nonetheless, Bayesian logic can provide valuable reasoning support within a CR system by tempering the predicted outcome based on prior experience.

12.3.4 Decision Trees

A decision tree, as the name implies, is a directed graph consisting of a hierarchical set of nodes connected via arcs, where each node represents a choice or decision, and the arcs leading from that node to the next decision node represent the set of possible choices for a given node. A decision tree can be visualized as a sequence of choices in which the path taken through the choices from the starting point to an endpoint is governed by the choices made at the starting node and each successive node. The root node provides the starting point and the leaf nodes contain the decisions or actions to be taken. Decision trees may be simplistic in nature, representing a fixed procedural process, such as problem diagnosis, or may be more complex, applying Bayesian reasoning to the decision process governing the transition between the nodes of the graph.

This type of reasoning approach is referred to as a *solution space search problem*; that is, the system generates a set of possible alternative actions for a given state. Each of the generated alternatives is then explored, alternative actions are generated for each of those states, and so on. As each alternative is generated, its weight is calculated, subsequent alternatives are generated, and a total weight or probability is calculated. Thus, for any given node in a decision tree, the total weight for the path leading to that point is maintained.

The calculation of the weights can be dynamically adjusted based on observed state compared to anticipated state. Thus, the system can learn to respond differently to different scenarios and states. The learning aspect can be extended further by enabling the system to generate wholly new actions for a given node. This enables the radio system to not only learn new behavioral responses based on performance observations but also to extend the range of choices.

12.3.5 **Reinforcement-Based Learning**

Reinforcement-based learning provides a method for assessing the success of a particular action, where success is defined as the actual outcome of the event being the same as (or near to) the anticipated or desired outcome. Based on the degree of success, a reward weight is assigned to the action, thereby reinforcing the selection of that particular action if the system finds itself in the same state at some future time.

There have been a number of advances in reinforcement learning [17]. Reinforcement learning is represented by a series of states as a directed graph. As the system moves through the sequence of states by applying some policy for selecting a path or alternative, the degree of success attained in achieving the system's goals, both short-term and long-term, are measured. Based on the degree of success, the policy is reinforced.

At each point in time, t, the system is in some state, s. The system has a set of actions, A, from which it selects a specific action, a, where $a \in A$, to perform in the given situation with the objective of transitioning to some goal state, s', at the completion of the action. To select an action, the system has a set of policies or rules for deciding which action to apply. The selection of a specific action, a, given a state, s, by a policy, π, can be represented as $a = \pi(s)$.

Based on the degree of success, the system assigns a reward to the action performed, which in effect is a reinforcement of the policy-selection rules. The reward function, $R(s,a,s')$,[1] provides immediate feedback to the selection process. So, as the system selects actions that achieve the goal state, the policy rules that contributed to that selection process are reinforced and the probability of selecting the same action again, if the same set of conditions are present, is increased.

Each correct selection of an action provides an immediate reward. Over the course of several actions, a cumulative reward value is calculated. This provides the mechanism for strengthening the selection of a sequence of actions. In essence, it is reinforcing the ordered set of actions for the attainment of a final objective or goal. Therefore, two methods influence the selection of an action. One is the selection of the current action based on the reinforcement of that action successfully achieving the desired objectives for the current situation and the other is the influence of the successful attainment of the end objective providing reinforcement of the action as a part of the ordered set of actions.

This attainment of a long-term goal is typically represented within the reinforcement paradigm as a cumulative value that is the sum of each of the immediate reward values. Each of the immediate reward values that are part of the sum is discounted by applying a factor, γ^t, to the reward value. The cumulative reward is then expressed as:

$$CumulativeReward = \sum_{t=0}^{\infty} \gamma^t R(s, a, s') \tag{12.3}$$

where $0 \le \gamma^t < 1$.

---

[1] s' is the future state after transitioning from state s.

The effect of the discount factor is to balance the reinforcement value between realization of a short-term goal (i.e., whether the selected action yielded the expected state, given the current state) versus the long-term goal (i.e., whether the selected action contributed positively to the attainment of the long-term goal).

Thus, for any given state and action pair, there is a computed value function for the policy based on the current state, denoted as $V^\pi(s)$. Given that there may be multiple states, s', that may be the result of performing the selected action, a probability assignment is made for each of the possible states. The probability of ending in state s', given a current state of s and that action $\pi(s)$ is taken, is represented as $P(s'|s, \pi(s))$. This probability is then combined with the sum of the short-term reward associated with the current action, $R(s, \pi(s), s')$, and the cumulative discounted reward, $\gamma V^\pi(s')$. This computed value for each possible path is summed together:

$$V^\pi(s) = \sum_{s'} P(s'|s, \pi(s)) \bullet \left[R(s, \pi(s), s') + \gamma V^\pi(s') \right] \qquad \textbf{(12.4)}$$

The overall objective of reinforcement learning is to compute the optimal policy, $\pi*$, for each iteration of the process that maximizes the total reward shown in Eq. (12.4).

Note that the term *policy rule* can be realized within the reinforcement-learning system in a wide variety of ways. The policy may start out as nothing more than a random selection process that, over time, evolves to select the appropriate action based solely on matching the pattern of current state variables to those patterns that were previously encountered, stored, and reinforced because the system chose the right action.

Figure 12.11 illustrates the application of this approach. The radio system is in some current state, s, and a policy, a, has been selected. There are four possible states that may be entered on the completion of the action. For each of these possible states, a probability, $P(s'|s, \pi(s))$, has been assigned. The immediate reward, $R(s, \pi(s), s')$, for each

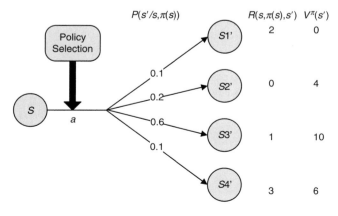

FIGURE 12.11

Reinforcement learning applied to policy selection.

of the subsequent states is shown, along with the estimated value of each of the states, $V^\pi(s')$.

Assume that the discount factor, γ, is equal to 0.6. This would yield a new estimated value of $V^\pi(s)$ as:

$$V^\pi(s) = 0.1(2 + 0.6(0)) + 0.2(0 + 0.6(4)) + 0.6(1 + 0.6(10)) + 0.1(3 + 0.6(6))$$

$$V^\pi(s) = 5.5 \tag{12.5}$$

This value becomes the new value of the selected policy for the current node, $V^\pi(s')$. This process of modifying the policy value of the current node based on the probable path and subsequent nodes is referred to as *backing up the policy value.*

Problems arise with reinforcement learning because, in order to function, a complete model of the system is required that provides the transition probability between states and the reward function. Defining the model to the level of detail required may not be feasible for a CR system.

Finally, in addition to the infeasibility of specifying the complete model description, there is the issue of the algorithm performance. Performing the policy iteration calculations requires $O(n^3)$ time, where n is the number of states in the system.

12.3.6 Temporal Difference

Temporal difference also applies a reinforcement algorithm to the policy selected based on the degree of success. However, the temporal difference algorithm does not require the a priori model of the sequence of possible states, as the standard reinforcement algorithm does. Thus, the temporal difference algorithm builds the state representation "on the fly" and does not require the backpropagation used in the reinforcement algorithm to update policy weights.

The general temporal difference algorithm, introduced by Sutton [18], is referred to as $TD(\lambda)$. There is a variant of temporal difference learning, $TD(0)$, that performs the computation of simple backups as the system transitions through its states. As with reinforcement learning, as described in Section 12.2.4, the system is in state s and, based on the selection of an action based on a policy associated for state s, $a = \pi(s)$, performs a probabilistic transition to a new state, s'. The transition produces the immediate reward, $R(s,a,s')$. The $TD(0)$ algorithm updates the value function by using this equation:

$$V^\pi(s) := (1 - \alpha)V^\pi(s) + \alpha\left[R(s, a, s') + \gamma V^\pi(s')\right] \tag{12.6}$$

In Eq. (12.6), α is a *learning rate parameter*. The initial value of this parameter is usually between 0.01 and 0.5 and it must, over time, reduce down to 0 to allow the $TD(0)$ algorithm to converge. As can be seen in the equation, as the learning parameter reaches 0, the latter portion of the function that incorporates the immediate reward and the discounted value of the next state is algebraically eliminated, and the system converges to the state value $V^\pi(s)$.

Thus, as a node is visited multiple times, the effect is the same as performing a simple backup, as with the reinforcement-learning algorithm in Section 12.2.4. The benefit of the temporal difference algorithm is that the policy value can be calculated without an explicit model. The set of states traversed as a result of the actions and the environment become the model. This can be accomplished because the empirical data gathered as the system traverses the set of states provide the basic data required to compute a probability function for each node traversed.

A second advancement of the temporal difference algorithm introduced by Sutton and Barto [17] over the reinforcement algorithm is that, rather than maintaining a separate value, $V^{\pi}(s)$, for each state s, the value function can be stored as a neural network or some other method for providing a differential approximation. Thus, the value function can be represented as $V(s,W)$, where W is a vector of weights modified based on observed state versus expected state. This prevents the ability to directly assign a value to a state. However, the weights specified in W may be adjusted such that $V(s,W)$ fits closely to the desired value through the use of an error function. The temporal difference error function is:

$$J(W) = \frac{1}{2}(V^{\pi}(s, W) - [R(s, a, s') + \gamma V^{\pi}(s', W)])^{2} \qquad (12.7)$$

This equation captures the temporal difference error as one-half the squared difference between the current estimated value of state s and the backed-up value. Thus, the objective is to modify the set of weights, W, such that the temporal difference error, $J(W)$, is reduced. By differentiating the equation with respect to W and limiting only the first occurrence as being adjustable, the general weight assignment function is obtained. This is shown as:

$$W = W - \alpha \nabla_{w} V(s, W)(V^{\pi}(s, W) - [R(s, a, s') + \gamma V^{\pi}(s', W)]) \qquad (12.8)$$

The term ∇ represents the gradient of V with respect to the weights W. The term α represents the step size to be taken to reduce the error term calculated as the temporal difference error.

12.3.7 Neural Networks

Neural networks and GAs also rely on the reinforcement of a decision or selection based on the actual result or outcome of a decision. (See Section 12.3.8 and Chapter 7 for a discussion about genetic algorithms.) The neural net typically has a vector of input values and a vector of output values. An intermediate layer connects the input and output values, propagating the input values along a set of connections from the input to the output.

Essentially, each node in the neural network takes an input vector, A, and applies a weight vector, W, to perform the propagation of the input value along the link to the next layer in the network. This propagation is typically tempered by a constant or bias value, b. Thus, for an individual node, the propagation output, A, of a given node, j, can be expressed as:

$$A_{j} = \left[\sum_{i=0}^{n} a_{i}w_{i} + b_{j} \right] \qquad (12.9)$$

When the output value, A_{j}, is greater than some threshold value, T_{j}, the value is propagated along the output, as illustrated in Figure 12.12. Learning, within the context of a

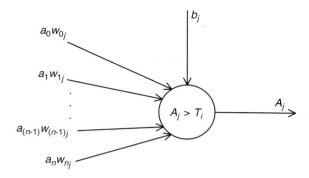

FIGURE 12.12

Neural network node illustration. Shown are the input vector A (with elements a_0, ..., w_n); weight matrix W (with elements w_{0j}, ..., w_{nj}); bias value, b; and output.

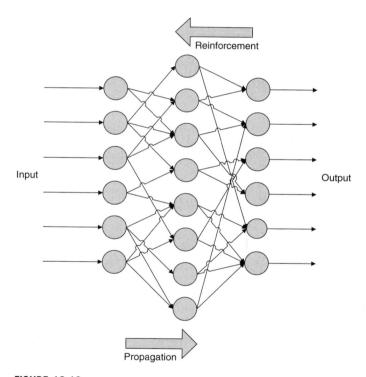

FIGURE 12.13

Neural network. Shown are forward propagation of input and backpropagation learning.

CR, involves the adjustment of the threshold value, the bias value, or the weight associated with a node.

A neural network is formed when a collection of individual nodes is organized in a multilayer fashion, as illustrated in Figure 12.13; the neural network has a set of input points. The values sensed or input through these points is propagated forward, through the middle layer (also referred to as the hidden layer), to a set of output points. The output points activated by the forward propagation are then compared with the actual value (i.e., anticipated versus actual). If there is a match, the path followed to arrive at the output point is followed through backpropagation, and the intervening nodes and paths are reinforced for that particular output point.

Learning within a neural network requires feedback from a training database that allows the network to compare the expected output value associated with a set of input data against the conclusion reached by the neural network. This forms the backpropagation illustrated in Figure 12.13. As the vectors of input values are applied to the system, the data propagate through the intermediate layers to the output. The expected output is mapped onto the output vectors. Those elements in an output vector with a value that matches the applied expected value are reinforced by backpropagation. Thus, the intermediate layers that contributed to the propagation of data resulting in the correct (i.e., expected) output values are reinforced. Reinforcement may be performed through increasing weights, or the bias value for each of the nodes involved in the propagation from the input data to the correct output or conclusion.

This reinforcement of neural links increases the probability that the same input values will result in the same propagation to the correct output values. Those output values that did not match the expected output value are weakened, thereby decreasing the probability that they would be applied again given the same set of conditions.

12.3.8 Genetic Algorithms

Genetic algorithms use a vector that represents a given condition or input stimuli and action. Based on the success of the action taken, the input vector is modified. In the case of GAs, however, the vector may undergo alteration, or genetic mutation, in a random fashion. Even though this may introduce potentially large vector sets, GAs may introduce new solutions through the random mutation process. This enables a system to generate new knowledge and evaluate its effectiveness by using empirical data collected through the operational environment.

The initial implementations of GAs use a bit sequence as the representation of the *gene*. The bit pattern essentially consists of two parts. The first part forms a pattern that is used to match a given set of conditions; the second part forms a bit pattern representing the action or actions to be taken as a result of the pattern match.

As the process of matching the input part and performing the corresponding actions is performed, two modifications of the gene occur. First is the strengthening of those genes that contributed to a successful set of actions (and a weakening of those genes that did not). This serves to enhance the survival rate of those genes that contributed positively, and effectively causes those genes that did not contribute positively to die off (i.e., they are removed from the set of bit sequences). The second action is a random mutation of the bit sequence. This provides a mechanism for generating new alternatives within the set of bit sequences. Thus, not only will genes live or die off through a form of natural selection, but also wholly new strains can be introduced, which may result in more efficient system behavior.

Implementation of GAs has continued to evolve to include the use of a vector of multivalued data items. In effect, each item in the sequence is represented as a discrete variable that can be modified as part of the process. (See Chapter 7 for a detailed discussion of genetic algorithms.)

12.3.9 Simulation and Gaming

The application of simulation and game theory to machine learning provides a method for developing and evaluating alternative actions in a given scenario, thereby proposing new paths or actions. At a fundamental level, game-theoretic learning uses a reinforcement mechanism similar to those previously discussed. Unlike reinforcement and temporal learning, which require a comprehensive model of states, game theory—due to its incorporation of strategy and probability—supports learning in systems that are not well defined or are changing. Game-theoretic learning uses game theory as a means for proposing actions to be taken, given a particular situation.

Game-theoretic learning introduces some features that are unique to this approach. The player chooses from a set of actions for any given state. The action may be selected based on a numerical probability, random generation, or any other selection method. As the player applies the selected strategy, the success of the strategy is recorded for future reference. Thus, as in reinforcement and temporal difference algorithms, there is a bias value applied to a set of actions for a given situation. Chapter 15 provides a comprehensive discussion of the application of game theory to CR systems.

12.4 IMPLEMENTATION CONSIDERATIONS

The previous sections discussed different approaches to knowledge representation as well as reasoning and learning approaches. This section discusses implementation and deployment considerations related to the use of this technology within a radio system. Several areas are related to the technological requirements necessary to realize a CR that learns, and some of the key issues that need to be addressed are related to operational and sociological issues raised when a learning system is deployed.

12.4.1 Computational Requirements

As with any type of work, some effort must be expended to produce the desired artifact or product. CR systems must expend effort in the form of computational resources and the power to drive them in order to reach conclusions or decisions concerning the system's operation. Similarly, computational effort must be expended by a learning system in order to monitor and analyze the results of the actions performed against the expected outcome, perform analysis on the differences between the two, and generate new or changed knowledge to be applied to the next iteration of the decision process.

It can be argued that, whereas the cognitive decision process is an integral part of the operational behavior of the radio system, the learning process is not. Consequently, relegating the learning process to noncritical performance times would be a reasonable operational assumption. However, the primary driver is the availability of computational resources that can be applied to the learning process without affecting the mission-critical performance of the radio system.

12.4.2 Brittleness and Edge Conditions

Learning has the potential to overcome the brittleness of cognitive implementations when faced with a new situation or scenario. However, the resources and time required to generate additional or new knowledge that can be applied to the situation may not be sufficient to result in a feasible alternative in a time frame that can be applied to the situation. Nonetheless, the ability of the system to extend its behavior provides unique opportunities for radio systems. Systems that are to be deployed for long-duration missions in locations that cannot be easily serviced would benefit from the capabilities afforded by learning algorithms.

12.4.3 Predictable Behavior

One of the critical aspects of any automated system is the tacit assumption that the system is deterministic in its behavior; that is, given a set of inputs, the output or action taken by the system can be determined. The problem that arises in any complex system is that each step taken to reach a decision may be individually predictable, but the end decision or action reached may not be what was expected on the part of the human operating or interacting with it. This trait of individually predictable steps leading to an unexpected or unforeseen outcome is one of the founding premises of *chaos* theory. The unpredictability of the endpoint action is further exacerbated by the introduction of learning to the system.

Enabling learning within a system is a powerful two-edged sword. On the one hand, it enables the radio system to adapt autonomously to new situations, change operational characteristics, make operational decisions based on newly learned knowledge, and provide more reliable and dependable service. On the other hand, as systems learn, there is no guarantee that the observed data used by the learning system to adapt or form new knowledge are accurate, nor is it assured that the methodology applied to form some new chunk of knowledge is without flaws.

Just as in the case of humans, the ability to extend the knowledge and behavior of the system through learning is assuredly not predictable. Yet, it is predictability that underlies current methodologies and approaches to radio system certification. Regulatory bodies, such as the Federal Communications Commission (FCC), are just beginning to consider the ramifications of a radio system that is largely or exclusively software based and may change its operational characteristics through changes in software. Regulatory methods and approaches, as discussed in previous chapters, will be stretched as CR technology becomes an integral part of radio systems, allowing them to autonomously change operational characteristics. However, implementation of a machine-executable version of regulatory policy constraints through a policy engine will minimize the extent of undesirable behaviors learned. Finally, as learning technologies are integrated within CR systems, the concept of radio system certification will again need to evolve as cognitive systems gain the ability not only to change operational behavior based on the operational environment, but also to modify the method by which the decisions to change operational behavior are reached.

12.5 SUMMARY

This chapter has reviewed several methods of knowledge representation and reasoning, approaches to computational learning, and possible applications of these methods within the domain of CR systems. An underlying assertion is that no single approach to knowledge representation and reasoning or learning will address all aspects of CR systems. The level of intelligence exhibited by a CR system will be dynamic and varied, depending on the implementation platform, the computational resources available, and the specific mission of the system.

Also, the type of knowledge representation and reasoning system employed, as well as the type of knowledge, may "prefer" a particular learning method. Radio systems may therefore employ a specific representation mechanism, and thus a specific learning implementation based on the operational constraints and objectives imposed. Some systems may host two or more methods, or may interact with other radios in a networked fashion, to implement a larger intelligent entity.

Learning within the context of a CR must encompass multiple representation paradigms, integrate multiple reasoning mechanisms, provide hybrid learning mechanisms, and span multiple functional layers within the CR. This concept is illustrated in Figure 12.14.

The CR can be viewed through three essential aspects or perspectives:

1. At the lowest layer is the physical set of hardware. This provides the physical infrastructure for the radio system.
2. The infrastructure aspect provides an abstraction layer that implements a logical set of interfaces to manage, control, configure, and operate the physical components, and it provides software views of these resources to the waveform implementation. This layer also manages the suite of components that implement a waveform.
3. At the application and services layer, waveforms and system services are viewed and manipulated as a logical entity.

A fourth perspective is the user's aspect. From that viewpoint, the user is either interacting with the physical radio (e.g., power up, diagnostic checks) or interacting with an instantiation of a waveform as a logical entity.

As the right side of Figure 12.14 illustrates, knowledge representation, reasoning, and learning must be implemented within and across each of the layers, yielding a hybrid system. Physical hardware will become smarter with low-level implementation of simple learning algorithms to adaptively modify the power usage or RF components. The software infrastructure will provide a basic reflexive behavioral and learning capability that includes, among other capabilities, safety limitations and operational constraints. At the upper levels of the CR, reasoning and learning become more abstract, addressing mission requirements and user needs.

Another basic tenet of this chapter is that the capabilities of CRs will vary depending on their mission, processing power, and connectivity to other CRs in a network. The radio will embody one or more of the levels of cognitive capability illustrated in Figure 12.15. Not all radios will embody all of the abilities identified. For example, a cognitive sensor network may be deployed as a set of intelligent sensing nodes. These

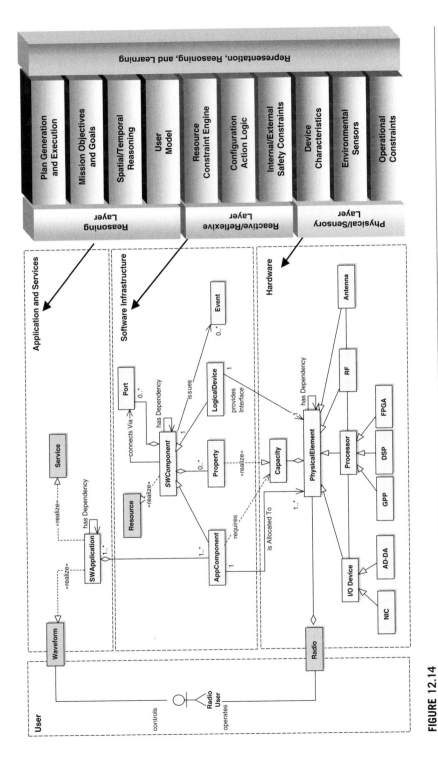

FIGURE 12.14

Multiple learning methods. Learning and reasoning span multiple architectural layers within the CR.

A Cognitive Radio:
- Is **aware** of its environment (*simple sensing*)
  - Space: Where it is
  - Time: When it is
  - Stimuli: What is happening
- **Reflects** on internal knowledge (*state interpretation*)
  - Goals and objectives (e.g., mission, communication plan)
  - Constraints (e.g., safety, physical, policy)
  - Internal state (e.g., current resource allocation)
- **Reasons** over state and knowledge (*plans and alternatives*)
  - Integration of internal state with external stimuli resulting in a set of decisions (i.e., a plan)
- **Acts** on a decision or plan (*autonomous behavior*)
  - Initiates actions following a plan and reacts to changes in state requiring plan modification
- **Explains** its behavior (*provides rationale*)
  - Radio actions may not correspond to user expectations (e.g., Why did it perform some action?)
- **Learns** from experience (*prevents system brittleness*)
  - Extends the range of situations over which the system may operate

Increasing
Cognitive
Ability

FIGURE 12.15

Layers of increasing cognitive capabilities.

in turn may feed one or more nodes that have the processing power to reflect and reason over the data collected. The sensory analysis could then be relayed to another node that analyzes the state information collected and evaluate courses of action.

Consequently, future research and development of computational learning applications for radio systems will need to address the hybrid nature of reasoning and learning within CR systems and between cognitive radios within a collaborative network.

EXERCISES

12.1 A detector used for sensing spectrum occupancy accurately senses energy 95 percent of the time. However, the detector is sensitive to ambient temperature and, when used in environments above 45°C, has a false-positive reading 8 percent of the time. Assuming that interference will be intermittently present 25 percent of the time for a given frequency, what is the Bayesian probability that a positive sensor output is correct?

12.2 Although CBR and rule-based systems have different representations and reasoning mechanisms, there are general similarities. Identify and describe three similarities.

12.3 In descriptive terms, how could the rules shown in Figure 12.9 be modified to incorporate an energy level as part of the spectrum-sensing test?

12.4 A neural network node has four input vectors with input values, a, of 1.1, 0.8, 2.0, 3.2, and weights, w, of 0.3, 0.18, 0.4, and 0.12, respectively; and it has a threshold value of 1.85. What is the propagated value, A? What is the minimum threshold value, T, that allows a positive propagation value to flow to connected nodes?

12.5 Identify and describe three key characteristics of machine learning as it applies to a CR.

12.6 How could the ontology shown in Figure 12.6 be extended to include the following devices?
   (a) Wireless microphone
   (b) Satellite internet connection
   (c) Remote temperature sensor
   (d) Controllable Web camera

12.7 You have a network of sensors providing data on RF activity for a given area to a central node for analysis. For the data collected, describe:
   (a) How the central node could use a neural net to identify a pattern of activity that is of interest.
   (b) How the data might be incorporated into a REM, as described in Chapter 11.

12.8 Describe how access policies, as discussed in Chapter 6, could be represented and implemented using a rule-based approach.

12.9 A specific state within a reinforcement-learning system has five policies with probabilities, immediate reward, and current estimated value for each of the policies shown here. Given a discount factor of 0.47, what is the updated reward value for the current node?

Policy	Probability	Immediate Reward	Current Value
1	0.14	3	1.2
2	0.2	1.5	3.1
3	0.06	0.7	4.5
4	0.4	3	2.7
5	0.2	2	1

REFERENCES

[1] Mitola, J. III, and G. Q. Maguire Jr., Cognitive Radio: Making Software Radios More Personal, *IEEE Personal Communications*, 6(4):13–18, 1999.

[2] Wang, J., D. Brady, K. Baclawski, and M. Kokar, The Use of Ontologies for the Self-Awareness of the Communications Nodes, *Proceedings Software Defined Radio Technical Conference*, Orlando, November 2003.

[3] Wang, J., M. Kokar, K. Baclawski, and D. Brady, Achieving Self-Awareness of SD Nodes Through Ontology-Based Reasoning and Reflection, *Proceedings Software Defined Radio Technical Conference*, Phoenix, November 2004.

[4] Kokar, M., D. Brady, and K. Baclawski, Roles of Ontologies in Cognitive Radios, *Cognitive Radio Technology*, B. Fette (ed.), pp. 401–434, Newens/Elsevier, 2006.

[5] Minsky, M., A Framework for Representing Knowledge, Massachusetts Institute of Technology, Technical Report, UMI Order Number: AIM-306, 1974.

[6] Bechhofer, S., et al., *OWL Web Ontology Language Reference*, M. Dean, G. Schreiber (eds.); available at *www.w3.org/TR/owl-ref/*, February 2004.

[7] Baclawski, K., D. Brady, and M. Kokar, Achieving Dynamic Interoperability of Communication at the Data Link Layer through Ontology Based Reasoning, *Proceedings Software Defined Radio Technical Conference*, Garden Grove, CA, November 2005.

[8] Schank, R., *Dynamic Memory: A Theory of Learning in Computers and People*, Cambridge University Press, 1982.

[9] Kolodner, J., Improving Human Decision Making through Case-Based Decision Aiding, *The AI Magazine*, 12(2):52–68, 1991.

[10] Forgy, C., and J. McDermott, OPS: A Domain-Independent Production System Language, *International Joint Conference on Artificial Intelligence*, pp. 933–939, 1977.

[11] Allen, J. F., Maintaining Knowledge about Temporal Intervals, *Communications of the ACM*, 26(11):832–843, 1983.

[12] Kovarik, V., and A. Gonzalez, An Interval-Based Temporal Algebra Based on Binary Encoding of Point Relations, *International Journal of Intelligent Systems*, 15(6):495–523, 2000.

[13] Allen, J. F., Time and Time Again: The Many Ways to Represent Time, *International Journal of Intelligent Systems*, 6(4):341–356, 1991.

[14] Kovarik, V., Temporal Reasoning in a Cognitive Radio System, *Proceedings Software Defined Radio Technical Conference*, Denver, November 2007.

[15] *www.en.wikipedia.ord/wiki/Ivan_Pavlov.*

[16] Tversky, A., Features of Similarity, *Psychological Review*, 84(4): 327–352, 1977.

[17] Sutton, R., and A. G. Barto, *Reinforcement Learning: An Introduction*, MIT Press, 1998.

[18] Sutton, R. S., Learning to Predict by the Methods of Temporal Differences, *Machine Learning*, 3(1):9–44, 1988.

The Role of Ontologies in Cognitive Radios

Mieczyslaw M. Kokar, David Brady, Kenneth Baclawski
Northeastern University
Boston, Massachusetts

13.1 OVERVIEW OF ONTOLOGY-BASED RADIOS

This chapter discusses the role of knowledge representation (ontologies) in cognitive radio (CR). The emphasis is on those capabilities of cognitive radio that are practical and amenable to ontological treatment. After a brief introduction to ontologies, the chapter develops an ontology for CR functionality that is organized by using the same layers as communication networks. Specific examples of the use of ontologies in CR are then developed in some detail. The examples show how ontologies can be the basis for achieving interoperability at the physical and data link layers. The chapter then outlines some of the major research issues for ontology-based CR.

13.2 KNOWLEDGE-INTENSE CHARACTERISTICS OF COGNITIVE RADIOS

The term *cognitive radio* is interpreted differently by different people. This chapter uses an interpretation that includes a *cognitive agent*, as well as standard radio functionality. Toward this aim we first discuss the features that a cognitive agent is expected to have, and then propose some features that a CR could have.

To put the discussion in context, a CR is viewed as being part of a larger functionality, that is, of an *intelligent agent* or *intelligent personal digital assistant* (PDA) that can support a mobile user. Such a PDA would not only have to advise the user, but it would also have to be connected essentially all the time. Although there are many definitions of a cognitive agent, all of them revolve around similar ideas. For instance, the Defense Advanced Research Projects Agency (DARPA) [1] provides the following definition of a cognitive system:

- Can reason, using substantial amounts of appropriately represented knowledge
- Can learn from its experience so that it performs better tomorrow than it did today

- Can explain itself and be told what to do
- Can be aware of its own capabilities and reflect on its own behavior
- Can respond robustly to surprise

The functionalities of a cognitive agent are often viewed according to Boyd's so-called OODA (Observe, Orient, Decide, Act) loop [2]. This approach is especially popular in the information fusion community where the OODA loop is used to model human behavior, which then serves as a pattern to be followed by a fusion system. In the intelligent control community, this loop is presented in a somewhat simpler form, called the perception–reasoning–action triad [3], used to represent an intelligent controller.

Following roughly the same line of reasoning, this chapter identifies the following basic functionalities that a CR should include:

- Information collection and fusion
- Self-awareness
- Awareness of constraints and requirements
- Query by user, self, or other radio
- Command execution
- Dynamic interoperability at any stack layer
- Situation awareness and advice
- Negotiation for resources

Each of these capabilities is discussed in this chapter from two points of view: (1) the user's point of view and (2) the radio's point of view. This reasoning is illustrated and explained in Figure 13.1. A refinement of the cognitive capabilities to the radio domain is provided in this book.

As shown in Figure 13.1, multiple conversations are taking place at several layers. At a higher layer, the two speakers pose and respond to queries through their radios; simultaneously, at a lower layer the radios converse in a similar fashion. In this case, one radio is trying to establish the multipath characteristics observed by the other with the intent of improving performance. To achieve this goal, the radios need to monitor the communication (indicated in the figure as dashed arrows between the radios and their users) as well as possibly some environmental characteristics, for instance, obtained from other sensors (indicated by the arrows from the radios to the "ancient astronomer" icon). The notion of simultaneous communications at different levels may be applied to each layer in the protocol stack, as discussed in Section 13.2.8. These conversations will be interrelated by the requirement that the "layer below" is supporting the "layer above."

13.2.1 Knowledge of Constraints and Requirements

At the base of any cognitive activity is knowledge. A cognitive agent must possess knowledge and must be able to make use of the knowledge for deriving its decisions and actions. Any software system contains some knowledge, but the real issue is how knowledge can be represented so it can be used in the most flexible way. This issue is

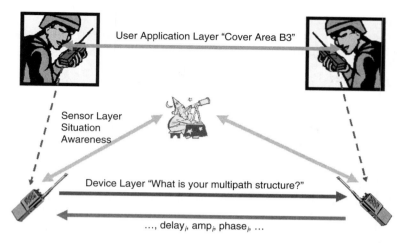

FIGURE 13.1

Three layers of CR conversation: user, sensor, and device. CRs permit conversations at all three layers. Conversation at the user layer is a standard feature of all radios. Communication at the sensor layer is required for self-awareness, situation awareness, and information collection. Communication at the radio layer enables resource negotiation, query, and response.

related to the requirement that a cognitive system "respond robustly to surprise." When knowledge is encoded in a purely procedural way, the use of that knowledge must be prescribed (encoded) during system development. Thus, it is the responsibility of the system developer not only to encode all the necessary *procedures*, but also to encode the *invocations* of the procedures. In contrast to this approach, the *declarative* programming approach requires only that knowledge fragments be represented in the system's data structures (e.g., rules); the selection of the knowledge is left to the system. Such decisions are made by the system at runtime based on *pattern matching*; that is, specific knowledge fragments are used when specific patterns are recognized by the system. This approach gives more flexibility to the use of stored knowledge and at least partially satisfies the requirement of the system responding to surprise.

13.2.2 Information Collection and Fusion

In addition to knowledge encoded in the cognitive agent, the agent must also have access to current information. Whereas knowledge is encoded by either the system designer or the user, the current information is collected from various sources at runtime. For instance, a PDA could have access to a global positioning system (GPS) receiver and thus obtain information of its current location. Additionally, a PDA could (potentially) have access to traffic reports, including the status of roads and bridges, as well as accident reports, weather conditions such as blizzards and black ice, and other pertinent details. Based on such information, the PDA could advise the user on efficient routing. Some of this information could be collected by subscribing to a service, whereas other information would require explicit requests. Because such advice would depend on various types of information from multiple sources, the information would have to be integrated and fused appropriately in order to be useful to the user.

13.2.3 Situation Awareness and Advice

Having knowledge and current information stored in the PDA's memory does not automatically guarantee that the PDA is aware of the current situation of the user (and, in effect, the PDA). According to Endsley and Garland [4], "*Situation Awareness* is the perception of the elements in the environment within a volume of time and space, the comprehension of their meaning, and the projection of their status in the near future." In other words, to understand a situation, the agent needs to know not only about all the objects of interest, but also about their relations to other objects and also possible future states of the objects and the relations. This in turn requires that knowledge of models of objects, including their dynamics, be known, so that future states could be predicted and rules for determining the relationships could be derived. With respect to the PDA's task as a travel advisor, an example of a relevant relation might be "on-the-path-from-to," meaning that a specific geographic feature (e.g., a bridge or a river) is on the path that leads "from-to" specific destinations.

Situation awareness has been recognized as one of the major features of a cognitive radio. Various chapters of this book discuss this issue. For instance, situation awareness has application in energy-efficient network routing or adjusting the end-user interface [29]. Marshall [33] discusses the spectrum awareness of CRs. Mitola [32] stresses the awareness of a CR itself. Situation awareness is also discussed in Zhao et al. [34], where three levels of awareness are identified, similarly as in Endsley and Garland [4].

13.2.4 Self-Awareness

To plan and schedule a task for execution, a PDA must know whether the task is within its own capabilities. It needs to understand what it does and does not know, as well as the limits of its capabilities. This is referred to as *self-awareness*. For instance, the radio should know its current performance, such as bit error rate (BER), signal-to-interference and noise ratio (SINR), multipath, and others. In a more advanced case, the agent might need to reflect on its previous actions and their results. For instance, for the radio to assess its travel speed a fortnight ago between locations A and B, it might be able to extract parameters from its log file and do the calculation. For the radio to decide whether it should search for the specific entries in the log and then perform appropriate calculations (or simply guess), it needs to know the effort required to perform such a task and the required accuracy of the estimate to its current task.

13.2.5 Query by User, Self, or Other Radio

Similar to humans, intelligent agents need to be able to answer queries from their users and other agents. Moreover, they should be able to query the user and other agents about relevant information whenever they cannot infer it from their own knowledge. For agents to formulate such queries, they must understand a common language, formally defined in syntax and semantics. Moreover, to formulate queries, agents must have a planning capability. Finally, they should be able to wait for an answer and incorporate the answer into their own knowledge structures. For instance, the PDA might ask the user, "Where do you go next?" or ask another agent, "What is the temperature at your location?" Another radio could ask, "What is your multipath structure?" or "What is your bit error rate?" Timely and accurate responses to these queries are key to the optimization of physical layer performance [30] and to the efficient formation of ad hoc networks [31].

13.2.6 Query Responsiveness and Command Execution

If two humans agree on a plan to perform a specific task, they then (normally) execute the plan by performing the actions they agreed to. For instance, if two people are driving around the neighborhood searching for a runaway dog, they communicate over cell phones and decide who will search which street and when. The implementation of such a feature in a software-defined radio (SDR) is not a simple matter. For instance, if two radios decide that one of them should adjust its communication protocol after exchanging information about the multipath structure, by selecting a different modulation scheme, the radio must be able to understand how to implement such a command.

A simple way would seem to be to provide a number of procedures and invoke one of them using an "if then else" programming construct. However, this would limit the flexibility of reacting to different multipath structures to only the procedures encoded at the development time of the software radio. Moreover, the invocation of the procedures would have to be decided at development time. At the other extreme, one might envision an approach in which code is automatically generated at runtime and then executed. A middle-ground solution would be to use the declarative programming approach in which knowledge fragments are developed and represented as rules at design time and then the runtime system selects, combines, and executes the rules as needed. This approach would provide more flexibility in terms of possible behaviors in response to various external conditions.

13.2.7 Negotiation for Resources

The human activity of planning involves, among other things, negotiation for resources. For instance, if two people decide that one of them will drive to a grocery store to get drinks and the other will drive to the hardware store to buy a ladder, and if they have a sedan and an SUV at their disposal, it is likely that through a rational negotiation process they will use the SUV for the hardware store and the sedan for the grocery store. Similarly, resource negotiation is very useful in software radios. Negotiation of the most important resource, the spectrum, is being investigated in the DARPA NeXt Generation (XG) program [5]. According to the XG vision, SDRs would be able to request unused spectrum that is allocated to another radio. This approach should lead to the achievement of a higher utilization of the spectrum resource.

13.2.8 Dynamic Interoperability at Any Stack Layer

The flexibility of ontology-based radios postulated in the preceding discussion can be extended to protocols at any layer of the protocol stack. This extension would enable intercommunication and negotiation among layers. Query and response among layers invites the possibility of cross-layer optimization of communication efficiency while maintaining the strict functional division between layers. For example, the medium access control (MAC) layer, which handles channel access and routing, might query the local physical (PHY) layer about the residual error in an equalizer, with the goal of preempting an outage. If the reported error is sufficiently large, the MAC might avoid an outage by seeking an alternative channel and route to the destination. As another example, MAC and transport layers at different nodes may also use query and response to reduce the frequency of broadcasting routing tables. Through the process of query and response, MAC layers could also provide a much more meaningful routing metric than hop count or aggregate delay because much more PHY layer information would be available. Data link (DL) layers, which manage Automatic Repeat ReQuest (ARQ) protocols, may avoid some of the rigidness of fixed protocols such as selective repeat go-back-in or stop-and-wait. Through the use of reasoning, ARQs may be tailored on-the-fly for individual links as the nodes learn about the link performance.

This chapter focuses on ontological representations for the physical and data link layers because these layers are the ones that were implemented in hardware or firmware prior to the introduction of SDRs. Higher layers are more likely to be already implemented in software. Furthermore, the lower layers present challenges that are unique to radio communication.

13.3 ONTOLOGIES AND THEIR ROLES IN COGNITIVE RADIO

An *ontology* is an explicit mechanism for capturing the basic terminology and knowledge (the concepts) of a domain of interest as well as the relationships among the concepts [6]. Ontologies are an increasingly important mechanism for the integration of disparate software systems. Indeed, a shared ontology is a fundamental prerequisite for meaningful communication between systems. The ontology can be hardcoded via shared data formats, database schemas, and procedures, but there are significant advantages for expressing the shared ontology by using a formal declarative ontology language that are either difficult to achieve or even impossible without it. The advantages include support for interoperability, flexible querying, runtime modifiability, validation against specifications, and consistency checking.

In the case of SDR, ontologies offer the additional advantage of *self-awareness*: Communication nodes can understand their own structure and can modify their functioning at runtime. Furthermore, nodes can query the capabilities and current state of other nodes, allowing them to modify the processing of packets during a communication session both at the source and the destination. The ontology specifies not only the structure of communication packets but also the processing of those packets according to the communication protocol. The use of ontologies adds flexibility, inferencing, and reasoning features that are not available with ad hoc data structures or database schemas [7, 8].

Systems that do not initially share an ontology might still be able to interoperate by mapping or merging ontologies to synthesize a shared ontology. It may be possible to construct systems that not only use ontologies, but also modify them or even learn from them dynamically.

13.3.1 Basics

An ontology specifies the concepts of a domain, attributes of the concepts, and relationships among the concepts. Each concept is expressed by using a *class*, which may be interpreted as a set of things. Anything that belongs to a class is called an *instance* of that class. Waveforms, packets, and symbol alphabets are all fundamental concepts in radio communication. In the ontology for SDRs, these concepts are expressed as classes. For example, Waveform is a class with instances that are particular waveforms. Classes are organized into a *hierarchy* of classes by the *subclass* relationship. For example, binary phase shift keying (BPSK) and quadrature phase shift keying (QPSK) are both special cases of *M*-ary phase shift keying (MPSK). This is expressed by specifying that BPSK and QPSK are subclasses of MPSK.

An *attribute* is a property that something has, such as the number of symbols in an alphabet or the carrier frequency of a waveform. An attribute is a characteristic of a single entity, where that characteristic is a data value such as a number. A *relationship* is an association among various entities. For example, a waveform is used to represent a sequence of symbols from an alphabet. This is expressed by linking the waveform to the symbol sequence. An ontology will generally have many different kinds of attributes and relationships. The number of symbols in an alphabet might be called *numberOfSymbols,* and the relationship of a waveform with the sequence of symbols it represents might be called the *usedToRepresent* relationship. The term *property* is used for either an attribute or a relationship. As with classes, one kind of property may be regarded as a set of instances, called *facts*. For example, when a particular waveform w is being used to represent a particular sequence S, this fact is the triple (w, *usedToRepresent*, S). Properties can be organized in a hierarchy by the *subproperty* relationship. For example, the *usedToRepresent* property is a subproperty of the more generic *used* property.

13.3.2 Ontology Languages

The rapid expansion of the World Wide Web (WWW) has had a profound impact on communication. One consequence of this expansion is the emergence of the eXtensible Markup Language (XML) as the most commonly supported format for data interchange [9]. This trend has also affected ontology and knowledge representation languages. In order to be interoperable, it is becoming essential that such languages be Web based and expressible in XML. Being Web based means that the ontology is concerned with *Web resources*. A Web resource is identified by a universal resource indicator (URI). In other words, anything being described by a Web-based ontology is a Web resource.

There are three major Web-based ontology languages: XML Topic Maps (XTM) [10], the Resource Description Framework (RDF) [11], and the Web Ontology Language (OWL) [12]. XTM is an International Organization for Standardization (ISO) standard (ISO13250), whereas RDF and OWL are standards of the World Wide Web Consortium (W3C). XTM allows one to specify relationships among any number of Web resources.

By contrast, RDF and OWL restrict all relationships to be binary. Restricting relationships to binary states simplifies the language and processors, but it makes it much more awkward to deal with relationships that involve more than two resources. When one restricts relationships to be binary, all facts are *triples*, consisting of the two entities being related (called the *subject* and *object*) and the relationship between them (called the *predicate*). RDF is an XML-based language for representing triples. The RDF Schema (RDFS) language is an extension of RDF that allows one to specify subclass and subproperty relationships. It also allows one to specify the domain and range of a property.

The OWL language has three levels: OWL Lite, OWL-DL, and OWL Full. They differ in the constructs that are allowed, with Lite being the most restrictive and Full being the least restrictive. OWL adds many new capabilities to RDF and RDFS, such as cardinality constraints, disjointness constraints, enumerations, and inverse properties. However, the most significant new feature is the ability to construct classes from other classes, such as defining a class to be the intersection or union of two or more other classes. For example, a BPSK is the special case of MPSK for which there are exactly two symbols in the waveform alphabet (each one being 180 degrees away from the other). Class constructors are the basis for a form of reasoning known as *description logic* [13].

OWL Lite differs from OWL-DL in the class constructors that are allowed. For example, in OWL-DL one can specify the *complement* of a class (i.e., all instances that are *not* in the class), but this specification is not allowed in OWL Lite. Although OWL-DL allows all class constructors, it does not allow one to cross *metalevels*. For example, in OWL-DL a class cannot be an instance of another class. By contrast, in OWL Full, a class can be an instance of another class, which itself is an instance of yet another class, and so on to any number of levels. Although OWL Full is a very rich ontology language, it is still not as rich as arbitrary first-order predicate logic. Table 13.1 summarizes the features and differences between the various Web-based ontology languages.

Table 13.1 Existing Web-Based Ontology Languages and Their Characteristics

Language/Organization	Features	Reasoning/Complexity
XTM/ISO	Higher-order relationships	None/linear
RDF/W3C	Binary relationships	Minimal
RDF Schema/W3C	RDF plus subclass, subproperty, domain, and range	Subsumption/polynomial
OWL Lite/ W3C	RDFS plus some class constructors, but no crossing of metalevels	Limited form of description logic/ exponential
OWL-DL/ W3C	All class constructors, but no crossing of metalevels	General description logic/decidable
OWL Full/ W3C	No restrictions	Limited form of first-order predicate logic/undecidable

Note: The last column describes the level of reasoning as well as the level of processing complexity as a function of the number of triples.

One might think that the richer the ontology language the better. However, richer ontology languages are also more difficult to process. RDF and RDF Schema are relatively easy to process. The time to process RDF is linear in the number of triples. RDF Schema is a little more difficult, being polynomial in the number of triples and NP-complete in the size of the query in the worst case. OWL Lite is much more difficult, requiring exponential time in the number of triples for the worst case. OWL-DL is *decidable*, meaning that the processing will finish in a finite amount of time, but the amount of time can be more than exponential. Finally, OWL Full is *undecidable*, which means that instance checking need not be answerable in a finite amount of time. The complexity of other tasks, such as concept subsumption and general query answering, has also been analyzed, and generally follows the same increase in complexity as just described for instance checking.

13.3.3 Querying

Given a database, one can extract information by using a *query*. The query language for RDF and OWL is called SPARQL [14]. The query language that has been proposed for OWL is called OWL-QL [15]. As is the case with most query languages, SPARQL is syntactically and semantically very similar to the Structured Query Language (SQL) for relational database systems. SPARQL differs primarily by allowing one to specify patterns. A *pattern* is a fact in which some of the components can be variables.

For example, the query in Figure 13.2 retrieves all fields of all data link frames. The query patterns in Figure 13.2 specify both variables, such as ?x, and constants, such as <http://ontoradio.org/2005/datalink#contains>. The constants are the Universal Resource Indicators (URIs) of Web resources. The ontoradio.org domain is the ontology-based radio (OBR) domain where the OBR ontology resources are defined.

The SPARQL query language differs from database query languages in several important ways. One important difference is that SPARQL is Web based. Whereas databases are generally restricted to a single server or at least one site, RDF and OWL effectively regard the entire Web as being a single database. OWL-QL has the additional feature of specifying a protocol for query requests and answers to support its use by Web services.

```
1.. SELECT ?x
2.. WHERE  (?x, <http://www.w3.org/1999/02/22-rdf-syntax-ns#type>,
                <http://ontoradio.org/2005/datalink#DataLinkFrame>)
3.. AND    (?y, <http://www.w3.org/1999/02/22-rdf-syntax-ns#type>,
                <http://ontoradio.org/2005/datalink#DataLinkField>)
4.. AND    (?x, <http://ontoradio.org/2005/datalink#contains>, ?y)
```

FIGURE 13.2

An example of SPARQL, a query language for RDF. The first line enables the retrieval of attributes of variable ?x that match patterns. Patterns are expressed as triples (subject, predicate, object) following the WHERE delimiter. The second line limits all ?x to be of type *DataLinkFrame*. The third line describes the attributes of interest (those ?y of type *DataLinkField*). The last line restricts the objects to be retrieved to *DataLinkFields* contained in *DataLinkFrames*.

The most important difference between database queries and SPARQL is their support for reasoning. In addition to facts that have been explicitly asserted, a query can also retrieve facts that have been inferred. It is for this reason that RDF and OWL databases are said to be knowledge bases. A *knowledge base* is the set of all currently known or inferred facts in a particular context. The reasoning capability of OWL is especially powerful. We elaborate on this feature in the following section.

Radio communication introduces an additional requirement on query languages. Unlike databases, which have explicitly stored data, and knowledge bases, which have a combination of explicitly asserted facts and inferred facts, an OBR knowledge base includes data that are embedded in the software that implement the communication protocols. Extracting such data requires a new software capability known as *self-awareness*, or *reflection*.

Reflection is a property that enables software to understand its own *runtime structure*. Reflection is a key feature of any software system that is expected to respond to unanticipated queries. When a program is compiled and executed, the structure of data in memory is highly optimized, and it need not be possible to relate it to the original source code. When reflection is supported, the relationship between data and source code is retained. This is essential for the software to be able to answer queries and execute methods dynamically. For more discussion of reflection, see Fette [35].

13.3.4 Reasoning

One of the important features of ontology languages that distinguishes them from databases is the ability to make logical deductions. In other words, one can *reason* about the information in a knowledge base. A fact is *deduced* if one can infer that it is true even though the fact has never been explicitly asserted to be true. One of the most important examples of deduction is *subsumption*. For example, if one knows that an analog signal uses BPSK, then one can deduce that it is also of type MPSK. In general, whenever something is an instance of a subclass, then it is also an instance of all superclasses of the subclass. Subsumption is the basis for reasoning in description logic. For example, all features and axioms applicable to MPSK signaling also apply to BPSK signaling.

While subsumption reasoning is useful, it is not sufficient for all reasoning tasks. When reasoning involves several linked facts, one cannot express the inference using subsumption alone. When database records are linked by common attributes, they are said to be *joined*. To express reasoning involving *joins* of facts, it is necessary to introduce *rules*. A rule is knowledge in the form of an "if-then" statement. If a *hypothesis* holds, then a *conclusion* must also hold. The hypothesis is also known as the *antecedent,* and the conclusion is also known as the *consequent.* The rule language that has been proposed for OWL is the Semantic Web Rule Language (SWRL) [16]. An example of an SWRL rule is shown in Section 13.4.2. A new standard for rules is under development now at the W3C. It is called the Rule Interchange Format (RIF) [36].

Ontologies can be used as the knowledge representation language for machine-learning and decision-making systems. However, this approach requires that the entities be organized into a hierarchy, at least initially (Section 12.2.2 of [24]). Rule-based systems have also been used for machine learning and decision making. However, rule-based systems usually lack self-awareness of their own structure (Section 12.2.5 of [24]).

Combining ontologies with a rule-based system allows one to have both self-awareness and the power of rules.

13.3.5 Role of Ontology in Knowledge-Intensive Applications

Two-way radio communication introduces a number of challenges not shared by most other ontology-based applications:

1. Real-time processing demands higher performance for inference and reasoning than an interactive application.
2. The knowledge base of a node includes state information that is continually varying, in contrast with the static knowledge bases required by most reasoning systems.
3. The facts are not explicitly stored in a knowledge base but rather are embedded in the software that implements the communication protocol [2].
4. The radios may not have access to the WWW or the Semantic Web for broad support.
5. Most radios are continually moving, and the link performance is bandwidth restrictive and time varying.

If these challenges can be overcome, ontologies can play a number of important roles in radio communication, such as the following:

Interoperability. Radios can use ontologies to deduce important information, such as the protocol being used by other radios.

Flexible querying. Information, such as multipath structures, can be queried. Furthermore, such queries can be answered without having any explicit preprogrammed monitoring capability.

Runtime modifiability. Protocols, packet structures, and even waveforms can be modified at runtime in response to environmental conditions and application requirements.

Validation. Formalization allows one to check the consistency of protocols and to validate the correctness of algorithms that implement the protocols.

Self-awareness. Communication nodes can understand their own structure and modify their functioning at runtime based on this understanding.

A promising application of ontologies is policy management. Ontologies make it possible to specify regulations for wireless communications, including complex, dynamic policies for spectrum management (Section 2.5.3 of [25]). A device that has the ability to encode and to enforce policies is said to be *policy enabled*. KAoS is an example of a policy management system based on OWL [26]. Specifying policies using Semantic Web languages, such as OWL, rather than compiled and interpreted programming languages has many advantages, including greater flexibility as well as more precise and reliable semantics (Section 6.5.2 of [27]). One possible shortcoming of this approach is the difficulty that reasoning systems may have as the complexity of the policies increase. However, Semantic Web technologies have steadily improved so this is becoming less of an issue over time.

13.4 A LAYERED ONTOLOGY AND REFERENCE MODEL

This section discusses two (partial) ontologies, for the PHY layer and for the data link layer. These ontologies are then used for the discussion of the realization of the cognitive aspects discussed in Section 13.5.

13.4.1 Physical Layer Ontology

A piece of the ontology for the PHY layer is shown in Figure 13.3. It is represented in the Unified Modeling Language (UML) notation. UML is used primarily in software engineering to represent software and systems. The boxes in UML represent *classes*, and

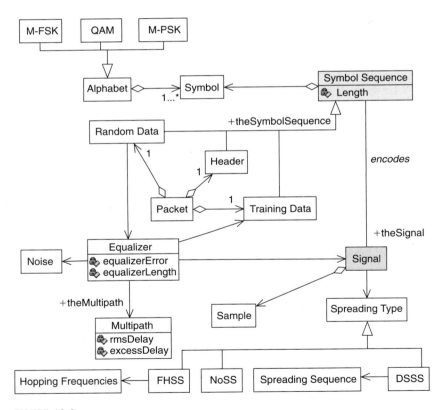

FIGURE 13.3

A partial ontology for the PHY layer in UML [8]. The boxes denote classes and the lines connecting them represent associations. The hollow arrowhead denotes the subclass relation and the diamond denotes aggregation.

the connecting lines represent *associations*. UML classes correspond to the classes used in object-oriented programming languages such as C++, Java, or C#. For programming languages such as C, which are not object-oriented, classes are implemented by using *structs*. Associations can be implemented in a variety of ways, depending on the kind of association. In OWL, associations are called *properties*.

Some associations have a predefined meaning. For instance, the hollow arrow at the end of the line in Figure 13.3 represents the *subclass* relation. In the figure, M-FSK, QAM, and M-PSK are all subclasses of `Alphabet`. The subclass relation is supported by all object-oriented programming languages. In Java and C# one specifies that a class is a subclass of another by using the "extends" keyword. C++ is more succinct, specifying a subclass by using the colon character (:). Diamonds represent *aggregation*. Thus, `Alphabet` is an aggregate of symbols (instances of class `Symbol`).

Aggregations are normally implemented by using arrays or linked lists, but other implementations are also used, depending on the number of elements in the aggregate and whether the aggregation has a variable number of elements. Because a particular `Alphabet` has a fixed number of `Symbol` instances, it would be implemented as an array. Programming languages that are not object-oriented, such as C, implement the subclass relation by using aggregation. Other associations are identified either by the names placed in the middle of the association line, such as `encodes`, or by the roles that elements of the related classes play (roles are attached to the line ends and are distinguished by the "+" symbol). So the association `encodes` in Figure 13.3 has two roles, `+theSymbolSequence` and `+theSignal`. Associations such as these are implemented by using pointers or references to the associated object. In some cases, associations have a prespecified multiplicity. For instance, an `Alphabet` must have at least one instance of `Symbol`. Multiplicity specifications affect how an association can be implemented. Because an `Alphabet` can have more than one instance of `Symbol`, it cannot be implemented using a single pointer or reference to a `Symbol`.

The ontology consists of two groups of classes. The upper part represents classes and properties related to symbols. We can see that `Packet` consists of one `Header`, one `Training Data`, and one `Random Data`. Each of them is an instance of `Symbol Sequence`.

The bottom part of Figure 13.3 represents classes and properties related to the signal domain. The main association between the two parts is `encodes`. It relates a `Symbol Sequence` to a `Signal`. `Signal` is an aggregate of instances of `Sample`. A signal can be modulated in different ways. This ontology example shows two of them, `DSSS` and `FHSS`. Another connection between the symbol and the signal domain occurs through `Equalizer`, which estimates the `Multipath` using the `Training Sequence` and then uses the result for signal interpretation.

13.4.2 Data Link Layer Ontology

The data link layer is responsible for transmitting frames and for error detection and correction in communication links. There are many *data link protocols*. Each protocol specifies the frame types and structure as well as how the communication link is controlled. Many of the data link protocols specialize and extend other protocols. Consequently, the data link protocols form a hierarchy as shown in Figure 13.4. Because of

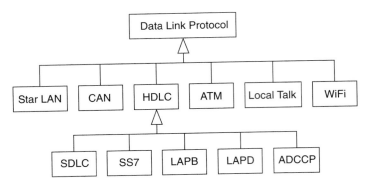

FIGURE 13.4

A partial hierarchy of data link protocols in UML.

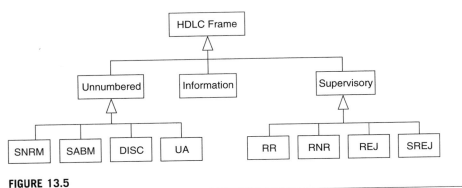

FIGURE 13.5

HDLC frame hierarchy.

the diversity and complexity of data link protocols, not all of the details of the data link ontology are shown in the figure.

There is a large variety of frame types. The names and semantics of the frame types depend on the protocol, but there is considerable overlap among the protocols. Consequently, although each protocol has its own frame type hierarchy, the frame types in different hierarchies can be related by the OWL *sameAs* relationship. The frame type hierarchy for High-Level Data Link Control (HDLC) protocols is shown in Figure 13.5 and the hierarchy for WiFi protocols is shown in Figure 13.6.

A frame consists of a sequence of *fields*. The frame structure is defined by the order of the field types, the number of bits allowed in each field, and the values (bit sequences) allowed in each field. The HDLC protocol has the field types shown in Figure 13.7.

The WiFi protocol has many of the same field types, except for changes in terminology. For example, the WiFi Frame Control Field is called the Control Field in the OWL syntax. Similarly, the WiFi Checksum Field is called the CRC Field in the OWL syntax. Using triples, this is written as: Frame Control Field *owl:sameAs* Control Field, and Checksum Field *owl:sameAs* CRC Field. The data link classes are related to one another, as shown in Figure 13.8.

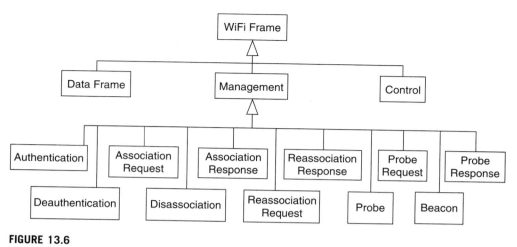

FIGURE 13.6

WiFi frame hierarchy.

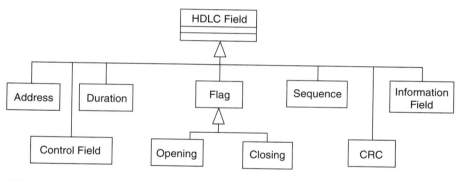

FIGURE 13.7

HDLC field hierarchy.

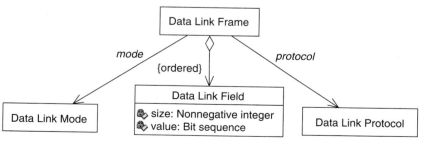

FIGURE 13.8

Data link layer ontology.

A data link frame contains an ordered sequence of fields, each of which has a size (in bits) and a value. A data link field belongs to a protocol and also has a *mode*. For example, the HDLC protocol has three operational modes: Normal Response (NR), Asynchronous Response (AR), and Asynchronous Balanced (AB). These modes are not shown in Figure 13.8.

In addition to the hierarchies and relationships already described, the data link ontology specifies a large number of rules that constrain data link fields within the same frame and in related frames. Examples of such rules shown informally (and implemented in Figures 13.9 and 13.10) include:

1. If a frame belongs to the HDLC protocol, then its opening flag field has value $0 \times 7E$ (i.e., the bit sequence 01111110).
2. If a frame belongs to the SDLC protocol, then the address field has 8 bits.
3. If a frame belongs to the SS7 protocol, then the address field has 0 bits.
4. If a frame belongs to the HDLC protocol, then the control field has 8 or 16 bits.
5. If a frame belongs to the HDLC protocol and the mode is ARM, then the address field has 0 bits.
6. If a frame belongs to the WiFi protocol, then the first 2 bits of the control field are zeroes. This subfield represents the version number.

All of these rules can be represented either in SWRL or in OWL. For instance, the first rule may be represented in OWL, as shown in Figure 13.9.

In XML, element tags, attribute names, and attribute values can belong to various *namespaces*. The namespace of a tag or an attribute name is specified with a colon.

```
1  <owl:Class>
2    <owl:intersectionOf rdf:parseType="Collection">
3      <owl:Class rdf:about="#OpeningFlagField"/>
4      <owl:Restriction>
5        <owl:onProperty rdf:resource="#containedIn"/>
6        <owl:allValuesFrom>
7          <owl:Restriction>
8            <owl:onProperty rdf:resource="#protocol"/>
9            <owl:allValuesFrom rdf:resource="#HDLCProtocol"/>
10         </owl:Restriction>
11       </owl:allValuesFrom>
12     </owl:Restriction>
13   </owl:intersectionOf>
14   <rdfs:subClassOf>
15     <owl:Restriction>
16       <owl:onProperty rdf:resource="#value"/>
17       <owl:hasValue rdf:datatype="&xsd;hexBinary">7E</owl:hasValue>
18     </owl:Restriction>
19   </rdfs:subClassOf>
20 </owl:Class>
```

FIGURE 13.9

An OWL implementation of the data link layer protocol, which categorizes the opening flag field of data link layer frames.

```
 1 <ruleml:Imp>
 2    <ruleml:body>
 3      <swrlx:classAtom>
 4        <owlx:Class owlx:name="#OpeningFlagField"/>
 5        <ruleml:var>x</ruleml:var>
 6      </swrlx:classAtom>
 7      <swrlx:individualPropertyAtom swrlx:property="#containedIn">
 8        <ruleml:var>x</ruleml:var>
 9        <ruleml:var>y</ruleml:var>
10      </swrlx:individualPropertyAtom>
11      <swrlx:individualPropertyAtom swrlx:property="#protocol">
12        <ruleml:var>y</ruleml:var>
13        <ruleml:var>z</ruleml:var>
14      </swrlx:individualPropertyAtom>
15      <swrlx:classAtom>
16        <owlx:Class owlx:name="#HDLCProtocol"/>
17        <ruleml:var>z</ruleml:var>
18      </swrlx:classAtom>
19    </ruleml:body>
20    <ruleml:head>
21      <swrlx:datavaluedPropertyAtomsw rlx:property="#value">
22        <ruleml:var>x</ruleml:var>
23        <owlx:DataValue
24          owlx:datatype="&xsd;hexBinary">7E</owlx:DataValue>
25      </swrlx:datavaluedPropertyAtom>
26    </ruleml:head>
27 </ruleml:imp>
```

FIGURE 13.10

An SWRL implementation of the rule from Figure 13.9. This rule is used to characterize the opening flag field of the data link layer frames.

For example, rdf: specifies the RDF namespace; rdfs: specifies the RDF Schema namespace; and owl: specifies the OWL namespace. Within an attribute value, a namespace is specified with an XML *entity*. For example, &xsd; specifies the XML Schema namespace. If no namespace is specified, the namespace is the current default namespace. In this case, the default namespace is obr, the namespace of OBR. For example, OpeningFlagField is in the obr namespace.

At the highest level, the rule in Figure 13.9 says that an intersection of two classes is a subclass of another class (lines 1, 2, 13, 14, 19, 20). This is expressed using the owl:intersectionOf (lines 2, 13) and the rdfs:subClassOf (lines 14, 19) properties. The intersection is used to represent the Boolean AND operator, and subclass is used to present the logical IMPLIES or IF-THEN operator. The two classes being intersected are the class of opening flag fields (line 3) and the class of HDLC fields (lines 4–12). The former class is part of the ontology and has a name (line 3). The latter class does not have a name. It is specified by two relationships in the ontology: containedIn (line 5) and protocol (line 8). An HDLC field is one that is contained in a frame with a protocol that is of type HDLC. The owl:Restriction is used for constructing classes of instances that satisfy a constraint for a particular property. The first restriction (line 4) specifies "a field that is contained in" (lines 5, 6), and the second restriction (line 7) specifies "a frame the protocol of which has type HDLC" (lines 8, 9). The last restriction (line 15) specifies a value that must be taken by the field (lines 16, 17). Putting all of these together, the rule can be expressed as "IF something is an opening flag field AND is a field that is contained in a frame the protocol of which has type HDLC, THEN it has value 7E." Alternatively, the same rule could be represented in SWRL, as shown in Figure 13.10.

At the highest level of the rule in Figure 13.10, there is an IF-THEN operator (lines 1 and 27). An IF-THEN operator has two parts: the *hypothesis* or *body* (lines 2–19), and the *conclusion* or *head* (lines 20–26). Within either body or head, one specifies a sequence of *atoms*. A *class atom* specifies an instance of a class. In this case the class atoms specify that x is an instance of the class `OpeningFlagField` (lines 3–6), and z is an instance of `HDLCProtocol` (lines 15–18). A *property atom* specifies a triple. Individual property atoms are for properties with values that are *individuals* (lines 7–10 and 11–14). Data-valued property atoms are for properties with values that are data values (lines 21–25). The second atom in the body specifies the triple (x obr:containedIn y) (lines 7–10) and the third atom in the body specifies the triple (y obr:protocol z) (lines 11–14). The head consists of a single atom, which specifies the triple (x obr:value "7E").

The full specification of all data link ontology rules in OWL is extremely large [17]. Rules dealing with multiple frames (e.g., a request and response to it) are especially complex because they are specifying functionality rather than just formats. The data link ontology provides for the following capabilities:

Self-awareness of data link layer functionality. The ontology specifies the protocols. More precisely, it specifies the frame structure and the functionality associated with the various fields.

Dynamic interoperability at the data link layer. An OBR can infer the data link protocol being used by another radio. A simple example of such a deduction was shown by the list of properties 1 through 6 earlier in this section.

Command capability. A radio can remotely request the use of a different protocol or modify the features of an existing protocol.

13.5 EXAMPLES

This section provide examples of how two radios can exchange information about their communications characteristics and parameters and then adapt the communications protocol so that the quality of communication is maintained or improved, demonstrating the cognitive aspects described in the preceding sections.

13.5.1 Responding to Delays and Errors

Wireless transmission requires a robust and efficient communication protocol. When the channel has been estimated and the estimation has been sent back to the transmitter, then the transmission can be adapted according to the channel characteristics. The basic idea behind adaptive transmission is to maintain a constant signal-to-noise ratio

```
1   PREFIX obr: <http://ontoradio.org/2006/obr#>
2   SELECT ?x ?y ?z
3   WHERE {
4     obr:multipath obr:rmsDelay ?x ; obr:excessDelay ?y .
5     obr:equalizer obr:equalizerError ?z
6   }
```

FIGURE 13.11

A SPARQL query. The query to radio B from radio A is requesting information concerning multipath parameters (rms delay and mean excess delay) as well as the mean-squared decision error at receiver B. The first line specifies the `obr` namespace prefix. The fourth line requests the `rmsDelay` (query variable x) and `excessDelay` (query variable y) of the multipath. The fifth line requests the `equalizerError` value (query variable z) of the equalizer.

(SNR) level, (E_b/N_0), by varying the transmission power level, symbol transmission rate, constellation size, and coding rate/scheme, or any combination of these parameters [7]. In experiments with protocol adaptation [7], radios monitored their excess delay (multipath delay spread), rms delay (root mean square delay spread) of the multipath structure of the channel, and the mean square root error of the equalizer. Here the mean square root error of the equalizer represents the average distance between the equalized data (the input data of the equalizer multiplied by the equalizer chips) and the output of the equalizer (the estimated symbol).

OBRs are able to query each other by using a query expressed in an appropriate query language (e.g., SPARQL). An example of such a query regarding the RMS delay, excess delay, and equalizer error is shown in Figures 13.11 and 13.12. In this example, radio A first sends the query shown in Figure 13.11 to radio B.

As is illustrated in Figure 13.11 for SPARQL, a query consists of a series of triples in which some of the slots contain variables. The first line specifies the `obr` namespace prefix. The fourth line requests the `rmsDelay` (query variable x) and `excessDelay` (query variable y) of the multipath. The fifth line requests the `equalizerError` value (query variable z) of the equalizer.

When radio B receives this query, it invokes its reasoner to derive an answer. The query was formulated in the ontology that the two radios share, so the other radio understands what `obr:rmsDelay`, `obr:excessDelay`, and `obr:equalizerError` mean. For instance, the ontology would include information on the units of measure for each of the parameters. To answer the query, the radio first searches its *annotation* (or *markup*) file to find or infer facts that can match the query pattern. However, this file contains only facts that have been explicitly asserted. It does not contain facts that are embedded in the software. Consequently, the reflection mechanism is invoked to extract facts that can be used to match the query pattern, either directly or by means of inference. The answers to the query are then transmitted to radio A. Such a response would look like the fragment in Figure 13.12.

Upon receiving this answer, radio A invokes its reasoner in order to decide how to adjust its protocol so that the communications quality is improved. For instance, the

```
<?xml version="1.0"?>
<!DOCTYPE sparql [
  <!ENTITY xsd "http://www.w3.org/2001/XMLSchema#">
]>
<sparql xmlns="http://www.w3.org/2005/sparql-results#">
 <head>
  <variable name="x"/>
  <variable name="y"/>
  <variable name="z"/>
 </head>
 <results distinct="false" ordered="false">
  <result>
   <binding name="x"><literal
datatype="&xsd;double">1.007837037250556</literal></binding>
   <binging name="y"><literal
datatype="&xsd;double">1.062759005498691</literal></binding>
   <binging name="z"><literal
datatype="&xsd;double">0.025987243652343</literal></binding>
  </result>
 </results>
</sparql>
```

FIGURE 13.12

A SPARQL response to radio A from radio B containing the values for query variables ?x, ?y, and ?z, which were requested by the query in Figure 13.11, namely rmsDelay, excessDelay, and equalizerError.

radio could do one of the following: select a different alphabet (e.g., 2-QAM, 4-QAM, or 16-QAM), increase the equalizer length, or increase the transmit power.

13.5.2 Example: Adaptation of Training Sequence Length

The following example considers the case of negotiating the length of the training data according to the channel dynamics and noise level [8]. The goal is to have a low packet overhead, when possible, and increase it in the case of high noise or high delay spread. In these experiments, the transmitter used the DSSS spreading type DSSS(2^7-1) for transmitting the header (i.e., each symbol is mapped to 127 chips), and a sequence of DSSS(2^5-1) symbols (31 chips for each symbol) for the training data. The number of symbols in the training data was changed gradually in response to the changing characteristics of the channel.

In the example described in Wang et al. [8], negotiation of the length of the training data was accomplished in six transmissions. First, node A sends data to node B. These data come to node B in an ontology-defined envelope (as is done in Figure 13.11) as *Data*. Node B then checks performance. For this purpose, nodes invoke their rules for performance checking. If performance is satisfactory, then node B returns a *Confirm* message with content "Continue" and node A continues to send data to B. If the performance is not satisfactory, a *Confirm* message with content "CommandReq" (request a command from the other node) is returned. Node A then generates a *Command* to change the communication protocol.

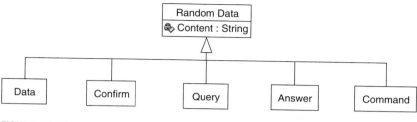

FIGURE 13.13

Classification of message types.

The command-generation rules first send a *Query* from node A to node B requesting the channel condition and the current protocol parameters (in a similar way as discussed in the previous example). When node B receives this query, it infers the answer and sends the *Answer* back to node A. When node A receives the answer, it generates the command. After the *Command* is generated, it is sent to node B, and executed on node A, thus changing the protocol at node A. When node B receives the *Command*, it executes the command, thus changing the protocol at node B. A *Confirm* message with content "Continue" is then sent to node A. The negotiation sequence involves five message types: data, confirmation, query, answer, and command, as shown in the inheritance diagram in Figure 13.13. These message types each have message contents, as described next.

To realize this kind of negotiation, the radios must not only share an ontology but also have means for inferring ways of behaving in particular situations. In pragmatic terms, this means that radios must have some domain knowledge of radio communication. The knowledge may be represented in many different ways, but as discussed in Section 13.4.1, the declarative representation is preferred to the procedural. Consequently, this knowledge should be encoded in the form of *rules*. For instance, the rule that is used in deriving a command concerning the number of symbols in the training sequence in response to this query might look like this:

IF
- the difference between the `rmsDelay` and the previous equalizer feedback coefficient vector length is larger than half of the previous equalizer feedback coefficient vector length

THEN
- construct a new equalizer with the length 3•*(integer(rmsDelay) + 1)* and construct new training data of length *3.rmsDelay.20 MOD symbol length*

ELSE IF
- the equalizer error is smaller than a predefined performance threshold

THEN
- decrease the length of the training data by a predefined value AND use the old equalizer

ELSE
- increase the length of the training data by a predefined value AND use the old equalizer

An informal textual representation of the rule is used here. The experiments reported by Wang et al. [8] used a version of Prolog (Kernel Prolog) for this purpose. In the future, the use of SWRL [16] is envisioned instead.

The experiments reported by Wang et al. [8] considered three different internode distances under different multipath conditions. It was shown that for the worst condition, about 20 symbols were selected according to the negotiation rules, whereas in other cases, only 5 symbols were needed to maintain a desired level of quality of communication.

13.5.3 Data Link Layer Protocol Consistency and Selection

The data link ontology specifies the formats and functionality of the data link protocols. This section presents a few examples of how the data link ontology might be used. The first example illustrates how the formal specification of protocols can uncover inconsistencies. This kind of reasoning would normally be performed offline. The second example shows how the ontology can be used for interoperability.

The first example uses the following two rules (2 and 5) from the data link ontology list presented in Section 13.3.2:

- If a frame belongs to the SDLC protocol, then the address field has 8 bits.
- If a frame belongs to the HDLC protocol and the mode is ARM, then the address field has 0 bits.

Because the SDLC protocol is a subclass of the HDLC protocol, the two rules imply that an ARM frame belonging to the SDLC protocol has an address field with both 8 and 0 bits. It follows that the SDLC protocol cannot have ARM frames.

We can address this inconsistency in several ways. We could remove the subclass relationship between the SDLC protocol and the HDLC protocol classes from the ontology. This has the disadvantage that the SDLC protocol would now have to be completely specified from scratch rather than being derived from a more general class. Alternatively, we could modify the first of these two rules to recognize that the address field for the SDLC protocol will have only 8 bits for modes other than ARM.

The second example concerns the issue of interoperability. Suppose that a radio receives a stream of packets from another radio with which it has not previously communicated. Can the receiving radio deduce the protocol being used by the transmitter? In the case of HDLC versus WiFi, the following two rules (1 and 6 from the list in Section 13.4.2) allow one to disambiguate the packets:

- If a frame belongs to the HDLC protocol, then its opening flag field has value $0 \times 7E$ (i.e., the bit sequence 01111110).
- If a frame belongs to the WiFi protocol, then the first 2 bits of the control field are zeroes. This subfield represents the version number.

From these two rules, we can deduce that the first 2 bits of the packet will distinguish these two protocols. If the first 2 bits are 01, then it is an HDLC frame. If the first 2 bits are 00, then it is a WiFi frame. If the bits have some other value, then the frame belongs to neither protocol. This example is somewhat artificial, and we could argue that it is easy to write a small procedure that could make this same determination. However, this is only one example among many possible inferences. The software would quickly become extremely unwieldy if we needed a separate procedure for every possible deduction. Indeed, we would need infinitely many procedures because we can make infinitely many deductions, and every time a new feature or protocol variation was added, we would need to revise all of the procedures.

13.6 OPEN RESEARCH ISSUES

Even though the concept of OBR can be implemented as discussed in this chapter, a number of open issues still need to be investigated. Additionally, some of the engineering issues need to be addressed before this concept finds its way into real radio applications. Several of the most important open issues are discussed in this section.

13.6.1 Ontology Development and Consensus

As discussed in this chapter, ontologies can bring many advantages to SDR communication. However, to use an ontology, one must first develop it. Building high-quality ontologies is a substantial task. It is further complicated by the diversity of Web-based ontology languages.

The first step in an ontology development project is to agree on the purpose and motivation for embarking on this activity. This step includes determining the following:

1. *Why* the ontology is being developed.
2. *What* will be covered by the ontology.
3. *Who* will be using the ontology.
4. *When* and for how long the ontology will be used.
5. *How* the ontology is intended to be used.

Once there is a clear understanding of the purpose of the ontology, four major activities must be undertaken: (1) choose an ontology language, (2) obtain a development tool, (3) acquire domain knowledge, and (4) reuse existing ontologies. None of these is especially easy, and it is important to reach consensus because the ontology forms the basis for communication within the community. Generally speaking, the larger the community, the more difficult it is to reach consensus. One can mitigate this difficulty to some degree by establishing a precise statement of the purpose of the ontology before starting the project. However, one cannot entirely eliminate it.

Ultimately, the success of ontologies for SDR and CR communications will depend not only on the quality of the ontology design but also on the extent to which the community accepts the ontology. The task is a challenging one, but experience in other communities suggests that it is achievable, nonetheless.

13.6.2 Ontology Mapping

At every level of the communication protocol stack there are many protocols currently in use. Each protocol has its own terminology and therefore its own ontology. Generally speaking, the higher in the stack, the more diverse the ontologies. To achieve interoperability between protocols and other applications, it is necessary to have mappings between the ontologies. Such mappings can be specified by subject-matter experts, but when the ontologies use the same concepts, it should be possible to automate the process of transforming from one ontology to the other. This problem has been studied for many years. Most of the work has been devoted to mapping relational database schemas, but there has also been some recent work on this problem for XML DTDs and even for the more sophisticated ontologies of the Semantic Web.

A survey of relational schema integration tools is presented by Rahm and Bernstein [18]. When the data from a variety of relational database sources are transformed to a single target database, the process is called *data warehousing*. Many data warehousing companies now also support XML. If a query using one vocabulary is rewritten so as to retrieve data from various sources, each of which uses its own vocabulary, it is called *virtual data integration* [19].

Ontology mapping depends on identifying semantically corresponding elements. This is a difficult problem to solve because different sources may use very different structural and naming conventions for the same entities. In addition, the same name can be used for elements having totally different meanings, such as different units, precision, resolution, measurement protocol, and so on. It is usually necessary to annotate an ontology with auxiliary information to assist one in determining the meaning of elements, but ontology mapping is difficult to automate even with this additional information. Furthermore, a single element in one ontology may correspond to multiple elements in another. In general, the correspondence between elements is many-to-many: many elements correspond to many other elements.

Tools for automating ontology mapping have been proposed, and some research prototypes exist. However, these tools mainly help to discover simple one-to-one matches, and they do not consider the meaning of the data or how the transformation will be used. Using such a tool requires significant manual effort to correct incorrect matches and to add missing matches. One of the few tools that can handle many-to-many mappings was developed by Goguen and his students [20].

In practice, ontology mapping is done manually by domain experts and is very time consuming when there are many data sources, or when ontologies are large or complex. Automated ontology mapping systems are designed to reduce manual effort. However, such systems require a substantial amount of time to prepare input to the system as well as to guide the matching process. This amount of time may easily exceed the time saved by using the system. Unfortunately, existing ontology-mapping prototypes focus on measuring accuracy and completeness rather than on whether they provide a net gain. Nevertheless, current systems can help to record and to manage the ontology matches that have been detected, by whatever means. For large, distributed ontology mapping projects this kind of support can have a considerable impact on productivity.

Many ontology mapping projects exist, and some have developed prototypes. Examples are PROMPT [21], from the Stanford Medical Informatics Laboratory, and the Semantic Knowledge Articulation Tool (SKAT) [22], also from Stanford. Considerable progress has been made on ontology-mapping tools, but ontology mapping remains a challenging field for development.

13.6.3 **Learning**

Ontology learning is an emerging field aimed at automated assistance in the construction of ontologies and semantic page annotation through the use of machine-learning techniques [23]. The relevance of this field is tied to the prohibitively large amount of manual ontological construction required to represent the vast amount of information in communication systems. The implementation of a fixed ontological system presumes that the knowledge engineer anticipated the extent of the language involved in queries, responses, and reasoning. Although this goal could be attained at some time instant, the continual evolution of radio standards and implementations mandates a requisite evolution in the ontology. For example, WiFi was not a recognized radio standard 15 years ago, IEEE 802.11g was not available prior to 2000, and the prerelease version of 802.11n was not available until 2004. Additionally, new variants and derivatives (subclasses) of the HDLC protocol class will continue to appear for some time. Without automated ontological learning, a CR that is expected to communicate using these emerging standards would require continuous updating of its ontological database system. Ontology learning is one mechanism to automate this evolution.

13.6.4 **Efficiency of Reasoning**

The implementation of the concept of OBR discussed in this chapter relies heavily on OWL, a language for representing ontologies. Another component of the OBR architecture is a generic reasoning engine. Even though a generic-reasoning mechanism satisfies the requirement of being able to respond to surprise by a cognitive agent, it also poses a very high demand on the reasoning engine. In general, logical reasoning is undecidable. In simple terms, this means that an inference engine might never terminate its process of answering a specific query.

With this concern in mind, the designers of OWL provided the language in three levels: OWL Full, OWL-DL, and OWL Lite, as discussed in Section 13.3.2. The three levels differ in both expressive power and computational complexity. OWL Lite is the simplest of the three, with the least expressive power of the three OWL languages. In spite of its relative simplicity, the computational complexity of OWL Lite is very high; it is in the EXPTIME complexity class, meaning that in the worst case a deduction will require time equal to $O(2^{p(n)})$, where $p(n)$ is a polynomial in the size n of the ontology. OWL-DL is more expressive than OWL Lite, but its complexity is even worse; it is in the NEXPTIME complexity class. Although the precise complexity of this class is not known, it is at least as high as EXPTIME, and it may be much higher still, possibly an exponential of an exponential of a polynomial—that is, $O(2^{f(n)})$, where $f(n) = 2^{p(n)}$.

Furthermore, there is no bound on how much memory may be required. OWL Full is the most expressive of the three OWL languages, but its complexity is the worst possible: it is undecidable. *Undecidability* means that in the worst case there is no bound on the amount of time or space that may be needed to perform a deduction. In practice, even OWL Full is not sufficiently expressive, and it is necessary to use rules to express some of the axioms of the ontology. Unfortunately, if rules are added to any of the three languages, including OWL Lite, then the computational complexity of deduction is undecidable.

To make systems based on such generic reasoners scalable, the issues of complexity of reasoning must be resolved so that reasoning conclusions are derived within the constraints of the radio domain. Various approaches to such a problem are known in the literature. In general, such approaches are based on a trade-off between the quality of the solution and the computation time. For instance, for algorithms known as *anytime algorithms*, the problem-solving activity can be stopped at any time, and the result will be the best one that has been found up to that time. Another approach is to combine logical reasoning with an uncertainty handling mechanism, like Bayesian networks, fuzzy logic, and others.

13.7 SUMMARY

This chapter presented an ontology-based approach to address some of the requirements of CR. Throughout this chapter we illustrated how to capture knowledge about the domain of radio communication using ontologies. All of the examples used a purely declarative representation of domain knowledge. The objective was to show that such a knowledge representation can be used by generic-reasoning tools both to formulate queries and to infer answers to such queries. Such a query-answering mechanism is generic and thus satisfies the requirement of being capable of "responding to surprise," one of the key features of a cognitive agent.

Another feature of a cognitive agent is the ability to collect and use information about the environment. The example in Section 13.6.1 showed how radios can request multipath information and then use it to modify the transmission protocol. Moreover, such radios are situation-aware because they know not only transmission conditions at other nodes, but also how such conditions relate to their own transmission protocol parameters. Section 13.3.3 also discussed how self-awareness can be implemented so that radios can know the values of their internal variables and how to change those values on demand. For instance, a radio can know its RMS delay, excess delay, or equalizer error. Another example of this capability, shown in Section 13.5.2, is the knowledge and use of the values of these internal variables such as the bit error rate and the multipath structure.

This chapter also showed examples of querying of radios by other radios. Although it did not show any queries by end users, such queries can be implemented in the same manner as queries by other radios. The chapter has also shown that self-aware radios can execute commands. For instance, if two radios agree to use a specific length of the training sequence, they first negotiate a contract, and then both update their internal structures. In general, any information expressible in terms of the ontology can be retrieved and modified.

EXERCISES

The following exercises are based on IEEE 802.15.4-2003 and thus require an ability to access this standard. The section of the standard from which the exercise arises is listed in brackets.

13.1 The coordinator on a PAN can optionally bound its channel time using a superframe structure. Define the components of a superframe structure using a property named `hasComponent`. [7.5.1.3]

13.2 Express the versions of the CSMA-CA algorithm as a class hierarchy. The version that is employed for transmissions in the CAP of the superframe is determined by whether beacons are being used in the PAN. Express this using class constructors based on a property named `beaconsBeingUsed`. [7.5.1.3]

13.3 Use a class hierarchy to express the various kinds of channel scans [7.5.2.1]. Use class constructors to specify:
   a. What each kind of scan can be used to accomplish
   b. When each kind of scan can be used
   c. Whether other commands must be transmitted

13.4 Use an OWL cardinality constraint to specify the number of GTSs that may be allocated by a PAN coordinator at one time. [7.5.7]

13.5 Define the information that must be maintained by the PAN coordinator for each GTS. Represent the definition in OWL. [7.5.7]

13.6 Use `owl:inverseFunctionalProperty` to specify how a GTS is uniquely determined within the PAN coordinator by GTS attributes. [7.5.7]

REFERENCES

[1] Brachman, R., A DARPA Information Processing Technology Renaissance: Developing Cognitive Systems; available at *www.darpa.mil/ipto/presentations/brachman.ppt*.

[2] Boyd, J., A Discourse on Winning and Losing, Technical Report, Maxwell AFB, 1987.

[3] Passino, K. M., and Antsaklis, P. J. (eds.), *Introduction to Intelligent and Autonomous Control*, Kluwer/Academic, 1992.

[4] Endsley, M., and D. Garland, *Situation Awareness, Analysis and Measurement*, Erlbaum, 2000.

[5] XG Working Group, The XG Vision, Request For Comments, Version 2.0.

[6] Guarino, N., Formal Ontology in Information Systems, *Proceedings of Formal Ontology in Information Systems*, pp. 3–15, Amsterdam, 1998.

[7] Wang, J., D. Brady, Baclawski K., M. Kokar, and L. Lechowicz, The Use of Ontologies for the Self-Awareness of the Communication Nodes, *Proceedings Software Defined Radio Technical Conference*, 2003.

[8] Wang, J., M. M. Kokar, K. Baclawski, and D. Brady, Achieving Self-Awareness of SDR Nodes Through Ontology-Based Reasoning and Reflection. *Proceedings Software Defined Radio Technical Conference*, Phoenix, November 2004.

[9] World Wide Web Consortium, eXtensible Markup Language; available at *www.w3.org/XML/*, 2001.

[10] TopicMaps.org, The XTM Web site (*topicmaps.org*), 2000.

[11] Lassila, O., and R. Swick, Resource Description Framework (RDF) Model and Syntax Specification, February 1999; available at *www.w3.org/TR/REC-rdf-syntax*.

[12] van Harmelen, F., J. Hendler, I. Horrocks, D. McGuinness, P. Patel-Schneider, and L. Stein, *OWL Web Ontology Language Reference*, M. Dean and G. Schreiber (eds.), March 2003; available at *www.w3.org/TR/owl-ref/*.

[13] Baader, F., D. Nardi, D. McGuinness, P. Patel-Schneider, and D. Calvanese. *The Description Logic Handbook: Theory, Implementation and Applications*, Cambridge University Press, 2003.

[14] Seaborne, A., SPARQL—A Query Language for RDF, January 2008; available at *www.w3.org/TR/rdf-spqrl-query*.

[15] Fikes, R., P. Hayes, and I. Horrocks, OWL-QL: A Language for Deductive Query Answering on the Semantic Web, *Journal of Web Semantics*, 2(2), 2005.

[16] Horrocks, I., P. Patel-Schneider, H. Boley, S. Tabet, B. Grosof, and M. Dean, SWRL: A Semantic Web Rule Language Combining OWL and RuleML, 2003; available at *www.daml.org/2003/11/swrl*.

[17] Brady, D., K. Baclawski, and M. Kokar, Achieving Dynamic Interoperability of Communication at the Data Link Layer through Ontology Based Reasoning, *Proceeding SDR Forum Technical Conference*, Anaheim, November 2005.

[18] Rahm, E., and P. A. Bernstein, On Matching Schemas Automatically, Department of Computer Science, University of Leipzig, 2001.

[19] Embley, D., D. Jackman, and L. Xu., Multifaceted Exploitation of Metadata for Attribute Match Discovery in Information *Proceedings Integration, International Workshop on Information Integration on the Web*, 2001.

[20] Nam, Y., J. Goguen, and G. Wang, A Metadata Integration Assistant Generator for Heterogeneous Distributed Databases, *Proceedings International Conference on Ontologies, Databases, and Applications of Semantics for Large-Scale Information Systems*, Lecture Notes in Computer Science, 2519:1332–1344, Springer-Verlag, 2002.

[21] Noy, N., and M. Musen, PROMPT: Algorithm and Tool for Automated Intelligence, Austin, TX, 2000.

[22] Mitra, P., G. Weiderhold, and J. Jannink, Semi-automatic Integration of Knowledge Sources, *Proceedings Second International Conference on Information Fusion*, 1999.

[23] Omelayenko, B., Learning of Ontologies for the Web: The Analysis of Existent Approaches, *Proceedings International Conference on Database Theory*, London, January 2001.

[24] Kovarik, V., Jr., Cognitive Research: Knowledge Representation and Learning, *Cognitive Radio Technology*, B. Fette (ed.), Newnes/Elsevier, 2006.

[25] Kolodzy, P., Communications Policy and Spectrum Management, *Cognitive Radio Technology*, B. Fette (ed.), Elsevier, 2006.

[26] Uszok, A., et al., KAoS Policy Management for Semantic Web Services, *IEEE Intelligent Systems*, 19(4):32–41, 2004.

[27] Wellington, R., Cognitive Policy Engines, *Cognitive Radio Technology*, B. Fette (ed.), Newnes/Elsevier, 2006.

[28] Polson, J., Position Awareness, *Cognitive Radio Technology*, B. Fette (ed.), Newnes/Elsevier, 2006.

[29] Polson, J., Technologies Required for Cognitive Radio, *Cognitive Radio Technology*, B. Fette (ed.), Newnes/Elsevier, 2006.

[30] Bostian, C., Cognitive Techniques—Physical and Link Layer, *Cognitive Radio Technology*, B. Fette (ed.), Newnes/Elsevier, 2006.

[31] Reed, J., Cognitive Radio Network Performance Analysis, *Cognitive Radio Technology*, B. Fette (ed.), Newnes/Elsevier, 2006.

[32] Mitola, J., III, Cognitive Radio Architecture, *Cognitive Radio Technology*, B. Fette (ed.), Newnes/Elsevier, 2006.

[33] Marshall, P., Spectrum Awareness, *Cognitive Radio Technology*, B. Fette (ed.), Elsevier, 2006.

[34] Zhao, Y., B. Le, and J. H. Reed, Network Support: The Radio Environment Map, *Cognitive Radio Technology*, B. Fette (ed.), Newnes/Elsevier, 2006.

[35] Fette, B. A., History and Background of Cognitive Radio Technology, *Cognitive Radio Technology*, B. Fette (ed.), Newnes/Elsevier, 2006.

[36] World Wide Web Consortium, *RIF Basic Logic Dialect*, October 2007; available at *www.w3.org/TR/2007/WD-rif-bld-20071030/*.

Cognitive Radio Architecture

14

Joseph Mitola III
Stevens Institute of Technology
Castle Point on Hudson, New Jersey

14.1 INTRODUCTION

Architecture is a comprehensive, consistent set of *design rules* by which a specified set of *components* achieves a specified set of *functions* in products and services that evolve through multiple design points over time [1].[1] This section introduces the fundamental principles by which software-defined radio (SDR), sensors, perception, and machine learning (ML) may be integrated into SDR and system-on-chip (SoC) cognitive radios (CRs). These CRs have better quality of information (QoI) through capabilities to observe (sense, perceive), orient, plan, decide, act, and learn (the so-called OODA loop) in radio frequency (RF), via network intelligence, and in the user domains.

This chapter develops five complementary perspectives of cognitive radio architecture (CRA), called CRA-I through CRA-V, each building on the previous in capability. Architecture is driven top-down by market needs and bottom-up by available, affordable technologies. Taking the top-down perspective requires some attention to the use cases that the functions are intended to realize. This chapter therefore begins with a review of the substantial changes in use cases that drive cognitive wireless architecture.

14.1.1 Use Case Evolution

The use cases that have captured market share and propelled radio engineering to its current levels of success have been based on the proliferation of cellular wireless networks. In addition, the affordability of fiber-optic core networks and short-range wireless local area networks (WLAN) of the Internet are propelling markets toward convergence. Thus, the achievement of ubiquity brings with it a shift of use case from mere ubiquity toward differentiated multimedia services. There also is greater integration of historically distinct market segments such as commercial and public safety wireless (e.g., Block D in the US 700 MHz wireless auctions). Block D challenges have been characterized by the SDR Forum (reported online in *Communications Daily*, December 11, 2007) as "meeting the divergent needs of commercial and public safety users, cov-

---

[1]This chapter is adapted from the text *Cognitive Radio Architecture: The Engineering Foundations of Radio XML*, Wiley, 2006, with permission.

Table 14.1 Internet to Wireless Use Case Evolution

Use Case Parameters	Foundation Era (1990–2005)	Evolution Era (2005–2020)
Core wireless use cases	Toward ubiquitous access	Toward integrated services
Profit margins	High (handsets-infrastructure) then handset profits declining	Low (handsets-infrastructure) to high for differentiated services
Value proposition	*QoS (Connectivity, data rate)*	*QoI (User experience)*
PSTN integration	SS7[0], SDH[0][GDCS2]	IP-SIP[0], Mobile IP, or IPv6[0]
Reconfigurable HW	*Not worth the cost versus chipset*	*Transitioning to mainstream?*
Location awareness	Niche applications	Ubiquitous
Multimedia	*Infeasible to feasible*	*Strong differentiator*
Spectrum awareness	Within allocated band	Across multiple bands
Spectrum Auctions	*Large blocks for long term*	*Small space–time holes short term*
Public safety	Distinct markets	Integration with agility
Data rate framework	*Stationary, walking, vehicle*	*Hot spot, traveling, emergency*
Sentient Spaces	*Video surveillance markets*	*Elder care and home robotics*

The lines of the table without italics have been well established during the past few years and thus need little elaboration, but set the stage for the more speculative use case projections in italics. (Source: © Dr. Joseph Mitola III, used with permission.)

erage, shared operational control, robustness, adaptability, and spectrum use in the absence of network buildout." Such public commentary reflects an evolution of use case that drives wireless architectures from the relatively monolithic cellular radio networks (with gateways to the public switched telephone network (PSTN)) toward greater integration with the Internet as characterized in Table 14.1.

14.1.2 Organization of the Chapter

These use cases drive the evolution of architecture from RF awareness (e.g., for dynamic spectrum access) to include user awareness for high QoI.[2] The CRA-I perspective facilitates this evolution via six functional components, black boxes to which are ascribed a first-level decomposition of CR functions and among which important interfaces are defined. One of these boxes is SDR, a proper subset of CR. One of these boxes performs cognition via the <Self/>,[3] a self-referential software architecture that strictly embodies finite computing (e.g., no `While` or `Until` loops) mitigating the Gödel-Turing paradox[4] of self-referential systems.

The CRA-II perspective examines the flow of inference through a cognition cycle that arranges the core capabilities of *ideal* CR (iCR) in temporal sequence for a logical flow and circadian rhythm for the CRA. The CRA-III perspective examines the related

---

[2]High QoI means the information is closely matched to the information required by the user.

[3]Note that <Self/> is how the CR will refer to itself, so "<Self/>" can also be read as "the radio." This chapter will uses brackets (< >) to indicate items about which the radio can reason in Radio XML (RXML).

[4]This refers to the classic Turing machine, which can fall into unbounded analysis when attempting to reason about itself. One solution to this problem is a timer that kills any task that isn't completed within a specified time.

levels of abstraction for CR to sense elementary sensory stimuli and to perceive QoI relevant aspects of a <Scene/> consisting of the <User/> in an <Environment/> that includes RF.[5] The CRA-IV perspective examines the mathematical structure of this architecture, identifying mappings among topological spaces represented and manipulated to preserve set-theoretic properties. Finally, the CRA-V perspective reviews SDR architecture, sketching an evolutionary path from the Software Communications Architecture (SCA)/Software Radio Architecture (SRA) to the CRA. The CRA is expressed in Radio eXtensible Markup Language (RXML). The CRA is introduced in this chapter and developed in [12]. The chapter ends with a discussion of making the CR architecture sufficiently robust to be useful under diverse real-world conditions.

14.2 CRA-I: FUNCTIONS, COMPONENTS, AND DESIGN RULES

The *functions* of CR exceed those of SDR. Reformulating the CR <Self/> as a *peer* of its own <User/> establishes the need for added functions by which the <Self/> accurately perceives the local scene including the <User/> and autonomously learns to tailor the information services to the specific <User/> in the current RF and physical <Scene/>. Scaling up such tailoring requires machine learning for affordability.

14.2.1 Cognitive Radio Functional Component Architecture

The SDR components and the related cognitive components of iCR appear in Figure 14.1. The cognition components describe the SDR in RXML [12] so that the <Self/> can know that it is a radio and that its goal is to achieve high QoI tailored to its own users. RXML intelligence includes a priori radio background and user stereotypes as well as knowledge of RF and space–time <Scenes/> perceived and experienced. This includes both structured reasoning with iCR peers and cognitive wireless networks (CWNs), and ad hoc reasoning with users, all the while learning from experience.

The detailed allocation of functions to components, with interfaces among the components, requires closer consideration of the SDR component as the foundation of CRA. In addition, SoC architectures may employ the CRA for agility in the management of RF chip sets.

SDR Components

SDRs include a hardware platform with RF access and computational resources plus at least one software-defined personality. The SDR Forum has defined its Software Communication Architecture (SCA) [3] and the Object Management Group (OMG) has defined its Software Radio Architecture (SRA) [4]. There are similar fine-grained architecture intended to enable wireless connectivity with greater flexibility at lower cost. These SDR architectures are defined in Unified Modeling Language (UML) object models [5], Common Object Request Broker Architecture (CORBA) [6], Interface Design Language (IDL) [7], and XML descriptions of the UML models. The SDR Forum and OMG standards describe the technical details of SDR both for radio engineering and for an initial level of wireless air interface ("waveform") plug-and-play. The SCA/SRA

---

[5]The use of Semantic Web technology (e.g., referring to the <User/> in an <Environment/>) is addressed by Mahonen [2].

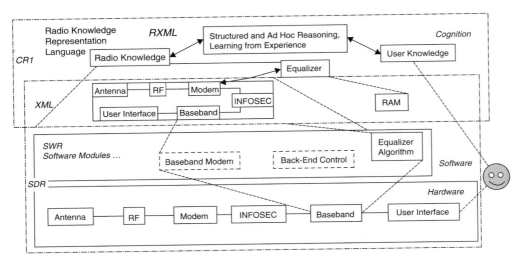

FIGURE 14.1

The CRA augments SDR with computational intelligence and learning capacity. (*Source:* © 1998 Dr. Joseph Mitola III, used with permission.)

was sketched in 1996 at the first US Department of Defense (DoD) inspired Modular Multifunctional Information Transfer System (MMITS) Forum. It was developed by the DoD in the 1990s and the architecture is now in use by the US military [7]. This architecture emphasizes radio personalities on computationally capable mobile nodes where network connectivity is often intermittent.

The commercial wireless community [8], in contrast, led by cell phone giants Motorola, Samsung, Ericsson, and Nokia, envisioned a much simpler architecture for mobile wireless devices, consisting of two application programming interfaces (APIs)—one for the service provider and another for the network operator. Those users define a knowledge plane in future intelligent wireless networks that is not dissimilar from a distributed CWN. That community promotes the business model of the user (user → service provider → network operator → large manufacturer → device), in which the user buys mobile devices consistent with services from a service provider, and the technical emphasis is on *intelligence in the network*. This perspective no doubt will yield computationally intelligent networks in the near to mid-term.

The CRA developed in this text, however, envisions the computational intelligence to create ad hoc and flexible networks with the *intelligence in the mobile device* as well as the network. This technical perspective enables the business model of user → device → heterogeneous networks, typical of the Internet where a user device (e.g., a wireless laptop) connects to the Internet via any Internet service provider (ISP). Current multiplay integrated markets accommodate both business models with consumer-purchased femtocells augmenting cellular networks. The CRA builds on the SCA/SRA, commercial APIs, and the Semantic Web for more of an Internet business model. Yet, SoC, SDR, CR, and iCR form a continuum facilitated by radio exchange languages such as RXML (Radio XML [12]).

CR Node Functional Components

A basic CRA includes the functional components shown in Figure 14.2. A functional component is a black box to which functions have been allocated, but for which implementation is not specified. Thus, while the applications component is likely to be primarily software, the nature of those software components is yet to be determined. User-interface functions, however, may include optimized hardware (e.g., for computing video flow vectors in real time to assist scene perception). At the level of abstraction of this figure, the components are functional, not physical, as follows:

- The user sensory perception (SP), which includes haptic, acoustic, and video sensing and perception functions.
- The local environment sensors (location, temperature, accelerometer, compass, etc.).
- The system application's (sys-apps) media-independent services (e.g., playing a network game).
- The SDR functions, which include RF sensing and SDR applications.
- The cognition functions (symbol grounding for system control, planning, and learning).
- The local effector functions (speech synthesis, text, graphics, and multimedia displays).

These functional components are embodied on an iCR platform, a hardware realization of the six functions. To support the capabilities described in the prior chapters, these components go beyond SDR in critical ways. First, the user interface goes well beyond buttons and displays. The traditional user interface has been partitioned into a substantial user sensory subsystem and a set of local effectors. The user sensory interface includes touch sensor buttons (the haptic interface); and microphones (the audio interface) with acoustic sensing that is directional, capable of handling multiple speakers simultaneously, and able to include full motion video with visual scene perception. In addition, the audio subsystem does not just encode audio for (possible) transmission; it also parses and interprets the audio from designated speakers, such as the <User/>, for a high-performance spoken natural language (NL) interface via which to perceive

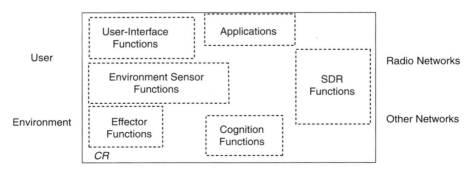

FIGURE 14.2

Minimal CR node architecture. (*Source:* © Dr. Joseph Mitola III, used with permission.)

user context. Similarly, the text subsystem parses, and interprets the language, to track the user's information states, detecting plans and potential communications and information needs unobtrusively as the user conducts normal activities. The local effectors synthesize speech along with traditional text, graphics, and multimedia displays.

Sys-apps are those *information services* that generate value for the user. Historically, voice communications with a phone book, text messaging, and the exchange of images or video clips have comprised the core value proposition of sys-apps for SDR. These applications are generally integral to the SDR application, such as data services via GSM packet radio service (GPRS), which can be considered a wireless SDR personality as well as an information service. CR sys-apps break the service out of the SDR waveform, so that the user need not be concerned with details of wireless connectivity unless that is of particular interest. Should the user care whether he or she plays the distributed video game via 802.11 or Bluetooth over the last 3 months? Probably not. The typical user might care if the CR wants to switch to beyond third generation (B3G) at $5 per minute, but a particularly affluent user might not care and would leave all that up to the CR.

The cognition component provides all the cognition functions—from the semantic grounding of the <Self/> with respect to entities in the perception system to the control of the overall system through planning and initiating actions—learning user preferences and RF support needed in particular situations in the process. Each of these subsystems contains its own processing, local memory, integral power conversion, built-in test (BIT), and related technical features.

The Ontological <Self/>

The CRA consists of six functional components: user SP, environment, effectors, SDR, sys-apps, and cognition. Those components of the <Self/> enable external communications and internal reasoning about the <Self/> by using the RXML syntax. Given the top-level outline of these functional components along with the requirement that they be embodied in physical hardware and software (the "platform"), the six functional components are defined ontologically in the equation in Figure 14.3. In part, this equation expresses the design premise that the hardware–software platform and the functional components of the CR are independent. Platform-independent computer languages (e.g., Java) are well understood. This ontological perspective formulates platform independence as an architecture design principle for CR. In other words, the burden is on the (software) functional components to adapt to whatever RF-hardware-operating system platform might be available.

```
<Self>
<ICR-Platform/>
<Functional-Components>
          <User SP/><Environment/><Effectors/><SDR/><Sys Apps/><Cognition/>
</Functional-Components>
</Self>
```

FIGURE 14.3

Components of the CR <Self/>. The CR <Self/> is defined to be an iCR platform, consisting of six functional components using the RXML syntax.

14.2.2 Design Rules Include Functional Component Interfaces

The six functional components (see Table 14.2(a) and (b)) imply associated functional interfaces. In architecture, design rules may constrain the quantities and types of components as well as the interfaces among those components. This section addresses the interfaces among the functional components.

The CR N-squared diagram of Table 14.2(a) characterizes CR interfaces. These constitute an initial set of CR APIs, augmenting the established SDR APIs. This enables basic CRs to accommodate the dynamic spectrum behavior of the Defense Advanced Research Projects Agency (DARPA) NeXt-Generation (XG), and Wireless Network after Next (WNaN) radio communication programs.

In other ways, these APIs supersede the existing SDR APIs. In particular, the SDR user interface (GUI) becomes the user sensory and effector API. User sensory APIs include acoustics, voice, and video, and the effector APIs include speech synthesis to give the CR <Self/> its own voice. In addition, wireless applications are growing rapidly. Voice and short-message service provide an ability to exchange images and video clips with semantic tags among wireless users. The distinctions between cell phone, personal digital assistant (PDA), and game box continue to disappear.

These interface changes enable the CR to sense the situation, to interact with the user, and to access radio networks on behalf of the user according to its situational assessment. This matrix characterizes internal interfaces between functional processes. Interface notes 1 to 36 are explained in Table 14.2(b).

The preceding information flows, aggregated into an initial set of CR APIs, define an information services API (ISAPI) by which an information service accesses the other five components (interfaces 13–18, 21, 27, and 33 in Table 14.2(a)). They would also define a CAPI by which the cognition system obtains status and exerts control over the rest of the system (interfaces 5, 11, 17, 23, 25–30, and 35 in Table 14.2(a)). Although the constituent interfaces of these APIs are suggested in this table, it would be premature to define these APIs without first developing detailed information flows and

Table 14.2(a) CR N-Squared Diagram

From/To	User SP	Environment	Sys-Apps	SDR	Cognition	Effectors
User SP	1	7	13 PA[a]	19	25 PA[b]	31
Environment	2	8	14 SA[a]	20	26 PA[b]	32
Sys-Apps	3	9	15 SCM[a]	21 SD[a]	27 PDC[a,b]	33 PEM[a]
SDR	4	10	16 PD[a]	22 SD	28 PC[b]	34 SD
Cognition	5 PEC[b]	11 PEC[b]	17 PC[a,b]	23 PAE[b]	29 SC[b]	35 PE[b]
Effectors	6 SC	12	18a	24	30 PCD[b]	36

Note: This matrix characterizes internal interfaces between functional processes. Interface notes 1–36 are explained in Table 14.2(b).   P = primary; A = afferent; E = efferent; C = control; M = multimedia; D = data; S = secondary; others not designated P or S are ancillary.

[a]Information services API.

[b]CAPI.

Table 14.2(b) Explanations of Interface Notes for Functional Processes Shown in Table 14.2(a)

Note No.	Process Interface	Explanation
1	User SP–User SP	Cross-media correlation interfaces (video-acoustic, haptic-speech, etc.) to limit search and reduce uncertainty (e.g., if video indicates user is not talking, acoustics may be ignored or processed less aggressively for command inputs than if user is speaking).
2	Environment–User SP	Environment sensors parameterize user sensor-perception. Temperature below freezing may limit video.
3	Sys-Apps–User SP	Sys-apps may focus scene perception by identifying entities, range, expected sounds for video, audio, and spatial perception processing.
4	SDR–User SP	SDR applications may provide expectations of user input to the perception system to improve probability of detection and correct classification of perceived inputs.
5	Cognition–User SP	This is the *primary control efferent* path from cognition to the control of the user SP subsystem, controlling speech recognition, acoustic signal processing, video processing, and related SP. Plans from cognition may set expectations for user scene perception, improving perception.
6	Effectors–User SP	Effectors may supply a replica of the effect to user perception so that self-generated effects (e.g., synthesized speech) may be accurately attributed to the <Self/>, validated as having been expressed, and/or cancelled from the scene perception to limit search.
7	User SP–Environment	Perception of rain, buildings, indoor/outdoor can set GPS integration parameters.
8	Environment–Environment	Environment sensors would consist of location sensing such as GPS or Galileo; ambient temperature; light level to detect inside versus outside locations; possibly smell sensors to detect spoiled food, fire, etc. There seems to be little benefit in enabling interfaces among these elements directly.
9	Sys-Apps–Environment	Data from the sys-apps to environment sensors would also be minimal.
10	SDR–Environment	Data from the SDR personalities to the environment sensors would be minimal.
11	Cognition–Environment (primary control path)	Data from the cognition system to the environment sensors control those sensors, turning them on and off, setting control parameters, and establishing internal paths from the environment sensors.
12	Effectors–Environment	Data from effectors directly to environment sensors would be minimal.
13	UserSP–Sys-Apps	Data from the user SP system to sys-apps is a *primary afferent path* for multimedia streams and entity states that effect information services implemented as sys-apps. Speech, images, and video to be transmitted move along this path for delivery by the relevant sys-apps or information service to the relevant wired or SDR communications path. Sys-apps overcomes the limitations of individual paths by maintaining continuity of conversations, data integrity, and application coherence (e.g., for multimedia games). Whereas the cognition function sets up, tears down, and orchestrates the sys-apps, the primary API between the user scene and the information service consists of this interface and its companions—the environment afferent path, the effector efferent path, and the SDR afferent and efferent paths.
14	Environment–Sys-Apps	Data on this path assist sys-apps in providing location awareness to services.

Table 14.2(b)　*Cont'd*

Note No.	Process Interface	Explanation
15	Sys-Apps–Sys-Apps	Different information services interoperate by passing control information through the cognition interfaces and by passing domain multimedia flows through this interface. The cognition system sets up and tears down these interfaces.
16	SDR–Sys-Apps	This is the primary afferent path from external communications to the AACR. It includes control and multimedia information flows for all the information services. Following the SDR Forum's SCA, this path embraces wired as well as wireless interfaces.
17	Cognition–Sys-Apps	Through this path, the AACR <Self/> exerts control over the information services provided to the <User/>.
18	Effectors–Sys-Apps	Effectors may provide incidental feedback to information services through this afferent path, but the use of this path is deprecated. Information services are supposed to control and obtain feedback through the mediation of the cognition subsystem.
19	User SP–SDR	Although the SP system may send data directly to the SDR subsystem (e.g., to satisfy security rules that user biometrics be provided directly to the wireless security subsystem), the use of this path is deprecated. Perception subsystem information is supposed to be interpreted by the cognition system so that accurate information, not raw data, can be conveyed to other subsystems.
20	Environment–SDR	Environment sensors such as GPS historically have accessed SDR waveforms directly (e.g., providing timing data for air-interface signal generation). The cognition system may establish such paths in cases where cognition provides little or no value added, such as providing a precise timing reference from GPS to an SDR waveform. The use of this path is deprecated because all of the environment sensors, including GPS, are unreliable. Cognition has the capability to "deglitch" GPS (e.g., recognize from video that the <Self/> is in an urban canyon and therefore not allow GPS to report directly, but report to the GPS subscribers, on behalf of GPS, location estimates based perhaps on landmark correlation, dead reckoning, etc.).
21	Sys-apps–SDR	This is the primary efferent path from information services to SDR through the services API.
22	SDR–SDR	The linking of different wireless services directly to each other is deprecated. If an incoming voice service needs to be connected to an outgoing voice service, there should be a bridging service in sys-apps through which the SDR waveforms communicate with each other. That service should be set up and taken down by the cognition system.
23	Cognition–SDR	This is the primary control interface, replacing the control interface of the SDR SCA and the OMG SRA.
24	Effectors–SDR	Effectors such as speech synthesis and displays should not need to provide state information directly to SDR waveforms, but if needed, the cognition function should set up and tear down these interfaces.
25	User SP–Cognition	This is the primary afferent flow for the results from acoustics, speech, images, video, video flow, and other sensor-perception subsystems. The primary results passed across this interface should be the specific states of <Entities/> in the scene, which would include scene characteristics such as the recognition of landmarks, known vehicles, furniture, and the like. In other words, this is the interface by which the presence of <Entities/> in the local scene is established and their characteristics are made known to the cognition system.

Table 14.2(b) Explanations of Interface Notes for Functional Processes Shown in Table 14.2(a)—*Cont'd*

Note No.	Process Interface	Explanation
26	Environment–Cognition	This is the primary afferent flow for environment sensors.
27	Sys-Apps–Cognition	This is the interface through which information services request services and receive support from the AACR platform. This is also the control interface by which cognition sets up, monitors, and tears down information services.
28	SDR–Cognition	This is the primary afferent interface by which the state of waveforms, including a distinguished RF-sensor waveform, is made known to the cognition system. The cognition system can establish primary and backup waveforms for information services, enabling the services to select paths in real time for low-latency services. Those paths are set up and monitored for quality and validity (e.g., obeying XG rules) by the cognition system, however.
29	Cognition–Cognition	The cognition system as defined in this six-component architecture entails the following: (1) orienting to information from RF sensors in the SDR subsystem and from scene sensors in the user SP and environment sensors; (2) planning; (3) making decisions; and (4) initiating actions, including the control over all of the cognition resources of the <Self/>. The <User/> may directly control any of the elements of the systems via paths through the cognition system that enable it to monitor what the user is doing in order to learn from a user's direct actions, such as manually tuning in the user's favorite radio station when the <Self/> either failed to do so properly or was not asked.
30	Effectors–Cognition	This is the primary afferent flow for status information from the effector subsystem, including speech synthesis, displays, and the like.
31	User SP–Effectors	In general, the user SP system should not interface directly to the effectors, but should be routed through the cognition system for observation.
32	Environment–Effectors	The environment system should not interface directly to the effectors. This path is deprecated.
33	Sys-Apps–Effectors	Sys-apps may display streams, generate speech, and otherwise directly control any effectors once the paths and constraints have been established by the cognition subsystem.
34	SDR–Effectors	This path may be used if the cognition system establishes a path, such as from an SDR's voice track to a speaker. Generally, however, the SDR should provide streams to the information services of the sys-apps. This path may be necessary for legacy compatibility during the migration from SDR through AACR to iCR, but it is deprecated.
35	Cognition–Effectors	This is the primary efferent path for the control of effectors. Information services provide the streams to the effectors, but cognition sets them up, establishes paths, and monitors the information flows for support to the user's <Need/> or intent.
36	Effectors–Effectors	These paths are deprecated, but may be needed for legacy compatibility.

interdependencies. We will define and analyze these APIs in this chapter. It would also be premature to develop such APIs without a clear idea of the kinds of RF and user domain knowledge and performance expected of the CR architecture over time. These aspects are developed in the balance of this chapter, enabling one to draw some conclusions about these APIs in the final part of this chapter.

A fully defined set of interfaces and APIs would be circumscribed in RXML.

14.2.3 Near-Term Implementations

One way to implement this set of functions is to embed a reasoning engine into an SDR, such as a rule base with an associated inference engine, as the cognition function. If the effector functions control parts of the radio, then we have the simplest CR based on the simple six-component architecture of Figure 14.2. Such an approach may be sufficient to expand the control paradigm from today's state machines with limited flexibility to tomorrow's CR control based on reasoning over more complex RF states and user situations.

This incremental step doesn't suggest how to mediate the interfaces between multisensory perception, situation-sensitive prior experience, and a priori knowledge to achieve situation-dependent radio control. Such radio control enables more sophisticated information services. A simple architecture does not proactively allocate machine-learning (ML) functions to fully understood components. For example, will AML require an embedded radio propagation modeling tool? If so, then what is the division of function between a rule base that knows about radio propagation and a propagation tool that can predict values such as the received signal-strength indicator (RSSI)? Similarly, in the user domain, some aspects of user behavior, such as movement by foot and in vehicles, may be modeled in detail based on physics. Will movement modeling be a separate subsystem based on physics and GPS? How will this work inside of buildings? How is the knowledge and skill in tracking user movements divided between physics-based computational modeling and the symbolic inference of a rule base or set of Horn clauses[6] [34]. with a PROLOG engine? For that matter, how will the learning architecture accommodate a variety of learning methods such as neural networks, PROLOG, forward chaining, or support vector machines (SVMs) if learning occurs entirely in a cognition subsystem?

Although hiding such details may be a good thing for CR in the near term, it may severely limit the mass customization needed for CRs to learn user patterns and thus to deliver RF services dramatically better than mere SDRs. Thus, we need to go further "inside" the cognition and perception subsystems to establish more of a fine-grained architecture. This enables one to structure the data sets and functions that mediate multisensory domain perception of complex scenes and related learning technologies that can autonomously adapt to user needs and preferences. The sequel [12] thus proactively addresses the embedding of ML technology into the radio architecture.

Next, consider the networks. Network-independent SDRs retain multiple personalities in local storage, whereas network-dependent SDRs receive alternate personalities

---

[6]A Horn clause is a Boolean expression in which no more than one of the Boolean variables is positive (not negated). Horn clauses are used in artificial intelligence (AI) systems to prove theorems [9].

from a supporting network infrastructure—CWNs. High-end SDRs both retain alternate personalities locally, and have the ability to validate and accept personalities by download from trusted sources. Whatever architecture emerges must be consistent with the distribution of RXML knowledge aggregated in a variety of networks from a tightly coupled CWN to the Internet, with a degree of <Authority/> and trust reflecting the pragmatics of such different repositories.

Thus, the stage is set for the development of CRA. The following sections address the cognition cycle, the inference hierarchies, and the SDR architecture embedded into the CRA.

14.2.4 Cognition Components

Figure 14.1 shows three computational intelligence aspects of CR:

1. Radio knowledge—RXML:RF
2. User knowledge—RXML:User
3. The capacity to learn

The minimalist architecture of Figure 14.2 and the functional interfaces of Table 14.2(a) and (b) do not assist the radio engineer in structuring knowledge, nor do they assist much in integrating ML into the system. Rather, the fine-grained architecture developed in this chapter is derived from the functional requirements to fully develop these three core capabilities.

Radio Knowledge in the Architecture

Radio knowledge has to be translated from the classroom and engineering teams into a body of computationally accessible, structured technical knowledge about radio. RXML is the primary enabler and product of this foray into formalization of radio knowledge. This text starts a process of RXML definition and development that can be brought to fruition only by industry over time. This process is similar to the evolution of the SCA of the SDR Forum [3]. The SCA structures the technical knowledge of the radio components into UML and XML. RXML will enable the structuring of sufficient RF and user world knowledge to build advanced wireless-enabled or enhanced information services. Thus, whereas the SRA and SCA focus on building radios, RXML focuses on using radios.

The World Wide Web (WWW) is now sprouting with computational ontologies, some of which are nontechnical but include radio, such as the open Cyc[7] ontology. They bring the radio domain into the Semantic Web, which helps people know about radio. This informal knowledge lacks the technical scope, precision, and accuracy of authoritative radio references such as the European Telecommunications Standards Institute (ETSI) documents defining the Global System for Mobile Communications

---

[7]Cyc is an AI project that attempts to assemble an encyclopedic comprehensive ontology and database of everyday commonsense knowledge, with the goal of enabling AI applications to perform humanlike reasoning [9].

(GSM) and the International Telecommunication Union (ITU) definitions of, for example, 3GPP.[8]

Not only must radio knowledge be precise, it must be stated at a useful level of abstraction, yet with the level of detail appropriate to the use case. Thus, ETSI GSM, in most cases, would overkill the level of detail without providing sufficient knowledge of the user-centric functionality of GSM. In addition, CR is multiband, multimode radio (MBMMR), so the knowledge must be comprehensive, addressing the majority of radio bands and modes available to an MBMMR. This knowledge is formalized with precision that should be acceptable to ETSI, the ITU, and regulatory authorities (RAs), yet also be at a level of abstraction appropriate to internal reasoning, formal dialog with a CWN, or informal dialog with users.

The capabilities required for a CR node to be a cognitive entity are to sense, perceive, orient, plan, decide, act, and learn. To relate ITU standards to these required capabilities is a process of extracting content from highly formalized knowledge bases that exist in a unique place and that bear substantial authority, encapsulating that knowledge in less complete and therefore somewhat approximate form that can be reasoned with on the CR node and in real time to support RF-related use cases. Table 14.3 illustrates this process.

Table 14.3 is illustrative, not comprehensive, but it characterizes the technical issues that drive an information-oriented CR node architecture. Where ITU, ETSI, . . . (meaning

Table 14.3 Radio Knowledge in the Node Architecture

Need	Source Knowledge	AACR Internalization
Sense RF	RF platform	Calibration of RF, noise floor, antennas, direction
Perceive RF	ITU, ETSI, ARIB, RAs	Location-based table of radio spectrum allocation
Observe RF (sense and perceive)	Unknown RF	RF sensor measurements and knowledge of basic types (AM, FM, simple digital channel symbols, typical TDMA, FDMA, CDMA signal structures)
Orient	XG-like policy	Receive, parse, and interpret policy language
	Known waveform	Measure parameters in RF, space, and time
Plan	Known waveform	Enable SDR for which licensing is current
	Restrictive policy	Optimize transmitted waveform, space–time plan
Decide	Legacy waveform, policy	Defer spectrum use to legacy users per policy
Act	Applications layer	Query for available services (white/yellow pages)
	ITU, ETSI, … CWN	Obtain new skills encapsulated as download
Learn	Unknown RF	Remember space–time–RF signatures; discover spectrum use norms and exceptions
	ITU, ETSI, … CWN	Extract relevant aspects (e.g., new feature)

---

[8]3GPP is a collaboration agreement for the 3G portable phone among ETSI (Europe), the Association of Radio Industries and Businesses/Telecommunication Technology Committee (ARIB/TTC; Japan), the China Communications Standards Association (CCSA; China), the Alliance for Telecommunications Industry Solutions (ATIS; North America), and the Telecommunications Technology Association (TTA; South Korea).

other regional and local standards bodies), and CWN supply source knowledge, the CWN is the repository for authoritative knowledge derived from the standards bodies and RAs, the <Authorities/>. A user-oriented CR may note differences in the interpretation of source knowledge from <Authorities/> between alternate CWNs, precipitating further knowledge exchanges.

User Knowledge in the Architecture

Next, user knowledge is formalized at the level of abstraction and degree of detail necessary to give the CR the ability to acquire, from its owner and other designated users, the user knowledge relevant to information services incrementally. Incremental knowledge acquisition was motivated in the introduction to ML by describing how frequent occurrences with similar activity sequences identify learning opportunities. ML machines may recognize these opportunities for learning through joint probability statistics <Histogram/>. Effective use cases clearly identify the classes of user and the specific knowledge learned to customize envisioned services. Use cases may also supply sufficient initial knowledge to render incremental ML not only effective, but also—if possible—enjoyable to the user.

This knowledge is defined in RXML:User. As with RF knowledge, the capabilities required for a CR node to be a cognitive entity are to OOPDAL.[9] To relate a use case to these capabilities, one extracts specific and easily recognizable <Anchors/> for stereotypical situations observable in diverse times, places, and situations. One expresses the anchor knowledge in RXML for use on the CR node.

Cross-Domain Grounding for Flexible Information Services

The knowledge about radio, and about user needs for wireless services, must be expressed internally in a consistent form so that information service relationships may be autonomously discovered and maintained by the <Self/> on behalf of the <User/>. Figure 14.4 shows relationships among user and RF domains.

Staying better connected requires the normalization of knowledge between <User/> and <RF/> domains. If, for example, the <User/> says, "What's on one oh seven, seven," near the Washington, DC, area, then the dynamic <User/> ontology should enable the CR to infer that the user is talking about the current frequency modulation (FM) radio broadcast, the units are in MHz, and the user wants to know what is on WTOP. If it can't infer this, then it should ask the user or discover by first dialing a reasonable default (e.g., 107.7 FM, a broadcast radio station) and asking, "Is this the radio station you want?" Steps 4, 5, and 6 in Figure 14.4 all benefit from agreement across domains on how to refer to radio services. Optimizing behavior to best support the user requires continually adapting the <User/> ontology with repeated regrounding of terms in the <User/> domain to conceptual primitives and actions in the <RF/> domain.

The CRA facilitates this by seeding the speech-recognition subsystem with the most likely expressions a particular <User/> employs when referring to information services. These would be acquired from the specific users via text and speech recognition, with dialogs oriented toward continual grounding by posing yes–no questions, either verbally

---

[9]OOPDAL—Observe, Orient, Plan, Decide, Act, Learn; these steps in reasoning will be discussed in Section 14.3.1.

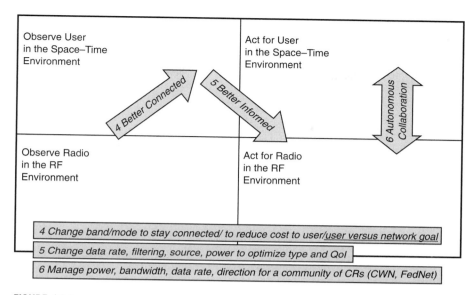

FIGURE 14.4

Discovering and maintaining services. (*Source:* © Dr. Joseph Mitola III, used with permission.)

or in displays or both, and obtaining reinforcement either verbally or via haptic interaction or both. The required degree of mutual grounding would benefit from specific grounding-oriented features in the CR information architecture [12].

The process of linking user expressions of interest to the appropriate radio technical operations sometimes may be extremely difficult. Military radios, for example, have many technical parameters: a "channel" in SINCGARS[10] consists of dehopped digital voice in one context (voice communications) or a 25 kHz band of spectrum in another context (and that may be either an FM channel into which its frequency hop waveform has hopped or a frequency division multiple access (FDMA) channel when in single-channel mode). If the user says, "Give me the commander's channel," the SINCGARS user is talking about a "dehopped CVSD voice stream."[11] If the same user a few seconds later says, "This sounds awful. Who else is in this channel?" the user is referring to interference with a collection of hop sets. If the CR observes, "There is strong interference in almost half of your assigned channels," then the CR is referring to a related set of 25 kHz channels. If the user then says, "OK, notch the strongest three interference channels," the user is talking about a different subset of the channels. If in the next breath the user says, "Is anything on our emergency channel?" then the user has switched from SINCGARS context to <Self/> context, asking about one of the cognitive military radio's physical RF access channels. The complexity of such exchanges demands cross-domain grounding; and the necessity of communicating accurately but quickly

---

[10]SINCGARS is the Single-Channel Ground and Airborne Radio System used by the US DoD (see Chapter 4).

[11]CVSD stands for *continuously variable slope delta modulation*, a voice coding method.

under stress motivates a structured NL and rich radio ontology aspects of the architecture, developed further in this section.

Thus, both commercial and military information services entail cross-domain grounding with ontology oriented to NL in the <User/> domain and oriented to RXML formalized, a priori knowledge in the <RF/> domain. Specific methods of cross-domain grounding with associated architectural features include:

1. *<RF/> to <User/> shaping dialog* to express precise <RF/> concepts to nonexpert users in an intuitive way, such as:
   (a) *Grounding:* "If you move the speaker box by a few inches, it can make a big difference in how well the remote speaker is connected to the wireless transmitter on the television (TV)."
   (b) *CR information architecture:* Include facility for *rich set of synonyms* to mediate cognition–NL–synthesis interface (<Antenna/> ≅ <Wireless-remote-speaker/> ≅ "Speaker box").
2. *<RF/> to <User/> learning jargon* to express <RF/> connectivity opportunities in <User/> terms:
   (a) *Grounding:* "Tee oh pee" for "WTOP," "Hot ninety-two" for "FM 92.3."
   (b) *CR information architecture:* NL-visual facility for single-instance update of user jargon.
3. *<User/> to <RF/> relating values to actions:* Relate <User> expression of values ("low-cost") to features of situations ("normal") that are computable (<NOT> (<CONTAINS> <Situation> <Unusual/> </..>)) and that relate directly to <RF/> domain decisions, for example:
   (a) *Grounding:* Normally wait for free WLAN for big attachment; if situation is <Unusual/>, ask if user wants to pay for 3G.
   (b) *CR information architecture:* Associative inference hierarchy that relates observable features of a <Scene/> to user sensitivities, such as <Late-for-work/> ⇒ <Unusual/>; "The President of the company needs this" => <Unusual/> because "President" => <VIP/> and such a <VIP/> is not present in most scenes.

14.2.5 Self-Referential Components

The cognition component must assess, manage, and control all of its own resources, including validating downloads. Thus, in addition to <RF/> and <User> domains, RXML must describe the <Self/>, defining the CR architecture to the CR itself in RXML.

Self-Referential Inconsistency

This class of self-referential reasoning is well known in the theory of computing to be a potential black hole for computational resources. Specifically, any Turing-capable (TC) computational entity that reasons about itself can encounter unexpected Gödel-Turing situations from which it cannot recover. Thus, TC systems are known to be "partial"—only partially defined because the results obtained when attempting to execute certain classes of procedure are not definable (the computing procedure will never terminate).

To avoid this paradox, CR architecture mandates the use of only "total" functions, typically restricted to bounded minimalization [10]. Watchdog "step-counting" functions [11] or timers must be in place in all its self-referential reasoning and radio functions. The timer and related computationally indivisible control constructs are equivalent to the computer-theoretic construct of a step counting function over "finite minimalization." It has been proven that computations that are limited with certain classes of reliable watchdog timers on finite computing resources can avoid the Gödel-Turing paradox or at least reduce it to the reliability of the timer. This proof is the fundamental theorem for practical self-modifying systems.

In brief, if a system can compute in advance the amount of time or the number of instructions that any given computation should take, then if that time or step count is exceeded, the procedure returns a fixed result such as "Unreachable in Time T." As long as the algorithm does not explicitly or implicitly restart itself on the same problem, then with the associated invocation of a tightly time-constrained and computationally constrained alternative tantamount to giving up, it:

a. is not TC, but
b. is sufficiently computationally capable to perform real-time communications tasks such as transmitting and receiving data as well as bounded user-interface functions, and
c. is not susceptible to the Gödel-Turing incompleteness dilemma, and thus
d. will not crash because of consuming unbounded or unpredictable resources in unpredictable self-referential loops.

This is not a general result. This is a highly radio domain-specific result that has been established only for isochronous communications domains in which

- Processes are defined in terms of a priori tightly bounded time epochs such as code division multiple access (CDMA) frames and Signaling System 7 (SS7) timeouts.[12]
- For every situation, there is a default action that has been identified in advance that consumes $O(1)$ resources.
- The watchdog timer or step-counting function is reliable. There are no external loops that invalidate the preceding prohibitions and controls.

Because radio air interfaces transmit and receive data, there are always defaults, such as "repeat the last packet" or "clear the buffer," that may degrade the performance of the overall communications system. A default has $O(1)$ complexity and the layers of the protocol stack can implement the default without using unbounded computing resources.

---

[12]SS7 is the seventh and most stable software update of the long-distance telephony software. *Timeout* here means that the software expected tasks to complete by a certain timeout; if not, it determined that something had gone awry, and either tried a different route or delivered a tone indicating that a system fault had occurred.

Watchdog Timer

Without the reliable watchdog timer in the architecture and without this proof to establish the rules for acceptable computing constructs on CRs, engineers and computer programmers would build CRs that would crash in extremely unpredictable ways as their adaptation algorithms got trapped in unpredictable unbounded self-referential loops. Because planning problems exist that cannot be solved with algorithms so constrained, either an unbounded community of CRs must cooperatively work on the more general problems or the cognitive network (CN) must employ a TC algorithm to solve the more difficult problems (e.g., *NP*-hard with large *N*) offline. There is also the interesting possibility of trading off space and time by remembering partial solutions and restarting *NP*-hard problems with these subproblems already solved. Although it doesn't actually avoid any necessary calculations, with $O(N)$ pattern matching for solved subproblems, it may reduce the total computational burden, somewhat.[13] This class of approach to parallel problem solving is similar to the use of pheromones by ants to solve the traveling salesman problem in less than $(2^N)/M$ time with M ants. Such theoretical issues are not developed further in this text. It suffices to show the predictable finiteness and proof that the approach is boundable and hence compatible with the real-time performance needs of CR.

This timer-based finite computing regime also works for user interfaces because users will not wait forever before changing the situation (e.g., by shutting off the radio or hitting another key); and the CR can always ask the user to take over.

Thus, with a proof of stability based on the theory of computing, the CRA structures systems can not only modify themselves, but can also do it in such a way that they are not likely to induce nonrecoverable errors because of the "partial" property of self-referential computing.

14.2.6 Flexible Functions of the Component Architecture

Although this chapter develops the six-element component architecture of one particular information architecture and one reference implementation, many possible architectures exist. The purpose is not to try to sell a particular architecture, but to illustrate the architecture principles. The CRA and research implementation CR1[14] therefore offer open-source licensing for noncommercial educational purposes. Table 14.4 further differentiates architectural features.

The functions of the architecture shown in the table are not different from those of the six-component architecture, but represent varying degrees of instantiation of the six components. Consider the following degrees of architecture instantiations:

Cognition functions of radio entail the monitoring and structuring knowledge of the behavior patterns of the <Self/>, the <User>, and the environment (physical, user

---

[13]For example, the fast Fourier transform (FFT) converts $O(N^2)$ steps to $O(N\log N)$ by avoiding the recomputation of already computed partial products.
[14]CR1 is a CR research prototype developed by Joseph Mitola III [12]. CRII, CRIII, CRIV, and CRV are successive extensions of this architecture.

Table 14.4 Features of CR to be Organized via Architecture

Feature	Function	Examples (RF, vision, speech, location, motion)
Cognition	Monitor and learn	Get to know user's daily patterns and model the local RF scene over space, time, and situations
Adaptation	Respond to changing environment	Use unused RF, protect owner's data
Awareness	Extract information from sensor domain	Sense or perceive
Perception	Continuously identify knowns, unknowns, and backgrounds in the sensor domain	TV channel; depth of visual scene, identity of objects; location of user, movement and speed of <Self/>
Sensing	Continuously sense and preprocess single-sensor field in single-sensory domain	RF FFT; binary vision; binaural acoustics; GPS; accelerometer; etc.

situation, and radio) to provide information services, learning from experience to tailor services to user preferences and differing radio environments.

Adaptation functions of radio respond to a changing environment, but can be achieved without learning if the adaptation is preprogrammed.

Awareness functions of radio extract usable information from a sensor domain. Awareness stops short of perception. Awareness is required for adaptation, but awareness does not guarantee adaptation. For example, embedding a GPS receiver into a cell phone makes the phone more location aware, but unless the value of the current location is actually used by the phone to do something that is location dependent, the phone is not location adaptive, only location aware. These functions are a subset of the CRA that enable adaptation.

Perception functions of radio continuously identify and track knowns, unknowns, and backgrounds in a given sensor domain. Backgrounds are subsets of a sensory domain that share common features that entail no particular relevance to the functions of the radio. For a CR that learns initially to be a single-owner radio, in a crowd, the owner is the object that the radio continuously tracks in order to interact when needed. Worn from a belt as a cognitive wireless PDA (CWPDA), the iCR perception functions may track the entities in the scene. The nonowner entities comprise mostly irrelevant background because no matter what interactions may be offered by these entities, the CR will not obey them—only the interactions of the perceived owner. These functions are a subset of the CRA that enable cognition.

The sensory functions of radio entail those hardware and/or software capabilities that enable a radio to measure features of a sensory domain. Sensory domains include anything that can be sensed, such as audio, video, vibration, temperature, time, power, fuel level, ambient light level, sun angle (e.g., through polarization), barometric pressure, smell, and anything else imaginable. Sensory domains for vehicular radios may be much richer, if less personal, than those of wearable radios. Sensory domains for fixed infrastructure could include weather features such as ultraviolet sunlight, wind direction and speed, humidity, traffic flow rate, or rain rate. These functions are a subset of the CRA that enable perception.

The platform-independent model (PIM) in the UML of SDR [13] provides a convenient, industry-standard computational model that a CR can use to describe the SDR and computational-resource aspects of its own internal structure, as well as facilities that enable radio functions. The general structure of hardware and software by which a CR reasons about the <Self/> in its world is also part of its architecture defined in the SDR SCA/SRA as resources.

14.3 CRA-II: THE COGNITION CYCLE

The CRA comprises a set of design rules by which the cognitive level of information services may be achieved by a specified set of components in a way that supports the cost-effective evolution of increasingly capable implementations over time [1]. The cognition subsystem of the architecture includes an inference hierarchy and the temporal organization and flow of inferences and control states—the cognition cycle.

14.3.1 Cognition Cycle

The cognition cycle developed for CR1 [14] is illustrated in Figure 14.5. This cycle implements the capabilities required of iCR in a reactive sequence. Stimuli enter the CR as sensory interrupts, dispatched to the cognition cycle for a response. Such an iCR continually observes (senses and perceives) the environment, orients itself, creates plans, decides, and then acts. In a single-processor inference system, the CR's flow of control may also move in the cycle from observation to action. In a multiprocessor system, temporal structures of sensing, preprocessing, reasoning, and acting may be parallel and complex. Special features synchronize the inferences of each phase. The tutorial code all works on a single processor in a rigid inference sequence defined in Figure 14.5.

This process is called the "wake epoch" because the primary reasoning activities during this large epoch of time are reactive to the environment. We will refer to "sleep epochs" for power-down conditions, "dream epochs" for performing computationally intensive pattern recognition and learning, and "prayer epochs" for interacting with a higher authority (e.g., network infrastructure). See Section 14.5.4 for further discussion of behavioral epochs.

During the wake epoch, the receipt of a new stimulus on any of a CR's sensors or the completion of a prior cognition cycle initiates a new primary cognition cycle. The CR observes its environment by parsing incoming information streams. These can include monitoring and speech-to-text conversion of radio broadcasts (e.g., the Weather Channel, stock ticker tapes, etc.). Any RF-LAN or other short-range wireless broadcasts that provide service awareness information may be also parsed. In the observation phase, a CR also reads location, temperature, and light-level sensors, among other parameters, to infer the user's communications context.

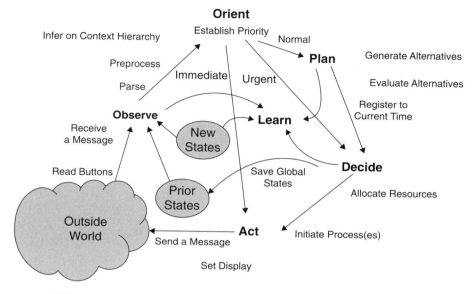

FIGURE 14.5

Simplified cognition cycle. The OOPDA loop is a primary cycle; learning, planning, and sensing the outside world are crucial phases of the larger OOPDA loop. (*Source:* © Dr. Joseph Mitola III, used with permission.)

14.3.2 Observe (Sense and Perceive)

The iCR senses and perceives the environment (via "observation phase" code) by accepting multiple stimuli in many dimensions simultaneously and by binding these stimuli—all together or more typically in subsets—to prior experience so that it can subsequently detect time-sensitive stimuli and ultimately generate plans for action.

Thus, iCR continuously aggregates experience and compares prior aggregates to the current situation. A CR may aggregate experience by remembering everything. This may not seem like a very smart thing to do until you calculate that all the audio, unique images, and emails the radio might experience in a year takes up only a few hundred gigabytes of memory, depending on image detail. So the computational architecture for remembering and rapidly correlating current experience against everything known previously is a core capability of the CRA. A *novelty* detector identifies new stimuli, using the new aspects of partially familiar stimuli to identify incremental-learning primitives.

In the six-component (user SP, environment, effectors, SDR, sys-apps, and cognition) functional view of the architecture defined in the CR Node Functional Components subsection (see p. 433), the observe phase comprises both the user SP and the environment (RF and physical) sensor subsystems. The subsequent orient phase is part of the cognition component in this model of architecture.

14.3.3 Orient

The orient phase determines the significance of an observation by binding the observation to a previously known set of stimuli of a "scene." The orient phase contains the internal data structures that constitute the equivalent of the short-term memory (STM) that people use to engage in a dialog without necessarily remembering everything with the same degree of long-term memory (LTM). Typically, people need repetition to retain information over the long term. The natural environment supplies the information redundancy needed to instigate transfer from STM to LTM. In the CRA, the transfer from STM to LTM is mediated by the sleep cycle in which the contents of STM since the last sleep cycle are analyzed both internally and with respect to existing LTM. How to do this robustly remains an important CR research topic, but the overall framework is defined in Section 14.5.4.

Matching of current stimuli to stored experience may be achieved by "stimulus recognition" or by "binding." The orient phase is the first collection of activity in the cognition component.

Stimulus Recognition

Stimulus recognition occurs when there is an exact match between a current stimulus and a prior experience. The CR1 prototype is continually recognizing exact matches and recording the number of exact matches that occurred along with the time measured in the number of cognition cycles between the last exact match. By default, the response to a given stimulus is to merely repeat that stimulus to the next layer up the inference hierarchy for aggregation of the raw stimuli. But if the system has been trained to respond to a location, a word, an RF condition, a signal on the power bus, or some other parameter, it may either react immediately or plan a task in reaction to the detected stimulus. If that reaction were in error, then it may be trained to ignore the stimulus, given the larger context, which consists of all the stimuli and relevant internal states, including time.

Sometimes the orient phase causes an action to be initiated immediately as a "reactive" stimulus–response behavior. A power failure, for example, might directly invoke an act that saves the data (the "immediate" path to the act phase in Figure 14.5). A nonrecoverable loss of signal on a network might invoke reallocation of resources (e.g., from parsing input to searching for alternative RF channels). This may be accomplished via the path labeled "urgent" in Figure 14.5.

Binding

Binding occurs when there is a nearly exact match between a current stimulus and a prior experience and very general criteria for applying the prior experience to the current situation are met. One such criterion is the number of unmatched features of the current scene. If only one feature is unmatched and the scene occurs at a high level, such as the phrase or dialog level of the inference hierarchy, then binding is the first step in generating a plan for behaving in the given state similar to the last occurrence of the stimuli. In addition to number of features that match exactly, which is a kind of hamming code, instance-based learning (IBL) supports inexact matching and binding. Binding also determines the priority associated with the stimuli. Better binding yields higher priority for autonomous learning, whereas less effective binding yields lower priority for the incipient plan.

14.3.4 Plan

Most stimuli are dealt with "deliberatively" rather than "reactively." An incoming network message would normally be dealt with by generating a plan (in the plan phase, the "normal" path). Such planning includes plan generation. In research quality or industrial-strength CRs, formal models of causality must be embedded into planning tools. The plan phase should also include reasoning about time. Typically, reactive responses are preprogrammed or defined by a network (i.e., the CR is "told" what to do), whereas other behaviors might be planned. A stimulus may be associated with a simple plan as a function of planning parameters with a simple planning system. Open-source planning tools enable the embedding of planning subsystems into the CRA, enhancing the plan component. Such tools enable the synthesis of RF and information access behaviors in a goal-oriented way based on perceptions from the visual, audio, text, and RF domains, as well as RA rules and previously learned user preferences.

14.3.5 Decide

The decide phase selects among the candidate plans. The radio might have the choice to alert the user to an incoming message (e.g., behave like a pager), or to defer the interruption until later (e.g., behave like a secretary who is screening calls during an important meeting).

14.3.6 Act

Acting initiates the selected processes using effector modules. Effectors may access the external world or the CR's internal states.

Externally Oriented Actions

Access to the external world consists primarily of composing messages to be spoken into the local environment or expressed in text form locally or to another CR or CN using the Knowledge Query and Manipulation Language (KQML), Radio Knowledge Representation Language (RKRL), Web Ontology Language (OWL), Radio eXtensible Markup Language (RXML), or some other appropriate knowledge interchange standard.

Internally Oriented Actions

Actions on internal states include controlling machine-controllable resources (e.g., radio channels). The CR can also affect the contents of existing internal models, such as adding a model of stimulus–experience–response (serModel) to an existing internal model structure [12]. The new concept itself may assert related concepts into the scene. Multiple independent sources of the same concept in a scene reinforce that concept for that scene. These models may be asserted by the <Self/> to encapsulate experience. The experience may be reactively integrated into RXML knowledge structures as well, provided the reactive response encodes them properly.

14.3.7 **Learning**

Learning is a function of perception, observations, decisions, and actions. Initial learning is mediated by the observe phase perception hierarchy in which all SP are continuously matched against all prior stimuli to continually count occurrences and to remember time since the last occurrence of the stimuli from primitives to aggregates.

Learning also occurs through the introduction of new internal models in response to existing models and case-based reasoning (CBR) bindings. In general, there are many opportunities to integrate ML into CR. Each of the phases of the cognition cycle offers multiple opportunities for discovery processes, such as <Histogram/>, as well as many other ML approaches. The architecture includes internal reinforcement via counting occurrences and via serModels, so ML with uncertainty is also supported [12, 31].

Finally, a learning mechanism occurs when a new type of serModel is created in response to an action to instantiate an internally generated serModel. For example, prior and current internal states may be compared with expectations to learn about the effectiveness of a communications mode, instantiating a new mode-specific serModel.

14.3.8 **Self-Monitoring**

Each of the prior phases must consist of computational structures for which the execution time may be computed in advance. In addition, each phase must restrict its computations to not consume more resources (time x allocated processing capacity) than the precomputed upper bound. Therefore, the architecture has some prohibitions and some data set requirements needed to obtain an acceptable degree of stability of behavior for CRs as self-referential, self-modifying systems.

Since first-order predicate calculus (FOPC) used in some reasoning systems is not decidable, one cannot in general compute in advance how much time an FOPC expression will take to run to completion. There may be loops that will preclude this, and even with loop detection, the time to resolve an expression may be only loosely approximated as an exponential function of some parameters (e.g., the number of statements in the FOPC database of assertions and rules). Therefore, unrestricted FOPC is not allowed.

Similarly, unrestricted `For`, `Until`, and `While` loops are prohibited. In place of such loops are bounded iterations in which the time required for the loop to execute is computed or supplied independent of the computations that determine the iteration control of the loop. This seemingly unnatural act can be facilitated by next-generation compilers and computer-aided software engineering (CASE) tools. Because self-referential, self-modifying code is prohibited by structured design and programming practices, no such tools are available on the market today. But since CR is inherently self-referential and self-modifying, such tools most likely will emerge, perhaps assisted by the needs of CR and the architecture framework of the cognition cycle.

Finally, the cognition cycle itself cannot contain internal loops. Each iteration of the cycle must take a defined amount of time, just as each frame of a 3G air interface takes 10 milliseconds. As CR computational platforms continue to progress, the amount of computational work done within the cycle will increase, but under no conditions should explicit or implicit loops be introduced into the cognition cycle that would extend it beyond a given cycle time.

Retrospection

The assimilation of knowledge by ML can be computationally intensive, so, as previously stated in Section 14.3.1 and further discussed in Section 14.5.4, CR has sleep and prayer epochs that support ML. A sleep epoch is a relatively long period of time (e.g., minutes to hours) during which the radio will not be in use, but has sufficient electrical power for processing. During the sleep epoch, the radio can run ML algorithms without detracting from its ability to support its user's needs. ML algorithms may integrate experience by aggregating statistical parameters. The sleep epoch may rerun stimulus-response sequences with new learning parameters in the way that people dream. The sleep cycle could be less anthropomorphic, however, employing a genetic algorithm to explore a rugged fitness landscape, potentially improving the decision parameters from recent experience.

Reaching Out

Learning opportunities not resolved in the sleep epoch can be brought to the attention of the user, the host network, or a designer. We refer to elevating complex problems to an infrastructure support as a prayer epoch.

14.4 CRA-III: THE INFERENCE HIERARCHY

The phases of inference from observation to action show the flow of inference, a top-down view of how cognition is implemented algorithmically. The inference hierarchy is the part of the algorithm architecture that organizes the data structures. Inference hierarchies have been in use since Hearsay II in the 1970s [15], but the CR hierarchy is unique in its method of integrating ML with real-time performance during the wake epochs. An illustrative inference hierarchy includes layers from atomic stimuli at the bottom to information clusters that define action contexts, as shown in Figure 14.6.

The pattern of accumulating elements into sequences begins at the bottom of the hierarchy. Atomic stimuli originate in the external environment, including RF, acoustic, image, and location domains, among others. The atomic symbols extracted from them

Sequence	Level of Abstraction
Context Cluster	**Scenes** in a Play, Session
Sequence Clusters	**Dialogs**, Paragraphs, Protocol
Basic Sequences	**Phrases**, Video Clip, Message
Primitive Sequences	**Words**, Token, Image
Atomic Symbols	**Raw Data**, Phoneme, Pixel,
Atomic Stimuli	External Phenomena

FIGURE 14.6

Standard inference hierarchy. (*Source:* © Dr. Joseph Mitola III, used with permission.)

are the most primitive symbolic units in the domain. In speech, the most primitive elements are the phonemes. In the exchange of textual data (e.g., email), the symbols are the typed characters. In images, the atomic symbols may be the individual picture elements (pixels) or they may be small groups of pixels with similar hue, intensity, texture, and so forth.

A related set of atomic symbols forms a primitive sequence. Words in text, tokens from a speech "tokenizer," and objects in images (or individual image regions in a video flow) are primitive sequences. Primitive sequences have spatial and/or temporal coincidence, standing out against the background (or noise), but there may be no particular meaning in that pattern of coincidence. Basic sequences, in contrast, are space–time–spectrum sequences that entail the communication of discrete messages.

These discrete messages (e.g., phrases) are typically defined with respect to an ontology of the primitive sequences (e.g., definitions of words). Sequences cluster together because of shared properties. For example, phrases that include words, such as "hit," "pitch," "ball," and "out," may be associated with a discussion of a baseball game. Knowledge Discovery in Databases (KDD) and the Semantic Web offer approaches for defining, or inferring, the presence of such clusters from primitive and basic sequences.

A scene is a context cluster, a multidimensional space–time–frequency association, such as a discussion of a baseball game in the living room on a Sunday afternoon. Such clusters may be inferred from unsupervised ML (e.g., using statistical methods or nonlinear approaches such as SVMs).

Although presented here in a bottom-up fashion, there is no reason to limit multidimensional inference to the top layers of the inference hierarchy. The lower levels of the inference hierarchy may include correlated multisensor data. For example, a word may be characterized as a primitive acoustic sequence coupled to a primitive sequence of images of a person speaking that word. In fact, taking the cue that infants seem to thrive on multisensory stimulation, the key to reliable ML may be the use of multiple sensors with multisensor correlation at the lowest levels of abstraction. Each of these levels of the inference hierarchy is discussed in more detail in the following sections.

14.4.1 Atomic Stimuli

Atomic stimuli originate in the external environment and are sensed and preprocessed by the sensory subsystems, which include sensors of the RF environment (e.g., radio receiver and related data and information processing) and of the local physical environment, including acoustic, video, and location sensors. Atomic symbols are the elementary stimuli extracted from the atomic stimuli. Atomic symbols may result from a simple noise-riding threshold algorithm, such as the squelch circuit in RF that differentiates signal from noise. Acoustic signals may be differentiated from simple background noise this way, but generally the result is the detection of a relatively large speech epoch that contains various kinds of speech energy. Thus, further signal processing is typically required in a preprocessing subsystem to isolate atomic symbols.

The transformation from atomic stimuli to atomic symbols is the job of the sensory preprocessing system. Thus, for example, acoustic signals may be transformed into phoneme hypotheses by an acoustic signal preprocessor. However, some software tools may not enable this level of interface via an API. Advanced speech-recognition and video-processing software tools are needed to develop industrial-strength CR. Speech tools yield an errorful transcript in response to an acoustic signal. Thus, it is preferable for speech tools to map from stimuli to basic sequences. One of the important contributions of architecture is to identify such maps and to define the role and the level of the mapping tools.

There is nothing about the inference hierarchy, however, that forces data from a preprocessing system to be entered at the lowest level. For the more primitive symbolic abstractions (e.g., atomic symbols) to be related to more aggregate abstractions, one may either build up the aggregates from the primitive abstractions or derive the primitive abstractions from the aggregates. People are used to being exposed to "the whole thing" by immersion in the full experience of life—touch, sight, sound, taste, and balance—all at once; therefore, it seems possible (even likely) that the more primitive abstractions are somehow derived through the analysis of aggregates, perhaps by cross-correlation. This can be accomplished in a CRA sleep cycle. The idea is that the wake cycle is optimized for immediate reaction to stimuli, similar to what our ancestors needed to avoid predation, whereas the sleep cycle is optimized for introspection, for analyzing the day's stimuli to derive those objects that should be recognized and acted on in the next cycle.

Stimuli are each counted. When an iCR that conforms to this architecture encounters a stimulus, it both counts how many such stimuli have been encountered and resets to zero a timer that keeps track of the time since the last occurrence of the stimulus.

14.4.2 Primitive Sequences: Words and Dead Time

The accumulation of sequences of atomic symbols forms primitive sequences. The key question at this level of the data structure hierarchy is the sequence boundary. The simplest situation is one in which a distinguished atomic symbol separates primitive sequences, which is exactly the case with white space between words in typed text. A text-based ML system may need white space to separate a text stream into primitive sequences.

14.4.3 Basic Sequences

The pattern of aggregation is repeated vertically at the levels corresponding to words, phrases, dialogs, and scenes. The data structures generated by processing nodes create the concept hierarchy of Figure 14.6. These are the reinforced hierarchical sequences. They are reinforced by the inherent counting of the number of times each atomic or aggregated stimulus occurs. The phrase level typically contains or implies a verb (the verb "to be" may be implied if no other verb is implicit).

Unless digested (e.g., by a sleep process), the observation phase hierarchy accumulates all the sensor data, parsed and distributed among processing nodes for fast parallel retrieval. Because the hierarchy saves everything and compares new data to memories, it is a kind of memory-based learning approach, which takes a lot of space. When the stimuli retained are limited to atomic symbols and their aggregates, however, the total amount of data that need to be stored is relatively modest. In addition, recent research shows the negative effects of discarding cases in word pronunciation. In word pronunciation, no example can be discarded even if it is "disruptive" to a well-developed model. Each exception has to be followed. Thus, in the CR1 prototype, when multiple memories match partially, the nearest match informs the orientation, planning, and action.

14.4.4 NL in the CRA Inference Hierarchy

In speech, words spoken in a phrase may be coarticulated with no distinct boundary between the primitive acoustic symbols in a basic acoustic sequence. Therefore, speech-detection algorithms may extract an acoustic sequence, but the interpretation of that sequence as constituent primitive sequences may be much less reliable. Typically, the correct parse is within the top ten candidates for contemporary speech tools. The flow of speech signal processing may be similar to the following:

- Isolate a basic sequence (phrase) from background and noise by using an acoustic analysis to determine speech versus background.
- Analyze the basic sequence to identify candidate primitive sequence boundaries (words).
- Analyze the primitive sequences statistically using Hidden Markov sequence Modeling (HMM).
- Evaluate primitive and basic sequence hypotheses based on a statistical model of language to rank-order alternative interpretations of the basic sequence.

So a practical speech-processing algorithm may yield alternative strings of phonemes and candidate parses "all at once." NL-processing (NLP) tool sets may be embedded into the CRA inference hierarchy, as illustrated in Figure 14.7. Speech and/or text channels may be processed via such NL facilities with substantial a priori models of language and discourse. The use of those models entail mappings among the word, phrase, dialog, and scene levels of the observation phase hierarchy and the encapsulated component(s).

It is tempting to expect CR to integrate a commercial NLP system such as IBM's ViaVoice or a derivative of an NLP research system such as SNePS [16], AGFL [17], or XTAG [18] perhaps using a morphological analyzer such as PCKIMMO [19]. These tools go too far and yet not far enough in the direction needed for CRA. One might like to employ existing tools by using a workable interface between the domain of radio engineering and some of these NL tool sets. The definition of such cross-discipline interfaces is in its infancy. At present, one cannot just express a radio ontology in Interlingua and neatly plug it into XTAG to get a working CR.

The internal data structures that are used in radio mediate the performance of radio tasks (e.g., "transmit a waveform"). The data structures of XTAG, AGFL, and so forth mediate the conversion of language from one form to another. Thus, XTAG wants to know that *transmit* is a verb and *waveform* is a noun. The CR needs to know that if the user says "transmit" and a message has been defined, then the CR should call the SDR function `transmit()`. NLP systems also need scoping rules for transformations on the linguistic data structures. The way in which domain knowledge is integrated in linguistic structures of these tools tends to obscure the radio engineering aspects.

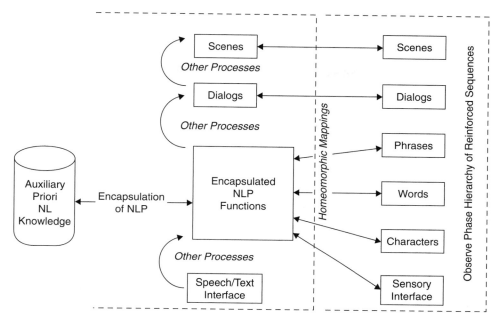

FIGURE 14.7

NL encapsulation in the observation hierarchy. (*Source:* © Dr. Joseph Mitola III, used with permission.)

NLP systems work well on well-structured speech and text, such as the prepared text of a newsanchor. But they do not yet work well on noisy, nongrammatical data structures encountered, for example, when a user is trying to order a cab in a crowded bar. Thus, less-linguistic or metalinguistic data structures may be needed to integrate core CR reasoning with speech and/or text-processing front ends. The CRA has the flexibility illustrated in Figure 14.7 for the subsequent integration of evolved NLP tools. The emphasis of this version of the CRA is a structure of sets and maps required to create a viable CRA. Although introducing the issues required to integrate existing NLP tools, the discussion does not pretend to present a complete solution to this problem.

14.4.5 Observe–Orient Links for Scene Interpretation

CR may use an algorithm-generating language with which one may define self-similar inference processes. In one example, the first process (Proc1) partitions characters into words, detecting novel characters and phrase boundaries as well. Proc2 detects novel words and aggregates known words into phrases. Proc3 detects novel phrases, aggregating known phrases into dialogs. Proc4 aggregates dialogs into scenes, and Proc5 detects known scenes. In each case, a novel entity at level N will be bound in the context of the surrounding known entities at that level to the closest match at the next highest level, $N + 1$. For example, the word–phrase intersection of Proc2 would map the following phrases:

- "Let me introduce Joe."
- "Let me introduce Chip."

Since "Chip" is unknown, and "Joe" is known from a prior dialog, integrated CBR matches the phrases, binding <Chip/> = <Joe/>. In other words, it will try to act with respect to Chip in the way it was previously trained (at the dialog level) to interact with Joe. In response to the introduction, the system may say, "Hello, Chip. How are you?" mimicking the behavior that it had previously learned with respect to Joe. Not too bright, but not all that bad either for a relatively simple ML algorithm.

There is a particular kind of dialog that is characterized by reactive world knowledge in which there is some standard way of reacting to given speech-act inputs. For example, when someone says, "Hello," you may typically reply with "Hello" or some other greeting. The capability to generate such rote responses may be preprogrammed into a lateral component as Hearsay knowledge source (KS). The responses are not preprogrammed, but the general tendency to imitate phrase-level dialogs is a preprogrammed tendency that can be overruled by plan generation—but that is present in the orient phase, which is Proc6 in Figure 14.8.

Words may evoke a similar tendency toward immediate action. What do you do when you hear the words "Help!" or "Fire, fire! Get out, get out!"? The CR programmer can capture reactive tendencies in a CR by preprogramming an ability to detect these kinds of situations in the word-sense KS, as implied by Figure 14.8. When confronted with such wording (which is preferred), a CR should react appropriately if properly trained. To cheat, one can preprogram a wider array of stimulus–response pairs so that the CR has more a priori knowledge, but some of it may not be appropriate. Some

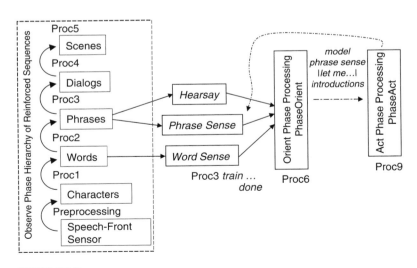

FIGURE 14.8

The inference hierarchy supports lateral KS. (*Source:* © Dr. Joseph Mitola III, used with permission.)

responses are culturally conditioned. Will the CR be too rigid? If it has too much a priori knowledge, it will be perceived by its users as too rigid. If it doesn't have enough, it will be perceived as uninformed.

14.4.6 Observe-Oriented Links for Radio Skill Sets

Radio knowledge may be embodied in components called radio skills. Declarative radio knowledge is static, requiring interpretation by an algorithm (e.g., an inference engine) to accomplish anything. Radio skills, on the other hand, are knowledge embedded in serModels through the process of training or sleeping/dreaming. This knowledge is continually pattern matched against all stimuli in parallel. That is, there are few a priori logical dependencies among knowledge components that mediate the application of the knowledge. With FOPC, the theorem prover must reach a defined state in the resolution of multiple axioms to initiate action. In contrast, serModels are continually compared to the level of the hierarchy to which they are attached, so their immediate responses are always cascading toward action. Organized as maps primarily among the wake-cycle phases "observe" and "orient," the radio procedure skill sets (SSs) control radio personalities, as illustrated in Figure 14.9.

These SSs may either be reformatted into serModels directly from the a priori knowledge of an RKRL frame, or they may be acquired from training or sleep/dreaming. Each skill set may also save the knowledge it learns into an RKRL frame.

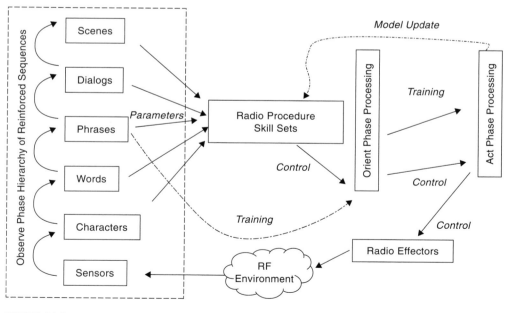

FIGURE 14.9

Radio skills respond to observations. (*Source:* © 1999, Dr. Joseph Mitola III, used with permission.)

14.4.7 General World Knowledge

A CR needs substantial knowledge embedded in the inference hierarchies. It needs both external RF knowledge and internal radio knowledge. Internal knowledge enables it to reason about itself as a radio. External radio knowledge enables it to reason about the role of the <Self/> in the world, such as respecting rights of other cognitive and not-so-cognitive radios.

Figure 14.10 illustrates the classes of knowledge a CR needs to employ in the inference hierarchies and cognition cycle. It is one thing to write down that the universe includes a physical world (there could also be a spiritual world, and that might be very important in some cultures); it is quite another thing to express that knowledge in a way that the CR will be able to use it effectively. Symbols (e.g., universe) take on meaning by their relationships to other symbols and to external stimuli. In this ontology, metalevel knowledge consists of *abstractions*, distinct from existential knowledge of the physical universe. In RXML, this ontological perspective includes all in a universe of discourse, <Universe>, expressed as shown by Figure 14.11.

Abstractions include informal and formal metalevel knowledge from unstructured knowledge of concepts to the more mathematically structured models of space, time, RF, and entities that exist in space-time. To differentiate "now" as a temporal concept from *Now*, which is the Chinese name of a plant, the CRA includes both the a priori knowledge of "now" as a space-time locus, <Now/>, as well as functions that access and manipulate instances of the concept <Now/>. <Now/> is axiomatic in the CRA, for temporal reference in planning actions. The architecture allows an algorithm to return the date-time code (e.g., from the operating system) to define instances of <Now/>.

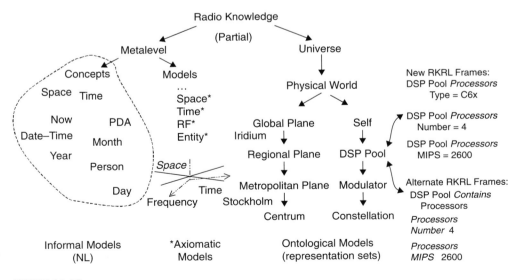

FIGURE 14.10

External radio knowledge includes concrete and abstract knowledge. (*Source:* © 1999, 2000, Dr. Joseph Mitola III, used with permission.)

```
<Universe>
<Abstractions> <Time> <Now/> </Time> <Space> <Here/> </Space> ... <RF/> ...
<Intelligent-Entities/> ... </Abstractions>
<Physical-universe> ... <Instaces/> of Abstraction ... </Physical-universe>
</Universe>
```

FIGURE 14.11

Equation for the universe of discourse of CR. This <Universe/> consists of abstractions plus the physical universe.

Definition-by-algorithm permits an inference system like the cognition subsystem to reason about whether a given event is in the past, present, or future. What is the present? The present is some region of time between "now" and the immediate past and future. If the user is a paleontologist, "now" may consist of the million-year epoch in which it is thought that humans evolved from apes. To a rock star, "now" is probably a lot shorter than that. How will a CR learn its user's concept of now? The CRA design offers an axiomatic treatment of time, but the axioms do not reflect such subjective reality from the <User/> perspective. The CR1 [12] aggregates knowledge of time by a temporal CBR that illustrates the key principles. The CR1 [12] does not fix the definition of <Now/>, but enables the <Self/> to define the details in an <Instance/> in the physical world about which it can learn from the user, whether a paleontologist or a rock star.

Given the complexity of a system that includes both a multitiered inference hierarchy and the cognition cycle's observe, orient, plan, decide, act sequence with AML throughout, it is helpful to consider the mathematical structure of these information elements, processes, and flows. The mathematical treatment is the subject of the next section.

14.5 CRA-IV: ARCHITECTURE MAPS

Cognition functions are implemented via cognition elements consisting of data structures, processes, and flows, which may be modeled as topological maps over the abstract domains identified in Figure 14.12.

The <Self/> is an entity in the world, whereas the internal organization of the <Self/> (annotated PDA in the figure) is an abstraction that models the <Self/>. The hierarchy of words, phrases, and dialogs from sensory data to scenes is not inconsistent with visual perception. Words correspond to visual entities; phrases to detectable movement and juxtaposition of entities in a scene. Dialogs correspond to a coherent sequence of movement within the scope of a scene, such as walking across the room. Occlusion may be thought of as a dialog in which the room asserts itself in part of the scene while observable walking corresponds to assertion of the object. The model data structures may be read as generalized words, phrases, dialogs, and scenes that may be acoustic, visual, or perceived in other sensory domains (e.g., infrared). These structures refer to set-theoretic spaces consisting of a set X and a family of subsets Ox that contain $\{X\}$

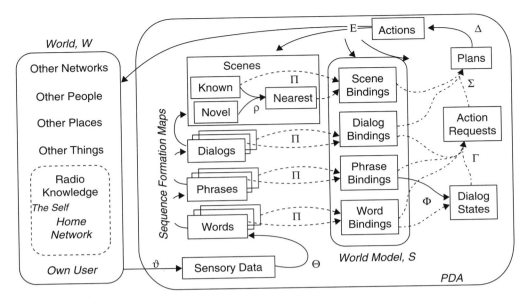

FIGURE 14.12

Architecture based on the cognition cycle. (*Source:* © 2000, Dr. Joseph Mitola III, used with permission.)

and {}, the null set, and that are closed under union and countable intersection. In other words, each is a topological space induced over the domain. Proceeding up the hierarchy, the scope of the space (X, Ox) increases. A <Scene/> is a subset of space–time that is circumscribed by the entity by sensory limits.

The cognition functions modeled in these spaces are topology-preserving maps (see Figure 14.12). Data and knowledge storage spaces are shown as rectangles (e.g., dialog states, plans), whereas processing elements that transform sets are modeled as homeomorphisms, or topology-preserving maps, shown as directed graphs (e.g., Π) in the figure.

14.5.1 CRA Topological Maps

The processing elements of the architecture are modeled topological maps, as shown in Figure 14.12:

1. The input map ϑ consists of components that transform external stimuli to the internal data structure sensory data.

2. The transformation Θ consists of entity recognition (via acoustic, optical, and other sensors), lower-level software radio (SWR) waveform interface components, and so forth, that create streams of primitive-reinforced sequences. The model includes maps that form successively higher-level sequences from the data on the immediately lower level.

3. Reasoning components include the map ρ that identifies the best match of known sequences to novel sequences. These are bound to scene variables by projection components, Π. The maps ϑ, Θ, ρ, and Π constitute observe-phase processing.

4. Generalized word- and phrase-level bindings are interpreted by the components Φ to form dialog states. Train, for example, is the dialog state of a training experience in the CRA.

5. The components of Γ create action requests from bindings and dialog states. The maps Φ and Γ constitute orient-phase processing.

6. Scene bindings include user communications context. Context-sensitive plans are created by the component Σ that evaluates action requests in the plan phase.

7. The decision-phase processing consists of map Δ that maps plans and scene context to actions.

8. Finally, the map E consists of the effector components that change the PDA's internal states, change displays, synthesize speech, and transmit information on wireless networks using the SWR personalities.

14.5.2 CRA Identifies Self, Owner, and Home Network

The sets of entities in the world that are known to the CR are modeled graphically as rounded rectangles in Figure 14.12. These include the self that is grounded in the outside world ("self"), as well as its knowledge of the self as radio self (e.g., as "PDA"). The critical entities are world, W; the PDA; and the PDA's World Model, S. (In the CRA, S includes the orient phase data structures and processes.) Entities in the world include the differentiated entities "Own User," or owner, and "Home Network." The architecture requires that the PDA be able to identify these entities so that it may treat them differentially. Other networks, people, places, and things may be identified in support of the primary cognition functions, but the architecture does not depend on such a capability.

14.5.3 CRA-Reinforced Hierarchical Sequences

The data structures for perception include the reinforced hierarchical sequences: words, phrases, dialogs, and scenes of the observe phase. Within each of these sequences, the novel sequences represent the current stimulus–response cases of the cognitive behavior model. The known sequences represent the integrated knowledge of the cognitive behavior model. Known sequences may consist of a priori RXML statements embedded in the PDA or of knowledge acquired through independent ML. The nearest sequence is the known sequence that is closest in some sense to the novel sequence. The World Model, W, consists primarily of bindings between a priori data structures and the current scene. Such structures are also associated with the observe phase. Dialog states, action requests, plans, and actions are additional data structures needed for the observe, orient, plan, and act phases, respectively. Each internal data structure maps to an RXML frame consisting of element (e.g., set or stimulus); model (e.g., embedded procedure, parameter values); content, typically a structure of elements terminating in either primitive concepts <concept/> (e.g., subset or response) or instance data; and associated resources. Context is defined as the RXML URL or root from <Universe/>, to include source, time, and place of the <Scene/>.

14.5.4 Behaviors in the CRA

CRA entails three modes of behavior: waking, sleeping, and praying. Behavior that lasts for a specific time interval is called a behavioral epoch. The axiomatic relationships among these behaviors are expressed in the topological maps of Figure 14.13.

Waking Behavior

Waking behavior is optimized for real-time interaction with the user, isochronous control of SWR assets, and real-time sensing of the environment. The conduct of the waking behavior is informally referred to as the *awake state*, although it is not a specific system state but a set of behaviors. Thus, referring to Figure 14.13, the awake state cognition actions (α) map the environment interactions to the current stimulus–response cases. These cases are the dynamic subset of the embedded serModels. Incremental ML (δ) maps these interactions to integrated knowledge, the persistent subset of the serModels.

Sleeping and Dreaming Behaviors

Cognitive PDAs (CPDAs) detect conditions that permit or require sleep and dreaming. For example, if the PDA predicts or becomes aware of a long epoch of low utilization (e.g., overnight hours), then the CPDA may autonomously initiate sleeping behavior. Sleep occurs during planned inactivity (e.g., to recharge batteries). Dreaming behavior

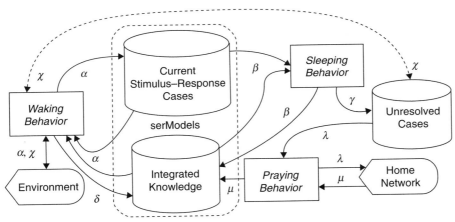

α: action cycle; δ: incremental machine learning; β: nonincremental machine learning; γ: learning conflicts; λ: RKRL/KQML requests for external assistance; μ: authoritative assistance; χ: attempt to resolve problems

FIGURE 14.13

Cognitive behavior model consists of domains and topological maps. (*Source:* © 2000, Dr. Joseph Mitola III, used with permission.)

employs energy to retrospectively examine experience since the last period of sleep. In the CRA, all sleep includes dreaming. In some situations, the CPDA may request permission to enter sleeping/dreaming behavior from the user (e.g., if predefined limits of aggregate experience are reached). Regular sleeping/dreaming limits the combinatorial explosion of the process of assimilating aggregated experience into the serModels needed for real-time behavior during the waking behaviors. During the dreaming epochs, the CPDA processes experiences from the waking behavior using nonincremental ML algorithms. These algorithms map current cases and new knowledge into integrated knowledge (β).

A conflict is a context in which the user overrode a CPDA decision about which the CPDA had little or no uncertainty. Map β may resolve the conflict. If not, it will place the conflict on a list of unresolved conflicts (map γ).

Prayer Behavior

Attempts to resolve unresolved conflicts via the mediation of the PDA's home network may be called prayer behavior, referring the issue to a completely trusted source with substantially superior capabilities. The unresolved-conflicts list γ is mapped (λ) to RXML queries to the PDA's home CN expressed in XML, OWL, KQML, RKRL, RXML, or a mix of declared knowledge types. Successful resolution maps network responses to integrated knowledge (μ). Many research issues surround the successful download of such knowledge, including the set of support for referents in the unresolved-conflicts lists and the updating of knowledge in the CPDA needed for full assimilation of the new knowledge or procedural fix to the unresolved conflict. The prayer behavior may not be reducible to finite-resource introspection, and thus may be susceptible to the "partialness" of Turing Capable (TC), even though the CPDA and CWN enforce watchdog timers.

Alternatively, the PDA may present the conflict sequence to the user, requesting the user's advice during the wake cycle (map π).

14.5.5 From Maps to APIs

Each of these maps has a domain and a range. Axiomatically, the domain is the set of subsets of internal data structures over which the map is defined, and the range is the set of subsets onto which the map projects its effects. Thus, for each map, $M{:}D \Rightarrow R$, there is an associated API or API component.

$$\text{API-}M{:}\{m \in M{:}\{d \in D \Rightarrow r \in R\}\}.$$

In other words, the API for map M specifies methods, or attached procedures, defined over subsets d of domain D that map onto subsets r of range R. So each map can be interpreted as a generalized API. Some APIs may entail more than one map. A planning API, for example, might include the maps that generate the plans and the maps that select among plan components and schedule plans for actions. In fact, APIs for many CR functions from perception to planning and action include more functionality (e.g., visualization tools and user interfaces) than is needed for embedding into a CR. Therefore, the representation of API components as maps establishes the foundations of the API without overconstraining the definition of APIs for a given CR design. The evolution of the CRA from this set of maps to a set of APIs with broad industry support may be facilitated by the framework of the maps.

14.5.6 Industrial-Strength Inference Hierarchy

Although the CRA provides a framework for APIs, it doesn't specify the details of the data structures or of the maps. The CRA research prototype emphasizes ubiquitous learning via serModels and CBR, but it doesn't implement critical features that would be required in deployable CRs. Other critical aspects of such industrial-strength architectures include more capable scene perception and situation interpretation, specifically addressing the following:

Noise: In utterances, images, objects, location estimates, and the like. Noise sources include thermal noise; conversion error introduced by the process of converting analog signals (audio, video, accelerometers, temperature, etc.) to digital form; error in converting from digital-to-analog form; preprocessing algorithm biases and random errors, such as the accumulation of error in a digital filter; or the truncation of a low-energy signal by threshold logic. Dealing effectively with noise differentiates a tutorial demonstration from a useful product.

Hypothesis management: Keeping track of more than one possible binding of stimuli to response, dialog sense, scene, etc. Hypotheses may be managed by keeping the *N*-best hypotheses (with an associated degree of belief), by estimating the prior probability or other degree of belief in a hypothesis, and by keeping a sufficient number of hypotheses to exceed a threshold (e.g., 90 percent or 99 percent of all the possibilities) or by keeping multiple hypotheses until the probability for the next most likely (second) hypothesis is less than some threshold. The estimation of probability requires a measurable space, a sigma-algebra[15] that defines how to accumulate probability on that space, proof that the space obeys the axioms of probability, and a certainty calculus that defines how to combine degrees of belief in events as a function of the measures assigned to the probability of each event.

Training interfaces: The reverse flow of knowledge from the inference hierarchy back to the perception subsystems. The recognition of the user by a combination of face and voice could be more reliable than single-domain recognition either by voice or by vision. In addition, the location, temperature, and other aspects of the scene may influence object identification. Visual recognition of the owner outdoors in a snowstorm, for example, is more difficult than indoors in an office. Even though the CR might learn to recognize the user based on weaker cues outdoors, access to private data might be constrained until the quality of the recognition exceeds some learned threshold.

Nonlinear flows: Although the cognition cycle emphasizes the forward flow of perception enabling action, it is crucial to realize that actions may be internal, such as advising the vision subsystem that its recognition of the user is in error because

---

[15]In mathematics, a *sigma-algebra* over a set is a functional correspondence between subsets of that set and operations that measure (e.g., the size of the subset). The sigma-algebra is fundamental to probability theory [9].

the voice does not match and the location is wrong. Due to the way the cognition cycle operates on the self, these reverse flows from perception to training are implemented as forward flows from the perception system to the self, directed toward a specific subsystem (e.g., vision or audition). There may also be direct interfaces from the CWN to the CR to upload data structures representing a priori knowledge integrated into the UCBR learning framework.

14.6 CRA-V: BUILDING THE CRA ON SDR ARCHITECTURES

A CR is an SWR or SDR with flexible formal semantics-based entity-to-entity messaging via RXML and integrated ML of the self, the user, the RF environment, and the "situation." This section reviews SWR, SDR, and the SCA, or SRA, as they relate to the SRA. Although it is not necessary for a CR to use the SCA/SRA as its internal model of itself, it certainly must have some model, or it will be incapable of reasoning about its own internal structure and adapting or modifying its radio functionality autonomously.

14.6.1 Review of SWR and SDR Principles

Hardware-defined radios, such as the typical amplitude/frequency modulation (AM/FM) broadcast receiver, convert radio to audio using such radio hardware as antennas, filters, analog demodulators, and the like. SWR is the ideal digital radio in which the analog-to-digital converter (ADC) and digital-to-analog converter (DAC) convert digital signals to and from RF directly, and all RF channel modulation, demodulation, frequency translation, and filtering are accomplished digitally. For example, modulation may be accomplished digitally by multiplying sine and cosine components of a digitally sampled audio signal (called the "baseband" signal, to be transmitted) by the sampled digital values of a higher-frequency sine wave to up-convert it, ultimately to the RF spectrum.

Figure 14.14 shows how SDR principles apply to a cellular radio basestation. The ideal SWR would have essentially no RF conversion, just ADC/DAC blocks accessing the full RF spectrum available to the (wideband) antenna elements. Today's SDR basestations approach this ideal by digital access (DAC and ADC) to a band of spectrum allocations, such as 75 MHz allocated to uplink and downlink frequencies for 3G services. In this architecture, RF conversion can be a substantial system component, sometimes 60 percent of the cost of the hardware, and not amenable to cost improvements through Moore's law. The ideal SDR would access more like 2.5 GHz from, say 30 MHz to around 2.5 GHz, supporting all kinds of services in TV bands, police bands, air traffic control bands, and other bands. Although this concept was considered radical when introduced in 1991 [20] and popularized in 1995 [21], recent regulatory rulings are encouraging the deployment of such "flexible spectrum" use architectures.

This ideal SWR may not be practical or affordable, so it is important for the radio engineer to understand the trade-offs (see Mitola [1] for SDR architecture trade-offs). In particular, the physics of RF devices (e.g., antennas, inductors, filters) makes it easier to synthesize narrowband RF and intervening analog RF conversion and intermediate frequency (IF) conversion. Given narrowband RF, the hardware-defined radio might employ baseband (e.g., voice frequency) ADC, DAC, and digital signal processing. The

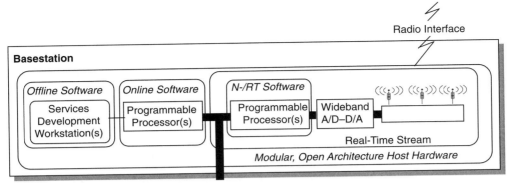

N-/RT = Near Real Time and Real Time

FIGURE 14.14

SWR principle applied to cellular basestation. (*Source:* © 1992, Dr. Joseph Mitola III, used with permission.)

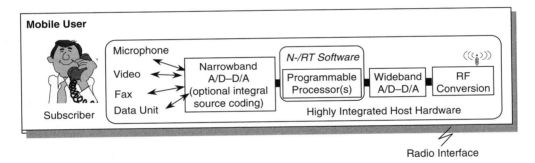

FIGURE 14.15

SWR principle: "ADC and DAC at the antenna" may not apply. (*Source:* © 1992, Dr. Joseph Mitola III, used with permission.)

programmable digital radios (PDRs) of the 1980s and 1990s used this approach. Historically, this approach has not been as expensive as wideband RF (i.e., the cost of antennas, conversion), ADCs, and DACs. Handsets are less amenable to SWR principles than the basestation (Figure 14.15). Basestations access the power grid. Thus, the fact that wideband ADCs, DACs, and DSP (digital signal processor) consume many watts of power is not a major design driver. Conservation of battery life, however, *is* a major design driver in the handset.

Thus, insertion of SWR technology into handsets has been relatively slow. Instead, the major handset manufacturers include multiple single-band RF chip sets into a given handset. This has been called the "Velcro" radio or "slice" radio.

The ideal SWR is not readily approached in many cases, so the SDR has comprised a sequence of practical steps from the baseband DSP of the 1990s toward the ideal

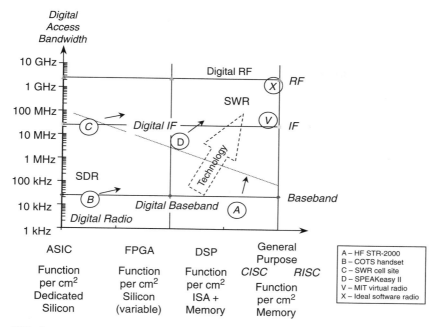

CISC: Complex instruction set computer; ISA: Instruction set architecture; RISC: Reduced instruction set computer

FIGURE 14.16

SDR design space. This figure shows how designs approach the ideal SWR. (*Source:* © 1996–2008, Dr. Joseph Mitola III, used with permission.)

SWR. As the economics of Moore's law and of increasingly wideband RF and IF devices allow, implementations move upward and to the right in the SDR design space (Figure 14.16).

This space consists of the combination of digital access bandwidth and programmability. Access bandwidth consists of ADC/DAC sampling rates consistent with the Nyquist criterion[16] and effective bandwidth. Programmability of the digital subsystems is defined by the ease with which logic and interconnect may be changed after deployment. Application-specific integrated circuits (ASICs) cannot be changed at all, so the functions are "dedicated" in silicon. Field-programmable gate arrays (FPGAs) can be changed in the field, but if the new function exceeds some performance parameter of the chip, which is not uncommon, then one must upgrade the hardware to change the function. DSPs are typically easier or less expensive to program and are more efficient in power use than FPGAs. Memory limits and instruction set architecture (ISA)

---

[16]The Nyquist criterion is that a signal must be sampled at more than twice the highest-frequency component present in the signal. Failure to do so will result in an alias, in which interference will be shifted from out of band to within band. This is generally a distortion that most systems cannot tolerate. In other words, the Nyquist frequency is half the sample rate of the ADCs or DACs.

complexity can drive up costs of reprogramming the DSP. Finally, general-purpose processors, particularly reduced instruction set computers (RISCs), are the most cost effective for accommodating software changes. To characterize a cell phone with a CDMA-ASIC, DSP speech codec, and RISC microcontroller, we weight the point in the design space by equivalent-processing capacity.

Where should one place an SDR design within this space? The quick answer, along a migration path of radio technology from the lower left toward the upper right, benefiting from lessons learned in the early migration projects, is captured in *SRA* [1].

14.6.2 Radio Architecture

The discussion of the SWR design space contains the first elements of radio architecture. It defines a mix of critical components for the radio. For the SWR, the critical hardware components are the ADC, DAC, and processor suite. The critical software components are the user interface, the networking software, the information security (INFOSEC) capability (hardware and/or software), the RF media access software, including the physical (PHY) layer modulator and demodulator (modem) and medium access control (MAC), and any antenna-related software (e.g., antenna selection, beam forming, pointing). INFOSEC consists of transmission security, such as the frequency-hopping spreading code selection, plus communications security encryption.

The SDR Forum defined a very simple, helpful model of an SDR in 1997, which is shown in Figure 14.17. This model highlights the relationships among radio functions at a tutorial level. The CR has to "know" about these functions, so this model is a good start because it shows both the relationships among the functions and the typical flow-of-signal transformations from analog RF to analog or (with SDR) digital modems, and on to other digital processing, including system control of which the user interface is a part.

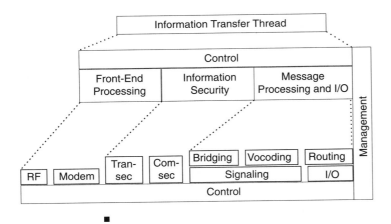

FIGURE 14.17

SDR Forum (formerly MMITS) information-transfer thread architecture. (*Source:* © 1997, SDR Forum, used with permission.)

This model, the techniques for implementing an SWR, and the various degrees of SDR capability are addressed in depth in the various texts about SDR [22–25].

14.6.3 The SCA

The US DoD developed the SCA for its Joint Tactical Radio System (JTRS) family of radios. The SCA identifies the components and interfaces shown in Figure 14.18. The APIs define access to the PHY layer, to the MAC layer, to the logical link control (LLC) layer, to security features, and to the input/output (I/O) of the physical radio device. The physical components consist of antennas and RF conversion hardware that are mostly analog and that typically lack the ability to declare or describe themselves to the system. Most other SCA-compliant components are capable of describing themselves to the system to enable and facilitate plug-and-play among hardware and software components. In addition, the SCA embraces the portable operating system interface (POSIX) and CORBA.

The model evolved through several stages of work in the SDR Forum and OMG into a UML-based object-oriented model of SDR (Figure 14.19). Waveforms are collections of load modules that provide wireless services, so from a radio designer's perspective, the waveform is the key application in a radio. From a user's perspective of a wireless PDA (WPDA), the radio waveform is just a means to an end, and the user doesn't want to know or have to care about waveforms. Today, the cellular service providers hide this detail to some degree, but consumers sometimes know the difference between CDMA and GSM—for example, because first-generation CDMA works in the United States but not in Europe. With the deployment of the 3G of cellular technology, the

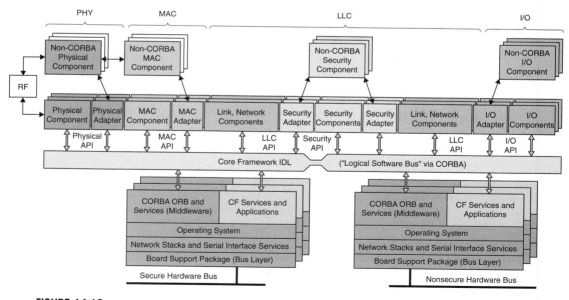

FIGURE 14.18

TRS SCA Version 1.0. (*Source:* © 2004 SDR Forum, used with permission.)

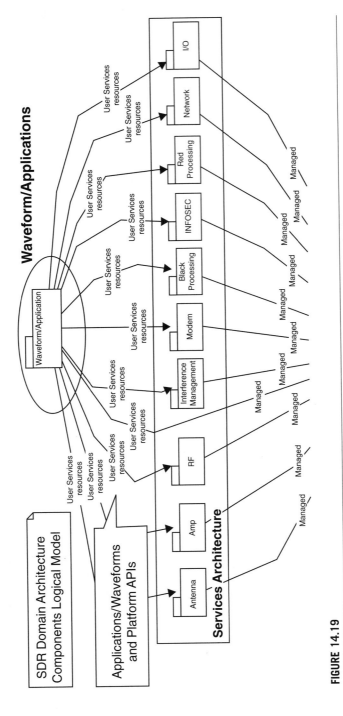

FIGURE 14.19

SDR Forum UML model of radio services. (*Source:* © 2004 SDR Forum, used with permission.)

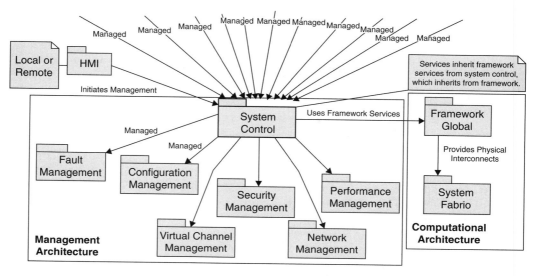

FIGURE 14.20

SDR Forum UML management and computational architectures. (*Source:* © 2004 SDR Forum, used with permission.)

amount of technical jargon consumers will need to know is increasing. So the CRA insulates the user from those details, unless the user really wants to know.

In the UML model shown in Figure 14.19, amp refers to amplification services, RF refers to RF conversion, and interference management refers to both avoiding interference and filtering it out of one's band of operation. In addition, the jargon for US military radios is that the "red" side contains the user's secret information, but when it is encrypted, it becomes "black" (protected) so that it can be transmitted. Black processing occurs between the antenna and the decryption process. Notice also that Figure 14.19 has no user interface. The UML model contains a sophisticated set of management facilities, illustrated further in Figure 14.20, to which the human–machine interface (HMI) or user interface is closely related.

Systems control is based on a framework that includes very generic functions (e.g., event logging) organized into a computational architecture, heavily influenced by CORBA. The management features are needed to control radios of the complexity of 3G and of the current generation of military radios. Although civil sector radios for police, fire, and aircraft lag these two sectors in complexity and are more cost sensitive, baseband SDRs are beginning to insert themselves even into these historically less technology-driven markets.

Fault management features are needed to deal with the loss of a radio's processors, memory, or antenna channels. CR therefore interacts with fault management to determine which facilities may be available to the radio given recovery from hardware and/or software faults (e.g., error in a download). Security management is increasingly important in the protection of the user's data by the CR; balancing convenience and security can be very tedious and time consuming. The CR will direct virtual channel management (VCM) and will learn from the VCM function which radio resources are available, such as which bands the radio can listen to and transmit on, and how many bands it can use at once. Network management does for the digital paths what VCM does for the radio paths. Finally, SDR performance depends on the availability of analog and digital resources, such as linearity in the antenna, millions of instructions per second (MIPS) in a processor, and the like.

14.6.4 Functions–Transforms Model of Radio

The CRA uses a self-referential model of a wireless device, the functions–transforms model, to define the RKRL and to train the CRA. In this model, illustrated in Figure 14.21, the radio knows about sources, source coding, networks, INFOSEC, and the collection of front-end services needed to access RF channels. Its knowledge also extends to the idea of multiple channels and their characteristics (the channel set), and the radio part may have many alternative personalities at a given point in time. Through evolution support, those alternatives change over time.

CR reasons about all of its internal resources via a computational model of analog and digital performance parameters, and how they are related to features it can measure or control. MIPS, for example, may be controlled by setting the clock speed. A high clock speed generally uses more total power than a lower clock speed, and this tends to reduce battery life. The same is true for the brightness of a display. The CR only "knows" this to the degree that it has a data structure that captures this information and algorithms, preprogrammed and/or learned, that deal with these relationships to the benefit of the user. Constraint languages may be used to express interdependencies, such as how many channels of a given personality are supported by a given hardware suite, particularly in failure modes. CR algorithms may employ this kind of structured reasoning as a specialized KS when using case-based learning to extend its ability to

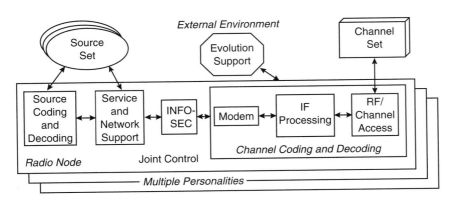

FIGURE 14.21

Functions–transforms model of a wireless node. (*Source:* © 1996, Dr. Joseph Mitola III, used with permission.)

```
<SDR>
        <Sources/> <Channels/> < Personality>
                <Source-Coding-Decoding/> <Networking/> <INFOSEC/>
<Channel-Codec><Modem/> <IF-Processing/> <RF-Access/> <Channel-Codec>
</Personality>
        <SDR-Platform/> <Evolution-Support/>
</SDR>
```

FIGURE 14.22

Equation that defines SDR subsystem components.

cope with internal changes. The ontological structure of the preceding may be formalized as shown by the equation in Figure 14.22.

Although this text does not offer a comprehensive computational ontology of SDR, semantically based dialogs among CRs about internal issues (e.g., downloads) may be mediated by RXML with the necessary ontological structures.

14.6.5 Architecture Migration: From SDR to CR

Given the CRA and contemporary SDR architecture, one must address the transition of SDR through a phase of CRs, toward the iCR. As the complexities of handheld, wearable, and vehicular wireless systems increase, the likelihood that the user or network will have the skill necessary to do the optimal thing in any given circumstance is reduced. Today's cellular networks manage the complexity of individual wireless protocols for the user, but the emergence of multiband, multimode CR moves the burden for complexity management toward the PDA. Likewise, the optimization of the choice of wireless service between the "free" home WLAN and the "for-sale" cellular equivalent moves the burden of radio resource management from the network to the WPDA.

14.6.6 Cognitive Electronics

The increasing complexity of the PDA–user interface also accelerates the trend toward increasing the computational intelligence of personal electronics. CR is in some sense just an example of a computationally intelligent personal electronics system. For example, using a laptop computer in the bright display mode uses up the battery power faster than when the display is set to minimum brightness. A cognitive laptop could offer to further reduce the brightness level when only half of the charge remains and it is operating in battery-powered mode. It would be even nicer if it would recognize operation aboard a commercial aircraft and know that the user's preference is to set the brightness low on an aircraft to conserve the battery, and automatically do so. A cognitive laptop shouldn't make a big deal over that, and it should let the user turn up the brightness without complaining. If it had an ambient light sensor or ambient light algorithm for an embedded camera, it also could tell that a window shade was open and that the user has to deal with the brightness. By sensing the brightness of the *on-board aircraft* scene and associating the user's control of the brightness of the display with the brightness of the environment, a hypothetical cognitive laptop could learn to advise the user to do the right thing in the right situation (pull down the shade).

How does this relate to the CRA? For one thing, the CRA could be used as-is to increase the computational intelligence of the laptop. In this case, the self is the laptop and the PDA knows about itself as a laptop, not as a WPDA. It knows about its sensors suite, which includes at least a light-level sensor if not a camera through the data structures that define the <Self/>. It knows about the user by observing keystrokes and mouse action as well as by interpreting the images on the camera; it verifies that the user is still the owner because that is important to building user-specific models. It might build a space–time behavior model of any user or it might be a one-user laptop. Its actions include the setting of the display intensity level. In short, the CRA accommodates the cognitive laptop with suitable laptop knowledge and functions implemented in the CRA map sets.

14.6.7 When Should a Radio Transition Toward Cognition?

If a wireless device accesses only a single-RF band and mode, then it is not a very good starting point for CR—it's just too simple. Even as complexity increases, as long as the user's needs are met by wireless devices managed by the network(s), then embedding computational intelligence in the device has limited benefits. In 1999, Mitsubishi and AT&T announced the first "four-mode handset." The T250 operated in time division multiple access (TDMA) mode on 850 or 1900 MHz, in first-generation advanced mobile phone system (AMPS) mode on 850 MHz, and in cellular digital packet data (CDPD) mode on 1900 MHz. This illustrates the early development of multiband, multimode, multimedia (M3) wireless. These radios enhanced the service provider's ability to offer national roaming, but the complexity was not apparent to the user because the network managed the radio resources of the handset.

Even as device complexity increases in ways that the network does not manage, there may be no need for cognition. There are several examples of capabilities embedded in electronics that are not heavily used. For example, how many people use their laptop's speech-recognition system? What about its Infrared Data Association (IrDA) port? The typical users in 2004 didn't use either capability of their Windows XP laptop all that much. So complexity can increase without putting a burden on the user to manage that complexity if the capability isn't central to the way in which the user employs the system.

For the radio, as the number of bands and modes increases, the SDR becomes a better candidate for the insertion of cognition technology. But it is not until the radio or the wireless part of the PDA has the capacity to access multiple RF bands that cognition technology begins to pay off. With the liberalization of RF spectrum use rules, the early evolution of CR may be driven by RF spectrum-use etiquette for ad hoc bands such as the FCC use case. In the not-too-distant future, SDR PDAs could access satellite mobile services, cordless telephone, WLAN, GSM, and 3G bands. An ideal SDR device with these capabilities might affordably access three octave bands, from 0.4 to 0.96 GHz (skipping the air navigation and GPS band from 0.96 to 1.2 GHz), from 1.3 to 2.5 GHz, and from 2.5 to 5.9 GHz (Figure 14.23). Not counting satellite mobile and radio navigation bands, such radios would have access to more than 30 mobile subbands in 1463 MHz of potentially sharable outdoor mobile spectrum. The upper band provides another 1.07 GHz of sharable indoor and RF-LAN spectrum.

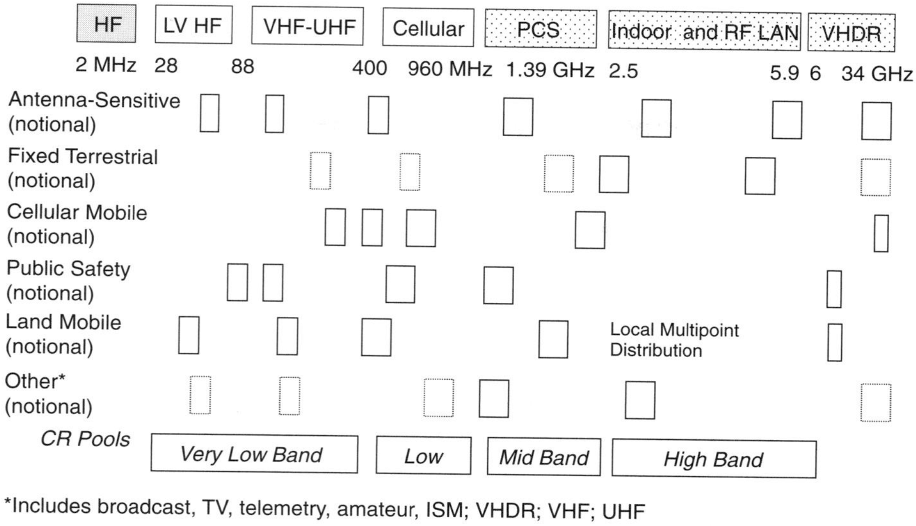

FIGURE 14.23

Fixed spectrum allocations versus pooling with CR. (*Source:* © 1997, Dr. Joseph Mitola III, used with permission.)

This wideband radio technology will be affordable first for military applications, next for basestation infrastructure, then for mobile vehicular radios, and later for handsets and PDAs. When a radio device accesses more RF bands than the host network controls, it is time for CR technology to mediate the dynamic sharing of spectrum. It is the well-heeled conformance to the radio etiquettes afforded by CR that makes such sharing practical.

14.6.8 Radio Evolution Toward the CRA

Various protocols have been proposed by which radio devices may share the radio spectrum. The FCC Part 15 rules permit low-power devices to operate in some bands. In 2003, an FCC Report and Order (R&O) made unused TV spectrum available in the United States for low-power RF-LAN applications, making the manufacturer responsible for ensuring that the radios obey this simple constraint. DARPA's XG program developed a language for expressing spectrum use policy [26]. Other more general protocols based on peek-through to legacy users have also been proposed [33].

Does this mean that a radio must transition instantaneously from the SCA to the CRA? Probably not, because the six-component CRA may be implemented with minimal SP, minimal learning, and no autonomous ability to modify itself. Regulators hold manufacturers responsible for the behavior of such radios. The simpler the architecture, the simpler the problem of explaining it to regulators, and of getting concurrence among manufacturers regarding open-architecture interfaces that will facilitate technology insertion and teaming. Manufacturers who fully understand the level to which a highly autonomous CR might unintentionally reprogram itself might be able to violate regulatory constraints. Therefore, the manufacturer may decide it wants to field aware-adaptive (AA) radios, but may not want to take the risks associated with self-modifying CRs yet.[17]

Thus, one can envision a gradual evolution toward the CRA beginning initially with a minimal set of functions mutually agreeable among the growing community of CR stakeholders. Subsequently, the introduction of new services will drive the introduction of new capabilities and additional APIs, perhaps informed by the CRA.

14.7 COGNITION ARCHITECTURE RESEARCH TOPICS

The cognition cycle and related inference hierarchy imply a large scope of hard research problems for CR, including reliable inference of the situation from vision, speech, and text sensory subsystem; methods that integrate temporal calculus [27]; constraint-based scheduling [28]; task planning [29]; and modeling [30]. Resources may include algebraic methods for wait-free scheduling protocols [31], open distributed processing (ODP), and parallel virtual machines (PVMs). Finally, ML remains one of the core challenges in artificial intelligence research [32] applicable to CR. The focus of CRA research, then, may not be on the development of any one of these technologies per se. Rather, it is on the organization of cognition tasks and on the development of cognition data structures needed to integrate contributions from these diverse disciplines for the context-sensitive delivery of wireless services by SDR.

Learning the difference between situations in which a reactive response is needed versus those in which deliberate planning is more appropriate is a key challenge in ML for CR. The CRA framed the issues. The CR1 goes further [12], providing useful KS and related ML so that the CR designer can start there in developing good engineering solutions to this problem for evolving CR applications domains. Finally, policy languages are needed that not only define constraints but also that empower proactive behavior in novel circumstances [34].

14.8 INDUSTRIAL-STRENGTH CR DESIGN RULES

The CRA allocates functions to components based on design rules. Typically design rules are captured in various interface specifications, including APIs, and object interfaces, such as Java's JINI/JADE structure of intelligent agents. This chapter so far has introduced the CRA; this section suggests additional design rules by which user domains, sensory domains, and radio RF band knowledge may be integrated into industrial-strength CR products and systems.

The following design rules further circumscribe the integration of cognitive functions with the other components of a WPDA within the CRA:

1. The cognition function should maintain an explicit 3D behavioral model of space-time of the user, the physical environment, the radio networks, and the internal states of the radio (the <Self/>).

---

[17]It has been recognized that CRs will not learn unacceptable behaviors because the policy component will not allow such modes and thus successful use of them and subsequent learning of such modes.

2. The CRA requires each CR to predict, in advance, an upper bound on the amount of computational resources (e.g., time) required for each cognition cycle. The CR must set a trusted (hardware) watchdog (e.g., a timer) before entering a cognition cycle. If the watchdog is violated, the system must detect that event, log that event, and mark the components invoked in that event as nondeterministic.

3. The CRA should internalize knowledge as procedural skills (e.g., serModels):
   (a) The CRA requires each CR to maintain a trusted index to internal models and related experience.
   (b) Each CR must preclude unbounded iterative cycles from its internal models and skills graph. A CRA conformance requires reliable detection of unbounded cycles (e.g., via timer) to avoid Gödel-Turing unbounded resource use endemic to self-referential TC computational entities such as CRs.

4. Context that references space, time, RF, the <User/>, and the <Self/> for every external and internal event shall be represented formally using a topologically valid and logically sound model of space–time–context.

5. Each CR conforming to the CRA shall include an explicit grounding map, M, that maps its internal data structures onto elements sensed in the external world represented in its sensory domains, including itself. If the CR cannot map a sensed entity to a space–time–context entity with specified time allocated to attempt that map, then the entity should be designated "ungroundable."

6. The model of the world shall follow a formal treatment of time, space, RF, radio propagation, and the grounding of entities in the environment.

7. Models shall be represented in an open-architecture RKRL suited to the representation of radio knowledge (e.g., a Semantic Web derivative of RKRL). That language shall support topological properties and inference (e.g., forward chaining), but must not include unconstrained axiomatic FOPC, which per force violates the Gödel-Turing constraint.

8. The cognition functions shall maintain location awareness, including (a) the sensing of location from global positioning satellites, (b) sensing position from local wireless sensors and networks, and (c) sensing precise position visually:
   (a) Location shall be an element of all contexts.
   (b) The cognition functions shall estimate time to the accuracy necessary to support the user and radio functions.
   (c) The cognition functions shall maintain an awareness of the identity of the PDA, of its owner, of its primary user, and of other legitimate users designated by the owner or primary user.

9. The cognition functions shall reliably infer the user's communications context and apply that knowledge to the provisioning of wireless access by the SDR function.

10. The cognition functions shall model the propagation of the user's radio signals with sufficient fidelity to estimate interference to other spectrum users:
    (a) The cognition function shall also ensure that interference is within limits specified by the spectrum use protocols in effect in its location (e.g., in spectrum rental protocols).
    (b) The cognitive function shall defer control of the <Self/> to the wireless network in contexts where a trusted network manages interference.

11. The cognition functions shall model the domain of applications running on the host platform, sufficient to infer the parameters needed to support the application. Parameters modeled include quality of service (QoS), data rate, probability of link closure (grade of service), and the space–time–context domain within which wireless support is needed.

12. The cognition functions shall configure and manage the SDR assets to include hardware resources, software personalities, and functional capabilities as a function of network constraints and use context.

13. The cognition functions shall administer the computational resources of the platform. The management of SWR resources may be delegated to an appropriate SDR function (e.g., the SDR Forum domain manager). Constraints and parameters of those SDR assets shall be modeled by the cognition functions. The cognition functions shall ensure that the computational resources allocated to applications, interfaces, cognition, and SDR functions are consistent with the user communications context.

14. The cognition functions shall represent the degree of certainty of understanding in external stimuli and in inferences. A certainty calculus shall be employed consistently in reasoning about uncertain information.

15. The cognition functions shall recognize preemptive actions taken by the network and/or the user. In case of conflict, the cognition functions shall defer the control of applications, interfaces, and/or SDR assets to the owner, to the network, or to the primary user, according to the appropriate priority and operations assurance protocol.

14.9 SUMMARY AND FUTURE DIRECTIONS

Often technical architectures of the kind presented in this chapter accelerate the state of practice by catalyzing work across the industry on plug-and-play, teaming, and collaboration. The thought is that to propel wireless technology from limited spectrum awareness toward valuable user awareness, an architecture (e.g., the CRA) will be needed. In short, the CRA articulates the functions, components, and design rules of next-generation stand-alone and embedded wireless devices and networks. Each of the different aspects of the CRA contributes to the government, academic, and industry dialog:

1. The functional architecture identifies components and interfaces for CRs with sensory and perception capabilities in the user domain, not just the radio domain.

2. The cognition cycle identifies the processing structures for the integration of sensing and perception into radio: observe (sense and perceive), orient (react if necessary), plan, decide, act, and learn.

3. The inference hierarchies suggest levels of abstraction helpful in the integration of radio and user domains into the synthesis of services tailored to the specific user's current state of affairs given the corresponding state of affairs of the radio spectrum in space and time.

4. The introduction to ontology suggests an increasing role for Semantic Web technologies in making the radios smarter—initially, about radio; about time, and about the user (see [12]). However, cognitive linguistics and behavior modeling appear to

be necessary to overcome the limitations of logic and rules in empowering appropriate behavior in novel circumstances [34].

5. Although not strictly necessary for CR, SDR provides a very flexible platform for the regular enhancement of both computational intelligence and radio capability, particularly with each additional Moore's law cycle.

6. Finally, this chapter has introduced the CRA to the reader interested in the cutting edge, but it has not defined the CRA. The previous list suggested a few of the many aspects of the embryonic CRA that must be addressed by researchers, developers, and markets in the continuing evolution of SDR toward ubiquitous and "really fun" CRs.

In conclusion, CR seems headed for behavior modeling, but the markets for services layered on practical radio networks will shape that evolution. Although many information-processing technologies from e-Business Solutions to the Semantic Web are relevant to CR, the integration of audio and visual SP into SDR with suitable cognition architectures remains both a research challenge and a series of increasingly interesting radio systems design opportunities. A CRA that is broadly supported by industry could accelerate such an evolution.

14.10 EXERCISES

14.1 What is iCR and how does it differ from software radio, SDR and AAR?

14.2 Which advanced services would be enabled by iCR?

14.3 How will emerging CR services differentiate products and benefit users on the way to such a vision of the future?

14.4 What is the simplest CRA? How could the architecture evolve through initiatives such as the SDR Forum's CR special-interest group? Where would QoI drive the PHY–MAC technology? Network layers? Applications? User as the eighth layer of the protocol stack?

14.5 Which radio knowledge embedded into SDR enables migration toward iCR? How could computational ontology be used to represent this knowledge, and how is the technology drawn from yet distinct from the Semantic Web? What are the shortfalls of behavior modeling both technically and institutionally?

14.6 Which new sensors are needed for migration toward iCR? When does AAR have sufficient sensor perception for dynamic spectrum? For user-situation awareness? For machine learning of user preferences in the CR?

14.7 Which SDR platforms are compatible and incompatible with CR evolution? How does a radio engineering group get started with CR? Which skills must a radio systems organization add to its workforce for CR? Machine vision? Natural language processing (NLP)? Machine learning (ML)? Ontologists?

14.8 How is regulatory rule making shaping CR markets?

14.9 Search the Web for CR and describe the main thrusts in US, European, and Asian R&D and productization. Be sure to include E3.

14.10 How will today's discrete cell phone, PDA, and laptops merge into the iCR wardrobe?

REFERENCES

[1] Mitola, J., III, *Software Radio Architecture*, Wiley-Interscience, 2000.

[2] Mahonen, P., *Cognitive Wireless Networks*, RWTH Aachen, 2004.

[3] *www.sdrforum.org*.

[4] *www.omg.org*.

[5] Eriksson H.-E., and M. Penker, *UML Toolkit*, Wiley, 1998.

[6] Mowbray, T., and R. Malveau, *CORBA Design Patterns*, Wiley, 1997.

[7] *www.jtrs.mil*.

[8] Wireless World Research Forum (*www.wwrf.com*), 2004.

[9] *www.wikipedia.org*.

[10] Mitola, J., III, Software Radio Architecture: A Mathematical Perspective, *IEEE JSAC*, IEEE Press, 1998.

[11] Hennie, R,. *Introduction to Computability*, Addison-Wesley, 1997.

[12] Mitola, J., III, *Cognitive Radio Architecture*, Wiley, 2006.

[13] *www.omg.org/UML*.

[14] Mitola, J., III, *Cognitive Radio: An Integrated Agent Architecture for Software Defined Radio*, KTH, The Royal Institute of Technology, Stockholm, June 2000.

[15] Erman, L., F. Hayes-Roth, V. Lesser, and R. Reddy, The Hearsay-II Speech Understanding System: Integrating Knowledge to Resolve Uncertainty, *ACM Computing Surveys*, 12(2):213–253, 1980.

[16] SNePS, 1998; available at *ftp.cs.buffalo.edu:/pub/sneps/*.

[17] Koser, et al., *read.me*, University of Nijmegen, The Netherlands, March 1999; available at *www.cs.kun.nl*.

[18] The XTAG Research Group, A Lexicalized Tree Adjoining Grammar for English, Institute for Research in Cognitive Science, University of Pennsylvania, Philadelphia, 1999.

[19] PC-KIMMO Version 1.0.8 for IBM PC, February 18, 1992.

[20] Mitola, J., III, Software Radio: Survey, Critical Evaluation and Future Directions, *Proceedings of the National Telesystems Conference*, May 1992.

[21] Mitola, J., III, Software Radio Architecture, *IEEE Communications Magazine*, May 1995.

[22] Tuttlebee, W., *Software Defined Radio Enabling Technologies*, Wiley, 2002.

[23] Reed, J. H., *Software Radio: A Modern Approach to Radio Engineering*, Prentice Hall, 2002.

[24] Mitola, J., III, and Z. Zvonar (eds.), *Software Radio Technologies*, IEEE Press, 1999.

[25] Jondral, F., *Software Radio*, Universität Karlsruhe, Karlsruhe, Germany, 1999.

[26] Marshall, P., *Remarks to the SDR Forum*, September 2003.

[27] Phillips, C., Optimal Time-Critical Scheduling, *STOC*, 1997; available at *www.acm.org*.

[28] Esmahi, L., et al., Mediating Conflicts in a Virtual Market Place for Telecommunications Network Services, *Proceedings of the Fifth Baiona Workshop on Emerging Technologies in Telecommunications*, Vigo, Spain, 1999.

[29] Das, S. K., et al., Decision Making and Plan Management by Intelligent Agents: Theory, Implementation, and Applications, *Proceedings of Autonomous Agents*, 1997; available at *www.acm.org*.

[30] Pearl, J., *Causality: Models, Reasoning, and Inference*, Morgan Kaufmann, 2000.

[31] Michalski, R., I. Bratko, and M. Kubat, *Machine Learning and Data Mining*, Wiley, 1998.

[32] Clark, K., and S. A. Tarnlund, *Logic Programming*, Academic Press, 1982.

[33] Mitola, J., III, Cognitive Radio for Flexible Mobile Multimedia Communications, *Mobile Multimedia Communications*, New York, November 1999.

[34] Mitola, J., III, Cognitive Radio Policy Languages, *Proceedings ICC*, 2009.

Cognitive Radio Performance Analysis

James O. Neel
Cognitive Radio Technologies, LLC
Lynchburg, Virginia

Jeffrey H. Reed, Allen B. MacKenzie
Wireless @ Virginia Tech, Bradley Department
of Electrical and Computer Engineering
Blacksburg, Virginia

15.1 INTRODUCTION

Unlike traditional radios, cognitive radios (CRs) both react to and alter their operating environment. In isolation, this is not much of a problem, but when multiple CRs are present, each radio's adaptations change the observed state of the spectrum for the other radios, which in turn influences the adaptations of every other spectrum-agile radio. Such interactions could spawn infinite adaptation loops, drive the networks to unexpected self-jamming states, and lead to brittle unstable networks.

To illustrate this phenomenon, consider the network depicted in Figure 15.1, which consists of three coexisting, but uncoordinated, wireless links (1,2,3) connecting three access nodes (ANs) to three clients. Each link implements a dynamic spectrum access (DSA) algorithm based on Digital European Cordless Telephone (DECT) where each link chooses between two orthogonal channels, {0,1} (for point of this illustration other channels are presumed offlimits due to policy considerations), to minimize the interference experienced by the link's client. For simplicity, we assume that observed interference from other clients is negligible so observed interference is dominated by AN-to-client interference terms and that all ANs transmit at the same power level. Now suppose that $g_{31} > g_{21}$, $g_{12} > g_{32}$, $g_{23} > g_{13}$, where g_{ij} is the link budget gain (pathloss) from the AN of link i to the client of link j. In other words, client 1 is interfered with more by AN 3 than by AN 1; client 2 is interfered with more by AN 1 than by AN 2; and client 3 is interfered with more by AN 2 than by AN 1. Without a loss of generality, assume $g_{31} = g_{12} = g_{23} = 1.0$ and $g_{21} = g_{32} = g_{13} = 0.5$ and a transmit power of 2 so that the observed interference at each client is equal to twice the sum of path gains from the other links' ANs to the client.

In this two-channel system, there are eight (2^3) different channel allocations that could be made by the independent choices of the three links. For these eight combinations, the interference levels experienced by each client are shown in Table 15.1, where

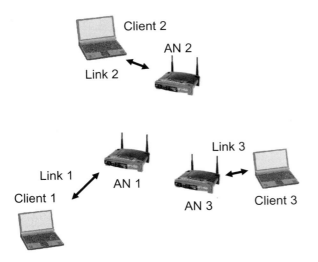

FIGURE 15.1

Three coexisting uncoordinated wireless links.

Table 15.1 Interference Levels for Example DFS Algorithms			
Channel	(0,0,0)	(0,0,1)	(0,1,0)	(0,1,1)
Interference	(3,3,3)	(1,2,0)	(2,0,1)	(0,1,2)
Channel	(1,0,0)	(1,0,1)	(1,1,0)	(1,1,1)
Interference	(0,1,2)	(2,0,1)	(1,2,0)	(3,3,3)

the entries on the channel-labeled rows specify the choice of channels by each AN (1,2,3) and inference-labeled entries specify the interference levels seen by the client of links (1,2,3). If each AN chooses the channel with the least amount of interference at its client, the system will enter into an infinite loop—(1,0,0), (1,0,1), (0,0,1), (0,1,1), (0,1,0), (1,1,0), (1,0,0),...—from any initial channel allocation. So while each AN's adaptation process would be attractive in isolation, in a network it is decidedly undesirable.

Unfortunately, this link gain pattern is not a special case as one out of every four deployments of this system will enter into an infinite loop! In fact, as the number of coexisting links increases, the probability of an infinite loop rapidly approaches 1. Even increasing the number of channels does not eliminate this problem as long as the number of channels is less than the number of adapting links—a seemingly assured situation for most deployments.

This example illustrates a critical challenge to designing CRs—the interactions of cognitive radios must be considered when evaluating performance and the following questions should be answered:

- Will the CRs interactions have a steady state (or steady states) that we can identify so we can anticipate performance?
- Will that performance be desirable or will the adaptations result in a vicious cycle of deteriorating network performance?
- Which conditions will be necessary to ensure these adaptations arrive at a desirable steady state?
- Will the steady states be stable or will the inherent variations of the wireless medium make the system unpredictable?

The remainder of this chapter presents techniques for answering these questions and presents principles for ensuring desirable network operation.

Section 15.2 provides an overview and formalization of the CR network analysis problem. Section 15.3 addresses traditional engineering techniques that are useful for analyzing distributed adaptive radio systems, including dynamical systems theory, contraction mappings, and Markov models. To introduce a foundation for addressing a broader range of applications, Section 15.4 defines the basic elements of game theory and describes how this tool, which was originally used in the field of economics, can be applied to the analysis of CR networks. Building on this foundation, Section 15.5 presents two important game models that facilitate rapid analysis of many CR network algorithms.

15.2 THE ANALYSIS PROBLEM

Throughout this chapter, we take a slightly different view of how CRs will operate. In the preceding chapters, we frequently made use of the cognition cycle to understand the operation of CRs. However, this notional representation has an incomplete view of the *outside world* (the environment to which the CRs are observing, learning, and reacting).

A more complete view of the operation of CRs would depict an outside world with a state that is jointly determined by the adaptations of many cognitive radios. Thus, for the purposes of understanding how cognitive radios will behave, we should envision CRs operating as shown in Figure 15.2, where they react to an outside world determined both by other CRs and by non-CRs.

15.2.1 A Formal Model of a CR Network

Building on the conceptual model of CR interactions shown in Figure 15.2, the following symbols and conventions will be used to facilitate a formal discussion of modeling elements. Elements needed to model algorithm-specific variables will be defined as they are introduced later in this chapter.

- N—The (finite) set of CRs under study. For convenience, we say that n represents the number of elements in N (i.e., $n = |N|$).
- i, j—Particular CRs in N.
- A_j—The set of actions available to CR j.

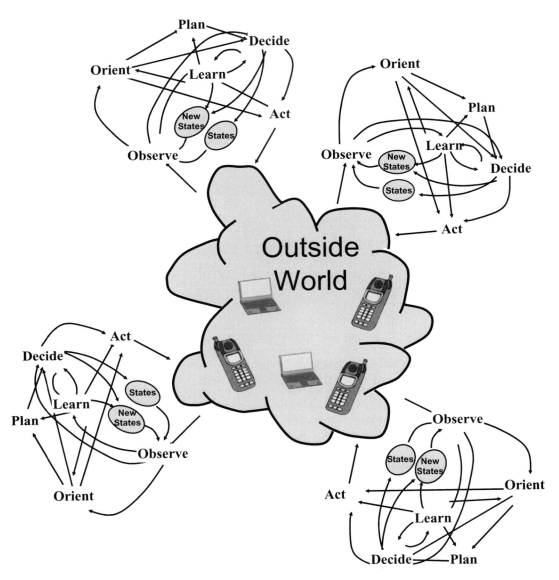

FIGURE 15-2

The interactive CR problem. In addition to the interference profile produced by the static devices in the outside world, cognitive radios also respond to the adaptations of other CRs. (*Source:* This figure is modified from Mitola [1], Figure 4.2.)

The available actions are intended to model all adaptations available to each radio, and since the adaptations can include a number of independent types of adaptations (e.g., power levels, modulations, channel and source coding schemes, encryption algorithms, MAC algorithms, center frequencies, bandwidths, and routing algorithms), A_j will generally be a multidimensional set. Throughout this chapter we assume that we are analyzing adaptations only over short time intervals so that A_j will not be a function of time. However, for longer time intervals, A_j could be expected to grow as CR j learns new waveforms or A_j could change as j moves through different policy regimes that may constrain the adaptations available to the radio. Additionally, different CRs may have different capabilities perhaps due to having additional radio frequency (RF) chains or additional processing capabilities so that in general $A_j \neq A_i$.

- A—The *action space*, that is, the set of all possible combinations of actions by all the CRs. Throughout this chapter we assume that A is formed by the Cartesian product of each radio's action sets (i.e., $A = A_1 \times A_2 \times A_3 \times \cdots \times A_n$). For some algorithms, it is convenient to think of A as a vector space with orthogonal bases A_1 through A_n.

- a—An *action tuple*, that is, a particular combination of actions where each CR in N has implemented a particular action or waveform (equivalently a is a point in A). CR j's contribution to a is written as a_j, and the choice of actions by all CRs other than j is written as a_{-j}.

- O—The *observed outcome space*, that is, the set of all possible realizations of the outside world as determined by the choice of actions available to each CR and its operating environment (outside world) as observed by the radios.

- o—An *observation tuple*, that is, a particular combination of observations where each CR in N has made a particular observation (equivalently o is a point in O). CR j's contribution to o is written as o_j, and the observations made by all CRs other than j is written as o_{-j}. Frequently, we also refer to o as an *outcome* or observed outcome. The importance in distinguishing between actions and observations can be seen in the following example.

Consider two CRs, {1,2}, with actions (waveforms) $\{\omega_{1_a}, \omega_{1_b}\}$ and $\{\omega_{2_a}, \omega_{2_b}\}$, respectively, that are communicating with a common receiver that reports back to each CR its *signal-to-interference ratio* (SIR). Assuming fixed transmission powers for simplicity, the actual outcomes the radios observe can be described by:

$$o_j = \gamma_j = \frac{g_j}{g_{-j} \left| \rho(\omega_{j_k}, \omega_{-j_m}) \right|} \tag{15.1}$$

where $j \in \{1,2\}$, γ_j is the observed SIR for radio j, g_j is the link budget gain of radio j to the common receiver, g_{-j} is the gain of the radio other than j to the common receiver, and $\left| \rho(\omega_{j_k}, \omega_{-j_m}) \right|$ is the absolute value of the statistical correlation between the waveforms chosen by the radios.

In this scenario, there are four different possible elements in A, which form the set $\{(\omega_{1_a}, \omega_{2_a}), (\omega_{1_a}, \omega_{2_b}), (\omega_{1_b}, \omega_{2_a}), (\omega_{1_b}, \omega_{2_b})\}$. However, there is an infinite number of possible observations due to the infinite number of possible channel realizations between the radios and their common receiver. Throughout this chapter we assume that traditional analysis techniques have been used to supply an expression that relates the observed outcome with the implemented waveforms (action tuple).

- $f_j(o)$—The *decision update rule* that describes how radio j updates its decisions given the observation, o, that is, $f_j : O \rightarrow A_j$. When there is a clear and well-defined mapping between A and O, we treat f_j as a function mapping a to an a_j, that is, $f_j : A \rightarrow A_j$. When appropriate, we also use the notation f^t to denote the network update function at time t where in general f^t captures the adaptations of the subset $M \subset N$ of radios that update their decisions at time t (i.e., $f^t = \underset{j \in M}{\times} f_j$). While it is also possible that a radio bases its decisions on past observations and predictions about the future state of the network, for simplicity in this chapter we assume that f_j^t is only a function of CR j's most recent observation.

For compactness, we will also say that the network is updating *synchronously* when all radios adapt at the same time; is updating its decisions in *round-robin* order when radios take turns adapting; is updating its decisions in *random* order when one randomly chosen radio adapts at each instance; and is updating *asynchronously* when no structure can be inferred about the updating process. We say that the network is updating its decisions in an *asynchronous* manner. For asynchronously updating networks, there may be some instances where multiple radios adapt.

Systems with synchronous timings are most frequently encountered in centralized systems and will be rarely encountered in an interactive CR decision process, which implies some degree of distributed decision timings. A round-robin scheme can occur in centralized systems with distributed decision making with scheduling (as might occur in a hybrid Automatic Repeat ReQuest (ARQ) scheme). Without a synchronizing agent, and assuming an arbitrary fineness in the time scale, every distributed CR algorithm is a randomly time-ordered process. However, because of the coarseness of the timing of observation processes, many systems with random timings can be expected to behave in a manner that is indistinguishable from an asynchronous system. As we show in subsequent sections, these different decision update timings—synchronous, round-robin, random, and asynchronous—can have a significant impact on the analysis of our CR network.

15.2.2 Analysis Objectives

Before formally presenting the analysis objectives, consider a network of three radios, $N = \{1,2,3\}$, where each radio, j, can choose an action, a_j, which is drawn from a convex action set according to a decision update rule, f_j. Starting at any initial action vector, repeated applications of the radios' decision update rules trace out paths in the action space. Sometimes these paths terminate in a stable point; under different conditions the paths may enter into an infinite loop. There also may be points in the action space that the decision update rules would not independently adapt away from, but after a small external perturbation (perhaps from noise or channel fluctuations), the decision rules drive the network away from these points. Each of these concepts is illustrated in the example interaction diagram shown in Figure 15.3 where paths are shown by arrows and steady states are labeled as Nash Equilibrium (NE) for reasons that will be explained in Section 15.4.3.

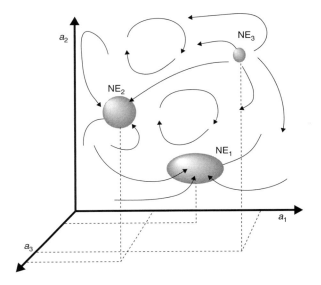

FIGURE 15.3

A three-radio interaction diagram with three steady states (NE_1, NE_2, and NE_3) and adaptation paths.

This conceptual interaction diagram illustrates the four different analysis questions that we would like to answer when analyzing a network of CRs.

- What is the expected behavior of the network?
- Does this behavior yield desirable performance?
- What conditions must be satisfied to ensure that adaptations converge to this behavior?
- Is the network stable?

The following formalizes these questions into specific analysis objectives.

Establishing Expected Behavior

As is the case for many systems, the analysis in this chapter assumes that expected behavior of a CR network is equivalent to its steady-state behavior. Accordingly, to establish expected behavior we analyze the following steady-state issues:

- *Existence*—Does the system have a steady state?
- *Identification*—What are the specific steady states for the system?

The following subsections introduce a number of different techniques to address these issues, including showing that f^t is a variant of a *contraction mapping*, that the network can be modeled as an *absorbing Markov chain*, and that f^t has a fixed point—that is, an action vector a^* such that $a^* = f^t(a^*) \forall t \in T$.

Desirability of Expected Behavior

Of course, determining a CR network's steady states tells us little about the desirability of the algorithm under study. We also need to address whether those steady states are "good" steady states or "bad" steady states and whether there are other action vectors that would be preferable from a network designer's perspective. Again, there are two specific issues we would like to address:

- *Desirability*—How "good" are the steady states of the algorithm?
- *Optimality*—Does an optimal action vector exist and how close do the steady states come to achieving optimal performance?

There are many different ways of identifying whether an action vector is a "good" steady state, but we assume in this text that there is some global objective or cost function, $J:A \to \Re$, that we wish to max(min)imize (perhaps total system goodput or total network interference).[1] Assuming we wish to maximize J, we'll treat action vector a^2 as more desirable than a^1 if $J(a^2) > J(a^1)$. To determine whether an optimal action vector exists and whether our steady states are indeed optimal, we introduce gradient techniques and *Pareto optimality* criteria. These concepts are discussed in greater detail later in the Establishing Optimality and Desirability subsection (pp. 493 and 508, respectively).

Convergence Conditions

Along with identifying that a CR algorithm has desirable steady states, it is important to identify the conditions under which the decision update algorithms *converge* to a steady state. Formally, given a sequence of action vectors, $\{a(t)\}$ formed from the decision update rules as $a(t^{k+1}) = f^t(a(t^k))$ and an initial action vector $a(0)$, we'll say that $\{a(t)\}$ *converges* to action vector a^* if for every $\varepsilon > 0$, there is some time $t^* \in T$ such that $t \geq t^*$ implies $\|a(t), a^*\| < \varepsilon$. In other words, $\{a(t)\}$ converges to a^* if for every arbitrarily small *neighborhood* of a^* (a neighborhood of a^* is the set $S \subset A$ such that $\|a, a^*\| < \varepsilon \; \forall a \in S$), there is a time after which $\{a(t)\}$ remains "trapped" in that neighborhood of a^*.

For convergence, we'll be interested in addressing the following issues:

- *Rate*—Given $\{a(t)\}$, a^*, f^t, and a neighborhood (e.g., a specified ε), what is the value or expected value of t^*?
- *Sensitivity*—Do changes in the value of $a(0)$ or a different realization of f^t (perhaps modeling an asynchronous system instead of a round-robin system) impact convergence, and if so, how?

For the various analysis techniques that we discuss in the following subsections, we highlight how those techniques provide insight into convergence rate and how different decision update rules impact convergence.

Network Stability

While the preceding analysis objectives implicitly assume that radios are behaving in a deterministic manner, wireless networks are stochastic, which means the CRs are

---

[1] J is a global objective, or cost function, that maps from the action space to the set of real numbers.

responding to estimates of their operating environment. Accordingly, the radios will frequently make mistakes in their adaptations due to this imperfect information.

For this chapter, we assume the radios' estimates are unbiased and their errors random. However, this does lead to stability concerns as small perturbations could potentially lead to undesirable behavior. Accordingly, this chapter addresses the following analytical issues with respect to an algorithm's steady state(s):

- *Lyapunov stability*—After a small perturbation, will the system stay within a neighborhood of the steady state?
- *Attractivity*—After a small perturbation, will the network converge back to the steady state?

Section 15.3.3 defines Lyapunov stability and attractivity with greater rigor and introduces techniques for determining whether a network has Lyapunov stable and attractive steady states.

15.3 TRADITIONAL ENGINEERING ANALYSIS TECHNIQUES

This section reviews some traditional engineering techniques from dynamical systems, optimization theory, parallel processing, and Markov chain theory that can be leveraged to analyze CR networks.

15.3.1 A Dynamical Systems Approach

Dynamical systems theory is concerned with analyzing the behavior of dynamical systems and designing mechanisms so that the systems act in a desirable manner. Typical analysis goals of dynamical systems theory are similar to the ones that we set out in Section 15.1: determining the expected behavior, convergence, and stability of the system.

Formally, a dynamical system is a system with a change state that is determined by a function of the current state and time. In other words, a dynamical system is any system of the form given by

$$\dot{a} = g(a, t) \qquad (15.2)$$

which describes the change in the state of a system as a function of the current system state, a, and current time, t.[2] Implicitly the system is assumed to be at state $a(0)$ at time $t = 0$.

When Eq. (15.2) is not directly dependent on t (i.e., $\dot{a} = g(a)$), the system is said to be *autonomous*. For our purposes, it makes sense to treat synchronous systems as autonomous; but for random and asynchronous systems, it is difficult to eliminate the time dependency.

The first step in a traditional dynamical systems analysis is to solve Eq. (15.2) to yield the *evolution function* that describes the state of the system as a function of time. This typically involves solving an ordinary differential equation—a task we would prefer to

---

[2] The change in system state is a function of the current state and time.

not undertake without knowing that a solution exists. In general, this solution would take the form we supposed existed for the decision update rule, f^t, in our model in Section 15.2.1.

Given a dynamical systems model, we can be assured that such a solution exists by the Picard-Lindelöf theorem [4], which states that given an open set $D \subset A \times T$, if g is continuous on D and *locally Lipschitz* with respect to a for every $a \in D$, then there is a unique solution, f^t, to the dynamical system for every $a(0)$ while f^t remains in D.

A function, $f^t : A \times T \rightarrow A$, $A \subset \mathfrak{R}^n$,[3] is said to be *Lipschitz continuous*[4] at (a, t) if there exists a $K < \infty$ such that $\|f^t(a^1, t) - f^t(a^2, t)\| \leq K\|a^1 - a^2\|$ for all $a^1, a^2 \in A$; f^t would be *locally Lipschitz continuous* if this condition were only satisfied for some open set $D \subset A \times T$. Similarly, the function f is Lipschitz continuous if it is Lipschitz continuous for all $(a, t) \in A \times T$. Any function that is Lipschitz continuous is also continuous.

Fixed Points and Solutions to CR Networks

A solution for the evolution function f^t may imply a system that is changing states over time, perhaps bounded within a certain region or wandering around the entire action space. For some systems, continual adaptations may not be an issue and may even be desirable. However, continual adaptations for a CR network imply that significant bandwidth is being consumed to support the signaling overhead required to support these adaptations.

For a CR network, we would prefer that the network settle down to a particular steady state and only adapt as the environment changes. Identifying these steady states also allows a CR designer to predict network performance. In the context of our state equation, such a steady state is a *fixed point* of f^t—a point $a^* \in A$ such that $a^* = f^t(a^*) \forall t \geq t^*$. For one-dimensional sets, it is convenient to envision a fixed point of a function as a point where the function intersects the line $x = f(x)$. Figure 15.4 illustrates a function, $f(x)$, that has three fixed points.

Solving for fixed points can be tedious because it may involve a search over the entire action space, so we would like to know whether a fixed point exists before we begin our search. Fortunately, this can be readily established by the Leray-Schauder-Tychonoff fixed-point theorem, given by Proposition 1.3 in Chapter 3 of Bertsekas and Tsitsiklis [5], which states that if $A \subset \mathfrak{R}^n$ is *nonempty*, *convex*, and *compact*, and if $f^t : A \rightarrow A$ is a *continuous* function, then there exists some $a^* \in A$ such that $a^* = f^t(a^*)$.[5] Note that this definition is inappropriate for finite action sets, which while compact, are not convex.

However, actually solving for a fixed point under such general conditions can be much more difficult. Subsequent subsections refine these conditions to facilitate fixed-point identification.

---

[3] f^t is a function that maps from the Cartesian product of the action space with the set of all update times to the action space, where the action space is a subset of all real n tuples; that is, given an initial action state, the function describes how the network state changes over time.

[4] A function is Lipschitz continuous if there exists a finite real K, such that for all action states a^1 and a^2 in the action space, the Euclidean distance between their next action state in time is less than K times the distance between their current action states.

[5] Leray-Schauder-Tychonoff actually considers continuous mappings instead of continuous functions, but a continuous function is a continuous mapping.

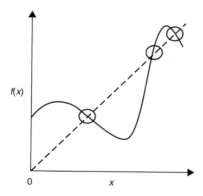

FIGURE 15.4

A function with three fixed points (*circled*). For functions on one-dimensional sets, the points at which the function intersect the line $f(x) = x$ (*dashed*) are fixed points.

Establishing Optimality

Perhaps the easiest way to establish that a solution to a CR network is optimal is to show that it maximizes (or minimizes) some objective function $J{:}A \rightarrow \mathfrak{R}$. For networks with a finite action space, we can perform an exhaustive search and evaluate J at each point in A.

However, this approach is impractical for infinite action spaces. When J is differentiable and A is a compact interval of $\mathfrak{R}^n$, however, we can reduce the search space by noting that if a particular action vector, a^*, is optimal, then a^* must either be a boundary point or $\nabla J(a^*) = 0$. Recall that

$$\nabla J(a) = \frac{\partial J(a)}{\partial a_1}\hat{a}_1 + \frac{\partial J(a)}{\partial a_2}\hat{a}_2 + \cdots + \frac{\partial J(a)}{\partial a_n}\hat{a}_n$$

where each $\hat{a}_j$ is a dimension of A.[6] So in effect, this condition says that for a^* to optimize J, there must be no direction that can be followed from a^* that increases J. If J is *pseudo-concave*, we can change this to a sufficient condition; that is, if there exists some point such that $\nabla J(a^*) = 0$, then it is optimal. J is said to be pseudo-concave if $\nabla J(a'') \cdot (a'-a'') \leq 0 \Rightarrow J(a') \leq J(a'')$ for all points $a',a'' \in A$ [6]. More familiarly, a function that is *concave* is also pseudo-concave. Formally, a function, $J{:}A \rightarrow \mathfrak{R}$, is concave on the set A if for all $a_1,a_2 \in A$, $J(\lambda a_1 + (1 - \lambda)a_2) \geq \lambda J(a_1) + (1 - \lambda)J(a_2)$ for all $\lambda \in [0,1]$. Equivalently, a function is concave if it is impossible to join two points in the function with a line that contains points above the function.

Figure 15.5 shows an example of a function that is pseudo-concave, but not concave. This function can be verified to not be concave by considering a line joining the points

---

[6]The gradient of the cost function J, is in general a vector-valued function that, when evaluated at a particular point, a, indicates the magnitude and direction of greatest increase for J at a. When J is a function of a single dimension, then the gradient of J is equivalent to the slope of J.

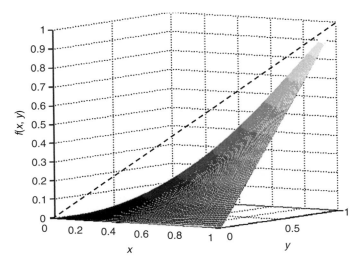

FIGURE 15.5

$f(x, y) = xy$, x, $y > 0$—A function that is pseudo-concave, but not concave.

(0, 0) and (1, 1) (shown as a *dashed line*); except for the endpoints, all of the points in this line lie above the function.

Convergence and Stability

When discussing convergence and stability of a decision rule's fixed point, it is convenient to make use of two forms of stability: *Lyapunov stability* and *attractivity*.

Formally, we say that an action vector, a^*, is *Lyapunov stable* if for every $\varepsilon > 0$ there is a $\delta > 0$ such that for all $t \geq t^0$, $\|a(t^0), a^*\| < \delta \Rightarrow \|a(t), a^*\| < \varepsilon$.[7] While no particular relation between δ and ε can be inferred from this definition, an engineer may be more comfortable thinking of Lyapunov stability as akin to Bounded-Input-Bounded-Output (BIBO) stability wherein after a bounded "stimulus" of δ is added to a system operating at a^*, the system remains within a bounded distance ε of a^*.

The action vector a^* is said to be *attractive* over the region $S \subset A$, $S = \{a \in A | \|a, a^*\| < M\}$, if given any $a(t_0) \in S$, the sequence $\{a(t)\}$ converges to a^* for $t \geq t_0$. We say that a^* is *asymptotically stable* if it is both Lyapunov stable and attractive.

Note that Lyapunov stability does not imply attractivity nor does attractivity imply Lyapunov stability. For instance, the fixed point (0, 0) in Figure 15.6 is Lyapunov stable, but not attractive; meanwhile the fixed point (0, 0) in Figure 15.7 is attractive, but not Lyapunov stable. However, a method does exist for simultaneously establishing both stability and attractivity—Lyapunov's direct method.

---

[7] Equivalently, the action vector a^* is said to be Lyapunov stable if for every arbitrarily sized $\varepsilon > 0$, it is possible to identify $\delta > 0$ such that after a perturbation to any point $a(t^0)$, all subsequent action vectors are no more than a Euclidean distance of ε away from a^*.

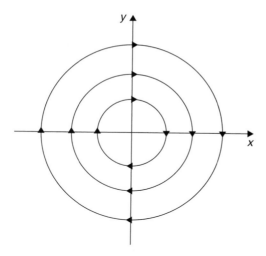

FIGURE 15.6

Paths (formed by iterative application of f^t and shown with arrows) for a system that is Lyapunov stable, but not attractive.

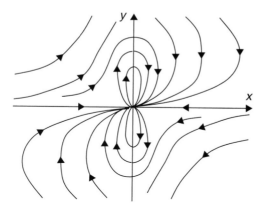

FIGURE 15.7

Paths for a fixed point that is attractive, but not Lyapunov stable.

Lyapunov's Direct Method for Discrete Time Systems

Instead of attempting to directly apply the definitions of Lyapunov stability and asymptotic stability, we can use *Lyapunov's direct method*, which says that if we can find a function that strictly decreases along all paths created by the adaptations of a CR network, then that cognitive radio network is asymptotically stable.[8]

---

[8] A more rigorous definition is given by Medio and Lines [7], Theorem 3.4, with a longer discussion in the context of CR networks in Neel [29].

The existence of a Lyapunov function can be used to establish the existence and identify the network's steady states, namely, all points where $L(a^*) = 0$. Further, unlike the Picard-Lindelöf equation, Lyapunov's second method can be readily applied to both synchronous and asynchronous CR networks—the only requirement being that each adaptation must decrease the value of the Lyapunov function.

15.3.2 Contraction Mappings and the General Convergence Theorem

The preceding discussion assumed a closed-form expression for the next network state as a function of current network state. Now suppose that after one recursion of the network update rule we are unable to precisely predict the next network state. However, we are able to bound the network state within a particular set of states $A(t^1)$. Then suppose that armed with the knowledge that the network starts in $A(t^1)$, we could say that after the second iteration the network state would have to be within another set $A(t^2)$, which is a subset of $A(t^1)$. Extending this concept, suppose that given any set of network states, $A(t^k)$, we know that the decision update rule always results in a network state in the set, $A(t^{k+1})$, which is a subset of $A(t^k)$.

In effect, this process is saying that as the recursion continues, finer and finer approximations on the operating point of the network are possible, perhaps resulting in a prediction of a specific steady state for the network. This iterative restriction on a recursion's possible points forms the basis of numerous valuable algorithms and is a characteristic of a special class of algorithms known as *contraction mappings*.

Contraction Mappings
Given a recursion $a(t^{k+1}) = f(a(t^k))$, f is said to be a *contraction mapping* with modulus α if there is an $\alpha \in [0,1)$ such that $\|f(a), f(b)\| \leq \alpha \|a, b\| \ \forall \ b, a \in A$. While applying this definition to a decision rule can be difficult, it has been shown that an arbitrary recursion, f, is a contraction mapping if the following two conditions[9] are satisfied [8]:

- *Monotonicity*—Given bounded functions g_1, g_2: $A \rightarrow \Re$, where $g_1(a) \leq g_2(a)$ $\forall \ a \in A$, f must satisfy $f(g_1(a)) \leq f(g_2(a)) \ \forall \ a \in A$.
- *Discounting*—There exists a $\beta \in (0,1)$ such that $f(g_1(a) + c) = f(g_1(a)) + \beta c$ for all bounded, $g_1: A \rightarrow \Re$, $c \geq 0$, $a \in A$.

Analysis Insights
Knowing that our decision rule constitutes a contraction mapping immediately provides us with several valuable insights. From Banach's contraction mapping theorem [9], we know that f has a unique fixed point to which the recursion f converges from any starting point. After k iterations, a bound on the distance of the current state from the fixed point is given by:

$$\|a(t^k), a^*\| \leq \frac{\alpha^k}{1 - \alpha}\|a(t^1), a(t^0)\| \tag{15.3}$$

Eq. (15.3) is also useful for bounding the error in estimating f's fixed point by recursively evaluating f. Additionally, a Lyapunov function for any contraction mapping with fixed point a^* is given by the following equation:

---

[9] These conditions are known as *Blackwell's conditions*.

$$L(a) = \|a, a^*\| \qquad (15.4)$$

Thus, every contraction mapping, f, has a unique stable fixed point to which f converges at a predictable rate.

Pseudo-contractions

A pseudo-contraction eliminates the contraction mapping's requirement that all points move closer to each other after each iteration, but still requires that after each iteration all points move closer to a unique fixed point. Formally, given mapping $f: A \rightarrow A$ with fixed point, a^*, we say f is a pseudo-contraction if there is an $\alpha \in [0,1)$ such that $\|f(a), f(a^*)\| \leq \alpha \|a,a^*\| \ \forall a \in A$. By definition, f has a unique fixed point, a^*, and the distance to a^* at time t^k is given by:

$$\|a(t^k), a^*\| \leq \alpha^k \|a(0), a^*\| \qquad (15.5)$$

Note that evaluation of Eq. (15.5) requires knowledge of the fixed point, so, unlike Eq. (15.3), it is not appropriate for bounding the error on an estimate of the system's fixed point while iterating to solve for the fixed point.

General Convergence Theorem

Most contraction mappings assume that the updating process occurs synchronously (recall the discussion of decision timings in Section 15.2.1). We can relax this assumption by introducing the general convergence theorem presented by Bertsekas and Tsitsiklis [5, Proposition 2.1 of Chapter 6].

Suppose we know that $\cdots \subset A(t^{k11}) \subset A(t^k) \subset \cdots \subset A(t^0)$, where $A(t^k)$ represents the possible states of the network after k iterations and $A(t^0)$ represents all possible initial states for the network. Then if the following two conditions hold, f also converges asynchronously.

1. *Synchronous convergence condition*
   (a) $f(a) \in A(t^{k+1}) \ \forall k, a \in A(t^k)$
   (b) If $\{a(t^k)\}$ is a sequence such that $a(t^k) \in A(t^k)$ for every k, then every limit point of $\{a(t^k)\}$ is a fixed point of f.
2. *Box condition*—For every k, there exist sets $A_j(t^k) \subset A_j$ such that $A(t^k) = A_1(t^k) \times \cdots A_n(t^k)$.

For our purposes, the general convergence theorem states that under our regular action sets assumption, any contraction or pseudo-contraction mapping that converges synchronously also converges asynchronously. However, we can also apply the general convergence theorem to algorithms that are not obviously contraction mappings, as seen in the extended example presented in the next subsection.

Standard Interference Function Model

Many traditional analyses consider specific decision rules that model specific applications. The following discussion presents such an analysis that is also an example of a nonobvious contraction mapping. Yates [10] considers a power control algorithm operating on the uplink of a cellular system. For this algorithm, there is a set of N mobiles where each mobile, j, attempts to achieve a target-received signal interference-to-noise ratio (SINR), $\hat{\gamma}_j$. The development of this algorithm assumes that each mobile is capable of observing its received SINR (perhaps via feedback from a basestation) which is generally given by:

$$\gamma_j = \frac{g_{jj} p_j}{\sum_{k \in N} g_{kj} p_k + N_j} \tag{15.6}$$

where g_{kj} can be the link budget gain from mobile k to the basestation (BS) of mobile j, p_k is the transmit power of mobile k, and N_j is the noise power at the BS that is receiving mobile j's signal.

Based on observations of Eq. (15.6), the mobiles compute a scenario-dependent *interference function*, $I_j(\mathbf{p})$, which is formed as the ratio of the target SINR, $\hat{\gamma}_j$, and the effective SINR, λ_j, or $I_j(\mathbf{p}) = \hat{\gamma}_j / \gamma_j$ where $\mathbf{p}$ is the vector of transmit powers, $\mathbf{p} = (p_1, p_2, \ldots, p_n)$, drawn from the power vector space $\mathbf{P}$.

Generalizing beyond this ratio formalization, Yates [10] defines any interference function to be *standard* if it satisfies the following three conditions.

1. *Positivity*—$I(\mathbf{p}) > 0$
2. *Monotonicity*—If $\mathbf{p}^1 \geq \mathbf{p}^2$, then $I(\mathbf{p}^1) \geq I(\mathbf{p}^2)$
3. *Scalability*—For all a > 1, $\alpha I(\mathbf{p}) > I(\alpha \mathbf{p})$

Here we write $\mathbf{p}^1 > \mathbf{p}^2$ if $p_j^1 > p_j^2 \ \forall j \in N$ and $I(\mathbf{p})$ is the synchronous evaluation of all $I_j(\mathbf{p})$.

Assuming the existence of a standard interference function, Yates [10] defines a synchronous updating process of the form $\mathbf{p}(t^{k+1}) = f(\mathbf{p}(t^k))$ where $f(\mathbf{p}) = f_1(\mathbf{p}) \times \cdots \times f_n(\mathbf{p})$ is given by:

$$f_j(\mathbf{p}(t^k)) = p_j(t^k) I_j(\mathbf{p}(t^k)) \tag{15.7}$$

When the target SINR vector, $\hat{\mathbf{g}}$, is *feasible*,[10] Yates [10] is able to show that an algorithm updating the power vector according to Eq. (15.7) has the following properties:

1. A fixed point exists; that is, there is some $\mathbf{p}^*$ such that $\mathbf{p}^* = f(\mathbf{p}^*)$.
2. This fixed point is unique.
3. Starting from any initial power vector, f converges to $\mathbf{p}^*$.

Whereas Yates [10] shows these results in an ad hoc manner, Berggren [11] shows that this updating process constitutes a pseudo-contraction that could be used to establish these same results in a different manner.

The fact that f constitutes a pseudo-contraction implies that $\cdots \subset \mathbf{P}(t^{k11}) \subset \mathbf{P}(t^k) \subset \cdots \subset \mathbf{P}(t^0)$, where $\mathbf{P}(t^k)$ is the power vector space for iteration k. Coupled with the just-established synchronous convergence of f and implicit satisfaction of the box condition, this means that f has satisfied the conditions for the general convergence theorem. Thus, it is known that f converges both synchronously and asynchronously.

---

[10]An SINR vector is feasible if there exists a $\mathbf{p} \in \mathbf{P}$ such that $\gamma_j \geq \hat{\gamma}_j \forall j \in N$. Neel [29] presents a method for calculating the required $\mathbf{p}$ to show feasibility.

Further Standard Interference Function Analysis Insights

Assuming the SINR feasibility criterion is satisfied, Yates [10] also shows that the following target SINR arrangements of BSs and mobiles have standard interference functions and thus converge synchronously and asynchronously to a unique power vector when the decision update rule is given, as in Eq. (15.7).

- *Fixed assignment*—each mobile is assigned to a particular BS
- *Minimum power assignment*—each mobile is assigned to the BS in the network where the mobile's SINR is maximized
- *Macro diversity*—all BSs in the network combine the signals of the mobiles
- *Limited diversity*—a subset of the BSs combine the signals of the mobiles
- *Multiple connection reception*—the target SINR must be maintained at a number of BSs

15.3.3 Markov Models

Perhaps due to uncertainty in the order of adaptation (as would be the case for a randomly or asynchronously timed process) or due to uncertainties in the decision rules, it may be impossible to derive a closed-form expression for an evolution equation or to even bound the adaptations into sequential subsets. Instead, suppose we can model the changes of the CR network from one state to another as a sequence of probabilistic events conditioned on past states that the system may have passed through. The probability distribution for the next state in time, $a(t^{k+1})$, is conditioned solely on the most recent state, as shown in the following equation:

$$P\left(a\left(t^{k+1}\right) = a^k \middle| a(0), \dots, a(t)\right) = P\left(a\left(t^{k+1}\right) = a^k \middle| a\left(t^k\right)\right) \tag{15.8}$$

The random sequence of states, $\{a(t)\}$, is said to be a *Markov chain*. A model of a system with states that form a Markov chain is said to be a *Markov model*. Throughout the remainder of this section, we use these two terms interchangeably.

Formalizing this model, assume that the state space is finite. This is not a requirement for a Markov chain, but the assumption is useful for the subsequent discussion. Further, assume that if the network is in state $a^m \in A$ at time t^k, then at time t^{k+1}, the network transitions to state $a^n \in A$ with probability p_{mn}, where

$$p_{mm} \geq 0 \ \forall \ a^m, \ a^n \in A \ \text{ and } \ \sum_{j \leq |A|} p_{mj} = 1.$$

Of course, it is also permitted that the system remains in state a^m, which it does with probability p_{mm}. To simplify notation, we make use of a *transition matrix*, which we represent with the symbol $\mathbf{P}$. The transition matrix is formed by assigning p_{mn} to the entry corresponding to the mth row and nth column.

Markov Model Analysis Insights

From $\mathbf{P}$ we can then form $\mathbf{P}^2$ as the matrix product $\mathbf{PP}$. Now entry p_{mn}^2 in the mth row and nth column of $\mathbf{P}^2$ represents the probability that a system is in state a^n two iterations after being in state a^m. Similarly, if we consider the matrix $\mathbf{P}^k$ formed as $\mathbf{P}^k = \mathbf{PP}^{k-1}$ (an example of a Chapman-Kolmogorov equation for a Markov chain [13]), then entry

p_{mn}^k in the mth row and nth column of $\mathbf{P}^k$ represents the probability that a system is in state a^n k iterations after being in state a^m.

A similar relationship can be found when the initial state is specified by a random probability distribution arranged as a column vector π where

$$\pi_m \in [0, 1] \; \pi_m \in [0, 1] \text{ and } \sum_{m=1}^{|A|} \pi_m = 1$$

where π_m represents the probability of starting in state a^m. For such a situation, the state probability distribution after k iterations is given by $\pi^T \mathbf{P}^k$.

There may also be some distribution π^* such that $\pi^{*T} \mathbf{P} = \pi^{*T}$. Such a distribution π^* is said to be a *stationary distribution* for the Markov chain defined by $\mathbf{P}$. Note that solving for a stationary distribution is equivalent to solving the eigenvector equation $\pi^{*T} \mathbf{P} = \lambda \pi^{*T}$ where $\lambda = 1$. A related concept is the *limiting distribution* which is the distribution that results from evaluating $\lim_{k \to \infty} \pi^{0T} \mathbf{P}^k$. Although the network is not generally a steady state as considered in the previous discussion, showing that a Markov chain has a unique distribution that is both stationary and limiting would permit us to characterize the behavior of the network. Specifically, given the unique stationary limiting distribution π^*, we could predict that at a particular instance in time and after a sufficient number of iterations, the network would be in state a^m with probability p_m. Thus, it is desirable to be able to identify when such a unique stationary limiting distribution exists.

The ergodicity theorem [14] states that if a Markov chain is *ergodic*, then there exists a unique limiting and stationary distribution for all initial distributions π^0. A Markov chain is ergodic if it is: (1) *irreducible*, (2) *positive recurrent*, and (3) *aperiodic*. A Markov chain is *irreducible* if $\forall a \in A$, there exist sequences of state transitions with nonzero probability that lead to every state. A Markov chain is *positive recurrent* if the expected time to return to every state is finite. Finally, a Markov chain is *aperiodic* if for each state in the chain there is no integer, $m > 1$, such that once the system leaves the state, it can only return to the state in multiples of m iterations.

Identifying that these three conditions are satisfied can be a rather daunting task. However, Kemeny and Snell's Theorem 4.1.2 [15] shows that for finite Markov chains, $\mathbf{P}$ is ergodic if and only if there is some k such that $\mathbf{P}^k$ has no zero entries. Thus, by identifying this simple condition, we know that a unique stationary limiting distribution exists.

Absorbing Markov Chains

Unfortunately, the limiting distribution for an ergodic matrix is rather unsatisfying because all states will have nonzero probability of being occupied. However, this is not a problem for absorbing Markov chains. A state a^k is said to be an *absorbing state* if there are no paths that leave a^k—that is, $p_{km} = 0$ $\forall k \neq m$ and $p_{kk} = 1$. A Markov chain is said to be an *absorbing Markov chain* if:

- It has at least one absorbing state.
- From every state in the Markov chain there exists a sequence of state transitions with nonzero probability that lead to an absorbing state. These nonabsorbing states are called *transient states*.

Eq. (15.9) is an example of a transition matrix for an absorbing Markov chain, where a^4 is the absorbing state and a^1, a^2, and a^3 are the transient states:

$$
\mathbf{P} = \begin{array}{c} \\ a^1 \\ a^2 \\ a^3 \\ a^4 \end{array} \begin{array}{cccc} a^1 & a^2 & a^3 & a^4 \\ \left[\begin{array}{cccc} 0.1 & 0.3 & 0.1 & 0.5 \\ 0.4 & 0.0 & 0.3 & 0.3 \\ 0.4 & 0.1 & 0.3 & 0.2 \\ 0 & 0 & 0 & 1 \end{array} \right] \end{array} \qquad (15.9)
$$

Note that when represented as a transition matrix, state a^m is an absorbing state if and only if $p_{mm} = 1$.

Absorbing Markov Chains Analysis Insights

Within the context of our analysis objectives, an absorbing state is a fixed point or steady state that, once reached, the system never leaves. Similarly, valuable convergence insights can also be gained when the system can be modeled as an absorbing Markov chain as follows.

First, form the modified transition matrix, $\mathbf{P}'$, as shown in the equation:

$$
\mathbf{P}' = \left[\begin{array}{c|c} \mathbf{Q} & \mathbf{R} \\ \hline \mathbf{0} & \mathbf{I}^{ab} \end{array} \right] \qquad (15.10)
$$

where $\mathbf{I}^{ab}$ is the identity matrix corresponding to the state transitions between the absorbing states of the chain, $\mathbf{Q}$ represents the state transitions between the nonabsorbing states of the chain, $\mathbf{0}$ is a rectangular matrix filled with all zeros representing the probability of transition from absorbing states to nonabsorbing states, and $\mathbf{R}$ represents the rectangular matrix of state transition probabilities from nonabsorbing states to absorbing states.

Given $\mathbf{P}'$, Markov theory provides us with information on convergence and the expected frequency that the system visits a transitory state. First, $\lim_{k \to \infty} \mathbf{Q}^k \to \mathbf{0}$ implies that the probability of the system not being "absorbed" (i.e., not terminating in one of the absorbing states of the chain) goes to zero.

Beyond this basic result, more specific convergence results can be stated by introducing the *canonical form* for the absorbing chain. Given an absorbing chain with a modified transition matrix, as in Eq. (15.10), the *fundamental matrix* is given by:

$$
\mathbf{N} = (\mathbf{I} - \mathbf{Q})^{-1} \qquad (15.11)
$$

Solving for the fundamental matrix $\mathbf{N}$ permits a number of valuable analytic insights. First, Kemeny and Snell's Theorem 3.2.4 [15] states that the entry n_{km} gives the expected number of times that the system will pass through state a_m given that the system starts in state a_k. Second, their Theorem 3.3.5 [15] states that if we evaluate $\mathbf{t} = \mathbf{N1}$ where $\mathbf{1}$ is a column vector of all ones, then t_k gives the expected number of iterations before the state is absorbed when the system starts in state a_k. Finally, their Theorem 3.3.7 [15] states that if we evaluate $\mathbf{B} = \mathbf{NR}$ where $\mathbf{R}$ is as given the preceding above Eq. (15.10), then entry b_{km} in $\mathbf{B}$ specifies the probability the system ends up in absorbing state a_m if the system starts in state a_k.

Thus, once we show that a Markov model for a network of CRs with transition matrix **P** is an absorbing Markov chain, the following insights are readily gained:

- Steady states for the system can be identified by finding those states a^m for which $p_{mm} = 1$.
- Convergence to one of these steady states is assured, and the expected distribution of states can be found by solving for **B**.
- Given an initial state, a^k, convergence rate information is given by solving for **t**.

15.4 APPLYING GAME THEORY TO THE ANALYSIS PROBLEM

The techniques presented in Section 15.3 provide us with tools that allow us to answer many of our analysis questions, but there are still some noticeable limitations. Establishing the existence and uniqueness of a fixed point for an evolution function says little about convergence or stability. Finding an appropriate Lyapunov function can provide valuable convergence and stability information, but finding that Lyapunov function is largely a hit-or-miss affair. Contraction mappings provide many of the results that we desire, but are encountered infrequently. Although Markov models can do an excellent job of handling the nondeterministic nature of one of the most promising CR adaptation algorithms—genetic algorithms (GAs)—solving for a network's transition matrix can be a daunting task. Finally, for all of these approaches, analyzing one decision rule says little about the performance of related decision rules. It would be nice if, given the application-specific goal of a CR, we were able to immediately predict what decision processes would be required to achieve the desired level of network performance.

Frequently, progress in analysis occurs by introducing additional information. In this case, introducing CRs' goals allows us to apply techniques from game theory to gain additional insights. As we will see in the remainder of this chapter, game theory has a number of advantages over more traditional techniques. A number of readily identifiable game models allow us to simultaneously identify the existence of steady states, convergence criteria, and stability of CR algorithms. Further, these game models provide the capacity to analyze broad classes of decision rules, such as the random better response dynamic considered in the Identification of Exact and Ordinal Potential Games section, which encompasses the GAs in Chapter 7. Further, game theory provides a means to analyze the interactions of the ontologically defined CRs in Chapter 13 for which no predefined decision rules exist—an impossible analysis problem for more conventional techniques.

It is assumed that game theory is a new concept to most readers, so this section provides an extended discussion of the basic elements of game theory, how game theory can be applied to CR networks, basic game models, and the analytic insights that can be gained by applying the game models. Section 15.5 goes into greater depth by describing two game models that permit rapid analysis of CR networks in terms of our analysis objectives.

15.4.1 Basic Elements of Game Theory

Game theory is a collection of models and analytic tools used to study *interactive decision-making processes*. In brief, an interactive decision process is a process with an outcome that is a function of the inputs (actions) from several different decision makers (players) who may have conflicting goals with regard to the outcome of the process.

The fundamental modeling tool of game theory is the *game*. Whether explicitly or implicitly, every game includes the following components:

- A set of *players*
- *Actions* for each of the players
- Some method for determining *outcomes* according to the actions chosen by the players
- *Preferences* for each of the players defined over all the possible outcomes
- *Rules* governing the order of play

The following subsections provide brief descriptions of these components in light of the CR network model introduced in Section 15.2.1.

Players

The players in a game are the decision-making entities in the modeled interactive process. In our case, the players are the CRs in the network. For notational continuity, we refer to the set of players (CRs) as N and individual players as i or j. As a rule, games only consider situations of two or more players because a single-player game would by definition not be an interactive process.

Actions and Outcomes

For our purposes, we continue to use actions and outcomes (or observations of the outcomes) in the same manner as introduced in Section 15.2. The actions are the adaptations (waveforms) available to the radio, and the outcomes are the observations of the network.

Utility Functions

Game theorists frequently employ *utility functions* (sometimes called *objective functions*), which assign a real number to each outcome for a particular player such that higher values indicate that the player (the cognitive radio) achieves a more desirable outcome.

Frequently, we write the utility functions as functions of the action vectors that yield the outcomes, $u_j{:}A \to \Re$, instead of as functions of the outcome space, $u_j{:}O \to \Re$. As was the case for the model in Section 15.2, this simplification is appropriate as long as the mapping between A and O is clear. Because different players generally ascribe different valuations to the same action vector (or outcome), we sometimes make use of a *payoff vector* that lists the utility that each player assigns to a particular action vector. For example, rather than writing $u_1(a) = 1$, $u_2(a) = -3$, and $u_3(a) = 4$, we could write $u(a) = (1,-3,4)$. With this notation, it also sometimes makes sense to describe a single utility function that maps A into $\Re^n$ where n is the number of players in the game.

Rules Governing the Order of Play

Different game models assume different rules for when the players are allowed to "play" (choose an action). As with the CR network model introduced in Section 15.2, for our games we concern ourselves with games that adopt synchronous, asynchronous, round-robin, or randomly ordered timings for decisions.

15.4.2 Mapping a Game's Basic Elements to the Cognition Cycle

Fundamentally, game theory can be applied to the analysis of the adaptations of any set of intelligent agents, and the cognition cycle represents the processes that go on in any intelligent being—including humans. So it is not surprising that we can establish connections between the components of a game and the cognition cycle.

A CR network described via interactive cognition cycles can be modeled by using the elements of a game as follows. First, the players of the game are all the nodes in a network adapting their waveforms, N. Each player's action set is formed from the various adaptations available to the radio, A_j, and the action space for the game is simply formed from the Cartesian product of the radios' available adaptations A. A utility function for each player is provided by the CR's goal, and the arguments and valuation for this utility function are taken from the outputs of the CR's observation and orientation steps. Loosely, the observation step provides the player with the arguments to evaluate its utility function, and the orientation step determines the valuation of the utility function.

Note that we have ignored the learning step of the cognition cycle. This is neither an oversight, nor indicative, of a limitation of game theory. Subsequent sections show that some simple models can also be used to handle many situations in which the CRs must learn how their operating environment impacts their goals.

15.4.3 Basic Game Models

Different game models incorporate these basic game elements in different ways. Some game models, such as an "extensive form game model," have complex iteration-varying rules governing the actions and order of play. Other game models (e.g., the "normal form game model") are much simpler. For the purposes here, we are particularly interested in the normal form and repeated game models.

Normal Form Game Model

The simplest and most frequently encountered game model used to describe an interactive decision-making process is the *normal form game*. In addition to the basic elements, a normal form game adds the following rules:

- *Synchronous single-shot play*—All players make their decisions simultaneously and only make a single decision. Thus, T has a single element.
- *Perfect information*—The players know their own utility functions and the utility functions for all the other players in the game.
- *Perfect implementation*—All players exhibit perfect implementation[11] (i.e., no player accidentally implements action a_j^1 instead of a_j^2).

---

[11] The *trembling hand* model is an example where perfect implementation is not assumed.

With these rules governing the timing and implementation of play, a normal form game is defined by the 3-tuple, $\Gamma = \langle N, A, \{u_j\}_{j \in N} \rangle$, where N is the set of players, A is the action space, and u_j is the utility function for player j (each player in N has its own utility function).

Particularly for two-player games, it is convenient to represent a normal form game in *matrix form*. In a matrix form representation of a two-player normal form game, all possible action vectors are arrayed in a matrix such that player 1's actions (the first component of the action vector) are given by the rows of the matrix, and player 2's actions (the second component of the action vector) are given by the columns of the matrix. Each cell in this matrix is thus determined by a unique action vector (row, column) and is filled with the payoff vector associated with that action vector.

EXAMPLE 15.1

The Cognitive Radio's Dilemma

Two CRs are operating in the same environment and are attempting to maximize their throughput. Each radio can implement two different waveforms—one a low-power narrowband waveform, the other a higher-power wideband waveform. If both radios choose to implement their narrowband waveforms [action vector (n, N)], the signals will be separated in frequency and each radio will achieve a throughput of 9.6 Kbps. If one of the radios implements its wideband waveform while the other implements a narrowband waveform [action vectors (n, W) or (w, N)], then interference will occur. In this event, the narrowband signal will achieve a throughput of 3.2 Kbps and the wideband signal will yield a throughput of 21 kbps. In the event that both radios choose to implement wideband waveforms, then each radio will achieve a throughput of 7 Kbps.

These waveforms can be visualized in the frequency domain as shown in Figure 15.8 and represented in matrix form as shown in Figure 15.9. Without going into the analysis of this game

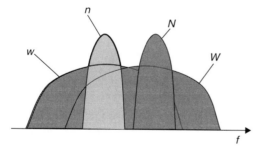

FIGURE 15.8

Frequency domain representation of waveforms in CR's dilemma.

Γ	N	W
n	(9.6, 9.6)	(3.2, 21)
w	(21, 3.2)	(7, 7)

FIGURE 15.9

The cognitive radio's dilemma in matrix form.

(presented in Example 15.2), the insightful reader may already anticipate that the design of this algorithm will tend to lead to less than optimal performance.

Repeated Game Model

A repeated game is sequence of "stage games" in which each stage game is the same normal form game. Based on their knowledge of the game (e.g., past actions, current observations, and future expectations), players choose *strategies* (choices of actions over subsequent stages). These strategies may be fixed (not responsive to choices by other players in later stages) or adapt in response to the actions of other players. Further, these strategies can be designed to punish players who deviate from agreed-on behavior. When punishment occurs, players choose their actions to minimize the payoff of the offending player.

For our purposes, we consider repeated games defined by the 4-tuple,

$$\Gamma^R = \left\langle N, A, \{u_j\}_{j \in N}, \{T_j\}_{j \in N} \right\rangle$$

where N, A, and $\{u_j\}_{j \in N}$ are specified by the normal form stage game

$$\Gamma = \left\langle N, A, \{u_j\}_{j \in N} \right\rangle$$

and T_j represents the times (or with a simple notational change, the stages) at which player j can change its decisions. For modeling purposes, this is the same T_j that we used in the model of Section 15.2.1.

15.4.4 Fundamental Game Theory Analysis Techniques

Unlike in Section 15.3, where we analyzed predefined decision rules, game theory permits us to predict the behavior of large classes of decision rules by considering the goals and preferences of the radios. The following discussion shows how game theory addresses the general issues of identifying steady states, measuring optimality, and determining convergence. Stability is not particularly well addressed for the general game models presented in Section 15.4.3, but stability is considered in the discussion of the game models presented in Section 15.5.

Steady States

In game theory, the typically discussed steady-state concept is the Nash Equilibrium (NE). An action vector, a^*, is said to be an NE if and only if Eq. (15.12) is satisfied $\forall i \in N$, $b_i \in A_i$:

$$u_i(a^*) \geq u_i\left(b_i, a^*_{-i}\right) \tag{15.12}$$

In other words, an action vector is an NE if no player (radio) can improve its performance by itself. When only one player changes its action, the adaptation is called a *unilateral deviation*. In a normal form game, where each player has knowledge of the other players' utility functions, an NE is predicted as the most likely action tuple that the players will choose; in a repeated game without complex punishment strategies, an NE is similarly predicted to be like a traditional steady state, in which once that action vector is reached, adaptations cease because no player can improve its performance.

Both of these assertions can be seen to hold as long as each player's decision rule results in adaptations that further the players' own self-interest (i.e., the player acts in a way that increases its own utility or goal). In game theory parlance, a player acting in its own interest is said to be *rational*. Now, consider a repeated game the previous iteration of which resulted in an action vector that satisfies the conditions for an NE. Assuming no coordination between groups of players, then if the players are following decision rules that adapt in the direction of improving performance, there is no directional adaptation that increases any player's payoff, and thus the next iteration of the game will again be in the NE. If each player at every iteration chooses the action that maximizes its performance, then the locally optimal choice is still the NE, and play does not change in the next iteration. If possible, new actions to be considered for implementation are generated randomly (perhaps via a GA), therefore the NE will remain inescapable if each player's decision rule requires the new action to improve performance. So, by analyzing the utility functions instead of a specific decision rule, an NE identifies the steady state for all decision rules that a game theorist would say are rational.

This rationality requirement readily extends to CR networks because every cognitive radio can be considered to be acting in its own interest, in its user's interests, or in the interests of its network. In all these cases, the CR is observing, orienting, deciding, learning, and acting in a way that maximizes an objective function. And if the network arrives in a state from which no radio in the network can find a profitable adaptation that increases its objective, then a game theorist would identify this state as an NE, and for any rational decision rule the network would remain in the NE.

EXAMPLE 15.2

Identifying the NE of the Cognitive Radio's Dilemma

Consider the CR's dilemma of Example 15.1 again. This game has a unique NE of (w, W), which is circled in Figure 15.10. Note that although (n, N) would actually yield superior performance for both radios, neither radio would unilaterally deviate from (w, W) to improve its performance, given performance would drop from (7, 7) to (3.2, 7).

Γ	N	W
n	(9.6, 9.6)	(3.2, 21)
w	(21, 3.2)	(7, 7)

FIGURE 15.10

The CR's dilemma. This game has a unique NE at (w, W).

NE Existence

Now that we have seen the power of the NE concept, how can we know that our network of CRs has an NE? As we did in the discussion of dynamical systems, we turn to "fixed-point" theorems,[12] but now as applied to the goals of the players, not the decision rules.

---

[12] Fixed-point concepts of game theory are more fully explained in Section 15.5.2.

The most important fixed-point theorem for normal form and repeated games is the Glicksberg-Fan fixed-point theorem ([16], Theorem 1.2):

> Given a normal form game $\Gamma = \langle N, A, \{u_i\} \rangle$, where all A_i are nonempty *compact convex* subsets of $\Re^M \forall i \in N$, if $\forall i \in N$ u_i is *continuous* in a and *quasi-concave* in a_i, then Γ has a pure strategy NE.

Note that unlike the Leray-Schauder-Tychonoff fixed-point theorem (see the Fixed Points and Solutions to CR Networks, section, p. 492), which needs a specific decision rule, the Glicksberg-Fan theorem considers utility functions on which any self-interested decision rule could be implemented.

The previously undefined term from the Glicksberg-Fan theorem, *quasiconcave*, is used for a function if all of its *upper-level sets* are convex. Given a point a^* and a function, $f:A \to \Re$, the upper-level set for a^* is given by $U(a^*) = \{a \in A : f(a) \geq a^*\}$. Contrasting quasi-concavity to the previously considered concepts of concavity and pseudo-concavity, Neel [29] provides an example of a function that is quasi-concave, but neither concave nor pseudo-concave.

It is also important to note that if A is finite, then Glicksberg-Fan cannot be applied to show that the system has a fixed point because it is not convex. For instance, a game model of the network described in Section 15.1 (see Figure 15.1) would have no NE. However, if the radios are permitted to mix their strategies (i.e., if a radio is permitted to randomly alternate between playing actions a_i and b_i), then even games with nominally finite action spaces will have an NE. The existence of an NE under mixed strategies is a result of Nash's fixed-point theorem.

NE Identification

As was the case for the dynamical systems approach, identifying the steady states of a general normal form or repeated game can be quite difficult. Because the only generally applicable approach is to perform an exhaustive search with repeated application of Eq. (15.12), NE identification for a game is an NP-complete problem [17]. When attempting to identify all of the NE in a game, some analysts are forced to turn to simulations—the very step we are intent on minimizing. For example, Ginde et al. [3] used an exhaustive simulation that ran for days to show that a GPRS network employing joint rate-power adaptations had four NE, even though the modeled system included only seven players. Fortunately, the potential game model presented in Section 15.5.1 provides further information that can be used to simplify the NE identification process.

Desirability

The most typically encountered criterion in the game theory literature that demonstrates that an NE is desirable is *Pareto optimality* [18–20]. Formally, an action vector, a^*, is said to be Pareto optimal if there exists no other action vector, $a \in A$ such that $u_i(a) \geq u_i(a^*)$ $\forall i \in N$ with at least one player strictly greater.

Unfortunately, Pareto optimality is a weak concept because a very large number of states may be Pareto optimal, and some Pareto optimal states are neither desirable nor fair, as Example 15.3 demonstrates. Accordingly, it is preferable to adopt the optimality approach of Section 15.3 wherein steady states are evaluated via some network objective function that is appropriate to the CR engineer's (and hopefully the user's) objectives such as Erlang capacity.[13]

EXAMPLE 15.3

SINR Maximizing Power Control

Neel et al. [17] consider a single-cluster direct spread spectrum (DS-SS) network with a centralized receiver in which all of the radios are running power-control algorithms in an attempt to maximize their signals' SINR at the receiver.

A normal form game, $\Gamma = \langle N, A, \{u_i\} \rangle$, for this network can be formed with the CRs as the players, the available power levels as the action sets, and the utility functions given by:

$$u_i(\mathbf{p}) = h_i p_i \Big/ \left((1/K) \sum_{k \in N \setminus i} h_k p_k + \sigma \right) \tag{15.13}$$

where K is the statistical cross-correlation of the signals.

As might be expected, the unique NE for this game is the power vector where all radios transmit at maximum power. This outcome can be verified to be Pareto optimal because any more equitable power allocation will reduce the utility of the radio closest to the receiver, and any less equitable allocation will reduce the utility of the disadvantaged nodes.

However, this is not a network we would want to implement because:

1. This state greatly reduces capacity from its potential maximum due to near–far problems (unless our network is in the unlikely configuration of having all radios at the same radius from the receiver).
2. The resulting SINRs are unfairly distributed (the closest node will have a far superior SINR to the farthest node).
3. Battery life is greatly shortened.

Convergence

It makes little sense to speak of convergence of a normal form game because it is defined as having only a single iteration. Accordingly, convergence is more frequently discussed in the context of repeated games. However, the properties of repeated games are largely defined by their stage game, which we are assuming to be a normal form game. So, to analyze the convergence of repeated games, we must identify properties of normal form games that lead to convergent behavior when the normal form game is played repeatedly.

In particular, there are two normal form game properties for games with finite action spaces that can be used for establishing some conditions for convergence: (1) the *finite improvement path* (FIP) property and (2) the *weak finite improvement path* (weak FIP) property.

A normal form game is said to have the FIP property if every sequence of profitable unilateral deviations (called an *improvement path*) is finite. Implicit in this property are the results that every game with FIP has at least one NE, that there are no

---

[13] *Erlang capacity* refers to the call arrival rate that a network can accommodate without the network's quality-of-service (QoS) dropping below some threshold, frequently measured via an inverse relationship with the probability of a call being blocked.

improvement cycles, and that play converges to an NE (Neel [29]). Accordingly, any network that can be modeled as a repeated game with a stage game that has FIP will converge if the following two conditions are met: (1) the radios implement rational decision rules, and (2) the network has either random or round-robin decision timings.

A normal form game is said to have the weak FIP property if, from every action vector, there exists at least one FIP that leads to an NE. By this definition, every game that has the FIP property also has weak FIP. Weak FIP implies the existence of at least one NE, but permits the existence of improvement cycles.

There are two different scenarios for which play converges when the normal form game has weak FIP. First, if the decision rules are designed such that the adaptations always follow the FIP, then play converges. Although this may seem like a difficult condition to achieve, supermodular games, which are discussed in Section 15.5.2, have a well-defined algorithm for achieving this result.

Second, if we consider decision rules by which adaptations are chosen randomly from the set of possible actions that improve performance, where all elements in the set have nonzero probability of being selected, such as would be the case for a genetic algorithm, then a repeated game with weak FIP converges to an NE for any decision update timing. The convergence of this process can be ensured by noting that the process can be modeled as an absorbing Markov chain, where the game's NE forms the absorbing states for the game.

An example of a normal form game with FIP is shown in the CR's dilemma of Example 15.1. An example of a game with weak FIP but not FIP is shown in Figure 15.11. Here an improvement cycle [(a, A), (a, B), (b, B), (b, A), (a, A)] is labeled with the circular arrow, and an NE is circled at (c, C). However, starting from any action vector, it is possible to find an FIP that terminates in the NE. An example of a normal form game that has neither FIP nor weak FIP is given by Exercise 15.7 in Section 15.7.

For an arbitrary normal form game, there is no generalizable technique for establishing FIP or weak FIP outside of an exhaustive search of a game's improvement paths. Due to this difficulty, many authors have considered convergence separately from the game theoretic analysis [20] or through an exhaustive simulation [3].

Fortunately, more powerful game models exist that not only permit establishment of FIP and weak FIP, but also perform NE identification and establish stability criteria. Two such game models—potential and supermodular—are discussed in Sections 15.5.1 and 15.5.2, respectively.

Γ	A	B	C
a	(1, −1)	(−1, 1)	(0, 2)
b	(−1, 1)	(1, −1)	(1, 2)
c	(2, 0)	(2, 1)	(2, 2)

FIGURE 15.11

A game with weak FIP but not FIP. The game has an improvement cycle (shown by the circular arrow) and an NE (*circled*). (*Source:* Reproduced from Neel et al. [17], Figure 2.)

15.5 RELEVANT GAME MODELS

This section presents two readily identified normal form game models—*potential games* and *supermodular games*—that will enable us to immediately establish the following results for decision update rules:

- Existence and sometimes identification of a Nash equilibrium/steady state
- Convergence conditions
- Stability conditions

15.5.1 Potential Games

As formalized by Monderer and Shapley [21], a potential game is a normal form game that has the property that there exists a function known as the *potential function*, $V{:}A \to \Re$, that reflects the change in value accrued by a unilaterally deviating player. In many ways, the concept of a potential function is identical to that of a Lyapunov function—a topic discussed in the Convergence and Stability subsection (p. 494) and again in the Stability subsection (p. 516).

Given an arbitrary unilateral deviation by player j from a_j to b_j (the actions of the other players (a_{-j}) remain fixed), several different types of potential games can be defined by the relationship between the value accrued by the unilateral deviation, $\Delta u_i(a,b_i) = u_i(b_i,a_{-i}) - u_i(a_i,a_{-i})$, and the change in value of the potential function, $\Delta V(a,b_i) = V(b_i,a_{-i}) - V(a_i,a_{-i})$:

- If there exists a function, V, such that $\Delta u_i(a,b_i)$ and $\Delta V(a,b_i)$ are *exactly* equal for all unilateral deviations, then the game is said to be an *exact potential game*.
- And if V preserves the *ordinal* relations of u_i for all unilateral deviations and for all $i \in N$, that is, $u_i(b_i,a_{-i}) > u_i(a_i,a_{-i}) \Leftrightarrow u_i(b_i,a_{-i}) > u_i(a_i,a_{-i}) \Leftrightarrow V(b_i,a_{-i}) > V(a_i,a_{-i})$, then the game is said to be an *ordinal potential game*. Other forms of potential games are defined and compared in Chapter 5 of Neel [29] but not used in this chapter.

These relationships are summarized later in Table 15.2. For purposes of shorthand notation, we refer to the potential function of an exact potential game as an *exact potential function* and the potential function of a weighted potential game as a *weighted potential function* and so on.

By these definitions, every exact potential game is an ordinal potential game such that the set of possible exact potential games is a subset of the set of ordinal potential games. As all exact potential games are ordinal potential games, this implies that exact potential games preserve the properties of its larger parent set.

EXAMPLE 15.4

Potential Game Examples

To illustrate these games in a more concrete manner, we present examples of exact and ordinal potential games taken from Chapter 5 of Neel [29]. Throughout this example, we make use of a two-player normal form game, Γ, with action sets {a, b} and {A, B} portrayed in matrix representation. Each game is accompanied by an associated potential function, also presented in matrix representation.

Γ	A	B
a	(3, 3)	(0, 5)
b	(5, 0)	(1, 1)

V(·)	A	B
a	0	2
b	2	3

FIGURE 15.12

An exact potential game and its associated exact potential function.

Γ	A	B
a	(3, 3)	(0, 5)
b	(7, 0)	(2, 1)

V(·)	A	B
a	0	2
b	2	3

FIGURE 15.13

An ordinal potential game and its associated ordinal potential function.

Figure 15.12 depicts an exact potential game and its associated exact potential. Note that $u_1(b,A) - u_1(a,A) = V(b,A) - V(a,A) = 2$, and that $u_2(b,A) - u_2(b,B) = V(b,A) - V(b,B) = 1$. Similar relationships hold for the other two possible unilateral deviations.

Figure 15.13 shows an ordinal potential game and its associated ordinal potential function. Consider the normal form game, Γ, shown in the figure's matrix representation. This game is an ordinal potential game with a potential function given by V. Note that this potential function is neither a weighted nor an exact potential function.

Identification of Exact and Ordinal Potential Games

At this point, potential games may appear as difficult to identify as Lyapunov functions. Fortunately, some techniques have been developed for identifying potential games.

Exact Potential Games

First, Monderer and Shapley [21] state that when all $u_k \in \{u_i\}$ are everywhere differentiable and V is an exact potential function, Eq. (15.14) must hold $\forall i, j \in N, \forall a \in A$. They also state that when all $u_k \in \{u_i\}$ are everywhere twice differentiable, Eq. (15.15) is a sufficient condition for the existence of an exact potential function:

$$\frac{\partial^2 u_i(a)}{\partial a_i \partial a_j} = \frac{\partial^2 u_j(a)}{\partial a_j \partial a_i} = \frac{\partial^2 V(a)}{\partial a_i \partial a_j} \tag{15.14}$$

$$\frac{\partial^2 u_i(a)}{\partial a_i \partial a_j} = \frac{\partial^2 u_j(a)}{\partial a_j \partial a_i} \tag{15.15}$$

When these equations are satisfied (Monderer), they give Eq. (15.16) for finding the potential function:

$$V(a) = \sum_{i \in N} \int_0^1 \frac{\partial u_i}{\partial a}(x(t)) x_i'(t) \, dt \tag{15.16}$$

Table 15.2 Common Exact Potential Game Forms

Utility Function Form	Potential Function	Game
$u_i(a) = C(a)$	$V(a) = C(a)$	Coordination Game
$u_i(a) = D_i(a_{-i})$	$V(a) = c,\ c \in \Re$	Dummy Game
$u_i(a) = C(a) + D_i(a_{-i})$	$V(a) = C(a)$	Coordination-Dummy Game
$u_i(a) = S_i(a_i)$	$V(a) = \sum_{i \in N} S_i(a_i)$	Self-Motivated Game
$u_i(a) = \sum_{j \in N \setminus \{i\}} w_{ij}(a_i, a_j) - S_i(a_i)$ where $w_{ij}(a_i,a_j) = w_{ji}(a_j,a_i)$	$V(a) = \sum_{i \in N} \sum_{j=1}^{i-1} w_{ij}(a_i, a_j) - \sum_{i \in N} S_i(a_i)$	Bilateral Symmetric Interaction (BSI) Game
$u_i(a) = \sum_{\{S \in 2^N; j \in S\}} w_{S,j}(a_S) + D_i(a_{-i})$ where $w_{s,i}(a_s) = w_{s,j}(a_s) \forall i,j \in S$	$V(a) = \sum_{S \in 2^N} w_S(a_S)$	Multilateral Symmetric Interaction (MSI) Game

Source: Adapted from table in Chapter 5 of Neel [29].

where x is a piecewise continuously differentiable path that connects some fixed-action tuple b to some other action tuple a such that $x:[0,1] \to A$ ($x(0) = b$, $x(1) = a$).

Even when the utility functions are twice differentiable, the evaluation of Eq. (15.16) can be quite tedious. Instead, the solution of a potential function is more readily accomplished by demonstrating that the game satisfies the conditions of one of a handful of common exact potential game forms and then applying its associated equation to find its exact potential function. Chapter 5 of Neel [29] provides a listing of types of exact potential games and their associated exact potential functions (Table 15.2). For our purposes, when the goals or utility functions of all radios in the network take the form shown in the first column of Table 15.2, then the network has an exact potential function given by the corresponding entry in the second column and is said to be the type of game listed in the third column.

When reading this table, note that a function that is not subscripted (e.g., $C(a)$) is an arbitrary function that all players use as part of their utility function. When a function is subscripted with i (e.g., $D_i(a_{-i})$), then each player has its own arbitrary function using the same kind of argument. Specifically, when $D_i(a_{-i})$ is present, each player has its own "dummy" function with a valuation that is only a function of the actions of the other players. For an entry that includes $S_i(a_i)$, the player has a function with a valuation that is only a function of its own action. These sort of utility function components can be encountered when interference ($D_i(a_{-i})$) can be separated from the signal component ($S_i(a_i)$) of a player's goal (e.g., an SINR goal measured in dB). When a function is subscripted with multiple players (e.g., ij or S), then all players listed receive the same payoff. For example, assuming that the players' goals are functions of the correlations between waveforms (as occurs when measuring interference), then given choices of waveforms a_i and a_j, both players would measure the same correlation between a_i and a_j, w_{ij}.

EXAMPLE 15.5

A BSI Interference Avoidance Game

Consider a network with a frequency reuse scheme such that cross-cluster interference is negligible. Each cluster is power controlled so that received power at the cluster head for all radios is constant. However, each radio that is communicating with the cluster head is also attempting to minimize the interference its signal experiences at the receiver by adapting its waveform.

We can consider this as a repeated game with the normal form stage game modeled as follows. Each adaptive radio in a cluster is a player, the actions for each radio are its available waveforms, and utility functions are given as Eq. (15.17), where $\rho(a_j, a_k)$ is the statistical correlation between waveforms a_j and a_k with the assumption that $\rho(a_j, a_k) = \rho(a_k, a_j)$:

$$u_j(a) = -\sum_{k \in N \setminus j} \rho(a_j, a_k) \tag{15.17}$$

From Table 15.2, we can see that Eq. (15.17) satisfies the conditions for a bilateral symmetric interaction (BSI) game, where $S_j(a_j) = 0 \; \forall j \in N$. The table indicates that Eq. (15.18) is an exact potential function for this game:

$$V(a) = \sum_{i \in N} \sum_{j=1}^{i-1} \rho(a_i, a_j) \tag{15.18}$$

Ordinal Potential Games

Monderer and Shapley [21] state that if a game has FIP and if $\forall a_{-i} \in A_{-i}$ and all $i \in N$, $u_i(a_i, a_{-i}) \neq u_i(b_i, a_{-i}) \forall a_i, b_i \in A_i$, then the game is an ordinal potential game. A simpler, though less systematic approach is to apply the concept of *better response equivalence*. A game $\Gamma = \langle N, A, \{u_i\} \rangle$ is said to be *better response equivalent* to game $\Gamma' = \langle N, A, \{v_i\} \rangle$ if $\forall i \in N$, $\forall a \in A$, $u_i(a_i, a_{-i}) \geq u_i(b_i, a_{-i}) \Leftrightarrow v_i(a_i, a_{-i}) \geq v_i(b_i, a_{-i})$. For notational simplicity, if Γ is better response equivalent to Γ' we write $\Gamma \sim \Gamma'$.

An example of better response equivalence is given in the games presented in Example 15.4. These games also imply a few important properties that are preserved by better response equivalence, namely NE, improvement path properties, and ordinal potential functions.

These preservation properties imply a different approach to identifying ordinal potential games—identifying a *better response transformation* (a reformulation of a game's utility functions in a way that the resulting game is a better response equivalent to the original game) that yields an identifiable exact potential game. Examples of better response transformations include scalar multiplication and logarithmic transformations.

EXAMPLE 15.6

An Ordinal Potential Interference Avoidance Game

Suppose we modify the network of Example 15.5 so that instead of minimizing interference, the radios adapt their choice of operating frequencies to maximize their throughput. In this context, throughput is a monotonic function of interference in that any decrease in interference results in an increase in throughput. Accordingly, these two games are a better response equivalent and we can conclude that this modified game is an ordinal potential game.

Fixed Points and Steady States for Potential Games

It is relatively easy to show that every potential game where a maximum value of V exists (e.g., games with a finite action space and games where V is bounded and A is compact) has an NE. Specifically, if a^* is a global maximizer of V, then there can be no a' such that $u_j(a') > u_j(a^*)$ where a' and a^* differ only in the jth component. Otherwise, $V(a') > V(a^*)$ and a^* is not a global maximizer of V.

Thus, by showing that each iteration of a CR algorithm can be modeled by the same potential game and that the associated potential function has a maximum, we also show that a steady state exists for the algorithm. Further, the preceding discussion also shows that we can identify the steady states for the algorithm by solving for the global maximizers of V. Although other NEs may exist for a potential game, as we show in the Stability subsection (p. 516), only those NEs that are maximizers of V are stable.

Desirability

In general, little can be said about the optimality, or desirability, of the steady states of a cognitive potential game. They need not be Pareto efficient, and they are not generally maximizers of a design objective function. However, when the potential function is also the network objective function, that is, when $V = J$, then if V admits a global maximum, there exists an NE that is optimal. Further, because deterministic play increases the value of V with each iteration, it is safe to say that the stable steady state of the network will give better performance than the initial state of the network.

As an example, consider the adaptive interference avoidance game of Example 15.5. This game necessarily has numerous steady states, but if the network's designer is attempting to minimize total network interference, then the steady states will generally be desirable and performance will improve with each adaptation.

Convergence

Up to this point we have considered only the goals of the radios. Let us now consider the set of possible decision rules, f^t, that could be implemented to achieve convergence. Specifically, let us consider rational adaptations in which each radio, j, chooses to change its action from a_j to b_j if and only if $u_j(a_j,a_{-j}) > u_j(b_j,a_{-j})$. Further, let us restrict ourselves to those algorithms with either round-robin or random decision timings. With these assumptions, the sequence $\{V(a(t))\}$ where $a(t^{k+1}) = f^{t^k}(a(t^k))$ is monotonically increasing.

If we consider the situation in which the potential functions are bounded, both $\{V(a(t))\}$ and the sequence $\{a(t)\}$ must converge. For finite games implementing random or round-robin self-interested algorithms, $\{a^t\}$ converges to an NE. However, this need not be the case for games with infinite action spaces because infinitesimally small steps could be taken to cause convergence to an action vector other than an NE.

However, for games that are generalized ε-potential games (recall that all exact potential games are generalized ε-potential games; Neel [29]), we can introduce a different class of algorithms that can be shown to converge, called *ε-self-interested updates*. For an ε-self-interested update, each radio, j, chooses to change its action from a_j to b_j if and only if $u_j(a_j,a_{-j}) + \varepsilon < u_j(b_j,a_{-j})$. For a generalized ε-potential game with a bounded potential function, this implies that there is an $\varepsilon_V > 0$ such that for every round-robin or random ε-self-interested adaptation, V increases by at least an ε_V. In this case, the

Γ	A	B
a	(0, 0)	(2, 2)
b	(2, 2)	(0, 0)

FIGURE 15.14

An exact potential game that could oscillate between (a, A) and (b, B) for synchronous decision timings.

adaptations are guaranteed to converge to an ε-NE. An action vector a^* is said to be an ε-NE if $\forall j \in N$ when there is no $b_j \in A_j$ such that $u_j(a_j^*, a_{-j}^*) + \varepsilon < u_j(b_j, a_{-j}^*)$.

For synchronous algorithms, note that little can be said about the convergence of a traditional potential game. For instance, consider the exact potential game shown in Figure 15.14. Starting from (a, A), it is possible to enter into the oscillation $(a, A) \rightarrow (b, B)$ for a deterministic self-interested algorithm with synchronous timing. However, it is readily observed that such an oscillation is broken when random decision timing is assumed.

Convergence Rate

An infinite generalized ε-potential game with a potential function bound by $|V| \leq K$ and step size of ε_V with round-robin decision timing cannot have more than $2|N|K/\varepsilon_V$ iterations. This particular bound on the number of iterations comes from an assumption of an initial action vector such that $V(a(0)) = -K$ and the requirement that for every $|N|$ iterations at least one player must be able to improve its payoff by at least an ε (otherwise an ε-NE has been reached).

Stability

As pointed out earlier in this section, the definition of a potential function is similar to that of a Lyapunov function. As shown in Neel [29], if a CR network has any rational decision rule with random or round-robin timing with iterations that can be modeled as a potential game with potential function V, then it has a Lyapunov function related to the potential function, such that all maximizers of V (our method for finding NE in a potential game) are also Lyapunov stable. Also note that if the game has a unique NE, then a^* is globally Lyapunov stable. And if the decision rule fits into one of the classes of algorithms that deterministically converge, as discussed in the Convergence subsection, then the algorithm is asymptotically stable as well.

So for traditional analysis techniques, Lyapunov stability has to be considered for a specific decision rule, but for potential games, we can show that an entire class of decision rules is stable by examining the goals of the radios (the radios are the players in a game model of a CR network).

Designing Potential Game Networks

In general, the actions, observations, decision rules, utility functions, and operating context are all highly interdependent. But if we start from the premise that our utility functions will satisfy the conditions of a potential game, we can significantly relax the constraints on our selection of decision rules. However, the assurance of convergence and stability is not accompanied by an assurance of optimality, so when adopting this approach, care must still be given to the design of the observation processes which can depend on the choice of objective functions if we do not assume the existence of a common knowledge database such as the REM (see Chapter 11).

To address this issue, Chapter 6 of Neel [29] introduces the concept of the *Interference Reducing Network* (IRN)—a CR network with utility functions designed such that the network satisfies the conditions of an exact potential game where the potential function is a negative scalar multiple of the sum of interference terms observed by the radios in the network. In such a setting, every selfish adaptation increases the potential function, which decreases the sum of observed interference—thus providing the implied convergence and stability of a potential game with the added benefit that actions that correspond to interference minimizers will create an NE.

Chapter 6 of Neel [29] introduces various different methods by which an IRN can be achieved. Some of these require significant collaboration or networkwide omniscience. However, Neel also shows that an IRN can be realized by modifying the radios' observation processes to induce a symmetry condition called *Bilateral Symmetric Interference* (BSI) between the interference measurements used by all pairs of spectrum management processes. Consciously modeled on BSI games (see Table 15.2), BSI is satisfied when Eq. (15.19) holds for all possible pairwise subnets formed by only considering decision processes j and k (and their controlled radios):

$$I_j(\omega_j, \omega_k) = I_k(\omega_k, \omega_j) \forall (\omega_j, \omega_k) \in \Omega_j \times \Omega_k \qquad (15.19)$$

where Ω_j is the set of waveforms that j can implement as defined by policy or by device capabilities and $I_j(\omega_j, \omega_k)$ is the interference observed by j when j and k implement waveforms ω_j and ω_k, respectively.

This condition is accompanied by a requirement that each decision process is attempting to minimize its own observed interference as shown by:

$$u_j(\omega) = -I_j(\omega) = -\sum_{j \in N \setminus j} g_{kj} p_k \rho(\omega_j, \omega_k) \qquad (15.20)$$

where p_k is the transmission power of radio k, g_{kj} is the link gain (pathloss) from k to where j's interference observation is made (e.g., a client device in the introductory example), and $\rho(\omega_j, \omega_k)$ is the absolute value of the correlation between the basis functions of waveforms ω_j and ω_k. In such a context, (Eq. 15.19) is equivalent to satisfying:

$$g_{kj} p_k \rho(\omega_k, \omega_j) = g_{jk} p_j \rho(\omega_j, \omega_k) \forall j, k \in N, \forall (\omega_j, \omega_k) \in \Omega_j \times \Omega_k \qquad (15.21)$$

That this induces a potential function can be readily verified by examining Table 15.2 and substituting $w_{jk}(a_j, a_k) = -I_j(\omega)$ and $S_i(a) = 0$. Table 15.2 gives the resulting potential function as:

$$V(\omega) = \sum_{i \in N} \sum_{k=1}^{j-1} I_{jk}(\omega_j, \omega_k)$$

which is exactly one-half of the negation of all interference observations, or:

$$V(\omega) = -\sum_{j \in N} \sum_{k \in N} I_{jk}(\omega_j, \omega_k)/2$$

Rather simply, BSI would hold in a network consisting of two CRs, j and k, each transmitting at the same power level, p, with a collection of mutually orthogonal waveforms (i.e., $\rho(\omega_j, \omega_k) = 1$ if $\omega_j = \omega_k$ and $\rho(\omega_j, \omega_k) = 0$ otherwise). This can be readily verified as channel reciprocity implies that the link gain from j to k is the link gain from k to j so that $I_j(\omega) = g_{kj} \, p_k \, \rho(\omega_j, \omega_k) = I_k(\omega)$. Neel and Reed [30] and Neel [31] introduce several more practical examples where waveform knowledge is leveraged to design cognitive observation processes to induce the BSI condition for infrastructure and ad hoc networks.

Because these networks satisfy the BSI condition, relatively unsophisticated distributed selfish decision rules will converge to stable optimal radio resource allocations. In effect, BSI enables CR networks to achieve the performance of a centralized algorithm with the simplicity of a distributed, selfish algorithm.

EXAMPLE 15.7

A BSI DFS Algorithm for 802.11 Networks in Infrastructure Mode

Neel [31] presents the embedded DFS algorithm for fixed 802.11 networks, illustrated in Figure 15.15. In this system, each AN chooses an operating channel to minimize its observed interference. Critically, interference levels observed by the clients are not included in the interference observations and each $I_j(\omega)$ is instead based only on the ANs' observations of the receive signal power levels of request to send (RTS) and clear to send (CTS) signals broadcast by other ANs.

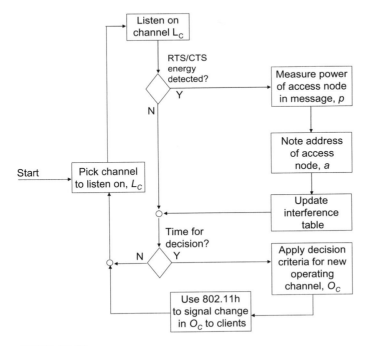

FIGURE 15.15

A simple noncooperative algorithm that achieves optimal frequency reuse patterns. (*Source:* Reproduced from Neel [31], Figure 2.)

This combination of conditions has the following effects on the components of $I_j(\omega)$:

- $p_k = p_j \ \forall \ j, \ k \in N$, because all RTS/CTS messages are generally broadcast at the same power level
- $\rho(\omega_j, \ \omega_k) = \rho(\omega_k, \ \omega_j)$, because as for channel selection, ρ assumes values of 1 ($\omega_k = \omega_j$) and 0 ($\omega_k \neq \omega_j$) for orthogonal channels and is also symmetric for nonorthogonal channels [8]
- $g_{kj} = g_{jk}$, because for any given frequency in a fixed network, the link gain from device j to device k is the link gain from device k to device j

So by careful construction of an observation process we achieve a network where BSI holds. This then assures the convergence of all selfish decision processes guided by $I_j(\omega)$ to a channel allocation that minimizes sum network interference.

The behavior of such a network is illustrated in a simulation of 30 ANs randomly distributed over 1 km^2 operating in an environment with a pathloss exponent of 3, random device placements, and randomly assigned initial channels. The radios are constrained to operate in the 11-channel 5.47 to 5.725 GHz European upper UNII band (channels 100–140) and to transmit at a common 1 W and assumed to have noise floors of –90 dBm. Figure 15.16 depicts the transient behavior

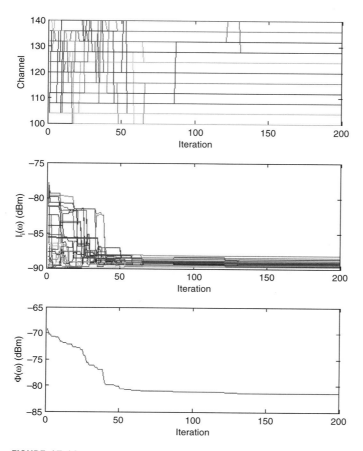

FIGURE 15.16

Transient behavior for the network-implementing algorithm shown in Figure 15.15. (*Source:* Reproduced from figure in Neel [31].)

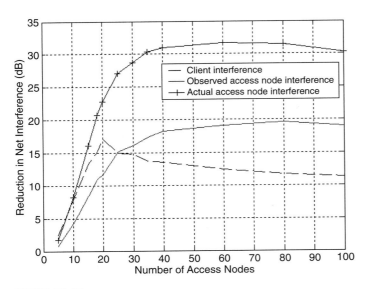

FIGURE 15.17

Average reduction in interference levels. (*Source:* Reproduced from figure in Neel [31].)

of the network with the operational channels for each AN (*top*), perceived interference levels by the ANs (*middle*), and the sum of observed interference levels (*bottom*) for the simulated network. Note that sum-observed interference is a monotonically decreasing function as predicted by potential game theory.

While the adaptations intend only to minimize AN-to-AN interference, the performance of client devices also dramatically improves with this algorithm. Figure 15.17 depicts the average reduction in interference levels seen by the clients and access nodes and the average aggregate reduction in AN-to-AN interference (−2V) when twice as many clients are present as ANs and 100 trials performed per number of ANs in the sweep. The ANs were permitted to operate over all 18 channels in the European UNII band and controlled the nearest client devices.

For each trial, all positions and all initial channel assignments were randomly assigned. The algorithm yields a greater reduction in interference for the client devices than for the ANs' observed interference for low-density deployments with the situation reversed for high-density deployments. In all cases, the reduction in the ANs' actual interference (~30 dB for high density) is more than the clients' reduction (~12 dB for high density) and more than the reduction in AN-to-AN interference (~19 dB for high density). In general, similar performance improvements would be expected in a centrally planned network, but now we get the benefit at runtime (postdeployment) without using any bandwidth to coordinate decisions or transmit information between ANs.

15.5.2 Supermodular Games

Supermodular games are encountered frequently in algorithms in which an increase in a_{-i} results in a corresponding increase in a_i—a concept known as *increasing differences*. This section presents a method for determining if a cognitive radio network satisfies the modeling condition of a *smooth supermodular game* and describes what analytical insights can be gained by demonstrating that the iterations of a CR algorithm can be modeled as a smooth supermodular game. A more general discussion of supermodular games is presented in Neel [29].

EXAMPLE 15.8

Supermodular Game Example

Let us define a supermodular game with the following components: player set $N = \{1, 2, \ldots, n\}$, action sets given by $A_i = [0,1] \forall i \in N$, and utilities given by $u_i(a) = \sum_{j \in N} a_i a_j$:

$$u_i(a) = \sum_{j \in N} a_i a_j$$

This game can be shown to be a smooth supermodular game. Condition (1) is satisfied as A_i is a compact subset of $\Re(k_i = 1 \forall i \in N)$. Condition (2) holds as u_i is twice differentiable. Condition (3) does not apply as the action sets are single dimensions. Condition (4) can be verified as $\partial^2 u_i(a)/\partial a_i \partial a_{-i} = 1$. Note that this game is also a BSI game and thus an exact potential game.

Properties of Supermodular Games

This subsection describes some of the properties of supermodular games that are relevant to the analysis of CR networks. In particular, supermodular games are useful for CR that implement decision rules that perform local optimizations, or in game theory parlance, best response decision rules.

Given a_{-j}, player j's *best response* to a_{-j} is given by the set of action tuples for which no higher utility can be found. This set can be formally specified as:

$$BestResponse = \{b_j \in A_j : u_j(b_j, a_{-j}) \geq u_j(a_j, a_{-j}) \forall a_j \in A_j\} \qquad \textbf{(15.22)}$$

We can define a best response function for each player $j \in N$, $\hat{B}_j(a)$ where $\hat{B}_j(a)$ returns an action that satisfies Eq. (15.22). We can also define a joint best response to a as $\hat{B}(a)$, where $\hat{B}(a)$ returns an action vector formed by the synchronous application of $\hat{B}_j(a) \forall j \in N$.

Fixed Points in Supermodular Games

Before considering NE and fixed points in supermodular games, we need to introduce Tarski's fixed-point theorem:

> Let A be a nonempty compact sublattice of $\Re^n$. Let $f: A \to A$ be an increasing function. Then f has a fixed point.

Unlike the previous fixed-point theorems considered in this chapter, there is no requirement that f be continuous, nor is there a requirement that A be a convex set. The only requirement is that A is a lattice and f is an increasing function.

Now consider a CR network with iterations that can be modeled as a repeated smooth supermodular game. Then assume that the decision rule is synchronous and of the form $f^t(a) = \hat{B}(a)$. By property (6) of Lemma 4.2.2 of Topkis [23], $f^t(a)$ is then an increasing function, and by virtue of being smooth and by Tarski's fixed-point theorem, $f^t(a)$ must have at least one fixed point (NE).

For supermodular games, identifying the NE is not as easy as it is for potential games. However, by leveraging property (2) of Lemma 4.2.2 of Topkis [23], we see that the set of NE must form a complete lattice. Thus, once we identify a pair of NE (a^*, a^{**}) we know that their joins and meets must also be an NE.

Desirability

Little can be said about the desirability of the fixed points in a supermodular game in general. As is the case for many systems, the fixed points of the game should be evaluated via some networkwide cost function to determine desirability.

Convergence

Consider a synchronous locally optimal decision rule, $f^t(a) = \hat{B}(a)$, a round-robin locally optimal decision rule, $f^{t_k}(a) = \hat{B}_{\mod(k,|N|)}(a)$, or a randomly timed locally optimal decision rule, $f^{t_k}(a) = \hat{B}_{rand(k)}(a)$. Because $\hat{B}(a)$ and $\hat{B}_j(a)$ are increasing functions, for all $a^0 \leq \inf(\{a \in A : a \in \hat{B}(a)\})$, the sequence formed by f^t is nondecreasing. Similarly, for all $a^0 \geq \sup(\{a \in A : a \in \hat{B}(a)\})$, the sequence formed by f^t is nonincreasing. As these two sequences squeeze together, the recursive calculation of f^t must converge to a region bounded by the greatest and least NE. Of course, if the NE lattice has only one unique element, then these adaptations must converge to this unique element.

Theorem 4.3.1 in Topkis [23] states that when the round-robin best response algorithm is modified so that the least element in $\hat{B}(a)$ is selected at each stage and the algorithm is started at the least element in A, then the number of iterations this algorithm will take to reach an NE (in particular, the least NE in the NE lattice) is bounded by:

$$(|N| - 1)\left(\sum_{j \in N} |A_i|\right) - |N|^2 + |N| + 1 \qquad (15.23)$$

Adaptive Dynamic Process

A variation on the simultaneous best response algorithm is presented by Milgrom and Roberts [24], wherein the players follow what is termed an *adaptive dynamic process*. In an adaptive dynamic process, all players play a best response to some arbitrary weighting of actions played by other players in the recent past (i.e., not just the most recently observed actions).

Formally, a process is defined [24, (A6)] as an adaptive dynamic process, $\forall T \exists T'$, such that $\forall t \geq T'$, $a^t \in \bar{U}([\inf(P(T, t)), \sup(P(T, t))])$, where $P(T, t)$ denotes the actions played between times T and t, $U(a)$ is the list of undominated responses to a for each player, and $\bar{U}(a) = [\inf(U(a)), \sup(U(a))]$.

The corollaries to Theorem 8 in Milgrom and Roberts [24] show that a smooth supermodular game following an adaptive dynamic process converges to a region bounded by the NE lattice and that iterative elimination of dominated strategies converges to a region defined by the NE lattice. Note that when the NE is unique, the adaptive dynamic process converges to NE.

Random Sampling

For supermodular games, there exists an improvement path that terminates in an NE starting from every action tuple. Accordingly, all smooth supermodular games have the weak FIP property described in the Convergence subsection (p. 509).

Friedman and Mezzetti [25] present a modified better response dynamic, here called a *random better response dynamic*, that assumes the normal form game has weak FIP. In the random better response dynamic, the player that adapts its behavior is chosen at random, and the choice of adaptation by that player is chosen randomly and implemented if the adaptation would improve performance. Specifically, when deciding on its action, player i randomly samples action a_i' from $A_i \setminus \{a_i\}$, where all actions in $A_i \setminus \{a_i\}$ have equal probability of being chosen. If $u_i(a_i', a_{-j}) > u_i(a_i, a_{-j})$, then player i will implement a_i'. Friedman and Mezzetti show in that any finite supermodular game, (finite A) converges according to random better response dynamics. Based on the discussion in this chapter, we can see that such a process on a game with weak FIP forms an absorbing chain and thus must end in an absorbing state, in this case, an NE.

In the context of cognitive radios, a CR network following random sampling would permit the radios to "try out" a particular waveform and then continue using that waveform if it improved its performance. In effect, algorithms that converge via random sampling are algorithms that permit CRs to learn about their operating environment, specifically learning how the operating environment impacts performance even when they start with no knowledge of the environment.

Because this result requires only that the stage game has weak FIP, all potential games with finite action spaces (which have FIP and thus also weak FIP) also converge under random sampling.

Stability

For any finite best response convergent normal form game following any one of the best response algorithms discussed in the preceding subsection, a Lyapunov function exists and is given by:

$$V(a) = |\sigma(a)| + \sum_{a' \in \sigma(a)} V(a') \qquad (15.24)$$

where $\sigma(a) = \{a' \in A : a \in \hat{B}(a')\}$.

For supermodular games with infinite action sets (for which the concept of stability is better defined), additional information must be introduced to determine stability, perhaps by leveraging whatever information can be gleaned from traditional analysis techniques. An example analysis for which additional information can be used to determine the stability of a supermodular game is given in the Analysis subsection (p. 524).

EXAMPLE 15.9

Distributed Power Control

Neel et al. [26] present an analysis of a distributed power control algorithm on an ad hoc network in which each link, j, varies its transmit power in an attempt to achieve a target SINR, γ_j, measured in dB at the receiving end of the link. This scenario can be thought of as analogous to the fixed assignment scenario presented by Yates [10] and presented in the Standard Interference Function Model subsection (p. 497). Indeed, this analysis can be considered an extension of that scenario to ad hoc networks with additional consideration given to stability.

Using the notation presented in the Standard Interference Function Model subsection, in a network, N, of CRs the SINR of the signal transmitted by j and received by its node of interest measured in dB is given by:

$$\gamma_j = 10\log_{10}(g_{jj}p_j) - 10\log_{10}\left(\sum_{k \in N\setminus j} g_{kj}p_k + N_j\right)(\text{dB}) \qquad (15.25)$$

where g_{kj} is the effective fraction of power transmitted by node k that is received at j's node of interest (receiving end of j's link) and N_j is the noise at the receiving end of link j.

Stage Game Model

Based on the preceding discussion, a normal form stage game can be formulated as follows:

- *Player Set, N*—Set of decision-making links
- *Player Action Set, A_j*—The real convex, compact set of powers, $[0, p_j^{\max}]$, where $p_j^{\max}$ is the maximum transmit power of CR j. The action space, $A \subset \mathfrak{R}^n$, is given by $A = A_1 \times A_2 \times \ldots \times A_n$.
- *Utility*—An appropriate utility function for a target SINR (dB) algorithm given by:

$$u_j(p) = -\left(\hat{\gamma}_j - 10\log_{10}(g_{jj}p_j) + 10\log_{10}\left(\sum_{k \in N\setminus j} g_{kj}p_k + N_j\right)\right)^2 \qquad (15.26)$$

where $\hat{\gamma}_j$ is the SINR target of CR j.

Analysis

Altman and Altman [27] claim that the cellular fixed-assignment scenario of Yates [10], on which this ad hoc network model is based, is supermodular. The following analysis parallels that given by Neel et al. [26], which showed that this stage game constitutes a smooth supermodular game for an ad hoc network.

A stage game can be shown to be a smooth supermodular game by applying the second-order conditions presented in Section 15.5.2. First, notice that the action space forms a complete lattice because it is a compact interval of Euclidean space. Then, evaluating the second derivative with respect to p_j and p_k, where k is any CR $k \in N\setminus j$, yields:

$$\frac{\partial^2 u_j(p)}{\partial p_j \partial p_k} = \frac{200 g_{kj}}{p_j\left(\sum_{k \in N\setminus j} g_{kj}p_k + N_j\right)\ln(20)} \qquad (15.27)$$

Because Eq. (15.27) is strictly positive, the last condition for a smooth supermodular game is satisfied. Accordingly, we know the following about the network:

- It has at least one steady state.
- The network converges for synchronous and asynchronous best response algorithms (local optimization), even with adaptation errors to a region bound

by the greatest and least NE. Thus, convergence to a bounded region is assured even as the radios learn their operating environment.

Further, because the best response algorithm given by:

$$p_j^{t_{k+1}} = p_j^{t_k} \frac{\overline{\gamma}_j}{\gamma_j} \tag{15.28}$$

is a known standard interference function, we know the following:

- The network adaptation process constitutes a pseudo-contraction and thus has a unique fixed point. Thus, the greatest and least NE are the same point and play must converge to a unique point.
- The network achieves the target SINR vector with the smallest possible power vector (when the SINR vector is feasible) implying the algorithm is optimal in terms of minimizing power consumption.
- A Lyapunov function is given by the distance between the current power vector and the fixed-point power vector.

Finally, for a feasible SINR target vector (see Standard Interference Function Model subsection, p. 497), the unique steady state for this game can be found by solving the linear program $\mathbf{Z}\vec{\mathbf{p}} = \vec{\mathbf{Y}}$ where:

$$\mathbf{Z} = \begin{bmatrix} b_{1v_1} & -\hat{\gamma}_1 b_{1v_2} & \cdots & -\hat{\gamma}_1 b_{1v_n} \\ -\hat{\gamma}_2 b_{2v_1} & b_{2v_2} & \cdots & -\hat{\gamma}_2 b_{2v_n} \\ \vdots & \vdots & \ddots & \vdots \\ -\hat{\gamma}_n b_{nv_1} & -\hat{\gamma}_n b_{nv_2} & \cdots & b_{nv_n} \end{bmatrix},$$

$$\mathbf{Y} = \begin{bmatrix} \hat{\gamma}_1 N_1 & \hat{\gamma}_2 N_n & \cdots & \hat{\gamma}_n N_n \end{bmatrix}^T$$

$$\mathbf{Y} = \begin{bmatrix} \hat{\gamma}_1 N_1 & \hat{\gamma}_2 N_n & \cdots & \hat{\gamma}_n N_n \end{bmatrix}^T$$

and

$$\mathbf{p} = \begin{bmatrix} p_1 & p_2 & \cdots & p_n \end{bmatrix}^T$$

$$\mathbf{p} = \begin{bmatrix} p_1 & p_2 & \cdots & p_n \end{bmatrix}^T$$

where the b variables as are in the Standard Interference Function Model subsection.

Validation

Consider the ad hoc network shown in Figure 15.18, which is operating at a single frequency, where each terminal is attempting to maintain a target SINR at a cluster head and where each cluster head is maintaining a target SINR at the gateway node. The signals employed by the radios have a statistical spreading factor of K. Here, $b_{jk}=g_{jk}/K$ for $j \neq k$ and $b_{jj} = g_{jj}$.

Assuming these devices implement decision rules that are locally optimal, the network is implementing a decision rule that we know converges to the unique steady state. Accordingly, we would expect that any initial power vector would converge to the unique power vector, and that even when corrupted by noise, the system would remain in a region near this steady state as the analysis predicts Lyapunov stability.

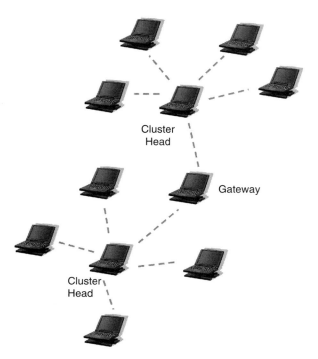

FIGURE 15.18

Simulation scenario for an ad hoc power control example.

To confirm these predictions, a synchronous simulation was constructed for deterministic and stochastic simulation scenarios. The simulation results for these scenarios are shown in Figures 15.19 and 15.20, respectively, where the upper graphs (a) plot objective function values as a function of iteration, and the lower graphs (b) plot power level as a function of iteration. Note that the locally optimal algorithm rapidly converges to the steady state in both of the scenarios and that, even in the presence of random noise-induced perturbations, the network remains in a region that is around the deterministic steady state.

15.5.3 The Value of Game Theory to CR Networks

The combination of traditional and game theoretic analysis and design techniques presented in this chapter permit us to address a wide array of CR networking problems. But what is game theory adding to our capabilities? After all, the supermodular power control example was shown to also conform to the standard interference model, and while the DFS examples were designed using potential game techniques, it is not inconceivable that similar algorithms could have been derived without the benefit of game theory. The following highlights some key capabilities that will be critical to the success of cognitive radio that can only be provided via approaches grounded in game theory.

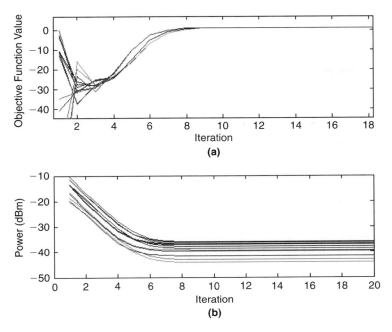

FIGURE 15.19

Deterministic simulation of an ad hoc network of CRs with synchronous adaptations and utility functions given by Eq. (15.26). The graph in (a) plots the value of each radio's utility function versus the iteration; the plot in (b) shows the power levels for each radio versus iteration.

Suitability for Imprecisely or Undefined Decision Rules

By far the greatest unique benefit of game theory is that game theory does not require closed-form expressions for decision rules to yield valuable analysis insights. Just by examining the radios' objectives or preferences, we can predict network steady states and make convergence and stability characterizations. Thus, networks of radios controlled by ontological reasoning engines (see Chapter 13) or genetic algorithms (see Chapter 7), for which it is not generally possible to express the network behavior in terms of an evolution function, can still be analyzed using game theory. For radios controlled by genetic algorithms, the network could be modeled and analyzed using Markov models. However, any useful transition matrix would have to be determined empirically—the very process we are seeking to avoid. So if we were limited to traditional engineering analysis techniques, modeling and analyzing of most CR network behavior would be impractical.

Thus, the techniques presented in this chapter permit us to analyze problems that we could not handle with traditional approaches.

Simplified Compatibility Analysis

If we solve for a fixed point of an evolution function, we have identified a steady state for a particular combination of decision rules. If a different combination of decision rules is deployed, then the analysis will need to be repeated.

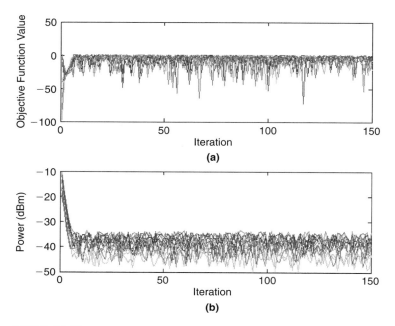

FIGURE 15.20

Simulation of an ad hoc network of CRs with synchronous adaptations, utility functions given by Eq. (15.26), and stochastic channel models. The graph in (a) plots the value of each radio's utility function versus the iteration; the plot in (b) shows the power levels for each radio versus iteration.

On its own, this does not seem like a significant burden. But consider the deployment of CRs in unlicensed bands—the location where CR (in the form of 802.11h) is already being deployed. One of the benefits of opening up unlicensed spectrum is that it permits the fielding of numerous different devices from various vendors, thus driving prices for consumers down. To differentiate their products, vendors typically employ different algorithms, which is permissible because wireless standards frequently do not specify radio resource management algorithms. For instance, while 802.22 specifies times within which a radio has to vacate a band when a primary user is detected, it does not currently specify the algorithm by which a new band is selected. Likewise, 802.11h mandates dynamic frequency selection (DFS) and transmit power control (TPC) and specifies messages to support these operations, but the algorithms by which frequency and power are adjusted are not specified.

Considering just 802.11h, in May 2008, the WiFi alliance listed 29 different vendors and 2332 different products for 802.11h. If the WiFi alliance, the Federal Communications Commission (FCC), or any radio designer wanted to ensure that the DFS algorithms in all WiFi-certified 802.11h products will not negatively interact (only radar avoidance is currently addressed), they would have some $2^{29} - 1$ combinations of decision rules to analyze if every vendor used their own algorithm.

Using game theoretic techniques, we only need to know the goals of the radios and the permissible actions, with the latter almost certainly defined as part of any standard, in order to determine the steady states using the NE concept. Thus, instead of performing $2^{29} - 1$ analyses for 802.11h, only a single analysis would need to be performed!

Simplified Spectrum Management of CRs

The preceding implies another advantage of a game theoretic approach to analysis—simplified spectrum management. If the FCC or some primary spectrum holder specifies a particular combination of goals and allowable actions as part of a licensing agreement, then device testing can be simplified to merely verifying that the radio's algorithms act to improve its performance according to one of the allowable goals.

Currently, spectrum policy focuses solely on a specification of permissible actions (e.g., a spectral mask), but it seems likely that an additional specification of permissible goals would be sufficient to ensure acceptable performance while still permitting vendors and secondary users to employ varying decision rules to differentiate their products.

In general, the actions, observations, decision rules, utility functions, and operating context are all highly interdependent when examining the performance of a CR network. But if we start from the premise that our utility functions will satisfy the conditions of a potential game, then we can significantly relax the constraints on our selection of decision rules. However, the assurance of convergence and stability is not accompanied by an assurance of optimality, so when adopting this approach care must still be given to the design of the observation processes; this can depend on the choice of objective functions if we do not assume the existence of a common knowledge database such as the REM.

While promising tremendous improvement in network and mission performance, the use of cognitive principles with any distribution of decision making will result in a complex systems-of-systems problem as the adaptations of a single cognitive engine influence the actions taken by other cognitive engines. For example, the choice of one radio to operate in a band makes that band less desirable for other radios due to the increased interference; increasing the power budget to improve the performance of one cluster reduces the performance of other clusters. The competing demands present in a network necessarily lead to resource competition, and these interactions make the system's equilibriums, convergence, and stability difficult to ascertain.

Within a closed centralized or collaborative system, where all devices are working toward the same goal and there are no interactions spawned by adaptations, any real-time evaluatable function can serve as the objective for the system without negatively impacting considerations (e.g., performance, convergence, or stability). However, the scale of the envisioned network implies that no centralized or collaborative system will be able to be closed as processing and communication time constraints will limit the applicability of a purely centralized approach.

15.6 SUMMARY

This chapter has presented several modeling approaches for describing the interactive decision processes that occur in a network of CRs, specifically covering dynamical systems, contraction mappings, standard interference functions, Markov models, games, potential games, and supermodular games.

For these game models, this chapter presented analysis insights that can be gleaned by demonstrating that a CR network satisfies the modeling conditions for one of these models, covering the steady-state properties, the convergence properties, and the stability properties for each of these models. As Example 15.9 showed, sometimes CR networks satisfy the conditions of multiple models. In these cases, the analytic insights from each of the applicable multiple models are available.

Two different model-independent approaches to determining the desirability of network behavior were also presented: Pareto optimality and evaluation of a network objective function. Demonstrating that a network state is Pareto optimal was shown to be of less value than demonstrating that the state maximized a network objective function.

In addition, this chapter presented a significant number of useful analytic results, but note that it was able to include only a brief treatment of these extensive models. Many of the models have entire disciplines dedicated to their analyses and applications. Accordingly, the interested reader is encouraged to explore the texts listed in the references.

Certain analytical difficulties arise when information is limited—a situation that will become increasingly prevalent as the technologies of the preceding chapters are realized. We may not be able to precisely describe a radio's future decision update rule as the decision processes and goals evolve to better reflect a user's preferences. A radio's available actions may also evolve in time to incorporate new waveforms that could not be anticipated ahead of time.

From an analysis perspective, this situation is analogous to attempting to solve a system of equations of unknown order with unknown coefficients and an unknown number of variables. Perhaps it will be possible to broadly classify the decision update processes and action sets, in which case a game theoretic preference approach should be able to address this situation. Perhaps different models will be needed, or perhaps a completely different approach should be taken.

Although this chapter describes methodologies suitable for analyzing the cognitive radios of today, analyzing those radios of the future will continue to be an active area of research and a remaining "hard problem."

EXERCISES

15.1 Prove that every contraction mapping satisfies the synchronous convergence condition of the asynchronous convergence theorem.

15.2 Given that the SIF is a pseudo-contraction, prove the following:
   (a) That it has a unique fixed point.
   (b) Synchronous adaptations converge to the fixed point.
   (c) Synchronous adaptations are stable.

15.3 Consider a network consisting of three terminals and an access node with noise power of −80 dBm implementing TPC with a statistical spreading gain of 64. Suppose gains to the access node 1, AN 2, and AN 3 are −10, −15 and −20 dB, respectively, and each node would like to achieve a target SINR of 8 dB. Assume each radio's set of transmit power levels is convex:

(a) Determine whether these target SINRs are feasible.

(b) If these SINRs are feasible, solve for the operating power vector.

(c) Based on the discussion in this chapter, what conditions are necessary to ensure convergence?

(d) Is this operating power vector stable? How do you know?

15.4 Repeat exercise 15.3, assuming target SINR for the AN 1, AN 2, and AN 3 are 6, 8, and 10 dB, respectively.

15.5 Repeat exercise 15.3, assuming the radios operate with discrete power levels.

15.6 Consider a pair of CR LANs, where each LAN is attempting to maximize its network capacity. Each LAN must make a one-time choice of frequencies $\{f_1, f_2, f_3\}$. If the two LANs choose the same frequency, the capacity of both LANs is 1/2 (reflecting a probability of collision of 1/2); if they choose different frequencies, each LAN has a capacity of 1 (reflecting a probability of collision of 1/2). Model this situation as a normal form game in matrix representation and solve for any NE.

15.7 Consider the network introduced in Section 15.1. Construct a game model of this network. Show that the game has no NE. Now suppose the links can employ mixed strategies, perhaps by employing frequency hopping and varying the ratio of time spent in each channel. Model this new network as a game and show that it has an NE in mixed strategies. Finally, propose a modification of the observation processes that would induce the BSI condition.

15.8 For the scenario in exercise 15.6, suppose the LANs are allowed to repeatedly adapt their decisions. Model this situation as a repeated game and describe what conditions need to be met to ensure convergence. How might a random timer, such as the one used in a random back-off scheme, be useful here?

15.9 Suppose a CR network consists of a set of ontologically defined radios where each radio is attempting to minimize its observed interference by adapting its waveform. Model this network as a game and identify conditions for which this network could be modeled as a potential game. (*Hint:* Suppose the radios have access to a radio environment map.)

15.10 Consider a set of 802.11 devices implementing DFS and TPC in Europe. Model this network as a normal form game by specifying the players, action sets, and appropriate goals. Identify an appropriate network objective function. Under what conditions could we expect uncoordinated CRs implementing GAs to converge to a desirable steady state for the network?

15.11 Suppose each link in the network in Example 15.7 transmits at a different power level. Why does the network not satisfy BSI? What are some issues introduced by violating the BSI condition? In light of these power variations, define new observation and/or utility functions that induce the BSI condition.

REFERENCES

[1] Mitola, J., *Cognitive Radio: An Integrated Agent Architecture for Software Defined Radio*, PhD Dissertation, Royal Institute of Technology, Stockholm, May 2000.

[2] IEEE Std. 802.11h™-2003 Part 11: Wireless LAN Medium Access Control (MAC) and Physical Layer (PHY) Specifications Amendment 5: Spectrum and Transmit Power Management Extensions in the 5 GHz Band in Europe, *Proceedings IEEE*, New York, October 2003.

[3] Ginde, S., R. M. Buehrer, and J. Neel, A Game Theoretic Analysis of the Joint Link Adaptation and Distributed Power Control in GPRS, *VTC Fall*, 2:732-736, 2003.

[4] Walker, J., *Dynamical Systems and Evolution Equations: Theory and Applications*, Plenum Press, 1980.

[5] Bertsekas, D., and J. Tsitsiklis, *Parallel and Distributed Computation: Numerical Methods*, Athena Scientific, 1997.

[6] Zangwill, W., *Nonlinear Programming: A Unified Approach*, Prentice-Hall, 1969.

[7] Medio, A., and M. Lines, *Nonlinear Dynamics: A Primer*, Cambridge University Press, 2001.

[8] Blackwell, D., Discounted Dynamic Programming, *The Annals of Mathematical Statistics*, 36(1):226-235, 1965.

[9] Sundaram, R., *A First Course in Optimization*, Cambridge University Press, 1999.

[10] Yates, R., A Framework for Uplink Power Control in Cellular Radio Systems, *IEEE Journal on Selected Areas in Communications*, 13(7):1341-1347, 1995.

[11] Berggren, F., Power Control, Transmission Rate Control and Scheduling in Cellular Radio Systems, PhD Dissertation, Royal Institute of Technology, Stockholm, May 2001.

[12] Zander, J., and S. Kim, *Radio Resource Management for Wireless Networks*, Artech House, 2001.

[13] Stewart, W. J., *Introduction to the Numerical Solution of Markov Chains*, Princeton University Press, 1994.

[14] Turin, W., *Digital Transmission Systems: Performance Analysis and Modeling*, McGraw-Hill, 1999.

[15] Kemeny, J., and J. Snell, *Finite Markov Chains*, Van Nostrand Company, 1960.

[16] Fudenberg, D., and J. Tirole, *Game Theory*, MIT Press, 1991.

[17] Neel, J., J. Reed, and R. Gilles, Game Models for Cognitive Radio Analysis, *SDR Forum Technical Conference*, November 2004.

[18] Sung, C., and W. Wong, A Noncooperative Power Control Game for Multirate CDMA Data Networks, *IEEE Transactions on Wireless Communications*, 2(1):186-190, 2003.

[19] Krishnaswamy, D., Game Theoretic Formulations for Network-Assisted Resource Management in Wireless Networks, *VTC Fall*, 3:1312-1316, 2002.

[20] Hayajneh, M., and C. Abdallah, Distributed Joint Rate and Power Control Game-Theoretic Algorithms for Wireless Data, *IEEE Communications Letters*, 8(8):511-513, 2004.

[21] Monderer, D., and L. Shapley, Potential Games, *Games and Economic Behavior*, 14:124-143, 1996.

[22] Voorneveld, M., and H. Norde, A Characterization of Ordinal Potential Games, *Games and Economic Behavior*, 19:235-242, 1997.

[23] Topkis, D., *Supermodularity and Complementarity*, Princeton University Press, 1998.

[24] Milgrom, P., and J. Roberts, Rationalizability, Learning, and Equilibrium in Games with Strategic Complementarities, *Econometrica*, 58(6):1255-1277, 1990.

[25] Friedman, J., and C. Mezzetti, Learning in Games by Random Sampling, *Journal of Economic Theory*, 98:55-84, 2001.

[26] Neel, J., R. Menon, A. MacKenzie, and J. Reed, Using Game Theory to Aid the Design of Physical Layer Cognitive Radio Algorithms, *Conference on Economics, Technology and Policy of Unlicensed Spectrum*, Lansing, MI, May 2005.

[27] Altman, E., and Z. Altman, S-Modular Games and Power Control in Wireless Networks, *IEEE Transactions on Automatic Control*, 48:839-842, 2003.

[28] Sung, C. W., and K. K. Leung, On the Stability of Distributed Sequence Adaptation for Cellular Asynchronous DS-CDMA Systems, *IEEE Transactions on Information Theory*, 49(7):1828–1831, 2003.

[29] Neel, J., Analysis and Design of Cognitive Radio and Distributed Radio Resource Management Algorithms, PhD Dissertation, Virginia Tech, September 2006.

[30] Neel, J., and J. Reed, Performance of Distributed Dynamic Frequency Selection Schemes for Interference Reducing Networks, *Proceedings Milcom*, Washington, DC, October 2006.

[31] Neel, J., Synthetic Symmetry in Cognitive Radio Networks, *SDR Forum Technical Conference*, November 2007.

Cognitive Radio in Multiple-Antenna Systems

16

Jae Moung Kim, Sung Hwan Sohn, Ning Han
INHA University, Incheon, Korea

Seungwon Choi, Chiyoung Ahn, Gyeonghua Hong, Yusuk Yun
Hanyang University, Seoul, Korea

16.1 INTRODUCTION

Since first proposed by Dr. Joseph Mitola in his dissertation, cognitive radio (CR) technology has drawn attention in the research community as a means for secondary systems to share frequency bands with primary systems. We expect that its implementation will become more common in the wireless systems of the future, as spectrum sharing can be better achieved by implementing a CR approach in multiple-antenna systems to increase spectrum efficiency.

Multiple-antenna techniques such as beamforming and spatial-domain multiple access (SDMA) can provide substantial performance enhancements depending on the users' requirements and the operating environment. An advantage of cognitive radio is that it can learn from its environment, allowing it to provide a better solution than fixed-choice multiple-antenna systems.

Multiple antennas can provide better support for such core techniques as spectrum sensing. By using multiple antennas, spectrum sensing can be carried out in more dimensions, increasing the number of degrees of freedom available to the radio for learning the spectrum environment.

Adapting CR to multiple-antenna systems requires several changes in conventional cognitive radio functionalities. For radio environment observation, CR is required to provide more parameters, such as the direction of arrival (DOA) of the primary user and channel correlation. The cognitive engine (CE) also uses a more complex parameter optimization process.

This chapter presents an overview of the techniques used in a multiple-antenna (MA)[1] system and discusses the key issues in introducing CR technology. This discussion is followed by an explanation of the applications of the new system, which combines CR and MA techniques.

---

[1] Whereas other parts of this book refer to multiple access as MA or MAC, in this chapter we refer to multiple antenna as MA.

16.2 MULTIPLE-ANTENNA TECHNIQUES

Adapting CR to the MA system helps the system learn from the environment and could provide better results than a noncognitive multiple-antenna system alone. In this chapter, we introduce the principles behind various MA techniques and discuss the critical criteria in selecting the optimum MA scheme.

16.2.1 Multiple-Antenna Systems

In this section we will discuss beamforming, diversity, and spatial multiplexing.

Beamforming System

Beamforming is a signal-processing technique used in sensor arrays for directional signal transmission (TX) or reception (RX). Spatial selectivity is achieved by using an adaptive, or fixed, receive/transmit beam pattern. The improvement provided over an omnidirectional RX/TX setup is the enhancement of communication capacity and quality both for the intended network nodes and for other nodes in the vicinity where interference levels are reduced.

When the user is moving or the DOAs vary during receiving, the effects of interference can be decreased by using an array antenna. An array antenna provides the ability to optimize communications for signals from a specific DOA or to transmit signals in a specific direction. When transmitting, a beamformer controls the phase and relative amplitude of the signal at each transmitter antenna element to create a pattern of constructive and destructive interference in the wavefront.

An example of a beamforming antenna is the uniform linear array antenna in which the distance between each array element is half of the carrier wavelength. Each element is assumed to be an omnidirectional antenna having fixed gains in all directions. Another assumption is that the transmitted/received signals from each array are plane waves [1, 2].

Figure 16.1 is a conceptual description of the transmission array antenna system. This array antenna transmits the signal in the direction defined by the control phases $w[n]$. Every modulated symbol, $s[n]$, is weighted differently at each antenna. The weighted signal at each antenna is expressed as

$$w_l[n] = e^{j(l-1)\pi \sin \theta} \quad l = 1, 2, \ldots, M \tag{16.1}$$

where n is the index of the snapshot, l is the index of the antenna array, M is the number of the antenna array, and θ is the DOA estimated in the uplink.

The transmit power at each antenna array is p_l, and the transmitted signal is

$$x_l[n] = \alpha_l s[n] e^{-j(l-1)\pi \sin \theta} \tag{16.2}$$

where $\alpha_l = \sqrt{p_l}$. The signal transmitted from the M-array antenna is received by the mobile station. The received signal is

$$y[n] = \sum_{l=1}^{M} \alpha_l s[n] e^{-j(l-1)\pi \sin \theta} h_l[n] + z[n] \tag{16.3}$$

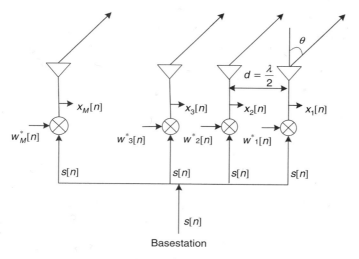

FIGURE 16.1

Conceptual description of the transmission antenna array system.

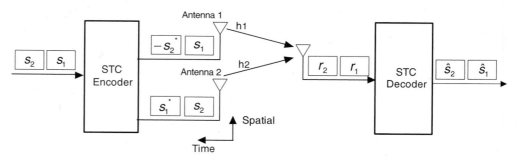

FIGURE 16.2

A two-branch transmission diversity scheme with one receiver.

where $h_l[n]$ is the channel between the l antenna and the receiver, and $z[n]$ is the additive white Gaussian noise (AWGN) at the receiver.

Transmit Diversity System

Space–time coding (STC) is a scheme that can achieve transmission diversity. The encoding is done in space and time. Figure 16.2 shows the baseband representation of a two-branch transmission diversity scheme. This scheme uses two transmission antennas and one receiving antenna and may be considered a 2×1 STC [3].[2] Table 16.1 shows this encoding approach.

---

[2] Throughout this chapter, we will refer to multiple-antenna systems as $n \times m$, where n is the number of transmitting antennas and m is the number of receiving antennas.

Table 16.1 Two-Branch Transmit Diversity Encoding and Transmission Sequence Scheme

	Antenna 1	Antenna 2
Time 0	s_1	s_2
Time T	$-s_2^*$	s_1^*

The coding rate of the 2×1 STC is 1, since two symbols are transmitted during each two-symbol duration. The signals received during the first symbol duration, r_1, and during the second symbol duration, r_2, are as follows:

$$r_1 = h_1 s_1 + h_2 s_2 + n_1$$
$$r_2 = -h_1 s_2^* + h_2 s_1^* + n_2$$

(16.4)

Assume that the channels, h_1, h_2, remain constant over the two-symbol duration, and n_1, n_2 are AWGN.

Using a matrix expression, the received symbol vector can be expressed as

$$\begin{bmatrix} r_1 \\ r_2 \end{bmatrix} = \begin{bmatrix} s_1 & s_2 \\ -s_2^* & s_1^* \end{bmatrix} \begin{bmatrix} h_1 \\ h_2 \end{bmatrix} + \begin{bmatrix} n_1 \\ n_2 \end{bmatrix}$$

(16.5)

The transmitted symbols, $\hat{s}_1$, $\hat{s}_2$, are estimated from the received symbols, r_1, r_2, as

$$\begin{bmatrix} \hat{s}_1 \\ \hat{s}_2 \end{bmatrix} = \begin{bmatrix} h_1^* & h_2 \\ h_2^* & -h_1 \end{bmatrix} \begin{bmatrix} r_1 \\ r_2^* \end{bmatrix} = \begin{bmatrix} \left(|h_1|^2 + |h_2|^2\right) \cdot s_1 + \tilde{n}_1 \\ \left(|h_1|^2 + |h_2|^2\right) \cdot s_2 + \tilde{n}_2 \end{bmatrix}$$

(16.6)

As for Eq. (16.6), we can see that the 2×1 STC obtains a diversity gain similar to the maximum ratio combining (MRC) of the 1×2 reception diversity system [4].

This scheme increases information reliability by obtaining a diversity gain similar to the maximum ratio combining in receiving antennas, but the data rate is not increased compared to the single-input, single-output (SISO) system.

Spatial Multiplexing

The spatial multiplexing (SM) scheme transmits different signals from each transmission antenna, so that the data rate is increased without any change in frequency band or transmission power. Figure 16.3 shows a representation of an SM transceiver with two transmission antennas and two receiving antennas. Two independent data symbols are transmitted by each transmission antenna. The signals received by the antennas consist of the sum of the transmitted data. Received signals can be expressed as

$$\begin{bmatrix} r_1 \\ r_2 \end{bmatrix} = \begin{bmatrix} h_{11} & h_{12} \\ h_{21} & h_{22} \end{bmatrix} \begin{bmatrix} s_1 \\ s_2 \end{bmatrix} + \begin{bmatrix} n_1 \\ n_2 \end{bmatrix}$$

(16.7)

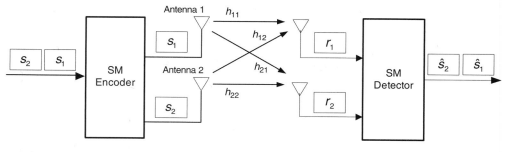

FIGURE 16.3

Structure of the 2×2 SM transceiver.

Using a matrix expression, Eq. (16.7) can be written as

$$\underline{r} = \underline{\underline{H}}\,\underline{s} + \underline{n} \tag{16.8}$$

where $\underline{r}$ is the received symbol vector, $\underline{\underline{H}}$ is the channel matrix, $\underline{s}$ is the transmitted symbol vector, and $\underline{n}$ is AWGN.

To detect received symbols, detection processing is required. There are various methods currently in use for detection processing, including maximum likelihood (ML), zero forcing (ZF), minimum mean-square error (MMSE), successive interference cancellation (SIC), and ordered SIC (OSIC) [5].

16.2.2 Critical Criteria

To provide criteria for the radio to select the optimal MA schemes based on observation of its current environment, we investigated the performance of the different MA schemes in the IEEE 802.16e standard system. For comparison, we chose 2×2 SM, 2×2 STC, 2×1 STC, and selected the ML method from among the various SM detection algorithms.

Criteria Used for Equivalent Data Modulation

Figure 16.4 shows the bit error rate (BER) performance of an MA system in a Rayleigh frequency-flat fading channel [6] environment. A 1/2 rate Convolutional Turbo Code (CTC) [7] and 16-QAM (quadrature amplitude modulation) data modulation were adopted.

If there is only one receiving antenna, we choose beamforming, or 2×1 STC. When the signal-to-noise ratio (SNR) is less than 8 dB, the BER of beamforming is lower than that of 2×1 STC at the same SNR; however, the BER of 2×1 STC is less than that of beamforming when the SNR is greater than 8 dB. So it is reasonable to choose beamforming when SNR is less than 8 dB, and 2×1 STC when the SNR is greater than 8 dB. If there are two receiving antennas, 2×2 STC is a good selection for retrenching transmitting power, and 2×2 SM provides a high data rate.

Figure 16.5 shows the BER performance of an MA system in a Rayleigh frequency-selective fading channel environment [6]. In this case, a 1/2 rate CTC and 16-QAM data modulation were adopted. The figure shows that the BER of the beamforming is always lower than that of 2×1 STC, so it is reasonable to choose beamforming for a

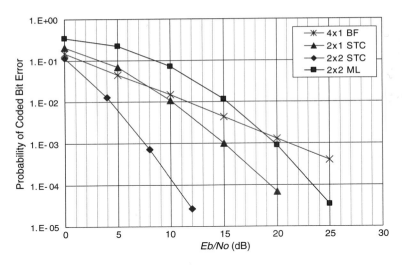

FIGURE 16.4

BER performance in a Rayleigh frequency-flat fading channel environment (16-QAM).

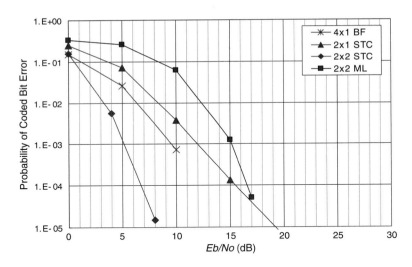

FIGURE 16.5

BER performance in a Rayleigh frequency-selective fading channel environment (16-QAM).

one-receiving antenna system. Similarly, we should choose 2×2 STC for a two-receiving antenna system.

Criteria Used for Equivalent Frequency Efficiency

In the preceding discussion for equivalent data modification, we investigated the criteria used for MA schemes in the same modulation (16-QAM) environment. In this section,

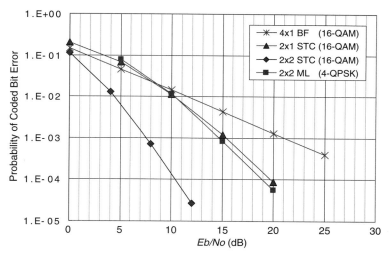

FIGURE 16.6

BER performance in a Rayleigh frequency-flat fading channel environment.

to determine the criteria used for MA schemes in the same frequency–efficiency environment, we adopted quadrature phase shift keying (QPSK) for ML (SM) and 16-QAM for STC. The channel coding was CTC with an $R = 1/2$ coding rate.

Figure 16.6 shows the BER performance of an MA system in a Rayleigh frequency-flat fading channel environment. If there is only one receiving antenna, it is reasonable to choose beamforming when the SNR is less than 8 dB, and 2×1 STC when the SNR is greater than 8 dB. If there are two receiving antennas, 2×2 STC is a better choice than 2×2 SM, even though the frequency efficiencies of the two systems are equivalent.

Figure 16.7 shows the BER performance of an MA system in a Rayleigh frequency-selective fading channel environment. The figure shows that it is reasonable to choose beamforming for a one-receiving-antenna system, and 2×2 STC for a two-receiving-antenna system. In the common channel circumstance, or in a cell edge circumstance that has poor channel quality, STC is the conventional selection; however, SM may be selected in the case of good channel quality.

The analysis described in this section provides the criteria by which the CR engine chooses the best MA scheme according to the current channel environment. The criteria take into account the SNR and fading environment.

16.3 COGNITIVE CAPABILITY IN AN MA SYSTEM

A system that combines CR and MA techniques is able to increase the spectrum efficiency by taking advantage of two functionalities: radio environment observation and the CE (Figure 16.8). The environmental information provided by radio environment observation includes information related to the primary system (e.g., its type and DOA)

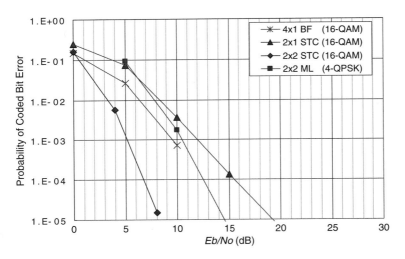

FIGURE 16.7

BER performance in a Rayleigh frequency-selective fading channel environment.

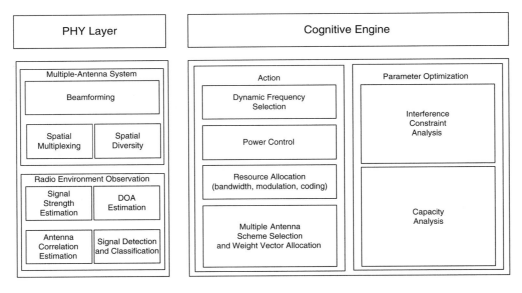

FIGURE 16.8

Structure of the CR in an MA system.

and channel information for the link between a secondary transmitter and receiver. The CE, in contrast, takes advantage of cognitive techniques to generate the optimum transmission parameters for the current environment. The spectrum efficiency is increased in two ways: (1) the radio can use the spectrum occupied by the primary system by avoiding transmission in the directions in which the primary system is operating, and

(2) the radio can select the optimum MA technique for operation in the channels in which the primary system is not operating.

16.3.1 Structure of the CR in an MA System

The smart antenna system is equipped with a radio environment observation block in its physical (PHY) layer in addition to the conventional MA system block, which includes three kinds of implementation techniques: beamforming, spatial multiplexing, and spatial diversity, the selection of which is determined by the CE. The radio environment observation block, shown in Figure 16.8, monitors the radio environment and keeps a record of the current environment in memory.

The "Signal Detection and Classification" is used to detect and identify the specific type of signal operating in the band. If the signal is identified as a primary signal, related information can be obtained from the database in which the specifications of primary systems are stored. The "Signal Strength Estimation" is used to estimate the received signal strength (RSS) of the primary signal. The "DOA Estimation" calculates and stores the direction of arrival of the received signal. RSS and DOA information are able to assist the MA system to estimate the position of the primary user. Thus, the MA system is able to manage its transmission and keep interference to the primary system lower than the limit. The "Antenna Correlation Estimation" computes the correlation of the current antenna through the received signal in each antenna mainly during the operation period without the primary user (PU) in the frequency band. The correlation information is important for choosing a different MA scheme.

Additionally, the information via radio environment observation is periodically sent to the CE to be further analyzed and subsequently used to change the parameters of the MA system to achieve optimum performance. The capacity of the multiple-antenna system is improved with the help of the CE, which contains the current status of the system and is able to combine information about the current environment to dynamically select the optimum transmission parameters. The purpose of the CE is to fully utilize its knowledge of the radio environment and the current status of the MA system to take advantage of reasoning techniques to generate the optimum operation parameters that allow for the best system performance. There are two major blocks, "Action" and "Parameter Optimization," that interact to obtain the system's optimum parameters.

The Action block is the set of actions available to the CE ordered by how they serve the radio's objectives. Generally, five possible actions can be explored: dynamic frequency selection, power control, resource allocation, and MA scheme selection and weight vector allocation. In the Parameter Optimization block, both the interference constraint and the capacity are analyzed for optimization. The "Interference Constraint Analysis" block analyzes the available interference level periodically to keep interference under a level that can be tolerated by the primary system. The "Capacity Analysis" block monitors the system capacity to maximize the spectrum efficiency.

This completes the introduction to all the function blocks in the MA system that use the cognitive concept; the operational procedure is illustrated in Figure 16.9. The operation begins by requesting a frequency band. Next, a radio environment observation is performed to first detect and identify the signal in the frequency band, and then

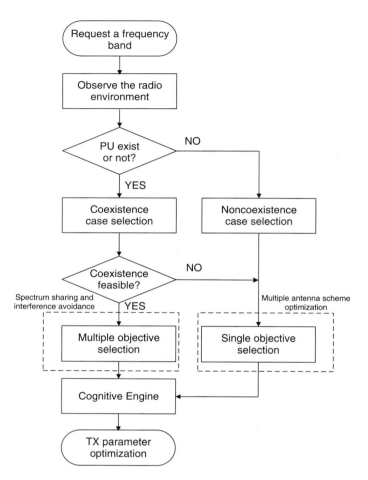

FIGURE 16.9

The operational procedure of the CR in an MA system.

to estimate such additional information as signal strength and DOA if a primary signal is detected. Usually, the system operates in two modes: the PU present in the current channel and the PU absent in the current channel.

Primary User Present in the Current Channel

The case in which the PU is present in the current channel is shown in Figure 16.10, where a pair of secondary transmissions interfere with one PU because they are all operating in the same frequency band. In a single-antenna system, the secondary transmission has to change its operation frequency band to avoid interfering with the PU's operation. However, by using multiple antennas with beamforming technology, the secondary system can construct its transmission beam pattern such that its interference with the primary system is either minimized or constrained, as long as the PU is sepa-

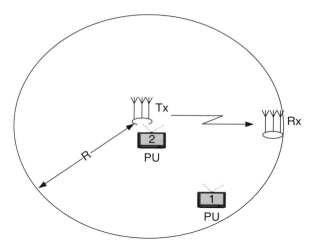

FIGURE 16.10

Presence of PU in the same channel.

rated from the secondary transmitter as shown in position 1 in Figure 16.10. Thus, the secondary system must be aware of the approximate location of the primary receiver. This knowledge may be acquired in various ways, as described in the following sections. If the primary user exists in position 2, however, the secondary system has to avoid using the same channel to protect the primary operation.

Primary User Absent in the Current Channel

When there is no primary user in the observed frequency band, then the "non-coexistence case" will be chosen. Since the PU is absent in the frequency band, related information (e.g., signal strength and DOA) is not required. However, the antenna correlation information is still required to compute the correlation of the current antenna. By using this information, the CE must determine the best possible parameters for the MA system. The best scheme, with the most suitable parameters, is the one that maximizes its spectral efficiency or throughput, or minimizes interference while maintaining transmission power at or below a certain level. The relevant parameters are transmission power, carrier frequency, MA scheme, modulation, and coding rate. The relevant MA schemes are beamforming, spatial multiplexing, and spatial diversity.

To optimize the parameters based on current information from observation of the radio environment, CR techniques are used to formulate the problem of searching for optimal parameters. This technique has the ability to search parameter spaces despite having little or no information.

16.3.2 Radio Environment Observation

Radio environment observation is a new functionality that results from introducing the cognitive concept to the MA system. It has two tasks: (1) to detect the existence of a primary system and identify the characteristics of its signal, and (2) to estimate the response of the channels between the secondary basestation and secondary users (SUs). Several methods used in a conventional CR system (e.g., energy detection and feature detection) can be improved in a smart antenna system.

Spectrum-Sensing Method

The optimal approach to signal detection is a matched filter because it maximizes the received SNR [8]. However, a priori knowledge of the detected signal is required when using a matched filter because it requires demodulation of the received signal. Moreover, timing and carrier synchronization information are required to achieve the expected performance.

Energy detection [9] is a suboptimal and simple method that can be applied to any signal type without requiring any information about the received signal. In determining whether a signal of a certain bandwidth is present in a spectral region of interest, energy detection using a single antenna suffers from poor performance in low SNR regions. After receiving and downconverting, the received signal, $y(t)$, is sent to the energy detector. The energy, E, is calculated using

$$E = \int_0^T y^2(t)\, dt \qquad (16.9)$$

In an MA system, $y(t)$ could be formed by using different selection processes. It could be determined by linearly combining the signals received from each antenna. It could also be formed by selecting the signal with the highest SNR among all signals.

As a method for signal detection, energy detection suffers from several drawbacks. First, to improve detection reliability, increased sensing time is required. Moreover, there is a minimum SNR below which no signal can be detected. Noise uncertainty, caused by various factors (e.g., temperature changes, ambient interference, and filtering), is unavoidable and leads to errors when setting the threshold for signal detection.

In addition to detecting the signal, feature detection can identify the received signal among all possible signals. Signals used in different wireless systems exhibit different features. Through recognition of these characteristics, the signal can be identified. Generally, cyclostationarity is considered to be a key feature in the separation of wireless communication signals. The goal of cyclostationary detection is to determine the cycle frequencies, α_n, included in the spectral correlation function of the signal given by [10]:

$$S_x^{\alpha_n}(f) = \frac{1}{T}\int_{-T/2}^{T/2} Y(t, f + \alpha_n/2)\cdot Y^*(t, f - \alpha_n/2)\, dt \qquad (16.10)$$

where the spectral function, $Y(t)$, is the Fourier transform of $y(t)$ over the time interval $[t - T/2, t + T/2]$. Similar to the method used for energy detection in an MA system, the two spectral terms can be replaced by the signal received from different antennas by some selection process. An obstacle to the implementation of this detection method is the complexity of extracting the cyclostationary features.

To ensure that an SU will not interfere with primary users, reliable sensing techniques should be employed. However, spectrum sensing is compromised when a CR user is subjected to an environment in which fading or shadowing occurs. Therefore, collaboration among SUs should be implemented by the secondary system. Using the simple "or" rule [11] or weighted-collaborative [12] criteria, the collaborative sensing method can significantly improve sensing performance over individual sensing.

DOA Estimation

When using multiple antennas, the spatial information of the PU's transmitter is available to the SU through DOA estimation. By knowing the DOA of the primary signal, the secondary user is able to avoid directly transmitting toward the primary user's transmitter while carrying out transmissions in other directions. If the direction of the receivers in the PU's network is also known, then the antenna pattern may be able to synthesize a beam shape that does not transmit toward receivers in the primary user network. This increases the efficiency of spectrum utilization.

The received signal, $\mathbf{x}(n)$, at the SU that was transmitted by q sources is modeled as

$$\underline{\mathbf{x}}(n) = \sum_{k=1}^{q} \underline{\mathbf{a}}(\theta_k) s_k(n) + \underline{\mathbf{n}}(n) \tag{16.11}$$

$$\underline{\mathbf{x}}(n) = [x_1(n), x_2(n), \ldots, x_M(n)]^T \tag{16.12}$$

$$\underline{\mathbf{a}}(\theta_k) = \left[1, e^{-j2\pi d \sin(\theta_k)/\lambda}, \ldots, e^{-j(M-1)2\pi d \sin(\theta_k)/\lambda}\right]^T, \tag{16.13}$$

where $\underline{\mathbf{a}}(\theta_k)$ is the steering vector containing the phase information of the kth ($k = 1$, $2, \ldots, q$) signal that arrives at the SU (Eq. 16.13), θ_k is the direction of arrival of the kth source, d is the distance between two adjacent array elements, λ indicates the wavelength, and $\underline{\mathbf{n}}(n)$ represents AWGN. The task of DOA estimation is to calculate θ_k for each transmitted signal. There are several methods that can be implemented in this estimation.

The likelihood function of the received signal is expressed as

$$f(\underline{\mathbf{x}}) = \frac{1}{(\pi\sigma^2)^M} \cdot \exp\left(-\frac{1}{\sigma^2}|\underline{\mathbf{x}}(n) - \underline{\mathbf{a}}(\theta)s(n)|^2\right) \tag{16.14}$$

To estimate the DOA, we find the θ_k that maximizes this function. Thus, the estimation problem is equivalent to the minimization problem shown in Eq. (16.15). This method accurately estimates the DOA in regions with a low SNR using a small number of samples, but it is computationally intense. Moreover, it is not statistically efficient for a limited number of receiving antennas.

$$\min_{\theta}\left\{\left|\underline{\mathbf{x}}(n) - \underline{\mathbf{a}}(\theta)\left(\underline{\mathbf{a}}^H(\theta)\cdot\underline{\mathbf{a}}(\theta)\right)^{-1}\underline{\mathbf{a}}^H(\theta)\cdot\underline{\mathbf{x}}(n)\right|^2\right\} \tag{16.15}$$

Another method is to exploit the eigenstructure of the covariance matrix expressed in Eq. (16.16). The orthogonality of the steering vectors corresponding to the signal components and the noise subspace eigenvectors gives rise to peaks in the MUSIC spectrum, as shown in Eq. (16.17).

$$\hat{\underline{\underline{\mathbf{R}}}}_{xx} = \frac{1}{k}\sum_{k=o}^{K} \underline{\mathbf{x}}_k \underline{\mathbf{x}}_k^H \tag{16.16}$$

$$\hat{p}_{MUSIC}(\phi) = \frac{\underline{\mathbf{a}}^H(\phi)\underline{\mathbf{a}}(\phi)}{\underline{\mathbf{a}}^H(\phi)\underline{\underline{\mathbf{V}}}_n \underline{\underline{\mathbf{V}}}_n^H \underline{\mathbf{a}}(\phi)} \tag{16.17}$$

where $\underline{\underline{V}}_n = \left[\underline{q}_D, \underline{q}_{D+1}, \ldots, \underline{q}_{M-1} \right]$ is the eigenvector of $\hat{\underline{\underline{R}}}_{xx}$. The main advantage of this method is that it is able to distinguish between arbitrarily closely spaced signals. Modified versions of this method can also be used to interpret the DOA estimation for the PU.

The DOA information can also be obtained from a database. For example, PUs may be static computer stations with fixed locations. The method of cooperation among several secondary users is also suitable. In this method, each SU estimates the distance to and arrival angle of the primary signal by monitoring the uplink communication of the primary users. Then the SUs combine their information in a centralized or a decentralized manner to determine the DOA of the primary signal. A number of measurement sensors can also be placed in the coverage area to estimate the DOA information of the primary signal.

Antenna Correlation Measurement

Most of the adaptive antenna array-processing techniques depend on a correlation of the signals received at the array. It has been shown [13] that the real and imaginary parts of $R(m, n)$, the signal correlation between the mth and nth antenna elements, are given by

$$\operatorname{Re}\{R(m, n)\} = J_0(z_{mn}) + 2\sum_{l=1}^{\infty} J_{2l}(z_{mn})\cos(2l\theta)\operatorname{sinc}(l\Delta) \tag{16.18a}$$

$$\operatorname{Im}\{R(m, n)\} = 2\sum_{l=1}^{\infty} J_{2l+1}(z_{mn})\sin((2l+1)\theta)\operatorname{sinc}\left(\left(l+\frac{1}{2}\right)\Delta\right) \tag{16.18b}$$

where θ is the mean angle of arrival measured with respect to the normal line joining the two antennas (Figure 16.11), and $z_{mn} = \omega d_{mn}/c$, where d_{mn} is the distance between the mth and nth antennas. We assume that all signals from the arriving signal at the antenna array arrive within $\pm\Delta/2$ of the mean angle of arrival, θ.

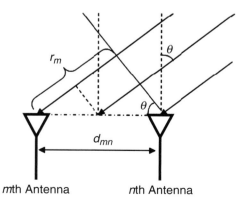

r_m

θ

θ

d_{mn}

mth Antenna nth Antenna

FIGURE 16.11

Model geometry.

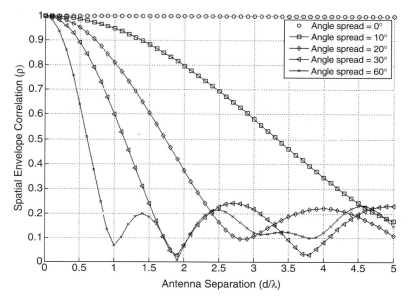

FIGURE 16.12

Spatial envelope correlation versus antenna spacing: DOA $\theta = 0°$.

The spatial envelope correlation is given by $\rho = |R(m, n)|$. For $\theta = 0°$ and various angle spreads, Δ, Figure 16.12 shows that as Δ decreases, the first zero in the correlation occurs at a larger antenna spacing. When the signal arrives from a direction other than broadside, the antenna spacing for low correlation increases, and the envelope correlation is never zero for all values of $\theta \neq 0°$ and $\Delta < 2\pi$ (ρ is zero when Re$\{R(m, n)\}$ and Im$\{R(m, n)\}$ have zero crossings at exactly the same spacing).

We can estimate the spatial envelope correlation by using

$$\rho = E\left[w_1 w_2^* \right] \tag{16.19}$$

where $\rho(n) = f\rho(n-1) + (1-f)\left(w_1(n)w_2^*(n)\right)$, $E[\cdot]$ means the expected value, f is the forgetting factor with [0, 1], and $w = [w_1 w_2, \ldots, w_n]$ is an array response vector with n antennas.

Eigenvalue-Based Detection Method Using Multiple Antennas

Using multiple antennas for signal detection enables the SU to receive multiple replicas of the primary signal. The eigenvalue-based detection method can be used to determine whether the primary signal exists [14].

By collecting enough samples $(L + N_s)$ of the received signal, the sample covariance matrix, S, of the signals received from M antennas can be built. L indicates the first L samples of the output signal and N_s is the number of samples after the L samples to form the covariance matrix, S. Based on the eigenvalue, λ_i, of S, signal detection can be carried out.

One approach is to find the maximum and minimum of the eigenvalues. The test statistic is defined as the ratio of these two eigenvalues:

$$Z = \lambda_{max} \Big/ \lambda_{min} \qquad (16.20)$$

The threshold is determined by exploiting the distributions of both λ_{max} and λ_{min} as discussed previously [14], and the corresponding threshold for a specific false-alarm probability, P_{fa}, is derived as

$$\gamma = \frac{\left(\sqrt{N_s} + \sqrt{ML}\right)^2}{\left(\sqrt{N_s} - \sqrt{ML}\right)^2} \cdot \left(1 + \frac{\left(\sqrt{N_s} + \sqrt{ML}\right)^{-2/3}}{(N_s ML)^{1/6}} F_1^{-1}(1 - P_{fa})\right) \qquad (16.21)$$

where F_1^{-1} is the inverse cumulative distribution function of the first-order Tracy-Widom distribution [15].

16.3.3 Cognitive Engine in Multiple-Antenna Systems

One feature of CR is its ability to adaptively determine appropriate operational parameters in a dynamic wireless channel environment while meeting both the system's operational requirements and the users' requirements. The CE process involves three categories of parameters: environmental parameters, operational parameters, and optimization objectives. The CE determines the operational parameters of the radio according to environmental parameters it has obtained for the purpose of achieving optimized objectives. This process involves a cognitive system's decision about the values of the radio's operational parameters [16]. Therefore, the reasoning system is the main controller of the CR, and the reasoning selection substantially affects system performance.

Several genetic algorithm (GA) based CEs have been proposed by groups at Virginia Tech [17] and University of Kansas [16]. Both have demonstrated that, in a single-antenna system, their GA implementation is able to change the operational parameters based on a set of objectives. Furthermore, the CE designed by University of Kansas provides a numerical analysis of the relationships between the environmental and transmission parameters.

In an MA system, the cognitive engine is a controller for transmission parameter optimization, as illustrated in Figure 16.13. The difference between the two systems is in the choice of optimization objectives that depend on the presence or absence of a cochannel PU. The radio must determine if the PU is on the same channel, and then select suitable optimization objectives for the radio to optimize its operational parameters. For instance, if a cochannel exists, the radio must select the transmission parameters that protect the operation of the PU system as well as maximize the capacity of the CRs.

Environmental Parameters
Environmental parameters provide information regarding the system's internal state and surroundings. Information on the internal state includes the transmission rate, SNR, and BER. Information on the system's surroundings includes the channel occupation and channel response. Table 16.2 shows the additional parameters that must be included in MA systems. PU information must be included because SUs are allowed to operate in the same channel through separation in the spatial domain. This information contains

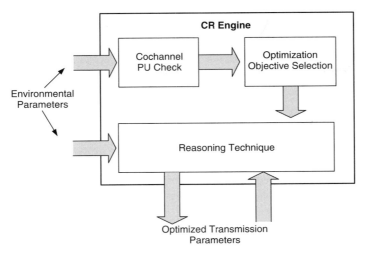

FIGURE 16.13

Structure of the cognitive engine.

Table 16.2 Additional Environmental Parameters

Parameter Name	Description
DOA of PU	The direction of PU signal's arrival
Transmission power of PU	Magnitude in decibels of PU's transmission power
Antenna correlation factor	Multipath delay profile, angle spread and correlation of MA
Number of antenna of SU, CR-BS	Available user array antenna
SM, STC capability of PU	PU capability of the SM and STC
MIMO channel	Conditional number of matrix

the DOA of the PU signal and its transmitted signal power. Moreover, the channel correlation between each antenna should be measured to provide reliable information for the selection of different multiple-input, multiple-output (MIMO) techniques. All of these parameters can assist the CE in making a selection of transmission parameters. Because they are the input to the fitness function, environmental parameters are linked to the objective of optimization.

Transmission Control Parameters

With the aid of the cognitive reasoning engine, the transmission control parameters are optimized to achieve the maximum spectral efficiency while allowing minimal interference with the primary system. Table 16.3 shows that the operational parameters that provide information regarding system operations including the transmission power, antenna scheme selection, and weight vector allocation.

Table 16.3 Major Operational Parameters

Parameter Name	Description
Transmission power	Raw transmission power
Antenna scheme selection	The antenna scheme selected for transmission or reception
Weight vector allocation	The allocated weight vector for antenna system

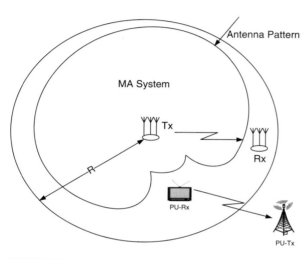

FIGURE 16.14

Primary user present in the same channel with beamforming.

Objective Selection

In MA systems, the objective selection is performed differently than in a single-antenna system due to the reuse of the frequency band in the spatial domain. The criterion for objective selection is the existence or absence of a PU in the same channel. Therefore, two scenarios with different objectives are considered in the following subsections. Multiple-antenna techniques (e.g., STC and beamforming) are selected to illustrate the different scenarios.

1. PU present in the same channel

In this case, the PU is present in the same frequency band of the secondary system. Depending on the relative location of the PU and the secondary system, as shown in Figures 16.14 and 16.15, we must select different objectives with which the CE can perform optimization.

In Figure 16.14, the secondary system partially overlaps the PU, but we can still reuse a fraction of the frequency resource. The optimization objectives must limit the interference with the PU to a constraint level. Meanwhile, optimizing the system capacity and transmission power can still be an objective. Thus, the CE can probably select

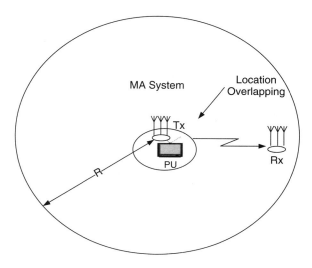

FIGURE 16.15

Primary user in the same channel with channel selection.

the beamforming technique to avoid transmission in the direction of the PU, and calculate the corresponding operational parameters based on the environment parameters through the artificial intelligence (AI) technique.

In Figure 16.15, because the PU is within the communications range of the SUs, it is impossible to reuse any part of the frequency band. In this case, the secondary user must switch operation to other frequency bands, and the same optimization process as described in Figure 16.14 can be implemented to select the best operational parameters except that, in this case, interference with the PU is not a concern.

2. PU absent in the same channel

In this case, the frequency band of interest is determined to be empty within the operating range (Figure 16.16); therefore, the secondary system can operate in this frequency band as licensed systems do. The objective is to increase system efficiency by increasing capacity while keeping the transmission power as low as possible. The key is in the selection of the proper MA technique.

16.4 APPLICATION TO NEXT-GENERATION WIRELESS COMMUNICATIONS

The possible role of cognition in wireless communication systems increases significantly when looking at the several international standards currently under development. WiFi, or IEEE 802.11, is working on opening a new frequency band (3600–3700 MHz in the United States) to operate on a coprimary basis with higher power, allowing it to extend its coverage to 5 km [18].

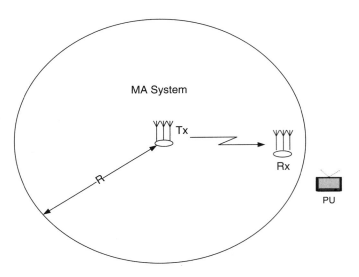

FIGURE 16.16

No PU in the neighborhood.

WiMax, also known as IEEE 802.16, has been working on improving the coexistence mechanisms for license-exempt operation through the License-Exempt Task Group (IEEE 802.16h) since 2004 [19]. Wireless broadband (WiBro) Evolution of IEEE 802.16m is a good candidate for next-generation wireless communications with a target of providing 100 Mbps service for mobile terminals, and 1 Gbps service for fixed terminals. The major technology trends seen in WiBro Evolution are summarized in Figure 16.17 [20]. To increase capacity, technologies such as MIMO and self-configuration are good solutions. MIMO technology includes STC, SM, collaborative SM (CSM), beamforming, and multiuser (MU) MIMO. In addition, various downlink MU-MIMO technologies (e.g., dirty paper coding (DPC))[3] will also be adopted in WiBro Evolution. A high average throughput is expected with the help of channel state information (CSI) feedback from mobile stations. CSI from mobile stations to the basestation provides a large amount of feedback information, which will require research into the reduction of information content. Software-defined radio (SDR) and CR can add the ability of self-configuration to these systems.

IEEE 802.22 is developing a new system that provides broadband wireless service in rural areas through TV channels with the help of CR techniques [21]. The CR-multiantenna system (CR-MAS) described in this chapter has proved itself to be a good candidate for spectrum sharing. It could be implemented in the newly developed systems to improve their feasibility and capacity.

From the network point of view, CR-MAS is a strong candidate for interference cancellation and capacity improvement. Very high-speed wireless mesh networks have

---

[3] Dirty paper coding is a method to counteract interference in the communication channel by encoding the data in a manner to cancel the interference.

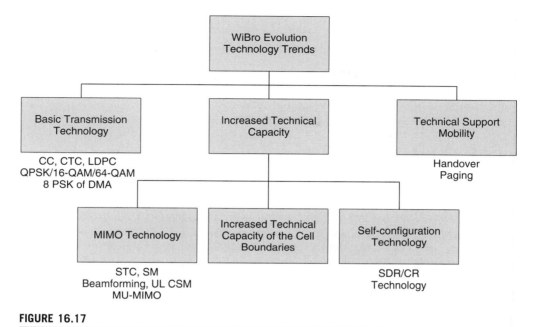

FIGURE 16.17

WiBro technical evolution.

been created over the primary wireless communication system by using CR and MIMO technologies [22].

CR-MAS could be considered a part of IEEE SCC41 [23], which aims to address issues related to the deployment of next-generation radio systems and advanced spectrum management.

16.5 SUMMARY

In this chapter, we discussed the issue that combines the MA system with CR technology to increase the utilization of spectrum by sharing the spectrum among different systems as well as to improve the transmission efficiency by selecting the optimum transmission schemes under certain operational environments. We introduced the new functionalities, namely, radio environment observation and cognitive engine, in the MA system by adding cognitive capability. In such a system, besides signal detection and identification, the DOA and antenna correlation information should be obtained with the help of multiple antennas. By combining the environmental information, CE is able to select an optimum operation mode for the radio, depending on the related position between the PUs and SUs. On the one hand, spectral efficiency is increased by sharing the frequency bands between primary and secondary users via transmission beamforming. On the other hand, based on the criteria discussed in Section 16.2.2, the SU system is able to select the optimum transmission scheme to increase the system capacity. Finally, possible implementations of CR-MAS have been discussed.

REFERENCES

[1] Litva, J., and T. Kwok-Yeung Lo, *Digital Beamforming in Wireless Communications*, Artech House, 1996.

[2] Stutzman, W. L., and G. A. Thiele, *Antenna Theory and Design*, John Wiley, 1981.

[3] Alamouti, S. M., A Simple Transmit Diversity Technique for Wireless Communications, *IEEE Journal on Selected Areas in Communications*, 16(8):1451–1458, 1998.

[4] Proakis, J. G. *Digital Communications*, Fourth Edition, McGraw-Hill, 2001.

[5] Larsson, E. G., and P. Stoica, *Space Time Block Coding for Wireless Communications*, Cambridge University Press, 2003.

[6] International Telecommunication Union/ITU Radiocommunication Sector, *Guidelines for Evaluation of Radio Transmission Technologies for IMT-2000*, IHS, January 1, 1997.

[7] IEEE Std., DRAFT Standard for Local and Metropolitan Area Networks, Part 16: Air Interface for Broadband Wireless Access Systems (revision of IEEE Std 802.16-2004 and amended by Std. 802.16f-2005 and IEEE Std. 802.16e-2005), pp. 1069–1089, October 2007.

[8] Cabric, A., D. Tkachenko, and R. W. Brodersen, Spectrum Sensing Measurements of Pilot, Energy, and Collaborative Detection, *Proceedings Military Communications Conference*, pp. 1–7, October 2006.

[9] Urkowitz, H., Energy Detection of Unknown Deterministic Signals, *Proceedings IEEE*, 55:523–531, April 1967.

[10] Gardner, W. A., *Cyclostationarity in Communications and Signal Processing*, IEEE, 1994.

[11] Mishra, S. M., A. Sahai, and R. Brodersen, Cooperative Sensing among Cognitive Radios, *Proceedings International Conference on Communications*, June 2006.

[12] Huang, X., N. Han, G. Zheng, S. H. Sohn, and J. M. Kim, Weighted-Collaborative Spectrum Sensing in Cognitive Radio, *Proceedings IEEE International Conference on Communications and Networking in China*, August 2007.

[13] Naguib, A. F., Adaptive Antennas for CDMA Wireless Networks, Ph.D. Dissertation, Stanford University, August 1996.

[14] Zeng, Y., and Y. C. Liang, Maximum-Minimum Eigenvalue Detection for Cognitive Radio, *IEEE Personal, Indoor and Mobile Radio Communications*, September 2007.

[15] Tracy, C. A., and H. Widom, On Orthogonal and Symplectic Matrix Ensembles, *Comm. Math. Phys.*, 177:727–754, 1996.

[16] Newman, T. R., B. A. Barker, A. M. Wyglinski, A. Agah, J. B. Evans, and G. J. Minden, Cognitive Engine Implementation for Wireless Multicarrier Transceivers, *Wiley Journal on Wireless Communications and Mobile Computing*, 7(9):1129–1142, 2007.

[17] Rieser, C. J., Biologically Inspired Cognitive Radio Engine Model Utilizing Distributed Genetic Algorithms for Secure and Robust Wireless Communications and Networking, Ph.D. Dissertation, Virginia Polytechnic Institute and State University, Blacksburg, April 2004.

[18] 802.11y Meeting Reports, IEEE, May 2007; available at *www.grouper.ieee.org/groups /802/11/Reports/tgy_update.htm*.

[19] IEEE 802.16's License-Exempt Task Group; available at *www.wirelessman.org/le/*.

[20] IEEE 802.16 Task Group m (TGm); available at *www.wirelessman.org/tgm*.

[21] IEEE 802.22 WRAN; available at *www.ieee802.org/22/*.

[22] Sakaguchi, K., and T. Fujii, Cognitive MIMO Mesh Network for Spectrum Sharing, *Tutorial in Crowncom*, May 2008.

[23] *www.scc41.org/*.

Cognitive Radio Policy Language and Policy Engine

17

Grit Denker, Daniel Elenius, David Wilkins
SRI International, Menlo Park, California

17.1 INTRODUCTION

Wireless communication is facing several challenges, including spectrum scarcity, deployment delays, and formation and management of dynamic networks. Cognitive radios (CRs) will help address these challenges, but only if they reason about *policies* to guide their behavior. Policies are sets of declarative statements with unambiguous semantics. *Policy engines* (PEs) are software components that reason with policies so that a particular communication device, or network of devices, obeys a given set of policies during its operation.

There are many advantages in using a policy-based approach to CRs. Deployment delays are drastically reduced because a policy-based architecture enables policies, policy reasoners, and radio devices to be accredited separately. Policies can be used to describe preferences and constraints on parameters, such as priority of traffic, security, quality of service, and probability of interference. Radio behavior can be changed in flexible ways at runtime by changing such policies. Finally, by having policies with clear, easily understood semantics, we can coordinate a variety of organizational entities. For example, regulators can specify admissible transmission behavior in a policy, or network managers with proper authority can activate, or deactivate, policies to flexibly control the network. Detailed arguments discussing the benefits of policy-based CRs are provided in Section 17.2.

To satisfy any such policy, radio engineers need only design their devices to correctly interact with a certified PE. Of course, radios may need to add capabilities to take advantage of all opportunities provided by a set of policies.

We focus this vision by describing a specific policy language (PL) and engine that were developed at SRI to use policies that provide opportunistic spectrum access, thus addressing the challenge of spectrum scarcity. Our PL is sufficiently general and expressive to be useful for the variety of policies mentioned above. However, our PE was implemented to meet the operational constraints of our spectrum-sharing environment. For example, per our current requirements, radios must be able to abandon channels within 500 ms, which makes it necessary for our policy reasoner to evaluate transmission requests in less than 500 ms.

The design challenge was to provide a language rich enough to express numerical constraints (e.g., frequency ranges, power limits, temporal intervals) and extensible enough to support unanticipated future policies, while at the same time supporting efficient reasoning by the engine. In the remainder of this chapter, we describe a policy language and engine that meet these challenges for radios that are exploring a new, policy-based approach to spectrum sharing. The language and engine were demonstrated on actual radio hardware used in field experiments.

17.1.1 Opportunistic Spectrum Access Using Policies

Currently, static, centralized policies are developed by regulatory bodies and hardcoded into the radio fabric. Such policies are used to control frequency allocation [1]. This regime cannot adapt to the rapidly changing spectrum needs of users from the government, military, public safety, and commercial worlds. DARPA's NeXt Generation (XG) radio program[1] envisions opportunistic spectrum access [2], which can be realized by achieving the following capabilities:

- Sensing over a wide-frequency band and identifying primary users (PUs)
- Characterizing available transmission opportunities
- Communicating among devices to coordinate the use of identified opportunities
- Expressing and applying policies
- Enforcing behaviors consistent with applicable policies while using identified opportunities

The behavior of a radio is constrained along many operating dimensions (e.g., frequencies, waveforms, and power levels). Because the regulatory environment and application requirements frequently change, it is not realistic to expect a fixed algorithm, written when the radio was built, to be able to constrain the radio's behavior correctly while making efficient use of available spectrum over the life of the radio. Instead, a flexible mechanism must support spectrum sharing, while ensuring that radios will adhere to regulatory policies. It must be able to adapt to changes in regulatory policies, applications, and radio technology. The solution embraced by the XG program is to write formal, machine-processable policies for interference-limiting spectrum sharing in a device-independent PL. A PE in each device enforces the policies. For the remainder of this section, we use *policy, formal policy*, or *declarative policy* to refer to these formal policies. *Regulatory policy* refers to policies typically expressed using natural-language or other semiformal techniques.

This chapter focuses on PL and PE because they constitute major technological challenges. However, other components are necessary for a policy-based CR system: policy authoring and administration, policy analysis and validation tools, and policy dissemination mechanisms. Solutions toward these components have been shown in various XG demonstrations [3, 4]. We also do not attempt to give an overview of existing approaches to policy-based radios. Other approaches are also found in [3, 5–8].

---

[1]The DARPA NeXt Generation radio program uses specific requirements and terminology, which are used extensively in this chapter as an example of CRs. Even though some terminology may differ from that elsewhere in this book, the corresponding terminology should be clear to the reader.

Our approach allows policies to be dynamically changed and ensures that radio behavior is compliant with these policies. The former is achieved by expressing regulatory policies in a declarative language based on formal logic, and allowing devices to load and change policies at runtime. Instead of embedding policies in hardware, firmware, or device-specific software, our approach enables devices to adapt their behavior in a flexible way by changing declarative policies.

Two capabilities supported in the PE ensure that radios behave in accordance with policies: (1) the capability to check transmission requests for policy conformance and (2) the capability to search for desirable transmission opportunities.

For the PE to grant or deny access to spectrum, the radio must submit a request consisting of a set of values that fully characterize the current situation of the radio (sensed data, radio characteristics, and so on) that is sufficient for the PE to decide whether the request is in conformance with the policies. The PE can thus reply to the radio either *yes* (meaning "transmission with parameter values as specified in the request is allowed") or *no* (meaning "transmission with parameter values as specified in the request is not allowed").

There might be cases where a radio cannot, or does not form such a fully specified request, and thus the PE cannot come to a conclusive decision about granting or denying access to the requested spectrum. A basic radio that requests authorization to transmit may not identify all the requirements it needs to fulfill prior to making a request. This may occur because the device is not aware of all the applicable policies, or because of a strategy employed by the radio to not initiate costly sensing operations unless required. For example, suppose a radio requests transmission at 100 mW, which the policies permit only if the transmission time is less than 2 ms. Because the radio is not aware of this restriction, the request submitted to the PE will not match the policy rule. This is an example of a transmission request that is underspecified. It is useful if the PE accepts such underspecified requests and is able to determine further policy constraints that must be satisfied. In this example, the engine determines that the request would be valid if the radio limits the transmission to 2 ms. This capability is referred to as a *search for transmission opportunities*.

Details about the advantages of a policy-based approach to radio operation are presented in Section 17.2. The PE architecture that realizes policy-based radio operation is introduced in Section 17.3. A broad range of technology is available when designing and implementing a PE. Section 17.4 reviews the design considerations for languages and reasoning about dynamic spectrum access (DSA). This leads to the design decisions of our PL and PE, presented in Sections 17.5 and 17.6, respectively. We report on field experiments in Section 17.7. We conclude with a brief discussion of lessons learned from our approach and future challenges in Section 17.8. Section 17.9 summarizes the chapter.

17.2 BENEFITS OF A POLICY-BASED APPROACH

In current radios, regulatory policies are programmed, or hardwired, into the radio and form an inseparable part of the radio's firmware. Typically, radio engineers use *imperative* (procedural) languages such as C for radio software. One can envision implementing spectrum-sharing algorithms, and behaviors on radios, using imperative languages such as C, C++, or Java. Complex, imperative programs do not have an easily understood semantics. Although these programs could be used to implement some policies, they are too machine oriented and not expressive enough to generally specify regulatory policies, nor do they have the reasoning techniques required to enforce radio behavior that is compliant with loaded policies.

The obvious drawback of using imperative languages is that any change in regulatory policies requires reimplementation (and reaccreditation) in the firmware of every radio that might operate in bands affected by the changes. Clearly, this approach does not scale well as technological advances lead to an increasing number of radio designs. Further, it is not sufficiently scalable or flexible to deal with frequently changing regulatory policies or those that are written by many different authorities in many countries. Further compounding the problem is the fact that spectrum-sharing regulatory policies will most likely have a larger number of operating dimensions (e.g., the state of sensed spectrum), and may initially change frequently as best practice is discovered or additional opportunities are exploited.

The key difference in our approach is that *declarative* policies are expressed in terms of *what* should be protected or made available rather than *how* spectrum is protected or made available. Such policies are higher level than typical radio code and free from implementation details. This often makes it easier to intuitively grasp the meaning of a policy.

Declarative languages based on logic are expressive enough to specify not only policies, but also define the reasoning techniques. The advantage is that solutions are not given explicitly, and especially not as a sequence of imperative instructions. Instead, solutions are characterized in a more abstract way by using sentences in some logic. Examples of declarative languages are languages for logic programming (e.g., Prolog), constraint logic programming (e.g., Eclipse), functional programming (e.g., ML, Haskell), equational logic programming/specification (e.g., Curry, Maude), first-order logic reasoning (e.g., Otter), higher-order logic reasoning (e.g., HOL, PVS), and description logics (e.g., OWL). All these languages have different expressive power, and hence the implementations of such languages have very different capabilities. In Section 17.4, we provide insights on our design decisions for choosing a PL and implementing a reasoning engine for it.

Several considerations argue for this declarative policy-based approach over encoding spectrum-sharing algorithms directly in radios:

Adaptability of Radio Behavior. Radio behavior can quickly adapt to a changing situation. Whereas policies themselves can be written to behave differently in different situations, declarative policies can also be dynamically loaded without the need to recompile any software on the radio. For example, a policy might be loaded to more aggressively exploit spectrum-sharing opportunities in emergencies.

Ease of Policy Changes. Policy changes can be limited to certain regions, frequencies, times of day, or any other relevant parameters. In addition, regulators can more easily modify policies. Because policies are platform independent, they can be loaded on different types of radios. Each radio runs the PE on the currently loaded policies, so a new policy only has to be loaded to take effect.

Reduced Certification Effort. Our approach decouples policy definition, loading, and enforcement from device-specific implementations and optimizations. One advantage is a reduced certification effort. When policies, PEs, and devices can be accredited separately, accreditation becomes a simpler task for each component. Changes to a component can be certified without accrediting the entire system. In particular, certification amounts to certifying the PE and each policy once, independent of the radio, and then verifying that device configurations correctly interpret the results from the PE. As a consequence, a change in the device results in only one new check. In effect, the cost of accrediting the policies and PE is shared across all radio platforms.

Separation of Radio Technology and Regulatory Policy. An advantage of decoupling policies from radio implementation is that devices and policies can evolve independently over time. If a radio does not understand a policy and cannot fulfill its requirements, it will not have transmissions approved by this policy, thus missing opportunities, but avoiding potentially creating interference. In contrast, if a radio has more capabilities than required by a certain policy, it can use just what is required. Thus, new policies do not require changes in radio software or hardware, and existing policies will work on new radio hardware. Today, a cyclic dependency exists in which regulatory bodies must wait for technology, and technology must wait to see what the policies look like. The policy-based approach allows radio technology to develop in advance of policies and vice versa.

Extensibility. A policy-based approach is extensible with respect to the kinds of regulatory policies that can be expressed. We already know many relevant parameters and the interrelationships among various categories of regulatory policies, including structural relations such as hierarchies. However, we cannot predict the degrees of freedom in policy definition that may be required in the future. Our approach allows for the definition of new policy parameters. For example, future policies may include functional allocation of spectrum; geographic or temporal restrictions; or host nations, authorities, or services restrictions. In addition, extensibility is crucial when policies are used more broadly (e.g., for dynamic networks).

Fine-Grained Control of Spectrum. An unprecedented amount of use and control of spectrum is possible, allowing stakeholders to design spectrum policies (as allowed by regulations) that best fit their objectives.

The basis for policy-defined radios is a PL that acts as an interface between regulators and radio engineers by providing the means to formally express regulatory policies as well as the methods for interpreting policies and determining efficient, automated policy enforcement. Before we dive into the specific design decisions for SRI's PL and PE, we describe how these components fit into the overall architecture of a policy-based CR.

17.3 NEXT-GENERATION SPECTRUM POLICY ARCHITECTURE

At the highest level of abstraction, an XG radio has four main components, as shown in Figure 17.1.

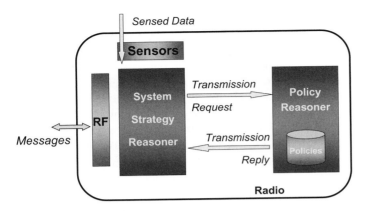

FIGURE 17.1

XG architecture. The small boxes are hardware components. The SSE (also known as SSR) is a radio component that exploits transmission opportunities.

1. *Sensors.* XG radios need sensors to discover available spectrum and transmission opportunities.

2. *Radio Frequency (RF).* The RF component transmits and receives.

3. *System Strategy Engine (SSE).* The SSE controls the radio's transmissions, and can transmit only when the PE has approved transmission. It builds transmission requests based on sensor data received from the sensors and its current strategies. Replies to its requests from the PE may affect strategy.

4. *Policy Engine (PE).* The PE accepts transmission requests from the SSE and checks policy conformance. The PE has all active policies loaded. The PE is platform independent. The PE runs on the radio—as opposed to other possible architectures, including distributed models—and it must be compiled for the target hardware.

Several different types of messages can be sent between the different components of the architecture:

RF–SSE. All incoming messages to the XG radio arrive at the RF unit, and end up in the SSE. These can be control messages (e.g., updates to system strategies, updates to policies, or messages controlling the coordination with other radios). Similarly, all messages going out from the XG radio originate in the SSE and are passed through the RF component. Outgoing messages can also be control messages or data messages.

Sensors–SSE. The details of this interface will be determined by the radio designer. We assume that the sensors send their received data (or conclusions drawn from it) to the SSE. The analysis of sensor data, sensor data aggregation, signal detection, and other such processing could happen in the sensor component(s), in the SSE, or in a dedicated component (not shown). The SSE may send control messages to the sensor components.

SSE–PE. Several types of messages exist in the interface between the SSE and the PE.

- *Transmission requests:* The SSE builds a transmission request, and sends it to the PE. The PE reasons about the request and the active policies, and responds by sending one of three types of replies: (1) the transmission is allowed, (2) the transmission is not allowed, or (3) specified additional constraints must be satisfied. Given acceptable values of the underspecified request parameters, the transmission will be allowed.
- *Policy updates:* The SSE can send policy-update messages to the PE, to update the PE's policy base, by adding or removing policies and activating or deactivating policies.
- *Policy information:* The SSE can request information regarding which policies are loaded or active. The PE never initiates a message exchange. It should be noted that this interface minimizes the amount of "state information" that the PE needs to keep. The only persistent state is the set of loaded policies. Other than that, each message and reply is independent from previous messages and replies.

Given the open-ended nature of the SSE–PE interaction, how should the SSE behave? What should it request? The XG architecture does not offer any answers to these questions. We envision a broad range of radio devices, some of which may have a sophisticated SSE component, and others that may have a simpler SSE.

A simple SSE might put only the requested frequency and maximum power in the request and hope for approval. If additional constraints are returned, it might just try another frequency from a table until the request is approved—or the SSE may just give up. A cognitive SSE, on the other hand, might exploit spectrum opportunities by constructing a request that satisfies the additional constraints returned by the PE. For example, the SSE might have to perform sensing actions and include information about the sensed spectrum in its requests in order for them to succeed.

The specifics of request formation and returned constraints depend on the PL and the domain concepts that are shared between the SSE and PE. We provide such details in Sections 17.5 and 17.6.

17.4 POLICY LANGUAGE AND ENGINE DESIGN

Various design considerations apply to any PL and PE for CRs and DSA. Different choices lead to different expressiveness of language and varying reasoning capabilities. This section provides an overview of the requirements for the PL and PE that inform our design.

17.4.1 Policy Language Design Considerations

In this section we will cover the basic properties of the policy language and what it is intended to specify and accomplish.

Declarative Policy Language

The PL serves as an interface between at least two different viewpoints, namely, between regulators and radio engineers.

The main goal of regulators is the specification of permissible transmission behavior. Regulators are usually not interested in how policy conformance is checked as long as the check is correctly implemented and the radio enforces the policy. This is referred to as the *soundness* of the check. Regulators are not interested in the strategy used to discover opportunities, assuming that policy conformance is ultimately enforced. Furthermore, they are not interested in verifying if a radio's strategy engine can exploit all transmission opportunities. Various trade-offs (e.g., cost of sensing versus need for spectrum), radio capabilities (e.g., ability to sense the spectrum), and the quality (degree of completeness) of the strategy engine itself will all affect which opportunities are exploited.

The main interest of radio engineers, in contrast, is to exploit as many policy-conforming transmission opportunities as possible. Thus, they have an incentive to enhance capabilities of both the strategy engine and the policy-conformance engine.

To support both of these views, a PL with a simple and unambiguous semantics is needed. The foremost objective is to specify—as opposed to implement—policy-conforming behavior. Thus, a declarative language is a considerably better fit than an imperative language such as C.

Permissive and Restrictive Policies

The PL should provide linguistic constructs for both permissive and restrictive policies. Permissive policies describe conditions under which transmission is allowed, and restrictive policies describe conditions under which transmission is not allowed. The conditions in permissive policies impose constraints on a possible transmission, whereas the conditions in a restrictive policy define situations in which all transmission requests will be denied. The XG approach is based on the assumption that XG devices will be designed to operate on a do-not-interfere basis. That is, the device will not transmit if not explicitly authorized to do so by a policy.

Ontology of Domain Concepts

In general, before a request for a transmission can be authorized, three types of information need to be available: the capabilities of the radio, the current environment of the radio, and the characteristics of the requested transmission. Such information is heavily dependent on domain concepts. There is a need for ontologies of domain concepts that are common to the SSE maker, the PE implementor, and the regulatory policy authors.

17.4.2 Policy Engine Design Considerations

The PE must be able to efficiently reason about the policies to perform policy-conformance checks of transmission requests or to search for transmission opportunities. When we investigated the existing declarative languages and their reasoning technologies, we found that they are either not sufficient to address both policy conformance and search, or not efficient enough to meet operational constraints.

Logic Programming Rules

Features of the PL, such as those for defining permissive and restrictive policies, also have implications on the reasoning system used to process policies. It is intuitive to write policies as *rules*—for example, of the form Allow ⟸ *Constraints* (for permissive policies), in which we try to prove Allow by solving the constraints. (Replace Allow by Disallow for restrictive policies.) It is also useful to define auxiliary predicates using rules for modularity, reusability, and convenience of specification. This speaks in favor of a logic programming type of language.

Reasoning about Numerical Constraints

Policies often define conditions about transmission opportunities in terms of numerical constraints. For example, a policy may restrict transmissions to carrier frequencies within a certain range, or to situations in which the peak transmitted power is lower than a given threshold. These conditions are defined as numerical constraints that the PL must be able to represent and about which the PE has to adequately reason. For example, if a policy defines the allowable range of carrier frequencies to be between 3100 and 3300 MHz, then the PE must be able to recognize whether a given carrier frequency is within this range. This problem is made more complex by combinations of interacting constraints from multiple conditions and policies.

Returning Constraints for Underspecified Requests

A policy conformance engine can determine whether a request satisfies a policy base so it can return "yes" or "no" answers. Searching for transmission candidates requires an additional reasoning feature, that of returning "yes, if . . ." answers—answers giving additional constraints that, when fulfilled, will allow the request to take place. For example, a radio might request a frequency band from 1000 to 1002 MHz, and the engine might answer "yes, if the transmitted power is less than 10 dBm," or "yes, if you have sensed that there is no other signal detected over -90 dBm within 300 KHz of the carrier center frequency." This capability is important for several reasons:

- In our architecture, the *cognitive* elements of the radio are the SSE and PE. The SSE determines what transmission properties can be used, sends a request to the PE, and receives a reply. Because the SSE must be agnostic as to which policies are loaded, it cannot know in advance which requests will be approved. The only information it has about the policy constraints must come from the PE. With only "yes" or "no" answers, there is very little information coming from the PE, and the SSE has to search blindly for transmission opportunities. If, however, the PE can return "yes, if . . ." answers, the SSE can use the returned constraints to guide its search for transmission opportunities. This will become increasingly important as spectrum is more fully utilized and opportunities are rarer.

- Policies may rely on a large variety of information that the SSE can provide. It is not realistic for the SSE to provide all this information in every request, because there is often a cost associated with gathering information. For example, some policies may have geolocation constraints, or constraints on sensed power in order to avoid primary users. A radio may have a GPS and power sensors, but not use them all the time (e.g., to preserve battery life). With a PE that returns constraints, the SSE can choose to activate its sensors only when there is a constraint that requires it. Furthermore, the constraint may include information on which band to sense, so that the SSE can perform a focused sensing operation on a small band.

- In our vision of CRs, policies and radio capabilities are described using ontologies. It is unreasonable to assume that every radio manufacturer and every policy author will agree on one ontology that will remain unchanged indefinitely. Certainly, some similarities are necessary if there is to be any hope of interoperability, but we expect that different stakeholders will introduce specific ontological extensions to suit their needs. This extensibility means that the PE may comprise policies containing ontological elements that the SSE does not understand. The SSE will never put those elements into its requests, and will, with a simple yes/no engine, get "no" answers back, without knowing why. With a constraint capability, however, the SSE will see that there is a constraint involving some unknown element, and when these conditions occur, it may steer its transmission requests toward opportunities that do not involve this unknown element.

- An engine that can return constraints is needed for policy analysis. For example, policy authors can examine the results of combining different policy sets. If the engine returns "yes" when given an empty request, we know that the policy base is too permissive—it will allow every request. In contrast, if it returns "no", we know that it is too restrictive—it will deny every request. If there exists any chance of getting approval from the engine, a constraint would have been returned. Thus, we can make sure that a set of policies is consistent, and this cannot be done with a simple yes/no engine.

- The capability to return constraints is important for other types of policies beyond spectrum policies. For example, when evaluating routing policies, a set of allowable routes as constraints may be returned. For network management policies, the constraints may represent network configurations that satisfy a mission goal.

Stateless Policy Engine

It is imperative that the first generation of engines used in this framework be easy to verify and implement. The PE should, therefore, be designed as a stateless system that does not maintain the state of the SSE and PE interaction, and thus every request can be viewed as an atomic request. A stateless PE is much easier to verify, and is also simpler to design and implement. A stateful engine might be nearly impossible for governing authorities to verify because of the potentially unbounded number of states and the unpredictability of the quality and timing of state updates.

In summary, the core reasoning problem in XG is to infer, from a given set of policies and a given set of facts asserted by the radio, whether a transmission is permitted or not. Transmission is not allowed if permission cannot be inferred in the time allowed. In such a case, the PE should conclude what other constraints must be fulfilled to approve transmission. The engine may return a number of solutions, each representing one choice of parameters that would lead to a valid transmission.

Thus, we need a declarative language that allows rules in the logic programming style, functions defined using equations, and flexible handling of numerical constraints to meet the requirements on the language used to define the policies and the reasoning system used to process them. However, no declarative language and reasoning system exists that satisfies all requirements. In the next two sections we describe the design of our PL and the implementation of our policy reasoner.

17.5 SRI SPECTRUM POLICY LANGUAGE

Based on the design considerations discussed in Section 17.4, SRI developed a language called CoRaL [9–12], Cognitive Radio Language, based on typed first-order logic with equality. CoRaL was given model-theoretic and operational semantics in terms of our Universal Policy Logic (UPL) [13]. Subsequently, we moved to Semantic Web Rule Languages for our syntax, due to a need to use standard languages. In particular, we use Web Ontology Language (OWL) [14], Semantic Web Rule Language (SWRL) [15], and Semantic Web Rule Language First-Order Logic (SWRL FOL) [16]. We also defined a few extensions to SWRL FOL that proved useful, and in some cases necessary,[2] but we will not discuss those extensions in detail here (see [17, 18]). We also mapped these Semantic Web languages to UPL [19], which means that there is a well-defined mapping between CoRaL and OWL+SWRL. Here, we give a high-level overview of the capabilities of our PL, mostly through examples.

Typical spectrum policies restrict or permit transmissions on the basis of frequency, location, time, device capability, node identity, or sensed data. Other conditions may also be used as needed. Some simple example policies, encodable in CoRaL, follow.

- *Frequency Band.* "Allow transmission between 5180 MHz and 5250 MHz."
- *Time.* "Disallow transmission between 00:00 and 5:30 GMT time and between 22:00 and 23:59 GMT."
- *Location.* "Allow transmission if radio is at most 30 miles away from the geographic coordinates (39°10′ 30″ N, 75°01′ 42″)."
- *Node Identity.* "Allow transmission if radio belongs to the Red Cross."
- *Sensed Data—Listen before Talk.* "Allow transmission if radio's peak sensed power in the band is at most −80 dBm within 10 MHz of carrier center frequency and its EIRP for transmission is at most 10 dBm."

Our PE (see Section 17.6) tries to prove a special implication. By defining this predicate in terms of other predicates, we can achieve a very flexible policy structure. Typically, this top-level predicate is:

```
Permit iff Allow and not Disallow
```

This lets us define permissive and restrictive policies, as discussed in Section 17.4. A policy is composed of several *rules*. A policy must have at least one Allow or Disallow rule. To support permissive as well as restrictive requirements, rules use either the Allow or Disallow predicate, respectively. Policy rules are logical axioms that express which *conditions* these predicates hold. These axioms can involve any

---

[2]The extensions are inductive predicates and function applications.

declared parameters, which represent capabilities of the radio and the results of sensing actions (among other things).

Conditions can also use predicates, which express modes of operation, locations, and so forth. Thus, conditions allow for dynamic adjustment of policies to the current situation. For example, a rule could allow military radios to use the Global System for Mobile (GSM) band when a conflict starts, but not earlier. Such context-sensitive policies can respond to the situation in various ways, invoking either restrictive or permissive rules.

Numerical constraints are often used in policy specifications and can be directly expressed by using built-in predicates (e.g., $<$, $\leq$). For example, a policy might require that for frequencies of 5000 to 5500 MHz, the transmission power should be at most 10 dBm.

A policy can also be extended by rules in another policy without causing logical inconsistencies. For example, one policy may have a rule allowing the use of frequencies of 5000 to 5500 MHz, whereas another policy might disallow the use of a carrier frequency of 5250 MHz,[3] but allow frequencies of 5200 to 6000 MHz. Thus, the combination of policies will allow the use of frequencies of 5000 to 6000 MHz, with the exception of a frequency of 5250 MHz. The rule above has the effect that restrictive (`Disallow`) rules take precedence over permissive (`Allow`) rules.

Policies and ontologies can refer to concepts defined in other ontologies. This capability supports modular specification and reuse of policies and ontologies.

17.5.1 Ontologies

The SSE and the PE share knowledge of domain concepts such as frequency, power, location, and signal characteristics. Shared concepts are defined in common ontologies so that the SSE and the PE, as well as various policymaking bodies and SSE implementors, can consistently and unambiguously refer to radios, radio capabilities and parameters, and the relevant properties of the current radio environment. However, because not all uses of the PL can be foreseen, and not all policies or radio capabilities can be known in advance, the language must support user-defined concepts that can be added as needed and used in policy specification.

The ontologies developed so far deal with common entities such as bandwidth, frequency, power, radio capabilities, evidence, signals, time, powermasks, and transmissions. The structure of the ontologies, in terms of class–subclass, property–subproperty, and class–instance relationships, as well as property domains and ranges, is shown in Figure 17.2. The ontologies also contain additional axioms (not shown). These ontologies give an extensible base for the parameters over which policies can be formulated, but should not be regarded as the only possibility.

As an example, we consider a fragment of the ontology for transmitters.[4] One property of the `Transmitter` class is the `centerFrequency` it is set to use. This request

---

[3]Actual requirements are more likely to specify not only frequency but also bandwidth, such that interference is not produced at 5250 MHz regardless of carrier frequency and modulation chosen. Throughout this chapter, simple examples have been chosen.

[4]Because the application is the XG program and is restricted in frequency and bandwidth, it is possible to simplify coding of frequency to integers.

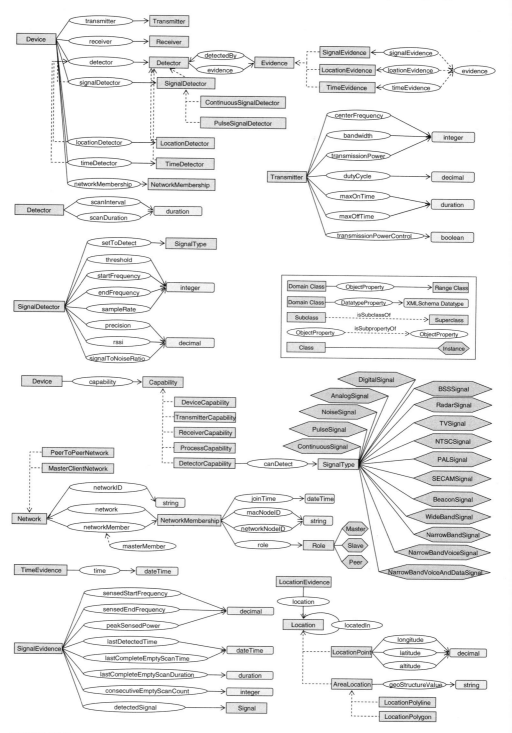

FIGURE 17.2

Example ontologies.

parameter, together with the `bandwidth`, determines the band in which the radio wants to operate (namely, `centerFrequency +/- bandwidth/2`).

```
Class(xg:Transmitter)
DatatypeProperty(xg:centerFrequency Functional
   domain(xg:Transmitter) range(xsd:integer))
DatatypeProperty(xg:bandwidth Functional
   domain(xg:Transmitter) range(xsd:integer))
DatatypeProperty(xg:transmissionPower Functional
   domain(xg:Transmitter) range(xsd:integer))
DatatypeProperty(xg:dutyCycle Functional
   domain(xg:Transmitter) range(xsd:decimal))
DatatypeProperty(xg:transmissionPowerControl Functional
   domain(xg:Transmitter) range(xsd:boolean))
DatatypeProperty(xg:maxOnTime Functional
   domain(xg:Transmitter) range(xsd:duration))
DatatypeProperty(xg:minOffTime Functional
   domain(xg:Transmitter) range(xsd:duration))
```

17.5.2 A Note on Notation

In the remainder of this section, and in Section 17.6, we make use of some common mathematical/logical notations. Details can be found in any introductory textbook on logic. Here, we have space for only a brief reminder.

Terms are variables (x, y), constants (c, d), or function applications. A function application consists of a function symbol (f, g) followed by a number of terms, in parentheses (e.g., $f(x, g(y, c))$). Note that this is a recursive definition. Every term has a *denotation*, or value.

Formulas are built from *atomic* formulas, *logical connectives*, and *quantifiers*. Atomic formulas are the special formulas `True` and `False`, and predicate applications. A predicate application consists of a predicate symbol (P, Q) followed by a number of terms in parentheses, for example $(P(x, f(x, y, c)))$. The logical connectives combine atomic formulas into larger formulas: $\wedge$ for "and" (conjunction), $\vee$ for "or" (disjunction), $\Rightarrow$ for "implies," $\Leftarrow$ for "implied by" (backwards implication), $\neg$ for "not" (negation), and $\Leftrightarrow$ for "if and only if" (equivalence). The *quantifiers* are $\exists$ for "exists" (existential quantification) and $\forall$ for "for all" (universal quantification). Formulas are either true or false.

We use `this` font to indicate concrete identifiers in the language, and *this* font as "metavariables"—to stand for *any* formula. In Section 17.6, we use $\rightarrow$ to stand for the rewrite relation, and this should not be confused with $\Rightarrow$ (implication). We also use the "proves" relation, which is written $\vdash$. For example, $A \vdash B$ means that "A proves B," or B is a logical consequence of premise A, given the proof system in question.

17.5.3 **Policy Examples**

Because the XML (eXtensible Markup Language) syntax for policies gets rather verbose, we use a syntax similar to the "human-readable syntax" suggested by the SWRL specification[5] for some policy examples. This syntax uses a subset of the notation discussed earlier. We make a few modifications for readability:

- We write the implications backward, with the consequent before the antecedent, as in Prolog (e.g., a⇐b∧c instead of b∧c⇒a).
- We write some of the SWRL built-in language functions in infix notation (e.g., a<b instead of `swrlb:lessThan(a,b)`).
- We use a functional notation for all functional properties, not just for functional built-ins (e.g., `transmitter(?d)=?tr` instead of `transmitter(?d;?tr)`).
- We omit the `xg:` prefix everywhere for brevity (consider it the default name space).

By using our XG ontology, we can express the policy "Allow a device with bandwidth less than 25,000 Hz and power less than 10 dBm to transmit," as a single permissive rule (policy 1).

```
Allow ⇐(Device(?d) ∧ transmitter(?d) = ?tr ∧
bandwidth(?tr) < 25000 ∧ transmissionPower(?tr) <10)            policy (1)
```

The next example (policy 2) is a so-called "listen before talk" policy, which requires the radio to provide signal evidence showing that there is no other signal in the spectrum over a certain threshold.

```
Allow ⇐(Device(?d) ∧ signalDetector(?d, ?sd) ∧
signalEvidence(?sd, ?se) ∧ peakSensedPower(?se) < −100)         policy (2)
```

Another common type of policy constraint is time of day, shown here (policy 3) using `swrlb:dateTime`, a SWRL built-in predicate supported by the PL and PE. This policy allows transmissions except between noon and 1 p.m.

```
Allow ⇐(Device(?d) ∧ timeDetector(?d, ?td) ∧
timeEvidence(?td, ?te) ∧ time(?te, ?t) ∧ swrlb :
dateTime(?t, ?year, ?month, ?day, ?hour, ?min, ?sec) ∧
?hour ≠ 12)                                                     policy (3)
```

Finally, we show a slightly more complex policy that has a constraint on the total emissions of the radio. A policy can define the maximum emissions as a function, using conditional equations such as the following.

```
maxPower(?f) = 0.0 ⇐ ?f < 10.0
maxPower(?f) = 10.0 ⇐ 10.0 ≤ ?f ∧ ?f < 30.0
maxPower(?f) = 20.0 ⇐ 30.0 ≤ ?f ∧ ?f < 40.0
```

---

[5]*www.w3.org/Submission/SWRL/#2.2.*

```
maxPower(?f) = 10.0 ⟸ 40.0 ≤ ?f ∧ ?f < 50.0
maxPower(?f) = 0.0 ⟸ 50.0 ≤ ?f
```

The policy is written as a rule (policy 4) that refers to the function defined above and is illustrated as a powermask in Figure 17.3 with a dotted line.

```
Allow(?f) ⟸(Device(?d) ∧ transmitter(?d) = ?tr ∧
emissions(?tr, ?f) < maxPower(?f))                                policy (4)
```

17.5.4 Requests and Replies

A transmission request is a set of facts (ground atoms or equations) that uses the concepts in the ontology, for example:

```
reqEmissions(?f) = 0.0 ⟸ ?f < 15.0
reqEmissions(?f) = 8.0 ⟸ 15.0 ≤ ?f ∧ ?f < 25.0
reqEmissions(?f) = 15.0 ⟸ 25.0 ≤ ?f ∧ ?f < 35.0
reqEmissions(?f) = 8.0 ⟸ 35.0 ≤ ?f ∧ ?f < 40.0
reqEmissions(?f) = 0.0 ⟸ 40.0 ≤ ?f
Device(d1)
transmitter(d1, t1)
emissions(t1, ?f) = reqEmissions(?f)
timeDetector(d1, td1)
timeEvidence(td1, te1)
time(te1, t)
swrlb : dateTime(t, 2008, 4, 4, 11, 30, 0, 0)
signalDetector(d1, sd1)
```

This request defines a function reqEmissions, and states that this function is the emissions of the radio's transmitter, along with a number of facts concerning time and signal detection. The emission mask is depicted in Figure 17.3 with a dashed line.

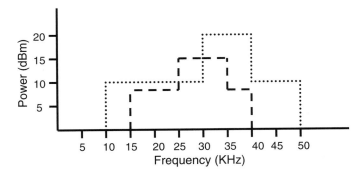

FIGURE 17.3

Powermasks defined as functions from frequency to power. The powermask depicted using the dotted line defines maximum emission for a policy, and the powermask depicted using a dashed line defines the emission of a proposed radio waveform used in a request.

If this request was submitted to a PE that had loaded only policy (1), the reply would be a constraint:

```
bandwidth(t1) < 25000 ∧ transmissionPower(t1) < 10
```

For policy (2), the reply would be:

```
signalEvidence(sd1, ?se) ∧ peakSensedPower(?se) < -100
```

For policy (3), we get the reply `True` because all the constraints are satisfied. For policy (4) we get a constraint:

```
?f < 25 ∨ 30 ≤ ?f
```

This is the frequency range for which the emissions in the request are less than the maximum power defined in the policy.

17.5.5 Spectrum Policy Language Summary

The PL has been designed to be extensible and customizable. There are many ways to take advantage of these functionalities. Policy authors can specify additional (customized) domain knowledge as ontologies, which are imported into their policies and are reusable by other policies. Libraries of customized policies and concepts specific to a particular regime can be developed, accredited, and used by other authors. Such libraries further raise the level of abstraction as seen by authors and ease certification burdens.

In the realm of usability, policy authors will not have to see the PL syntax if they use an authoring tool. A first prototype of such a tool, providing a selection of appropriate abstractions for creating policies, was presented during a demonstration in March 2008 (see Section 17.7).

17.6 SRI POLICY ENGINE

The XG architecture allows for different PEs to be plugged in, as long as they follow the common protocol described in Section 17.3. The Stanford Research Institute (SRI) has developed two policy engines for the XG program. The first one was described by Denker et al. and Jondral and Marshall [10, 20] and demonstrated to the general public at the First and Third *IEEE International Symposia on New Frontiers in Dynamic Spectrum Access Networks (DySPAN)* in 2005 and 2007, respectively. It was also shown to military personnel in August 2006 at Fort A. P. Hill in Virginia [4, 20]. This engine was based on the CoRaL language (see Section 17.5) and used Prolog as the underlying reasoning engine. This engine was able to efficiently determine whether a request satisfied a policy base, that is, to return "yes" and "no" answers. However, it lacked an important feature: That of returning "yes, if . . ." answers giving additional constraints that, when fulfilled, would allow the transmission to take place. We therefore developed another PE based on the Maude system [21] that is able to return constraints. This section describes the reasoning principles underlying our new reasoner, and then goes into the implementation in some detail.

17.6.1 Reasoning with Constraints

We can view the reasoning problem as that of proving

Facts, policies ⊢ `Permit`

where ⊢ is the *proves* relation. Using rules and definitions from the policies, we can replace `Permit` with one big constraint containing the meaning of all the policies combined:

Facts ⊢ *Constraint*

The preceding is equivalent to

`True` ⊢ *Facts* ⇒ *Constraints*

and, to complete the proof, we now have to reduce the right side of the rule to `True`. If we can do that, then we have shown that the facts and policies permit transmission, and the PE will return a "yes" result. If we can reduce the right side to `False`, then we have proven that the policies and facts do not permit transmission, and return a "no" result. If the right side reduces to neither `True` nor `False`, we return the remaining constraint in a "yes, if..." result. The remaining constraint is what remains to be proven in order to transmit. With this reasoning principle, the yes/no reasoning is a special case of the reasoning required to return constraints. We *always* operate on constraints, and sometimes they reduce all the way to `True` or `False`.

As a simplistic example, we consider the following policies:

```
Permit   ⟺ Allow ∧ ¬Disallow
Allow    ⟸ 500 < freq ∧ freq < 1000
Allow    ⟸ 1200 < freq ∧ freq < 1400
Disallow ⟸ 900 < freq ∧ freq < 1300
```

The first policy is a "top-level" policy, which relates permissive and restrictive policies. The given top-level policy just says that we need to find *some* policy that allows, and *no* policy can disallow. Other top-level rules can be used to account for priorities and other relationships between policies. With these policies, after expanding the definition of `Permit`, we get the proof obligation

Facts ⊢ (500 < freq ∧ freq < 1000 ∨ 1200 < freq ∧ freq < 1400)
∧ ¬(900 < freq ∧ freq < 1300)

which can be further simplified to

Facts ⊢ 500 < freq ∧ freq ≤ 900 ∨ 1300 ≤ freq ∧ freq < 1400

If *Facts* contains, for instance, freq = 800, the whole constraint simplifies to `True`, which means that the proof is complete and the radio can transmit. However, we are also interested in the case where such facts are *not* provided because radios can make underspecified requests as a way of querying for available transmission opportunities. Whatever remains of the constraint after simplification has been performed should be returned as a result to the radio.

Thus, constraint simplification is the main reasoning technique used. Simplification is performed by using proof rules that encode valid simplification steps. For example,

A ∧ (B ∨ C) → (A ∧ B) ∨ (A ∧ C)

(distributivity of conjunction) is a valid proof rule in this system. More interesting are the rules that have to do with the combination of ordering constraints and Boolean constraints; for example,

¬(a < b) → b ≥ a

Because we want the constraint returned to contain *all* the opportunities allowed by the policy base (a form of *completeness*), all the proof rules are actually *equivalences*. In other words, whatever is on the left side of a proof rule is equivalent to what is on the right side.

So why do we simplify the constraint at all, if all we are doing is rewriting it in a different form? The answer is that we want to modify the constraint to the "simplest" form. It is not obvious what the simplest form is, but in this case we have some guidance from the domain. The PE should return something that can be recognized by the SSE as a set of opportunities, where each opportunity is straightforward to understand. We can achieve this by using the *disjunctive normal form* (DNF) as the target of our constraint simplification. A DNF formula is a disjunction of conjunctions, that is, it has the form

(A ∧ B ∧ C) ∨ (D ∧ E) ∨ (F ∧ G) ∨ ...

We can think of each of the conjuncts, (A ∧ B ∧ C), (D ∧ E), or (F ∧ G), as an opportunity because it is enough for the SSE to satisfy *one* of the disjunctions in order to satisfy the entire constraint. For example, if the radio provides the facts D and E, then formula (2) simplifies to

(A ∧ B ∧ C) ∨ True ∨ (F ∧ G) ∨ ... → True

by using the rule

True ∨ *P* → True

Furthermore, for the chosen opportunity, all the constraints have to be satisfied.

Simplifying to DNF, however, is not enough. For example, the following constraint is in DNF:

$$(\text{freq} < 500 \land \text{power} < 10) \lor (\text{freq} > 400) \tag{17.1}$$

but it can be simplified further to

$$\text{power} < 10 \lor \text{freq} > 400 \tag{17.2}$$

Constraint (17.2) is still in DNF, and is equivalent to, but in some sense simpler than, constraint (17.1). Our proof system takes care of such cases.

17.6.2 Implementation in Maude

We chose to implement a PE based on the principles previously discussed in the Maude system [21]. We briefly describe Maude, and then discuss why this was an appropriate choice. Maude can be thought of as containing two layers. The first layer is an *equational logic* [22]. This logic consists of three main elements: sorts, operators, and equations. The second layer is a *rewrite logic*, which is discussed in this section.

Maude has a `reduce` command for equational reduction or simplification in functional modules (modules with equations only, no rewrite rules), and a `search` command for a breadth-first search in the state space of system modules (system modules can have rewrite rules and equations). The search mechanism allows searching for the first answer, all answers, or only answers matching some goal term. The search mechanism encompasses the reduction mechanism, as equational reduction is performed between each application of rewrite rules.

Maude also has a `narrow` command. Narrowing in Maude is similar in many ways to the `search` mechanism mentioned above. Like search, narrowing nondeterministically selects rewrite rules, generates choice points, and can return answers in the same ways. There are two differences: (1) new variables are allowed on the right side of rewrite rules, and (2) when there are uninstantiated variables in a reducible expression (redex), unification is used instead of matching.

The addition of narrowing to a functional language gives us the ability to subsume logic programming. We encode relations and logical connectives as functions operating on the sort `Trm`, which is the sort we use for all constraint terms, since our proof system operates only on constraints.

```
sort Trm .
```

The logical connectives are defined through statements such as:

```
op _and_ : Trm Trm -> Trm [assoc comm prec 55] .
op _or_  : Trm Trm -> Trm [assoc comm prec 59] .
op not_  : Trm -> Trm [prec 53] .
vars A B C : Trm .
eq true and A = A .
eq false and A = false .
eq A and A = A .
```

The `not` operator takes one `Trm` as its argument and produces another `Trm`, the negation of the argument; `or` and `and` take two `Trm`s as arguments and produce a `Trm` as the disjunction and conjunction, respectively, of the arguments. The `assoc` and `comm` attributes declare the associativity and commutativity of the operators. The `prec` attributed set the precedence of an operator, so that the mixfix notation is parsed correctly without a proliferation of parentheses. We use *equations* to encode our proof rules, such as the distributive rule:

```
A and (B or C) = (A and B) or (A and C)
```

where A, B, and C are declared as variables of the `Trm` sort and the equation directly rejects the rule.

Similarly, an *n*-ary user-defined predicate P is encoded as

```
op P : Trm ... Trm -> Trm .
```

Rules are encoded as rewrite rules. For example,

$\forall x,y,z : uncle(x,y) \Leftarrow (father(x,z) \wedge brother(z,y))$

is encoded in Maude as

```
vars x,y,z : Trm .
rl uncle(x,y) => father(x,z) and brother(z,y) .        (17.3)
```

Facts are rules without bodies. For example, *father (John; Bob)* is encoded as

```
rl father(John,Bob) => true .                          (17.4)
```

Note the use of rewrite rules (`rl`) in codes (17.3) and (17.4) rather than equations (`eq`) for user-defined facts and rules. The reason for this is that we want to be able to use narrowing on these facts and rules.

One detail worth mentioning at this point is the treatment of negation in this encoding of Boolean logic. Maude does not have a built-in notion of negation-as-failure. The `not` operator is simply a truth function that negates the truth value of its operand. This corresponds to classical negation, and is similar to how negation is treated in other narrowing systems.

Advantages of Using Maude

Using equations as proof rules is especially appropriate for several reasons. First, we want our proof rules to have the form of equations so that we maintain completeness. Second, we read the proof rules as transformations from the left side to the right side. This happens to be precisely how Maude treats equations when we apply its *reduction* mechanism to an equational specification. When we ask Maude to *reduce* a term, it applies the given equations to the term in order to rewrite it, as long as any equations apply (Maude specifications are supposed to be *terminating*, so at some point, no equations will apply).

This reduction in Maude is highly optimized, which allows us to use Maude as a practical implementation language for our PE. Another feature of Maude worth mentioning is that it takes care of *associativity* and *commutativity* of operators in an efficient, built-in way. For example, for `and`, we have that

```
A and B = B and A                                      (17.5)
```

This equivalence is needed because we have other proof rules, such as

```
A and True = A
```

that will not recognize a constraint of the form `True and A`. However, if we include the commutativity Eq. (17.5), our proof system will not terminate because this equation can be used to move between the two equivalent forms back and forth forever. Instead, we mark the `and` operator as commutative with a special keyword that Maude recognizes:

```
op _ and _ : Trm Trm -> Trm [comm] .
```

Now, Maude will recognize that `True and A` matches the left side of Eq. (17.5), and will rewrite it accordingly. It is also clear that encoding the proof rules as Maude equations gives us a very direct implementation of the PE. If we were to implement the proof system in a common programming language such as C, it would be very difficult to understand and certify the engine. With our solution, if we trust that Maude itself does the right thing, it is easy to check that the implementation of the proof rules is correct with regard to a specification of them, because the *implementation is essentially the same as the specification!* In other words, Maude allows us to write *efficiently executable specifications*.

Encoding Policies in Maude

To use Maude as our reasoning engine, we need to encode our policies in Maude. The policies are written as SWRL rules (i.e., Horn clauses[6]) and refer to OWL ontologies. We can also translate a significant portion of axioms from OWL ontologies into Horn logic. Once we have all our statements in Horn clause form, it is straightforward to encode them in a very direct way in Maude, using the scheme described in the previous paragraph. Some specifics of the encoding follow.

First, we note that OWL does not have types or sorts in the sense that Maude or other programming languages do. However, Maude operators need to be declared with sorted signatures. We introduce one sort `Trm` and translate OWL individuals, classes, and properties as follows (*** denotes a comment in Maude):

```
op Radio1 : -> Trm . *** an individual
op Radio : Trm -> Trm . *** a class
op detector : Trm Trm -> Trm . *** a property
```

The signatures here are perhaps best understood as follows: An individual is an operator with no argument that returns itself (a constant). A class is an operator that takes an individual as an argument, and returns `True` or `False`, respectively, depending on whether or not the individual is a member of the class. A property is an operator that takes a subject and an object as arguments and returns `True` or `False`, respectively, depending on whether that subject has that object as a property value for the property in question.

We treat *functional* properties separately, translating them as

```
op role : Trm -> Trm . *** a functional property
```

where the operator works as a function; that is, it takes the subject as an argument and returns the object. This encoding makes reasoning more efficient in Maude, since it is essentially a functional language. We can now translate facts from the ontology into Maude *rewrite laws*, which constitute Maude's *rewrite logic* mentioned before. For the time being, we can think of rewrite laws as being similar to equations. The differences, and the reason why we do not use equations to encode facts, will become apparent shortly. Some examples are

```
rl Radio(Radio1) => True . *** class-instance fact
rl detector(Radio1,Detector1) => True . *** property-value fact
rl role(Radio1) => Slave . *** functional property-value fact
```

---

[6]A Horn clause is a disjunction of literals with at most one positive literal.

Finally, we can also encode Horn clauses into rewrite laws. For example,

$$\forall x,y : P(x,y) \Leftarrow Q(x) \wedge R(y)$$

is encoded in Maude as

```
vars x,y : Trm .
rl P(x,y) => Q(x) and R(y) .                                    (17.6)
```

(Recall that and is an operator, and : Trm Trm -> Trm.) The reason we cannot use an equation, such as eq P(x,y) = Q(x) and R(y), is that the left side is *not* equivalent to the right side. We do not, in general, have definitions of our predicates, and therefore, we cannot use Maude's equational logic and its associated *reduction* mechanism for these rules. There could be additional rules with P(x,y) on the left side, and the result of the reasoning would depend on which rule was selected. Fortunately, Maude provides a different solution: its rewrite logic and the associated search mechanism. With search, when Maude encounters several rewrite laws that apply to the term being operated on, Maude splits into several cases, one for each rewrite law. For example, if we have rule (17.6) and an additional rule,

```
rl P(x,y) => P'(x,y) .
```

and execute

```
search P(x,y) => T:Trm .
```

then we get back two answers,

```
Q(x) and R(y)
P'(x,y)
```

We have to deal with one more complication. The Horn clauses can contain new variables on the right side, for example,

```
vars x,y : Trm .
rl P(x) => Q(x) and R(y) [non-exec].
```

This is not supported by Maude's search mechanism, and such rules have to be marked as nonexecutable as shown. The new variables are *existentially quantified*, and we would like Maude to speculatively instantiate them with matching individuals, and then backtrack and try other matches. Fortunately, Maude also has support for this type of behavior, with its *narrowing* mechanism. For example, given the partial specification

```
rl R(foo) => True .
vars x,y : Trm .
rl P(x) => Q(x) and R(y) and P'(y) [non-exec] .                (17.7)
```

executing

```
narrow P(bar) => T:Trm .
```

will yield

```
Q(bar) and P'(foo)                                            (17.8)
```

as one solution. Narrowing augments the search mechanism with a way to instantiate existential variables.

This solution was produced as follows: First Maude matches P(bar) with P(x) in the rule, producing the solution Q(bar) and R(y) and P'(y). Because the solution contains variables, Maude tries to eliminate them by narrowing. This works by unifying the terms containing variables with the left sides of other rewrite laws. In this case, R(y) unifies with R(foo), which is rewritten to True, and yields the substitution y\foo. The resulting substitution is performed in the source term, giving the solution Q(bar) and True and P'(foo). This solution reduces, using the equational proof rules to Q(bar) and P'(foo) (search and narrow both include the reduction mechanism). This is similar to how Prolog operates.

In summary, we make use of two reasoning mechanisms in Maude, reduction and narrowing, both used for different purposes. Narrowing is used to find a set of solutions to the policy rules. Reduction is used to simplify those solutions to a usable form by using the proof rules.

Components

In the preceding discussion, we addressed the *principles* of our engine. Here, we describe the main components of our implementation. The engine needed to be implemented in C/C++ in order to run on resource-constrained radios. Maude is implemented in C++ and highly optimized, so this posed no particular problem. The policies are written in OWL, SWRL, and SWRL FOL, using the XML presentation syntax. There was no existing C/C++ parser for OWL, and no software support at all for SWRL FOL. Thus, we implemented our own parser/writer for OWL+SWRL+SWRL FOL, using the XML presentation syntax. The parser generates an internal representation based on our UPL [13], implementing the translation detailed in Patel-Schneider [19]. This representation can be used with different back ends, such as the Maude back end, which SRI implemented. Thus, the engine and the syntax are independent of each other, as long as they can both be converted to and from the UPL representation.

The Maude reasoner back end consists of three main parts: the UPL-to-Maude converter, the Maude reasoner specification, and the Maude-to-UPL converter. The UPL-to-Maude converter translates UPL elements to a Maude encoding, as described earlier. The Maude reasoner specification contains Maude equations representing the proof rules. The Maude-to-UPL converter is used to translate the result of the Maude reasoning back to UPL. All these components and their relationships are shown in Figure 17.4.

On a more detailed level, the Maude reasoner specification consists of a number of Maude modules (see Figure 17.5). A Maude module is a collection of Maude statements (sort and operator declarations, equations, and so on). The modules are TRM, TIME, GEO, POLICY, SIMP, CNF, DNF, and REASONER. The TRM module contains basic definitions of the Boolean algebra, arithmetic, and ordering constraints of the UPL language. The TIME and GEO modules contain *built-in* functions for temporal and geospatial reasoning, which can be used as SWRL built-ins in policies. The POLICY module contains the translation of all the policies and facts that are currently loaded. When the reasoner first starts, the POLICY module is empty. Whenever a new policy is loaded, the module gets overwritten with a new version. SIMP, CNF, and DNF contain different parts of the proof system. SIMP does a number of simplifications, such as eliminating negation (as far as possible) and implication. CNF converts to conjunctive normal form and does some simplifications that can be done only in this form. Similarly, DNF converts to disjunctive

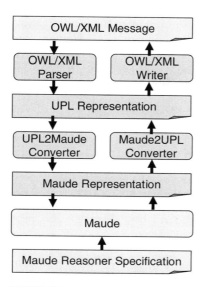

FIGURE 17.4

Components of the SRI policy engine implementation.

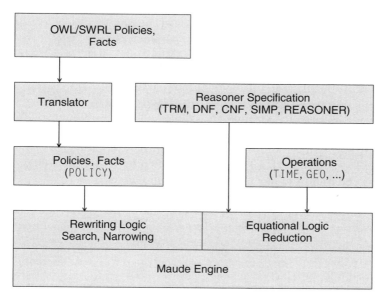

FIGURE 17.5

Maude reasoner modules.

normal form, and does some simplifications that can be done only in this form. If these three modules were combined into one, the reasoning would never terminate. For example, it could transform between CNF and DNF back and forth in all eternity. The REASONER module controls the ordered execution of the reasoning modules by using the Maude *metalevel*. First, narrowing is done in the POLICY module, then reduction in SIMP, CNF, and DNF, in that order.

17.7 SRI POLICY ENGINE DEMONSTRATION

In this section we will describe a policy engine developed at SRI including the authoring tool with which we can specify a policy and check its validity, and the results of experiments executing the policy on a modern PC.

17.7.1 SRI Policy Tool

We have developed a prototype graphical tool for defining policies and evaluating requests. The Maude-based reasoner described in Section 17.6 is used to evaluate requests. This policy tool is limited in scope to certain fixed policy parameters, but shows the principles of policies and policy reasoning. The tool could inspire a more complete future version. The policy tool has several parts that can all be started from the XG launchbar (see Figure 17.6).

Figure 17.7 shows the facilities for editing policies, in terms of the constraints on certain parameters. The parameters correspond to entities in the ontology (see Section 17.5), but the details are hidden from the user. Some parameters (e.g., "Role" and "Netword ID") are simply entered as text strings. The "Min Time" and "Max Time" calendar fields allow specifying the date and time range for which the policy applies.

Other parameters (e.g., powermasks and regions) have dedicated editing dialogs. For example, the "Region" dropdown list allows the user to select a geographic region within which this policy is applicable. Regions are defined using the Region Editor (Figure 17.8). Two types of regions can be created: circular and rectangular. A rectangular region is defined by specifying the top left and bottom right coordinates of the region in latitude/longitude values. A circular region is specified by giving the latitude/longitude values for the center and the radius of the region in meters.

The "Emissions" dropdown list allows the user to select from a predefined set of powermasks that define the maximum emissions allowed under this policy, and the "Sensed Power" dropdown list allows the user to select from a predefined set of powermasks that define peak received power allowed under this policy. Powermasks are

FIGURE 17.6

XG launchbar.

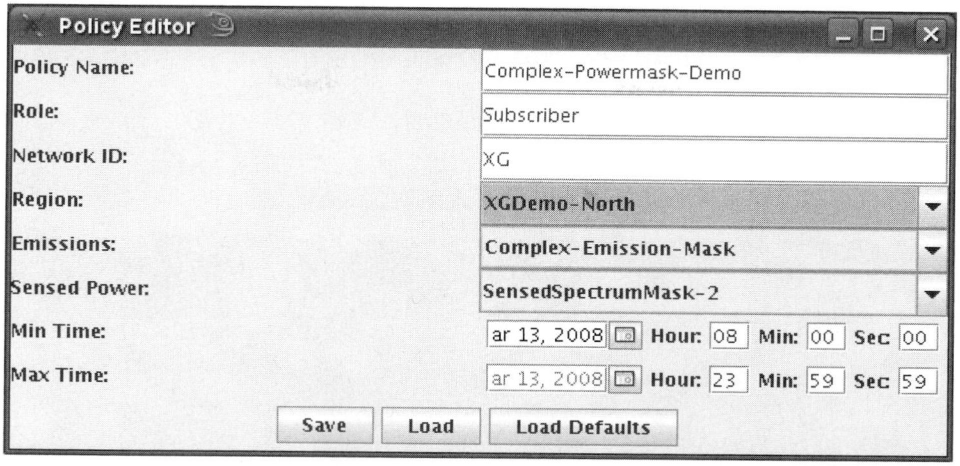

FIGURE 17.7

XG Policy Editor.

specified using the Powermask Editor (Figure 17.9). The table on the left of the figure is used to specify frequency and power values that are displayed in the central part. Each pair of such values is interpreted as the range up to which the given power value is true. Therefore, as an example, the two frequency/power value sets (232,0) and (234,40) in conjunction would be interpreted as any frequency less than 232 has a power value at most of 0 and any frequency between 232 and 234 has a power value of less than 40, respectively. The "Hi Val" field is used to specify the final power value that is applicable to any frequency greater than the highest frequency specified in the table on the left side of the figure.

Once policies have been edited, they are stored for later use, and loaded into the reasoner using the Policy Manager (Figure 17.10). The "Add" button is used to load policies to the PE. Pressing this button displays a list of policies that have been defined using the Policy Editor and that are not yet loaded into the PE. Once a policy is successfully loaded into engine, it appears on the "Active Policies" list in the Policy Manager. A particular policy cannot be loaded twice into the PE. The Policy Manager displays a message at the bottom of the panel indicating whether a policy loaded successfully. Policies are unloaded using the "Remove" button. To unload a policy, it must be selected from the "Active Policies" list in the Policy Manager. The system that is hosting the PE TCP/IP server needs to be provided to the Policy Manager. One can use either a fully qualified domain name of a system, or the IP address of the system that hosts the PE, or "localhost" if the PE is in the same system as the XG graphical user interface (GUI). The entry under the server port is the default port used by the PE for communication; this should *not* be changed.

The SSE panel is the main interaction method for sending requests to the PE and displaying the corresponding replies. The SSE consists of two tabular windows and two buttons ("Manual Request" and "Launch SSR"). Requests can be run manually or by

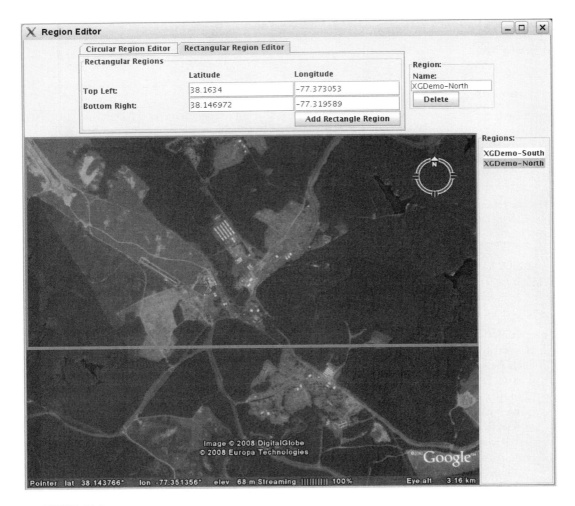

FIGURE 17.8

XG Region Editor.

using a strategy. The request editor shown in Figure 17.11 is used for manual requests. Running a request using a strategy uses a different window, which we will describe further down. The request editor is similar to the policy editor, and shares some of the specialized editors (e.g., for regions and powermasks), except for "Location", which is specified as a latitude/longitude pair instead of a region. Recall, however, that requests are sets of *facts*, whereas policies are sets of *constraints* that those facts must satisfy. Any values specified in the request editor are used in constructing the request that will be sent to the PE (using either "Manual Request" or "Launch SSE").

Reasoner replies are displayed as shown in Figure 17.12. A graphical indicator using color coding shows whether transmission was allowed (corresponding to a "true" reply indicated by green), denied (corresponding to a "false" reply indicated by red), or

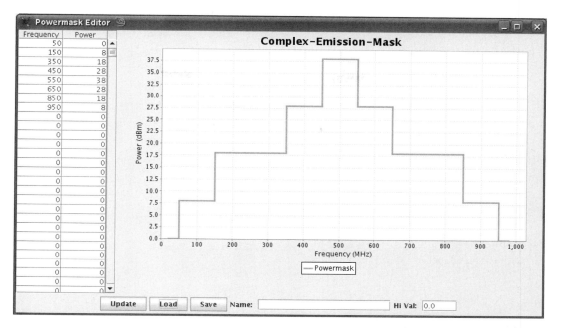

FIGURE 17.9

XG Powermask Editor.

FIGURE 17.10

XG Policy Manager.

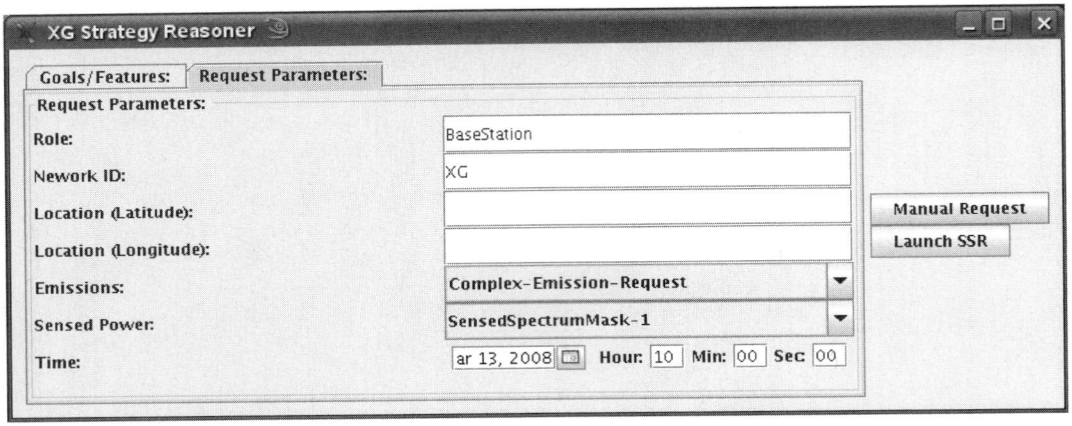

FIGURE 17.11

SSE Request panel.

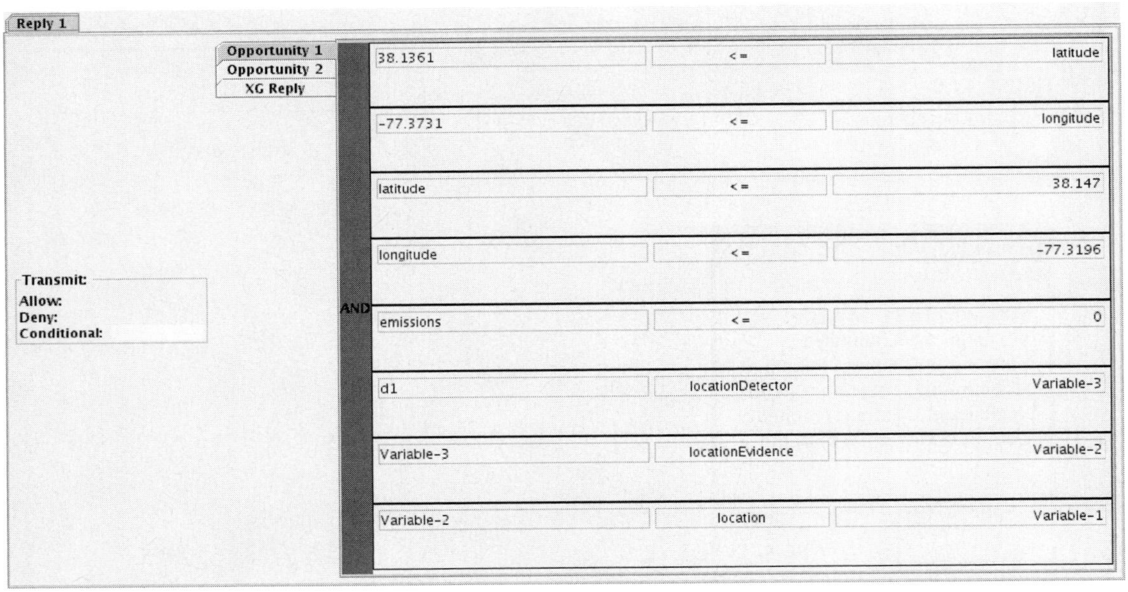

FIGURE 17.12

PE Reply view.

approved conditionally (corresponding to a set of constraints indicated by yellow). The constraints are in the form of a disjunction of conjunctions. Each disjunct is shown in the policy tool as an "Opportunity." Figure 17.12 shows two opportunities. The "XG Reply" tab that is beneath the "Opportunity" tabs displays the complete response as an XML document object model.

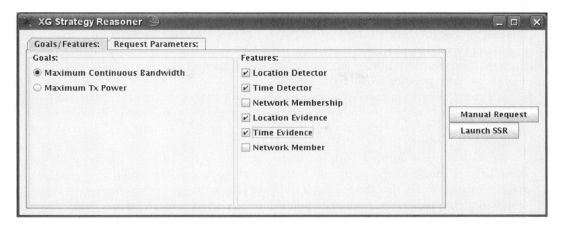

FIGURE 17.13

XG Strategy Reasoner—Goals/Features tab.

Figure 17.13 shows the "Goals/Features" tabular window. One panel allows selecting the goal of an interaction (either "Maximum Continuous Bandwidth" or "Maximum Tx Power"), and the other allows specifying the features supported by the radio. The shown strategies are just an example of the possible predefined strategies. They are used to demonstrate some rudimentary forms of reasoning that the SSE could perform. The user can also select whether the radio supports certain features (e.g., location detection or time detection), which will influence the requests that are generated. The "Goals" and "Features" specified in this panel are used as strategies when the SSE is launched by using the "Launch SSR" button. "Features" are used to prune PE replies to those that have constraints involving only radio features that were indicated in the "Features" panel, and "Goals" are used to further select among PE replies only those that satisfy the specified goals. As a result, SSE tries to reach transmission in a sequence of interactions with the PE.

The strategies built into the policy tool work as follows. First, an underspecified request is generated (using the input from the request panel of the SSE) and submitted to the PE. Then the tool (simulating an SSE) looks at the opportunities in the reply, and selects the one that best matches the chosen strategy. Next, a second request is generated, which will "fit" the constraints of the chosen opportunity. For this SSE, the second request is generated in such a way that the reply after submitting it to the PE will always be "True." In general, an SSE could submit any number of requests, successively increasing the amount of information put into them, and get back replies with fewer and fewer remaining constraints. However, as we discussed in Section 17.6, from the PE's point of view, every message from the SSE is independent from earlier and later messages. In other words, each request–reply interaction is an atomic transaction.

17.7.2 Experimental Results

On a modern PC, policy reasoning with our Maude-based PE is very fast. Typical response times are in the 50 ms range, with a few policies loaded, on a 2 GHz 64-bit machine. We ported our PE to a 32-bit PowerPC-based embedded system with 128 MB random access memory (RAM) and 390 MHz central processing unit (CPU), and got reasoning times that were about 30 times slower (i.e., about 1.5 seconds). In our experiences, these reasoning times are rather robust to changes in policies and go up only slightly as more policies are added.

Our previous, Prolog-based reasoner achieved times roughly one order of magnitude faster. However, as we discussed in Section 17.6, the Prolog-based reasoner does not return constraints, only "yes" or "no" answers, and is therefore of much more limited applicability.

We should point out that both reasoners are prototypes and have not been highly optimized or tuned for performance with typical workloads. The policies should be written in such a way that the SSE does not have to ask the PE for permission for every single outgoing packet, but rather is permitted transmission for a certain time, under certain constraints. Furthermore, today's high-end PC is tomorrow's embedded system. With a dual-core system, a radio would experience no interruption due to policy reasoning, as one processor core could be dedicated to this task.

17.8 LESSONS LEARNED AND FUTURE WORK

In this section, we discuss what we have learned about designing PLs and PEs to control CRs. We will describe how what has been learned affects the selection of policy language and the way in which it is encoded, as well as necessary extensions.

17.8.1 Lessons Learned

This section discusses what we have learned about temporal and geospatial reasoning. We will then discuss rules that specifically disallow certain behaviors, and how to structure these specifications.

Operations

A major challenge in designing both the PL and the PE was to express and reason about certain operations that involve complicated mathematical calculations—in particular, temporal reasoning, such as adding and subtracting time instants expressed in a human-friendly "calendar" format, and geospatial reasoning, such as finding the distance between two latitude/longitude points. SWRL does have a number of built-in operations for common operations, such as arithmetic and temporal reasoning. To meet the operational constraints of the hardware on which our PE was demonstrated, we decided to follow this approach, and implemented these built-ins in Maude as part of our PE. We also added built-ins for geospatial reasoning, which SWRL does not support. Unfortunately, using built-ins comes with problems because the operations do not have well-defined semantics and one cannot *dynamically*—without recompilation—add new operations. We decided to use built-ins for the field-test version of our engine to achieve response times of 50 ms on the experimentation hardware. However, our PL is expressive enough to define their semantics, and thus a complete Maude reasoner would have a well-defined semantics, although the resulting implementation would be less efficient.

One cannot define these operations in OWL and SWRL, a fact that was one of our motivations in designing a PL that goes beyond OWL and SWRL. The reason that one cannot define common operations such as for arithmetic or geospatial reasoning has to do with how these ontology languages represent structured data. In *functional* languages such as functional programming or equational logic, data structures are represented by using nested functions and constructors. It is easy to define operations over such structures using equations or other axioms. In OWL and SWRL, there are no functions, so everything has to be represented using classes and properties, that is, in a *relational* way. This relational representation is not amenable to be used as a basis for the aforementioned complex operations.

A possible solution would be to augment OWL and SWRL with a functional sublanguage for data representation, in which one could then define the operations. With our PL we have gone a step in this direction and we hope to do further research into how to efficiently implement such more expressive languages.

Disallowing Policies

An interesting observation concerns the applicability and limitations of restrictive policies. The main insight is that the use of restrictive policies has implications on how to handle policy changes. In particular, adding a restrictive policy to the policy base also requires extending the metarule with this new policy. We give a detailed example here. Even though this can be automated, it still influences the overall policy architecture and how policy administration is handled.

Consider the top-level policy rule we mentioned previously,

```
Permit iff Allow and not Disallow
```

This rule has the effect that we need *at least one* policy to allow, and *no* policy can disallow. Consider now a disallowing policy such as

```
Disallow ⇐(Device(?d) ∧ signalDetector(?d, ?sd) ∧
signalEvidence(?sd, ?se) ∧ peakSensedPower(?se) > −100)
```

The intended meaning is "disallow if the peak sensed power is greater than −100." If a radio submits a request,

```
Device(d1)
signalDetector(d1, sd1)
signalEvidence(sd1 se1)
peakSensedPower(se1, −110)
```

then we cannot prove `Disallow` by using this rule, and all is well. However, the same is true if the radio had submitted an *empty* request!

The root cause is that the negation in `not Disallow` is in the wrong place. The previous version says roughly "if there is *not* any signal evidence such that the peak sensed power is greater than −100, then disallow." What we want to say is "there must be signal evidence showing that peak sensed power is *not* greater than −100." In other words, we need to push this inside the `Disallow` rule to the actual constraint that we need to check; that is,

```
peakSensedPower(?se) > −100,

d1 ⇐(Device(?d) ∧ signalDetector(?d, ?sd) ∧
signalEvidence(?sd, ?se) ∧¬peakSensedPower(?se) > −100)
```

Then we must change the top-level rule so that conformance to this rule is still mandatory:

```
Permit iff Allow and d1
```

This top-level rule has to explicitly mention all such mandatory rules.

17.8.2 Future Work

Because of the expected complexity of future policies, a PL should allow for advanced forms of *policy analysis*, such as detection of logical inconsistencies caused by the combination of several policies. We believe that our existing reasoner could be used as the back end for an analysis toolkit, but needs to be augmented with a user-friendly graphical front end designed for this purpose.

A multitude of units (e.g., Hz, MHz, mW, dBm) are used for measuring various entities in radio communication. Currently, we have some hardcoded assumptions about the units used. For example, all frequencies are in Hz. With support for *dimension types* [23], one could ensure things such as distance divided by time gives a speed, and that the units for the different entities are correct and consistent.

Policy management requires support for version control and author identification, among other things. This is not a language or reasoning issue so much as an architecture issue, but some interesting challenges are associated with this.

Strategy reasoning is another open issue. In a CR, both the PE and the SSE perform reasoning, but the reasoning done in the SSE is quite different from what the PE does. The PE's job is conceptually rather simple: answer requests in a verifiably correct way. The SSE, on the other hand, can utilize any kind of heuristics and learning techniques to make better decisions on what kind of requests to make and when to make them. So far, we have only begun to experiment with very rudimentary system strategy reasoning, but this area needs to be explored in depth.

Although the PL has been developed for spectrum access policies, most of the language features are general and allow for specifying policies in many other domains. Ontologies are the vehicle to define domain concepts for use in formulating policies. We are interested in investigating the applicability of the PL and the reasoning techniques implemented in our Maude prototype to other domains, such as networking, routing, or security policies.

17.9 SUMMARY

As technologies evolve at ever-faster rates, the ability to more easily field and test new capabilities and paradigms of use is crucial. Policy-based CRs make it possible to upload new capabilities, tools, and policies to radios without redesigning or recertifying the radios. Policy reasoning is a key ingredient for achieving such a flexible, adaptive solution for CRs.

Declarative policies are a powerful mechanism. Benefits include coordination across a variety of organizational entities, reduced deployment delays, flexibly changing the behavior of wireless devices at runtime, and ease of certification.

We showed the feasibility of our vision by defining a specific PL and reasoning engine that were used to provide opportunistic spectrum access, thus addressing one of the main challenges of wireless networks: spectrum scarcity. We defined various types of notional spectrum-sharing policies and our engine reasoned with them while running on radio hardware.

The challenge was to design a language rich enough to express numerical constraints and extensible enough to support unanticipated future policies, while at the same time supporting efficient reasoning. Our PL is sufficiently expressive to be useful for policies in many domains besides spectrum sharing, such as security, quality of service, and dynamic network management. Our policy engine was implemented to meet the operational and radio hardware constraints of our spectrum-sharing environment. Thus, our engine might have to be adapted for other domains.

REFERENCES

[1] Stine, J. A., and D. L. Portigal, Spectrum 101: An Introduction to Spectrum Management, Technical Report MTR 04W0000048, MITRE, Bedford, MA, 2004.

[2] XG Working Group, The XG Vision, Request for Comments, Version 2.0, Technical Report, BBN Technologies, Cambridge, MA, 2005.

[3] Perich, F., Policy-based Network Management for Next Generation Spectrum Access Control, in F. Jondral and P. Marshall, 2007; available at *www.ieee-dyspan.org*.

[4] Denker, G., E. Elenius, R. Senanayake, M.-O. Stehr, D. Wilkins, C. M. Conway, and R. A. Newell, *Demonstration of a Policy Engine for Spectrum Sharing, Paper accompanying system demonstration*, in F. Jondrall and P. Marshall (eds.), 2007; available at *www.ieee-dyspan.org*.

[5] Berlemann, L., S. Mangold, G. Hiertz, and B. Walke, Policy-Defined Spectrum Sharing and Medium Access for Cognitive Radios, *Journal of Communications*, 1(1):1–12, 2006.

[6] Berlemann, L., S. Mangold, and B. Walke, Policy-based Reasoning for Spectrum Sharing in Cognitive Radio Networks, *IEEE First International Symposium on New Frontiers in Dynamic Spectrum Access*, pp. 1–10, Baltimore, November 2005.

[7] Kokar, M., M. D. Brady, and K. Baclawski, Roles of Ontologies in Cognitive Radios, *Cognitive Radio Technology*, First Edition, B. Fette (ed.), pp. 401–433, Newens/Elsevier, 2006.

[8] Baclawski, K., D. Brady, and M. M. Kokar, Interoperability Communication at the Data Link Layer through Ontology-Based Reasoning, *Proceedings of Software Defined Radio Conference*, 2005.

[9] Wilkins, D. E., G. Denker, M.-O. Stehr, D. Elenius, R. Senanayake, and C. Talcott, Policy-Based Cognitive Radios, *IEEE Wireless Communications*, 14(4):41–46, 2007.

[10] Denker, G., E. Elenius, R. Senanayake, M.-O. Stehr, and D. Wilkins, *A Policy Engine for Spectrum Sharing*, in F. Jondrall and P. Marshall (eds.), pp. 55–65, 2007; available at *www. ieee-dyspan.org*.

[11] Elenius, D., G. Denker, M.-O. Stehr, R. Senanayake, C. L. Talcott, and D. Wilkins, CoRaL—Policy Language and Reasoning Techniques for Spectrum Policies, *Eighth IEEE International Workshop on Policies for Distributed Systems and Networks* (Policy 07), pp. 261–265, 2007.

[12] Denker, G., D. Elenius, R. Senanayake, M.-O. Stehr, C. Talcott, and D. Wilkins,, Cognitive Policy Radio Language (CoRaL)—A Language for Spectrum Policies, XG Policy Language, Version 0.1, Technical Report ICS-16763-TR-07-001, SRI International, Menlo Park, CA, April 2007.

[13] Stehr, M.-O., Toward a Universal Policy Logic, Technical Report ICS-16763-TR-07-003, SRI International, Menlo Park, CA, April 2007.

[14] McGuiness, D. L., and F. van Harmelen, OWL Web Ontology Language Overview, August 2003; available at *www.w3.org/TR/owl-features/*.

[15] Horrocks, I., P. F. Patel-Schneider, H. Boley, S. Tabet, B. Grosof, and M. Dean, SWRL: A Semantic Web Rule Language Combining OWL and RuleML, W3C Member Submission 21, May 2004; available at *www.w3.org/Submission/SWRL/*.

[16] Patel-Schneider, P. F., A Proposal for a SWRL Extension to First-Order Logic, November 2004; available at *www.daml.org/2004/11/fol/proposal*.

[17] Elenius, D., and M.-O. Stehr, Rules and Computation on the Semantic Web, Technical Report, SRI International, 2007; available at *http://xg.csl.sri.com/technical_reports.php*.

[18] Elenius, D., Extensions to OWL—Functions and Inductive Predicates, Technical Report ICS-16763-TR-07-005, SRI International, Menlo Park, CA, April 2007.

[19] Elenius, D., and M.-O. Stehr, Translating SWRL FOL, SWRL, and OWL DL to the Universal Policy Logic, Version 0.1, Technical Report ICS-16763-TR-07-004, SRI International, Menlo Park, CA, April 2007.

[20] Jondral, F., and P. Marshall (eds.), *Second IEEE International Symposium on Dynamic Spectrum Access Networks*, Dublin, April 2007; available at *www.ieee-dyspan.org*.

[21] Clavel, M., F. Duran, S. Eker, J. Meseguer, and M.-O. Stehr, Maude as a Formal Meta-tool, *FM '99—Formal Methods*, J. Wing and J. Woodcock (eds.), Springer-Verlag (LNCS 1709, pp. 1684–1703), 1999.

[22] Meseguer, J., Conditional Rewriting Logic as a Unified Model of Concurrency, *Theoretical Computer Science*, 96(1):73–155, 1992.

[23] Kennedy, A., Dimension Types, *ESOP '94: Proceedings Fifth European Symposium on Programming*, pp. 348–362, London, 1994.

Spectrum Sensing Based on Spectral Correlation

Chad M. Spooner
NorthWest Research Associates, Monterey, California

Richard B. Nicholls
Tektronix Incorporated, Beaverton Oregon

18.1 INTRODUCTION

Within the animal kingdom, the eye is an excellent sensor of the visible-light portion of the electromagnetic spectrum—it is sensitive to total received energy, the relative contributions from distinct spectral subbands, and temporal variations indicative of motion. The eye evolved in the context of the Earth's atmosphere and the Sun's radiation, which combine to provide a window of relative transparency called the visible spectrum. The eye operates within its external constraints to perform sensing that is used by the brain to build a picture of the physical environment in which the animal exists.

A cognitive radio (CR) must also sense the electromagnetic spectrum. The constraints on its evolving design arise from the intrinsic properties of the radio signals that it must detect in order to function. An important component of a radio signal is its spectral location, which can be determined by detecting the presence of energy in a band of frequencies. But this is not the only fundamental attribute of radio signals that can be advantageously adapted for spectrum sensing. Virtually all human-made radio signals exhibit a fundamental property known as *spectral correlation*, which is a measure of the redundancy in the information contained in pairs of signal subbands. Therefore, like the eye, a CR's spectrum sensor can conceivably take advantage of the total signal energy, as well as the relative contributions from distinct subbands and their temporal variation to build a picture of the electromagnetic environment in which it must operate.

In this chapter, we explore spectrum sensing based on exploitation of spectral correlation for use in CR systems. Along the way, we contrast this approach with conventional sensing approaches based on matched filtering and energy detection. We provide several generic algorithmic approaches to signal detection based on spectral correlation that are applicable to a wide variety of modern communication signals. To validate the overall claims, the spectral-correlation properties of signals arising from systems such as the Global System for Mobile Communication/Advanced Data for GSM Evolution (GSM/EDGE), Advanced Television Systems Committee digital television (ATSC DTV), code division multiple access/wideband CDMA (CDMA/WCDMA), WiFi (802.11a/b/g), and NADC (IS-54/136) are studied by using both mathematical models and captured radio fraquency (RF) signals.

18.1.1 Cognitive Radio: Aware and Adaptive

Cognitive radios are transceivers that are aware of their physical, operational, and electromagnetic environments. They can also autonomously modify their operating parameters to enhance performance or utility [1–18]. The radio can be aware of many aspects of the physical environment, including position in space, proximity to various networks, and weather, among others. The radio might also be aware of its user's usage patterns and operating preferences. In this chapter, however, we are concerned with the CR's awareness of the local radio spectrum. In particular, we focus on the means by which a CR can become aware of signals in the local spectrum using only its own resources. This capability is typically called *spectrum sensing*, and the means we advocate involve the exploitation of spectral correlation.

18.1.2 Spectral Correlation in a Nutshell

Spectral correlation is a statistical property belonging to *cyclostationary (CS) signals* [25–28]. CS signals possess one or more probabilistic parameters (e.g., mean, autocorrelation, probability density function, nth-order moment, or nth-order cumulant) that are periodically time variant. One consequence of this fundamental statistical structure is spectral correlation, the existence of distinct subbands with contents that are temporally correlated. Another is the appearance of finite-strength additive sine-wave components in the outputs of quadratic and higher-order nonlinearities. That is, the squaring of a CS signal usually produces one or more sine waves in the output. Some signals will require a more general quadratic functional such as a delay-and-multiply (DM) operation to produce sine waves. The various facets of cyclostationarity are shown in Figure 18.1.

A key fact is that the separations between correlated signal subbands and the frequencies of generated sine waves are given by the same set of numbers: the *cycle frequencies* [25, 28]. These are also the periodicities in the time-variant moment and cumulant functions, as illustrated in Figure 18.1. This fact can be used to develop multiple distinct interpretations of a given signal-processing algorithm and can also suggest distinct approaches to algorithm invention.

Spectral correlation is directly exploited in frequency-shift filters (linear periodically time-varying filters) [39], and nonlinearly generated sine-wave components are exploited by synchronizers [40] and the DM and chip-rate detectors [24].

We describe spectral correlation in mathematical terms in Section 18.2, but this overview is sufficient to continue our introduction to spectrum sensing in a CR, and to begin to understand why spectral correlation may offer advantages over other sensing methods.

18.1.3 Spectrum-Sensing Considerations

The core idea of spectrum sensing is to determine whether a particular band of radio frequencies (i.e., a channel) is occupied. In general settings, a CR might view any unoc-

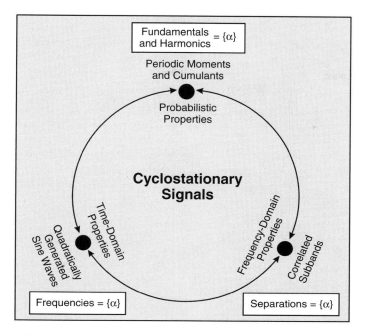

FIGURE 18.1

Facets of a CS signal. The set of CFs {α} characterizes the periodicities in the autocorrelation, the subbands' correlation, and the frequencies of quadratically generated sine waves.

cupied band as a potential communication channel, and therefore might survey large swaths of the spectrum to find exploitable *spectrum holes* [4, 19]. Occupied portions of the spectrum are often referred to as *black space*, whereas unoccupied bands—the spectrum holes—are *white space*. The concept of partially used bands has also been introduced [12] to indicate, for example, the presence of multiple CDMA signals to which the CR might add one more without causing significant interference. Such bands are called *gray space*. Key attributes of the spectrum-sensing problem are considered in the following subsections.

The Hidden-Node Problem

In most models of CR communication, the CR can use a particular RF band once it is known to be unoccupied. However, the CR may declare a channel unoccupied while at the same time a nearby receiver declares it occupied and can in fact successfully demodulate the signal. This may be because of inadequate detection sensitivity on the part of the CR, or it may be due to the *hidden-node problem*, in which the signal is severely attenuated during its propagation to the CR, but not during its propagation to its intended receiver.

Figure 18.2 illustrates the hidden-node problem. In this figure, the CR is attempting to detect a primary user's signal in a band centered at f_1. The propagation channel experienced by the primary signal en route to the CR is characterized by strong fading

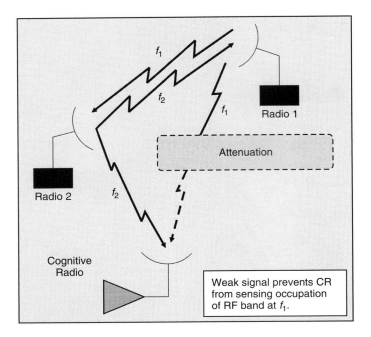

FIGURE 18.2

The hidden-node problem in CR. The CR is attempting to detect if a primary user's signal exists in a band centered at f_1, but the signal is shadowed by an obstruction.

or shadowing, whereas the primary receiver experiences a benign channel and can successfully demodulate its signal. The CR makes an error if it declares the primary signal at f_1 absent and commences its own transmission.

The hidden-node problem can be mitigated by using a spatially dense collection of CR spectrum sensors so that, on average, propagation conditions to at least one of the CRs will allow detection. However, due to the possibility of time-variant multipath conditions, this does not completely solve the hidden-node problem. No matter how many CR sensors are employed, the hidden-node problem suggests a maximum-sensitivity requirement on the CR spectrum sensor.

General Spectrum Sensing

A CR regularly attempts to survey the spectrum through which it can transmit and receive. If it can find white spaces, it can potentially rendezvous with another CR and communication can commence. For this general sensing problem, the CR must be able to determine the white spaces in complex RF environments characterized by high dynamic ranges in several variables, including power, spectral density, number of black spaces, and number of modulation types. In addition, there may be substantial temporal variation with which to contend. Through use of a database that contains the sensing results obtained over time, the CR can predict the appearance of white spaces and also select the white spaces that are most likely to remain white over a time interval extending into the future [5].

An example of general spectrum sensing that involves only a form of multiresolution spectral analysis is shown in Figure 18.3 [10]. For this synthetic RF scene, there are literally hundreds of bands of interest (BOIs), and therefore hundreds of potentially exploitable white spaces. Note the very large dynamic range in power and bandwidth exhibited by the signals. The method is good at determining the black spaces, but since it is fundamentally spectral analysis, it cannot reliably determine whether or not the white spaces are truly white—they could contain weak signals due to the hidden-node problem.

Figure 18.3 shows that it is possible to simultaneously detect many kinds of black spaces, whether narrow or wide, closely spaced or isolated, weak or strong. For example, many narrow occupied signal bands (black spaces) between 696 and 697 MHz are detected by the algorithm, but are not labeled due to insufficient space for the labels. Close examination of the figure also reveals several subbands that probably contain a signal, but the algorithm did not detect this fact. In particular, there is a signal near 697.5 MHz and several between detected bands 171 and 172. These are *potential white spaces*, for which further signal processing must be performed to ensure that they are truly unoccupied.

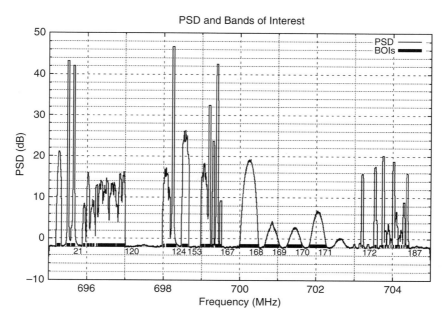

FIGURE 18.3

Illustration of general wideband spectrum sensing, in which no assumptions are made regarding the signals in the environment. Multiresolution spectral analysis can still yield an adequate spectrum sensor [10].

Constrained Spectrum Sensing

If a CR is to be part of a designed network, or will be capable of entering one or more existing networks [6], the spectrum-sensing task may be constrained relative to the general sensing problem just discussed. In particular, the bandwidth that must be scanned can be small, the signal or channel bandwidths may be elements of a small set, and the number of modulation types could be small. In this case, it is much easier to design spectrum analysis methods that are optimized because the resolution bandwidth of the core spectrum estimates can be well matched to the expected signal bandwidths, and optimized modulation classifiers can be employed to recognize signals from relevant networks. For example, it is feasible in such problems to simply implement the few full demodulators needed for accessing the available communication systems.

An example of constrained spectrum sensing is shown in Figure 18.4. In this figure, a set of IS-54/136 $\pi/4$-DQPSK signals are generated with random power levels over a 30-dB range. The generated signals have the IS-54/136 modulation type, transmit filtering, and symbol rate. Three sets of signals are generated. The first set ends near 498.8 MHz, the second ends near 500.2 MHz, and the third ends near 501.6 MHz. In each set, all but five available IS-54 channels are occupied. The figure shows the automatically detected white spaces. In the first and second sets, none of the five white IS-54 channels are adjacent, whereas in the third set, two are adjacent. Therefore, we see in the first set five detected white spaces (2–6, not all labeled due to limited space);

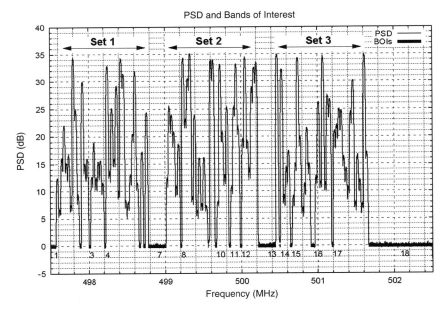

FIGURE 18.4

Illustration of constrained spectrum sensing. Here the sensing problem is severely constrained due to the small number of possible signal bandwidths and separations between signals. This kind of constraint eases the difficulties related to sensing methods based on spectral analysis because an optimum spectral resolution can be chosen in advance of processing [10].

in the second set, five white spaces (8–12); and in the third set, only four because two are adjacent (14–17). The white space detection algorithm [10] also finds the white spaces between the sets (labeled 1, 7, 13, and 18 in Figure 18.4).

Constrained Sensing Resources

In most cases, the CR will have strong constraints on resources brought to bear on the sensing problem. Examples include limited processing-block length, memory size, processor type, speed, and available cycles. In such cases, effective wideband sensing may require the collection and processing of many small blocks, greatly increasing the latency in declaring white space as exploitable. However, we are primarily concerned in this chapter with technology for future CR systems, where we can confidently expect to benefit from technological advances and Moore's Law [52].

The Role of Signal Detection

At a minimum, the spectrum-sensing function of the CR must be able to distinguish between noise-only bands and signal-plus-noise bands. However, relatively simple spectral analysis can fail to detect a weak signal. When it is vital to detect the presence of signals that must not be interfered with (e.g., IEEE 802.22), then spectral analysis can usefully partition the subbands into black spaces and potential white spaces. The latter can then be subjected to far more sensitive—and expensive—signal detectors, such as those based on exploiting spectral correlation.

The Role of Signal Classification

In some cases, the decision to transmit in a particular band of frequencies will depend on whether a specific kind of signal exists in the band or in a nearby band. In these situations, a signal must be detected and classified in order to supply the CR with enough information to make a correct decision.

An example of the role of signal classification relates to the notion of gray space [12]. If a CDMA-enabled CR senses a potential channel is underutilized by a CDMA system, it can add its transmission to the channel with negligible effect on the incumbent users. This kind of decision making requires that the CR be able to count the number of CDMA signals in the channel, which is a joint detection and classification problem.

Another example of the role of signal classification involves CR policies in which the cognitive radio is not allowed to transmit in an occupied channel, nor in channels adjacent to it when the channel is occupied by a signal of a particular type (e.g., IEEE 802.22). Therefore, the CR must be able to classify the modulation type of signals in detected black spaces in order to comply with mandatory policy.

Now that the basic idea and some of the nuances of the spectrum-sensing problem have been presented, we turn to the main topic of this chapter: sensing based on exploiting spectral correlation. We first present some classical alternatives that represent the main competitors to spectral correlation.

18.1.4 Spectrum-Sensing Solutions

In this section, we present an overview of some candidate solutions for spectrum sensing in CR. The idea is to provide a conceptual understanding of their applicability, strengths, and weaknesses, rather than a complete treatment.

Matched Filtering

A filter that is matched to a signal $x_0(t)$ has an impulse response equal to a conjugated and time-reversed version of $x_0(t)$. This filter provides a maximum signal-to-noise ratio (SNR) output when $x_0(t)$ is embedded in white Gaussian noise (WGN). This *matched filtering* is applicable to random communication signals when they contain some periodically repeated known component—the matched filter can be matched to the relatively short-duration known component(s).

Some examples of the possible application of matched filtering to spectrum sensing in CR include the known components of GSM, ATSC DTV, IS-54/136, 802.11a/g orthogonal frequency-division multiplexing (OFDM), and many others. In GSM, the filter can be matched to the 26-bit midamble in the center of each 156-bit traffic time slot; in ATSC DTV, it can be matched to the rather long pseudonoise synchronization sequence that is repeated every 24 ms; and in 802.11a/g OFDM, it can be matched to the 127-bit repeated pilot subcarrier synchronization sequence.

The strengths of matched filtering include computational simplicity, optimality for additive WGN (AWGN) channels, and reasonably general applicability. The most serious drawbacks are poor performance in non-AWGN channels and sensitivity to imperfect synchronization. This last weakness of matched filtering means that if the residual carrier offset after downconversion is not zero, the output of the filter can be zero even at the correct optimal delay at which the signal component is perfectly time aligned with the filter impulse response. In this situation, the application of the matched filter will require a search over the carrier offset parameter as well as the delay parameter.

A variant of matched-filter detection is *tone detection*, in which the presence of a finite-strength tone is detected, and the presence of the tone implies the presence of a particular signal associated with that tone's frequency. This can be done by using a matched filter or through Fourier analysis of a small band of frequencies near the putative tone's frequency. Examples of signals amenable to tone detection are broadcast analog modulation (AM), ATSC DTV, conventional analog TV, some forms of frequency shift keying (FSK), and on–off keying (OOK).

Energy Detection

For weak random signals in AWGN channels, an optimal detection scheme is to measure the received energy and compare it to a precomputed threshold [21–23]. The energy can be measured in several ways, depending on how much is known about the signal's power spectral density (PSD).

When the PSD is known, the optimum scheme correlates the measured periodogram with the ideal PSD to obtain the optimal energy detection statistic,

$$Y_{oed}(t) = \int S_s^0(f) I_{x_T}^0(t, f) df \qquad (18.1)$$

where

$$I_{x_T}^0(f) = \frac{1}{T} |X_T(t, f)|^2 \qquad (18.2)$$

is the periodogram, and

$$X_T(t, f) = \int_{t-T/2}^{t+T/2} x(u) e^{-i2\pi f u} du \qquad (18.3)$$

is the time-variant finite-time Fourier transform of the data $x(t)$, T is the data observation length, and f represents frequency.

When the exact shape of the signal's PSD is not known, a suboptimal version of the energy detector simply integrates the squared periodogram over some range of frequencies,

$$Y_{sed}(t) = \int_B \left| I_{x_t}^0(t, f) \right|^2 df \tag{18.4}$$

In either case, the detection variable is compared to a threshold, η, that takes into account the noise spectral density height, N_0, to achieve a given probability of detection, P_D, for a specified probability of false alarm, P_{FA}.

The strengths of the energy-detection method of spectrum sensing include universal applicability, relative computational simplicity, and reduced amount of required prior signal knowledge. Its weaknesses are similar to those of matched filtering. In particular, the energy detector's performance is quite sensitive to uncertainty in the background noise spectral density N_0 and to the presence of in-band interference [33, 34].

Delay-and-Multiply Detection

Another sensing technique is the delay-and-multiply signal detector [24], which multiplies the collected data block with a delayed and conjugated version of itself in order to generate an additive sine-wave component the presence of which can be detected by using Fourier methods. The presence of the tone implies the presence of the signal, and the exact frequency of the tone provides a parameter estimate for the signal, usually equal to the symbol rate (chip rate for direct sequence spread spectrum (DSSS) signals).

The DM detector is a simple exploitation of cyclostationarity in that it employs a quadratic data transformation to generate a spectral line. This is only possible for CS signals. That is, the DM detector exploits spectral correlation.

The strengths of the DM detector are that it can provide superior sensitivity relative to the energy detectors, is robust to uncertainties in the noise power and interference parameters, and is computationally less expensive than more thorough methods of exploiting the cyclostationarity property. Its main weaknesses are that it is not applicable to a large number of signals and that optimum performance requires knowledge of the optimum delay, which in turn requires knowledge of the transmitter filtering applied to the signal to be detected. That is, the optimum delay for rectangular-pulse signals is half the symbol interval, but for signals that have been filtered with a square-root raised-cosine filter, the optimum delay is zero.

Cycle Detection

The topic of cycle detection will be explored in greater detail in subsequent sections of this chapter, but it is helpful for us to present it early in this section along with the other alternative spectrum-sensing methods.

A *single-cycle detector* [33–35] matches the ideal spectral-correlation function (SCF) for a single value of the cycle frequency (CF) α with a measured version. It is computationally similar to the optimum energy detector, except that the ideal PSD is replaced by the ideal SCF, and the measured periodogram is replaced by the measured *cyclic periodogram*. A suboptimal version replaces the ideal SCF with a rectangular window over a band of frequencies in which the signal's SCF is expected to reside if present.

A *multicycle detector* [33, 35] coherently combines the statistics from two or more single-cycle detectors. A suboptimal version combines the magnitudes of the single-cycle detector outputs, since their phases depend on unknowns (e.g., symbol-clock and carrier phases). A further retreat from optimality noncoherently combines the outputs of the suboptimal single-cycle detectors.

Swiss Army Knife Solutions

It is possible to create a spectrum-sensing function that contains a highly specialized detector for each signal that must be detected: a matched filter for ATSC DTV, a DM detector for DSSS, an energy detector for GSM, etc. We call this kind of sensing strategy a *Swiss Army knife* (SAK) solution because of the disparate nature, computational requirements, and achievable performance of the various signal-specific sensors. In this chapter, we argue that an alternate sensing architecture is preferable in many respects, and it is an architecture based on the SCF.

Radio Frequency Situation Awareness

An alternative to SAK architectures is *radio frequency situation awareness* (RFSA), which is based on the fundamental statistical nature of all communication signals and uses a foundation of spectral-correlation analysis. This style of architecture leads to a set of signal-specific sensing algorithms that are more similar than they are disparate. It can also be used to create a completely blind spectrum-sensing algorithm that can, in principle, automatically detect and classify all signals in all RF bands with no prior knowledge [14, 45, 46]. Such an algorithm is said to provide blind RFSA.

As the ultimate in spectrum sensing, blind RFSA is not necessarily appropriate for all cognitive radio applications, but the basic approach can be employed to yield less capable algorithms that meet the requirements at hand. Moreover, the spectral-correlation-based RFSA approach allows a valuable kind of extensibility: new signals are easily added because we don't need to find yet another blade to add to a crowded SAK.

The foundation of general blind RFSA must consist of the fundamental statistical nature of the objects encountered in the RF spectrum: cyclostationarity. Although it has been demonstrated that complete blind RFSA requires the use of higher-order statistics [44–46, 14], an excellent first step consists only of first- and second-order statistics, which means first estimating and using the cyclic autocorrelation and/or SCFs before any higher-order statistics are estimated.

18.1.5 Archetypal Example

In this section, we present simulation results that illustrate the idea that the best-performing spectrum-sensing method depends strongly on the signal type, collection limitations, and RF environment. The simulation involves ATSC DTV and binary phase shift keying (BPSK) signals, and is meant to be illustrative rather than definitive; the full details of the algorithms and simulation parameters are not listed in favor of stressing the main results.

In the first part of the example, we consider detecting the presence of an ATSC DTV signal in WGN. The results for an in-band SNR of −10 dB are shown in Figure 18.5(a)

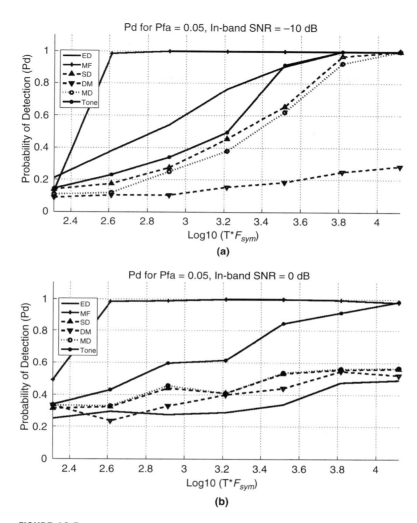

FIGURE 18.5

Sensing algorithm performance for an ATSC DTV signal in a benign environment (a) and difficult environment (b). *Note:* ED = energy detector, MF = matched filter, SD = single-cycle spectral correlation detector, MD = multiple-cycle SD, DM = delay-and-multiply detector, Tone = sine-wave detector.

for a range of observation interval lengths. The figure shows that the matched filter (MF) has the best performance, followed by energy detection (ED) and tone detection (Tone). The spectral-correlation detectors (SDs) are competitive with the energy detector, and the DM detector has the worst performance. Here, the matched filter is matched to the long pseudorandom synchronization sequence used in ATSC DTV, which repeats every 24 ms.

In the second part of the example, we repeat the ATSC DTV experiment with the added impairments of a variable noise power, in-band interference with variable parameters, and variable signal power. The results are shown in Figure 18.5(b). Again we see that the matched filter is superior. The tone detector is second, as before, but has a decreased performance level. The spectral correlation and DM detectors come next, and the energy detector has the worst performance. This confirms that the energy detector has the largest sensitivity to unknown or time-varying parameters (e.g., noise or interference levels). The matched filter continues to deliver excellent performance, although the addition of a multipath channel and/or unknown carrier frequency offset will degrade its performance significantly.

In the third part of the archetypal example, the signal to be detected is a generic BPSK signal in WGN. This signal has no known components and no additive sine-wave components. Therefore, the matched filter and tone detector are not applicable. The results for an in-band SNR of −10 dB are shown in Figure 18.6(a). Energy detection and spectral-correlation detection are competitive, and the DM detector delivers the worst performance, but lags the others by a relatively small amount.

In the fourth and final part of the example, we repeat the third part with the added impairments discussed in the second part: variable noise level, variable signal power, and variable-parameter in-band interference. The results are shown in Figure 18.6(b). Here we see that the DM detector is best, followed by the SDs, and the energy detector delivers the worst performance.

The lesson from the archetypal example is that the worst detector for a particular signal in a particular environment may be the best detector for a different signal or different environment. The specific details of the signal, noise, and interference structures must inform the design process: there is no single "best spectrum sensor." Nevertheless, building a spectrum sensor on the foundation of energy detection appears to be unwise. A foundation of spectral correlation, on the other hand, appears to provide a great deal of flexibility and robustness to changing and unknown environments.

18.1.6 Organization of the Chapter

The remainder of this chapter is organized as follows. The fundamental statistical nature of human-made communication and radar signals—cyclostationarity—is reviewed in Section 18.2. Methods of spectrum sensing based on exploitation of cyclostationarity (spectral correlation) are presented in Section 18.3 and are applied to modern communication signals in Section 18.4. Concluding remarks are provided in Section 18.5.

18.2 THE STATISTICAL NATURE OF COMMUNICATION SIGNALS

Human-made communication signals are commonly modeled as stationary random processes or stationary time series [29], but this requires willfully ignoring their true statistical nature by inserting phase-randomizing variables. In this section, we present the statistical parameters of CS random processes or time series, which vary periodically (or almost periodically [30]) over time.

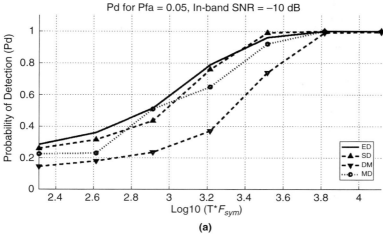

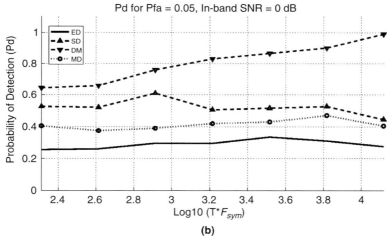

FIGURE 18.6

Sensing algorithm performance for a BPSK signal in a benign environment (a) and a difficult environment (b).

18.2.1 Stationary and Nonstationary Signals

For a wide-sense stationary signal, the mean and autocorrelation are time invariant. Let $x(t)$ represent a complex-valued stationary signal over all time, t, and assume that $x(t)$ is a power signal.[1] The mean, $m_x(t)$, and autocorrelation, $r_x(t_1, t_2)$, are defined by the following equations.

---

[1]A power signal is one for which the mean-square value is finite and nonzero. This is in contrast to an energy signal, which when averaged over all time has an average mean-square value approaching zero.

$$m_x(t) = E[x(t)] \tag{18.5}$$

$$r_x(t_1, t_2) = E[x(t_1)x*(t_2)] \tag{18.6}$$

where $E[\cdot]$ denotes expectation and $*$ denotes conjugation. Wide-sense stationarity of $x(t)$ implies that the mean and autocorrelation (first- and second-order moments) are constant and a function of the time difference $t_1 - t_2$, respectively,

$$m_x(t) = M_0 \tag{18.7}$$

$$r_x(t_1, t_2) = R_x(t_2 - t_1) \tag{18.8}$$

If we reparameterize the second-order moment using $t_2 = t + \tau/2$ and $t_1 = t - \tau/2$, we obtain the more familiar result

$$r_x(t_1, t_2) = R_x(\tau) \tag{18.9}$$

The spectral density of time-averaged power, or power spectral density, is given by the Fourier transform of the time-invariant autocorrelation,

$$S_x(f) = \int_{-\infty}^{\infty} R_x(\tau)e^{-i2\pi f\tau} d\tau \tag{18.10}$$

For nonstationary signals, the time invariance of the mean and autocorrelation does not hold. For almost all such signals, the structure of these two functions is such that it cannot be measured by using arbitrarily long observation blocks: the time-varying nature of the statistics is averaged away. However, there is one type of time-varying behavior that is persistent and regular: periodic time variation, which is exactly what is exhibited by the class of nonstationary signals known as CS signals.

18.2.2 The Cyclic Autocorrelation Function

For CS signals, the second-order moment can be represented as a Fourier series,

$$R_x(t, \tau) = \sum_{\alpha} R_x^{\alpha}(\tau)e^{i2\pi\alpha t} \tag{18.11}$$

where α is called a CF. If the autocorrelation is periodic, then the sum is over $\alpha = k/T_0$ for all integers k, where T_0 is the period. If the autocorrelation is almost periodic, then it is the sum of two or more periodic functions with incommensurate periods. In this case, the sum over the CFs includes all harmonics of each fundamental period. To accommodate this ambiguity, the sum over CF is left unspecified in most cases, as in Eq. (18.11).

The Fourier coefficient $R_x^{\alpha}(\tau)$ is called the cyclic autocorrelation, and it is equal to the conventional autocorrelation Eq. (18.9) for $\alpha = 0$. It is given by the usual Fourier coefficient expression,

$$R_x^{\alpha}(\tau) = \lim_{T\to\infty} \frac{1}{T} \int_{-T/2}^{T/2} R_x(t, \tau)e^{-i2\pi\alpha t} dt \tag{18.12}$$

where the limit is required for an almost periodic CS signal, but can be replaced by an integration over a single period for all other CS signals. The cyclic autocorrelation can also be computed from the data itself,

$$R_x^{\alpha}(\tau) = \lim_{T\to\infty} \frac{1}{T} \int_{-T/2}^{T/2} x(t + \tau/2)x^*(t - \tau/2)e^{-i2\pi\alpha t} dt \tag{18.13}$$

18.2.3 **The Spectral Correlation Function**

By analogy with the Fourier relationship between the autocorrelation and PSD for stationary signals, the SCF can be defined as the Fourier transform of the cyclic autocorrelation,

$$S_x^\alpha(f) = \int_{-\infty}^{\infty} R_x^\alpha(\tau) e^{-i2\pi\alpha\tau} d\tau \qquad (18.14)$$

However, it can also be derived by taking into account the density of temporal correlation between the complex envelopes of two narrowband components of the signal, where the separation between the center frequencies of the bands is given by α and the center point by f [25]. This gives rise to the name *spectral correlation*.

18.2.4 **Extensions for Complex-Valued Signals**

Working with complex-valued data has many advantages, but it also imposes a significant complication when exploiting cyclostationarity. The idea is that there are two kinds of cyclic autocorrelation functions and two kinds of SCFs to deal with, instead of one. This complication turns out to be quite important when applying the theory to real-world signals.

Consider for a moment the analytic-signal representation of an RF signal,

$$s_r(t) = \Re[s_a(t)] \qquad (18.15)$$

where

$$s_a(t) = s_c(t) e^{i2\pi f_c t + i\varphi_0} \qquad (18.16)$$

is the analytic signal, $s_c(t)$ is the complex envelope, and α denotes the real-part operation. The statistical nature of $s_a(t)$ must be identical to the statistical nature of

$$s_a(t) + s_a^*(t) = \left[s_c(t) e^{i2\pi f_c t + i\varphi_0} + s_c^*(t) e^{-i2\pi f_c t - i\varphi_0} \right] \qquad (18.17)$$

The second-order moment for $s_a(t)$ must therefore involve the following expectations $E[s_c(t + \tau/2) s_c(t - \tau/2)]$, $E\left[s_c(t + \tau/2) s_c^*(t - \tau/2) \right]$, $E\left[s_c^*(t + \tau/2) s_c(t - \tau/2) \right]$, and $E\left[s_c^*(t + \tau/2) s_c^*(t - \tau/2) \right]$. Due to symmetry considerations, we need only concern ourselves with the following two second-order moments:

$$R_{s_c}(t, \tau) = E\left[s_c(t + \tau/2) s_c^*(t - \tau/2) \right] \qquad (18.18)$$

$$R_{s_c s_c^*}(t, \tau) = E[s_c(t + \tau/2) s_c(t - \tau/2)] \qquad (18.19)$$

Because the standard correlation function involves one conjugated and one non-conjugated factor, the correlation in Eq. (18.19) is commonly called the *conjugate autocorrelation function*, even though it does not contain any conjugated factors.

For many modulation types [36, 37], both the non-conjugate and conjugate correlation functions are periodic or almost periodic functions. This leads to the necessary consideration [20, 28] of both the conjugate cyclic autocorrelation function:

$$R_{xx^*}^\beta(\tau) = \lim_{T\to\infty} \frac{1}{T} \int_{-T/2}^{T/2} R_{xx^*}(t, \tau) e^{-i2\pi\beta t} dt \qquad (18.20)$$

$$= \lim_{T\to\infty} \frac{1}{T} \int_{-T/2}^{T/2} x(t + \tau/2) x(t - \tau/2) e^{-i2\pi\beta t} dt \qquad (18.21)$$

and the conjugate SCF:

$$S_{xx^*}^{\beta}(f) = \int_{-\infty}^{\infty} R_{xx^*}^{\beta}(\tau) e^{-i2\pi f\tau} d\tau \qquad (18.22)$$

18.2.5 Spectral Coherence

The SCF is a correlation function, and if the two involved quantities have zero means, it is also a covariance function. As such, it can be normalized to yield a correlation coefficient. For the non-conjugate SCF, the two quantities undergoing correlation are narrowband spectral components at frequencies $f \pm \alpha/2$. Any covariance can be normalized by the geometric mean of the variance of the two variables in question. Since the variances of the two spectral components are equal to the PSD values for the two frequencies $f \pm \alpha/2$, we have the following spectral coherence:

$$C_x^{\alpha}(f) = \frac{S_x^{\alpha}(f)}{\left[S_x^0(f + \alpha/2) S_x^0(f - \alpha/2) \right]^{1/2}} \qquad (18.23)$$

For the conjugate SCF, we obtain the following conjugate spectral coherence:

$$C_{xx^*}^{\alpha}(f) = \frac{S_{xx^*}^{\alpha}(f)}{\left[S_x^0(f + \alpha/2) S_x^0(\alpha/2 - f) \right]^{1/2}} \qquad (18.24)$$

The two coherence functions are particularly useful for signal detection via cycle-frequency detection because the value of the coherence is independent of the absolute power levels of the signals in the data $x(t)$, which simplifies threshold setting.

18.2.6 Key Statistical Properties

In this section, we briefly review and demonstrate the key properties of the SCF that render it useful in difficult signal-processing situations, such as spectrum sensing for CR.

Noise and Interference Tolerance

The most important feature of the SCF is its insensitivity to noise and co-channel interference. By this we mean that the contributions to the SCF from different signals can appear in disjoint regions of the f-α plane. Mathematically, this can be seen by first creating a multiple-signal model such as

$$x(t) = \sum_{j=1}^{M} s_j(t) + w(t) \qquad (18.25)$$

where $s_j(t)$ are M statistically independent signals and $w(t)$ is stationary noise (e.g., WGN). The signals' power levels, relative delays, phases, and modulation types are subsumed into the notation $s_j(t)$. For this model, it can be shown that the SCF for the observable data $x(t)$ is the sum of SCFs for the component signals,

$$S_x^{\alpha}(f) = \sum_{j=1}^{M} S_{s_j}^{\alpha}(f) + S_w^{\alpha}(f) \qquad (18.26)$$

$$S_{xx^*}^{\alpha}(f) = \sum_{j=1}^{M} S_{s_j s_j^*}^{\alpha}(f) + S_{ww^*}^{\alpha}(f) \qquad (18.27)$$

Choosing $\alpha \neq 0$ in the non-conjugate SCF forces $S_w^\alpha(f)$ to vanish due to its assumed stationarity, leaving us with

$$S_x^\alpha(f) = \sum_{j=1}^{M} S_{s_j}^\alpha(f) \tag{18.28}$$

and choosing $\alpha = \alpha_{j_0} \neq 0$, which is unique to signal $j = j_0$, yields

$$S_x^{\alpha_{j_0}}(f) = S_{s_{j_0}}^{\alpha_{j_0}}(f) \tag{18.29}$$

Similar statements hold for the conjugate SCF, except that $\alpha = 0$ has no special significance there and $S_{ww^*}^\alpha(f)$ is typically zero for all α and f. These results reveal that the SCF for observed data that are corrupted by interference and noise can reflect only the spectral-correlation properties of a single signal if the CF is unique to that signal.

An illustrative example relevant to CR is provided in Figure 18.7, in which an 802.11 OFDM signal is added to a Bluetooth signal and stationary noise. The sampling rate is 20 MHz.

In Figure 18.7, the Bluetooth hops have varying power and so are more or less obvious in the PSD (non-conjugate SCF with $\alpha = 0$). The noise floor is clearly visible in the PSD but does not visibly influence the SCF in other regions of the planes. Note that the Bluetooth signal is an 8-PSK (phase shift key) signal with square-root raised-cosine transmit filtering and a symbol rate of 1.0 MHz. The features for the Bluetooth hops appear at the CF of $\alpha = 1.0$ MHz and are centered at each of the observed hop frequencies. The OFDM signal has many features in both planes, but the dominant features are due to the presence of the four perfectly correlated BPSK pilot subcarriers, which are separated by 4.375 MHz. This gives rise (in a nonobvious way) to dominant non-conjugate CFs of 4.25, 4.5, 8.75, 13.0, and 13.25 MHz.

Signal-Parameter Dependence

We have seen that the SCF for a particular signal can be separated from the SCF for an observed data set by selecting only that signal's CFs for analysis. It follows that any modulation-specific parameters on which the SCF depends can be estimated from the SCF for the observable data. The SCF depends on a variety of parameters, including the basic modulation parameters of bit, symbol, chip, hop, slot, and frame rates, as well as carrier-offset frequency, average power, symbol-clock phase, and carrier phase. These latter two are reflected in the complex-valued SCF, whereas all the others are reflected in the SCF magnitude.

This observation implies that the SCF can be used for symbol and carrier synchronization, transmit-filter estimation, hop-rate estimation, signal detection, channel estimation, and automatic modulation classification. Moreover, if the signal's CFs are unique, these signal-processing tasks can be performed in the presence of strong noise and/or co-channel interference [40].

Near-Universal Applicability

A further advantage of the SCF as a spectrum-sensing tool is that it is applicable to nearly all human-made communication signals. The set of basic modulation types that exhibit nonzero SCFs includes MPAM, MPSK, MQAM, CPM, CPFSK, AM-DSB, FH, DSSS, hybrid FH-DSSS, $\pi/4$-DQPSK, MSK, OQPSK, GMSK, and many others.

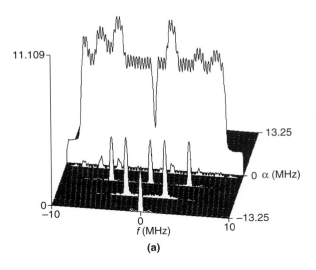

(a)

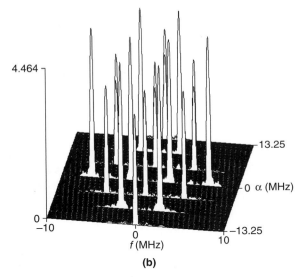

(b)

FIGURE 18.7

Non-conjugate (a) and conjugate (b) SCFs for 802.11 OFDM with in-band Bluetooth interferers. The stronger Bluetooth signals can be seen in the PSD ($\alpha = 0$ in (a)), although this may not be true when the OFDM signal experiences multipath propagation. Their contribution to the SCF is seen in (a) for $\alpha = 1.0$ MHz, which is the Bluetooth symbol rate. The 802.11a/g OFDM signal possesses many strong features in both (a) and (b). The features will still be clearly visible and will correspond to the correct values of α even for multipath propagation channels.

The multiple-access method of a radio network also has an influence on the SCF; in some cases, this is a very strong influence. For example, the spectral-correlation properties of GSM traffic signals are quite different from those for the base Gaussian minimum shift keying (GMSK) modulation type (see Section 18.4.5). In some cases, the base modulation has relatively weak SCF features, but the transmitted waveform has much stronger features due to the addition of the access layer components of the physical layer signal. In the case of OFDM, for example, the quadrature amplitude modulation (QAM) subcarriers combine to yield relatively weak non-conjugate features [6], but when some subcarriers are assigned periodically repeating synchronization sequences, the resulting signal has quite strong features, as seen in Figure 18.7.

Influence on Signal Design

The spectral-correlation property can influence various aspects of the modulation-design stage of communication-system design. One obvious area of application is to covert communications. In this situation, the signal can be designed to thwart the various spectral-correlation-based signal detectors described in this chapter by limiting the sets of correlated spectral components, which can be accomplished, for example, by severe transmit filtering such as seen in duobinary signaling [51].

A recent proposal in the field of CR is to modify the physical layer signal on a per-user basis to allow fast and robust user or network identification via SD [13]. In this proposed OFDM scheme, selected data subcarriers are used to transmit identical data streams, which introduces spectral correlation with CFs that depend on the frequency separation between the selected subcarriers.

18.2.7 Efficient Statistical Estimators

In this section, we describe two useful and quite distinct spectral-correlation estimators. The first, the frequency-smoothed cyclic periodogram, is highly efficient when the number of CFs that require analysis is small compared to the number of samples to be processed. The second method is a time-smoothing method called the strip spectral-correlation analyzer (SSCA), which is efficient when an estimate of the SCF is desired over all frequencies and CFs. The frequency smoothing method (FSM) is often used for focused analysis for a known (or previously estimated) CF, whereas the SSCA is used for search applications when little is known a priori or when the number of CFs that require detection is large [38]. We apply both the FSM and SSCA estimators to various sets of collected data in Section 18.4.

The Frequency Smoothing Method

The SCF for a single value of α and all frequencies f can be estimated by Fourier transformation of an estimate of the cyclic autocorrelation, but a simpler and more efficient method is to frequency smooth the cyclic periodogram, which is defined by

$$I_T^\alpha(t, f) = \frac{1}{T} X_T(t, f + \alpha/2) X_T^*(t, f - \alpha/2) \tag{18.30}$$

and is equal to the conventional periodogram when $\alpha = 0$. The frequency smoothing is simply a convolution in frequency with some pulselike window function $h(f)$,

$$\hat{S}_x^\alpha(t, f) = b(f) \otimes I_T^\alpha(t, f)$$

$$= \frac{1}{T} \sum_{n=0}^{T-1} b(f - n/T) X_T(t, n/T + a/T) X_T^*(t, n/T - a/T) \tag{18.31}$$

for $f = m/T$, where a is such that $|\alpha/2 - a/T|$ is minimized with respect to a. It can be shown that this estimate converges to the ideal SCF when the block length, T, is first increased without bound and then the width of $b(f)$ (the spectral resolution) is decreased to zero [25]. Similar statements apply to the conjugate SCF. The non-conjugate and conjugate spectral coherence functions can be estimated by combining FSM estimates of the SCF and PSD.

For many applications, only $b(f)$ width, not its exact form, is important. In these situations, the smoothing can be efficiently accomplished by using a rectangular $b(f)$.

The Strip Spectral-Correlation Analyzer

The SSCA has been defined and analyzed by Brown and Loomis and Roberts et al. [31, 32]. In Brown and Loomis, the definition is restricted to real-valued signals, whereas in Roberts et al., complex-valued signals are also considered. The differences are minimal, and we take our notation from Roberts et al. [32], where the SCF point estimate is defined by

$$\hat{S}_{xy_T}^{f_k+q\Delta\alpha}\left(n, \frac{f_k}{2} - \frac{q\Delta\alpha}{2}\right)_{\Delta t} = \sum_r X_T(r, f_k) y^*(r) g(n-r) e^{-i2\pi qr/N} \tag{18.32}$$

where $\alpha = f_k + q\Delta\alpha$, $f_k = kf_s/N'$, q ranges from $-N/2$ to $N/2 - 1$, and k ranges from $-N'/2$ to $N'/2 - 1$. Here, the integer N denotes the total number of samples processed by the estimator and it determines the ultimate cycle resolution $\Delta\alpha$. N' is a channelization parameter, discussed next, that determines the spectral resolution Δf. Eq. (18.32) is the cross (non-conjugate) SCF estimate between the two signals $x(t)$ and $y(t)$. The demodulates $X_T(\cdot,\cdot)$ are defined by

$$X_T(n, f) = \sum_{r=-N'/2}^{N'/2-1} a(r) x(n-r) e^{-i2\pi f(n-r)} \tag{18.33}$$

and consist of downconverted channelized versions of $x(t)$. The real-valued function $a(\cdot)$ is the *data-tapering window*, N' is a power of two (dyadic number) representing the number of channels into which $x(t)$ is divided (also the number of *strips*), and $T = N'T_s = N'$ is the length of data in the channelizer fast Fourier transform (FFT). Note that $a(\cdot)$ has width N' with $\sum_n a^2(n) = 1$. The real-valued function $g(\cdot)$ in Eq. (18.32) is the time-smoothing window with width $\Delta t = NT_s = N$ and $\sum_n g(n) = 1$. The cycle resolution is $\Delta\alpha = 1/\Delta t = 1/(NT_s) = 1/N$. The channelizer frequencies are denoted by $f_k = kf_s/N' = k/N'$, for $-N'/2 \le k \le N'/2 - 1$. In terms of the usual SCF parameterization of (f, α), the point estimates from the SSCA Eq. (18.32) correspond to

$$f = (f_k - q\Delta\alpha)/2 \tag{18.34}$$

$$\alpha = f_k + q\Delta\alpha \tag{18.35}$$

Reexpressing leads to

$$\alpha = 2f_k - 2f \tag{18.36}$$

So, for a fixed value of f_k, the point estimates lie along a line parallel to $\alpha = -2f$, which is called a strip.

There are three basic computational elements to the SSCA. The first is the computation of the demodulates $X_T(r, f_k)$, which is also referred to as channelization. When N' is a dyadic number, this can be done efficiently with the FFT. The second computational step is the computation of the signal product, which is the product of the demodulate, the time-domain waveform, and the time-averaging window. The third computational step is the correlation of the signal product with the complex exponential $e^{-i2\pi qr/N}$, which, for dyadic N, can also be performed efficiently with the FFT.

The spectral resolution of the estimate is determined by the channelization step, and so is approximately $\Delta f = 1/N'$. The temporal resolution is determined by the final smoothing operation, which averages over the entire block of available data; therefore, $\Delta t = N$. The resolution product is therefore

$$\Delta f \Delta t \approx N/N' \qquad \qquad (18.37)$$

To meet the resolution-product requirement for a reliable estimate [25], we need $\Delta f \Delta t \gg 1 \Rightarrow N \gg N'$. So this tends to favor a small number of channels in the channelizer. To resolve the detailed features of the SCF (and PSD), however, we desire small spectral resolution, $\Delta f \ll 1 \Rightarrow N' \gg 1$, which favors a large number of channels. In general, we want to choose N' just large enough to allow resolution of the expected spectral features in the PSD and SCF. This will allow the largest-resolution product and adequate spectral resolution. A similar analysis holds for the FSM, where the frequency resolution is equal to the width of $h(f)$ and the temporal resolution is equal to the FFT size T.

18.3 SPECTRUM SENSING BASED ON SPECTRAL CORRELATION

In this section, we provide examples of sensing strategies that are based on the SCF. In most cases, optimum exploitation [33, 34] is not practical, so suboptimum or ad hoc methods are used. In spite of their suboptimality, they can be preferred over other sensing methods because they inherit the desirable statistical properties of spectral correlation.

18.3.1 Cycle-Frequency Detection

In many sensing situations, it is enough to determine the presence of a particular signal type the parameters of which are known in advance. For example, in the IEEE 802.22 context, we want to detect the presence of an ATSC DTV signal, about which much is known, including modulation type, symbol rate, relative spectral location, and relative power of the pilot tone, as well as the exact nature of the periodically repeated long synchronization sequence. Similarly, for a cellular-radio CR, the radio needs to detect the presence of only one of a handful of well-known standards, such as GSM, EDGE, CDMA, or time division multiple access (TDMA) (IS-54/136) [6].

The following is the traditional formulation of the binary-hypothesis random-signal detection problem,

$$H_1: x(t) = s(t) + w(t)$$
$$H_0: x(t) = w(t)$$
(18.38)

where $s(t)$ is the signal to be detected and $w(t)$ is noise, is replaced by the SCF-based cycle-frequency detection problem,

$$H_1: S_x^\alpha(f) \not\equiv 0$$
$$H_0: S_x^\alpha(f) \equiv 0$$
(18.39)

or the coherence-based CF detection problem,

$$H_1: C_x^\alpha(f) \not\equiv 0$$
$$H_0: C_x^\alpha(f) \equiv 0$$
(18.40)

In these latter two problems, we explicitly test for the presence of nonzero SCF or coherence at a particular CF as a way of indirectly determining the presence of the signal itself.

As a practical matter, the specified CF α should be considered only nominally known because the transmitter is allowed to produce a waveform that deviates from nominal specifications in small ways. This leads to signal-presence tests that search for significant values of the coherence or spectral correlation in a small band of CFs near the nominal one. Or, in mathematical terms, the coherence-based detection statistic is

$$Y_c(\alpha) = \max_{f \in F, a \in A} \left| \hat{C}_{x_T}^a(f) \right|$$
(18.41)

where F is some set of spectral frequencies and A is an interval of CFs centered at α. The signal-presence detection test is then

$$Y_c(\alpha) > \eta_c \Rightarrow H_1; \text{ else } H_0$$
(18.42)

The threshold η_c can be chosen empirically or from considerations of the statistical properties of the spectral coherence [7].

18.3.2 Joint Cycle-Frequency Detection

Many signal types exhibit more than one CF. For such signals, multiple single-CF detection statistics could be combined in various ways to achieve better performance. This can be particularly useful in frequency-selective fading channels, where the SCF for one CF is affected less than for another by the particular frequency dependence of the fade.

As an example, we can find the average coherence over multiple CFs and compare to a threshold,

$$Y_m = \sum_{j=1}^{L} Y_c(\alpha_j) > \eta_m \Rightarrow H_1; \text{ else } H_0$$
(18.43)

An alternative that emphasizes low probability of false alarm is to insist that each of the separate CFs are present with sufficient strength,

$$(Y_c(\alpha_1) > \eta_{c,1}) \text{ AND } (Y_c(\alpha_2) > \eta_{c,2}) \ldots \text{ AND } (Y_c(\alpha_N) > \eta_{c,N}) \Rightarrow H_1; \text{ else } H_0 \quad (18.44)$$

18.3.3 Spectral-Correlation Matching: The Cycle Detectors

When the full SCF for the signal of interest is known in advance, the optimum detector for a weak version of the signal is the multicycle detector [33, 34]. This detector correlates the cyclic periodogram for each known CF with the ideal SCF for that CF and coherently adds the result to obtain the optimal multicycle detection statistic,

$$Y_{omd} = \sum_{j=1}^{L} \int S_x^{\alpha_j}(f) I_T^{\alpha_j}(t, f) df \qquad (18.45)$$

This statistic is complex-valued, and the test consists of comparing its real part to a threshold. The drawback of this statistic is that it requires that the phases (e.g., symbol-clock and carrier) of the signal be known so that the coherent addition of the complex numbers in the sum over j will be constructive. This requirement is difficult to meet for a signal with a presence that is not yet known.

An individual term in the multicycle detector is the optimal single-cycle detector [42, 43],

$$Y_{osd}(\alpha) = \int S_x^{\alpha}(f) I_T^{\alpha}(t, f) df \qquad (18.46)$$

for which the magnitude can be taken at a small loss of optimality. This notion leads to the suboptimal noncoherent multicycle detector given by

$$Y_{smd} = \sum_{j=1}^{L} |Y_{osd}(\alpha)| \qquad (18.47)$$

Finally, in the case for which the ideal SCF is not completely known—for example, the exact transmitter pulse shaping is unknown or could be corrupted by the propagation channel—the ideal SCFs can be replaced by rectangular windows centered at the nominal feature center frequency and the magnitude of the cyclic periodogram summed over the interval. For example, the suboptimal single-cycle detector is

$$Y_{ssd}(\alpha) = \int_B \left| I_T^{\alpha}(t, f) \right|^2 df \qquad (18.48)$$

where B represents all frequencies for which the ideal SCF is known to be nonzero. As seen from the IEEE 802.11a/g OFDM results (see Section 18.4.7), this does not necessarily mean a single contiguous frequency interval.

18.3.4 Connections to Conventional Processors

In this section, we point out the mathematical connections between the sensing methods based on spectral correlation and more traditional sensing methods. Finally, we outline the ways to extend the spectral correlation methods to higher-order nonlinear processing, and describe some situations in which it is desirable to do so.

Energy Detection

The optimal energy detector (radiometer) for a weak random signal in WGN is given in Section 18.1.4 (Eq. (18.1)) and can now be seen as the special case of the optimum single-cycle detector Y_{osd} (Eq. (18.46)) with CF set to zero. It is also, trivially, the optimum multicycle detector for the special case in which the signal has no second-order cyclostationarity.

Delay-and-Multiply Detection

The DM detector is a reduced-dimension version of the suboptimum single-cycle detector. The latter can be reexpressed in the time domain as the correlation between the measured cyclic autocorrelation and the ideal cyclic autocorrelation for the signal of interest. The DM detector is proportional to this detector when the range of correlation is severely restricted to a single delay.

Cyclic-Cumulant Detection

Many possible detection schemes could employ cyclic moments or cumulants [44–50]. When the highest order of moment or cumulant employed is two, then the method is exploiting only spectral correlation. When the method combines moments or cumulants of more than one order (e.g., two and four), then the method exploits spectral correlation and higher-order statistics. In most cases, the method involves correlating measured cumulant functions with ideal ones for the set of modulation types of interest in the sensing problem. Because the spectral correlation is a function of two variables and the higher-order statistics are multidimensional functions with dimensions higher than two, only a very small subset of the chosen functions can be used in the correlation. In this sense, most higher-order methods are similar to extensions of the DM detector from second-order to joint use of multiple orders.

The use of higher-order statistics (e.g., cyclic cumulants and cyclic moments) can be justified when the number of possible signal types is large or includes QAM and PSK. For digital QAM and PSK, only the BPSK (2-QAM) signal can be distinguished from therest. All of the higher-alphabet signals possess the same SCF (to within a single scale factor) provided the transmitter pulse-shaping filter is the same for all of them. However, they possess distinct higher-order statistics, and in most cases they possess distinct fourth-order statistics, which implies that there is a large gain in classification resolution even if the maximum employed order is restricted to four [14].

18.4 APPLICATION TO MODERN COMMUNICATION SIGNALS

In this section we explain the potential of spectral-correlation-based spectrum-sensing methods for modern signal types by using analysis and collected data sets.

18.4.1 General Approach to Sensing Algorithm Development

Our approach consists of the following elements:

Mathematical Modeling. Using standards documents and application notes, determine a mathematical model for the transmitted waveform, including the effects of the access method and the use of any known components (e.g., tones) or symbol sequences (e.g., synchronization bursts).

Spectral-Correlation Analysis. Using the obtained model, derive the SCF or the cyclic autocorrelation function. In some cases, this may be intractable, but determination of the signal's CFs can still be mathematically feasible.

Spectral-Correlation Verification. Develop a simulator for the signal with sufficient fidelity to represent the signal at the level of detail used in the mathematical modeling step. Estimate the SCF for the simulated signal using the SSCA to determine all exhibited non-conjugate and conjugate CFs. Compare them to the set of cycle frequencies from the analysis. Use the FSM to determine the exact shape of the SCF for each detected CF and compare them to the analysis. If an agreement is not obtained, refine both the mathematical model and simulator, as needed, until agreement is reached.

Identification of Low-Cost Algorithm. Using the knowledge gained from modeling, analysis, simulation, and verification, develop the simplest possible algorithms for determining the presence of the signal based on measuring some portion of the spectral-correlation plane. This can include CF detection, single-cycle detection, DM detection, or multicycle detection.

Validation Using Collected Data. Apply the SSCA and FSM to collected data for the signal type under study. Compare the detected CFs and measured SCFs to those from the analysis and simulation. If an agreement is obtained, test the identification algorithms against collected data sets. Adjust the thresholds as needed based on the empirical data results.

In this section, we provide only a summary of the steps and results needed to obtain algorithms that simultaneously achieve low computational cost and high detection performance.

We consider ATSC DTV, CDMA, North American digital cellular (IS-54/136), the GSM family of signals, and the several kinds of signals used in the IEEE 802.11a/b/g standards. The ATSC DTV signal appears in the IEEE 802.22 wireless regional area network CR context, where 802.22 transceivers may be allowed to transmit on fallow 6 MHz wide broadcast TV channels. The CDMA, TDMA, and GSM signal types are candidates for detection by CRs if the radio is to autonomously determine its options for cellular telephony or data connection at any given time and place. Finally, the WiFi IEEE 802.11a/b/g signals must coexist in unlicensed bands, together with other signals (e.g., Bluetooth) and are therefore candidates for sensing by advanced radios that may operate in such bands.

Before moving to the collected data results, we point out that our focus on modern wireless signals bridges a perceived gap between the well-established theory and algorithms of CS signal processing—typically illustrated with basic modulations such as BPSK—and the complicated spectral-correlation properties of modern RF signals.

18.4.2 ATSC DTV

The ATSC DTV signal is a heavily filtered vestigial sideband (VSB) 8-level (or 16-level) signal with an added pilot tone. The symbol rate is 10.7622 MHz, and the signal is filtered to a bandwidth of 6 MHz. This removes all non-conjugate CFs, but because the filtering is asymmetrical with respect to the baseband center frequency, and the modulation is real valued, there remain two conjugate CFs equal to the original doubled carrier offset, $2f_0$ and $2f_0 + F_{sym}$. Since the pilot tone is added to the signal at the location of the original baseband carrier, and then the signal is upconverted to its RF center

Table 18.1 ATSC DTV Physical Layer Parameters

Parameter	Value	Comment
Modulation	8 PAM or 16 PAM VSB	
Symbol rate	10.7622 MHz	F_{sym}
RF carrier frequency	f_c	Various
Pilot tone frequency	$f_c - F_{sym} / 4$	F_{pilot}
Pilot tone power	−11.3 dB	Relative to total power
Occupied bandwidth	≈ 6.0 MHz	
Synch sequence length	832 symbols	
Synch repetition interval	24 ms	Every 260,000 symbols
Non-conjugate CFs	0.0	
Conjugate CFs	$2F_{pilot}, 2F_{pilot} + F_{sym}$	

frequency, the two conjugate CFs for the transmitted signal are equal to $2F_{pilot}$ and $2F_{pilot} + F_{sym}$. More details on the modulated signal are shown in Table 18.1.

Matched filtering is applicable to this signal in at least two ways. First, the pilot tone can be detected by using Fourier methods; and second, the synchronization sequence can be detected by using conventional finite impulse response (FIR) filtering. Spectral correlation can be exploited by joint CF detection by using the two conjugate features. For moderate-to-high SNR, we have verified that the signal can be detected by using joint CF detection—even when subjected to frequency-selective fading—for captured block lengths of less than 10 ms. This gives spectral-correlation methods an advantage over filtering matched to the synchronization sequence in terms of time-to-detect performance, since the long synchronization sequence repeats only every 24 ms.

We have found that a high-performance algorithm for situations involving strict limits on the captured data block length involves joint detection of the pilot tone and the two conjugate CFs.

To validate the approach, we present the measured conjugate SCF for simulated and captured ATSC DTV signals in Figure 18.8. The signals are normalized to have unit power, are sampled at a rate of 7.0 MHz, and 65,536 samples are processed to obtain each estimate. For the conjugate feature with CF of $\alpha = 2F_{pilot}$, the pilot tone is first excised from the data before the SCF is estimated. This procedure reveals the contribution to the feature that is due only to the second-order statistical nature of the signal.

The captured ATSC DTV signals shown in Figure 18.8 correspond to RF center frequencies of 551, 569, 629, 647, and 659 MHz. The figure reveals that the features are reliably estimated and are easily detectable in spite of the wide variety of channel-distortion functions. This can be seen by first examining (a), where the measured PSDs are plotted with thin dotted lines and the ideal PSD is plotted with a thick solid line. The effects of the propagation channel on the collected signals are evident as the measured PSDs randomly fluctuate across the 6 MHz passband. Nevertheless, the measured SCFs for the two conjugate CFs, shown by the thin dotted lines in the lower two graphs in Figure 18.8, are all of a similar shape and also rise many decibels above the sur-

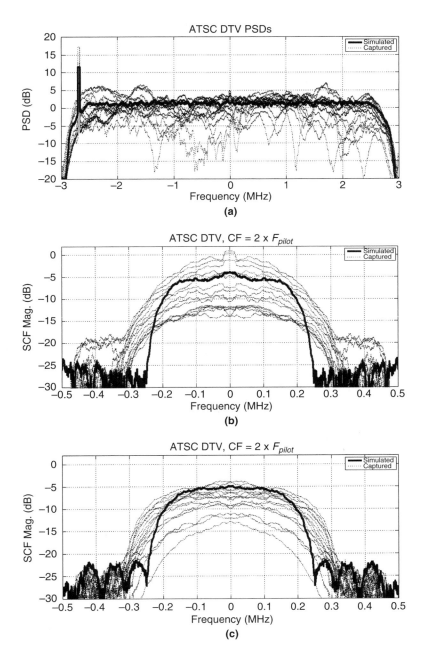

FIGURE 18.8

PSD and SCF estimates for simulated and captured ATSC DTV signals. Note the regularity and prominence of the measured SCFs in spite of the wide variety of channel distortion functions that are evident in the measured PSDs.

rounding measurement noise. The shapes of the measured SCFs also compare well with the ideal SCFs, plotted with thick solid lines in the two lower graphs.

18.4.3 The CDMA Family of Signals

The CDMA family consists of DSSS signals using BPSK and QPSK modulation formats. The second-generation mobile telephony version of CDMA is based on the IS-95 standard and uses a 1.2288 MHz chip rate DSSS QPSK signal in a 1.25 MHz wide channel. The transmitter filters the outgoing signal so that its excess bandwidth (filter roll-off in excess of the Nyquist rate) is limited to 17 percent. A 64-chip Walsh code is used together with a 32,768 chip long pseudorandom (PN) code to scramble and spread the data symbols.

A DSSS QPSK signal exhibits no conjugate CFs and a set of non-conjugate CFs that are harmonics of the code-repetition rate. Typically, the strongest feature is for the chip rate.

Because the PN sequence is so long (26.7 ms), many samples must be collected to exploit cyclostationarity related to the code-repetition rate. However, the non-conjugate SCF feature for the chip-rate CF can be exploited as long as a sufficient number of chips are processed, which is the case even for data blocks with lengths much smaller than 26.7 ms.

An effective detection algorithm for CDMA consists of CF detection to ensure that the chip-rate CF is accurately estimated, followed by the single-cycle detector, which performs a correlation between the measured and ideal SCFs for the chip-rate CF.

To validate the approach, we present in Figure 18.9 the ideal PSD and SCF for a CDMA signal modeled as a QPSK signal with a symbol rate of 1.2288 MHz, together with measured PSDs and SCFs for a set of collected CDMA signals. In all cases, the functions and CFs are measured by using 65,536 samples at a sampling rate of 2.8 MHz. The variation in the measured functions is due to several possible causes, including multiple-access interference and transmitted data that might not be independent and identically distributed. The CDMA signals were captured at RF center frequencies of 1.955, 1.9575, 1.96875, and 1.98625 GHz.

Figure 18.9 shows that it is possible to reliably detect the presence of the CDMA signal by detecting the presence of the chip-rate feature. This can be seen by noting that the measured SCFs are quite similar to each other, with only a few exceptions, and that they rise several to many decibels above the surrounding measurement noise floor. The features are somewhat narrower in frequency than the ideal, plotted as a solid black line, because the ideal signal does not fully capture all the CDMA physical layer attributes.

18.4.4 The Cellular TDMA Family of Signals

The TDMA family of signals comprises those signals used in North American digital cellular (NADC) systems that adhere to standards IS-54 or IS-136. These signals use a six-slot TDMA mechanism together with the $\pi/4$-DQPSK modulation type and a symbol rate of 24.3 kHz. Although this system is inherently narrowband and is being replaced by third-generation systems, it is instructive to include it here for completeness: all the wireless standards exhibit exploitable spectral correlation regardless of the modulation type and the access method.

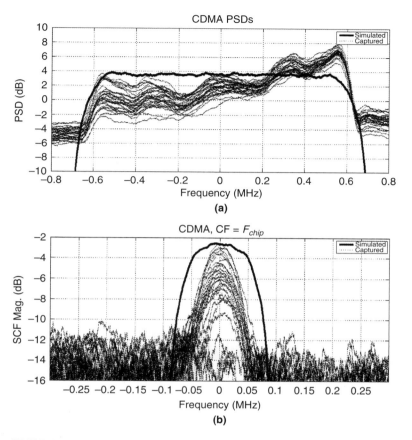

FIGURE 18.9

PSD and SCF estimates for simulated and captured CDMA signals. Note the measured SCFs' consistency in spite of the variation of the captured-signal PSD. CDMA can be reliably detected through detection of the presence of this single, relatively weak, spectral correlation feature.

The $\pi/4$-DQPSK modulation method uses one four-point QAM constellation for its even-indexed symbols and a second constellation, rotated by $\pi/4$ radians with respect to the first, for its odd-indexed symbols. This small change results in little effect on the SCF (it does, however, have a major affect on the higher-order statistics of the signal) so that the signal has a single non-conjugate CF equal to the symbol rate and no conjugate CFs. When combined with the slotting access method, further cycle frequencies are created, but are reduced in strength relative to the symbol-rate feature.

To verify that NADC signals exhibit exploitable spectral correlation at the 24.3 symbol-rate CF, we plot the measured SCF for several captured[2] NADC TDMA signals as well as for a simulated $\pi/4$-DQPSK signal in Figure 18.10. The ideal signal does not

---

[2]Unlike the other signal types in this section, the NADC signals were generated by using a hardware simulator and captured over the air in a laboratory setting.

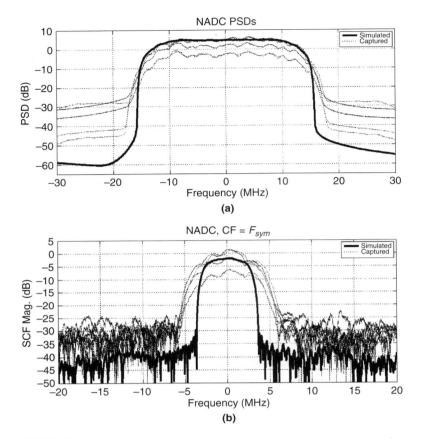

FIGURE 18.10

PSD and SCF estimates for simulated and captured TDMA (NADC) signals. Note that the measured SCF peaks rise many decibels above the measurement noise.

fully capture all the IS-54/136 physical layer attributes, but forms a reasonable baseline, especially for short capture lengths.

Figure 18.10 reveals, as in the cases of ATSC-DTV and CDMA signals, that the captured signals exhibit spectral correlation at the expected CF and with the expected shape and width. Note that the SCF extends several tens of decibels above the surrounding measurement noise. This validates the claim that IS-54/136 could be reliably detected by detecting the presence of this single SCF feature.

18.4.5 The GSM Family of Signals

The GSM family of cellular telephony signals includes GMSK-based GSM and 8PSK-based EDGE. These signals combine the basic modulation type with a TDMA-FDMA access method that uses eight slots per frame and 156.25 symbols per slot at a symbol rate of 270.833 kHz. The basic physical layer parameters are shown in Table 18.2.

Table 18.2 GSM Physical Layer Parameters

Parameter	Value	Units and Comments
Modulation	GMSK	GSM, $BT_b = 0.3$
	8PSK	EDGE, $3\pi/8$ version
Multiple access	TD-FD	
Slots/frame	8	
Information bits/slot	148	
Bits/guard	8.25	
Frame rate	216.6 Hz	F_{frame}
Slot length	577 μs	
Slot rate	1.733 kHz	F_{slot}
Bit rate	270.833 kHz	F_{bit}
Symbol rate	135.417 kHz	F_{sym}, SQPSK representation
Channel spacing	200.0 kHz	
Non-conjugate CFs	$k \times F_{frame}$	Subset of these
Conjugate CFs	$2f_c \pm F_{sym} \mp F_{frame}$	Dominant CFs

The base modulation type of GMSK (GSM and GPRS) exhibits detectable spectral correlation [41]. For a GMSK signal with symbol rate of F_{sym} and carrier offset of f_c, when viewed as an offset QPSK signal, the non-conjugate CFs have negligible-strength SCF, and the conjugate CFs are $2f_c \pm F_{sym}$. So for a GSM signal at complex baseband, we have $f_c = 0$ and the two CFs are ± 135.4 kHz.

For EDGE, the basic modulation type is a variant of 8PSK called $3\pi/8$-8PSK and the EDGE system uses a transmitter pulse-shaping filter that spectrally shapes the signal so that it can fit into the standard 200 kHz wide GSM channels. The transmitter filtering reduces the non-conjugate set of CFs to just {0}—the PSD. Moreover, 8PSK and $3\pi/8$-8PSK do not exhibit any conjugate spectral correlation. Therefore, the base modulation for EDGE is a second-order stationary signal.

In both GSM and EDGE, the base modulation is combined with the access mode, in this case an eight-slot TDMA scheme per RF carrier. It can be shown mathematically that this combination produces a great number of non-conjugate cyclic features even if the base modulation type is stationary. Moreover, for GSM, there is also a large number of conjugate features. When the transmitted bit stream includes periodically repeated components (e.g., the GSM/EDGE 26-symbol midamble), further cyclic features of both types are generated.

The set of non-conjugate CFs for a GSM traffic signal is a subset of the harmonics of the slot-loading repetition interval. By *slot-loading*, we mean the particular pattern of active and inactive slots within the eight-slot frame. For example, if only one slot is active, then the slot-loading repetition interval is equal to the frame length of 8×577 $= 4616$ μs. However, if the first and fifth slots are the active ones, then the slot-loading repetition interval is half the frame rate. Let the slot-loading repetition interval be represented by T_0. Then the non-conjugate CFs are a subset of the harmonics k/T_0, and the conjugate CFs are a subset of

$$\alpha = 2f_c \pm F_{sym} + k/T_0 \qquad\qquad (18.49)$$

It is generally true that the strongest conjugate GSM features for any loading pattern are $\alpha = 2f_c + F_{sym} - 1/T_0$ and $2f_c - F_{sym} + 1/T_0$, where T_0 represents the full eight-slot loading pattern. Thus, spectrum-sensing techniques for GSM can consist of finding chains of harmonically related CFs and detecting the presence of the two dominant conjugate CFs, which are separated by $2F_{sym} - 2/T_0 = 270.4$ kHz.

Examples of measured CFs together with SCF and coherence magnitudes are shown in Figure 18.11 for five pairs of captured GSM signals. The sample rate is 280 kHz, and 65,536 samples are processed in all cases. The strip spectral-correlation analyzer is used to efficiently estimate the SCF over its entire domain of definition, and the maximum value of the SCF and coherence is retained for each CF. The chains of equispaced CFs are evident, as is the variation in their spacing due to slot loading. In particular, the first set of non-conjugate CFs exhibits relatively wide spacing, the second slightly more narrow spacing, and so on until the fifth and final set of non-conjugate CFs, which are so closely spaced that they cannot be resolved in the plot. In contrast, the separation between the two dominant conjugate CFs is consistent throughout the five sets of measurements: 270.4 kHz.

18.4.6 IEEE 802.11b: DSSS and CCK

The various data rates available in WiFi are made possible in part by using different modulation types. The basic WiFi modulation types and their parameters are presented in Table 18.3.

In IEEE 802.11b, DSSS BPSK, DSSS QPSK, four-level complementary code keying (CCK), and eight-level CCK are used to enable rates of 1.0, 2.0, 5.5, and 11.0 Mbps, respectively. The DSSS signals operate at a symbol rate of 1.0 MHz and a chip rate of 11.0 MHz, whereas the CCK signals operate at a symbol rate of 1.375 MHz. The CCK-4 signal sends 4 bits per symbol, for a bit rate of 5.5 Mbps, and the CCK-8 signal uses 8 bits per symbol, achieving the 11.0 Mbps rate. The physical layer parameters for 802.11b/g signals are provided in Tables 18.4 and 18.5 (see p. 627) for DSSS and CCK signals, respectively, along with the CFs obtained from mathematical analysis of these signals.

802.11b DSSS

The spectral correlation of DSSS signals is well understood, because over short intervals (relative to the code-repetition interval), the signal is well modeled by PSK or QAM with a symbol rate equal to the DSSS chip rate, and over long intervals, the signal is well modeled as a PSK or QAM signal with a symbol rate equal to the code-repetition rate and a pulse function equal to the repeated chipping sequence. Over short intervals, then, the SCF for most DSSS BPSK signals is identical to that for a general BPSK signal. For longer intervals, there are many non-conjugate and conjugate CFs for DSSS BPSK. The non-conjugate CFs are harmonics of the code-repetition rate, which is equal to the data rate (bit rate) when the code is repeated for each bit (as it is in 802.11b DSSS BPSK). The conjugate CFs are equal to the non-conjugate CFs plus the doubled carrier frequency. The non-conjugate CFs for DSSS QPSK are identical to those for DSSS BPSK, and the conjugate CFs are not present.

The symbol rate (data rate) for the DSSS signals is 1.0 MHz, and there are 11 chips per symbol, which means the chip rate is 11.0 MHz. Therefore, we should see non-

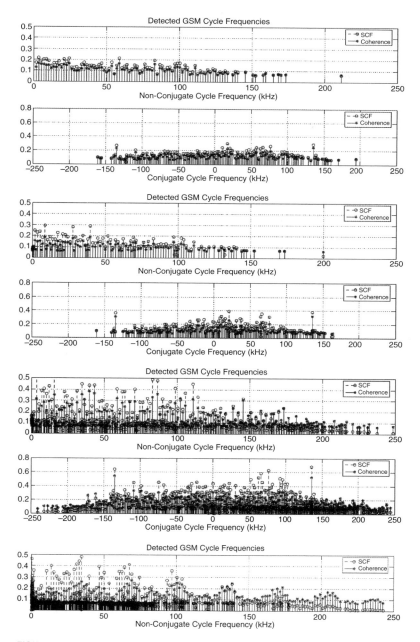

FIGURE 18.11

Detected CFs for five sets of captured GSM data. Non-conjugate CFs are shown in the upper graph in each pair, and conjugate CFs are shown in the lower graph in each pair. Note the decreasing separation between non-conjugate CFs in this sequence of measurements (upper graph in each pair), and the constant separation between the two dominant conjugate CFs (lower graph in each pair).

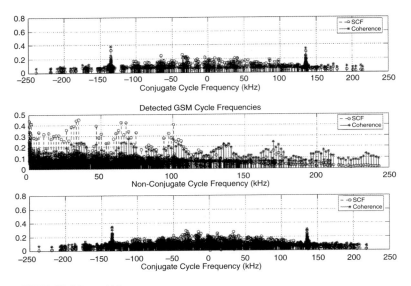

FIGURE 18.11 *cont'd*

Table 18.3 Overview of the 802.11a, b, and g WLAN Standards

| Modulation Type | Rate (Mbps) | Frequency (GHz) | | | RF BW (MHz) |
		11a	11b	11g	
DSSS BPSK, 11 chips/bit	1		2.4–2.5	2.4–2.5	22
DSSS QPSK, 11 chips/symbol	2		2.4–2.5	2.4–2.5	22
CCK, 4 bits/symbol	5.5		2.4–2.5	2.4–2.5	22
CCK, 8 bits/symbol	11		2.4–2.5	2.4–2.5	22
OFDM, 64 subcarriers, 4 pilot subcarriers, 13 null; BPSK, QPSK, 16-QAM, 64-QAM	6, 9, 12, 18, 24, 36, 48, 54	≈ 5.5		2.4–2.5	17

conjugate CFs of k MHz up to about $k = 2 \times 11 = 22$. Results for captured 802.11b DSSS BPSK and QPSK signals are shown in Figure 18.12, where it is evident that the modeling and mathematical predictions are correct (compare the CFs listed in Table 18.4 with the measurements in Figure 18.12).

Because 802.11b DSSS signals have many CFs and associated strong SCFs, many detection strategies are effective. A particularly effective and low-cost strategy is to jointly detect several of the cycle frequencies using CF detection. This can be done by using the SSCA and searching its output for "chains" of CFs separated by 1.0 MHz, or by using a set of FSM-based cycle-frequency searches in narrow bands of cycle frequencies near the nominal values of k MHz.

Table 18.4 Parameters for the 802.11b/g DSSS Waveform

Parameter	Value	Comments
Data rates	1.0, 2.0 Mbps	BPSK, QPSK
Symbol rate	1.0 MHz	
Chip rate	11.0 MHz	
Chips per symbol	11	
Chip sequence type	Barker	{1, −1, 1, 1, −1, 1, 1, 1, −1, −1, −1}
Transmit filtering	EBW = 100%	
Occupied bandwidth	22.0 MHz	
Non-conjugate CFs	k MHz, $k = 0, 1, \ldots, 22$	BPSK, QPSK
Conjugate CFs	$2f_c \pm k$ MHz, $k = 0, 1, \ldots, 20$	BPSK
	None	QPSK

Table 18.5 Parameters for the 802.11b/g CCK Waveform

Parameter	Value (CCK-4)	Value (CCK-8)
Symbol rate	1.375 MHz	1.375 MHz
Bits per symbol interval	4	8
Bits per pulse function	2	6
Bits per QPSK symbol	2	2
Chips per pulse	8	8
Transmit filtering	100% EBW	100% EBW
Occupied bandwidth	22 MHz	22 MHz
Non-conjugate CFs	$k \times 1.375$ MHz $k = 0, 1, 3, 5, 7, 8, 9, 11$	11.0 MHz
Conjugate CFs	None	None

802.11b CCK

The CCK signals in IEEE 802.11b are an unusual class of signals that can be viewed as QAM signals with random symbols and random pulse functions. For each symbol interval, a set of bits to be transmitted determines which symbol to send and which pulse to use. Together with the use of differential QPSK symbol alphabets, this signaling scheme can create CF patterns that are quite distinct from all other signals known to the authors. In particular, for four-level CCK, the set of non-conjugate CFs is given by $\alpha = kF_{\text{sym}}$ for $k = 1, 3, 5, 7, 8, 9, 11, 13$, and the set of conjugate CFs is empty. The CF pattern for eight-level CCK is quite a bit simpler, and consists of the single non-conjugate CF of $8f_{\text{sym}}$. For the 802.11b-CCK symbol rate of 1.375 MHz, these results imply that the four-level CCK signal possesses the non-conjugate CF set of $\alpha = \{1.375, 4.125, 6.875, 9.625, 11.0, 12.375, 15.125, 17.875\}$ MHz, and the eight-level CCK signal possesses only the non-conjugate CF of $\alpha = 11.0$ MHz. These theoretical predictions are confirmed by Figure 18.12.

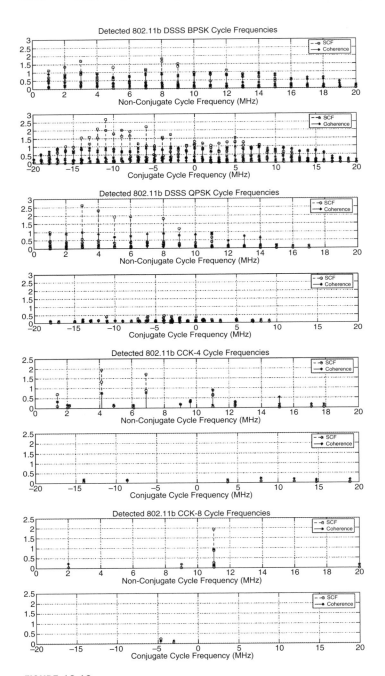

FIGURE 18.12

PSD and SCF estimates for captured 802.11b signals. From top in pairs: DSSS BPSK, DSSS QPSK, four-level CCK, and eight-level CCK.

All four 802.11b modulation types exhibit the 11.0 MHz non-conjugate CF, which could be used to detect the presence of "802.11b," but they all also exhibit distinct overall patterns of spectral correlation, which could be used to perform fine-grained classification if desired. Because the number of significant CFs is large (more than a few) for all but the CCK-8 signal type, the SSCA could be used to advantage here. After computing the SCF output using the SSCA, searches for conjugate and non-conjugate CF chains with separations of 1.0 and 1.375 MHz could be performed. The presence of the chains will then serve as the detection statistic. If no chains are found, then the presence of the 11.0 non-conjugate CF could be used to declare the presence of CCK-8.

18.4.7 IEEE 802.11a/g: OFDM

In the 802.11a/g standards, OFDM is used to obtain higher data rates relative to the 802.11b signals. This is possible due to the excellent multipath tolerance exhibited by OFDM. The signal can be viewed as a collection of low-rate QAM signals with center-frequency spacing equal to their symbol rate. Each individual QAM signal is referred to as a subcarrier. OFDM requires unusually good carrier and symbol synchronization, and therefore a subset of the subcarriers are reserved for transmitting a periodically repeated known binary sequence, which is used by the receiver to support its synchronization task. These synchronization subcarriers (also called pilots) produce significant spectral correlation, as we have already seen in Section 18.2.6.

The relevant physical layer parameters for 802.11a/g OFDM are provided in Table 18.6, along with the CFs predicted by mathematical analysis of the 802.11a/g OFDM signal model. The CFs are closely related to the frequency separations between the pilots, as expected from the spectral-correlation interpretation of cyclostationarity; these subcarriers are transmitting identical information, and so should be correlated over time.

To validate the mathematical predictions, the measured SCFs for several captured 802.11a/g OFDM signals are plotted together with ideal SCFs obtained from simulated OFDM signals in Figure 18.13. The PSDs (a) reveal that the captured signals are subjected to mild-to-severe channel effects and that the subcarriers are clearly visible as a periodic ripple along the signal passband. The measured SCFs for the two non-conjugate CFs of 4.25 and 8.75 MHz in (b) and (c) in Figure 18.13 are indicated by thin dotted lines, and the ideal SCFs are plotted using thick solid lines. There is some variation in the strength of the features, but in general, they rise above the surrounding measurement-noise floor by 10 decibels or more.

18.5 SUMMARY

This chapter provides an overview of detecting signals—sensing the electromagnetic spectrum—by exploiting spectral correlation. Spectral correlation is a statistical property possessed by nearly all communication signals, and is a consequence of their CS nature.

Spectrum sensing can be performed in many ways. The best ways depend on several factors, including the available computational resources, tolerable processing delays, and the signal environment—noise, interference, and propagation channel—in which the sensing will take place. Perhaps the most important factor, however, that enters the engineering trade-space for creating sensing methods is the fundamental shared statistical nature of the signals themselves. By taking advantage of the signals' common

Table 18.6 Parameters for the 802.11a/g OFDM Waveform

Parameter	Value	Comments
Subcarriers	64	FFT-IFFT size
Null subcarriers	13	One in band center, six on each edge
BPSK pilot subcarriers	4	8, 22, 36, and 50
Subcarrier separation	312.5 kHz	
Length of BPSK pilots	127 bits	
OFDM symbol rate	250 kHz	
Cyclic prefix	16 samples	16/64 = 1/4
Data constellations	BPSK, QPSK, 16-QAM, 64-QAM	All subcarriers have the same constellation in an OFDM symbol
Occupied bandwidth	17.0 MHz	
Non-conjugate CFs	±{0, 4.25, 4.5, 8.75, 13.0, 13.25} MHz	Dominant CFs
Conjugate CFs	$2f_c \pm$ {0, 4.25, 4.5, 8.75, 13.0, 13.25} MHz	Dominant CFs

spectral-correlation property, a general-purpose sensing platform can be constructed on the foundation of efficient spectral-correlation estimators.

The core technical idea of this chapter is that spectral-correlation detection is applicable to the transmitted waveforms of modern communication systems (e.g., ATSC DTV, CDMA, WCDMA, GSM/GPRS/EDGE, Bluetooth, WiMAX, and IEEE 802.11 WiFi), and that some care must be taken to determine the spectral-correlation properties of the transmitted signals. It is often the case that the spectral-correlation properties of the base modulation (e.g., BPSK, QPSK, or GMSK) interact with the multiple-access aspects of the system (e.g., TDMA, FDMA, FH, or CDMA) to yield surprisingly strong observable spectral correlation and robust high-performance detectors.

For CRs of the near future, spectrum-sensing requirements may involve only a few signal types, and the ranges of noise and interference environments can be narrow. For example, in IEEE 802.22, the number of signals that must be sensed is quite small, including voice FM, analog and digital broadcast TV, and the 802.22 signals themselves. In such cases, spectral correlation may or may not provide the best engineering solution, but is almost always applicable.

For an advanced far-future CR system in which the spectrum-sensing function must have the greatest capability, spectral correlation is a natural choice for building a suite of sensing algorithms due to its inherent noise and interference tolerance as well as its broad applicability. This idea has been established in other spectrum-sensing contexts where generality, accuracy, and noise tolerance are paramount.

EXERCISES

18.1 Consider the complex-valued pulse-amplitude-modulated (PAM) signal defined by

$$s(t) = \sum_{k=-\infty}^{\infty} a_k p(t - kT_0 - t_0),$$

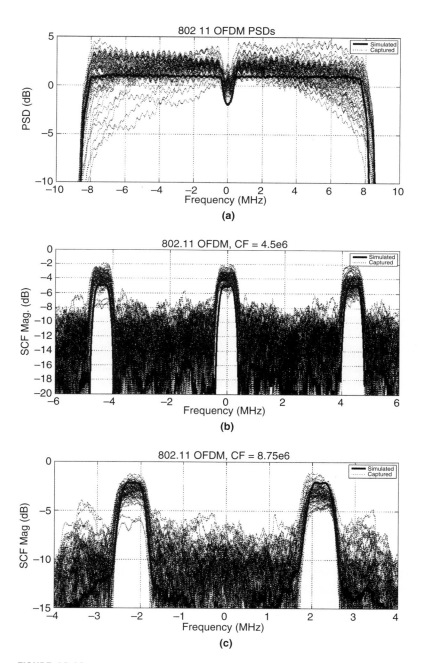

FIGURE 18.13

PSD (a) and SCF (b, 4.25 MHz; c, 8.75 MHz) estimates for simulated and captured 802.11a/g OFDM signals. Note the variation in the PSDs due to channel effects. The SCFs are reliably estimated for the two chosen CFs, since they almost all rise above the surrounding measurement noise.

where $\{a_k\}$ is a sequence of independent and identically distributed (IID) complex-valued symbols, $p(t)$ is the transmit pulse function, $1/T_0$ is the symbol rate, and t_0 is the symbol-clock phase. This signal models the complex envelope of M-ary PSK and QAM signals such as BPSK, QPSK, 16-QAM, and 64-QAM.

Assuming that the pulse function is a unit-height rectangle with width T_0, centered at $t = 0$, derive the following parameters:
(a) Autocorrelation function, $R_s^0(\tau)$.
(b) Power spectral density (PSD), $S_s^0(f)$.
(c) Cyclic autocorrelation function, $R_s^\alpha(\tau)$. Specify the values of the cycle frequency α for which the cyclic autocorrelation is not identically zero.
(d) Spectral correlation function (SCF), $S_s^\alpha(f)$.
(e) Repeat for the conjugate cyclic autocorrelation and conjugate SCF.

18.2 Use MATLAB to implement the frequency-smoothing method (FSM) of SCF estimation for a general smoothing window $h(f)$. Validate by estimating the PSD (non-conjugate SCF for $\alpha = 0$) for a signal consisting of unit-power white Gaussian noise (WGN). The value of the PSD should be near 1.0 for all frequencies except those close to the band edges (± 0.5). The non-conjugate SCF for $\alpha \neq 0$ and the conjugate SCF for all α should be small for all values of frequency f.

18.3 Modify the FSM code for the special case of a rectangular smoothing window function $h(f)$ so that it uses the efficient head-tail method of convolution. Compare the execution times for the general and efficient methods for various rectangle widths.

18.4 Compare the execution times for the general FSM using a half-cosine window and the efficient FSM using a rectangle width equal to the width of the half cosine. A half-cosine window is equal to a half period of a real-valued sine wave and is zero at its endpoints.

18.5 Use MATLAB to implement a signal generator that can produce rectangular-pulse complex PAM signals (Exercise 18.1). Validate by viewing real and imaginary components in the time domain.

18.6 (a) Express a generic OFDM signal as the sum of M evenly spaced (in frequency) complex-valued PAM signals. The frequency spacing is equal to the symbol rate. *Hint:* To frequency shift (to the right) a complex signal $x(t)$ by an amount $f_0 = 1/T_0$, multiply by a complex exponential to yield $y(t)$:

$$y(t) = x(t)e^{i2\pi f_0 t}$$

(b) Derive the cyclic autocorrelation and SCF assuming that the data used in all the constituent QAM signals (the subcarriers) is IID.
(c) Generalize the complex-PAM signal generator to allow production of such OFDM signals.
(d) Validate the derived formulas using the OFDM simulator and the MATLAB-based FSM.

REFERENCES

[1] Fette, B. (ed.), *Cognitive Radio Technology*, First Edition, Elsevier, pp. 1–28, 2006.
[2] *Notice of Proposed Rulemaking on Facilitating Opportunities for Cognitive Radio Technologies*, FCC ET Docket No. 03–108, December 2003.

[3] *Notice of Inquiry and Notice of Proposed Rulemaking for Establishment of an Interference Metric*, FCC ET Docket No. 03-237, November 2003.

[4] Haykin, S., Cognitive Radio: Brain-Empowered Wireless Communications, *IEEE Journal on Selected Areas in Communication*, 23(2):201-220, 2005.

[5] Jondral, F. K., Cognitive Radio: A Communications Engineering Perspective, *IEEE Wireless Communication*, 14(4):28-33, 2007.

[6] Oner, M. M., Air Interface Identification for Software Radio Systems, Doctoral Dissertation, Universitat Karlsruhe, 2004.

[7] Carter, G. C., Receiver Operating Characteristics for a Linearly Thresholded Coherence Estimation Detector, *IEEE Transactions on Acoustics, Speech, and Signal Processing*, pp. 90-92, 1977.

[8] Cabric, D., S. M. Mishra, and R. W. Brodersen, Implementation Issues in Spectrum Sensing for Cognitive Radio, *Conference Record of the 38th Asilomar Conference on Signals, Systems, and Computers*, pp. 772-776, November 2004.

[9] *Report of the Spectrum Policy Task Force*, FCC ET Docket No. 02-135, November 2002.

[10] Spooner, C. M., Multiresolution White-Space Detection for Cognitive Radio, *Proceedings of MILCOM*, pp. 1-9, October 2007.

[11] Spooner, C. M., Spectrum Sensing Alternatives for Cognitive Radio, Invited Presentation at the SDR Forum's Cognitive Radio Workshop, April 10, 2006.

[12] Mody A. N., et al., Recent Advances in Cognitive Communications, *IEEE Comunication Magazine*, 45(10): 54-61, 2007.

[13] Sutton, P. D., K. E. Nolan, and L. E. Doyle, Cyclostationary Signatures in Practical Cognitive Radio Applications, *IEEE Journal on Select Areas of Communication*, 26(1):13-24, 2008.

[14] Spooner, C. M., Cognitive Radio: Observing the Environment, chapter in *Designing Software and Cognitive Radios*, J. H. Reed (ed.), in preparation.

[15] Simeone O., et al., Spectrum Leasing to Cooperating Secondary Ad Hoc Networks, *IEEE Journal on Select Areas of Communication*, 28(1):203-213, 2008.

[16] Ganesan G., and Y. Li, Cooperative Spectrum Sensing in Cognitive Radio, Part I: Two User Networks, and Part II: Multiuser Networks, *IEEE Transactions on Wireless Communication*, 6(6):2204-2222, 2007.

[17] Gandetto, M., and C. Regazzoni, Spectrum Sensing: A Distributed Approach for Cognitive Terminals, *IEEE Journal on Select Areas of Communication*, 25(3):546-557, 2007.

[18] Nolan, K. E., and L. E. Doyle, Teamwork and Collaboration in Cognitive Wireless Networks, *IEEE Wireless Communication*, August 2007.

[19] Thomson, D. J., Spectrum Estimation and Harmonic Analysis, *Proceedings IEEE*, 70(9):1055-1096, 1982.

[20] Schreier, P. J., and L. L. Scharf, Second-Order Analysis of Improper Complex Random Vectors and Processes, *IEEE Transactions on Signal Processing*, 48(3):1055-1096, 2003.

[21] Proakis, J. G., *Digital Communications*, Fourth Edition, McGraw-Hill, 2001.

[22] Urkowitz, H., Energy Detection of Unknown Deterministic Signals, *Proceedings IEEE*, 55(4):523-531, 1967.

[23] Krasner, N. F., Optimal Detection of Digitally Modulated Signals, *IEEE Transactions on Communications*, 30(5):885-895, 1982.

[24] Spooner, C. M., and W. A. Gardner, Robust Feature Detection for Signal Interception, *IEEE Transactions on Communications*, 42:2165-2173, 1994.

[25] Gardner, W. A., *Statistical Spectral Analysis*, Prentice-Hall, 1987.

[26] Gardner, W. A. (ed.), *Cyclostationarity in Communications and Signal Processing*, IEEE Press, 1994.

[27] Gardner, W. A., Exploitation of Spectral Redundancy in Cyclostationary Signals, *IEEE Signal Processing Magazine*, April:14-36, 1991.

[28] Spooner, C. M., and W. A. Gardner, The Cumulant Theory of Cyclostationary Time-Series, Part I: Foundation, and Part II: Development and Applications, *IEEE Transactions on Signal Processing*, 42:3387-3429, 1994.

[29] Gardner, W. A., Two Alternative Philosophies for Estimation of the Parameters of Time-Series, *IEEE Transactions on Information Theory*, 37(1):216–218, 1991.

[30] Izzo, L., and A. Napolitano, The Higher-Order Theory of Generalized Almost-Cyclostationary Time Series, *IEEE Transactions on Signal Processing*, 46(11):2975–2989, 1998.

[31] Brown, W. A., and H. H. Loomis Jr., Digital Implementations of Spectral Correlation Analyzers, *IEEE Transactions on Signal Processing*, 41(2):703–720, 1993.

[32] Roberts, R. S., W. A. Brown, and H. H. Loomis, Jr., Computationally Efficient Algorithms for Cyclic Spectral Analysis, *IEEE Signal Processing Magazine*, April:38–49, 1991.

[33] Gardner, W. A., Signal Interception: A Unifying Theoretical Framework for Feature Detection, *IEEE Transactions on Communications*, 36(8):897–906, 1988.

[34] Gardner, W. A., and C. M. Spooner, Signal Interception: Performance Advantages of Cyclic-Feature Detectors, *IEEE Transactions on Communications*, 40(1):149–159, 1992.

[35] Gardner, W. A., and C. M. Spooner, Detection and Source Location of Weak Cyclostationary Signals: Simplifications of the Maximum-Likelihood Receiver, *IEEE Transactions on Communications*, 41:905–916, 1993.

[36] Gardner, W. A., Spectral Correlation of Modulated Signals: Part I—Analog Modulation, *IEEE Transactions on Communications*, 35(6):584–594, 1987.

[37] Gardner, W. A., W. A. Brown, and C.-K. Chen, Spectral Correlation of Modulated Signals: Part II—Digital Modulation, *IEEE Transactions on Communications*, 35(6):595–601, 1987.

[38] Gardner, W. A., Measurement of Spectral Correlation, *IEEE Transactions on Acoustics, Speech, and Signal Processing*, 34(5):1111–1123, 1986.

[39] Gardner, W. A., Cyclic Wiener Filtering: Theory and Method, *IEEE Transactions on Communications*, 40:151–163, 1992.

[40] Gardner, W. A., The Role of Spectral Correlation in Design and Performance Analysis of Synchronizers, *IEEE Transactions on Information Theory*, 34:1089–1095, 1986.

[41] Napolitano, A., and C. M. Spooner, Cyclic Spectral Analysis of Continuous-Phase Modulated Signals, *IEEE Transactions on Signal Processing*, 49(1):30–44, 2001.

[42] Rostaing, P., E. Thierry, and T. Pitarque, Asymptotic Performance Analysis of Cyclic Detectors, *IEEE Transactions on Communications*, 47(1):10–13, 1999.

[43] Izzo, L., L. Paura, and M. Tanda, Signal Interception in Non-Gaussian Noise, *IEEE Transactions on Communications*, 40(6):1030–1037, 1992.

[44] Spooner, C. M., Classification of Cochannel Communication Signals Using Cyclic Cumulants, *Proceedings 29th Asilomar Conference on Signals, Systems, and Computers,* pp. 531–536, October 1995.

[45] Spooner, C. M., W. A. Brown, and G. K. Yeung, Automatic Radio-Frequency Environment Analysis, *Proceedings 34th Asilomar Conference on Signals, Systems, and Computers,* pp. 1181–1186, October 2000.

[46] Spooner, C. M., On the Utility of Sixth-Order Cyclic Cumulants for RF Signal Classification, *Proceedings 35th Asilomar Conference on Signals, Systems, and Computers,* pp. 890–897, November 2001.

[47] Swami, A., and B. M. Sadler, Hierarchical Digital Modulation Classification Using Cumulants, *IEEE Transactions on Communications*, 48(3):416–429, 2000.

[48] Marchand, P., J.-L. Lacoume, and C. Le Martret, Multiple Hypothesis Modulation Classification Based on Cyclic Cumulants of Different Orders, *Proceedings ICASSP*, pp. 2157–2160, 1998.

[49] Reichert, J., Automatic Classification of Communication Signals Using Higher-Order Statistics, *Proceedings of ICASSP*, pp. V-221–V-224, April 1992.

[50] Schreyogg, C., K. Kittel, and U. Kressel, Robust Classification of Modulation Types Using Spectral Features Applied to HMM, *Proceedings of MILCOM*, pp. 1377–1381, 1997.

[51] Lender, A., The Duobinary Technique for High-Speed Data Transmission, *IEEE Transactions on Communications and Electronics*, 82(5):214–218, 1963.

[52] *http://en.wikipedia.org/wiki/Moore's_law*.

Rendezvous in Cognitive Radio Networks

19

Luiz A. DaSilva
Virginia Tech, Blacksburg, Virginia

Ryan W. Thomas
Air Force Institute of Technology,
Wright Pattersen AFB, Ohio

19.1 INTRODUCTION

According to the dictionary, a *rendezvous* is a meeting at an appointed time and place. The word comes from the French expression *rendez vous*, or "present yourself." In the context of dynamic spectrum access (DSA), it refers to the ability of two or more radios to meet and establish a link on a common channel. As we will discuss in this chapter, neither the time nor the "place" (the channel where the link is established) need necessarily be appointed. Cognitive radios (CRs) must have the ability to present themselves to the network and, conversely, to detect the presence of others to establish communication.[1]

Although rendezvous is a requirement of any multichannel system, it is particularly important (and, maybe, particularly tricky) in DSA. For opportunistic channel access, secondary users (SUs) must dynamically sense a potentially large number of channels; if a channel is not occupied by a primary user (PU), it is available for use by the SU, which may need to share the channel with other SUs. There can be a large number of channels where rendezvous can potentially occur, and at any given time a PU may be active in some of these, so the question of how two radios wishing to establish a link will manage to find each other is not trivial.

In a broader sense, rendezvous encompasses not only the establishment of network links, but also the maintenance of those links as channel availability changes. For instance, opportunistic spectrum access requires that, upon the appearance of a PU on the channel, the SUs must promptly vacate the channel. These SUs must now rendezvous again in an alternate channel.

In Figure 19.1, we establish a possible taxonomy of approaches for rendezvous. First, we classify the rendezvous approach as aided (or infrastructure-based) or unaided (or infrastructureless). Aided rendezvous is accomplished with help from a server, which periodically broadcasts information regarding available channels and may even serve as a clearinghouse for link establishment and the scheduling of transmissions, typically

---

[1]The views expressed in this chapter are those of the authors and do not reflect the official policy of the US Air Force, Department of Defense, or the US government.

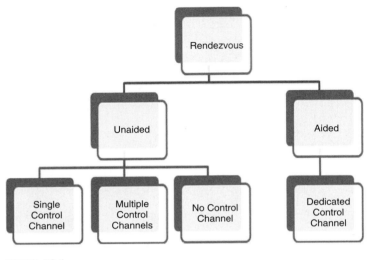

FIGURE 19.1

A classification of rendezvous solutions.

using a well-known control channel. In unaided rendezvous, each CR must find other nodes in the network on its own.

Another possible dimension for classification refers to the existence of a dedicated control channel. It is typical in aided rendezvous that the basestation uses a dedicated control channel to exchange channel availability information with mobile radios. An unaided rendezvous, in contrast, may or may not avail itself of a dedicated control channel. Even though using a single control channel simplifies the initial step of determining in which frequency to look for neighbors, it incurs additional overhead and creates a single point of failure; the common control channel may also become a bottleneck for communications as the network grows. To address some of these scalability issues, a distributed solution, in which different control channels are used for different clusters of nodes, can be adopted, at the cost of additional overhead to establish and maintain those clusters. A final approach refuses to dedicate any channel to control and link establishment, allowing all channels to be shared by control and data traffic. The process of establishing a link without the benefit of a control channel is sometimes referred to as a *blind rendezvous*.

The objective of this chapter is to discuss the main approaches to rendezvous proposed for CR networks. We start by discussing the trade-offs in the use of a common control channel, and then dedicate most of the chapter to describing and assessing approaches for rendezvous without the use of a control channel. Two approaches for blind rendezvous are explored: random rendezvous and sequence-based rendezvous. Rendezvous can also be accomplished through wideband sensing and signal classification approaches such as the detection of cyclostationary signatures proposed by Sutton et al. [1]. We focus, rather, on achieving rendezvous by sensing and communicating over one channel at a time.

19.2 THE USE OF CONTROL CHANNELS

In an infrastructure-based network, it is likely that basestations and mobile radios will all agree on a preset control channel (or channels) to be used to identify spectrum availability and request and schedule connections.

For example, Buddhikot et al. [2] have proposed an architecture in which some frequencies are set aside for use as *spectrum information channels*. Clients dedicate a wireless interface to scan these channels, where the basestations broadcast information regarding spectrum availability, interference conditions, and so on. Clients can use those same control channels to request the use of dedicated spectrum to their traffic (or, alternatively, clients may directly proceed to the data channels that they now know to be available).

The use of a common control channel simplifies the process of rendezvous and seems reasonable in an infrastructure-based wireless network. However, in a commercial DSA environment, some regulators worry about how to resolve the potential contention for the control channel. If multiple basestations and associated radios, all in close vicinity of one another, all wish to negotiate the use of available shared spectrum, policies must be in place to ensure that no one gets an unfair competitive advantage from being able to more efficiently access the control channel.

Dedicated control channels can also be adopted in decentralized networks (e.g., mobile ad hoc networks). Much of the work on multichannel medium access control (MAC) protocols for ad hoc networks assumes that each node in the network is equipped with two radios (as described by Mo et al. [3] and references therein). The idea is that one radio constantly monitors a dedicated control channel and utilizes that channel to reserve transmissions on one of multiple potential data channels, possibly through the use of request-to-send (RTS) and clear-to-send (CTS) frames. The second radio then tunes to the appropriate data channel for the exchange of data frames. The rendezvous process is thus simplified at the expense of dedicating one transceiver exclusively for the monitoring and transmission of control information. Also, the control channel acting as a bottleneck and single point of failure is still a concern.

It is possible to envision more flexible approaches that take advantage of multiple channels for exchange of control information. A variation is to have a dynamically changing control channel: for instance, radios may be programmed to always attempt to rendezvous in the lowest-number control channel that is currently not occupied by an incumbent. The potential problem with such a method is that two radios may not correctly sense the presence of a PU in a given channel (in fairness, this situation can also pose a challenge to the blind rendezvous methods discussed later). In the example illustrated in Figure 19.2, radios A and B are attempting to rendezvous. Suppose radio A is within the interference region of a PU operating in channel 1, whereas radio B is not. Radio A will be able to sense the primary and will know that the next channel to be used as a control channel will be channel 2; radio B, unaware of the primary, will continue to look for beacons in channel 1, and rendezvous will never occur. An important feature of any rendezvous solution is robustness to dynamic spectrum occupation as well as to differing views of current spectrum occupancy, due to differences in position, range, and sensing capabilities of different radios.

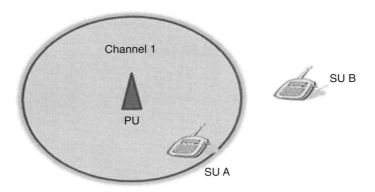

FIGURE 19.2

Secondary user A can detect the presence of a PU in channel 1 and will attempt to rendezvous in channel 2, while SU B attempts to rendezvous in channel 1.

The essential trade-off involved in the use of a control channel is between simplicity and flexibility. Dedicating one or more channels for the exchange of control information simplifies the handshakes needed for rendezvous, channel allocation, and reservation. In contrast, blind rendezvous, which does not rely on control channels, has the potential to achieve greater scalability and robustness to varying accuracy in sensing PUs, as well as improved efficiency in channel utilization. These benefits, as discussed in the next section, may come at the expense of greater complexity and time to complete rendezvous.

19.3 BLIND RENDEZVOUS

In *blind rendezvous*, all channels are potentially available for the exchange of control and data. Radios are responsible for determining which channels are available and then attempting to establish a link on one of those channels. Let us denote by $\{1, 2, \dots, N\}$ the set of channels that are potentially available for rendezvous. The radio must visit those available channels in random or preestablished order, alternatively transmitting beacons and listening for responses, until it is able to establish one or more links. An approach for emitting and scanning to establish a link is described in Horine and Turgut [4].

For simplicity of analysis, it is common to assume time to be slotted. A radio will change channels in consecutive timeslots, searching for others with which to communicate. During each timeslot, a radio senses the medium for the presence of a PU or other SUs in that channel. If it does not sense others in the vicinity, it will transmit a beacon or heartbeat, followed by another period of silence, while it waits for a response. Note that we need not assume that timeslots are aligned between two radios. In Figure 19.3, if all three radios are currently operating in the same channel, both radio B and radio C will be able to hear and respond to radio A's beacon, thereby completing the handshake needed for rendezvous, even though the slot boundaries are not aligned.

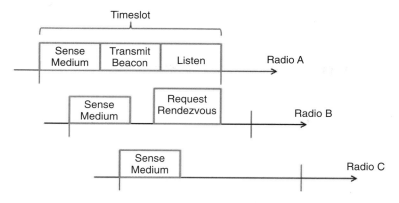

FIGURE 19.3

Blind rendezvous process. In this example, radios A and B successfully rendezvous with each other.

Of course, variations of the process depicted in Figure 19.2 and of the handshake method necessary to complete rendezvous are also possible. Regardless of the method, it is necessary that at some point in time the two radios will be operating in the same channel. In this section we discuss two alternative methods in which radios can, without assistance or coordination, eventually meet in the same channel. We call these methods *random rendezvous* and *sequence-based rendezvous*.

19.3.1 Random Rendezvous

In random rendezvous, a radio wishing to join a network visits the potential communications channels in random order. During each timeslot, the radio will select any of the N channels with probability $1/N$. For two radios following this procedure, rendezvous will be successful when two conditions occur: (1) the two nodes select the same channel; and (2) one of the radios is sensing the medium while the other is transmitting a beacon in such a way that the handshake required for rendezvous is possible, as in the example in Figure 19.3.

At time t, given that the two radios are operating in the same channel, the handshake is feasible if one of the radios is transmitting a beacon, while the other is sensing the medium, and it depends on the portion of the timeslot dedicated to each of these activities and to the amount of time that the channel-sensing mechanism requires to make a correct determination that there is another radio operating in the channel. We lump all of these conditions into the probability of a successful handshake p_H. The time-to-rendezvous (TTR) will depend on p_H, as well as on the expected time, until both radios select the same channel. For large N, the latter factor will dominate.

If we let E denote the event that, in a given timeslot, both radios select the same channel, we have:

$$P[\text{successful rendezvous}] = p_H P[E] \tag{19.1}$$

For random rendezvous, $P[E] = 1/N$, and the TTR is a geometrically distributed random variable representing the number of failures before the first success in a sequence of independent Bernoulli trials with probability of success equal to p_H/N. Thus, the probability mass function of the TTR can be expressed as:

$$P[TTR = k] = \left(\frac{p_H}{N}\right)\left(1 - \frac{p_H}{N}\right)^{k-1} \quad \text{for} \quad k = 1, 2, \ldots \tag{19.2}$$

The expected TTR is therefore equal to N/p_H. Without loss of generality, in subsequent discussion we will ignore the constant term $1/p_H$ and simply express $E[TTR] = N$ for random rendezvous. It is also simple to see that rendezvous will occur in any of the N channels with equal probability, and that the TTR is unbounded (i.e., arbitrarily large TTR occurs with positive probability).

19.3.2 Sequence-Based Rendezvous

An alternative approach to blind rendezvous, proposed by DaSilva and Guerreiro [5], is termed sequence-based rendezvous. This approach employs predefined sequences used by each radio to determine the order in which potential rendezvous channels are to be visited. The idea is that both transceivers follow the same sequence, albeit arbitrarily delayed with respect to each other. In Figure 19.4, we illustrate the process. In the example, secondary users A and B both visit channels according to the sequence (1, 1, 2, 3, 2, 1, 2, 3, 3, 1, 2, 3). Even though there is a lag (in this example, three timeslots) between the time that radio A and radio B start to look for others with which to rendezvous, they eventually occupy the same channel in the same timeslot, and therefore rendezvous is feasible.

These sequences must be constructed in such a way to minimize the maximum and/or expected TTR even when radios are not synchronized to each other. The properties of the time-to-rendezvous in this case depend on the selection of the sequence. By appropriately selecting the sequence, it is possible to: (1) establish a maximum TTR; (2) reduce the expected TTR as compared to random rendezvous; (3) establish a priority order for potential rendezvous channels.

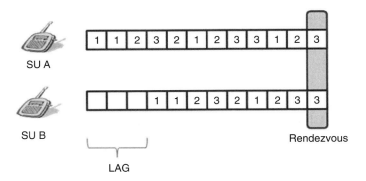

FIGURE 19.4

Secondary users A and B perform sequence-based blind rendezvous.

It should be clear that not any sequence will work. Consider again a set of N potentially available channels, numbered 1 through N. Now think of two radios visiting channels in ascending order of channel number: 1, 2, ... , N. If the two radios are synchronized with each other, they'll meet in the same channel immediately. However, in the vast majority of cases there is no practical way to synchronize multiple radios before they join the network. Therefore, following this simple sequence will result—with high probability—in the two radios never meeting, and the expected TTR is infinite. This will obviously not do. Thankfully, there are sequences that yield a finite expected TTR and other desirable properties.

The problem of selecting sequences for sequence-based rendezvous can be thought of as the dual of the problem of selecting hopping sequences for frequency-hopping spread spectrum (FHSS). In FHSS, desirable sequences are those that minimize the probability that, at a given time, multiple radios occupy the same frequency channel. In sequence-based rendezvous, desirable sequences are those that maximize that same probability. The former problem has been studied extensively (see, for instance, Sarwate [6]), whereas the latter is, to our knowledge, still open.

A visiting sequence $a = (a_1, a_2, a_3, \ldots)$ describes the order in which a radio visits channels in search of other radios with which to rendezvous. We have been investigating sequences that are infinite and periodic, and in our further discussion, we focus on one period of length M: $(a_1, a_2, \ldots, a_M)$. Further, for fairness reasons, it is desirable that each of the N channels appear the same number of times within one period, so M must be a multiple of N. Finally, it is desirable that the sequences be defined according to an algorithm that generalizes for any value of N.

One method for buiding such a sequence is to select a permutation $P(N)$ of the N channels (there are $N!$ such permutations) and building the sequence as illustrated in Figure 19.5. The selected permutation appears $(N + 1)$ times in the sequence: N times the permutation appears contiguously, and once the permutation appears interspersed with the other N permutations. The period of the sequence is $M = N(N + 1)$.

An example may make things more clear. Take $N = 5$, and select at random a permutation of these five channels, say the permutation (3, 2, 5, 1, 4). The method to form a sequence just described would yield:

<u>3</u>, 3, 2, 5, 1, 4, <u>2</u>, 3, 2, 5, 1, 4, <u>5</u>, 3, 2, 5, 1, 4, <u>1</u>, 3, 2, 5, 1, 4, <u>4</u>, 3, 2, 5, 1, 4

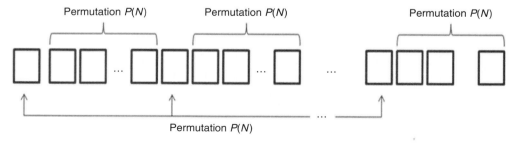

FIGURE 19.5

A method for creating a sequence for rendezvous.

More precisely, what we represented was one period of the sequence. This ordering of channels to be visited is then repeated indefinitely.

When radios visit potential rendezvous channels following such a sequence, the TTR can be shown to have some desirable properties. First of all, TTR can be shown to be bounded by N^2. This is an important property when trying to meet quality-of-service (QoS) objectives for link establishment in a DSA environment. Second, the sequence implicitly establishes a prioritization order for rendezvous channels. In other words, even though in each period of the sequence each channel appears the same number of times, rendezvous is more likely to occur in some channels than in others. This is useful whenever there is a reason to favor some channels over others—for instance, because some channels are less likely to have activity from incumbents than others, or due to propagation characteristics of a given set of channels. Further, the average TTR when rendezvous occurs in one of those "preferred" channels (i.e., the conditional expectation) is lower than for random rendezvous.

How about the average time to rendezvous? For this family of sequences, we have been able to derive the average TTR as (according to DaSilva and Guerreiro [5]):

$$E[TTR] = \frac{N^4 + 2N^2 + 6N - 3}{3N(N+1)} \qquad (19.3)$$

So while this sequence-based rendezvous outperforms random rendezvous in terms of maximum TTR and in its ability to support preferred rendezvous channels, it presents a higher average TTR.

It is, however, possible to select sequences that outperform the expected TTR achieved for random rendezvous, that is, for which $E[TTR] < N$. Some examples are presented in Figure 19.6 [7]. In all these sequences, each channel appears the same number of times over one period. The derivation of optimal sequences, defined as a sequence that, for a given N, minimizes either the maximum or the expected TTR (or, if we are lucky, both) is an area for further study.

We note that the sequence-based rendezvous method described here is completely different from MAC schemes, where hopping sequences are shared among nodes (see, e.g., So et al. [8]). These MAC protocols presuppose that nodes have already established communications, while the sequence-based rendezvous is used to bootstrap the entire process of establishing links. Further, the sequence-based rendezvous does not require that two radios follow the same sequence synchronously.

N	Sample sequence (one period)	Max TTR	$E[TTR]$
3	1 1 2 3 2 2 1 3 3 3 1 2	8	2.75
4	1 1 1 2 3 4 2 2 2 1 3 4 3 3 3 1 2 4 4 4 4 1 2 3	13	3.96
5	2 3 5 4 1 1 2 5 4 3 4 5 3 2 1 4 2 5 3 1 3 4 5 1 2 3 4 2 5 1	11	4.23

FIGURE 19.6

Sample sequences with bounded TTR and $E[TTR] < N$.

19.4 LINK MAINTENANCE AND THE EFFECT OF PRIMARY USERS

The presence of a PU in one or more channels where rendezvous would be attempted will impact the process. If all SUs correctly detect the presence of an incumbent, this may actually shorten the time required for blind rendezvous, as it decreases the number of channels to be visited to $N - K$, where K is the number of channels occupied by an incumbent. In both random and sequence-based rendezvous, once a channel is known to be occupied by a PU, and thus unavailable for rendezvous, it may make sense to remove it from the list of potential channels to be visited. This process, when applied to sequence-based rendezvous, is illustrated in Figure 19.7.

Even after a successful rendezvous, the appearance of a PU in the channel will require that the link be reestablished in an alternate channel. One obvious approach to this is to simply return to a rendezvous phase, and to repeat whatever process was originally used for rendezvous (random, sequence-based, common control channel-based, etc.).

A more efficient approach is for SUs, immediately after establishing the initial link, to decide on fallback channels in anticipation of the possible disruption of the link due to the appearance of an incumbent. In fact, instead of agreeing on one alternate channel in case they need to reestablish the link, SUs may exchange or negotiate a fallback dictionary, with an ordered list of such alternate channels. Fallback channels can be preselected or negotiated dynamically in the network [9]. The use of such a list is important in a DSA environment because no single channel can be guaranteed a priori to be available for use by SUs.

We note that a third alternative could be envisioned. Suppose that two SUs have established a link, and at some point in time they detect a PU. The SUs can then decide which channel to move to, inform each other, and then vacate the current channel. Although generally more efficient than both procedures described previously from the point of view of secondary users, this approach may not be practical when SUs have stringent requirements on the maximum time in which they are required to vacate a channel.

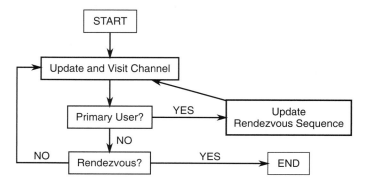

FIGURE 19.7

Sequence-based rendezvous, with updates to the sequence based on sensing of PUs.

19.5 SUMMARY

Radio rendezvous for DSA systems is still a largely open area for research. Much of the literature on multichannel communications simply assumes a common control channel (and, often, a radio dedicated to sensing and communicating over that control channel), thereby eliminating the rendezvous problem altogether. In the opportunistic access to spectrum, the use of a single control channel is not robust, as the appearance of an incumbent in that channel would be devastating to all communications among SUs. Also, the single control channel approach does not scale. Finally, even if nodes are equipped to multiple transceivers, it may be wasteful to dedicate one of them solely for monitoring a control channel.

A reasonable alternative is to use multiple channels, achieving varying trade-offs between robustness and TTR. In this chapter, we outlined some of those approaches and pointed out their main pros and cons. Surely, as more DSA systems are deployed, opportunistically able to access an increasing number of channels, other competing approaches will be suggested. Such approaches must be able to scale to a large number of channels and a large number of radios, achieving reasonable TTR and high probability of rendezvous as long as at least one channel is currently available.

REFERENCES

[1] Sutton, P. D., K. E. Nolan, and L. E. Doyle, Cyclostationary Signatures in Practical Cognitive Radio Applications, *IEEE JSAC*, 26(1):13–18, 2008.

[2] Buddhikot, M. M., P. Kolodzy, S. Miller, K. Ryan, and J. Evans, DIMSUMNet: New Directions in Wireless Networking Using Coordinated Dynamic Spectrum Access, *Proceedings Sixth IEEE International Symposium on a World of Wireless Mobile and Multimedia Networks*, pp. 78–85, June 2005.

[3] Mo, J., H.-S.W. So, and J. Walrand, Comparison of Multichannel MAC Protocols, *IEEE Transactions on Mobile Computing*, 7(1):50–65, 2008.

[4] Horine, B., and D. Turgut, Link Rendezvous Protocol for Cognitive Radio Networks, *Proceedings IEEE DySPAN*, pp. 444–447, April 2007.

[5] DaSilva, L. A., and I. Guerreiro, Sequence-Based Rendezvous for Dynamic Spectrum Access, *Third IEEE International Symposium on Dynamic Spectrum Access Networks*, pp. 1–7, October 2008.

[6] Sarwate, D. V., Optimum PN Sequences for CDMA Sequences, *IEEE Third International Symposium on Spread Spectrum Techniques and Applications*, 1:27–35, 1994.

[7] Martin, R. K., personal correspondence, June 24, 2008.

[8] So, H.-S., W. J. Walrand, and J. Mo, McMAC: A Multi-Channel MAC Proposal for Ad Hoc Wireless Networks, *Proceedings of IEEE Wireless Communication and Networking Conference*, pp. 209–218, March 2007.

[9] Silvius, M. D., R. Rangnekar, Y. Shi, A. B. MacKenzie, and C. W. Bostian, Channel Change: A Dynamic Spectrum Access Protocol for Spectrum Sharing in Smart Radios, under review, 2008.

Spectrum-Consumption Models

20

John A. Stine
MITRE Corporation
McLean, Virginia

20.1 INTRODUCTION

Spectrum management is the business of planning, managing, and coordinating the use of electromagnetic spectrum. By using operational, engineering, and administrative procedures a plurality of electronic systems can perform their functions without causing or suffering interference. As a process, spectrum management is very deliberative, usually taking years to resolve coexistence issues. This lack of agility in the spectrum management (SM) process and the observation that a lot of spectrum goes unused have motivated the promotion of cognitive radio (CR) and the use of dynamic spectrum access (DSA) as a technical solution.

Proponents of DSA claim that CR can resolve spectrum coexistence in a generic technical solution through the use of sensors, policy, and reasoning logic. Despite the hype, these systems have not yet proven themselves. Even if CRs can sense and use spectrum under the constraint of policy, it remains uncertain whether the results avoid causing harmful interference. Cognitive radio complicates, rather than simplifies, SM. Spectrum managers will need to certify whether sensor, reasoning, and policy combinations can adequately avoid harmful interference in every band and location in which their use is authorized. Not only the radios, but also the policy that is loaded on the radios must be managed. The prospect of this daunting task has brought the management of CR to the forefront of the world SM community, and it is an agenda item for the next World Radio Conference.

In this chapter, we describe a modeling approach to capture the consumption of spectrum that can be used as a comprehensive technique to manage all spectrum-consuming devices including CRs and DSA systems. We refer to these models as location-based spectrum rights (LBSRs). These models can convey spectrum consumption both as a definition of a system's current use of spectrum and as an authorization of an allowed use of spectrum. LBSRs combine a definition of harmful interference with models of location, transmission power, and signal attenuation. These provide an unambiguous means to compute allowed interference to primary systems and contributed interference from secondary systems based on location—therefore, regulators, systems, and even devices can compute whether their use of spectrum is compatible with another.[1] Thus, the models can be used for multiple purposes: to regulate and license spectrum use, to negotiate spectrum rights in secondary markets, to specify policy to CRs, to efficiently communicate policy over the air, and to identify reuse opportunities. A model created to articulate a primary use of spectrum can be a constraint and thus a policy to secondary users (SUs).

This chapter begins with an overview of the state of spectrum management with the intent of revealing the cause of the apparent underutilization of spectrum. Unless the deeply ingrained reasons for this cause are overcome, acceptance of DSA will face great resistance. Our motivation for putting forth this cause is to make the point that the spectrum-consumption models that we describe in this chapter provide a path to acceptance and a reasonable regulation of DSA technologies. We set out to describe the objectives of spectrum-consumption modeling: by defining spatial and spectral consumption, harmful interference, and time of applicability, as well as specifying special behavioral conditions for use. We describe the modeling components, the methods to build models using these components, the computations used with these models to assess compatible reuse, and the variety of applications for these models.

20.2 RECONCILING DSA AND SPECTRUM MANAGEMENT

The apparent underutilization of spectrum that is often used to motivate the need for DSA technologies is the result of an SM process that does not have the agility to capture temporal as well as spatial variations in spectrum use. The success of any particular DSA approach will depend on its acceptance by the SM community. Here we provide an overview of spectrum management and the challenge that confronts DSA as it tries to fit within these processes.

20.2.1 The Persistent Goal of Spectrum Management

Traditionally, spectrum management has been performed globally through international agreements and nationally by individual government administrations. Bands of spectrum are divided into allocations that are designated to support particular services. The allocations are subdivided into allotments that may be used by administrations in specified geographic areas. National administrations may further allot the spectrum into channels, specify the conditions of their use, and assign (or license) them to users. Historically, the growth in spectrum requirements was accommodated through technology that made the higher-frequency bands available for use. Little unassigned spectrum remains, so now spectrum management involves reallocating and reassigning spectrum.

Observations of inefficient spectrum use, the rise of new commercial applications for spectrum, and the opportunity to generate revenue for governments have motivated the reassessment of spectrum policy. In 2002, the US Federal Communications Commission (FCC) established a Spectrum Policy Task Force (SPTF) to provide recommendations on how to evolve spectrum policy into an "integrated, market oriented approach

---

[1]These models are regulatory in nature and are not intended to be an exact representation of the physical propagation condition. They attempt to capture the trend and also to bound the performance to ensure systems are protected from each other.

that provides greater regulatory certainty while minimizing regulatory intervention" [1]. In November of that year it produced a report [2] that, as its most significant recommendation, proposed that the FCC move more spectrum from the command and control management model to the exclusive use and commons models.

The command and control model is the legacy model by which an administration licenses spectrum to users under specific conditions. Changing uses of spectrum is a deliberative process that involves study and opportunities for public comment. The major complaints against this approach are that it is very slow to adapt, it is unfriendly to commercial interests, and it results in inefficient use of spectrum. Nevertheless, the command and control model is still necessary to protect public interests that are not market driven (e.g., public safety, scientific research, and government operations), as well as to conform to treaty obligations. Even with the use of the other SM models, the command and control model will remain the overarching SM model, the difference being that parts of the spectrum will have more liberal rules that allow commercial development and changing uses without the administrative proceedings.

The exclusive use model is a licensing approach in which the licensee has exclusive rights to a band of spectrum within a defined geographic region. The licensee has flexibility to implement different technologies and can transfer the use rights. The best example of the exclusive use model in practice is cellular telephony. The licensees develop the technologies, infrastructure, and services, and then transfer spectrum use to subscribers of those services. There are great incentives to promote this model, especially for the most desirable spectrum, because licensees bid for the spectrum, which brings revenue to governments and creates the incentive for the licensees to apply the spectrum for its best-valued use.

The commons model opens bands of spectrum for unlicensed use with etiquettes that allow as much coexistence among different applications and users as feasible. Examples of spectrum bands that are managed in this way are the industrial, scientific, and medical (ISM) bands. The 2.4 GHz ISM band has been very successfully used for wireless local area networks (LANs), personal area networks, microwave ovens, cordless telephones, and other consumer products. Harmful interference among devices (e.g., such as a cordless telephone with a wireless LAN) is tolerated or is resolved by the owners of the devices.

The FCC has also supported the sharing of spectrum between primary users (PUs) and SUs. In this approach, SU are allowed to use a specified band of spectrum as long as they do not cause harmful interference to the PUs. Meanwhile, the SUs must accept all interference from the PUs. In this dual approach to spectrum management, the command and control model designates who is in a primary or secondary status. The rules for secondary use are sufficiently restrictive to preclude harmful interference to the PU for most uses. An unfortunate outcome of such secondary access is the perception of users that when they buy a product that uses spectrum, they have the right to interference-free use. They are unaware of its secondary status. Politics can turn these expectations into pressure on the PUs to modify their behavior for the sake of SUs [3]. These experiences cause resistance to sharing.

In a recent ruling, the FCC opened the 3650–3700 MHz band of spectrum to still another approach to spectrum management [4]. The Commission decided to manage this band by using a nonexclusive licensing scheme coupled with a number of provisions designed to allow cooperative shared use of the band. These provisions include a streamlined licensing mechanism and the requirement that equipment that uses the band employs a contention-based protocol to minimize interference. Licenses are granted on a nationwide nonexclusive basis to licensees that demonstrate that the technology they will employ meets the contention-based protocol requirements. They then must obtain station authorization for each basestation they deploy.

Licensees and applicants are expected to cooperate in the selection and use of frequencies. To assist, the FCC will maintain a database identifying the locations of all registered stations. However, an existing station in an area does not preclude new stations from being deployed in its vicinity. These stations are expected to cooperate with each other to find a mutually satisfactory arrangement for shared use. Commercial standards are being developed for this model, most notably the IEEE 802.11y standard.

It is mistaken to classify the commons, exclusive rights, and nonexclusive rights management models of managing bands of spectrum as equivalents of the command and control model. The movement of radio frequency (RF) bands to these models is an outcome of the command and control process. These new management models enable the business models of service providers and product developers but cannot support many of the services that use RF spectrum (e.g., services where spectrum is not resold or where access must be protected such as global positioning system (GPS), air traffic control, radio location, and radio astronomy).

What should be apparent is that all of these approaches pursue a persistent assignment of spectrum to users and/or uses. The FCC attempts to avoid serving the role of a dynamic temporal arbitrator of spectrum use. It recognizes the need for cooperation, and, as the rules for the 3650 to 3700 MHz band indicate, it would prefer that the users cooperate with each other to reuse spectrum both temporally and spatially. Unfortunately, a number of uses of spectrum are sporadic yet intolerant of interference. Further, the users may be mobile and thus consume spectrum in only a small space at any time, but must be protected over larger spaces in anticipation of movement anywhere in those spaces. Persistent arrangements to protect these uses allow much of the potential spectrum utility in these bands to be lost.

The tendency of SM administrations to pursue persistent solutions sets the stage for DSA, using technology to fill the voids in spectrum use caused by the lack of temporal agility in spectrum management.

20.2.2 The Promise of Dynamic Spectrum Access

Although there are many visions for DSA, the dominant vision is that of CRs acting autonomously, searching for unused spectrum and using it when it is found. This type of opportunistic use of spectrum is based on the premise that pathloss is distance based, and if the detector is much more sensitive than primary receivers, then the CR can use spectrum without interfering. The measures of performance for CR concern the ability of its sensors to find vacant spectrum, and then the agility of the radio system to use these bands. In the ideal embodiment of a CR, it moves to and uses idle spectrum and then departs from the band without causing any interference to any legacy system.

The nature of the physics of propagation and many legacy uses of spectrum do not cooperate with the ideal DSA vision. Sensing can be unreliable because a host of other propagation effects create conditions similar to distance-based pathloss (e.g., destructive interference from reflected signals and shadowing [5]). Also, the components that need to be protected, the receivers, are passive and may not be detected by sensing. Figure 20.1 illustrates two scenarios in which differences in antenna heights cause a third radio to sense the same spectrum use but where its reuse of the same spectrum would be compatible in only the first scenario. Additionally, the absence of detection is not a conclusive indication that the spectrum is idle. Legacy use may be discontinuous where channel idleness contributes to spectrum use (e.g., arbitrating contention in a carrier-sensing medium access control (MAC) protocol) or allows for the reception of reflected radar signals or distant transmission by highly sensitive receivers. Further, sensing cannot indicate when one of these PUs will need the spectrum next. Using spectrum when a primary user next wants to use it is harmful interference. So, the use of spectrum during perceived spectrum idleness can harm legacy uses. These and many other scenarios give regulators cause to reject the ideal DSA vision.

The proposed solution to the limitations of the ideal DSA radio is to use policy to govern the behavior of CRs. Policy is written and loaded into a radio under the theory that a regulator would license equipment that can comply with policy written in this manner, and that the regulator then manages the policy used by the radios in their administrative region. Policy would handle the details of protecting legacy use by

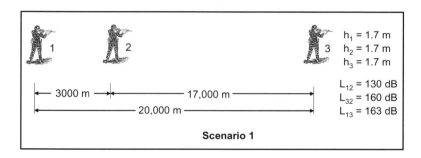

$h_1 = 1.7$ m
$h_2 = 1.7$ m
$h_3 = 1.7$ m

3000 m
17,000 m
20,000 m

$L_{12} = 130$ dB
$L_{32} = 160$ dB
$L_{13} = 163$ dB

Scenario 1

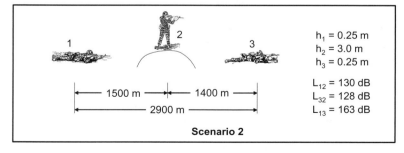

$h_1 = 0.25$ m
$h_2 = 3.0$ m
$h_3 = 0.25$ m

1500 m
1400 m
2900 m

$L_{12} = 130$ dB
$L_{32} = 128$ dB
$L_{13} = 163$ dB

Scenario 2

FIGURE 20.1

Sensing scenarios demonstrating the asymmetric effects of antenna height on sensing strength. The receivers at soldier 3 in each scenario sense the same signal strength from the transmitter at soldier 1. A protocol on the radio at soldier 3 then acts on this sensed state to successfully avoid interfering at soldier 2 in Scenario 1, but could cause harmful interference in Scenario 2.

regulating the requirements for sensing, the conditions for use, and the criteria for abandonment per channel or band of spectrum. And so the promise of DSA hinges on the ability of policy to manage the dynamic behavior of radios and to prevent those radios from causing harmful interference.

20.2.3 The Limitations of Policy

The vision for policy in CR is very broad, governing not only spectrum use but also all aspects of network operations and management. The lack of specifics leaves open the perceived possibilities. Figure 20.2 illustrates the Defense Advanced Research Projects Agency NeXt Generation (DARPA XG) vision of where policy reasoning falls within a CR [6], giving the impression that spectrum use decisions are occurring packet-by-packet. Further, the reasoning is informed solely by the options developed through spectrum sensing. The concerns are whether policy can do anything to overcome the limitations of sensing and whether the cognitive cycle of sensing, characterizing, reacting, and adapting can be quick enough for a networking application.

The answers to these questions will depend on the specific legacy use of spectrum and the capabilities of the components of the CR. The challenge varies from the easier coexistence among high-power broadcast channels with a low-powered cognitive use to the more challenging coexistence among low-powered distributed networking devices. In the former, a history of sensed silence would be a stronger indicator of sufficient displacement because there are no temporal changes in primary use. Further, the relative difference in transmit power and the nature of power-law attenuation gives the primary transmitter an advantage except in the very local region of the CR. In the

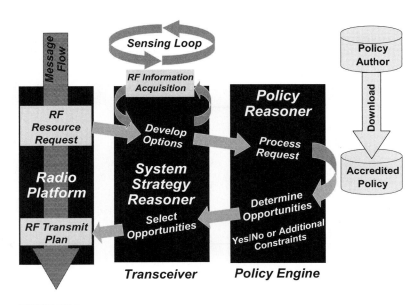

FIGURE 20.2

The DARPA XG model of the policy reasoning placement in a CR [6].

latter, the CR would not have the power advantage, so separation is not as easy to ensure, and errors in transmission would not be so forgiving. The temporal variation in the primary's use makes sensing strategies and policy reasoning difficult and puts a greater demand on the agility of a radio to maneuver in and out of spectrum.

Figure 20.2 gives an exaggerated view of cognitive agility, implying that a CR can sense and reason at a rate comparable to the rate of a standard access mechanism arbitrating per-packet access, but the duration of CR functions are orders of magnitude greater than those of access mechanisms. Sensing and reasoning can take tens to hundreds of milliseconds and a rendezvous with a distant radio is likely to take seconds, whereas the clear channel assessment used in access mechanisms of wireless networking modems takes just a few microseconds. With these differences in performance, the only policy that would likely protect the primary would be to prevent secondary use in its presence.

Thus, we have laid out the limitations of policy—it cannot overcome the shortcomings of sensing but only reduce the occurrence of errors, and its success depends largely on the nature of the PU. Most policy will need to ensure avoidance of primary systems rather than coexistence among them. While the promise of DSA is to provide the temporal agility that is not possible with standard SM processes, we see that in practice, policy will physically separate systems. Rather than coexisting among PUs, they would coexist adjacent to primary users, and the temporal agility in CR use would follow from the temporal use of the primary system in the macro sense of being used and not being used, rather than in the micro sense of the silences that naturally occur within the access mechanisms of the systems.

20.2.4 **The Challenges in Managing DSA**

The mechanics of certifying radios and of managing policy loaded on radios has not been defined. The key questions are who writes the policy and how is it certified to be compatible with any other user. Assuming there is a well-defined set of compatible policies, then the question concerns how users with policies are managed.

Using the certification of the dynamic frequency selection features in WiFi devices as an example, certification would begin with a definition of what sort of behavior respects the PUs and then testing to verify that the radio conforms to that behavior. So, the burden of proving a policy is compatible would rest with the developer of that policy through test procedures specified by the governing SM administration. The process of defining the behaviors that ensure compatibility with legacy users, and the tests that follow, will take a long time to execute. The costs of this process would require CR manufacturers and users to do the yeoman's work of developing policy, proving that the policy protects legacy users, and funding the testing of the CR-policy combination to demonstrate that the pair is effective. Then, assuming success, this would not mean carte blanche use of the radio.

An often overlooked detail of CR-policy combinations is that they are designed to be compatible with legacy users but not necessarily with peer users in separate networks. In application, CRs are likely to be used by small groups of users that desire to be isolated from others. As is demonstrated in tactical network design, subnetworking groups is a key component of network design necessary for the purposes of scalable routing. These subnets need to remain isolated from each other. CR technology may provide means to isolate subnets within a single system, so that the task previously performed by the system administrator is accomplished in a more automated fashion. Thus, in practice, spectrum management would treat each certified CR-policy combination in the same manner that it treats frequencies—they would be assigned to specific users in specific locations.

The conclusion is that CRs do not simplify spectrum management—they complicate it. CRs require a much more challenging process for certification. They add dimensions to the SM problem, such as capturing the pairwise compatibility of policy with legacy uses and then arbitrating the distribution of independent CR systems using the same policy.

20.2.5 The DSA Spectrum Management Alternative

Making spectrum management practical requires an approach to enabling DSA alternative to pursuing the ideal DSA vision. The problem with the ideal DSA vision is that policy is centered on spectrum sensing and what can be implied by sensing. The rich diversity of environments that drastically change propagation conditions and the variety of modulations and access schemes make the development of effective policy difficult if not impossible. Additionally, for the reasons described previously, if a set of policies were developed that worked with certain legacy spectrum users, the systems would remain difficult to manage. An additional cause for concern is that a predominant number of anticipated uses of CR involve hostile environments where there is an adversary. Hinging performance on sensing creates a vulnerability to denial-of-service attacks.

The purpose of this chapter is to describe an alternative approach to enabling DSA that places the SM task at the forefront of the solution. At its core is the definition of spectrum consumption. In this approach, particular uses of spectrum are captured using the LBSR models. We are unaware of any use of spectrum that cannot be bounded by these models. The utility of the models in DSA spectrum management is that a model of a primary use of spectrum is policy for secondary uses. SMs can manage spectrum by specifying that the potential spectrum CRs can use an LBSR, and then providing the LBSRs of primary users that are constraints to this use. Compatible reuse follows from the CR being cognizant of its location and then computing, based on these models, whether their reuse is compatible with the primary's rights. Cognitive radios that are designed to use this type of policy will be straightforward to certify because they will not require extensive testing in the presence of PUs. A certifiable CR would be one that accurately computes the criteria for compatible reuse based on its understanding of its location and the permissive and constraining rights it is given and then reliably complies with that computation. This approach to managing DSA systems provides an underlying mathematical basis for compatibility that removes the uncertainty of using policy to manage spectrum. Tools can be developed to simplify the task, and spectrum management can become dynamic and network-centric in solutions where rights to use spectrum can be conveyed dynamically.

For mobile uses, the CRs will need to compute location, but solutions for this task are well known. Concerns that location may be lost are legitimate, but there are many ways to mitigate this problem. Position location can be made more robust by using multiple systems in tandem (e.g., GPS, inertial systems, terrestrial reference signals, and so on). The criteria for using spectrum can be based on having a reliable location and, unlike spectrum-sensing systems, position location sensing systems can tell if there is problem—for example, GPS receivers know how many satellite signals they are receiving. And even without this constraint, if a system loses its capability to detect its location it may be possible to use a history of past locations to support radio use because locations do not change that quickly. The dynamic range in location errors is nowhere near that in spectrum-sensing errors. Finally, location-based systems are more difficult to attack than spectrum-sensing systems.

20.3 THE LOCATION-BASED METHOD TO SPECIFY RF SPECTRUM RIGHTS

Although the spectrum consumption of a system is a function of the location of the transmitters and the receivers of that system, the boundaries of use are less easy to define. Consumption depends on the signal space, the transmission power, the antennas used, the attenuation that occurs in propagation, and the susceptibility of the modulation to interference. Most of these cannot be modeled exactly. Variations in location and antenna orientation, variation of propagation effects that result from changes in the environment, and the imperfections in components make the consumption stochastic. Therefore, modeling consumption attempts to capture a bound on these effects—one that ultimately protects the users who need protecting. As we go about describing the LBSR, the reader should be attuned to the interaction of the components and how they are used collectively to bound consumption. There is an artful component to the modeling. Ultimately, we want the models to make the computation of compatible reuse unambiguous and tractable, and the uncertainty in what actually happens in spectrum use to be bounded.

20.3.1 The Ten Components

Stine [7] provides the first presentation of location-based spectrum rights; in that exposition, six modeling components were described. Since that time, we have expanded the definition to ten components to extend its versatility. These ten components are illustrated in Figure 20.3 and are briefly described in the following.

1. The maximum power density specifies that maximum power density at some designated distance toward any direction from a transmitting antenna for a transmitter right or at a receiving antenna from any direction for a receiver right.
2. The spectrum mask specifies the spectral power density relative to the maximum power density for all frequencies of a transmitter right.
3. The underlay mask specifies the spectral power density relative to a maximum power density at a receiving antenna of the maximum allowed interference of a remote interfering transmission.

1. Maximum power density: 20 dBW/m²

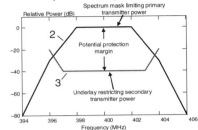

2. Spectrum mask: f_c: 400 MHz, f_i: 100 kHz
   (67, 80, 87, 30, 107, 0, 147, 0, 167, 30,
   187, 80, 255)
3. Underlay mask: f_c: 400 MHz, f_i: 100 kHz
   (97, 20, 102, 40, 152, 40, 157, 20, 255)

5
Propagation map: (10, 20, 220, 60,
125, 150, 60, 0)

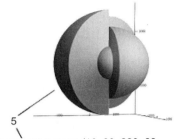

6

Location: 38.7486972, –90.3700289, 188.4

7. Minimum power density: –80 dBW/m²
8. Protocol or policy: TDMA2, 10 ms
9. Start time: 0700 12 Nov 2007
10. End time: 1200 12 Nov 2007

4
Power map: (15, 255, 50, 0, 25, 3, 40, 7, 92,
15, 251, 0, 0)

FIGURE 20.3

Examples of the ten components of a location-based spectrum right. These examples use the coded vector versions of the spectrum and underlay masks and the power and propagation maps.

4. The power map specifies the variation by direction of the maximum transmitted power density relative to the maximum power density of the right.

5. The propagation map specifies an attenuation model value by direction that indicates the rate of attenuation that should be used for computing compatible uses of spectrum.

6. The minimum power density specifies the attenuation level at which transmitted signals are no longer protected.

7. The location may be (a) points at which, (b) a volume within which, or (c) a track along which the components of the RF system receiving the RF spectrum rights may operate. The quantity of points and the specific methods for specifying volume and tracks vary.

8. The specified protocol or policy is a constraint that restricts the use of spectrum to a particular protocol or to a particular behavior.

9. The start time is the time that a right begins to apply.

10. The end time is the time that a right ends. Start and end times may be complemented with a function defining periodic changes in availability within the start and end times.

These components should be viewed as constructs from which rights are built. There is no requirement that each be used, and combining multiples of some components may be required to construct a right. Further, these components can have different meanings based on whether they provide permission to a transmitter to emit radiation, or they are used to define the protection for receivers. In the subsequent sections, we describe these components, their use, and their meaning in greater detail.

20.3.2 Maximum Power Density

The intent of LBSR is to create RF spectrum rights that have a geospatial limit. For such a system to work, the right must be decoupled from the antenna technology. So transmit power is defined as the effective power density at a specified distance away from the antenna. Transmitters with high gain antennas must still conform to these limits. The maximum power density specifies that maximum power density of any frequency in the right toward any direction from a transmitting antenna for a transmitter right or at a receiving antenna from any direction for a receiver right. All other transmitter power constraints in rights are relative to this maximum power density, so changing this value changes all power constraints in a right. The recommended practice, and the convention used in this chapter, is that transmitter power densities are specified for the 1-meter distance from the center of the transmitting antenna. These transmit power densities are equivalent to $RP(1\ m)$ in the log-distance power model, which is described later. Power densities in receiver rights are the power densities at the receiver antenna.

The maximum power density that governs an LBSR may be one of two types. The first is the *regulated* maximum, p_{RM}. The regulated maximum power density is the maximum allowed transmit power density permitted by an SM administration or a spectrum manager. In secondary use, a primary right may constrain the permitted transmission power of the secondary transmitter. In this case, we denote the *constraining* maximum transmit power density as p_{CM}.

20.3.3 Modeling Spectral Consumption and Signal Space

Spectral consumption is a function of the breadth of frequencies that a signal occupies and then its tolerance to other signals in the same spectrum. The definition of a signal's occupancy in spectrum is easily bounded by using a spectrum mask, discussed later. The more difficult part of the problem is defining tolerance of a signal to interference. The susceptibility of a signal to interference will depend on its modulation and that of the interfering signals and so there is not a single definition for this term. The general concept to define acceptable interference from an interfering signal on a protected signal is to use an underlay mask (discussed later) referenced to the spectrum mask of the protected signal. An underlay mask can be conservative and generic, specifying a constraint for all interfering signals, or be paired with a particular type of interfering signal with the intent of providing a more liberal reuse of spectrum by that signal. The underlay mask has a second purpose in defining rights. Its relative position to the spectrum mask of the protected signal can be shifted to account for the variations in power caused by fading and other randomly varying propagation effects.

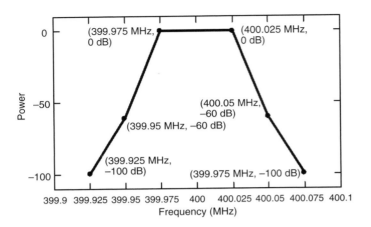

FIGURE 20.4

An example of a spectrum mask. The vector that defines this mask is (399.925 MHz, −100 dB, 399.95 MHz, −60 dB, 399.975 MHz, 0 dB, 400.025 MHz, 0 dB, 400.05 MHz, −60 dB, 400.075 MHz, −100 dB).

Spectrum Masks

A *spectrum mask* specifies the limit on the power over a band of spectrum that a transmitter may emit. It is typically presented as a piecewise linear graph of power versus frequency where power is the power density on a dB scale[2] and frequency has a linear scale. The recommended practice for specifying a spectrum mask is to use a vector of values alternating between frequency and power density of the form, $(f_0, p_0, f_1, p_1, \ldots, f_x, p_x)$, with each sequential pair specifying an inflection point in the mask. In keeping with our convention, the power density is the relative power density in dB from the specified maximum power density. The mask specifies the limit on emitted power by frequency. The emitted power for all frequencies within the mask may not exceed the mask's bound, and the emitted power for all frequencies outside the mask may not exceed the smallest value specified by the mask. Figure 20.4 illustrates an example of a spectrum mask and provides its vector. The highest power is 0 dB, which corresponds to the maximum power density that the transmitter can use.

Underlay Masks

An *underlay mask* is identical in structure to a spectrum mask but has a different function. Its purpose is to define the maximum interference a remote secondary system may cause at a primary receiver. It is selected to provide a margin of interference that is expected to protect the primary receivers from harmful interference. The piecewise linear graph specifies the limit to the power of interfering signals as a function of frequency. However, the restrictions of an underlay mask apply only to the limits of the

---

[2]Our intent is to create spectrum rights that have a geospatial limit. For such a system to work, the right must be decoupled from the antenna technology. So transmit power is defined as the effective power density at one meter from the antenna. Transmitters with high gain antennas must still conform to these limits in the rights. These transmit powers are equivalent to $RP(1\ m)$ in the log, distance pathloss model.

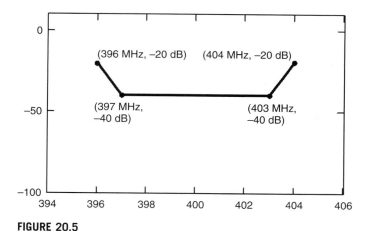

FIGURE 20.5

An example of an underlay mask. The vector that defines this mask is (396 MHz, −20 dB, 397 MHz, −40 dB, 403 MHz, −40 dB, 404 MHz, −20 dB).

mask. Identical to a spectrum mask, it consists of a vector of frequency and power density pairs. The power density in these pairs is the relative power density in dB from the specified maximum power density. Figure 20.5 illustrates an example of an underlay mask and provides its vector. Here the maximum power density of the mask is 20 dB below the maximum power density of the right.

20.3.4 Propagation Models

RF emissions attenuate as they propagate from their source. The quantity of attenuation is a function of frequency, distance, and the environment. Due to the rich diversity of environmental effects, there is no shortage of propagation models in the literature because no single model can do it all. Precise prediction of attenuation is usually untenable because total attenuation can vary significantly by slight movements and subtle changes in the environment. The model chosen in engineering is typically the one that best supports the specific task. In spectrum-consumption modeling, the propagation model needs to capture attenuation trends but result in tractable computations for spectrum reuse. The model chosen for this purpose is the log-distance pathloss model [8]. The log-distance pathloss model is a linear model in which pathloss is a function of distance related by $PL(d) = PL(1 \text{ m}) + 10n\log(d)$ on a logarithmic scale and is related by $PL = PL_{1m}d^n$ on a linear scale, where the parameters of the model are the pathloss PL_{1m} of the first meter and the pathloss exponent n. The PL_{1m} term accounts for the frequency effects on attenuation; the distance term accounts for the effect of distance; and the pathloss exponent, n, accounts for the environment. In the log-distance pathloss model, a pathloss exponent of 2 corresponds to the freespace pathloss model (i.e., Friis equation), and larger exponents are used in terrestrial models where reflected signals are likely to result in destructive interference and where foliage and atmospheric gases contribute to signal attenuation.

The log-distance model is generally considered to be an unreliable predictor of pathloss because of the wide variance in pathloss that occurs due to shadowing and multipath fading. Nevertheless, the log-distance pathloss model for spectrum rights possess advantages over comparable models, including its simplicity and that it leads to tractable computations of compatible reuse due to its monotonic trend. The variance in signal strength caused by fading and shadowing is accommodated by the protection margin of the LBSR created by choosing a larger 1-meter pathloss term, a larger exponent, or increasing the relative difference of the power levels in the underlay masks of the right.

20.3.5 Power Modeling

In spectrum-consumption modeling, the relative strength of signals, and not the amount of attenuation, determines whether two users can share the same spectrum. So it is necessary to convert the propagation model to predict power levels rather than pathloss. The log-distance pathloss model is easily converted into a received power-density model.

$$RP(d) = RP(1\text{ m}) - 10n \log(d). \tag{20.1}$$

Eq. (20.1) estimates the power density at distance, d. The term $RP(1\text{ m})$ specifies the power density at 1 meter and captures both the gain of the antenna and the pathloss of the first meter of propagation. This power density decreases by the second term that accounts for the log-distance pathloss. The terms $RP(1\text{ m})$ and $RP(d)$ are expressed in decibel units of power, dBm/m^2 or dBW/m^2, respectively. The 1-meter power density, $RP(1\text{ m})$, and the pathloss exponent, n, are specified in an LBSR. The 1-meter power density is referenced to the maximum power density, and the expectation is that the device using the spectrum will transmit within the bounds of this specification.

20.3.6 Directional Vectors Used for Power and Propagation Maps

The initial 1-meter power density of a signal can vary by direction due to the effects of the emitting antenna. Similarly, the attenuation of signals can vary by direction due to spatially dependent environmental effects. Thus, the LBSR can specify multiples of these parameters as a function of direction. The data structures that are used to specify parameters by direction are generally referred to as maps. The data structure to capture directional variation in 1-meter power density is referred to as a power map, and the data structure to capture directional variation in pathloss is referred to as a propagation map. Separate data structures are used for the two since these values are dependent on different external factors and can have different directional boundaries.

Both power maps and propagation maps use the same technique to specify model parameters, power densities, or pathloss exponents, by direction. A map is a vector that lists azimuths, elevations, and the model parameters in a prescribed order, so there is no ambiguity as to which elements in the vector represent what type of value. A complete map will specify a model parameter toward all directions. The vector of values starts with elevations (ϕ) from the vertical up direction 0° and reaching to the vertical down direction 180°, and azimuths (θ) reaching about the node on the horizon. The first and last azimuths point in the same direction, 0° and 360°, respectively. The vector

$(0°, 0°, n_{00}, \theta_{01}, n_{01}, \theta_{02}, ..., 360° \phi_1, 0°, n_{10}, \theta_{11}, ..., 360°, \phi_2, 0°, n_{20}, ..., n_{last}, 360°, 180°)$

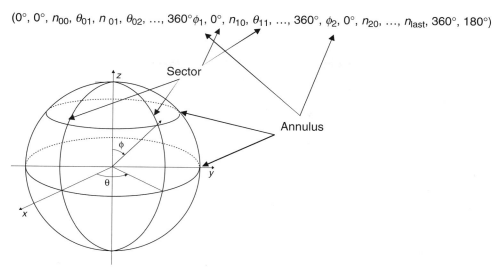

FIGURE 20.6

An example of the vector used in power and propagation maps and its translation to directions on a sphere.

uses two elevations to define a spherical annulus about a node and then a series of parameters and azimuths that specify different parameters on that annulus by sector. One or multiple annuli are ultimately defined to specify parameters for all directions.

The number and spacing of elevations and azimuths that are listed in the vector is arbitrary, used as necessary to provide resolution. The form of the vector is $(0°, 0°, n_{00}, \theta_{01}, n_{01}, \theta_{02}, ..., 360°, \phi_1, 0°, n_{10}, \theta_{11}, ..., 360°, \phi_2, 0°, n_{20}, ..., n_{last}, 360°, 180°)$. Figure 20.6 illustrates an interpretation of this vector. The vector starts in the 0° elevation and the 0° azimuth, and an annulus is specified for each pair of elevations, from 0° to ϕ_1, from ϕ_1 to ϕ_2, and so on. These elevations bound a series of values alternating between model parameters and azimuths. Each model parameter, n_{xy}, applies to the sector that reaches from elevation ϕ_x to ϕ_{x+1} and from azimuth θ_{xy} to $\theta_{x(y+1)}$. For example, in Figure 20.6, the model parameter n_{10} applies from elevation ϕ_1 to ϕ_2 and from azimuth 0° to θ_{11}.

Using maps in spectrum rights requires that they have a reference location and an orientation to the physical environment. Propagation maps are oriented based on the geographic location of the center. The horizon of the map is considered tangent to Earth's surface with the 0° azimuth pointing east and the 90° azimuth pointing north. The orientation of power maps may match that of the propagation map or be based on antenna orientation. By convention, we reference power maps to the coordinate system of the power map at its location.[3] Matching a power map's orientation to an antenna

---

[3]The orientation of an antenna may actually be referenced relative to a platform and then the platform has an orientation reference relative to its geographic location. The subsequent orientation of the power map is a function of the two. For purposes of our discussion we do not make this distinction and consider the power map orientation to be referenced directly to the geographic location.

is accomplished by shifting the vertical axis by a pitch (an elevation in the propagation map coordinate system) and a yaw (an azimuth in the propagation map coordinate system), and then rotating the power map coordinate system about the vertical axis by a roll. To simplify the discussion in the rest of this chapter, we assume the power maps are coincident to propagation maps.

20.3.7 Location Modeling

The purpose of location modeling is to specify the location of the components of an RF system and the relative location between the components of different RF systems. There are two aspects of modeling: capturing the shape of the space within which an RF system will be confined, and providing a point of reference for that space. Further, components of a model—specifically power and propagation maps—require a reference for orientation. The propagation map is referenced to a geographic location, and the orientation would depend where the location is on Earth. The power map, however, is more suitably referenced to the platform and antenna orientation, which in turn may be referenced to their geographic location. Maps are not perfectly symmetrical, and their orientation contributes to the effectiveness of the modeling. The horizon of a propagation map is best matched to the horizon of Earth's surface, whereas the axis of a power map may be more aptly made coincident with the boresight of an antenna. Meanwhile, all of these are referenced to a spherical Earth that is rotating, a fact that has special significance for systems in space.

Thus, location in spectrum-consumption modeling has three components: space, reference, and orientation. In the following subsections, we suggest techniques for modeling these components, recognizing that with experience additional techniques may be developed that may be more suitable to particular applications.

Position, Area, and Volume

A location in a spectrum right may be specified as a point, an area, a volume, or a movement track. A point would simply be a coordinate. Areas would be drawn on the surface of Earth with the implication that the components of the system are terrestrial and are used on the surface. Areas can be defined by using shape primitives (e.g., a circle), or be defined by a series of points that define the vertices of a polygon. Volumes may be specified by using a definition of a solid primitive oriented by one or more points; a series of points to define an irregular volume; or a set of solid primitives of which the intersection, union, difference, or some combination of these operations define the volume. Tracks convey the movement of systems and thus indicate the location of a component with a temporal reference (i.e., given a time, the track will reveal a point location for an object). There is usually uncertainty about the exact location of an object, and as such, the variance of this position may be specified by a solid primitive. That is, given a time, the track reveals the reference point of a solid primitive. A track may be periodic and be defined by an equation such as a satellite orbit or be composed of discrete trajectories. We have not yet defined these data structures and anticipate the use of conventions used in the applications that LBSRs ultimately support. In this chapter, we use points and simple solid primitives—cylinders and rectangular solids— as a generalization of location modeling.

Geographic Location and Orientation

Implementation of this technique requires a datum for location. The World Geodetic System (WGS 84) is well suited for this. It defines an Earth-centric ellipsoid to serve as a reference datum for location. It is a global system and the datum for GPS. The WGS 84 datum defines an ellipsoid that approximates Earth's surface. A WGS 84 coordinate consists of a latitude, φ, and a longitude, λ, which define a point on the surface of the ellipsoid, and then a height, h, which represents the distance above or below that point normal to the ellipsoid surface. These coordinates can be converted to Earth-centric Cartesian coordinates $\langle x, y, z \rangle$ or converted to coordinates to a propagation map or power map coordinate system. The horizon of a propagation map system would be parallel to the tangent plane to the ellipsoid at the $\langle \varphi, \lambda \rangle$ of the map's origin and would be oriented as described earlier.

20.3.8 Minimum Power Density

The minimum power density, p_{AM}, is used together with a maximum power density, a spectrum mask, a propagation map, and a power map as an alternative method to define the volume of a right. It is intended to be used with broadcaster rights to define the space where receivers should receive protection, making that protection contingent on the ability of the broadcaster to deliver a reasonably strong signal identified by the minimum power density. The surface of this volume is computed by identifying the range at which a signal attenuates to the minimum power density, using the transmission power specified by the maximum power density, the spectrum mask, and the power map, and then the attenuation specified by the propagation map.

20.3.9 Protocol and Policy

The protocol and policy component provides the means to specify behavioral guidance in a spectrum right. The distinction used to separate what is a protocol versus what is a policy concerns their scope and influence on the behaviors of a radio system. Policy refers to guidelines given to radios that specify the conditions those radios must assess to be present to consider spectrum available for those radios to use, and then the limitations on that use (e.g., power, bandwidth, etc.). Protocols are the detailed procedures used by radios to share information in order to assess spectrum availability and then to collaborate to use the spectrum. Thus, policy directs what spectrum a radio system is permitted to use and under what conditions it can use it, whereas a protocol defines the procedures that a radio or radio system follows to abide by the policy and to accomplish the radio system's purpose. An additional distinction in CR is that policy is executed in the reasoning part of the radio, where choices are made among alternatives, whereas a protocol is implemented in the mechanics of the radio, where specific actions follow an observed state. Radios are tuned to implement protocols, and thus protocol tasks are executed much faster than policy reasoning.

The use of LBSRs to specify which spectrum may be used is policy. Administrators can specify policy using just maximum power densities, spectrum masks, underlay masks, power maps, propagation maps, and locations. Additional sharing opportunities are possible by managing detailed behaviors of systems. The protocol or policy component of an LBSR exists to provide a means for SMs to convey those additional behaviors to systems. The LBSR remains a policy, and the guidance in this component provides additional criteria for spectrum use. At present, no specific methods to convey protocols or policies within the LBSR have been defined. These methods are likely to be defined when particular protocols or policies are found to provide compatible reuse. The creation or upgrade of radios to comply would follow. The remainder of this section provides additional details as to how protocols and policies can contribute to additional spectrum reuse.

Protocols make the temporal use of spectrum by systems predictable. In some primary uses of spectrum, it is possible for SUs to exploit the predictability of the primary's protocols and use the same spectrum in the same space. Examples of predictability in protocols that can be exploited are the chirping rates of radars that allow communication systems to communicate between chirps, and the slotting used in time division multiple access (TDMA) protocols, where secondary systems can use the space within unused slots to communicate. Further, protocols can be intentionally designed to allow multiple different users to share spectrum. It is anticipated that protocol compatibility will involve some certification process. Thus, the entity managing rights will know which protocols are compatible. Conveying the protocol to use involves identifying the protocol, and specifying key operating parameters of that protocol, which might include timing guidance and carrier-sensing requirements.

Policies, like protocols, specify behavior, but they differ in that their primary focus is to specify behaviors that prevent harmful interference as opposed to behaviors that describe how they use spectrum. Behavioral policies do not specify the details of how the system itself arbitrates the use of spectrum among system components. These details are left open for the user to choose. Policies can enable the use of sensing strategies, such as sensing threshold, sensing period, abandonment time, and disuse time. Through this component, the LBSR can replicate any policy that can be written using policy language techniques. Just as with protocols, it is anticipated that the entity managing rights will know which behavioral policies are compatible with other uses of spectrum. Conveying the behavioral policy to users involves identifying the set of behaviors, each by name, and specifying key operating parameters of those behaviors (e.g., sensing period = 1 second, abandonment = 10 minutes, and so on). The selection of behaviors, their names, and their parameters remain to be defined.

20.3.10 Time Models

Many standardized formats exist for specifying time, including, but not limited to, those within such standards as the International Organization for Standardization's ISO 8601. Similarly, many techniques exist to code time into one or more numbers. Currently, no specific format is recommended for the LBSR. Although a start and end time are listed, the intent is to allow any description of time. Periodic spectrum rights such as daily from a start-to-end time may be specified in lieu of a single start and end time (e.g., daily from 6:00–10:00 AM). The period and the duration of a right may be specified by providing a start and end time with a specified repeated duty cycle (e.g., three hours on and then four hours off). A temporal availability of spectrum may also be articulated by using a time bound together with a track. Spectrum may be used if within the time bound and the track indicates the primary system is out of range.

20.3.11 **Modeling Transmitter and Receiver Rights**

Spectrum rights may be articulated by using combinations of the maximum power density, spectrum masks, underlay masks, propagation maps, power maps, locations, protocol specifications, start times, end times, protocols, and policies. These parameters, either singly or in combination, may be used to convey transmitter rights and receiver rights. Transmitter rights specify the allowed power density of emissions from an RF system, and receiver rights specify the permitted interference.

A transmitter right includes (as a minimum) a maximum power density, a spectrum mask, a power map, a location, and optionally a propagation map, minimum power density, and a protocol specification. The assumption in a transmitter right is that the maximum power density and power map specify the maximum strength of the emission and that this strength attenuates as the signal propagates away from the emitting antenna. Figure 20.7 illustrates how the attenuation specified in propagation maps applies to transmitter rights. The strength of the signal is at its maximum at the transmitting antenna, and it attenuates linearly in dB with the logarithm of the distance it propagates.

A receiver right includes (as a minimum) a maximum power density, an underlay mask, a power map, a propagation map, and a location. The assumption in a receiver right is that the maximum power density and power map specify the maximum strength of an interfering signal at the receiver and that the strength of the interfering signal attenuates as it propagates away from the emitting antenna toward the receiving antenna. Figure 20.8 illustrates how the attenuation specified in propagation maps applies to receiver rights. The strength of the interfering signal is at its minimum at the receiving antenna, and the strength attenuates linearly in dB with the logarithm of the distance as the signal propagates toward the receiving antenna. The implication of the receiver right is that an interfering transmitter may transmit at a power no greater than that bounded by the receiver right based on the interfering transmitter's location. Figure 20.8 illustrates that authorized SUs can transmit more power the farther they are from the protected receiver. A receiver right is a constraint on distant transmitters and does not grant transmission rights.

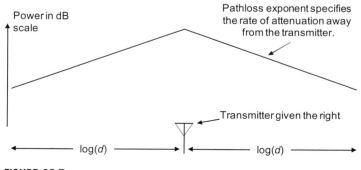

FIGURE 20.7

A transmitter spectrum (*right*), illustrating that the power bound attenuates with distance from the transmitter given the right.

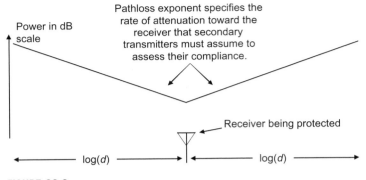

FIGURE 20.8

A receiver spectrum (*right*), illustrating that the power bound attenuates toward the receiver being protected.

The maximum power density in a right is the maximum power the right applies at the devices. All the parameters of the spectrum masks, underlay masks, and power map are referenced to this value. The attenuation that occurs away from a transmitter or toward a receiver adjusts the maximum power density based on the distance from the transmitter or receiver. The spectrum masks, underlay masks, and power map are all referenced to this adjusted maximum power density.

Timing parameters, start and end times, or definitions of periodic use are combined with transmitter and receiver rights to specify when they apply.

20.3.12 Compliance and Computing-Compatible Reuse

For any spectrum rights method to be effective, there must be a corresponding method to assess compliance. The compliance of primary transmitters (i.e., rights holders without any constraints from other users) to a transmitter right is generally straightforward. The transmitted signal should be within the spectrum mask and the power of emission at the transmitter should comply with the power map. Further, the transmitter should transmit only when it is in a location where it is authorized to use the spectrum. Systems that are intended to be cognizant must be able to identify their location and verify that they are within the location of the right. For noncognizant systems, the users must ensure they are operated in the authorized space.

The compliance of secondary transmitters (i.e., rights holders with restrictions placed on their use of spectrum by other users of spectrum) requires the SUs to compute the operating parameters that are compliant with the restrictions. Compliance is achieved by either of the following: (1) not interfering with a primary receiver by using different spectrum or a power that ensures the interference is below or outside the primary's underlay mask, (2) using a protocol that is compatible with a primary rights holder, or (3) using a policy that is compatible with the primary rights holder.

Noninterfering compliance requires computing the power that an SU may use in its transmissions to operate within the restrictions of a primary right. The computation may include four types of intermediate computations. The first type of intermediate computation determines the effect of the underlay masks of the constraining receiver rights on the transmit power of a secondary transmitter. The second type of computation is determining the difference from the maximum transmit power to the transmit power in the direction of the constraining point. The third type is the computation of attenuation based on the distance and the fourth type determines the location of an actual or notional receiver that would most constrain the transmit power of the SU, called a constraining point.

Figure 20.9 illustrates an example of how these computations might work together to compute the maximum power density allowed of an SU based on a primary right. Given the computed location of a constraining point, the process, as illustrated, begins by determining a minimum permissible difference or margin between the maximum power density of the primary right and the maximum level of the secondary signal, shown in the figure at (1), as constrained by the underlay mask of the primary right. The second step (2) is to determine the total attenuation of the transmitted signal at the constraining point (3) by using the propagation model specified for the primary right. The third step (4) is to determine the total attenuation of the transmitted signal from a secondary transmitter to the constraining point (3) by using the attenuation predicted by the governing propagation map. The constraining maximum power density at the SU (5) is the maximum power density of the primary transmitter less the margin caused by the underlay, less the directional difference determined from the power map (this is not illustrated), less the attenuation to the constraining point, and plus the attenuation of the secondary signal to the constraining point. This represents the

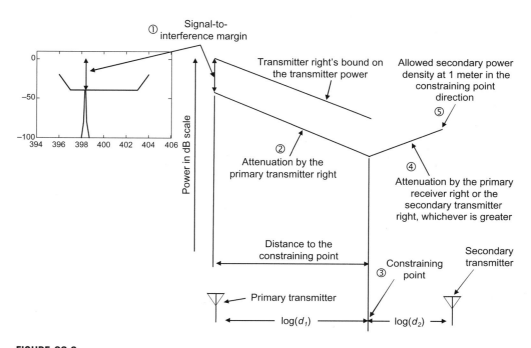

FIGURE 20.9

Exemplary process to compute the constraint of a primary right on the transmit power of a secondary transmitter involving constraining points.

maximum power that the SU may transmit in the direction of the constraining point. In implementation, multiples of these types of computations are made to determine the most restrictive constraint that ultimately governs the behavior of the secondary transmitter.

The computation of compatible reuse is at the heart of using the LBSR to manage spectrum. The variety of methods to specify rights, some of which are presented in Section 20.5, results in a large variety of scenarios in which the specific computations used to determine compatibility will be different. Here we describe the details of the types of computations that are required; in Section 20.5, we describe with exemplary rights the procedures that would be used by a secondary user to compute compatibility that incorporates the methods described here.

Underlay Margin Computations

The underlay margin is the minimum permissible difference, p_m, between the power density of the primary right and the power density of the secondary right at the locations of the primary right. One approach to determine this difference is to shift the transmitter spectrum mask in power to the point where the constraining mask first restricts the transmit power. Figure 20.10 illustrates the effect of a spectrum underlay of a primary receiver right on the allowed secondary power density of three secondary rights with identical spectrum masks but with shifted frequencies, a, b, and c. Each secondary right is constrained to a different power-density level.

The first channel, a, is allowed to operate at a power density 28 dB beneath the primary power density (i.e., $p_m = -28$ dB). The second channel, b, is allowed to operate at 40 dB beneath the primary power density (i.e., $p_m = -40$ dB), and the third channel, c, is allowed to operate at the same power density (i.e., $p_m = 0$ dB). Stine [10] provides an algorithm to compute the minimum permissible difference given a constraining underlay mask and a spectrum mask.

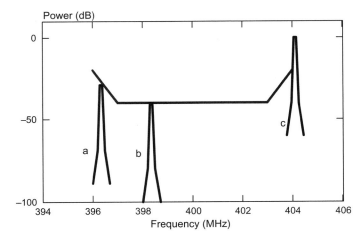

FIGURE 20.10

Three signals constrained to different power levels by the same constraining mask.

Power and Attenuation Computation

The reference power density in power computations is either the regulated maximum power density of a primary right transmitter, p_{RM}, the constaining power density that is computed as previously described, p_{CM}, or the regulated maximum power density of a receiver right, also denoted as p_{RM} in its LBSR data structure. Power and attenuation computations from a transmitter to a distant point may use propagation maps referenced to either the distant point or the transmitter and power maps referenced at either the transmitter or distant point or both. The maps that apply depend on how the rights are written and the purpose of the computation. If determining the power of a primary transmitter at a distant point, we would use the power map and the propagation map of the primary transmitter. If a receiver right applies to the distant point and both a propagation map and power map are specified and we want to determine the permitted transmit power from a secondary user, we would use the power and propagation map of that receiver right. It may occur that in providing protection to a receiver and a restriction to a secondary, two propagation maps apply. In this case, both maps would be checked and the most constraining exponent, the smallest, would be used in the computation. If a receiver right power map applies to the distant point, and a transmit power map applies to a secondary transmitter, and we want to determine the maximum power density at the secondary user, then the calculation would use the sum of the relative powers of the pair of power maps.

Computing the power at a point from a transmitter or the maximum power permitted by a secondary transmitter from a receiver right at a distant point requires determining the distance and directions between them. The directions are used to look up the relative powers in power maps and the pathloss exponent in the propagation maps. If a receiver right governs the computation, then the direction used to determine the power and pathloss exponent is from the distant point (i.e., the actual or notional receiver location) toward the transmitter; otherwise, the direction from the transmitter to the distant point is used.

Constraining Point Computations

Computation of the constraint placed on a secondary right by a primary right is dependent on the location of the primary receiver and the secondary transmitter. In rights where areas or volumes of operation are specified as opposed to single points, it is necessary to compute where those points are. The point in an area or volume of operation where a primary receiver most constrains a secondary transmitter is called a *constraining point*. The location of the constraining points depends on the relative distance between the primary receiver and the secondary transmitter, and the orientation of their power and propagation maps. Identification of constraining points is complicated by the changes in the relative orientation of the maps that occur with shifts in position that cause different power and pathloss parameters to apply in the computation of constraints. Identification of the pair of locations that most constrain secondary use will typically involve the application of heuristics in a search. Systems developed on this concept of location-based spectrum rights will differentiate themselves based on their heuristics to find spectrum they can compatibly reuse. We follow with some observation and theorems that should help in the development of those heuristics.

Observations on Volume Searches

The simplifying observation that will drive many heuristics is that if a single power and pathloss parameter apply to the volumes used by primary receivers and secondary transmitters, then the points that most constrain use are those that are closest to each other. A heuristic that is guaranteed to find a constraint that will prevent violation of the rights is to apply the most constraining pair of parameters in the maps to all directions from all locations in the areas or volumes in which the rights apply. The most constraining receiver parameters are the power map value that determines the lowest permitted interfering power density[4] and the smallest pathloss exponent. The most constraining transmitter parameters are the largest power and the smallest pathloss exponent. With these parameters, the closest points between the two volumes of operation are the most constraining. The path to greater resolution in these computations, and thus more opportunities for compliant reuse, is to identify subsets of volumes to which each combination of power and pathloss exponents apply and then to find the closest points between those subsets and the constraint made by those locations. The final constraint is the greatest constraint from the evaluation of all subsets of parameters.

Observations and Theorems for Protecting Broadcasts

When the locations of the primary receivers are specified through the use of a minimum power density, the computation of compatible reuse requires identifying the constraining points in the intersecting volumes of the primary and secondary power sectors. We define a power sector as the solid angle to which a single set of power gain and pathloss exponents applies. Figure 20.11 illustrates what is meant by an intersecting volume.

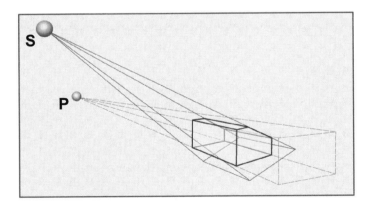

FIGURE 20.11

The intersecting volume that results from the intersection of a primary and a secondary power sector.

---

[4]Power maps in receiver rights can have different meanings. They can specify the allowed interference by direction or they can define the amount of gain of the receiving antenna. In the former, the smaller value is most constraining, and in the latter, the largest value is most constraining.

The figure shows a single power sector extending from a primary transmitter, P, a single power sector extending from a secondary transmitter, S, and the intersection of those sectors, which is the intersecting volume. Heuristics to find the constraint on a SU at a single point will involve considering the constraints of all intersecting volumes. The objective is to find the point in these volumes that most constrain the secondary transmitter and then use the constraints of the most constraining of these as the constraints that finally apply to the secondary user. Stine [7] provides several theorems that may be useful in developing algorithms to make these computations.

20.4 OPTIMIZED DATA STRUCTURES FOR THE LBSR

It is desirable to make LBSRs concise for over-the-air communication. RF systems are already constrained in their transport capacity, so it is beneficial to optimize the data structures within LBSRs to avoid unnecessary overhead. In this section, we describe techniques that improve the efficiency of spectrum masks and the directional vectors used in propagation and power maps.

20.4.1 Encoding Spectrum and Underlay Masks for Transmission

Spectrum masks can be made more concise, and therefore more suitable for over-the-air transmission, by encoding the vector into m-bit words, where m is an integer number greater than 1 but, for practical reasons, usually greater than or equal to 8. Each word corresponds to a unique frequency value or a unique power-density value. The concise spectrum mask data structure also alternates between the frequencies of the inflection points and their power-density levels, for example, $(f_0, p_0, f_1, p_1, \ldots, f_x, p_x, 2^m - 1)$. Three values orient the mask, the center frequency of the mask f_c, the reference transmission power of the mask p_c (which may be either the maximum power density of the right, p_{RM}, or the constraining maximum power density, p_{CM}), and the resolution of the frequency step f_i. There are 2^m frequency levels at which each subsequent value is separated by the specified frequency step resolution. The frequency $2^{m-1} - 1$ maps to the center frequency, and the value $2^m - 1$ is used only to denote the end of the mask. There are also 2^m power levels, where 0 represents the maximum power-density level of the mask, and each coded value maps directly to a decibel reduction in power from the maximum power.

Thus, the conversions between the frequency-coded values and their real values are

$$|f| = f_c + f_i(f - 2^{m-1} + 1)$$

$$f = \frac{|f| - f_c}{f_i} + 2^{m-1} - 1 \tag{20.2}$$

where f is the coded value and $|f|$ is the value that is coded. The conversions between the power values are

$$|p| = p_c - p$$
$$p = p_c - |p|$$

(20.3)

where all variables use the same decibel power units as p_c, p is the coded value, and $|p|$ is the real value that is coded.

Figure 20.12 illustrates the concise vector approach of specifying the spectrum and underlay masks previously depicted in Figures 20.4 and 20.5. The spectrum mask uses 8-bit words with a center frequency of 400 MHz, a frequency increment of 0.001 MHz, and a constraining power density of 20 dBm/m². The underlay mask also uses 8-bit words, and has a center frequency of 400 MHz, a frequency increment of 0.05 MHz, and a constraining power density of 10 dBW/m². The maximum power density of the

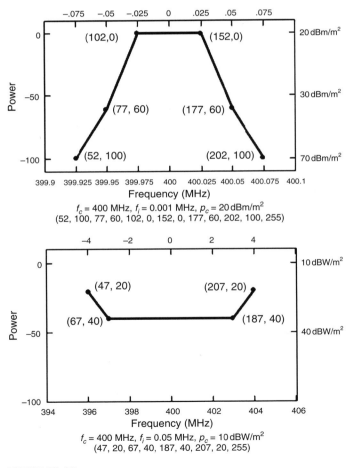

FIGURE 20.12

Examples of encoded spectrum and underlay masks.

concise mask is 0 dB and the positive power-density values are the number of dB below this maximum.

20.4.2 Encoding Directional Vectors for Transmission

The directional vectors used for maps may also be made more concise by encoding the vector into m-bit words, where m is an integer number greater than 1 and, for practical reasons, usually greater than or equal to 8. A concise map is identical in form to a regular map, except it uses a vector of m-bit words, where each word is coded and supports specifying up to 2^m model parameters mapped to values from some minimum to some maximum value: $2^m - 2$ elevations (ϕ), starting from the vertical up direction and reaching to the vertical down direction (an odd number of latitudes, so the middle latitude will point to the horizon), and $2^m - 1$ azimuths (θ), reaching about the node on the horizon (the first and last azimuths, 0 and $2^m - 1$ point in the same direction). If sectors were explicitly defined by the map, the map would have the form $(0, 0, n_{00}, \theta_{01}, n_{01}, \theta_{02}, \ldots, (2^m - 1), \phi_1, 0, n_{10}, \theta_{11}, \ldots, (2^m - 1), \phi_2, 0, n_{20}, \ldots, n_{\text{last}}, (2^m - 1), (2^m - 2))$. Since $\theta = 0$, $\theta = 2^m - 1$, $\phi = 0$, and $\phi = 2^m - 2$ appear predictably in the vector, most can be dropped. The reduced vector becomes $(n_{00}, \theta_{01}, n_{01}, \theta_{02}, \ldots, (2^m - 1), \phi_1, n_{10}, \theta_{11}, \ldots, (2^m - 1), \phi_2, n_{20}, \ldots, n_{\text{last}}, 0)$. The initial value of an unreduced vector is an elevation and the initial value is always 0 so it is not listed. The initial azimuth value of each annulus is also 0 and therefore is also not listed. Finally, a vector always ends in the combination $(2^m - 1)$, $(2^m - 2)$, and therefore these two terms are replaced with a single term 0 that would occur out of order for any other interpretation.

The discrete incremental values used to specify directions and exponents in propagation maps are mapped to values. Azimuth directions are evenly spaced about the map, with 0 and $2^m - 1$ values pointing in the same direction. The conversion from a map azimuth value to an angular direction is

$$|\theta| = \frac{\theta \cdot 360°}{2^m - 1} \tag{20.4}$$

where θ is the coded value and $|\theta|$ is the real azimuth that is coded. If an original vector azimuth direction is between the discrete values allowed by this encoding, then the discrete value that is used should be the one that enlarges the sector with the most conservative values—the values that cause the greatest separation of users.

Frequently, it is desirable to have greater elevation resolution near the horizon, especially for terrestrial applications where topology, human-made structures, and foliage can greatly affect propagation at relatively low elevations. Alternatively, it may be desirable to have a greater resolution near the axial directions, especially for airborne platforms with directional antennas oriented toward Earth. As a general method to provide the shifting of resolution, we incrementally scale subsequent elevation by some scaling factor moving from the axis to the horizon. Given a scaling factor of s, the relation of subsequent values are

$$
\begin{aligned}
(|\phi + 2| - |\phi + 1|) = s\,(|\phi + 1| - |\phi|) \quad & \phi \leq 2^{m-1} - 1 \\
(|\phi + 1| - |\phi|) = s\,(|\phi + 2| - |\phi + 1|) \quad & \phi > 2^{m-1} - 1
\end{aligned}
\tag{20.5}
$$

where ϕ is the coded value, $|\phi|$ is the real elevation that is coded, and where the elevation $\phi = (2^{m-1} - 1)$ points to the horizon. When the scaling achieves finer resolution at the horizon, $s < 1$, the conversion between values and coded values are

$$|\phi| = (1 - s^{\phi}) \frac{90°}{\left(1 - s^{2^{m-1}-1}\right)}, \quad 0 \le \phi \le 2^{m-1} - 1,$$

$$= 180° - \left(1 - s^{2^{m-2}-\phi}\right) \frac{90°}{\left(1 - s^{2^{m-1}-1}\right)}, \quad 2^{m-1} - 1 < \phi \le 2^m - 2; \tag{20.6}$$

$$\phi = \frac{\ln\left(1 - \dfrac{|\phi|\left(1 - s^{2^{m-1}-1}\right)}{90°}\right)}{\ln(s)}, \quad 0° \le |\phi| \le 90°,$$

$$= 2^m - 2 - \frac{\ln\left(1 - \dfrac{(180° - |\phi|)\left(1 - s^{2^{m-1}-1}\right)}{90°}\right)}{\ln(s)}, \quad 90° < |\phi| \le 180° \tag{20.7}$$

When there is no scaling, $s = 1$:

$$|\phi| = \frac{\phi}{2^m - 2} 180°, \tag{20.8}$$

and when finer resolution is used at the axes, $s > 1$:

$$|\phi| = 90° - \left(1 - \left(\frac{1}{s}\right)^{2^{m-1}-1-\phi}\right) \frac{90°}{\left(1 - \left(\dfrac{1}{s}\right)^{2^{m-1}-1}\right)}, \quad 0 \le \phi \le 2^{m-1} - 1,$$

$$= 90° + \left(1 - \left(\frac{1}{s}\right)^{\phi - 2^{m-1}-1}\right) \frac{90°}{\left(1 - \left(\dfrac{1}{s}\right)^{2^{m-1}-1}\right)}, \quad 2^{m-1} - 1 < \phi \le 2^m - 2; \tag{20.9}$$

$$\phi = 2^{m-1} - 1 - \frac{\ln\left(1 - \dfrac{(90° - |\phi|)\left(1 - \left(\dfrac{1}{s}\right)^{2^{m-1}-1}\right)}{90°}\right)}{\ln\left(\dfrac{1}{s}\right)}, \quad 0° \le |\phi| \le 90°,$$

$$= 2^{m-1} - 1 + \frac{\ln\left(1 - \dfrac{(|\phi| - 90°)\left(1 - \left(\dfrac{1}{s}\right)^{2^{m-1}-1}\right)}{90°}\right)}{\ln\left(\dfrac{1}{s}\right)}, \quad 90° < |\phi| \le 180°. \tag{20.10}$$

20.4.3 Encoding Power and Propagation Values for Transmission

The power-density model parameters are coded in a concise power map vector in the same manner as that used for concise spectrum masks. The pathloss exponent model parameters in concise propagation map vectors are coded such that subsequent coded exponents estimate nearly equidistant change in propagation range from the largest to the smallest exponent value. Range is the distance to the point where attenuation causes a signal to go below a threshold, RT, according to the model, and the smallest exponent value estimates the farthest range. The conversion equation may be created from a nominal $RP(1\text{ m})$ and RT and selected values for $|n_{\text{low}}|$ and $|n_{\text{high}}|$. A process to create the conversion equation first determines a maximum and minimum range predicted by the nominal $RP(1\text{ m})$ and RT and the selected values for $|n_{\text{low}}|$ and $|n_{\text{high}}|$. Further, the incremental distance, d_{inc} expressed by the exponents is determined by

$$d_{\text{low}} = 10^{\left(\frac{RP(1\text{ m})-RT}{10|n_{\text{low}}|}\right)} \tag{20.11}$$

$$d_{\text{high}} = 10^{\left(\frac{RP(1\text{ m})-RT}{10|n_{\text{high}}|}\right)} \tag{20.12}$$

$$d_{\text{inc}} = \frac{d_{\text{low}} - d_{\text{high}}}{2^m - 1} \tag{20.13}$$

The conversions between the coded exponents and the actual exponent values are

$$n = \frac{d_{\text{low}} - 10^{\left(\frac{RP(1\text{ m})-RT}{10|n|}\right)}}{d_{\text{inc}}} \tag{20.14}$$

$$|n| = \frac{RP(1\text{ m}) - RT}{10\log(d_{\text{low}} - n \cdot d_{\text{inc}})} \tag{20.15}$$

where n is the coded value and $|n|$ is the real elevation that is coded. As demonstrated, the interpretation of the concise propagation map is dependent on the method to code the exponents. In practice, the method may be explicitly defined by regulation or be arbitrary, dependent on the decision of the map's creator. In the case of the latter, the map must be accompanied with the values used in the conversion, specifically $|n_{\text{low}}|$, $|n_{\text{high}}|$, $RP(1\text{ m})$, and RT.

20.4.4 Concise Vector Examples

Figures 20.13 and 20.14 illustrate the interpretation of concise propagation maps. The parameters in Table 20.1 calibrate the maps, and the power density is assumed isotropic. The surface of these propagation maps identify the range from a transmitter where the signal strength threshold, RT, is reached.

Figure 20.13 illustrates a surface plot of the range predicted by a propagation map by using a single annulus. All values in the propagation map vector are coded, and the meanings of the values are known by their positions. For example, a 1-byte encoded propagation loss exponent 10 extends from an azimuth 0 to an azimuth 20; an exponent 220 extends from an azimuth 20 to an azimuth 60; an exponent 125 from an azimuth

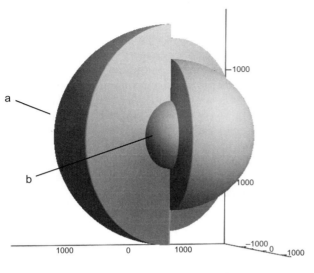

(10, 20, 220, 60, 125, 150, 60, 0)

FIGURE 20.13

Example propagation map demonstrating the definition of different pathloss exponents by direction by using the concise vector form with values encoded to 1 byte each.

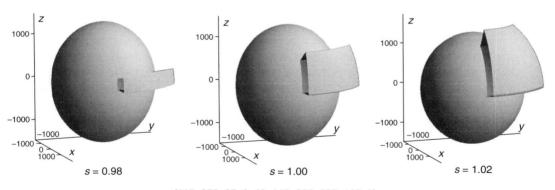

(115, 255, 85, 0, 40, 115, 255, 127, 115, 0)

FIGURE 20.14

Three illustrations of the same propagation map using different scaling factors, showing the ability to shift resolution toward the horizon or toward the axes.

60 to an azimuth 150; and an exponent 60 applies the rest of the way around the map. There are no elevation breaks in the example. In propagation maps, smaller exponents generally predict larger ranges, and therefore the surface labeled *a*, which corresponds to exponent 10, is further from the center than the surface labeled *b*, which corresponds to exponent 220.

Table 20.1 Propagation Map Parameters

Symbol	Description	Value
f_c	Center frequency	400 MHz
$P_c = (RP(1 \text{ m}))$	Maximum 1-meter power density	-24 dBm/m^2
RT	Receive power threshold	-80 dBm/m^2
n_{high}	Largest pathloss exponent	10
n_{low}	Smallest pathloss exponent	2
m	Number of bits per word	8

Note: These are example design parameters for the propagation map definitions used in this chapter.

These encoded pathloss exponents and azimuths are readily converted to their actual values using the parameters in Table 20.1 and the preceding equations. For pathloss conversion we can evaluate Eqs. (20.11) to (20.13) to determine d_{low}, d_{high}, and d_{inc}.

$$d_{low} = 10^{\left(\frac{RP(1 \text{ m}) - RT}{10|n_{low}|}\right)} = 10^{\left(\frac{-24 - (-80)}{10 \cdot 2}\right)} = 630.96$$

$$d_{high} = 10^{\left(\frac{RP(1 \text{ m}) - RT}{10|n_{high}|}\right)} = 10^{\left(\frac{-24 - (-80)}{10 \cdot 10}\right)} = 3.63$$

$$d_{inc} = \frac{d_{low} - d_{high}}{2^m - 1} = \frac{630.96 - 3.63}{2^8 - 1} = 2.46$$

So, for example, the conversion of the exponent $n = 10$ is determined using Eq. (20.15):

$$|n| = \frac{RP(1 \text{ m}) - RT}{10 \log(d_{low} - n \cdot d_{inc})} = \frac{-24 - (-80)}{10 \log(630.96 - 10 \cdot 2.46)} = 2.012.$$

Further, as an example, the azimunth $\theta = 20$ is determined using Eq. (20.4):

$$|\theta| = \frac{\theta \cdot 360°}{2^m - 1} = \frac{20 \cdot 360°}{2^8 - 1} = 28.24°$$

Figure 20.14 illustrates exemplary surface plots of the range predicted by a common propagation map vector by using different scaling factors. This map has two elevation breaks, one above the horizon and one at the horizon. An encoded propagation loss exponent encoded as 115 extends from azimuths 0 to 255, and the next value in the vector, 85, is the coded value of an elevation. In the second annulus, an exponent 0 applies to a sector from azimuths 0 to 40, and then an exponent 115 extends the rest of the way around to an azimuth 255. Since this is the end of the annulus, the next vector value, 127, is an elevation value that corresponds to the horizon. The last annulus has an exponent value of 115. We see that 0 follows 115 in the map vector and so the exponent 115 applies all the way around the annulus and down to the last elevation.

The solid angle projections differ because they use different scaling factors. With scaling factors of 0.98, 1.00, and 1.02, the actual value for the elevation 85 can be computed using Eqs. (20.6), (20.8), and (20.9), respectively.

$$|\phi| = \left(1 - s^\phi\right) \frac{90°}{\left(1 - s^{2^{m-1}-1}\right)} = \left(1 - 0.98^{85}\right) \frac{90°}{\left(1 - 0.98^{2^{8-1}-1}\right)} = 79.99°$$

$$|\phi| = \frac{\phi}{2^m - 2} 180° = \frac{85}{2^8 - 2} 180° = 60.24°$$

$$|\phi| = 90° - \left(1 - \left(\frac{1}{s}\right)^{2^{m-1}-1-\phi}\right) \frac{90°}{\left(1 - \left(\frac{1}{s}\right)^{2^{m-1}-1}\right)}$$

$$= 90° - \left(1 - \left(\frac{1}{1.02}\right)^{2^{8-1}-1-85}\right) \frac{90°}{\left(1 - \left(\frac{1}{1.02}\right)^{2^{8-1}-1}\right)} = 34.71°$$

Elevation 127 represents the horizon, so it has an actual value of 90° for all scaling factors. As described previously, the exponent 0 has a greater range than the exponent 115.

20.5 CONSTRUCTING RIGHTS

Constructing rights for uses of spectrum is an artful practice that attempts to bound the randomness that occurs in component performance, environmental effects, and operational use. The objective in constructing rights is to provide a sufficient level of protection to users while simultaneously trying to enable as much opportunity for reuse as possible. Modelers must balance the improved consumption properties that come with increased resolution models with the verbosity and complexity that can follow. Models should seek conciseness to avoid complicating compatible reuse computations and to reduce the attendant overhead required in communicating rights over the air.

20.5.1 Bounding Performance

Bounding performance involves specifying the borders of rights so that they contain the actual performance of the components of the system and the range of uses of the system throughout the duration of the right.

Signals

The spectral envelope of signals is one of the more predictable attributes of spectrum consumption. It is readily bounded with a spectrum mask. A spectrum mask is designed by selecting the points that define a piecewise linear bound on that envelope. Figure 20.15 illustrates an example of defining a mask on top of an envelope. Designing the mask with a small number of points reduces the size of the LBSR as well as the complexity of compatibility computations.

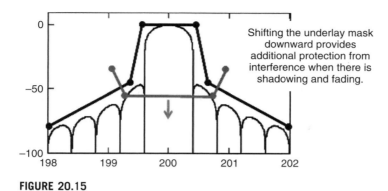

Shifting the underlay mask downward provides additional protection from interference when there is shadowing and fading.

FIGURE 20.15

Constructing spectrum masks to capture the spectral envelope and selecting an underlay mask that provides reliable protection from interference.

Defining the underlay mask is less deterministic. At the minimum, it should place a bound on the interfering signal strength that would interfere with the ideal signal, but the variations of propagation may require some additional margin. That margin is achieved by lowering the relative power levels with respect to the spectrum mask.

Intermodulation distortion may violate the limits of a spectrum mask that is modeled to contain the ideal signal. Although it is feasible for a modeler to explicitly call out this distortion, its existence is normally a function of happenstance and not something the modeler can reliably predict. Resolution of these types of issues is still likely to be handled like it is today, avoiding known situations that are likely to produce intermodulation distortion and reacting to them when they do occur.

Propagation

The modeling of propagation should be viewed as a technique to establish the spatial boundary of a use on account of propagation and not merely as an exercise to most accurately predict the strength of signals at a distance from a source. The more conservative model of propagation is one that predicts a lower rate of attenuation (i.e., a small pathloss exponent) because this will cause the greater separation between users. Square law attenuation in all directions can be a default when there is no knowledge of actual propagation conditions or when there is no evidence that a higher rate of attenuation will occur. Of course, the modeler can attempt to build a model that more accurately predicts performance, and this in turn will create boundaries that more accurately predict actual spectrum consumption.

Some phenomena of propagation cannot be modeled by using a pathloss exponent. The monotonic nature of this attenuation model does not capture the nonmonotonic effects of fading and shadowing in cases where signal strength alternates between strong and weak as the observation point moves along a direction. Modeling in this case would attempt to capture the more optimistic attenuation of the farthest region at which the signal is strong enough to be received. Similarly, the model cannot capture an abrupt change in propagation that might be caused by an obstacle or large terrain feature, for example a mountain. To benefit from the boundary that such a terrain feature provides requires using a pathloss exponent that overestimates attenuation to the obstacle so that the boundary of consumption matches the location of the obstacle. Thus, the model that most accurately captures the boundary of consumption does not necessarily provide the best estimate of signal strength.

Mobility contributes a special challenge to propagation modeling because propagation is so dependent on the environment and mobility continuously changes it. Mobility itself is accommodated by modeling a space that encompasses the range of movement; thus, a propagation model that accompanies such a space must account for the worst case in all of the space. Creating higher-resolution models of propagation when there is mobility requires dividing the regions of mobility into smaller areas, thus dividing up the use into temporal increments and then building separate propagation models for those increments.

Although the directional vector of propagation maps provide an unlimited ability to divide directions into different solid angles, providing the ability to fit a model to observations, doing so is usually not helpful. Increasing the number of directions and exponents used in a model increases the complexity of the computation of compatible reuse and decreases the efficiency of communicating the right. Modeling needs to weigh the benefit of having a higher-resolution model with these costs. It may also be reasonable for a regulatory right, say for a broadcaster, to be defined with a high-resolution model, but a system may reduce its resolution for more efficient computations and for less overhead.

Antennas

Unlike propagation modeling, power model values are not chosen to define a spatial boundary; rather, they are chosen to bound the actual power that is emitted from an antenna. The more conservative model overestimates the power transmitted from an antenna. Power maps can conform to a known antenna's power pattern, but in practice a lower-resolution model will typically suffice. As with propagation maps, it is desirable to avoid complexity and overhead.

Mobility, both the maneuvering of the antenna direction and the mobility of the antenna platform, contribute to the selection of a model. Angular displacement within the duration of a right is accommodated by a larger surface that captures the greatest gain that may be possible in a direction after an antenna is swept through its range of motion. Figure 20.16 shows a directional antenna that may be used by a ground control station of an unmanned aerial vehicle (UAV) and a hypothetical power pattern for that antenna. It then shows the power-density surface of a power map that would contain the antenna power pattern. Next, a mission volume for the UAV that the antenna would need to sweep and then the corresponding power map that contains the highest gain of the sweep are shown. A very simple vector with just a few values provides a map that bounds this operational use.

Power maps accommodate platform mobility in a manner similar to that shown for steering an antenna, capturing the range in orientations that follow from mobility. These changes might be caused by changing the directions that platforms move or be more unpredictable reorientations such as those caused by the undulation of ground vehicles that traverse rough terrain or by an aircraft maneuvering. The sectors with the

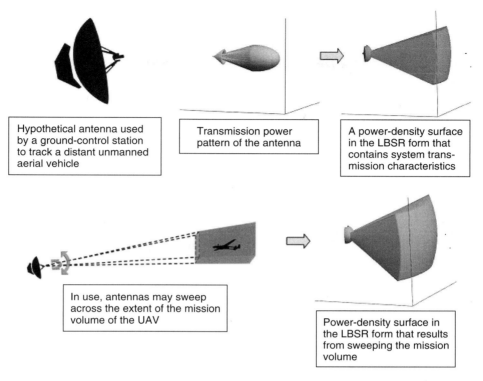

FIGURE 20.16

Example of creating power maps to contain the power patterns of antennas and the operational use of antennas.

largest power values become larger as the variability in orientation becomes larger, thus conveying a greater consumption of spectrum. Dividing a use into temporal or spatial segments in which there is less variability helps to reduce the consumption of the model.

Operating Regions

The goal in defining an operating region is to bound the likely location of system components. The selection of regions is based solely on the modeler's understanding of where components may be throughout an operational use. Uncertainty is accommodated by increasing the volume of use.

Subdividing regions into smaller spaces for different time increments of use is a technique to reduce volume. Figure 20.17 illustrates an example of subdividing a use into segments to reduce the volume at a point in time, which also allows the use of fewer spectrum-consuming power maps and propagation maps. In fact, as described previously, the selection of segments may seek to accomplish these improvements in the maps.

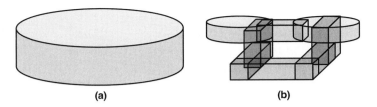

FIGURE 20.17

Options for modeling operating regions for spectrum use in a mission: (a) a large volume that captures a complete mission; (b) using smaller volumes to capture segments of a mission.

20.5.2 Example Models

The ten constructs described in Section 20.3.1 are all just constructs that are combined in various ways to build models of the spectrum consumption of a system. Considering the current interest in TV white space we provide an example of a broadcaster right used to compute the locations of compatible reuse.

Broadcaster

Broadcasters are single transmitters that transmit signals to a number of subscribing receivers. These communications are continuous and are only a downlink. Exemplary users of broadcasting include, but are not limited to, commercial television stations, commercial radio stations, and satellite television. Currently, broadcasters are regulated by placing limits on the amount of power they may use in their broadcasts and controlling where that broadcast might originate. One benefit of using an LBSR to define a broadcaster's rights is that the LBSR defines the geospatial limit on the broadcaster's right and reveals the conditions required for reuse of that spectrum.

In one approach to modeling a broadcaster's location-based RF spectrum right, three different rights tuples are defined. The first tuple is a transmission right that specifies the maximum power density, a power map, the spectrum mask, the right's propagation model and the location of the transmitter. The second tuple is a transmission right underlay referenced to the transmitter right spectrum mask. This underlay specifies a margin that quantifies the relative quality of reception that receivers must achieve and provides an opportunity for secondary spectrum users to use spectrum at a much reduced transmission power within the broadcaster's rights region. The third tuple is a receiver right and includes a power map, a propagation map that applies to interfering nodes, and a threshold receiver right power density. Times may be used to differentiate rights by time of day, such as specifying the use of different transmit powers between day and night. Protocols and policies are rarely given to broadcasters in their rights.

The three tuples work in concert with each other to define the opportunities for spectrum reuse, which can be both within the region that the broadcast subscribers operate as well as outside that region. The underlay works in concert with the transmitter right to identify the amount of interference that can be caused by an SU operating within the broadcast right. The receiver right works in concert with the underlay and the transmitter right to identify the amount of interference that can be caused by an SU outside the transmitter right. The boundary of the transmitter right is the point where the transmitter tuple's power map and propagation map predict that the transmitted signal attenuates to the threshold receiver right power density.

Figures 20.18 through 20.20 illustrate the creation of a broadcaster's spectrum right to conform to a particular operational requirement. Figure 20.18 illustrates an exemplary scenario for a broadcaster's spectrum right. The right needs to cover the shaded region, and a broadcast antenna is to be located at point *A*. In this scenario, it is assumed that an antenna can be built to any type of directionality. Figure 20.19 illustrates the use of a location, a maximum power density, a propagation map, a power map, and a threshold power density to build a right that covers the shaded region of Figure 20.20. The propagation map and the power map are coded into the concise versions of these masks. The shaded surface is the volume that this combination covers with a right. Figure 20.20 illustrates the spectrum mask and underlay mask for the right.

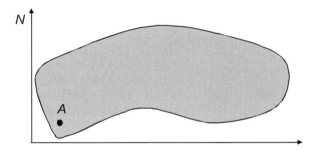

FIGURE 20.18

Sample scenario for demonstrating the construction of a broadcast transmission right. Point *A* marks the location of the transmitter and the shaded region is the space that its broadcast must cover.

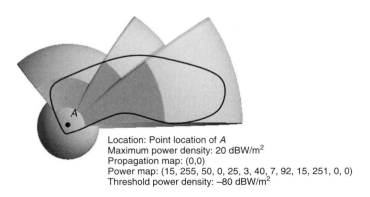

Location: Point location of *A*
Maximum power density: 20 dBW/m^2
Propagation map: (0,0)
Power map: (15, 255, 50, 0, 25, 3, 40, 7, 92, 15, 251, 0, 0)
Threshold power density: −80 dBW/m^2

FIGURE 20.19

Example of a broadcast right designed to conform to a space.

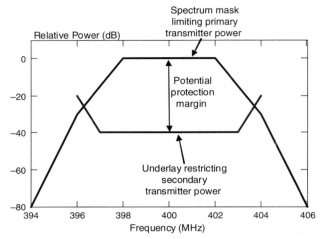

Center frequency, f_c: 400 MHz
Frequency increment, f_i: 100 kHz
Spectrum mask: (67, 80, 87, 30, 107, 0, 147, 0, 167, 30, 187, 80, 255)
Underlay mask: (97, 20, 102, 40, 152, 40, 157, 20, 255)

FIGURE 20.20

Example spectrum and underlay mask for a broadcaster right.

In contrast to this example propagation map that bases the receiver right on the combined use of a threshold power density and an underlay, another approach to model a broadcaster right may specify the spectrum rights with two tuples, a transmitter right and a receiver right. The transmitter right may include a location of the transmitter, a maximum power density, a power map, and the spectrum mask of the right. The receiver right may include a volume for the receiver right, an underlay mask, a power map, and a propagation map. The volume in the receiver right determines the spatial limits of the right. SUs would be able to use the spectrum as long as they respected the constraints of the receiver right portion of the broadcaster's right.

20.6 APPLICATIONS

The LBSR method of spectrum-consumption modeling has several features that make it attractive for new SM approaches: it is concise, it is generic, it is easily communicated, it eliminates the intermediate hurdle of converting a primary right to policy since the primary right is policy, and it provides a well-defined tractable approach to computing compatibility without dependence on external models and data sets. The significance of these features is that the SM task can be distributed throughout a system. Individual subsystems can compute the requirements for compatibility within their operating bands without having to have a comprehensive database of all users of spectrum. Next, we look at several applications that benefit.

20.6.1 Computing and Optimizing Spectrum Reuse

The LBSR was designed to simplify the computation of spectrum reuse. It provides a disciplined approach to capture the performance of RF components and their consumption of spectrum based on their operational use. It provides a model of spectrum consumption that makes the computation of reuse tractable. Thus, it is envisioned that the LBSR can be used as a tool to study and optimize the reuse of spectrum. Analysts seeking to determine the quantity of independent RF devices that can be deployed in a space would first model the component's use of spectrum by using the LBSR, and then compute reuse opportunities based on the consumption captured in the model. This general concept could find its way into mission-planning systems where missions of individual components are planned mindful of the spectrum they consume so that more missions can be executed concurrently. We use the unmanned aerial system (UAS) example, in which the modeling of specific flight plans of UAVs may allow the tighter packing of UAVs that use the same spectrum.

20.6.2 Conveying Machine-Readable Policy

All of the data structures used in the LBSR have an unambiguous meaning as they relate to the use and reuse of spectrum. Given a definitive syntax, it is possible to build machines to read and comply with the guidance in an LBSR. Its concise nature makes it easily communicated. The well-defined meaning of the LBSR eliminates the need to certify each policy. LBSRs that define legacy uses are all that must be communicated to protect these systems. Beyond the basic concept of complying with policy, the nature of the LBSR would allow radios to receive policy dynamically and for systems, radios, and users to arbitrate the use of spectrum with each other and with an SM. Managers can take on the role of designing policy based on the actual operational use on the ground, and radios can respond to these instructions.

20.6.3 Dynamic Spectrum Management

The culmination of the capabilities to compute and optimize the reuse of spectrum and to convey policy in a concise and machine-readable message makes the concept of dynamic spectrum management (DSM) possible. Figure 20.21 illustrates an example DSM system. The planning process in the DSM in the figure begins with (1) the mission-planning system requesting spectrum from the SM. The SM responds with rights for the UASs and offers relevant restrictions to spectrum use (2). Armed with these rights, and the planning tool's knowledge of spectrum use of its systems, the planner can consider available spectrum in planning and defining missions (3). Once the mission decides on the set of missions their unit will execute, they request the rights from the SM (4). If the requests have no errors, the SM loads the rights into the database and confirms the request (5). The SM can then search for reuse opportunities (6) and conveys these opportunities to SUs so that those systems can determine opportunities to reuse the same spectrum (7). In this case, the network manager (NM) identifies ways that it can use a portion of the available spectrum and responds to the SM with a request (8), which the SM subsequently grants (9). The NM requests spectrum for CRs and therefore, these radios are informed of the spectrum that they may try to use (10). Although the

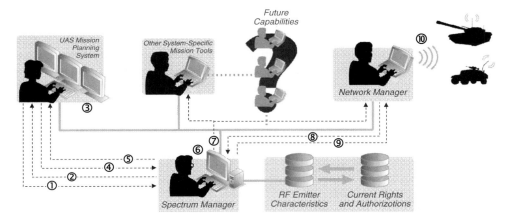

1. Request potential rights
2. SM sends relevant rights
3. Mission planner creates plan and appropriate spectrum rights
4. Mission-level spectrum rights requested
5. Mission-level spectrum rights granted
6. SM identifies reuse opportunities
7. Potential users of spectrum notified of opportunities
8. NM identifies reuse and requests rights
9. SM reviews NM's request and grants rights
10. NM informs CRs of rights

FIGURE 20.21

Example of a DSM system.

illustration in Figure 20.21 shows a man-in-the-loop, these processes could be partially, or even fully, automated.

Figure 20.21 also identifies future capabilities as part of the system. One future capability might be the use of a system of sensors to detect the uses of spectrum for the purposes of seeking reuse opportunities, identifying inappropriate use of spectrum, and calibrating rights. The same GUI of Figure 20.21 may also be used to display sensed use of spectrum. The display of the use of spectrum can be compared to dispensed rights to determine where rights are too liberal in their authorization of spectrum use or when spectrum is being used outside authorized areas.

20.6.4 Secondary Spectrum Markets

A common lament of manufacturers of RF products, of providers of services using RF spectrum, and of enterprises that use RF spectrum in their operations, is that there are many bands of spectrum that appear to be unused. However, users of that spectrum insist that their use is critical even if it is sporadic. A primary cause of the loss of use of spectrum is an artifact of spectrum management and administrative approaches that seek persistent assignments, and as such, sporadic users that have protection needs receive persistent protection. Spectrum administrations, such as the FCC, have clearly indicated through their policies that they do not intend to participate in the dynamic management and enforcement of spectrum use. Therefore, the detachment of the FCC provides the opportunity for nongovernment entities to take on the role of spectrum broker, with a net social benefit of making more of the RF spectrum available for use.

Assuming that a brokering activity can be self-supporting, the idea is to use all the same mechanisms used in DSM to control access to spectrum in a market-based approach. The existence of brokers is likely to increase the availability of spectrum. The opportunity to earn money on unused spectrum will serve as an incentive to sporadic users of spectrum to find ways that make their spectrum available on the secondary market, while also keeping it available for their use on amenable terms. If effective, governments may want to pool even more spectrum for brokers to manage as a means to raise revenue.

The LBSR and the SM capabilities it enables are well suited for brokering activities. It enables the combining of rights, or the subdivision of rights, into smaller amounts, dividing those smaller amounts by using any of the dimensions of time, space, or frequency.

20.7 FUTURE RESEARCH AND WORK

It should be clear that we envision a significant role for the LBSR in the future of CR and spectrum management. The LBSR has value within self-contained systems, but its larger purpose is the integration of all aspects of spectrum management across all systems that use spectrum. That vision can be achieved only through a collective effort to arrive at a standard specification for the components of the LBSR and its subsequent adoption by SM administrations across the world. The path to arrive at that vision is fraught with technical and political challenges. Focusing on the technical, the path to success is the incremental demonstration of its utility. Achieving utility hinges on the development of the algorithms used to compute compatible reuse and then the development of self-contained systems that use the LBSR as their technique for spectrum management.

The algorithms for computing compatible reuse must address three types of problems given the LBSR constraining rights and permissive rights: (1) determine the channels that are permitted to be used and the power limits on their use; (2) determine the spatial boundaries at which a particular band can be used at a specified power; and (3) search for the best band to support the requirement.

These algorithms need to be developed to be as efficient as possible, as they will affect the agility of the DSM technique. Most functions in these algorithms will need to be implemented in individual radios. They must be able to run in "near real time" for rapid adaptation to changes in the available spectrum and allocation of this spectrum. The seeking of this efficiency is likely to make algorithm development, similar to routing in wireless networks, an open area for research for years to come.

We see a utility for the LBSR in the better management of UASs. Each UAS is a self-contained system and the prize for better management is the flying of more missions with the same spectrum. Networks are another self-contained system that may benefit, but may be made difficult if the underlying protocols used by the network cannot exploit the additional spectrum that management provides. All too often, spectrum management in networks is nothing more than managing frequency assignment in the initial design of a network. The protocols do not support the dynamic reassignment of spectrum. Other networking technologies will need to be developed. For example, the networking concept described by Stine [9] is particularly well suited for DSM and the use of the LBSR.

The conclusion is that *the opportunities to research and exploit spectrum-consumption modeling are wide open*.

20.8 SUMMARY

We started this chapter by describing a significant challenge that confronts DSA technologies: its incompatibility with legacy spectrum management approaches that seek persistent but not dynamic solutions. We have explained why trying to place the solution entirely within a radio controlled by policy, specifically one contingent only on the sensing of spectrum use, is problematic. It is unlikely to prevent harmful interference and thus unlikely to be accepted by SMs. Additionally, it will be very difficult to write policy and to certify that it results in compatible reuse.

As an alternative, we have proposed a technique to model spectrum consumption called the location-based spectrum right. It is well suited for DSA management because the construction of a right for one user automatically becomes a constraint to the DSA user. This simplifies policy writing and system certification. Rather than trying to define a set of behaviors for cognitive systems to follow to protect secondary users, the task focuses on the much easier definition of a system's use of spectrum. Legacy uses of spectrum can be converted into these models, which then serve as policy constraints for dynamic access. Certification of systems would evaluate whether they can compute compatible reuse and whether they conform their use to these computations, rather than through expensive testing in operational scenarios. Initially, DSA based on using LBSRs would rely on the more reliable location sensing as opposed to spectrum sensing, but ultimately LBSRs can be used to implement any type of policy that is achievable by any other approach.

We have described that LBSRs consist of ten types of components, which are constructs that are assembled in many ways to articulate the consumption of spectrum. Each component has a specific meaning. Collectively, they allow an unambiguous computation of compatible reuse. Modeling the consumption of spectrum is an artful exercise that accounts for the capabilities of RF components and the operational use of systems. We describe what the modeler should consider in creating models for systems. Through this presentation we show that the LBSR can do much more than serve as a means to convey policy to CRs. Spectrum-consumption modeling can be the integrating concept that allows the creation of DSM processes and systems.

At present, the use of spectrum-consumption modeling is just a concept, so there is plenty of opportunity to influence its ultimate design and use.

REFERENCES

[1] *http://www.fcc.gov/sptf*.
[2] FCC, Spectrum Policy Task Force Report, ET Docket No. 02-135, November 2002.

[3] GAO-06-172R, Potential Spectrum Interference, December 2005.

[4] FCC 07-99, Memorandum Opinion and Order in the Matter of Wireless Operations in the 3650–3700 MHz Band, June 7, 2007.

[5] Sahai, A., N. Hoven, and R. Tandra, Some Fundamental Limits on Cognitive Radio, *Proceedings of Allerton Conference on Communication, Control, and Computing*, pp. 1–11, October 2004.

[6] Marshall, P., and D. Stewart, DARPA NeXt Generation (XG) Communications & Wireless Network after Next Information Brief, March 27, 2008. (Briefing given to the IEEE 1900.3 Working Group.)

[7] Stine, J. A., A Location-based Method for Specifying RF Spectrum Rights, *Proceedings IEEE DySPAN*, April 2007.

[8] Rappaport, T., *Wireless Communications, Principles and Practice*, Second Edition, Prentice Hall, 2002.

[9] Stine, J. A., Enabling Secondary Spectrum Markets Using Ad Hoc and Mesh Networking Protocols, *Academy Publisher Journal of Communication*, 1(1):26–37, 2006.

Protocols for Adaptation in Cognitive Radio

21

Michael B. Pursley, Thomas C. Royster IV
Clemson University, Clemson, South Carolina

21.1 INTRODUCTION

The protocols described and reviewed in this chapter are primarily for point-to-point, session-oriented traffic that is being sent over a communication link from one cognitive radio (CR), the *source*, to another cognitive radio, the *destination*. Our protocols are suitable for mobile ad hoc networks in which packets must be relayed to reach their destinations, but we focus on the transmissions between two radios that are within communication range of each other. These transmissions may take place on a link that is part of a route through the network. Two-way simultaneous transmissions are not required by any of the protocols, which permits each of them to be used in half-duplex radios. The only feedback that is needed by the source is provided in short acknowledgment packets that follow the successful reception of a sequence of one or more packets by the destination.

A typical session between the source and destination may involve the transfer of a file the size of which might range from about 50 KB to 10 MB or more. Depending on the code rates that are used for the session, a session for the transfer of files with sizes in this range requires the source to send between a few hundred and nearly a hundred thousand packets, each consisting of approximately 4000 binary code symbols. (In this chapter, the outputs of a binary encoder are referred to as *binary code symbols*, and the inputs are referred to as *information bits*.) The transmissions from the source to the destination may cause interference to other nearby radios operating in the same frequency band as the source. One goal of the protocol suite is to prevent such interference from disrupting the sessions of such radios.

Each CR has a family of codes and a set of modulation formats. The radio is permitted to change code-modulation combinations throughout the session, which may be necessary to compensate for changes in the propagation loss or interference environment. The radio is also permitted to change its transmitter power level, but power increases may be constrained by the need to avoid producing excessive interference to other sessions that are being conducted in the same frequency band. Our protocols govern the choice of the initial power level for each session and the adaptation of transmission parameters to accommodate variations in channel conditions from packet to packet throughout the session.

As a CR receives, demodulates, and decodes packets, it can extract information about its communication environment from the demodulation and decoding processes in its receiver. Cognitive radios exchange this information to permit the adaptation of future packet transmissions. The radios need not perform power measurements or obtain estimates of the channel gain, and the only feedback required for our protocols can be provided by the insertion of a few bits in each acknowledgment packet.

21.2 MODULATION

The three modulation formats that are used to illustrate our protocols are quadriphase shift key (QPSK) modulation; quadrature amplitude shift key (QASK) modulation, also known as quadrature amplitude modulation (QAM); and biorthogonal modulation. Each modulation format has its own set of *modulation symbols*, and the size of the set determines the number of binary code symbols that can be represented by each modulation symbol. For example, M-QASK is a QASK signal that has a set of $M = 2^m$ modulation symbols. Each such QASK symbol can represent $m = \log_2 M$ binary code symbols.

The signal for each of the modulation formats can be expressed as

$$s(t) = A\{a_1(t)\cos(\omega_c t + \varphi) - a_2(t)\sin(\omega_c t + \varphi)\} \qquad \textbf{(21.1)}$$

and the data modulation waveforms are given by

$$a_i(t) = \sum_{j=0}^{J-1} u_{i,j}\, p_\tau(t - j\tau) \qquad \textbf{(21.2)}$$

where $p_\tau(t)$ denotes the unit-amplitude rectangular pulse of duration $\tau > 0$ that begins at time $t = 0$. If $u_{i,j}$ takes values in the set $\{-1,+1\}$ for each $i(i = 1$ and $i = 2)$ and each $j(0 \leq j \leq J - 1)$, then the signal defined by Eqs. (21.1) and (21.2) represents a sequence of J modulation symbols, each of which uses QPSK modulation with symbol duration τ. If K_m is an odd positive integer, $u_{i,j}$ takes values in the set $\{-K_m, \ldots, -3, -1, +1, +3, \ldots + K_m\}$ for each i and j, and $M = (K_m + 1)^2$, then Eqs. (21.1) and (21.2) describe a sequence of J regular M-QASK modulation symbols with pulse duration τ.

Regular M-QASK [1] is a special case of M-QAM in which the pulses are rectangular and the points in the signal constellation are uniformly spaced on a square grid, so QPSK is the same as regular 4-QASK. In all that follows, we adopt the more popular term QAM rather than the more precise term QASK. If J is a multiple of the positive integer L and $a_2(t) = 0$ for each t, then we can use Eqs. (21.1) and (21.2) to represent a sequence of J/L pulse-coded modulation symbols with modulation symbol duration $T = L\tau$. We obtain L-ary orthogonal modulation if, for each j in the set $\{kL : 0 \leq k < J/L\}$, the L-tuple $(u_{1,j}, u_{1,j+1}, \ldots, u_{1,j+L-1})$ is from a fixed set of L orthogonal vectors of length L. For example, the set of orthogonal vectors could be the rows of an $L \times L$ Hadamard matrix [1], in which case $u_{1,j}$ takes values in the set $\{-1,+1\}$.

For $M = 2L$, a set of M biorthogonal vectors is obtained from a set of L orthogonal vectors by forming the union of the orthogonal vectors and their complements. If the L-tuples $(u_{1,j}, u_{1,j+1}, \ldots, u_{1,j+L-1})$, for $j \in \{kL : 0 \leq k < J/L\}$, are from a set of M

biorthogonal vectors, then Eqs. (21.1) and (21.2), represent a sequence of J/L pulse-coded modulation symbols that use the modulation known as M-biorthogonal modulation or M-biorthogonal keying (M-BOK). Orthogonal and biorthogonal modulation can be demodulated efficiently by using chip-matched filters and matrix multiplication (see [1], Appendix E), and the matrix multiplication can be accomplished with fast transform methods.

For pulse-coded modulation, the pulses $u_{1,j}p_\tau(t - j\tau)$ in Eq. (21.2) are referred to as *modulation chips* and τ is referred to as the *chip duration*. The modulation symbol duration for M-BOK is $T = L\tau$ and the modulation chip duration is τ, so M-BOK has $L = M/2$ modulation chips per modulation symbol. For each modulation format, we refer to the shortest pulse as a modulation chip; therefore, for QPSK and M-QAM, each modulation symbol consists of a single modulation chip. If the number of different modulation symbols is M, then each modulation symbol can represent $\log_2 M$ binary code symbols. Thus, each QPSK symbol can represent two binary code symbols, each 16-QAM symbol and each 16-BOK symbol can represent four binary code symbols, and each 64-BOK symbol can represent six binary code symbols. Notice that there is only 1 chip per modulation symbol for QPSK and QAM, but there are 8 chips per modulation symbol for 16-BOK and 32 chips per modulation symbol for 64-BOK. For each modulation format, the *chip rate* is $1/\tau$, the inverse of the chip duration. As discussed in Section 21.4, the chip rate should normally be held constant throughout each session, even if the system changes modulation formats during the session. If the chip rate is not changed, neither is the bandwidth.

The use of M-QAM, QPSK, and M-BOK here is for illustration only. Other modulation formats can be employed by our protocols. For example, a much larger signal set that has independent biorthogonal modulation on the inphase and quadrature components, referred to as inphase-quadrature biorthogonal (IQB) modulation, is included in the code-modulation library we previously employed [2]. In addition, any of the modulation formats can be used with multicarrier modulation (e.g., in orthogonal frequency-division multiplexing, or OFDM), spreading sequences can be applied to give direct-sequence spread-spectrum modulation, or the modulation formats can be used in frequency-hop spread spectrum.

21.3 ERROR-CONTROL CODES

For ease of implementation and to make it straightforward to use each error-control code in the code family with each of the three modulation formats, we focus on binary codes, as shown in Figure 21.1. To illustrate the performance of our protocols, we use a family of five turbo product codes that are denoted by C_1, C_2, C_3, C_4, and C_5, and have rates of $r_1 = 0.236$, $r_2 = 0.325$, $r_3 = 0.495$, $r_4 = 0.660$, and $r_5 = 0.793$, respectively. (The *rate* of a binary code is the number of information bits per binary code symbol.) A single hardware chip [3] can encode and decode each of these turbo product codes, and software decoders are also available. The block length is 4096 for C_2, C_3, and C_5, but it is only 2048 for C_1 and 1024 for C_4. For our numerical results, the interleaver in Figure 21.1 is an S-random interleaver [4], and each packet has 4096 binary code symbols. There is one code word per packet for C_2, C_3, and C_5; two code words per

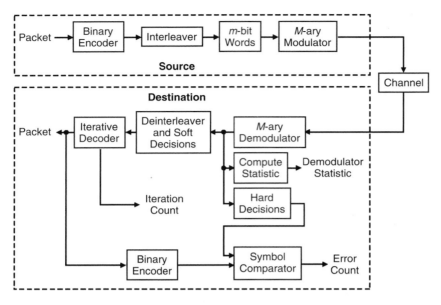

FIGURE 21.1

System model and receiver statistics.

packet for C_1; and four code words per packet for C_4. Soft-decision decoding is performed with quality information provided by the log-likelihood-ratio (LLR) metric [5]; however, a distance metric [6] is a good alternative for M-QAM.

21.4 PERFORMANCE MEASURES FOR A CODE-MODULATION LIBRARY

A *code-modulation library* for a CR is a collection of pairs $\mathcal{B} = (C, M)$ for which the first element is an error-control code and the second element is a modulation format. Each pair is referred to as a *code-modulation combination*. A radio is said to support the code-modulation combination $\mathcal{B} = (C, M)$ in the transmission mode if the radio can encode using code C and modulate using modulation format M. The radio is said to support $\mathcal{B}$ in the reception mode if it can demodulate M and decode C. A radio *fully supports* the code-modulation combination $\mathcal{B}$ if it supports $\mathcal{B}$ in both the transmission and reception modes. The code-modulation library that we use to illustrate the performance of our protocols is given in Table 21.1. The library in the table does not contain all 20 possible combinations of the five codes and four modulation formats. The process of selecting the subset of combinations for the library is described and illustrated in Section 21.5.

For digital communications, the appropriate measure of the signal-to-noise ratio (SNR) involves signal energy and noise power spectral density. For our purposes, the primary measure of signal energy is ε_c, the average energy per chip. Other measures of energy (e.g., ε_b, the average energy per information bit, and ε_s, the average energy per binary code symbol) can be obtained easily from ε_c. If the modulation is changed in a

Table 21.1 Code-Modulation Library with CENR Requirements for 10^{-2} Packet Error Probability on an AWGN Channel

Combination	Modulation	Code, Rate	CENR (dB)
$\mathcal{B}_1$	64-BOK	C_1, 0.236	−9.8
$\mathcal{B}_2$	64-BOK	C_3, 0.495	−8.0
$\mathcal{B}_3$	64-BOK	C_5, 0.793	−6.1
$\mathcal{B}_4$	16-BOK	C_3, 0.495	−3.8
$\mathcal{B}_5$	QPSK	C_1, 0.236	−1.7
$\mathcal{B}_6$	QPSK	C_2, 0.325	−0.8
$\mathcal{B}_7$	QPSK	C_3, 0.495	1.7
$\mathcal{B}_8$	QPSK	C_5, 0.793	4.9
$\mathcal{B}_9$	16-QAM	C_3, 0.495	7.2
$\mathcal{B}_{10}$	16-QAM	C_4, 0.660	10.0
$\mathcal{B}_{11}$	16-QAM	C_5, 0.793	11.2

way so that the bandwidth and average power are held constant, then the average energy per chip does not change, but ε_s and ε_b may change. If the one-sided power spectral density of the thermal noise is N_0, then the ratio of the energy per chip to the noise density is ε_c/N_0. To facilitate the specification of this ratio in decibels (dB), we define CENR = $10 \log_{10}(\varepsilon_c/N_0)$. Similarly, we define ENR = $10 \log_{10}(\varepsilon_b/N_0)$, which is the bit energy-to-noise density ratio (ENR) in dB, and SENR = $10 \log_{10}(\varepsilon_s/N_0)$, which is the ratio in dB of the energy per binary code symbol to the one-sided noise density.

The values of CENR required to give a packet error probability of 10^{-2} for an additive white Gaussian noise (AWGN) channel are listed in Table 21.1 for the code-modulation combinations in the library. The value of CENR required for code-modulation combination $\mathcal{B}_k$ is denoted by $CENR_k$ for $1 \leq k \leq K$. For an AWGN channel, the *span* of a code-modulation library is the difference between the largest of the required values of CENR and the smallest of these values. The span for the library of Table 21.1 is $CENR_{11} - CENR_1$ = 21 dB.

It is desirable for the library to have a large span so that the CR can adapt to large variations in the propagation loss. Suppose, for example, that the cognitive radio is transmitting packets with code-modulation combination $\mathcal{B}_{11}$, and then the shadow loss on the link increases by 15 dB. The radio can switch to combination $\mathcal{B}_4$ and continue communicating at a packet error probability of 10^{-2} without increasing its radiated power. We define the *information rate* of a code-modulation combination to be the number of information bits per chip provided by the combination. If the channel code is a binary code of rate r information bits-per binary code symbol, then the information rates for the three modulation formats used as examples in this chapter are as given in Table 21.2.

Although it is usually not possible to increase the spectral occupancy during a session, especially for dynamic spectrum access (DSA) systems, it is usually acceptable to decrease the bandwidth. For example, CRs could offset an increase in propagation

Table 21.2 Information Rates in Bits per Chip for Rate r Binary Codes with Three Modulation Formats

Modulation	Information Rate
QPSK	$2r$
M-QAM	$r \log_2 M$
M-BOK	$2rM^{-1} \log_2 M$

loss by increasing the chip duration, which corresponds to decreasing the chip rate and the bandwidth. This is not an efficient way to adapt to the increased loss, however, as can be seen from Table 21.1. For example, if a CR is using combination B_{11} when the propagation loss increases, it can decrease the bandwidth by a factor of 2 by doubling the chip duration.

This method of adaptation permits the system to offset an increase of 3 dB in the propagation loss, and continue to transmit at the same power level but with only half the information rate of B_{11}. Alternatively, the radio can change from B_{11} to B_8, which also gives half the original information rate; however, from Table 21.1 we see that the change from B_{11} to B_8 gains 6.3 dB. Even larger increases in propagation loss can be accommodated by switching to one of the combinations that uses 64-BOK, and such a switch provides considerably more improvement than a decrease in the bandwidth of the original modulation. In general, adapting the code-modulation combination is far superior to simply decreasing the bandwidth as a means of compensating for deterioration in the communication link.

The choice of 10^{-2} as the requirement for the packet error probability is reasonable for a wireless communication network, but all the concepts are the same for other error rates. For example, the required values of CENR increase by only a few tenths of a dB if the packet error probability requirement is reduced to 10^{-3}, and there is even less change in the span of the library. A packet error probability of 10^{-2} is low enough to provide acceptable retransmission requirements for most applications, and little or no benefit to the overall network performance is derived from a significant decrease in the packet error probability below 10^{-2}. If the packet error probability on a link is in the range from 10^{-2} to 10^{-3}, then the problems that arise at higher layers typically contribute more to the probability of an undelivered packet than decoding errors at the physical layer. Examples of such higher-layer problems result from attempts to send packets to busy radios at the medium access control (MAC) layer and routing errors at the network layer.

The information rate is an accurate measure of the rate at which information is delivered only if all packets are received correctly at the destination. Unless there are large differences in the block lengths of the codes in a family, the high-rate codes in the family typically cannot correct as many errors as lower-rate codes in the family. Therefore, maximization of the information rate does not necessarily result in the delivery of more information to the destination. It is usually necessary to perform a trade-off between the information rate and the packet success probability to maximize the rate at which information is delivered. The measure of delivered information that is used in this trade-off must count only the information bits that are contained in packets that are decoded correctly. Packets that are not decoded correctly must be retransmitted, so no credit is given for them. The throughput measure must also account for the amount of time required to send the information, so appropriate units are needed to measure the transmission time.

The choice of unit for measuring time depends on the characteristics of the protocol. We consider protocols for which the bandwidth is held constant for the duration of each session, even if the code-modulation combination is changed during the session. Such protocols are attractive for DSA networks because changes in the combination during a session do not require changes in spectral occupancy. If two code-modulation combinations are required to have the same bandwidth, then they must also have the same chip rate. If two combinations with the same chip rate do not have the same number of binary code symbols per chip, then their packet lengths are different, even if they have the same number of binary code symbols per packet. As a consequence, the throughput measure must be normalized in a way that permits an appropriate accounting of the information transfer per time unit. For our numerical results, we have selected the time required to transmit 2048 chips as the *time unit*, which corresponds to the transmission time for a packet of 4096 binary code symbols when QPSK is the modulation format.

A packet is accepted at the destination only if it is completely correct at the decoder output in the destination's receiver. Because only packets that are completely correct are accepted, we say that an information bit in a packet is *delivered* if every information bit in the packet is decoded correctly at the destination. A packet that has one or more erroneous information bits at the decoder output is not accepted, and it must be retransmitted in its entirety. For each code word in the transmitted packet, the receiver produces a corresponding *received word* at the output of the demodulator. The *packet success probability* is the probability that all the received words for the packet are decoded correctly. The *throughput* for a session is the number of information bits delivered to the destination per unit time that the source is transmitting during the session, including the time consumed by any retransmissions that are required.

Thus, the time spent transmitting a packet that is not decoded correctly counts in the denominator of the throughput measure, but such a packet contributes nothing to the numerator. The *maximum throughput* for a code-modulation combination is the throughput that is achieved if each packet in the session uses the code-modulation combination and each received word is decoded correctly (e.g., if the packet success probability is unity). If the packet success probability is unity when a particular code-modulation combination is used, then the throughput is proportional to the information rate of the combination, and it can be calculated as the number of information bits per chip times the number of chips per packet for the combination.

We define two special types of code-modulation combinations. A *robust* combination is one that provides a very high packet success probability, even if the SNR and signal-to-interference ratio (SIR) are low. The information rate is typically very low for a robust combination. A *high-throughput* combination has a high information rate and achieves an acceptable packet success probability for a practical range of channel parameters. The former condition means that the SNR requirement cannot be excessive for a high-rate code-modulation combination to be classified as a high-throughput combination. Even so, most high-throughput combinations require a much larger SNR than a robust combination, but typical high-throughput combinations have information rates more than an order of magnitude greater than the information rates of robust combinations. We assume that there is at least one robust code-modulation combination that is fully supported by every radio; we refer to this as the *initial code-modulation combination*. Normally, the initial combination, which we denote by B_1, is the library's most robust combination. In our protocol, it is always used at the beginning of a new session.

21.5 SPECIAL SUBSETS OF THE CODE-MODULATION LIBRARY

Before the transmission of the packets from the source to the destination can commence, the two radios must determine the session's *available library*, which is the set of all code-modulation combinations that are supported in the transmission mode by the source and in the reception mode by the destination. The available library, denoted by $\mathcal{L}$, can be determined from the exchange of two preliminary control messages between the source and destination. The control messages are sent with the initial code-modulation combination during the initial power-adjustment (P-ADJ) period, which is described in Section 21.7.

Because our protocols require only a few bits of feedback information for each packet that the destination receives from the source, the choice of the code-modulation combination for the reverse link from the destination to the source is straightforward in comparison with the choice for the forward link. In most applications, the most robust combination that is available for the reverse link should be used because this combination gives the maximum probability of correct reception of the feedback information. The feedback packets are very short, even if the lowest-rate combination is employed.

The next step in the process leading up to transmission of the packets is for the two CRs to determine the *active library* for the session, which is the subset of the available library that will actually be used for transmissions on the forward link throughout the session. The active library, $\mathcal{L}_a = \{B_k : 1 \le k \le K\}$, should have at least one combination that gives good performance for each communication environment. The available library for a session may contain one or more code-modulation combinations that do not provide significantly higher throughput than other combinations for a substantial range of channel parameters. Such combinations are excluded by the CRs when they choose the active library for the session. For instance, suppose that B_x and B_y are two code-modulation combinations, and further suppose that $v = (v_1, v_2, \ldots, v_q)$ is a vector that represents all the important parameters of the channel. For example v_1 might be the value of CENR at the destination, v_2 might represent the ratio of diffuse to specular power for Rician fading (see [1]), v_3 might represent the power in the interference on the channel, and so on. For an AWGN channel, the parameter CENR is sufficient to completely characterize the performance of each code-modulation combination. V should be set as all such vectors $\mathbf{v}$ that will occur for the intended application. Let $T_x(\mathbf{v})$ be the throughput that is achieved by combination B_x for a channel with parameter vector $\mathbf{v}$, and let $T_y(\mathbf{v})$ be the throughput that is achieved by combination B_y for the same channel. If $T_x(\mathbf{v}) \geq T_y(\mathbf{v})$ for all $\mathbf{v}$ in the set V, then we say that code-modulation combination $\mathcal{B}_x$ *dominates* code-modulation combination $\mathcal{B}_y$. If $\mathcal{B}_x$ dominates $\mathcal{B}_y$, then the CR should exclude $\mathcal{B}_y$ from the active library. Even if $\mathcal{B}_x$ does not dominate $\mathcal{B}_y$, it still may not be worth including $\mathcal{B}_y$ in the active library in some situations. For example, if $T_y(\mathbf{v})$ exceeds $T_x(\mathbf{v})$ by only a very small amount over a very limited range of values of $\mathbf{v}$, but $T_x(\mathbf{v})$ is much larger than $T_x(\mathbf{v})$ for all other values of $\mathbf{v}$, then very little is lost if code-modulation combination $\mathcal{B}_y$ is not in the active library.

The process of excluding code-modulation combinations from the available library to form the active library is illustrated by considering the throughput graphs in Figure 21.2 in which the channel parameter is CENR. The throughput curve for 16-QAM with the code of rate 0.236 is dominated by the throughput of QPSK with the code of rate 0.495 (and QPSK is easier to implement), so the code-modulation combination of 16-QAM and the turbo product code of rate 0.236 can be eliminated from the library with no reduction in throughput performance. The omission of the combination of 16-QAM and the code of rate 0.236 makes no change in the upper envelope of the throughput curves of Figure 21.2. When the code-modulation combinations are loaded into a radio, the radio is told that if the combinations (16-QAM, 0.236) and (QPSK, 0.495) are both in the available library for a session, then the former should be excluded from the active library. Of course, in the event that the destination does not support (QPSK, 0.495) but it does support (16-QAM, 0.236), then the latter should be included in the active library for the session.

For another example, we observe that although 16-QAM with the code of rate 0.325 is not dominated by other combinations, its omission changes very little in the upper

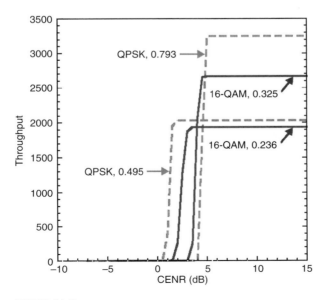

FIGURE 21.2

Selection of the library of code-modulation combinations.

envelope of the throughput curves in Figure 21.2. Hence, it can be eliminated from the library with only negligible reduction in throughput performance.

21.6 RECEIVER STATISTICS

A number of statistics can be generated in the receiver of a CR, but we focus on those that can be derived easily and require only minimal additional hardware or software beyond what is needed to demodulate and decode the received packets. Two of our statistics are derived primarily from the operation of the decoder: the error count and the iteration count. The receiver can determine each of these statistics for every decoded packet that is correct (e.g., by the use of a high-rate CRC code). The *error count* for such a packet is the number of binary symbol errors at the output of the demodulator. Many decoders can provide an error count as part of their normal decoding procedure. If the decoder does not provide the error count automatically, then the destination receiver can always encode the information bits from a correctly decoded packet and compare them with the binary representations of the demodulator hard decisions for the packet, as illustrated in Figure 21.1. Regardless of how the error count is provided, its purpose is to represent the number of binary symbol errors that would occur if there were no decoder in the receiver.

For receivers with iterative decoding, the number of iterations required to decode a received word is a measure of the quality of the channel. Normally, a packet that is received with a low SNR or SIR requires more iterations per received word than a packet that is received with high SNR and SIR. For a packet that is verified to be correct at the decoder output, the decoder reports the number of iterations that were required to decode each received word in the packet. The average number of decoder iterations per received word in a packet is the *iteration count* for the packet, which is illustrated in Figure 21.1.

The third statistic shown in Figure 21.1 is the *demodulator statistic*. Unlike the error count and the iteration count, the demodulator statistic is derived from the demodulation process rather than the decoding process. As a result, the demodulator statistic is obtained in different ways for different modulation formats, but it does not depend on the code used for the session. If the modulation is QPSK or M-QAM, then we consider the two-dimensional space in which one coordinate is for the inphase component of the received signal and the other coordinate is for the quadrature component. The distance between each received point and its closest point in the signal constellation is determined, and the *distance statistic* for a packet is the average of the distances for the modulation symbols in the packet.

The maximum-likelihood demodulator for M-BOK symbols produces $M/2$ decision statistics for each symbol in the packet. For each symbol, the receiver obtains the magnitude of each of the $M/2$ decision statistics and computes the ratio λ of the second largest magnitude to the largest magnitude. The *ratio statistic for the symbol* is $1 - \lambda$. Notice that $0 \leq \lambda \leq 1$, so $0 \leq 1 - \lambda \leq 1$. The average of the ratio statistics for all the symbols in a packet is the *ratio statistic for the packet*, denoted by γ, which also takes values between 0 and 1. If $1 - \lambda \approx 0$ for a symbol, then at least two of the decision statistics have magnitudes that are approximately the same, so the demodulated symbol is not very reliable. If $\gamma \approx 0$ for a packet, then the entire demodulated packet is not very reliable. In contrast, if $1 - \lambda \approx 1$ and $\gamma \approx 1$, then the symbol and the packet, respectively, are extremely reliable. Similar types of ratio statistics for individual symbols have been employed previously for mitigating jamming [7] and as a metric for soft-decision decoding [8]. The ratio statistic for a packet was employed as a demodulator statistic in [2].

21.7 INITIAL POWER ADJUSTMENT

The reason that initial power adjustment is required when a CR begins a session is that the propagation loss for the link to the destination is typically unknown, especially if the source and destination have not communicated recently in the frequency band selected for the session. The initial power level that the source uses for a new session may have to be selected without the benefit of recent experience in communicating with the destination in the designated frequency band. Without such experience, the propagation loss is very difficult to predict, even if the communication range is known. Variations in propagation loss of several decibels occur for both outdoor and indoor communications, even for a fixed distance between the source and destination (e.g., [9–12]). For example, the shadow losses that may be encountered in different urban environments can differ by 20 dB or more [9].

The *initial power-adjustment period* is the period at the beginning of each session in which the P-ADJ protocol compensates for the initial uncertainty in the propagation loss in the frequency band selected for the session. If the source's initial power level is much higher than is required to provide an acceptable packet error probability for the session, then it is important for the source to reduce its radiated power as quickly as possible during the initial P-ADJ period to save energy and permit frequency reuse in the network. If the initial power level is too low, then the transmitter power must be increased enough to provide an acceptable packet error probability, but not so much that the transmissions waste energy or cause unnecessary interference to other sessions that are using or wish to use the same frequency band. Elsewhere [2], we have provided additional discussion of initial P-ADJ as well as a detailed description of one initial P-ADJ protocol. In this chapter, we give an improved protocol that is more robust to variations in the channel that may occur during the initial P-ADJ period. The improved protocol also accommodates a limit on the amount of power that the source is allowed to transmit. For example, such a limit may be imposed by the spectrum etiquette that has been adopted in a DSA network.

We let P_{lim} denote the *power limit* for the session, which is defined as the maximum allowed transmitter power level that the source is permitted to use at any time during the session. For a DSA network, the power limit is determined when a CR decides to transmit in a particular frequency band. The purpose of the power limit may be to prevent the disruption of an ongoing session or to provide opportunities for new sessions to be started elsewhere in the network. If no limit has been imposed for a session, then P_{lim} is the maximum transmitted power that the radio can generate.

21.7.1 Overview of the Initial Power Adjustment

The protocol for adjustment of the initial power operates in two stages. The first stage establishes contact between the source and destination using B_1, the initial code-modulation combination defined in Section 21.4. In addition to the fact that each radio in the network fully supports the initial code-modulation combination, there are three other reasons for its use at the beginning of the initial P-ADJ period. First, it is normally the most robust combination in the library. Compared to higher-rate combinations, the initial code-modulation combination's energy requirements are lower, it produces less interference to other radios, and it usually has better interference-rejection capability. Second, typical initial code-modulation combinations use nonbinary orthogonal or biorthogonal modulation or a modulation format derived from them (e.g., IQB modulation). Such modulation techniques permit the use of the ratio statistic defined in Section 21.6, which is more accurate than the statistics used with higher-rate modulation formats. The ratio statistic has an especially large performance advantage when channel conditions are poor. Third, if the initial packet transmission does not produce an acknowledgment, then the P-ADJ protocol can be much more aggressive if it is using a low-rate combination rather than a high-rate combination. For example, we see from Table 21.1 that if the initial code-modulation combination is $\mathcal{B}_1$, then the initial P-ADJ protocol can increase the power by as much as 20 dB above the minimum requirement for $\mathcal{B}_1$ before it reaches the minimum power level needed by the higher-rate combination $\mathcal{B}_{11}$.

The goal of the first stage of initial P-ADJ is to obtain an estimate V_2 of the nominal power level P_1 for the initial combination $\mathcal{B}_1$. (This estimate is refined in the second stage.) The source then transmits a packet, referred to as the first P-ADJ packet, at power level V_2 using code-modulation combination $\mathcal{B}_1$. The first stage is finished when the source receives an acknowledgment of the first P-ADJ packet from the destination. The P-ADJ packets are just like any other packets (e.g., they carry information that the source needs to deliver to the destination), but their transmission represents the completion of a key step in the P-ADJ process.

During the second stage of the initial P-ADJ period, the source and destination determine the active library, and the source determines a *nominal power level* for each code-modulation combination in the active library. The intent is that the nominal power level P_k for code-modulation combination $\mathcal{B}_k$ should exceed the minimum required power level for the combination by approximately β_m dB, which is a margin selected by the system designer. The margin allows for fluctuations in the demodulator statistic and provides some compensation for changes in propagation loss that might occur between successive transmissions during initial P-ADJ. In their cooperative determination of nominal power levels, the source and destination do not use measurements of channel parameters or received power. Instead, the CRs apply a simple interval test to the demodulator statistic and then the source uses its list of required CENR values for the code-modulation combinations in the active library. Such a list is shown in Table 21.1. The maximum nominal power level that does not exceed the power limit is

$$P_l = \max\{P_k : 1 \le k \le K, P_k \le P_{\lim}\} \tag{21.3}$$

At the end of the P-ADJ phase of a new session, the source will use code-modulation combination B_ℓ to transmit a packet at power level P_ℓ. This packet, which is referred to as the second P-ADJ packet, has the maximum possible information rate subject to the constraint imposed by the power limit. If the power limit is sufficiently high, then $\ell = K$, and the source will use its highest-rate code-modulation combination for the packet that is sent at the end of the initial P-ADJ period. The power adjustment that is made for the transmission of the second PA packet is referred to as the *final power adjustment* for the initial power-adjustment protocol. If the second PA packet is decoded correctly at the destination, then the final power adjustment is the last function that is performed by the initial power-adjustment protocol and any changes that might occur in the channel after the final power adjustment is made are the responsibility of the adaptive transmission protocol.

21.7.2 Description of the Initial Power-Adjustment Protocol

In our description of the initial P-ADJ protocol, and in the subsequent discussion of performance results, all power increments and propagation losses are in decibels and all power levels are in decibel-watts (dBW). To simplify the presentation, we omit the phrase "subject to the power limit" for each increase in power. For example, the statement that the next power level is equal to the current power level plus the increment should be interpreted to mean that the next power level is equal to the power limit or the current power level plus the increment, whichever is smaller. The description includes parameters β_m, β_1, β_2, and β, with values that are determined by the system designer. The power margin β_m was mentioned previously, β_1 and β_2 are step sizes for increasing and decreasing power during the first stage of the initial P-ADJ period, and β is a step size that is used to respond to the results of the interval tests that are applied to the ratio statistic. For our numerical results, the parameter values are $\beta_m = 1$ dB, $\beta_1 = 10$ dB, $\beta_2 = 5$ dB, and $\beta = 0.5$ dB.

The system designer or the cognitive radio determines a set of nonoverlapping intervals that are defined by their endpoints μ_i, $1 \le i \le N$. The sequence $\mu_1, \mu_2, \ldots, \mu_N$ is an increasing sequence of numbers between 0 and 1, so the $N + 1$ intervals are $[0, \mu_1), [\mu_1, \mu_2), \ldots, [\mu_{N-1}, \mu_N)$, and $[\mu_N, 1]$, which cover the unit interval $[0, 1]$. The system designer can select the endpoints empirically (e.g., from simulations) by determining the relationship between the values of the ratio statistic γ and values of CENR for the code-modulation combinations. Alternatively, the CR can learn the values of the endpoints, as we discuss elsewhere [2]. An empirically determined set of endpoints is given in Table 21.3 for the code-modulation combinations of Table 21.1. We chose μ_1

Table 21.3 Endpoints for the Initial P-ADJ Protocol Interval Test			
μ_1 0.222	μ_{12} 0.504	μ_{23} 0.744	μ_{34} 0.865
μ_2 0.237	μ_{13} 0.533	μ_{24} 0.759	μ_{35} 0.873
μ_3 0.255	μ_{14} 0.560	μ_{25} 0.773	μ_{36} 0.880
μ_4 0.276	μ_{15} 0.587	μ_{26} 0.786	μ_{37} 0.887
μ_5 0.299	μ_{16} 0.611	μ_{27} 0.798	μ_{38} 0.893
μ_6 0.324	μ_{17} 0.634	μ_{28} 0.809	μ_{39} 0.899
μ_7 0.352	μ_{18} 0.655	μ_{29} 0.820	μ_{40} 0.905
μ_8 0.381	μ_{19} 0.676	μ_{30} 0.830	μ_{41} 0.910
μ_9 0.412	μ_{20} 0.694	μ_{31} 0.840	μ_{42} 0.915
μ_{10} 0.443	μ_{21} 0.712	μ_{32} 0.849	μ_{43} 0.920
μ_{11} 0.473	μ_{22} 0.729	μ_{33} 0.857	

so that if $\gamma = \mu_1$ then CENR $\approx$ CENR$_1$, the requirement for combination $\mathcal{B}_1$. Similarly, if $\gamma = \mu_N$, then CENR $\approx$ CENR$_K$, the requirement for combination $\mathcal{B}_K$. The remaining $N-2$ endpoints provide a convenient partition of the unit interval for use in the interval test.

The first stage of the initial P-ADJ begins with the transmission by the source of a packet at power level V_0 using code modulation $\mathcal{B}_1$. The value of V_0 is the cognitive radio's best estimate of the nominal power level for $\mathcal{B}_1$, but this estimate may be incorrect for the reasons described at the beginning of this section. If the packet is not acknowledged by the destination, then the next packet is transmitted at a power level equal to the previous power level plus the increment β_1. The process continues until the source receives an acknowledgment for the last packet it transmitted. The acknowledgment includes a value γ_0 for the ratio statistic from the destination's demodulator, and the source applies the first-stage interval test to γ_0.

If $\mu_i \leq \gamma_0 < \mu_{i+1}$ for some value of i in the range $1 \leq i \leq N-1$, then the first-stage interval test is complete. If $\gamma_0 < \mu_1$, the power is increased by β_1, another packet is transmitted, and the interval test is applied to the ratio statistic in the acknowledgment. The process is repeated until the ratio statistic for the most recent packet satisfies $\mu_i \leq \gamma < \mu_{i+1}$ for some value of i, which completes the first-stage interval test. If $\gamma_0 \geq \mu_N$, the power is decreased by β_2, another packet is transmitted, and the interval test is applied to the ratio statistic in the acknowledgment. The process is repeated until the ratio statistic for the most recent packet satisfies $\mu_i \leq \gamma < \mu_{i+1}$ for some value of i, which completes the first-stage interval test. In each case, if the transmitter power level is V_1 for the most recent packet, the power level is changed to $V_2 = V_1 + \beta_m - (i-1)\beta$, and the first P-ADJ packet is transmitted. For most values of i, the increment $\beta_m - (i-1)\beta$ is negative, so the power is decreased before the first P-ADJ packet is transmitted.

If an acknowledgment is received for the first P-ADJ packet, then the first stage of the initial P-ADJ is complete. In the very unlikely event that an acknowledgment is not received for the first P-ADJ packet, the power is increased by β_1 and the packet is retransmitted, and, if necessary, the process is repeated until an acknowledgment is received and the first stage of the initial P-ADJ is complete.

The second-stage of the initial P-ADJ begins with the application of the second-stage interval test to the ratio statistic γ_1 included in the acknowledgment of the first P-ADJ packet. Because the margin β_m is used in setting V_2, it is highly unlikely that the ratio statistic γ_1 from the first P-ADJ packet is less than μ_1; however, if it is, the power is increased by β_1 and the first stage is repeated. It is also very unlikely that $\gamma_1 \geq \mu_N$, but, if so, the power is decreased by β_2 and the first stage is repeated. The highly probable outcome is that the ratio statistic γ_1 satisfies $\mu_j \leq \gamma < \mu_{j+1}$ for some value of j, in which case the source determines the nominal power levels for each code-modulation combination $\mathcal{B}_k$, $1 \leq k \leq K$, from

$$P_k = V_2 - (j-1)\beta + CENR_k - CENR_1 + \beta_m \qquad (21.4)$$

The source then determines the maximum nominal power level from Eq. (21.3), which specifies the value of ℓ, and it transmits the second P-ADJ packet using code-modulation combination B_ℓ and power level P_ℓ. Once the second P-ADJ packet is transmitted, the initial P-ADJ period ends and the *adaptation period* begins. At this time, all future selections of code-modulation combinations and power levels are made by the Adaptive Transmission Protocol (ATP).

21.7.3 **Performance Evaluation of the Initial P-ADJ Protocol**

To test the initial P-ADJ protocol, we developed a simulation that models the initial propagation loss as a random variable. In the simulation, we normalize the propagation loss and power level as follows: When the source transmits at its *reference power level* $P_{t,0}$ and the channel's propagation loss is equal to its *nominal propagation loss* $L_{p,0}$, then CENR = 0 dB. The parameters $P_{t,0}$ and $L_{p,0}$ are used to facilitate the discussion of our performance results; they are not parameters that the radio should or even could determine. The reference power level and the nominal propagation loss are not unique, but their difference, $P_{t,0} - L_{p,0}$, is unique. (Recall that all power levels are in dBW and all losses are in dB.) From Table 21.1, we see that code-modulation combinations $B_1 - B_6$ can be used to transmit a packet of 4096 binary code symbols at a packet error of 10^{-2} or less at the reference power level over a channel with the nominal propagation loss, but code-modulation combinations $B_7 - B_{11}$ require a higher power level or lower propagation loss. If we desire to use B_{11} on a channel that has the nominal propagation loss, then the source must use a power level that is it at least 11.2 dB above the reference power level. If the source's maximum power level is 20 dB above the reference power level, then it cannot use B_{11} if the excess propagation loss is more than 8.8 dB.

If the source's transmitter power level is $P_{t,i}$ and the propagation loss is $L_{p,i}$ at the time that the ith packet is transmitted, then the value of CENR for the ith packet is

$$CENR(i) = (P_{t,i} - L_{p,i}) - (P_{t,0} - L_{p,0}). \qquad (21.5)$$

If $L_{p,i}$, $P_{t,0}$, and $L_{p,0}$ were known to the source radio, then it could simply choose $P_{t,i}$ to give the required value of CENR for the combination from Table 21.1 that it wants to use for the ith packet, provided, of course, that $P_{t,I} \leq P_{\lim}$. The source does not know $L_{p,i}$, $P_{t,0}$, or $L_{p,0}$, however, so it must rely on the receiver statistics supplied by the destination in the acknowledgments to decide what power level should be used. From Eq. (21.5), we see that the dependence of CENR(i) on the propagation loss is through the difference $L_{p,i} - L_{p,0}$ only; thus, the relative values of the propagation losses are important, not the individual values. For this reason, we define the *excess propagation loss* to be the amount by which the propagation loss exceeds the nominal propagation loss.

In our model, the excess propagation loss, $L_{p,i} - L_{p,0}$, for the ith packet in the session is the sum of two random components, a fixed component, L_R, and a state-dependent component $L_S(i)$ that can vary from packet to packet. Thus, the excess propagation loss for the ith packet is

$$L_{p,i} - L_{p,0} = L_R + L_S(i) \qquad (21.6)$$

A negative value for the excess propagation loss means that the actual propagation loss is less than the nominal propagation loss. It follows from Eqs. (21.5) and (21.6) that if the power level for the first packet is the reference power level, then the value of CENR for the first packet is

$$CENR(1) = -[L_{p,1} - L_{p,0}] = -[L_R + L_S(1)] \qquad (21.7)$$

In our simulations, the fixed propagation loss, L_R, is a random variable that is uniformly distributed over the interval $[-20, 20]$ dB, and it is the primary component of the initial uncertainty in the propagation loss for a source–destination pair that is

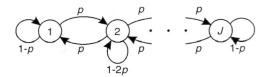

FIGURE 21.3

J-state Markov chain for modeling propagation losses.

starting a new session. We use a Markov model of the type shown in Figure 21.3 for the state-dependent component, which represents variations in propagation loss that might occur from one packet to the next during the P-ADJ period.

In this section, and in Sections 21.9.1 and 21.9.2, we employ finite-state Markov chains to provide the dynamical behavior of the channel for our analyses and simulations. Several general references on Markov models for channels with slow fading exist [13–16]. We have previously employed finite-state Markov models for other types of dynamic channels also, including those with time-varying partial band interference, time-varying multipath interference, and Rician fading with a time-varying fading parameter [17–20].

The state-dependent component of the excess propagation loss is governed by the *J*-state Markov chain shown in Figure 21.3. The state of the Markov chain and the propagation loss do not change while a packet is being transmitted, but with probability p the state changes between the completion of the reception of a packet and the start of the transmission of the next packet. If the channel is in state j at the time the *i*th packet is transmitted, then the state-dependent propagation loss experienced by the packet is $L_S(i) = L_j$, and the excess propagation loss is $L_R + L_j$. The initial state of the Markov chain in our simulations is selected at random with a uniform distribution over the J states, which is the steady-state distribution for the Markov chain of Figure 21.3. It follows that

$$P[L_S(i) = L_j] = P[L_S(1) = L_j] = 1/J \tag{21.8}$$

for each i and j.

The *propagation-loss increment* between state $j + 1$ and state j is $\Delta_j = L_{j+1} - L_j$. If the increment is the same for each pair of neighboring states, we let Δ denote the common value of Δ_j. For the numerical results on the performance of the initial P-ADJ protocol in this section, $J = 7$, $p = 0.1$, $\Delta = 2$ dB, and $L_j = 2(j - 4)$ for $1 \leq j \leq 7$. When the channel state is 4, the state-dependent propagation loss is 0 dB and the excess propagation loss is L_R, which is uniformly distributed from −20 dB to 20 dB. The state-dependent propagation loss ranges between $L_1 = -6$ dB and $L_7 = 6$ dB, so the excess propagation loss is between −26 dB and 26 dB, which gives an uncertainty range of 52 dB in the initial propagation loss. If the power level for the first packet is the reference power level, then it follows from Eq. (21.7) that CENR(1) is a random variable that takes values between −26 dB and 26 dB.

Let f denote the probability density function for $L_R + L_S(1)$, which is the initial excess propagation loss. Note that $f(n) = 0$ for $|u| > 26$. It is easy to show (see Problem 21.5 in Section 21.11) that f is the mixture of seven uniform density functions, each of which spans 40 dB. Let f_j denote the jth uniform density function, which is the conditional density of $L_R + L_S(1)$, given that j is the state of the Markov chain for the first packet. In other words, f_j is the density function for the random variable $L_R + L_j$. Because $L_j = 2(j - 4)$, the value of $f_j(u)$ is nonzero only if u is in the range from $-20 + 2(j - 4)$ to $20 + 2(j - 4)$. Notice that $f_4(u) = 1/40$ for $-20 \le u \le 20$ and $f_4(u) = 0$ for $|u| > 20$. More generally, for $1 \le j \le 7$, it is easy to show that if $f_j(u) = f_{8-j}(-u)$ for each u, so the densities satisfy $f_j(u) + f_{8-j}(u) = f_j(-u) + f_{8-j}(-u)$. Thus, $f_j(u) + f_{8-j}(u)$ is symmetrical about 0 (i.e., an even function) for each j, from which it follows that the density function for $L_R + L_S(1)$ is also symmetrical about 0. As a result, the density function for $L_R + L_S(1)$ is the same as the density function for $-[L_R + L_S(1)]$. In view of Eq. (21.7), this means that the density of CENR(1) is the same as the density function for $L_R + L_S(1)$, which is the function f.

21.7.4 Performance Results for Systems with Unlimited Power

Of course, any practical radio has an upper limit on its transmitter power. In some applications (e.g., transmission over a very short distance), the upper limit may be sufficiently high that the radio can transmit reliably at power levels well below its power limit for any excess propagation loss that will be encountered in the application, no matter what code-modulation combination is chosen. In such a situation, it is convenient to think of the radio as having unlimited power.

To be precise, we should always interpret "a radio with unlimited power" to mean a radio with a power limit that exceeds any power level that it will need to use in its intended application. For example, if the radio's power limit is 40 dB above the reference power level and the excess propagation loss will never exceed 26 dB in the intended application, then the radio's maximum power level always gives CENR $\ge$ 14 dB, which is more than enough for each code-modulation combination in Table 21.1.

We simulated the initial P-ADJ protocol for each of 100,000 sessions in a system with unlimited power for the random, time-varying excess propagation loss $LR + LS(i)$. The initial power level for each session is the reference power level, so Eq. (21.7) gives the value of CENR for the first packet in terms of the excess propagation loss for the first packet. The sessions were started with independent, random excess propagation losses. The propagation loss for the first packet in a simulated session was drawn at random according to the density function f, which is equivalent to drawing at random the value of L_R for the session and independently drawing at random the initial state of the Markov chain according to a uniform distribution on the set of states. As a result, the sessions started with independent, random values of CENR(1), each drawn at random according to the density function f.

The results of the simulation of the independently drawn random values for CENR(1) for 100,000 sessions are illustrated in Figure 21.4. Notice that part of the range of values for CENR(1) falls below the smallest value of CENR listed in Table 21.1. For such values, it is very unlikely that the first packet is received successfully by the destination, even though the first packet is always sent with combination B_1, the most robust code-modulation combination in the library. In this situation, it is necessary for the initial P-ADJ protocol to increase the power level in order to have reliable communications. Notice also that for a large range of values of CENR(1), the transmitted power level

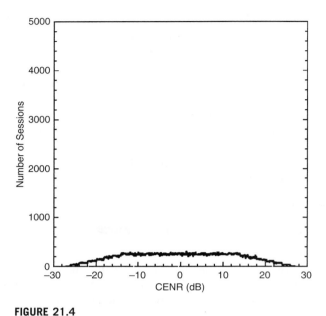

FIGURE 21.4

Simulated initial values of CENR for 100,000 sessions.

is several decibels higher than necessary, even for B_{11}, the code-modulation combination with the greatest power requirement. In this situation, it is important to decrease the power level to save energy and reduce interference to other radios.

The simulation results in Figure 21.5 show how much the range of values of CENR is reduced by the time that the power level is selected for the second P-ADJ packet. At that time, the initial P-ADJ protocol ceases operation and the control of transmission parameters is turned over to the ATP. The basis that we used for measuring the performance of the initial power-adjustment protocol is the power level that the protocol selects for the second PA packet (i.e., the final power level) rather than the signal-to-noise ratio at the destination when the second PA packet is demodulated and decoded. The difference is due to the occurrence of a transition time for the Markov channel between the time that destination obtains the receiver statistie that the source will use to select the final power level and the time that the destination receives the second PA packet. The initial power-adjustment protocol cannot respond to any changes in the propagation loss that occur after it makes its final power adjustment, so it should not be penalized for such changes. Therefore, we define the initial power-adjustment protocol's *final value* of CENR to be the value of CENR that the destination would see if there are no changes in the channel after the final power adjustment is made. It is the final value of CENR that is shown in Figure 21.5.

Recall that we want the initial P-ADJ protocol to send the second P-ADJ packet with the code-modulation combination of the highest rate, which is B_{11} in the library of Table 21.1. The target power level for sending the second P-ADJ packet using combination

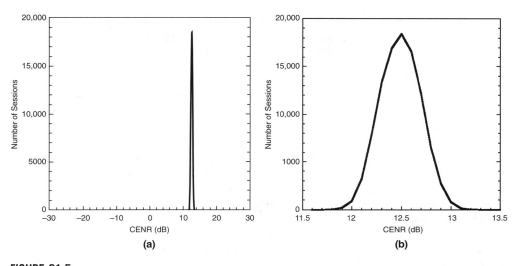

FIGURE 21.5

Two views of final values of CENR for a system with unlimited power.

β_{11} is the power level that gives CENR = $\text{CENR}_{11} + \beta_m$ = 12.2 dB, which is obtained from Table 21.1 and the specified margin β_m = 1 dB. As shown in Figure 21.5, the final values of CENR for the 100,000 sessions are all very close to the desired value.

In Figure 21.5(b), the abscissa of Figure 21.5(a) is truncated to the range 11.5 dB to 13.5 dB and the scale for the abscissa is expanded to provide a more detailed illustration of the simulation results. All 100,000 sessions had final values of CENR in the range from 11.8 to 13.3 dB. Nearly 90 percent of the sessions had final values within 0.5 dB of CENR = 12.2 dB, and more than 99 percent of the sessions had final values within 0.7 dB of CENR = 12.2 dB. Thus, the initial P-ADJ protocol is very successful in increasing power if necessary and decreasing power if conditions permit, yet it uses no power measurements or channel estimates and it is able to cope with unknown and time-varying propagation losses.

21.7.5 **Performance Results for Systems with Limited Power**

If a power limit is imposed on the source (e.g., by hardware constraints or by a DSA protocol) and the power limit is not large compared to the largest propagation losses expected in the application, then at times the excess propagation loss may be high enough so the source cannot use the higher-rate code-modulation combinations in Table 21.1. Therefore, for power limits below a certain level relative to the initial distribution of the excess propagation loss, we cannot expect to have a final distribution of CENR that is concentrated in the narrow range achieved when there is no power limit. Instead, some sessions will emerge from the initial P-ADJ period with a value of CENR that is below the 11.2 dB requirement for code-modulation combination β_{11} to achieve a packet error probability of 10^{-2}.

Recall that if the source transmits at its *reference power level* and the propagation loss on the channel is equal to its *nominal propagation loss*, then CENR = 0 dB. Recall also that the *excess propagation loss* is the amount in decibels by which the actual propagation loss exceeds the nominal propagation loss. For our test of the initial P-ADJ protocol in a radio with a power limit, we used a power limit P_{lim} that is 20 dB above the reference power level. For this limit, a source that transmits at its maximum power level, P_{lim}, over a channel with an excess propagation loss that is 0 dB, achieves CENR = 20 dB at the destination receiver. As before, the range of the random excess propagation loss is from −26 to 26 dB, so the range of values of CENR is from −6 to 46 dB if the source's power level is P_{lim}. Of course, we don't want the source to transmit at this power level if the excess propagation loss is −26 dB, because CENR = 46 dB is much higher than required for any of the code-modulation combinations in Table 21.1. (Recall that the excess propagation loss is negative if the actual propagation loss is less than the nominal propagation loss.) At the other extreme, the largest value of CENR that can be achieved if the excess propagation loss and transmitter power level are each at their maximum values is −6 dB, which is adequate for only code-modulation combinations B_1, B_2, and B_3. If a margin of more than 0.1 dB is desired, then only B_1 and B_2 can be used for transmission at the maximum power with the largest excess propagation loss. Even for an excess propagation loss of 20 dB (instead of maximum of 26 dB), the only combinations that can be used with a margin of 1 dB or more are $B_1 - B_5$; in particular, 16-QAM cannot be used with any code rate. For the source to transmit at its maximum

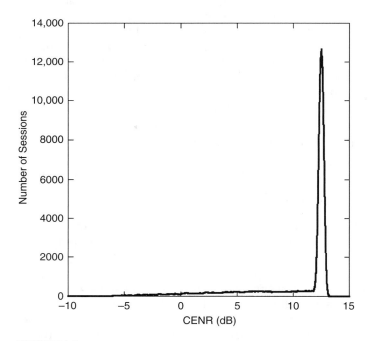

FIGURE 21.6

Two views of distribution of CENR for a power limit 20 dB above the reference level.

power limit, use 16-QAM, and achieve a packet error probability of 10^{-2} or less, the excess propagation loss cannot exceed 12.8 dB. If it is desired to use 16-QAM with the code of highest rate, then the excess propagation loss cannot exceed 8.8 dB.

For the simulation results in Figure 21.6, the power limit is 20 dB above the reference power level. Even with this limit, we see the range of values of CENR is greatly reduced from the initial distribution of Figure 21.4. The primary difference between the final distributions of Figures 21.5 and 21.6 is the spread of the final distribution below about 11.8 dB in Figure 21.6. This spread is caused by the power limit, and the tail of the distribution below 11.2 dB corresponds to final values of CENR that are too low to support the highest-rate combination.

For sessions with final values of CENR in the tail of the distribution, the ATP will be working initially with a second P-ADJ packet that uses one of the lower-rate code-modulation combinations.

21.7.6 Time Required for Initial Power Adjustment

The results in Figure 21.7 illustrate how quickly the protocol converges for a radio with unlimited power. For all 100,000 sessions, the initial P-ADJ protocol required no more than ten packet transmissions, and the last of these is the second P-ADJ packet, which is transmitted after the protocol has made its last adjustment.

For 98 percent of the sessions, six or fewer packet transmissions were needed, and nearly 80 percent of the sessions needed no more than four packet transmissions. From the description of our P-ADJ protocol, it is clear that the second P-ADJ packet can be no earlier than the third packet that is transmitted, and our results show that in fact it is the third packet in almost half the sessions. Therefore, in nearly half the sessions, only two packet transmissions were required before the protocol made its final power adjustment prior to sending the second P-ADJ packet.

The convergence results are approximately the same for systems that have a power limit.

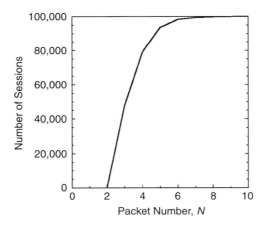

FIGURE 21.7

Number of sessions for which the P-ADJ period ends no later than the Nth packet.

21.8 ADAPTIVE TRANSMISSION

The ATP takes over after the initial power adjustment is complete. The goal of the ATP is to provide the maximum possible throughput for each set of channel conditions that arise during the session. It attempts to accomplish this goal without increasing the radiated power, because to do so would produce more interference elsewhere in the network. The consequences of an increase in power are illustrated in Figure 21.8. Prior to the power increase, several unintended receivers are beyond the interference range of the source. After the source increases its radiated power, perhaps because the destination moved farther away or the shadow loss increased, all the unintended receivers are within the source's interference range.

The primary adaptation mechanism for the ATP is to change the code-modulation combination as needed to compensate for changes in the channel, such as an increase in propagation loss. In our protocol, the adaptation that is performed to compensate for perturbations in the channel conditions does not increase the interference range of the source's transmission. Of course, the protocol has no direct information about the channel conditions, so it must rely on the receiver statistics that are described in Section 21.6.

For each packet that decodes correctly, the protocol uses one or more receiver statistics to decide which code-modulation combination to use for the next packet. For example, if a packet that does not decode correctly used combination B_k for $k > 1$, then the protocol designates B_{k-1}, which has a lower rate by one step, as the code-modulation combination for the next packet. If $k = 1$, then the protocol uses B_k again

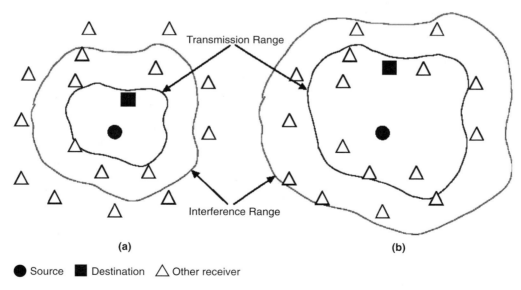

● Source   ■ Destination   △ Other receiver

FIGURE 21.8

The effects of an increase in radiated power: (a) before power increase and (b) after power increase.

for the next packet, but it may eventually have to increase the radiated power if doing so is permitted by the network's etiquette protocol.

The ATP may use one of the three types of receiver statistics described in Section 21.6 or it may use a combination of these statistics. For our illustrative performance results, only one receiver statistic is used, and it is either the iteration count or the error count. The receiver statistic is supplied by the destination in the acknowledgment packet. An interval test is applied to the statistic to decide what code-modulation combination to use for the next packet that the source sends to the destination. The intervals used for the interval test are $I_{-1} = (-\infty, \gamma_1)$, $I_0 = [\gamma_1, \gamma_2]$, and $I_1 = (\gamma_2, \infty)$. The intervals are defined by the endpoints γ_1 and γ_2, and our specifications of the intervals are given in terms of the endpoints.

The choice of which code-modulation combination to use for a packet is based on the combination that was used for the previous packet and the value of the receiver statistic obtained in the demodulation and decoding of the previous packet. If code-modulation combination B_k was used for the ith packet and the resulting statistic is z_i, then the code-modulation combination for the next packet is B_k if $z_i \in I_0$; B_{k+1} if $z_i \in I_{-1}$ and $k < K$; B_K if $z_i \in I_{-1}$ and $k = K$; B_{k-1} if $z_i \in I_1$ and $k > 1$, and B_1 if $z_i \in I_1$ and $k = 1$. The interval test, $z_i \in I_{-1}$, $z_i \in I_0$, or $z_i \in I_1$ is equivalent to two threshold tests, $z_i < \gamma_1$ and $z_i > \gamma_2$, because if neither of the inequalities is true, then we know $z_i \in I_0$.

In the description of the interval test, notice that if B_k was used for the previous packet, then we cannot use B_{k+1} for the next packet if $k = K$, and we cannot use B_{k-1} for the next packet if $k = 1$. If, for example, the source were to use B_K for a few consecutive packets and the receiver statistic is in I_{-1} each time, then it would be wise to reduce the transmitter power. Similarly, if the source were to use B_1 for a few consecutive packets, and for each packet the receiver statistic is in I_1 or there is a decoding failure, then the transmitter power should be increased if possible (i.e., if it is not at its maximum value). Such decreases or increases in the power level can be accomplished by employing a P-ADJ protocol similar to the one we described elsewhere [2].

Power increases are made only if the most robust code-modulation combination, which is B_1, cannot provide adequate performance and the current power level is below the maximum. If the ith packet is transmitted at power level $P_{t,i}$ using code-modulation combination B_1 and the source decides to increase the power level for packet $i + 1$, then it continues to use B_1 until the power can be reduced to $P_{t,i}$. Only then is it is permitted to change the code-modulation combination. Similarly, if the source transmits the ith packet at power level $P_{t,i}$ using code-modulation combination B_K and it finds that $z_i \in I_{-1}$, then it can decrease the power level. It continues using B_K until it has to increase the power level all the way to $P_{t,i}$, at which time it is permitted to change the code-modulation combination.

The endpoints that we use for our performance results in Section 21.9 are listed in Table 21.4. The endpoints for the error count are denoted by $\gamma_1(EC)$ and $\gamma_2(EC)$, and those for the iteration count are denoted by $\gamma_1(IC)$ and $\gamma_2(IC)$. The values in Table 21.4 are not optimized for a particular channel; instead, they were selected to give good performance over a wide range of channels. It is not necessary to determine the values for the endpoints in advance, however, because as the CR receives packets, it can learn the values for the endpoints or modify previously selected values. We give one possible algorithm for learning or modifying the endpoints elsewhere [2].

Table 21.4 Endpoints for the Interval Tests

Combination	γ_1(EC)	γ_2(EC)	γ_1(IC)	γ_2(IC)
$\mathcal{B}_1$	553	4096	2	32
$\mathcal{B}_2$	189	553	2	10
$\mathcal{B}_3$	21	189	2	10
$\mathcal{B}_4$	265	492	3	10
$\mathcal{B}_5$	763	901	3	10
$\mathcal{B}_6$	508	763	2	10
$\mathcal{B}_7$	172	508	2	18
$\mathcal{B}_8$	61	172	2	8
$\mathcal{B}_9$	300	512	3	8
$\mathcal{B}_{10}$	173	300	2	8
$\mathcal{B}_{11}$	0	173	0	6

21.9 PROTOCOL THROUGHPUT PERFORMANCE FOR DYNAMIC CHANNELS

Ideally, the ATP should select the code-modulation combination for a packet in a way that maximizes the throughput for the state that the channel will be in when the packet is transmitted. Because the channel state is not known to the protocol, it must rely on receiver statistics that are reported by the destination. We evaluate the throughput of our ATP, using the receiver statistic as the only source of information about the channel obtained by the destination node from the previous packet. We compare this throughput with that of two hypothetical protocols, each of which is given perfect channel-state information, and then make the ideal selection based on this information.

One protocol is told the state of the channel for the last packet that the source sent to the destination; the other is told the state that the channel will be in when the source sends its next packet to the destination. Upon receipt of this state information, each protocol makes the calculations necessary to determine which code-modulation combination maximizes the conditional expected throughput given the channel state. In addition to having perfect channel-state information, each hypothetical protocol knows the channel model and the value of each of the parameters of the model. For example, for the Markov channel model of Figure 21.3, the hypothetical protocols know the number of states, the channel parameters for each state (e.g., the propagation loss), and the state transition probabilities for the Markov chain. The hypothetical protocol that is told the previous channel state is referred to as the *perfect previous-state information* (PPSI) protocol and the hypothetical protocol that is told the state that the channel will be in for the next packet is called the *perfect next-state information* (PNSI) protocol.

The throughput obtained from the PPSI protocol is the best that can be obtained from any protocol that relies on feedback information from the previous packet to choose the code-modulation combination for the next packet. The PNSI protocol's throughput is the maximum possible throughput for any CR, including one that uses channel-state prediction techniques. The difference in the performance of the two hypothetical protocols is a measure of the potential performance improvement that can be obtained from channel-state prediction. In most situations, this measure over-estimates the potential gain because it is based on perfect prediction by the cognitive radio. The primary purpose of the comparison of our ATP with the two ideal protocols is to see whether the receiver statistics provide enough channel-state information for good performance.

Additional channel-state information can be obtained by the cognitive radio, but it may require the use of channel estimation, which adds complexity and may require the insertion of pilot symbols in each packet.

21.9.1 Time-Varying Propagation Loss

If the Markov chain in Figure 21.3 is used to model time-varying propagation loss, then each state in the chain corresponds to a different loss. Such a model can be used to model slow-fading or time-varying shadow loss in a packet communication system. For our purposes, the important feature of a Markov model is that it gives random changes in channel parameters from packet to packet, and we can test protocols to see if they respond appropriately to such changes. It is not very important that the Markov chain be a precise model for a specific channel because there are so many types of channels that a radio is likely to experience that it is unwise to focus on a particular type. As in Section 21.7.3, the excess propagation losses (in dB) are multiples of the propagation-loss increment Δ. For the results on the performance of ATPs, the excess propagation loss associated with state j is $L_j = (j - 1)\Delta$, so the excess propagation losses in decibels range from 0 for state 1 to $(J - 1)\Delta$ for state J.

The transmitter power is fixed for the entire session, so any change in the value of CENR is due to a change in the excess propagation loss as a result of a change in the state of the channel. We let $CENR_0$ be the value of CENR when there is no excess propagation loss. The value of CENR when the channel is in state j is $CENR = CENR_0 - L_j$, which ranges from $CENR_0$ for state 1 to $CENR_0 - (J - 1)\Delta$ for state J. For example, if $CENR_0 = 3$ dB and $\Delta = 2$ dB, then the value of CENR for a packet that is transmitted when the channel is in state 4 is $CENR = -3$ dB. From Table 21.1, we see that only the four code-modulation combinations that use BOK can provide a packet error probability of 10^{-2} or less for this value of CENR. For each set of performance results in this chapter, the state transition probability is $p = 0.1$.

We also obtained results for values of the transition probability up to $p = 0.2$ and found only minor differences in the results. For our performance results, the ATP does not know $CENR_0$, J, or p; however, the hypothetical protocols know each of these. The PPSI protocol knows the state that the channel was in for the previous transmission, and the PNSI protocol knows the state for the next transmission (the one for which the code modulation is being selected). The ATP knows neither of the states. The only information available to the ATP is the information it obtains from the receiver statistics derived by the destination from demodulation and decoding of the packet it received in the previous transmission.

We define a *transition time* for a Markov model to be a time at which a transition occurs. A transition may be a transition back to the same state, which occurs with probability $1 - p$ for states 1 and J and probability $1 - 2p$ for the other states. Therefore, there may not be a state change at each transition time. For all the results in this chapter, there are no transition times during the period in which an individual packet is being transmitted, so the channel state is fixed for the duration of each packet. For most of our performance results, there is a single transition time between each pair of consecutive packet transmissions; however, we also give performance results for Markov channels that have multiple transition times between consecutive packet transmissions. In all cases, the channel may change states at least once between any two consecutive packets, so the feedback information may be for a channel state that is different from the one that will be seen by the packet for which the code-modulation combination is being selected.

The performance of the hypothetical protocols can be determined analytically, once the error probability of each code-modulation combination is evaluated for each channel state. We explain the analytical method elsewhere [2]. The performance results for the ATP that relies only on receiver statistics from the previous packet are obtained by simulation. Throughput results are given in Figure 21.9 for a 12-state Markov channel model with $\Delta = 1$ dB, and in Figure 21.10 for a 6-state Markov channel model with $\Delta = 2$ dB. For the channel with $\Delta = 1$ dB, the curves are almost indistinguishable. Even

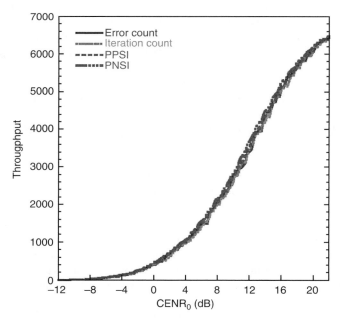

FIGURE 21.9

Throughput of the ATP and hypothetical protocols ($J = 12$, $\Delta = 1$ dB).

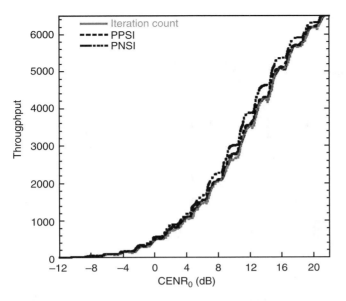

FIGURE 21.10

Throughput of the ATP and hypothetical protocols ($J = 6$, $\Delta = 2$ dB).

for $\Delta = 2$ dB, the performance of the ATP is nearly as good as that of the hypothetical PPSI protocol, which is the best that is possible for any protocol that uses only feedback from the previous packet. Consequently, there is very little reason to try to obtain additional channel-state information. By using only the error count, or the iteration count, the ATP achieves throughput levels that are nearly as good as those achieved by the hypothetical protocols that are given perfect channel-state information.

For the performance results in Figures 21.9 and 21.10, there is a single transition time between packets. In Figure 21.11, throughput curves are given for Markov models with n transition times between packets. As n increases, the feedback information is more likely to be obsolete. We see from Figure 21.11, however, that there is only a graceful degradation in throughput as n increases.

21.9.2 Time-Varying Interference

Especially for DSA networks, it is important to determine how the ATP deals with interference. Ideally, the network can employ spatially distributed reuse of frequency bands; that is, two or more source–destination pairs can use the same frequency band if each destination is sufficiently distant from the sources with which it is not paired. The transmissions from such sources cause interference in the destination radios for which those transmissions are not intended. The situation is depicted in Figure 21.12 for six source–destination pairs, two of which are operating in the same frequency band. Suppose the sources that are shaded are transmitting in the same frequency band,

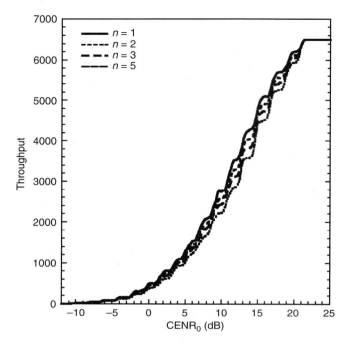

FIGURE 21.11

Throughput of the ATP for multiple transition times between packets ($J = 6$, $\Delta = 2$ dB).

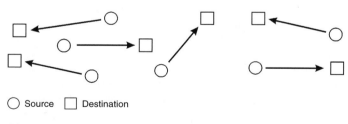

○ Source □ Destination

FIGURE 21.12

Source–destination pairs in a DSA network.

but most of the time the energy transmitted by a source does not interfere with the other source's destination, which we refer to as the unintended receiver. Occasionally, however, the power level, or the propagation loss, may change enough that interference is caused to the unintended receiver. For example, if a directional antenna is used by a source, the orientation of the source and its intended destination may change in a way that results in the antenna being pointed in the direction of the unintended receiver. Alternatively, there may be an obstacle that causes shadowing only part of the time (e.g., the obstacle is not always in the propagation path between the source and

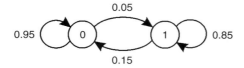

FIGURE 21.13

Two-state Markov chain for modeling interference.

the unintended receiver). A third possibility is that the source is always within range of the unintended receiver, but the source transmits only occasionally at the same time that the unintended receiver is attempting to demodulate a packet from its source. In such situations, there is intermittent interference in the unintended receiver as it attempts to receive packets from its source.

For a test of our ATP's ability to handle intermittent interference on a channel with time-varying propagation loss, we use two statistically independent Markov chains. The time-varying propagation loss is modeled by the Markov chain of Figure 21.3 with six states, propagation-loss increment 2 dB, and transition probability 0.1 (i.e., $J = 6$, $\Delta = 2$ dB, $p = 0.1$). The intermittent interference is modeled by the Markov chain of Figure 21.13, for which state 0 corresponds to a channel with no interference and state 1 corresponds to a channel with interference. The signal that is transmitted by the source is referred to as the *desired signal*; it is the signal that the destination wishes to demodulate and decode. The interference is modeled as a constant-envelope signal that has the same chip rate and carrier frequency as the desired signal. The intent is for the interference to model the type of signal that might be sent in the same frequency band by another radio in the network, as depicted in Figure 21.12. We assume that the interference signal arrives at the destination receiver in chip-and-phase synchronism with the desired signal, which typically represents the worst-case interference model. There is one transition time between each pair of consecutive packets for the Markov chain that models the propagation loss, and there is one transition time after every 500 packets for the Markov chain that models intermittent interference. We assume that the interference signal experiences the same propagation loss as the desired signal and is transmitted at the same power level.

As a result, the received power from the desired signal at the destination receiver is always the same as the received power from the interference signal, whenever the interference signal is being transmitted. Thus, the SIR is unity whenever the Markov chain for interference is in state 1.

The ability of the ATP to cope with intermittent interference is illustrated in Figure 21.14. As expected, the interference degrades the throughput for the entire range of values of $CENR_0$, especially when $CENR_0$ is large. For large values of $CENR_0$, the interference dominates over the thermal noise as a source of disturbance for the destination's demodulator and decoder. The additional disturbance caused by the interference forces the ATP to use a code-modulation combination of lower rate than it could use if there were no interference. The good news is that the performance degradation is graceful, and the ATP is able to compensate for the interference without increasing the source's radiated power. An increase in power would produce more interference in other

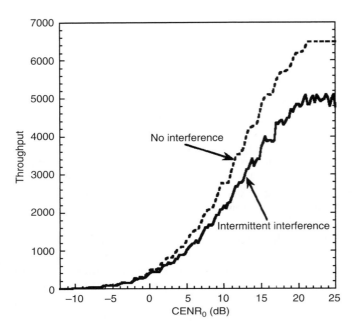

FIGURE 21.14

Throughput for time-varying interference and propagation loss.

receivers and perhaps prompt other transmitters to increase their power, thereby causing even more interference in the destination receiver. This ramping up of power by multiple transmitters in the network wastes energy. Notice from Figure 21.14 that no increase in throughput is obtained by increasing the power beyond a level that gives $CENR_0 = 21$ dB.

As mentioned, the results in Figure 21.14 are for an interference signal that experiences the same propagation loss as the desired signal. Other assumptions would give slightly different results. For example, we could have a third independent Markov chain that governs the propagation loss for the interference signal, which would cause the SINR to vary, even during the period that the Markov chain for interference stays in state 1.

21.10 SUMMARY

Cognitive radios are aware of their communication environments, and they can exchange the information that is needed to communicate with each other effectively and efficiently. Effective and efficient operation requires the radios to adapt their transmission parameters (e.g., code rate and modulation format) as changes occur in the network, including changes in propagation and interference. Because each transmission is a source of interference for each unintended receiver, it is important for the CR to transmit at the lowest power level that provides a satisfactory signal-to-noise ratio for the demodulators and decoders of the intended receivers. Our initial P-ADJ protocol finds this minimum power level at the start of a new session, even if the source and destination have not previously communicated in the frequency band that has been selected for the session. The initial P-ADJ protocol is used for only the first few packets of a session; after that, control of transmission parameters is maintained by the ATP. The goal of the ATP is to compensate for changes in the communication environment in a way that does not increase interference to unintended receivers. The ATP described and evaluated in this chapter accomplishes this objective and requires only a small amount of feedback information, which can be obtained easily by the demodulator and decoder in the destination's receiver.

EXERCISES

21.1. Suppose that the rate of the error-control code for a code-modulation combination is $r = 0.1$ and the energy per information bit is ε_b.
   a. How many binary channel symbols must be sent to represent 1000 information bits?
   b. Give an expression for ε_s, the energy per binary code symbol, in terms of ε_b.
   c. If the modulation is 16-BOK, then how many chips must be sent to represent 1000 information bits?
   d. For each of the three modulation formats in Table 21.1, give an expression for the ε_c, the energy per modulation chip, in terms of ε_b.

21.2. As discussed in Section 21.4, the chip rate is usually constant during a session, even if the code-modulation combination is changed. It is also desirable to hold the power constant because an increase in power results in an increase in interference to unintended receivers, and a decrease in power normally produces an increase in the error probability in the destination's receiver. Assume that the power and chip rate are held constant as the system changes from combination $\mathcal{B}_5$ to $\mathcal{B}_4$.
   a. Is the energy per information bit changed? If so, by how much?
   b. Is the energy per binary code symbol changed? If so, by how much?
   c. Is the energy per chip changed? If so, by how much?

21.3. For the code-modulation combinations in Table 21.1, what is the maximum throughput that can be obtained? What is the maximum throughput that can be obtained from the code-modulation combinations that use QPSK?

21.4. Suppose the reference power level for the source is 1W and the nominal propagation loss is 80 dB. What code-modulation combinations in Table 21.1 will provide a packet error probability of 10^{-2} or less for the transmission of a packet of 4096 binary code symbols when the transmitter's power level is 2W and the propagation loss is 85 dB?

21.5. For the model of the random initial excess propagation loss described in 21.7.3, derive the initial density function for CENR. *Hint:* First, determine the condition on the initial state of the Markov chain and find the conditional density given the state. Next, use the law of total probability for a continuous random variable X and a discrete random variable Y, which is

$$f_X(u) = \sum_{j=1}^{J} f_{X|Y}(u|j) P(Y = j)$$

21.6. The available library for a pair of CRs has three hypothetical modulation formats that are noted as M_4, M_8, and M_{16}. The modulation formats are used on an AWGN channel without error-control coding, so the information bits are mapped directly to the modulation symbols. The set of all modulation symbols for M_k has k symbols, so each symbol represents $\log_2 k$ bits of information. The modulation symbols for M_k consist of sequences of k fixed-magnitude modulation chips. The energy per modulation symbol is denoted by E_{ms}, and we define MSENR = 10 $\log_{10}(E_{ms}/N_0)$. It is known that, for each k, modulation format M_k achieves its maximum throughput if MSENR $\geq W_k$, and its throughput is zero for MSENR < W_k. It is also known that $W_4 = 10$ dB and $W_{16} = 0$ dB. Find the range of values of W_8 (in dB) for which it is beneficial for the pair of CRs to include all three modulation formats in their active library.

21.7. If modulation M is employed with a particular code of rate r_1, then the minimum requirement for a 10^{-2} packet error probability is CENR = −10 dB. If a particular code of rate r_2 is used instead, then CENR = 0 dB is the minimum requirement, and if a particular code of rate r_3 is used, then the minimum requirement is CENR = 10 dB. Assume that $r_1 < r_2 < r_3$. You have discovered a new receiver statistic G, which can be obtained easily in the demodulator and decoder. Analytical methods have shown that the relationship between G and CENR is $G = \exp\{CENR\}$. Your discovery of the receiver statistic G will allow you to remove an expensive channel estimator from your receiver implementation if you can design an interval test to use in the protocol for adaptive selection of the code rate. Find the three intervals of values of G that the protocol should employ, and specify which of the intervals corresponds to each rate.

REFERENCES

[1] Pursley, M. B., *Introduction to Digital Communications*, Prentice Hall, 2005.

[2] Pursley, M. B., and T. C. Royster IV, Low-complexity Adaptive Transmission for Cognitive Radios in Dynamic Spectrum Access Networks, *IEEE Journal on Selected Areas in Communications*, Jan(26):83–94, 2008.

[3] Advanced Hardware Architectures, Inc., Product Specification for AHA4501 Astro 36 Mbits/sec Turbo Product Code Encoder/Decoder; available at *http://www.aha.com*.

[4] Dolinar, S., and D. Divsalar, Weight Distributions for Turbo Codes Using Random and Non-random Permutations, *JPL TDA Progress Report 42-122*, pp. 56–65, August 1995.

[5] Le Goff, S., A. Glaviuex, and C. Berrou, Turbo-codes and High Spectral Efficiency Modulation, *Proceedings IEEE International Conference on Communications*, pp. 645–649, 1994.

[6] Phoel, W. G., J. A. Pursley, M. B. Pursley, and J. S. Skinner, Frequency-Hop Spread Spectrum with Quadrature Amplitude Modulation and Error-Control Coding, *Proceedings IEEE Military Communications Conference*, pp. 913–919, November 2004.

[7] Viterbi, A. J., A Robust Ratio-Threshold Technique to Mitigate Tone and Partial Band Jamming in Coded MFSK Systems, *Proceedings IEEE Military Communications Conference*, 2:22.4.1–22.4.5, October 1982.

[8] Pursley, M. B., and T. C. Royster, IV, High-rate Direct-sequence Spread Spectrum with Error-Control Coding, *IEEE Transactions on Communications*, 54(9):1693–1702, 2006.

[9] Jakes, W. C. (ed.), *Microwave Mobile Communications*, IEEE Press, 1974.

[10] Hata, M., Empirical Formula for Propagation Loss in Land Mobile Radio Services, *IEEE Transactions on Vehicular Technology*, 3(VT-29):317–325, 1980.

[11] Rappaport, T. S., *Wireless Communications: Principles and Practice*, Second Edition, Prentice Hall PTR, 2002.

[12] Goldsmith, A., *Wireless Communications*, Cambridge University Press, 2005.

[13] Wang, H. S., and N. Moayeri, Finite-state Markov Channel—A Useful Model for Radio Communication Channels, *IEEE Transactions on Vehicular Technology*, Feb(44):163–171, 1995.

[14] Zhang, Q., and S. A. Kassam, Finite-state Markov Model for Rayleigh Fading Channels, *IEEE Transactions on Communications*, Nov(47):1688–1692, 1999.

[15] Tan, C., and N. Beaulieu, On First-order Markov Modeling for the Rayleigh Fading Channel, *IEEE Transactions on Communications*, Dec(48):2032–2040, 2000.

[16] Hueda, M. R., and C. E. Rodriguez, On the Relationship between the Block Error and Channel-State Markov Models in Transmissions over Slow-fading Channels, *IEEE Transactions on Communications*, 52(8):269–1275, 2004.

[17] Pursley, M. B., and T. C. Royster IV, Adaptive Coding in Direct-sequence Spread-spectrum Systems for Channels with Time-varying Propagation Losses, *Proceedings IEEE Military Communications Conference*, 1:640–645, October 2005.

[18] Masse, M. R., M. B. Pursley, T. C. Royster IV, and J. S. Skinner, Adaptive Coding for Wireless Spread-spectrum Communication Systems, *Proceedings International Conference on Communications, Circuits, and Systems*, 1:1321–1326, June 2006.

[19] Pursley, M. B., and T. C. Royster IV, Adaptation of Modulation, Coding, and Power for High-rate Direct-sequence Spread Spectrum, *Proceedings IEEE Military Communications Conference*, pp. 1–6, October 2006.

[20] Pursley, M. B., and T. C. Royster IV, A Protocol Suite for Cognitive Radios in Dynamic Spectrum Access Networks, *Cognitive Wireless Communication Networks*, E. Hossain and V. Bhargava (eds.), Chapter 5, pp. 139–163, Springer, 2007.

Cognitive Networking

Ryan W. Thomas
Air Force Institute of Technology
Dayton Ohio

Luiz A. DaSilva
Virginia Tech
Blacksburg, Virginia

22.1 INTRODUCTION

Cognitive radio (CR) technology has emerged as an exciting field in the area of wireless communications research, allowing for the selection and optimization of radio parameters. However, in today's technological environment, radios rarely communicate in isolated pairs. Particularly for general-purpose data communications, individual radios are often part of a larger multihop network consisting of various other wired and wireless devices. For this reason, many of the philosophical underpinnings of CRs should be extended from the wireless link to encompass the entire network stack. Networks of nodes that intelligently select and optimize parameters based on the end-to-end requirements of the network are called cognitive networks (CNs).

The radio environment is clearly characterized by flexible operating parameters (e.g., frequency, modulation, power level) and dynamic behaviors (e.g., the intermittent and mobile presence of spectrum users). Thus, the motivation for CRs is clear: in the face of these challenges, CRs attempt to provide "highly reliable communications whenever and wherever needed [and] efficient utilization of the radio spectrum" [1]. The motivation for a CN operating with wireless components is extended from this motivation, because the wireless *network* environment is a superset of the radio environment, since the wireless network is built on the flexible and dynamic physical layer provided by the radios. All the complexities of the link layer exist in the network, with the additional problems of optimally coordinating and using the network's multiple connections and nodes.

Whereas CRs must coordinate radio parameters with, at the most, one other CR, CNs must coordinate network and radio parameters among multiple network nodes at all layers of the network stack. Working down the network protocol stack, examples of parameters (the "knobs and dials" of the network) that CNs could modify include multimedia codecs, buffer and window sizes for flow control and/or reliable transmission, routing metrics, network topologies, and medium access control (MAC) layer

---

Note: The views expressed in this chapter are those of the authors and do not reflect the official policy of the US Air Force, US Department of Defense, or the US government.

timings, in addition to the radio physical (PHY) layer parameters. The environment in which the CN makes its decisions goes beyond the dynamic physical environment of the CR because it includes the virtual environment of the network, which consists of many nodes and users running various applications with their own traffic and connectivity requirements.

The particular *framework*, *components*, and *mechanisms* of a network do not define it as cognitive. We define a CN as one that "has a cognitive process that can perceive current network conditions, and then plan, decide and act on those conditions. The network can learn from these adaptations and use them to make future decisions, all while taking into account end-to-end goals" [2].

This means that any network that uses some mechanism that provides a basic amount of cognition to achieve *end-to-end* or *networkwide* goals (as opposed to link or medium goals) can be considered a CN. Examples of possible end-to-end goals include quality-of-service (QoS) considerations (e.g., delay, jitter, and throughput), fault identification and repair, node connectivity, resource management, cost of operation, topology lifetime, authorization/trust, and security.

It becomes clear that CNs are not exclusively an extension of the CR concept. CNs have a scope that goes beyond the domain of cognitive radios. Although the most interesting CN research problems integrate well with the most interesting problems for CR, CNs are not restricted to this wireless domain. Completely wired networks, for instance, can be considered CNs if they meet the preceding definition. Similarly, wireless networks without a single CR can be considered CNs if they have a mechanism with a cognitive process in their framework.

A taxonomy can be developed that relates CRs to CNs, based on whether the CRs comprise a network, and whether that network meets the CN definition. A Venn diagram of this taxonomy is illustrated in Figure 22.1. Cognitive radios by themselves do not compose a network and thus are not considered to be a cognitive network. CRs that operate as a network, but pursue their own local, link-level objectives rather than their end-to-end objectives, are called networks of cognitive radios (NCRs). NCRs are not

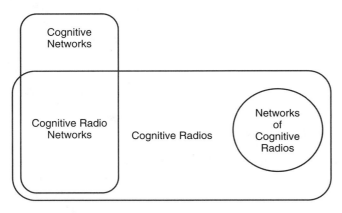

FIGURE 22.1

Venn diagram relating taxonomy of CN and CR concepts.

considered to be CNs because they are not actively pursuing any network-level objectives. NCRs that converge to a networkwide desirable state do so only when the component radios pursue individual goals that happen to be in alignment with the network goal. Neel et al. [3] investigated when NCRs achieve network objectives in this manner by using game theory to model a network of CRs as a large, multiplayer game, allowing for the determination of what conditions allow stable convergence of radio operation.

In contrast to NCRs, cognitive radio networks (CRNs) consist of multiple CRs that can take into account the network's end-to-end objectives. The radios may still pursue local, link-level objectives, but these objectives are aligned to the end-to-end objectives of the network. This is in contrast to the autonomous, link-based approach of the NCR. The focus of CRNs, as with cognitive radios, is primarily on MAC and PHY issues. Unlike CRs, the CRN attempts to manipulate these layers to achieve some end-to-end objective. In a CRN, the individual CRs comprise the distributed elements of the network's cognitive process.

Cognitive radio networks are described by other authors, although not necessarily by this name. Mitola described CRNs in his original thesis on CRs. He envisioned CRs operating and interacting within the system-level scope of a network [4]. Haykin [1] examines multiuser networks of CRs. He draws a functional distinction based on whether the CRs are using an *etiquette*, which he describes as a generalized protocol of behavior from a CR when dealing with other cognitive radios. Etiquette acts to enforce certain network objectives, so it can be used as a deciding factor to determine whether a system is an NCR (no etiquette) or a CRN (etiquette). Various authors have suggested applications for CRNs, including secondary-user (SU) networks, cooperative spectrum sensing [5, 6], and emergency radio networks [7].

The difference between NCRs and CRNs can seem simply semantic, but the underlying frameworks are very different. For example, take a group of CRs performing multihop routing and frequency selection. Assume the end-to-end objective is to minimize the spectral footprint of the network, meaning the total amount of unique spectrum the network requires to be fully connected should be minimized. An NCR may accomplish this goal, but it does so by the independent cognitive processes of each radio, with their decisions being made without any knowledge of this end-to-end goal. A CRN, in contrast, achieves this goal by explicitly pursuing it at the CR level. It does this by translating the network objective to local radio behaviors. In both scenarios, the network is achieving the end-to-end goal, but the CRN requires a mechanism for goal dissemination, translation, and agreement.

There are conceptual and system-level similarities between CNs and CRs, but cognitive networks also have similarities with the concept of cross-layer design, the approach of directly communicating or sharing internal information between adjacent and nonadjacent layers of the network stack. Particularly for wireless networks, where the interactions between nodes and layers are more pronounced than in wired networks, the violation of the traditional layered approach offers a tantalizing way to optimize performance. The ability of a CN's cognitive process to make decisions that can modify the parameters of any layers of the stack based on observations from across the stack places it squarely in the realm of cross-layer design.

An investigation into the evolution of cross-layer design illuminates this relationship with CNs. Initially, many designs involved simply merging two related layers (e.g., the

PHY and MAC layers) to accomplish a goal. ElBatt and Ephremides [8] provide an example of this kind of merging, where they combine scheduling with power control to increase throughput by reducing contention for the medium. The advantage to this kind of cross-layer design is that for all layers above and below the merged layers, the interfaces and operation of these merged layers appear the same as they did before the merge. The disadvantage of this kind of framework is that it typically optimizes for a single goal, at the expense of other objectives. The merged-layer scheme reduces contention, but it does not explicitly optimize any other objectives (e.g., fairness, node lifetime, or bandwidth allocation). If the goals of the network change, this design will have difficulty adapting to these new goals. Furthermore, if interactions from higher or lower layers are causing behaviors at the joint layer to be suboptimal, there is little the joint layer can do to prevent this because the scope of these adaptations is limited to the merged layers.

This "merged-layer" design was extended to allow uni- or bidirectional transfer of information between two nonadjacent layers. This is done by creating new interfaces at the selected layers beyond those used between layers. Chiang [9] provides an example of this kind of design in which information is shared between the transport layer and the physical layer to increase the throughput and energy efficiency of the network. Although this allows a more flexible pursuit of network goals, adding new interfaces to optimize specific metrics runs the risk of *interface creep*, in which the layered framework becomes meaningless as designers create and add interfaces without guidance. Furthermore, by opening more interfaces, designers are opening up more interactions, possibly making the network behavior more complex. Kawadia and Kumar [10] illustrate this point by showing the unintended interactions between the routing protocol and a MAC/PHY cross-layer design.

Recently, cross-layer designs have begun to use a different approach to avoid this problem. Instead of building communications between specific layers, proposals (e.g., CrossTalk [11], ECLAIR [12], CLD [13], and the framework of Gong et al. [14]) use a parallel structure that acts as a shared database of the system state accessible to whichever layers choose to use it. These kinds of frameworks are called *vertical calibrations* [15] because the system jointly tunes several parameters over the whole stack to achieve an application-level objective. The advantage to these vertical calibration frameworks is that they provide a structured method for accessing parameters and controlling and sensing the status of each layer. This addresses some of the concerns about cross-layer design decreasing the utility of the layered framework. To its disadvantage, even with its larger scope, the cross-layer interactions are stack-centric and do not incorporate properties outside the scope of the stack. For instance, PHY layer properties (e.g., battery life, bandwidth usage) of nodes not within range of one another cannot be jointly modified in a MAC/PHY cross-layer protocol. Internodal communication is limited to the scope of the layers being cross-designed.

It is here that the differences (and advantages) CNs provide over cross-layer design can be observed. A large difference is in their ability to support multiple network goals. All cross-layer designs have problems supporting trade-offs between multiple goals. Merged-layer designs are typically single-purpose in nature, which excludes the possibility of supporting multiple goals. Vertical calibrations have no mechanism for deconflicting multiple goals, since optimizations are performed independently and do not account for the set of performance goals as a whole. Trying to achieve each goal independently is likely to be suboptimal, and as the number of cross-layer designs within a node grows, conflicts between the independent adaptations may lead to adaptation loops [16]. Adaptation loops occur when conflicting goals cause a system to fail to converge to a stable operating point. CNs use a cognitive process that is not architecturally limited to considering network goals independently, potentially avoiding these pitfalls.

Furthermore, cross-layer designs are memoryless adaptations that will respond the same way when presented with the same set of inputs, regardless of how poorly the adaptation performed in the past. The ability to learn from past behavior is particularly important in light of the fact that understanding the interactions between layers is difficult. Several cross-layer designs incorporate adaptation, but often no discussion is given to intelligence, learning, or proactive adaptation. Because of the presence of a cognitive process that learns, these aspects are included, by definition, in a CN.

22.2 CURRENT CN RESEARCH

The definition of a CN does not specify the CN framework; however, several frameworks have been proposed in recent years. Not all frameworks use the term "cognitive network" to describe themselves, but they still meet the CN definition described earlier. These frameworks can be categorized by the overarching objectives they pursue: the first category of framework centers on using cognition to aid in the operation and maintenance of the network from an end-to-end perspective, whereas the second centers on cognition to solve "hard" end-to-end problems, where "hard" problems are those that require large amounts of resources—that is, a central processing unit (CPU), memory, bandwidth—to solve correctly. Examples of management problems include maintaining mobile network access or enforcing network policy. Management problems could be solved with noncognitive methods or by a system administrator, but the existing tools do not scale and are too time consuming, inflexible, or impractical. Examples of hard problems include creating optimal wireless topologies or estimating the radio frequency (RF) environment for selecting noninterfering communication channels. Hard problems are those that require nonpolynomial time for a correct solution, have a large solution space to search, or are not easily solved in a distributed fashion. Obviously, there is not a clean dichotomy between hard and management problems, and some management problems are hard.

Several of these frameworks are illustrated in Figure 22.2. Falling into the "management" category, the End-to-End Reconfigurability project (E^2R II) [17] is designing a complex, multifaceted framework that will allow the seamless reconfiguration of a network to achieve universal end-to-end connectivity. Although E^2R II is an ambitious project with many facets, the overarching end-to-end goal is one of maintaining a user's network connectivity. This is similar to the goal of the m@ANGEL platform from Motorola labs [18], which uses a CN-like framework for mobility management in a heterogeneous network. The m@ANGEL project consists of designing an intelligent management layer that sits between the reconfigurable radios and the network applications. These entities are hierarchically arranged so that lower-tier entities control

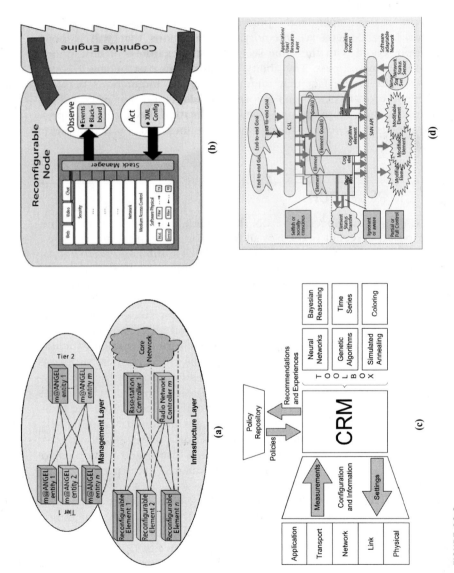

FIGURE 22.2

Illustrations of various CN-like frameworks: (a) m@ANGEL management framework, consisting of a tiered hierarchy of cognitive and reconfigurable elements [18]; (b) CTVR CRN framework, consisting of a CE that interacts with a "stack manager" on a CR, leveraging CRs for CN operation [19]; (c) the Mähönen et al. CN framework, using a cognitive "toolbox" to solve complex CN problems [20]; and (d) the three-layer Thomas et al. CN framework, providing a platform-agnostic three-layer approach to solving network problems in a distributed or centralized manner [2].

specific reconfigurable elements in the network, whereas higher-tier entities coordinate and assist the lower-tier entities.

In contrast to the management objectives of these two frameworks, the Centre for Telecommunications Value-Chain Research (CTVR) at Trinity College, Dublin [19], has presented a proposal for a general CRN framework to solve hard network problems. This framework consists of reconfigurable CRs working together to achieve network objectives. Although focused on problems associated with wireless connectivity, these nodes are able to solve a variety of network problems by modifying or changing the network stack parameters based on observed network behaviors. A "stack manager" is employed to implement and coordinate the modifications to the network that have been determined by a cognitive engine (CE), based on observations of the network environment. Changes in the stack may occur by modifying stack parameters (e.g., the radio transmit power), replacing individual layers of the stack (e.g., changing from one routing algorithm to another), or replacing the entire stack.

Mähönen et al. [20] propose a general framework of this category, using a collaborative cognitive resource manager (CRM) that provides cognitive behavior from a toolbox of machine-learning tools (e.g., neural networks, clustering, coloring, genetic algorithms, and simulated annealing). This toolbox allows for difficult problems in the network to be solved by applying policy decisions to the network stack based on measurements of the network environment.

Thomas's cognitive networking framework also addresses these difficult problems. It consists of three layers: at the top layer are the goals of the elements in the network that define the behavior of the system. These goals feed into the middle layer, the cognitive process, which computes the actions for the system to take. The bottom layer consists of the software-adaptable network (SAN), which is the physical control of the system that provides the action space for the cognitive process.

Expanding on each of the layers of Thomas's framework, the top-level component of the CN framework includes the end-to-end goals, Cognitive Specification Language (CSL),[1] and the resultant cognitive element goals. The end-to-end goals, which drive the behavior of the entire system, are put forth by the network users, applications, or resources. These end-to-end goals are interfaced to the cognitive process via the CSL, which provides behavioral guidance to the cognitive elements by translating the end-to-end goals to local element goals. The *cognition* associated with the cognitive process layer of the framework is said to be defined as any algorithm that "improves its performance through experience gained over a period of time without complete information about the environment in which it operates" [21]. The bottom-layer SAN consists of the application programming interface (API), modifiable network elements, and network status sensors.

The SAN is analogous to the software-defined radio (SDR) that CRs typically use. In a CRN, the SAN includes multiple SDRs. The SAN notifies the cognitive process of the status of the network. Many of the frameworks designed to solve difficult problems share the same basic design for the cognitive process. Either explicitly or implicitly, these frameworks reference Boyd's Observe-Orient-Decide-Act (OODA) loop. The

---

[1] Other chapters refer to the Cognitive Specification Language (CSL) as the System Strategy Reasoner (SSR).

OODA loop was originally used to help military officers understand the thought processes of their adversaries, but has since been adapted as a general decision-making cycle. The loop consists of four self-explanatory components that guide a decision maker through the process of choosing an appropriate action based on input from the environment.

22.2.1 Observe

The observations portion of the cognitive process consists of sensing the network environment and creating an internal model of it. Information can be either directly observed, via sensors in the SAN, or inferred from the sensed results of previous decisions. Possible information that is directly observed (the "meters" of the SAN) in the network includes the presence of primary and secondary spectrum user signals, received signal-to-interference and noise ratio (SINR), current connectivity, packet delays, and the state of other nodes' parameter choices (e.g., transmission power, location, or channel selection). This local information can be shared between elements of the cognitive process to create a more informed picture.

The degree of information that a cognitive process needs to make acceptable decisions is not a fixed parameter. The cognitive process can conceivably operate under some degree of ignorance. Ignorance in a CN can be categorized in three ways:

1. *Uncertain information*: The information, as measured, has some random, stochastic uncertainty in it.
2. *Missing information*: The information is missing the state of at least one other element.
3. *Indistinguishable information*: The information may indicate one of several states for the network, and it is impossible to distinguish among them.

The topic of storing and exchanging knowledge was examined early on in CRs, with Mitola describing a knowledge language called Radio Knowledge Representation Language (RKRL) [4]. RKRL is an instantiation of Knowledge Query Markup Language (KQML), an interaction language developed for communication between software agents [22]. In addition to exchanging information, KQML is designed to make and respond to requests for information, as well as to locate qualified agents. RKRL, according to Mitola, contains the following components:

- The *mappings* between the real world and the various models formed by the cognitive process
- A *syntax* defining the statements of the language
- *Models* of time, space, entities, and communications among entities (e.g., people, places, and things)
- An *initial set of knowledge*, including initial representation sets, definitions, conceptual models, and radio domain models
- Mechanisms for *modifying and extending* RKRL

KQML, and by association RKRL, makes some demanding assumptions as to the availability and reliability of communication channels. In particular, it is assumed that there are distinguishable connections among all agents, the connections are reliable, and they preserve the order of the messages [23]. These assumptions are appropriate for the local, point-to-point scope at which the CR operates, but are limiting for a dynamic, complex network of connections.

To attempt to address a language for higher-level goals, Mähönen et al. [20] suggest extending RKRL to encompass the high-level goals of the users of the network. They call this Network Knowledge Representation Language (NKRL). In this position paper, they did not fully describe nor develop the idea. However, based on the properties of the network environment, NKRL should capture the same core components as RKRL, but extend them to handle the complexities of the entire network environment. This means that NKRL should support the greater uncertainty and collaborative potential of the network environment.

Whereas RKRL focuses on modeling the knowledge that a radio has, NKRL should be capable of representing the knowledge that a network node does not have, including the uncertain, missing, and indeterminate information discussed previously. To this end, NKRL should also be able to represent compressed versions of the knowledge when sharing information between cognitive elements. This will avoid flooding other cognitive elements with too much information, including information that is either superfluous or repetitive. Furthermore, NKRL should be able to handle the dynamic, uncertain nature of the wireless, mobile network environment, making no assumptions about whether NKRL transactions will always be completed in a timely and reliable fashion.

22.2.2 Orient

To effectively use this information, the orient step must be implemented. For the cognitive process to effectively orient itself, the cognitive elements must have an interface to the sources of these network end-to-end objectives. In a CN, this role is performed by the CSL. The CSL describes, in a formal manner, the end-to-end goals for eventual use as cognitive element objectives. A CSL can be thought of as the architectural "blueprints" of the network behaviors. The CSL represents the end-to-end goals of the network in a standard, abstract fashion that various cognitive elements can interpret and act on. To understand the CSL, consider the following extension to the blueprint analogy: in a construction project, different contractors use the same blueprint to determine how to do their specific tasks, from correctly framing the building to running the venting system. In a similar fashion, the CSL provides guidance that different cognitive elements use to determine how to behave.

The CSL is distinctly different from the NKRL because it is not used to represent the observed network environment. Figure 22.3 illustrates the difference interfaces that the CSL and NKRL provide. The CSL is analogous in scope and intention to a QoS specification language. These languages are used to represent QoS requirements to the various mechanisms that the network offers to support them. There are already several different QoS specification paradigms in existence [24], and the concept of these languages— mapping requirements to underlying mechanisms—is the same here, except that the mechanisms are adaptive to the network capabilities as opposed to a fixed set of QoS capabilities.

In establishing criteria for a successful CSL, we adapt the criteria for a good QoS specification language, as described by Jin and Nahrstedt [24]. The following criteria represent design objectives for an effective CSL, representing measures of a successful

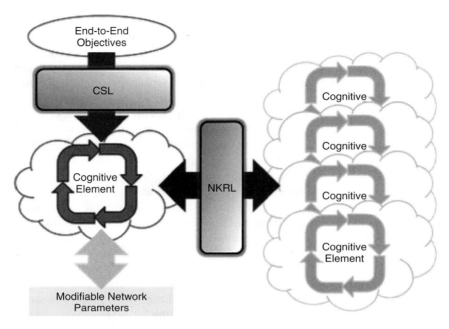

FIGURE 22.3

The CSL interface between other cognitive elements and the end to-end objectives. Unlike NKRL, which expresses and represents knowledge of the network state both internally and between cognitive elements, CSL represents the end-to-end objectives of the network to the cognitive elements.

CSL. A language that does not meet any or all of these requirements may still perform the role of a CSL, albeit less effectively.

Expressiveness: A CSL must be able to specify a wide variety of end-to-end goals. It should be able to express constraints, goals, priorities, and behaviors to the cognitive elements that make up the process. It should be able to express new goals without requiring a revision in the language.

Cognitive process independence: The cognitive process architecture and functionality should not dictate the CSL. Instead, the CSL should abstract away as much of the cognitive process as possible to the application, user, or resource. This allows a goal to be used with different cognitive processes with little modification and promotes reusability.

Interface independence: Whether the cognitive process is distributed or centralized in operation, autonomous or aggregated in architecture, the user should be presented as abstract an interface as possible. Like the previous criteria, this abstraction promotes reusability by allowing the reuse of goals over many different cognitive processes from the top layer with little effort.

Extensibility: The CSL should be extensible enough to adapt to new network elements, applications, and goals, some of which may not even be imagined yet.

The way the cognitive elements align themselves to the goals is a fundamental question. For cognitive processes with more than one cognitive element, the behaviors of the elements fall into a spectrum of behaviors ranging from purely selfish and individualistic, to socially conscious and altruistic. Even though altruistic behaviors may seem a natural method of accomplishing end-to-end goals, selfish behaviors sometimes lead to globally efficient adaptations. Furthermore, if the cognitive elements are autonomous and not under central control, selfish behaviors are a reasonable method of controlling the network because real-world systems often consist of unrelated nodes with varying degrees of internal selfish motivations. Additionally, as selfishness can require less coordination than altruism, it may lead to lower overhead.

22.2.3 Decide

Now that the cognitive process has observed the network environment and is oriented to the end-to-end objectives, it must make a decision. The cognitive process that makes the decisions can be implemented in two possible ways: either (1) a centralized decision-making unit that gathers network state data and distributes state information to the nodes of the network, or (2) a distributed process across the network nodes, with each node making decisions under some degree of autonomy. The first implementation suffers many of the weaknesses of centralized designs: poor scalability, a single point of failure, and high communication overhead. The second implementation is a good fit for CRNs, as the individual CRs act as a distributed cognitive process. Having the CRs exhibit the proper behaviors to achieve the end-to-end goals becomes a function of the interpretation of the CSL by the cognitive radios.

The cognitive process itself can take many forms, as long as it improves the network performance through experience. Several of the more common mechanisms for the cognitive process include metaheuristics, learning automata, and software agents. Metaheuristics consist of approaches such as neural networks, genetic algorithms (GAs), and evolutionary algorithms (EAs).

Neural networks use a bottom-up method of learning, simulating the biological neurons and pathways that the brain is believed to use. A series of these artificial neurons analyze different aspects of known inputs with some amount of unknown corruption. Pattern recognition is a common and straightforward application of neural networks. If network responses are modeled as a noisy pattern, a neural network could be used to categorize the pattern into predetermined responses.

GAs and EAs are used to optimize over large solution spaces where exhaustive searching would be too costly. By imitating the process of evolution (selection, recombination, and mutation), GAs are able to explore these large solution spaces for local optima. Genetic algorithms have many applications but work best for centralized problems for which the environment is well known. For this reason, if most of the current network state is known, GAs could be used to determine optimal behaviors.

Learning automata [21] are a distributed, adaptive control solution to identify the characteristics of an unknown feedback system. Learning automata maintain a probabilistic function for deciding what action to make. The function converges to decisions that generate desired responses in the system. Typical applications for learning automata include problems in which many dynamic elements interact with a complex system (e.g., intelligent vehicles [25], cross-layer optimization problems [26], or routing problems [21]). If the problem is distributed and requires little state information, learning automata can be a good approach.

Software agents are a general term given to processes that use some form of machine learning to observe and act on their environment. Depending on their functionality, software agents can perform tasks as diverse as winning auctions in an online bidding war or providing navigation control for a robot. In particular, the subfield of multiagent systems (MASs) shows promise for making distributed decisions. MASs are a catch-all model that includes variations on systems consisting of multiple software agents. MASs are a direct descendant of distributed artificial intelligence (DAI) [27], which pioneered research into such questions as task allocation, coordination, cooperation, and interaction languages between agents. MAS research goals are even broader than the artificial intelligence goals of DAI, consisting of almost any system with a distributed network of software agents. Although MASs have no universally accepted definition, we will adopt the definition of Jennings et al. [28], which relies on three concepts: situated, autonomous, and flexible.

> *Situated* means that agents are capable of sensing and acting on their environment. The agent is generally assumed to have incomplete knowledge and/or partial control of the environment, or both limitations [29].
>
> *Autonomous* means that agents have the freedom to act independently of humans or other agents, although there may be some constraints on the degree of autonomy for each agent.
>
> *Flexible* means that agents' responses to environmental changes are timely and proactive and that agents interact with each other and possibly humans as well in order to solve problems and assist other agents.

Dietterich and Langley [30] describe a standard agent model consisting of four primary components: (1) observations, (2) actions, (3) an inference engine, and (4) a knowledge base. In this agent model, reasoning and learning are a result of the combined operation of the inference engine and the knowledge base. Reasoning is the immediate process by which the inference engine gathers relevant information from the knowledge base and sensory inputs (observations) and decides on a set of actions. Learning is the longer-term process by which the inference engine evaluates relationships, such as between past actions and current observations or between different concurrent observations, and converts this to knowledge to be stored in the knowledge base. This model fits well within most of the cognitive architectures previously mentioned.

According to Wooldridge [31], there are two aspects of design that drive MAS research:

> *Agent design:* The task of creating software agents that are able to carry out tasks autonomously.
>
> *Society design:* The task of creating software agents that interact in a manner to carry out tasks in an uncertain environment.

These research areas integrate well with desired CN behavior: they can address complex systems, exhibit varying degrees of machine learning, have a naturally distributed nature, and can operate in less-than-reliable environments.

Pecarina [32] presents an example of using a MAS to solve network problems. This work presents an agent architecture called the Hybrid Agent for Network Control (HANC), which uses three layers of control. The first layer, called the deliberator, is a planner, which determines which long-term objectives to pursue to reach the end-to-end goals. The second layer, the sequencer, selects among a set of predefined behaviors to accomplish what the planner dictates. The bottom layer is the controller, which implements the sequencer behaviors. Information about the locally observed network state was exchanged between agents, allowing HANC to successfully route traffic according to flow priority in a dynamic, unreliable network.

Distributed cognitive process decision making can be implemented either synchronously or asynchronously. The details of making synchronous decisions across a large number of distributed elements with high reliability are likely to be complex. Trying to explicitly synchronize the decisions can be difficult because (due to the dynamic and unreliable nature of the network topology) not all nodes are guaranteed to receive notification of configuration changes at the same time. In contrast, the consequences of unsynchronized decision making may be worse than if simply no decision had been performed at all. One possible approach to ensuring synchronization is to require elements to be synchronized to some common time reference and have the SAN issue configuration to occur at some future timeslot, to allow synchronous adaptation. Unfortunately, this approach can delay network actions, resulting in lagging adaptations.

22.2.4 Act

Once decisions have been made by the cognitive process, they must be acted on. The process of acting involves the API portion of the SAN, which enables the cognitive process to implement its decisions into the network. For CRNs, the API that controls the radio parameters can use such software as the GNU Radio Project [33], the CTVR IRIS platform, or the DoD's Software Communications Architecture (SCA). Each of these software platforms provides, at the least, an interface for manipulating the radio waveform. IRIS and SCA also provide other aspects of the OODA loop functionality. Developing a general API for configuring the wireless parameters is relatively challenging because of the wide variety of radio functionality that may or may not be exposed by the various SDR hardware platforms. In contrast, the higher layers of the network stack are all software, meaning that differences between physical platforms have less impact on the API design. Simply replacing the standard network stack with one designed to work with the API can give additional control over needed network parameters.

The API acts as the control for the cognitive process, and raises the question of how much reach the cognitive process's control should have. If there is cognitive control over every parameter in the network, then the cognitive process could enact any particular solution to reach the end-to-end goals. Even if the cognitive process has less than full control over the action space of the network, in some cases it may still have enough to arrive at the desired system state. In this case, or the case where the cognitive process chooses not to exercise all of its control, the cognitive process has to use the functionality and interactions of the noncognitive aspects of the network to set the system state. Particularly if the desired network state is an attractor[2], the system may be pulled toward

---

[2] An attractor is a state of a dynamic system toward which the system state is drawn, even when disturbed or under dynamic conditions.

these states like water is drawn toward the hole at the bottom of a funnel. If a system has several attractors and some are more optimal or generally express broader influence than others, a few points of cognitive control may be enough to draw the system out of one attractor and into another.[3] An example of this can be seen in the Braess paradox, in which slightly changing the topology of a network to include an extra link can cause the network to perform significantly worse (in terms of average delay or cost per flow) than without the connector. Here, making a small change to the topology can push the network between two distinctly different routing performances.

22.3 RESEARCH HOLES AND FUTURE DIRECTIONS

The research area of cognitive networking is in its infancy, and there exist many open research questions that need investigation. In particular, the cognitive process, network architecture, and implementation present unsolved issues and open-ended questions.

We first examine the research holes of the cognitive process, looking at each stage of the OODA loop. The cognitive process observes the network environment via direct observations and shared information. This leads to the question of what and how much information should be observed locally versus gleaned from shared information. Shared information can come in smaller, more preprocessed and therefore information-rich chunks, saving the time and cost of locally observing the data. It has been shown that collaboration and information-sharing significantly improve the accuracy of spectrum sensing and modulation classification [34]. But these data can be outdated, inaccurate, or irrelevant by the time they reach the receiving node. There is some degree of tit-for-tat in information sharing—*all* nodes cannot depend entirely on shared information from other nodes, and every node can conceivably gather all its information locally (although not necessarily enough information). Given that some information is gathered through collaboration, there remains the question of how best to compress and represent the shared information, and then fuse it into other local and shared observations.

Regardless of the source of information, we still need to determine how much information the overall cognitive process requires to effectively operate the network toward the network goals. The cognitive process requires some level of information from the system to perform its computations. With more information, cognitive decisions should be more "correct" than those made with some degree of ignorance. There is often a high cost to acquire information, so the CN will have to be able to work with less than a full picture of the network status. Determining how little information the network can get by on and still have "good enough" performance is an open question, and probably depends heavily on the end-to-end objectives and the network architecture.

The orientation of the network to its environment requires that the end-to-end objectives be described by using the CSL. The sources of these goals are somewhat ambiguous. Often it is assumed that the goals are created by some centralized source. This is appropriate for small, homogeneous networks that operate under a single authority (e.g., a corporate, military, or public safety network). In larger networks of networks,

---

[3] This is analogous to a watershed, in which moving the source of water a few miles may be enough to change what river the water will finally flow into.

there may be many authorities or no authority at all. In these cases, there needs to be some manner of negotiating and selecting the end-to-end objectives that minimizes conflict and reduces the difficulty of the problems given to the cognitive process.

Determining exactly where the decision-making process resides in the CN may seem obvious for CRNs, as they use the CEs of the component CRs. For an objective that can be solved in a fully distributed manner, this decision-making process may be adequate, but for objectives that require more coordination, there may need to be some overarching process that efficiently and equitably manages the cognitive tasks given to the CRs. Whether this process itself operates in a centralized or distributed manner is an open question. For those objectives in which a centralized cognitive process is more effective than a distributed approach, determining which cognitive node will act as the "benevolent dictator" for the rest of the system is difficult, with selection depending on the scope of local network observations, connectivity to the rest of the network, and capabilities of the CE.

Most of the decision-making processes described to this point have been altruistic in nature, with collaboration and joint goal seeking assumed. As mentioned earlier, however, selfish behavior can often be an effective and efficient mechanism for solving problems. For problems in which selfishness does not lead to good solutions, using incentive or pricing mechanisms may be enough to allow selfishness to find stable, local optima for the network performance. Determining what mechanisms can implement these kinds of incentive-based direction is an open question. Furthermore, most analysis of selfish network operation has used a myopic viewpoint, meaning that the decisions made by network nodes are designed to maximally increase their progress toward their local objectives in the short term, based on current knowledge of what other nodes are doing. In essence, there is no long-term strategy or planning here, and the possible future actions of other nodes are not considered. Adding long-range planning capability to the selfish optimization process could lead to better equilibrium points, but it also results in more complex behaviors that are more difficult to analyze.

Determining the scope of available actions to the cognitive process is also a trade-off. This question is different than determining the amount of cognitive control in the network. Rather, it asks how many distinct network parameters (waveform, transmission power, buffer window) each cognitive node can change directly. Adding more action space requires software development time, creates additional complexity for the cognitive process (because there are now more variables), and will often correlate with the need for additional network status sensors to monitor the performance that these parameters effect. CNs must determine the appropriate amount of cognitive control that allows the network to achieve its objectives but minimize the cost to the developer, the cognitive process, and the equipment production cost.

The foundations of the CN architecture—the SAN and interface languages—are still not mature enough to support large-scale experimentation or deployment. The SAN, which makes up the virtual and physical foundation of a CN, does not exist as a widely available product. For publicly available SDR platforms that use a true general-purpose processor architecture, some limitations slow the development of a full SAN:

The MAC layer: Implementing random-access MACs (e.g., CSMA and its variants) is difficult because the long lag times between sense and transmit (due to running the receive and transmit chains in software on the general-purpose processor) decrease the channel throughput.

The "network wall": Due to academic and commercialization requirements, there exists a clear distinction between electrical engineers who work on physical layer issues and computer engineers and computer scientists who work on network issues. Until commercial opportunities open up for CNs, development on aspects of the SAN that integrate the two will be slow.

Similarly, although CN architectures may differ, all require some form of the interface languages discussed in this chapter. By developing a standardized language, cognitive processes of any architecture can represent knowledge, communicate among their distributed elements, and determine the end-to-end objectives. Standardization will allow some degree of interoperability between different CRs operating as a CRN and will allow the developers to focus on more functional issues, such as the cognitive process.

Researchers have just begun to study potential security attacks on CNs, such as denial-of-service attacks that seek to exploit coexistence between SUs and incumbent users of spectrum, as well as coexistence among secondary users. The increased flexibility and adaptability of CRs and networks may bring about new vulnerabilities and the need for new validation and verification methods. Conversely, the learning capabilities of CNs may prove useful in intrusion detection and mitigation of attacks.

Finally, there is a real need for additional implementation. Currently, most CN research is qualitative in nature. There are few theoretical frameworks available for analyzing CN architectures, and there has been little rigorous analysis to show that one given cognitive process is more effective than any other. By using architectural and design results from such related fields as robotics, game theory, and control theory, it may be possible to provide some formal design guidance. Furthermore, there exists very little measured, quantitative justification for engineering decisions made in terms of architecture, cognitive process, or applications. Some amount of experimental testing is needed to provide design tools for engineers architecting CNs. Although simulation and analysis is an important mechanism of determining performance bounds, experimental validation justifies design decisions and provides deeper insight.

Dynamic spectrum access is generally viewed as the first application area for cognitive networks and may provide the impetus to address some of these research issues. The United States has a strong interest in DSA for military use, and progress is being made in research on radios and networks that will support opportunistic spectrum access, through DARPA's NeXt Generation (XG) and Wireless Network after Next (WNaN) programs. In parallel, commercial applications of shared spectrum in the United States is signaled by recent Notices of Public Rule Making by the Federal Communications Commission (FCC), the dedication of 20 MHz of bandwidth for the creation of a spectrum sharing and innovation testbed, and the fact that spectrum sharing is a feature of the recent auction of spectrum vacated by the move to digital television.

Recent decisions by the European Commission show interest in increased flexibility in spectrum use and in the establishment of a spectrum market by 2010, and CNs are featured in the Seventh Framework Programme (FP-7), which funds research in the 2007 to 2013 time frame. Finally, the White Spaces Coalition and Wireless Innovation

Alliance, with participation from Google, Microsoft, and others, are strong advocates of broadband data delivery that opportunistically takes advantage of underutilized spectrum. There is reason to believe that DSA will play an important role in full interoperability and coexistence among diverse technologies for mobile networks, and that CNs will be the means to achieve this goal.

22.4 SUMMARY

This chapter provides a motivation for the cognitive network concept, arguing that CNs are a method of dealing with the complexity of the network parameters, wireless medium, and the demands of end-to-end objectives. In particular, we argue that complexity in wireless networks is a problem that cannot be solved or understood by using the local and reactive approach of networking protocols. Because of their similarities, we illustrate the relationship between CNs and two other tools for dealing with these complexities: cross-layer design and CRs. Their strengths and weaknesses when operating in a networking environment are illustrated, highlighting how many of the capabilities absent from their feature set exist in a CN. In particular, we show that to properly incorporate CRs into a CN requires having them act not just at the link level, but also at the network level, working toward a stated end-to-end objective.

Cogntive networking technology is rapidly advancing, with several competing visions for the technology's application and architecture. Regardless of which architecture and applications the research community coalesces on, there remain many open research questions that go beyond these problems. The cognitive decision-making process in particular has many fundamental questions related to how it observes, orients, decides, and acts. Whereas some of these questions are of immediate practical consideration, many are long-term questions that will require new frameworks for modeling and analyzing network behavior. Problems such as these can be understood only in the larger context of understanding the engineering trade-offs of knowledge, control, and node behavior.

REFERENCES

[1] Haykin, S., Cognitive Radio: Brain-Empowered Wireless Communication, *IEEE Journal on Selected Areas in Communication*, 23:201–220, 2005.

[2] Thomas, R. W., L. A. DaSilva, and A. B. Mackenzie, Cognitive Networks, *Proceedings of IEEE DySPAN*, pp. 352–360, November 2005.

[3] Neel, J. O., J. H. Reed, and R. P. Gilles, Convergence of Cognitive Radio Networks, *Proceedings of IEEE WCNC*, Vol. 4, pp. 2250–2255, 2004.

[4] Mitola, J., *Cognitive Radio: An Integrated Agent Architecture for Software Defined Radio*. PhD Thesis, Royal Institute of Technology (KTH), 2000.

[5] Ganesan, G., and Y. Li, Cooperative Spectrum Sensing in Cognitive Radio Networks, *Proceedings of IEEE DySPAN*, pp. 137–143, 2005.

[6] Mishra, S. M., A. Sahai, and R. W. Brodersen, Cooperative Sensing among Cognitive Radios, *Proceedings of IEEE ICC*, Vol. 4, pp. 1658–1663, 2006.

[7] Pawelczak, P., R. V. Prasad, L. Xia, and I. G. M. M. Niemegeers, Cognitive Radio Emergency Networks—Requirements and Design, *Proceedings of IEEE DySPAN*, pp. 601–606, 2005.

[8] ElBatt, T., and A. Ephremides, Joint Scheduling and Power Control for Wireless Ad-hoc Networks, *Proceedings of IEEE INFOCOM*, Vol. 2, pp. 976–984, 2002.

[9] Chiang, M., To Layer or Not to Layer: Balancing Transport and Physical Layers in Wireless Multihop Networks, *Proceedings of IEEE INFOCOM*, Vol. 4, pp. 2525–2536, 2004.

[10] Kawadia, V., and P. R. Kumar, A Cautionary Perspective on Cross-Layer Design, *IEEE Wireless Communications*, 12(1):3–11, 2005.

[11] Winter, R., J. H. Schiller, N. Nikaein, and C. Bonnet, CrossTalk: Cross-Layer Decision Support Based on Global Knowledge, *IEEE Communications Magazine*, 44(1):93–99, 2006.

[12] Raisinghani, V., and S. Iyer, Cross-Layer Feedback Architecture for Mobile Device Protocol Stacks, *IEEE Communications Magazine*, 44(1):85–92, 2006.

[13] Khan, S., Y. Peng, E. Steinbach, M. Sgroi, and W. Kellerer, Application-Driven Cross-Layer Optimization for Video Streaming over Wireless Networks, *IEEE Communications Magazine*, 44(1):122–130, 2006.

[14] Gong, M. X., S. E. Midkiff, and S. Mao, Design Principles for Distributed Channel Assignment in Wireless Ad Hoc Networks, *Proceedings of IEEE ICC*, Vol. 5, pp. 3401–3406, 2005.

[15] Srivastava, V., and M. Motani, Cross-layer Design: A Survey and the Road Ahead, *IEEE Communications Magazine*, 43(12):112–119, 2005.

[16] Kawadia, V., and P. R. Kumar, A Cautionary Perspective on Cross-layer Design, *IEEE Wireless Communications*, 12(1):3–11, 2005.

[17] Bourse, D., M. Muck, O. Simon, N. Alonistioti, K. Moessner, E. Nicollet, D. Bateman, E. Buracchini, G. Chengeleroyen, and P. Demestichas, End-to-End Reconfigurability (E2ER II): Management and Control of Adaptive Communication Systems. Presented at IST Mobile Summit, June 2006.

[18] Demestichas, P., V. Stavroulaki, D. Boscovic, A. Lee, and J. Strassner, m@ANGEL: Autonomic Management Platform for Seamless Cognitive Connectivity to the Mobile Internet, *IEEE Communications Magazine*, 44(6):118–127, 2006.

[19] Sutton, P. D., L. E. Doyle, and K. E. Nolan, A Reconfigurable Platform for Cognitive Networks, *Proceedings IEEE CROWNCOM*, Mykonos Island, Greece, June 2006.

[20] Mähönen, P., M. Petrova, J. Riihijärvi, and M. Wellens, Cognitive Wireless Networks: Your Network Just Became a Teenager, *Proceedings of IEEE INFOCOM*, 2006.

[21] Narendra, K. S., and M. A. Thathachar, *Learning Automata: An Introduction*, Prentice Hall, 1989.

[22] Labrou, Y., and T. Finin, A Proposal for a New KQML Specification, Technical Report CS-97-03, University of Maryland, 1997.

[23] Finin, T., J. Weber, G. Wiederhold, M. Genesereth, R. Fritzon, D. McKay, J. McGuire, R. Pelavin, S. Shapiro, and C. Beck, Draft Specification of the KQML Agent-Communication Language, Technical Report, The DARPA Knowledge Sharing Initiative, 1993.

[24] Jin, J., and K. Nahrstedt, QoS Specification Languages for Distributed Multimedia Applications: A Survey and Taxonomy, *IEEE Multimedia*, 11(3):74–87, 2004.

[25] Unsal, C., Intelligent Navigation of Autonomous Vehicles in an Automated Highway System: Learning Methods and Interacting Vehicles Approach, PhD Thesis, Virginia Polyechnic Institute and State University, January 1997.

[26] Haleem, M. A., and R. Chandramouli, Adaptive Downlink Scheduling and Rate Selection: A Cross-layer Design, *IEEE Journal on Selected Areas in Communications*, 23(6):1287–1297, 2005.

[27] Bond, A. H., and L. Gasser, *Readings in Distributed Artificial Intelligence*, Morgan Kaufmann, 1988.

[28] Jennings, N. R., K. Sycara, and M. Wooldridge, A Roadmap of Agent Research and Development, *Autonomous Agents and Multi-Agent Systems*, 1:7–38, 1998.

[29] Sycara, K. P., Multiagent Systems, *AI Magazine*, 19(2):79–92, 1998.

[30] Dietterich, T., and P. Langley, Cognitive Networks, M. Qusay, ed., *Machine Learning for Cognitive Networks: Technology Assessment and Research Challenges*, Wiley-Interscience, 2007.

[31] Wooldridge, M. J., *An Introduction to Multiagent Systems*, Wiley, 2002.

[32] Pecarina, J., Creating an Agent Based Framework for Maximizing Information Utility, Masters Thesis, AFIT/GCS/ENG/08-19, Air Force Institute of Technology, 2008.

[33] The GNU Radio Project, accessed April 2008; available at *http://www.gnu.org/software/gnuradio/*.

[34] da Silva, C. R., C. M., W. C. Headley, J. Reed, and Y. Zhao, The Application of Distributed Spectrum Sensing and Available Resources Maps to Cognitive Radio Systems, *Proceedings Information Theory and Applications Workshop*, La Jolla, CA, 2008.

The Role of IEEE Standardization in Next-Generation Radio and Dynamic Spectrum Access Developments

Ralph Martinez, Donya He

BAE Systems, Reston, Virginia

23.1 INTRODUCTION

IEEE standards have played an important role in the implementation of communications and network systems in the last four decades. IEEE standards for equipment and computers started with interfaces for serial and parallel communications methods. In the early 1980s, the IEEE standards for local area networks (LANs) defined the 802.3 Ethernet specification standards that helped shape the development of the Internet. Multiple technology variants of 802 standards were defined in the next 28 years, including 100 Mbps and 10 Gbps Ethernet, and 802.11 (WiFi) and 802.16 (WiMAX) for wireless networks. The current focus in international wireless standards is on dynamic spectrum access (DSA) methods and spectrum management in wireless radio access networks (RANs). DSA algorithms sense the immediate spectrum environment and determine which frequencies to use. Spectrum management provides the coordination and utilization of frequency bands in radio terminals that belong to a RAN. While there are several commercial implementations of DSA and spectrum management, the current state of the art continues to evolve, captured in the international standards for wireless RANs. The development of DSA methods and spectrum management has been a technology enabler for intelligent and adaptive radio terminals that are now called cognitive radios (CRs). There have been several definitions of cognitive radios in the last ten years [1, 2]. The IEEE SCC41 P1900.1 Working Group on Definitions and Terminology define a CR as follows [3]:

> a) *A type of* Radio *in which communication systems are aware of their environment and internal state and can make decisions about their radio operating behavior based on that information and predefined objectives.*
>
> Note—*The environmental information may or may not include location information related to communication systems.*

b) *Cognitive Radio (as defined in a) that utilizes radio, adaptive radio, and other technologies to automatically adjust its behavior or operations to achieve desired objectives.*

The promise of CRs is to provide intelligent use of limited spectrum in international communities that are heavily engaged in spectrum utilization. The US Department of Defense (DoD) is interested in improving spectrum utilization among legacy communications systems. Cognitive radios have been addressed by the Federal Communications Commission (FCC) for communications services in unlicensed very high frequency (VHF) and ultra high frequency (UHF) television bands. In software-defined radios (SDRs), many standards, such as WiFi (IEEE 802.11), Zigbee (IEEE 802.15.4), and WiMAX (IEEE 802.16), incorporate some form of CR technology for spectrum utilization and management.

This chapter contains a snapshot of the IEEE standards activities in 2008 for wireless environments. The authors summarize material from multiple sources, interpret the benefit of the IEEE standards, and give a brief analysis on CR technology through standards implementation. The future directions of IEEE standards are projected with respect to dynamic spectrum management (DSM) in heterogeneous wireless RANs. There are several other international standards activities taking place at the same time. Coordination of these activities is important to achieve international standards that can be realized in both the commercial and military sectors. Figure 23.1 shows a summary of international standards and some of the current IEEE standards [4].

Most of the international standards activities are focused on DSA and spectrum management. Table 23.1 summarizes the IEEE standard activities in the 802.11, 802.15, 802.16, and 802.22 working groups (WGs) that relate to CR technology [5]. Specifically,

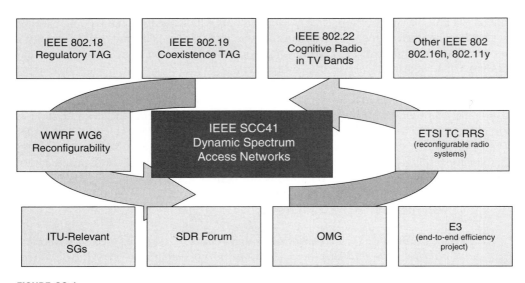

FIGURE 23.1

Summary of international standards in SDRs and cognitive technology [4].

Table 23.1 IEEE 802 Standards Activities Relating CR, DSA, and Coexistence Technologies

Standard	Scope
802.11h	This standard provides mechanisms for DFS and TPC that may be used to satisfy regulatory requirements for operation in the 5 GHz band in Europe. This standard can be applied to other regulatory domains. This document was superseded by IEEE standard 802.11-2007.
IEEE 802.11y	This amendment to the 802.11 standard allows application of 802.11 systems in the 3650–3700 MHz band in the United States. It standardizes mechanisms required to allow shared 802.11 operations with other users, such as specification of new regulatory classes (extending 802.11j), sensing of other transmitters (extending 802.11a), transmit power control (extending 802.11h), and dynamic frequency selection (extending 802.11h).
IEEE 802.15.2	This standard provides recommended practices for coexistence of IEEE 802.15 WPAN with other selected wireless devices operating in unlicensed frequency bands. The standard recommends practices for IEEE standard 802.11, 1999 edition devices to facilitate coexistence with IEEE 802.15 devices operating in unlicensed frequency bands. The standard suggests modifications to other IEEE 802.15 standards to enhance coexistence with other selected wireless devices operating in unlicensed frequency bands.
IEEE 802.15.4	This standard defines the protocol and interconnection of devices via radio communication in a PAN. The standard uses carrier sense multiple access with a collision avoidance medium access mechanism and supports star and peer-to-peer topologies. The standard defines DCS and operates at low power to support coexistence with other wireless devices.
802.16.2-2003	One of the first coexistence standards, this recommended practice provides guidelines for minimizing interference in fixed BWA systems. It addresses pertinent coexistence issues and recommended engineering practices, guidance for system design, deployment, coordination, and frequency usage. It covered frequencies of 10–66 GHz frequencies in general, but focuses on 23.5 to 43.5 GHz. This standard was superseded by the 802.16.2-2004 version.
802.16a	This amendment to the 802.16-2001 standard expands its scope by extending the WMAN air interface to address operational frequencies from 2–11 GHz. It adds DFS and TPC techniques. The standard includes an Annex (B.2) that discusses coexistence in license-exempt bands and provides interference analysis.
802.16.2	This revision of the 802.16.2-2001 added treatment of coexistence in the 2–11 GHz bands to the 802.16.2-2001 standard.
IEEE 802.16h	This amendment to the 802.16 standard specifies improved mechanisms (as policies and medium access control enhancements) to enable coexistence among license-exempt systems based on IEEE standard 802.16 and to facilitate the coexistence of such systems with primary users.
IEEE 802.16m	This amendment to the 802.16 standard provides an advanced air interface for operation in licensed bands. It meets the cellular layer requirements of IMT-advanced next-generation mobile networks while providing continuing support for legacy WMAN-OFDMA equipment.
802.19 WRAN	This standard recommends methods for assessing coexistence of WRANs. The document defines recommended coexistence metrics and methods for computing these coexistence metrics. The focus of the document is on IEEE 802 wireless networks, although the methods developed may be applicable to other standards development organizations and development communities.

this chapter describes the IEEE 802.22 and SCC41 WG activities, summarizing them and explaining how they are related to each other.

The IEEE SCC41 standard activities is one of the first standardization efforts that concentrates on system-level architectures and specifications. In previous standards, IEEE working groups have addressed specific technology functions and algorithms, mostly at the medium access control (MAC) and physical (PHY) layers. The SCC41 WGs, established as the P1900 standards activities in 2005, attempt to coordinate the IEEE CR technology activities at several levels. Table 23.2 summarizes the SCC41 WG activities.

Table 23.2 IEEE SCC41 Standards Activities Relating System-Level CR, DSA, Spectrum Management, and Coexistence Technologies

Standard	Scope
IEEE P1900.1: Terminology and Concepts for Next-Generation Radio Systems and Spectrum Management	This standard provides technically precise definitions and explanations of key concepts in the fields of spectrum management, cognitive radio, policy-defined radio, adaptive radio, radio, and related technologies. The standard describes how these technologies interrelate and create new capabilities while providing mechanisms for new spectrum management paradigms (e.g., DSA and dynamic spectrum assignment).
IEEE P1900.2: Recommended Practice for Interference and Coexistence Analysis	This standard describes the recommended practice for technical guidelines for analyzing the potential for coexistence or interference between radio systems operating in the same frequency band or between frequency bands.
IEEE P1900.3: Recommended Practice for Conformance Evaluation of Radio (SDR) Software Modules	This standard specifies techniques for testing and analysis to be used during compliance and evaluation of radio systems with DSA capability. The standard specifies radio system design features that simplify the evaluation challenge.
IEEE P1900.4: Architectural Building Blocks Enabling Network-Device Distributed Decision Making for Optimized Radio Resource Usage in Heterogeneous Wireless Access Networks	This standard defines the building blocks comprising (1) network resource managers, (2) device resource managers, and (3) the information to be exchanged between the building blocks, enabling coordinated network-device distributed decision making, which will aid in the optimization of radio resource usage, including spectrum access control, in heterogeneous wireless access networks. The standard will be limited to the architectural and functional definitions at the first stage. The corresponding protocol definitions related to information exchange will be addressed at a later stage.
IEEE P1900.5: Policy Language and Policy Architectures for Managing Cognitive Radio for Dynamic Spectrum Access Applications	This standard defines a policy language (or a set of policy languages or dialects) to specify interoperable, vendor-independent control of CR functionality and behavior for DSA resources and services. The standard defines a set of policy languages and their relation to policy architectures, for purpose of managing the features of CRs for DSA applications. Initial work concentrates on standardizing the features necessary for a policy language to be bound to one or more policy architectures to specify and orchestrate the functionality and behavior of CR features for DSA applications. Future work in this standard builds on this foundation to standardize how this is done in greater detail, paying special attention to interoperability concerns.

23.2 DEFINITIONS AND TERMINOLOGY

Each IEEE standards document includes a set of definitions, terminology, abbreviations, acronyms, and normative references. The IEEE 802.22 standard contains definitions and normative references that are specific to the definition of the spectrum-sensing functions, spectrum management, and geolocation service. These definitions are consistent with the 802.16 standard definitions and reflect new terms required in IEEE 802.22 that make the terminal cognitive. An example of a new term is the spectrum management (SM) module defined in the standard. The SM functionality is required at the basestation (BS) to control and coordinate the terminal operating modes and features. The SM functions include maintaining spectrum availability information, and the capability to make decisions on channel selection, channel management, and self-coexistence mechanisms. The interaction of the SM module in the BS is described in the standard at the MAC and PHY levels. Dynamic spectrum management is addressed at a higher level in the IEEE SCC41 (P1900) standards.

Definitions and terminology take on a new level in the IEEE SCC41 WGs. To establish a common level of understanding for new system-level architecture and functionality in heterogeneous wireless networks, the SCC41 Committee established the IEEE P1900.1 WG on "Standard Definitions and Concepts for Dynamic Spectrum Access: Terminology Relating to Emerging Wireless Networks, Spectrum Functionality, and Spectrum Management." As is demonstrated in the IEEE 802 set of standards, there is also considerable overlap in the definitions and terminology within the SCC41 WGs. In some instances, the definitions vary slightly within the context of the working group scope.

The P1900.1 standard attempts to coordinate a coherent view of the cognitive network (CN) technology area. The basic idea is to standardize and explain technically precise definitions related to CR technology. The standard describes definitions and key concepts in the fields of spectrum management, cognitive radio, policy-defined radio, adaptive radio, and related technologies. The standard includes basic definitions and terminology with amplifying text that explains the context of the technologies that use them. The standard describes how these technologies are interrelated and provides mechanisms supportive of new spectrum management paradigms (e.g., DSA). The P1900.1 standard goes beyond providing short definitions of terms. In addition to the short normative definitions, the standard provides informative sections that elaborate on the normative definitions. This approach is necessitated by the complexity of the terms being defined and their interrelationship. For example, informative tables and diagrams explain the relationships of CRs, software-controlled radios, intelligent radios, and adaptive radios. The definitions provided in this standard stem predominantly from a spectrum management point of view. Consequently, one section of the standard is devoted to normative spectrum management definitions. An informative section includes regulatory issues related to advanced radio systems such as CR and spectrum sharing.

The P1900.1 definitions and terminology are categorized into (1) definitions of advanced radio system concepts, (2) definitions of radio system functional capabilities, (3) definitions of network technologies that support advanced radio system technologies, (4) spectrum management definitions, and (5) a glossary of ancillary definitions. Some of the definitions in P1900.1 include the following [3].

Cognitive radio (defined in Section 23.1).

Intelligent radio: A type of CR that is capable of machine learning.

Machine learning: The capability to use experience and reasoning to adapt the decision-making process to improve subsequent performance relative to predefined objectives.

> *Note*—This notion of learning corresponds most closely to the subdiscipline of machine learning known as reinforcement learning. Learning in the intelligent radio context is meant to exclude learning "by being told" (e.g., acquiring information through environmental sensing, messages received from other systems, configuration files, and initialization parameters). Rather, learning implies the adaptation of decision making based on direct experience resulting from previous actions.

Policy-based radio: A type of **radio** in which the behavior of communications systems is governed by a policy-based control mechanism (see also *policy-based control mechanism*).

> *Note 1*—Policies may restrict behaviors (e.g., policies constraining time, power, or frequency use) associated with a specific set of radio functions, but do not necessarily change the functional capability of a radio. Because policies often do not change basic radio functionality, a policy-based radio need not also be a reconfigurable radio.

> *Note 2*—Because the definition for the term policy-based control mechanism considers radio policy to be a type of radio-control software, policy-based radio is considered a subset of software-controlled radio.

Cognitive control mechanism: A component of a *cognitive radio* that assesses inputs (e.g., environmental, spectral, and communications channel conditions) and predefined objectives to make decisions about radio operating behavior.

Cognitive radio network (CRN): A type of radio network in which the behavior of each radio is controlled by a cognitive control mechanism to adapt to changes in topology, operating conditions, or user needs.

> *Note*—Nodes in a cognitive wireless network do not have to be CRs. Rather, a CN is a network of radio nodes in which the nodes are subject to cognitive control mechanisms. Each node may have cognitive capabilities or it may receive instructions from another node with such capabilities. The cognitive capabilities potentially include awareness of the network environment, network state and topology, and shared awareness obtained by exchanging information with other nodes (typically neighboring nodes) or other network-accessible information sources. Cognitive decision making considers this collective information; this decision making may be performed in coordination or collaboration with other nodes. CNs are able to adjust radio behavior of each node to adapt to changes in topology, operating conditions, or user needs.

Dynamic spectrum access: The real-time adjustment of spectrum utilization in response to changing circumstances and objectives.

> *Note*—Changing circumstances and objectives include (but are not limited to) energy conservation, changes of the radio's state (e.g., operational mode, battery life, location, etc.), interference avoidance (either suffered or inflicted), changes in environmental/

external constraints (e.g., spectrum, propagation, operational policies, etc.), spectrum-usage efficiency targets, quality of service (QoS), graceful degradation guidelines, and maximization of radio lifetime.

Dynamic spectrum assignment: (a) The continuous update or assignment of specific frequencies or frequency bands within a wireless network operating in a given region and time to optimize spectrum usage. (b) The dynamic assignment of frequency bands to RANs within a composite wireless network operating in a given region and time to optimize spectrum usage.

> *Note*—The definition in (b) is specific to a class of network and device dynamic reconfiguration scenarios that enable coordinated network–device distributed decision making, including spectrum access control in heterogeneous wireless access networks as described in the draft standard for "Architectural Building Blocks Enabling Network–Device Distributed Decision Making for Optimized Radio Resource Usage in Heterogeneous Wireless Access Networks being developed by SCC41 project P1900.4."

Dynamic spectrum management: A system of spectrum management that dynamically adapts the use of spectrum in response of information about the use of that spectrum by its own nodes and other spectrum-dependent systems.

> *Note*—DSM helps to address the inherent inflexibility of static band allocations and the ability of future networks to simultaneously carry traffic corresponding to multiple radio communications services.

An example of the overlap in the IEEE SCC41 WG definitions is the use of DSA as a use case. The definition of DSA in IEEE P1900.4 [6] is:

> *Dynamic spectrum access:* The process and mechanisms for a type of spectrum access that occurs when different RANs and terminals dynamically access spectrum bands which are overlapping, in whole or in part, causing less than an admissible level of mutual interference, according to regulatory rules, and may be done with or without negotiation.

This definition overlaps with the P1900.1 definition for dynamic spectrum assignment, and is used in context with one of three use cases defined in P1900.4 that describes how frequency bands assigned to RANs are used. A frequency band can be shared by several RANs. The network reconfiguration managers in each RAN analyze the information from the spectrum-sensing algorithms and dynamically make spectrum access decisions according to composite wireless network policies.

In summary, the definitions and terminology in the IEEE 802.22 [7] and SCC41 WGs are mainly consistent; however, there are variances due the context of the specific standard. This condition is common among international standards that have similar and overlapping scopes and purposes.

23.3 OVERVIEW OF THE IEEE STANDARDS ACTIVITIES

The current evolution of the IEEE 802.22 and SCC41 standards enables commercial implementations of cognitive radio (terminals with spectrum-sensing and spectrum

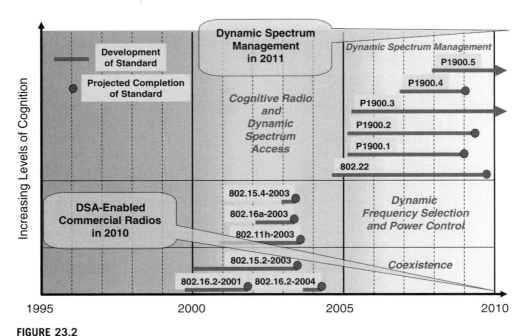

FIGURE 23.2

Projected evolution of IEEE standards that produce commercial products for CRs in 2009 and DSM in 2011 [5].

management functions) by 2009. The P1900.4 standard for DSM is on track for product deployment in 2011. Figure 23.2 shows the evolution of IEEE standards starting with coexistence standards development in 1999 (802.16.2). The early coexistence standards provided methods of measuring interference and mitigating interference through manual intervention. A second generation of standards includes such capabilities as DFS and TPC. Radios use coexistence to operate among other radios using different protocols in the same bands. Coexistence is demonstrated in unlicensed bands where a wide variety of unrelated protocols are used (e.g., IEEE standards IEEE 802.11, IEEE 802.15, and IEEE 802.16). The DSA standards in 802.22 and SCC41 address coexistence and have morphed from DFS work to provide a more intelligent and adaptive approach for CR implementations [8].

The CR and DSA standard timeline started with the IEEE 802.11h, 802.15.2, and 802.22 WGs. This work has been continued by the IEEE SCC41 WGs, as shown in Figure 23.2. The figure shows the relationship between the related IEEE 802.wiseless standards and SCC41 that deal with coexistence, dynamic frequency selection and power, CR, DSA, and DSA spectrum management. It is projected that DSA-enabled radios will appear commercially in 2009, and DSA spectrum management will start operation in 2011. This ambitious development schedule depends on the success of the SCC41 standards suite and the acceptance and subsequent use within the telecommunications and networking industries.

The standards development timeline shown in Figure 23.2 is a snapshot of the evolving standard at the time of this publication. Within the scope of definition of the CR in the P1900, standards activities have progressed at a rapid rate. This rapid rate of standards development has occurred at the expense of lack of measurement and prototypes that demonstrate the concepts and algorithms in the P1900 standards. The move toward standard implementation in the future by commercial and military markets will include testbeds for CR network measurement and demonstrations. The material in this chapter concentrates on the description of the evolving IEEE 802.22 wireless regional area network (WRAN) and SCC41 standards.

IEEE P1900 is a set of RAN standards established in 2005 jointly by the IEEE Communications Society and the IEEE Electromagnetic Compatibility Society. The P1900 committee objective is to develop supporting standards dealing with new technologies and techniques being developed for next-generation radio and advanced spectrum management. In March 2007, IEEE P1900 was reorganized as the Standards Coordination Committee 41 (SCC 41). The summary of the P1900 WGs is presented in Section 23.5. The current SCC41 WGs are:

- IEEE P1900.1: Terminology and Concepts for Next-Generation Radio Systems and Spectrum Management, *http://grouper.ieee.org/groups/scc41/1/index.htm.*
- IEEE P1900.2: Recommended Practice for Interference and Coexistence Analysis, *http://grouper.ieee.org/groups/scc41/2/index.htm.*
- IEEE P1900.3: Recommended Practice for Conformance Evaluation of Radio (SDR) Software Modules, *http://grouper.ieee.org/groups/scc41/3/index.htm.*
- IEEE P1900.4: Architectural Building Blocks Enabling Network–Device Distributed Decision Making for Optimized Radio Resource Usage in Heterogeneous Wireless Access Networks, *http://grouper.ieee.org/groups/scc41/4/index.htm.*
- IEEE P1900.5: Policy Language and Policy Architectures for Managing Cognitive Radio for Dynamic Spectrum Access Applications, *http://grouper.ieee.org/groups/scc41/5/index.htm.*

Access to WG information is available to the SCC41 entity and P1900 membership. A spectrum-sensing working group has been discussed within SCC41; however, the scope and purpose has not been approved at this time. The IEEE P1900.4 WG was formed in February 2007 and originally was the P1900.B study group, entitled "Architectural Building Blocks Enabling Network–Device Distributed Decision Making for Optimized Radio Resource Usage in Heterogeneous Wireless Access Networks." The P1900.4 objective is to standardize supporting procedures for flexible spectrum access in networks; it is active in defining RAN architectures and functions for distributed decision making to optimize radio resource usage.

23.4 IEEE 802 COGNITIVE RADIO-RELATED ACTIVITIES

The 802.22 WRAN standard was in its letter ballot phase at the time this chapter was written. Hence, the contents within this chapter include a snapshot of the proposed features and characteristics of the standard. The authors have interpreted the 802.22 WRAN standard and have summarized key features and highlights of the standard in the following sections.

The IEEE 802.22 WG was formed in 2004 in response to the FCC resolution entitled NPRM 04-186 that defines provisions that allow license-exempt devices to operate in the TV band providing they can coexist with existing services. The frequencies in the TV-band cover frequencies between 54 MHz and 862 MHz at 6, 7, or 8 MHz bandwidths. The standard defines a WRAN that contains CR devices that can sense the immediate spectrum. The 802.22 WRAN is the first IEEE standard to define how cognition in communications radios can be used in basestation and user terminals in a regional area network.

Specifically, the primary goal of the IEEE 802.22 WG on WRANs was to develop a standard for a CR-based PHY/MAC air interface for use in license-exempt wireless communication devices on a noninterfering basis with a TV broadcast spectrum. The 802.22 WG developed the specifications for a fixed point-to-multipoint WRAN environment that uses specific television channels and guard bands for communications in the UHF and VHF TV bands. The standard enables deployment of interoperable 802 multivendor WRAN products to facilitate competition in broadband access by providing alternatives to fixed infrastructure broadband access. The standard enables deployment of such systems into different geographic areas, including sparsely populated rural areas, while preventing harmful interference to incumbent licensed services in the TV broadcast bands. A secondary objective for the 802.22 standard is to serve dense population areas where spectrum is available.

The WRAN 802.22 describes the capability to support a mix of data, voice, and audio/video applications including Internet access, Voice over Internet Protocol (VoIP), and streaming video. The target markets and applications addressed by the 802.22 WRAN standard are single-family residential, multidwelling units, private enterprises, office and home locations, multitenant buildings, hotel complexes, shopping malls, small municipalities, and university and college campuses. In accordance with ITU-R definitions, the WRAN provides access to one or more core networks for Internet access, rather than forming an end-to-end communication system.

The 802.22 WG produced many working documents. One of the important normative documents is the Functional Requirements document [9]. These working documents served to scope and put the 802.22 WRAN standard into perspective with not only other IEEE standards, but also introducing new capability into network basestations and customer premise equipment (CPE). The Functional Requirements document includes operating conditions and performance parameters, such as the capacity of the user terminal to operate at 1.5 Mbps downlink and 384 Kbps uplink per user. The service availability due to radio frequency (RF) propagation is assumed to be at 50 percent of locations to allow the service provider to reach subscribers in fringe areas and 99.9 percent of the time to provide a reliable connection where it is possible. The average spectrum efficiency over the coverage area is expected to be around 2 bps/Hz given the adaptive modulation parameters and the operating constraints described later, assuming a 6 MHz TV channel bandwidth and a 40:1 oversubscription ratio resulting from the stochastic nature of the data network usage. This translates into a total of 255 user terminals that can be served by the basestation per TV channel. These parameters include the combination of coverage and capacity factors that affect access cost per user; the deployability, maintainability, and product costs associated with the customer premise installation; and spectrum reuse for serving the required number of customer locations with a minimum number of basestation locations and backhaul routes. Figure 23.3 shows the main characteristics of the 802.22 WRAN standard relative to the other existing wireless network standards. The figure depicts where 802.22 fits in with respect to the other similar wireless networking standards under the IEEE 802 umbrella. Typical IEEE 802.22 range is 17 to 30 km. Under special circumstances (e.g., the atmospheric tunneling), the BS may be able to serve CPEs located as far 100 km.

Many definitions are used in the 802.22 WRAN standard. One term that is used is regional area network (RAN). The same acronym is used in the IEEE P1900.4 standard, however it means *radio access network*. While Figure 23.3 could be used to describe similarities in the use of the two terms, the use of the RAN term must be considered in the corresponding context of each standard. Other definitions and terminology are included here for information [7]:

- *Customer premise equipment (CPE):* A generalized equipment set providing connectivity between a BS and a subscriber.

- *Channel selection function (CSF):* The function that combines the spectrum-sensing function's outputs from the BS and the CPE generate the available channel list.

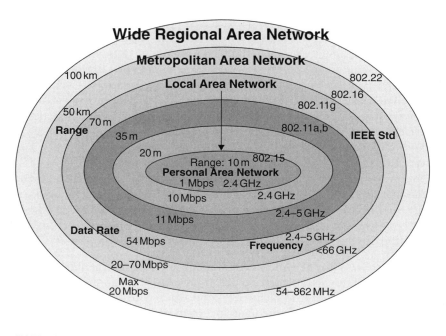

FIGURE 23.3

Frequency and range characteristics of the 802.22 WRAN standard relative to the other wireless network standards [10].

- *Spectrum-sensing function (SSF):* The function that observes the RF spectrum for a defined set of signal types and reports the results of its observations.

- *Waypoint:* A WRAN device the location of which is known and trustable and is to be used for geolocation purposes. A waypoint may be either a BS or a CPE that is located at a surveyed point and professionally installed.

Other definitions for terms, such as spectrum manager and geolocation, can be found in the 802.22 WRAN document sections.

23.4.1 802.22 Reference Architecture

In addition to the MAC/PHY layer descriptions and interfaces, the IEEE 802.22 standard describes a reference architecture for BS and CPE consideration. Several architectures were brought forth by industry in the WG. Almost every candidate proposed three major functions and several interfaces that make a BS or CPE DSA-enabled [6]. The functions include (1) spectrum-sensing function, (2) spectrum manager function, and (3) geolocation function. Specification of the three functions and describing how they provide cognitive features that integrate with an existing CPE architecture was a major activity of the 802.22 WRAN standard. Figure 23.4 shows an example of a candidate architecture proposed by BAE Systems in 2008 for the reference architecture, the three functions, and the required interfaces to MAC and PHY layers for existing terminals, either BSs or CPEs [11]. The figure also shows the feature contribution that includes a security sublayer with corresponding service access points (SAPs). SAPs are interfaces between sublayers and functions in the reference architecture. The ratified 802.22 WRAN standard may vary from this architecture contribution due to the standards process.

The rationale for this architecture contribution in the 802.22 WRAN standard is to enable a low level of cognition so that the MAC/PHY air interface is frequency agile. The frequency agility is used to change frequencies within fragmented TV bands and still avoid interference to the TV band incumbent services. Specifically, the capability is proposed for generating awareness of multiple TV channels used by incumbents to determine which channels can be used for WRAN dynamic-frequency selection algorithms that avoid the incumbent frequencies. The functionality for this capability is to provide this awareness in real time at the BS and CPE architecture.

This contribution includes the definition of a *cognitive plane* to the existing data and control plane that includes the spectrum-sensing, spectrum management, geolocation, and security sublayer functions. The Network Control and Management System (NCMS) in Figure 23.4 deals with the MAC/PHY layers and the upper protocol layers required to realize the application. The features of the proposed functions are:

Spectrum-sensing function: The SSF makes observations and measurements of the RF spectrum in the TV bands for a set of signal types, stores the observations, and reports them to the spectrum manager (SM). The SSF cooperates with the SM to observe, store, and learn of the terminal's immediate spectrum environment. The SSF may implement various classes of spectrum-sensing algorithms given that the interface and performance requirements are met. Inputs to the SSF may include channel bandwidth, signal type vector, signal type window specification vector, sensing mode, and maximum probability of false alarm. Outputs of the SSF may include sensing

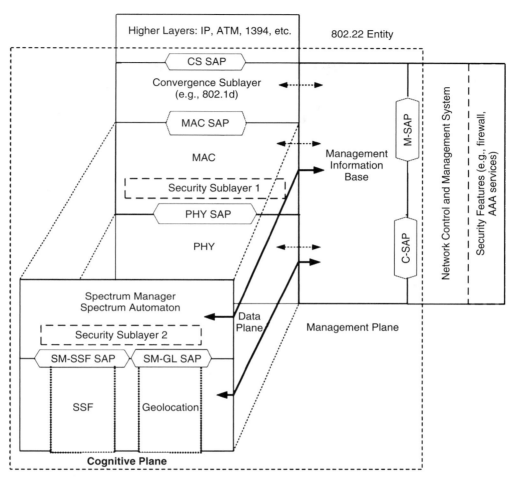

PHY: Physical layer
MAC: Medium access control layer
CS: Convergence sublayer
MIB: Management information base
PHY SAP: Physical layer service access point
MAC SAP: Medium access control service access point
CS SAP: Convergene sublayer service access point
SSF: Spectrum-sensing function
SM/SA: Spectrum manager/spectrum automaton
SM-SSF SAP: Spectrum manager, spectrum-sensing function service access point
SM-GL SAP: Spectrum manager, geolocation service access point
Security Sublayer: Security sublayers for the data/control, management, and cognitive planes
NCMS: Network control and management system
M-SAP: Management SAP
C-SAP: Control SAP

FIGURE 23.4

IEEE 802.22 defines the 802.22 WRAN reference architecture including functions for spectrum sensing, spectrum management, and geolocation [11].

mode and signal type, signal present, confidence, field strength estimate, error standard deviation vectors. The SSF candidate algorithms described in 802.22 take on variants of transmitter detection, cooperative detections, and interference-based detection methods [12].

Spectrum manager: The SM is embedded in every BS and CPE and takes on a corresponding role according to its location. The SM is responsible for the decision making for spectrum utilization that may be guided by an overall network policy and regulations. The SM performs operations to (1) maintain spectrum availability information, (2) channel classification and selection, (3) association control, (4) channel management, (5) interfaces to the SSF and geolocation function, and (6) self-coexistence with other WRANs. At the CPE, a slave function of the SM operates and is called the spectrum automation (SA) function. The SA includes functions for operation when the CPE is not under control of the BS, such as initialization and activation and responding to the BS commands. The SM uses the geolocation (GL) data obtained from the geolocation function to maintain a graphical representation of the CPEs in the vicinity of its host location. The SM includes all local spectrum knowledge at a BS and CPE and can be used to collaborate with other SM in other terminals to achieve networkwide spectrum management.

Geolocation function (GLF): The GLF provides the processes of acquiring the necessary location data, determining latitude and longitude, and producing the National Maritime Electronics Association (NMEA) string information to the SM. The GL has two modes of operation: (1) satellite-based (mandatory) and (2) terrestrial-based. The satellite-based GL mode determines the latitude and longitude of the BS transmitting antenna within a radius of 15 m and its altitude above mean sea level. The BS GL uses NMEA strings provided by the satellite-based GL at the CPE to determine the location of the CPE and the distance between the CPE location and each nearby incumbent protected contour. The terrestrial-based GL mode uses an abstract entity, called the geolocator, to send ranging requests (to the BS), receive the responses (from the BS), and derive the representations from a combination of ranging data and waypoints. The goal of this process is to allow the geolocator to compile a graphic representation of the set of CPEs that form a cell under a BS's control.

Security sublayer (SSL): Traditional broadband communications systems (e.g., 802.16) contain data, and control and management functions that require protection. However, due to the unique characteristics of the 802.22 systems, which include CR capability-enhanced security mechanisms are needed. These security features provide protection for the 802.22 users, service providers, and most importantly, the incumbents, who are the primary users of the spectrum. As a result, the protection mechanisms in 802.22 are divided into several security sublayers that target noncognitive, as well as cognitive functionality of the system and the interactions between the two. The SSL 1 provides subscribers with privacy, authentication, and confidentiality across the broadband wireless network. It does this by applying cryptographic transforms to MAC PDUs carried across connections between CPE and BS. To enhance the security for the cognitive functionality in 802.22, the SSL 2 is introduced. The security mechanisms for this layer include functions for authentication of the sensing information, ensuring spectrum availability, authorization to use the spectrum, authorization to configure the spectrum manager, confidentiality, and privacy of the competition-sensitive spectrum awareness information.

Management information base (MIB): In general, the MIB can be used to store the elements that are used to configure a communications device, or to pass on sensed/collected parameters to various entities, such as an Internet Service Provider (ISP) or a BS. MIBs make the system modular and well defined, and allow for relatively easy remote management. These MIB functions can be configured locally using a station management entity (SME), which may reside in the 802.22 Entity or remotely through the NCMS.

Network Control and Management System (NCMS): The NCMS abstraction allows the specified PHY and MAC layers to be independent of the network architecture, the transport network, and the protocols used at the back end, therefore allowing greater flexibility. The NCMS logically exists at the BS and CPE side of the radio interface and is termed NCMS (BS) and NCMS (CPE), respectively. Any necessary inter-BS coordination can be handled through the NCMS (BS). The NCMS allows remote management of the CPEs by the exchange of certain configuration parameters through an MIB. The NCMS may have an authenticator to ensure authentication, authorization, and accounting of any configuration information coming into the CR and its MIBs.

23.4.2 **802.22 PHY Layer Overview**

The IEEE 802.22 PHY protocol is based on the same technologies in the fixed-broadband 802.16 (2004) standard. Specifically, IEEE 802.22 uses Orthogonal Frequency-Division Multiple Access (OFDMA) as the PHY layer transport mechanism. The major differences between the IEEE 802.16 and IEEE 802.22 OFDMA protocols are the frequency ranges, channel sizes, special considerations, and options. Special considerations include better support that address the unique challenges associated with operating over the TV frequency band. The specification is designed for the 802.22 system using vacant TV channels to be able to provide wireless communication over a distance of up to 100 km under special circumstances. Fluctuating and diverse channel characteristics (e.g., multipath and atmospheric conditions) are a problem that the 802.22 WRAN WG must consider. Appendix D in the 802.22 standard [13] provides multipath and delay-spread profiles.

Spectral Efficiency

Spectral efficiency is an important performance parameter of any wireless access system. Spectral efficiency vary from location to location within the coverage area of the BS, depending on distance, propagation, channel impairments, interference, and their effect on the usable modulation and coding parameters over each given path. The 802.22 WRAN standard requires that the network must be capable providing spectral efficiencies in the range of 0.76-bit/sec/Hz to 3.78 bits/sec/Hz or better per transmission link depending on the transmission channel conditions.

Bandwidth Scalability

The current version of the IEEE 802.22 standard does not support bandwidth scalability. Hence, if an incumbent signal such as the TV or the wireless microphone is detected in Channel N, then 802.22 systems are supposed to vacate not just Channel N, but also Channels N + 1 and N − 1 to avoid intermodulation distortion. Based on the success and deployment of 802.22, future versions of the standard may allow operation in a fraction of a TV channel when interference is present with a requirement that the occupied bandwidth must be scalable. On the other hand, where spectrum is available, it may be useful for a WRAN system to use more than one TV channel (contiguous or not) to increase the capacity of the transmission link. However, from the modulation point of view, each TV channel must be able to free any one of these channels in case of interference. The 802.22 WRAN standard requires the minimum delivered peak data rate per subscriber to be 1.5 Mbps forward link and 384 Kbps return link.

Link Availability

An 802.22 WRAN system provides 99.9 percent of the link availability while supporting all intended services at the rated QoS levels. Graceful degradation through a reduction of the throughput rather than a complete loss of service is also required. The WRAN service coverage is based on 50 percent location availability at the edge of the coverage area for a median location. However, based on the preference of the system operator, the 802.22 WRAN system allows for the radio link to be engineered for different link availabilities.

Modulation and Coding

The IEEE 802.22 protocol uses OFDMA as the PHY transport mechanism with 2048 subcarriers. The subcarriers are classified as: (1) data subcarriers, (2) pilot subcarriers, (3) guard and null (including DC) subcarriers. The classification is based on the functionality of the subcarriers. Binary phase-shift keying (BPSK) up to 64-bit quadrature amplitude modulation (64-QAM) schemes are supported with the ability to dynamically adapt the method used as channel conditions change. Convolutional coding is mandatory with optional support for convolutional turbo codes (CTCs), low-density parity-check codes (LDPCs), and the shortened block turbo codes (STBCs).

Advanced Transmission Mechanisms

Much like the 802.16 and 802.20 standards, future versions of the 802.22 may support advanced transmission options—that is, STBC, adaptive beam forming, and various forms of multiple input, multiple output (MIMO) and space-division multiple access (SDMA). A feedback channel referred to as uplink channel sounding may be provided in the OFDMA framing structure to support the feedback paths required by mechanisms such as adaptive antenna system (AAS) and SDMA.

23.4.3 802.22 MAC Layer Overview

The 802.22 WRAN MAC protocols provide functions for coexistence and protection of TV bands incumbent services, as well as for self-coexistence. The central purpose of the MAC protocol is sharing of radio channel resources. The MAC protocol defines how and when a BS or subscriber station may initiate transmission on the channel. The 802.22 MAC is connection-oriented and provides flexibility in terms of QoS support. A BS manages all the activities within its 802.22 cell and all associated CPEs. The MAC supports unicast (addressed to a single CPE), multicast (addressed to a group of CPEs), and broadcast (addressed to all CPEs in a cell) services. The MAC layer implements a combination of access schemes that efficiently controls contention between users, while at the same time meeting the delay and bandwidth requirements. This is accomplished through four different types of uplink scheduling mechanisms using unsolicited bandwidth grants, polling, and contention procedures. The use of polling simplifies the access operation and guarantees that applications receive service on a deterministic basis.

Connection Establishment and Framing

Connection establishment in the 802.22 WRAN standard is different from that of 802.16 because other incumbent users may use the BS frequencies. As a result, the BS could be operating on any channel within the UHF/VHF band at any given time, making the task of performing user authentication and registration much more difficult. The CPEs account for this when they powerup by first searching all of the channels in the area to see if a BS is present. The presence of a BS is differentiated from other UHF/VHF users by the preamble sent at the start of each OFDMA frame. Once a user locates a BS, authentication and connection setup is done by the user injecting messages into the contention-based connection setup time allocated at the start of each frame.

Data in 802.22 networks are transmitted by an OFDMA frame structure where the BS controls all downlink traffic and users must request uplink slots before they transmit. Minor additions will be required in the future version of the standard to allow for channel bonding. If the channel bonding is used, a "super-frame header" is likely to contain information to indicate to the supporting users which channels they should look for data transmission. Furthermore, to provide support for users that do not support channel bonding a portion of the OFDMA super-frame may be reserved for such users to alert the BS of their restrictions so that the BS does not attempt to schedule transmissions and upload slots that the user cannot physically receive.

Incumbent Sensing Measurement and Detection

Incumbent sensing measurement and detection are essential for the 802.22 standard. To share the VHF/UHF bands with TV broadcasters or wireless microphones, a foolproof system of incumbent avoidance techniques needs to be implemented—a multitiered approach to sensing that aims to have a minimal impact on overall system performance. In the 802.22 standard, BSs control CPEs to conduct measurement activities and obtain measurement results. Scheduled quiet periods for sensing to take place may be required for this matter. Sensing should include signal signature identification and possible transmit unit identification.

At the time this chapter was written, it was proposed that both the BS and user terminals should participate in incumbent detection. This allows sensing tasks to be distributed. The BS reserves portions of the OFDMA frame to provide users with a list of channels to monitor. Users then perform sensing, collect these data, and report to the BS at prespecified intervals. Once all sensing information is collected, the BS then creates a revised list of occupied and unoccupied channel allocations. Depending on the required level of accuracy, sensing is done in either "fast" or "fine" mechanisms. Fast sensing is performed in-band during the guard interval portions of OFDMA frames. Fine sensing is done out-of-band during defined "quiet periods" when there is no network traffic. The BS schedules "quiet periods" ahead of time so that all user stations can synchronize their sensing with one another. This method of sensing provides a higher degree of accuracy, but at the cost of network overall throughput. Therefore, fine sensing should only be used when fast sensing did not detect any incumbents.

Self-Coexistence and Inter-BS Coordination

The issue of coexistence between competing 802.22 system operators becomes more significant when sharing spectrum. In areas with a significant number of incumbent users, open channels will be a scarce commodity. Therefore, when multiple 802.22 systems are located in the same region, a collaborative channel allocation and reservation method needs to be used so that no single system will occupy all available channels. The IEEE 802.22 WRAN protocols provide mechanisms for the exchange of information between BSs to allow for coexistence, interference avoidance, and sharing of radio resources among neighboring 802.22 basestations. CPEs and BSs will report interference received from other 802.22 basestations or CPEs. Neighboring BSs will cooperate to take measures to resolve any conflicts between competing 802.22 systems for channel allocations.

23.5 IEEE SCC41: DYNAMIC SPECTRUM ACCESS NETWORKS

In this section we will review the activities of the IEEE SCC41 standards coordinating committee. The work of this group is primarily involved in cognitive radio and dynamic spectrum access considerations. Included in this committee are five subcommittee activities labeled P1900.1, P1900.2, P1900.3, P1900.4, and P1900.5, all of which are explained in further detail in the following sections.

23.5.1 IEEE P1900.1: Terminology and Concepts for NG Radio Systems and Spectrum Management

The IEEE Standards Board approved the IEEE P1900.1 standard "Terminology and Concepts for Next-Generation Radio Systems and Spectrum Management" on June 12, 2008. The P1900.1 standard defines terms to establish common terminology for describing emerging networks and nodes employing radio devices characterized by cognition, adaptation, environment awareness, and policy-based adaptive techniques. Specifically, the definitions provided in this document are predominantly intended for spectrum management for RANs with cognitive terminals. It is anticipated that these definitions will ultimately mature and ideally achieve widespread acceptance among researchers, manufacturers, service providers, regulators, and operators.

The objective of the standard is to promote a common understanding of systems technology and spectrum management terms so that technologists in a variety of fields such as radio science (e.g., digital communications, computer science, and artificial intelligence) and regulators have a common understanding of the terminology [14]. The document provides concise definitions of key terms in advanced radio system technologies and in advanced spectrum management techniques. The focus of the standard is on terms and concepts relating to emerging wireless networks, radio, technology, system functionality, and spectrum management. It was agreed that, in some cases, multiple definitions for a specific term were appropriate. The development of this standard required: (1) the creation of new terminology (i.e., cognitive radio, CR network) and the development of definitions for these new terms, and (2) the development of alternative definitions for existing terms (e.g., dynamic frequency selection, dynamic frequency access) that have been defined by other international standards organizations. These alternative definitions were required for next-generation radio and spectrum management because the terms have new meanings when used in the context of discussing advanced radio systems.

To define all of the terms related to wireless heterogeneous radio networks is outside the scope of this standard. Appropriately, existing terms from the ITU-R, SDR Forum, and other IEEE documents are included for convenience, even though the terms and definitions are unchanged for NG radio and spectrum management. These terms are clearly identified in the text. The standard provides normative terms and definitions to support the research and deployment of DSM and DSA. Many factors are creating a need for DSM and DSA. Among the forces creating this need for change are:

- The increasing use of wireless services and their need for spectrum
- The increasing data load being transmitted wirelessly, requiring increasing spectrum bandwidth
- The emergence of multimode products, such as mobile, broadcast, and radiolocation, into single devices
- Increasing pressure to guarantee spectrum access for priority services (e.g., public safety) while allowing other uses for that same spectrum when not in use by those priority services

23.5.2 IEEE P1900.2: Recommended Practice for Interference and Coexistence Analysis

Due to CR technology in radio terminals, many radio systems coexist and attempt to optimize the utilization of spectrum in space and time simultaneously. The accurate measurement of interference has become a crucial requirement for the deployment of these CR technologies. The mandate of the 1900.2 WG is to recommend the interference analysis criteria and establish a well-thought-out framework for measuring and analyzing the interference between radio systems. New technologies, although attempting to improve spectral efficiency by being flexible, collaborative, and adaptive, can cause disputes. Therefore, this WG intends to establish a common standard platform by which the disputing parties can present their cases and resolve them amicably. The framework for interference analysis addresses the context of measurements and their purpose. Any adaptive system has a trade-off between cost and return on investment (ROI). Thus, the interference analysis should make the ROI explicit, together with the usage model used in the trade-off. Apart from the interference, power measurements and the context, impact, and remedies are important parts of the analysis and comparison. Finally, parameters for analysis are derived from scenarios, including the context and harmful interference thresholds. Uncertainty levels in measurements are compulsorily considered in the analysis. The P1900.2 standard describes the framework that measures and analyzes the interface between RANs with cognitive terminals.

23.5.3 IEEE P1900.3: Recommended Practice for Conformance Evaluation of SDR Software Modules

The P1900.3 standard describes techniques for testing and analysis to be used during compliance and evaluation of RANs that have DSA capability. The P1900.3 standard specifies radio system design features that simplify the evaluation phase and challenges. Although most of the current work has been performed by vigorous research on prototyping of RANs with DSA, the potential for deployment is affected by the ability of regulatory agencies and industry stakeholders to verify that a system conforms to applicable technical and policy requirements. The rationale for the project came from the need expressed by regulatory agencies and industry to have a standard for effective testing and analysis of the spectrum access behaviors and capabilities of DSA-enabled radio systems. The accepted DSA paradigm encompasses the idea that conformance assessment is more challenging than in static spectrum access radio systems. The spectrum access behavior of static systems is typically modifiable only through human-controlled interfaces in ways that have earned regulator and stakeholder trust. In a RAN with DSA-enabled radios, sophisticated control mechanisms may be used to decide when it is safe to transmit (e.g., frequency, power level, and bandwidth) based on a number of inputs (e.g., spectrum sensor readings, time and location information, and a database of rules that defines what spectrum resources may be used at any given time and place).

The complexity of the components involved can make it difficult to validate the spectrum access behavior of the radio system. Several stakeholders can benefit from the P1900.3 standard. Regulators can draw on the standard when setting radio system certification requirements. Manufacturers can design to the requirements of the standard and reduce certification risk. Other stakeholders (e.g., spectrum rights holders) can analyze the standard to determine whether radio systems tested and analyzed in this way are acceptable secondary users (SUs) of their spectrum. Finally, international harmonization of regulatory and other stakeholder requirements will promote market development by reducing barriers to the sale and use of radio systems with DSA.

23.5.4 IEEE P1900.4: Architectural Building Blocks Enabling Network–Device Distributed Decision Making for Optimized Radio Resource Usage

This standard activity defines the building blocks comprising (1) network resource managers, (2) device resource managers, and (3) the information to be exchanged between the building blocks, enabling coordinated network–device distributed decision making; this will aid in the optimization of radio resource usage, including spectrum access control in heterogeneous wireless access networks. The standard will be limited to the architectural and functional definitions at the first stage. The corresponding protocol definitions related to information exchange will be addressed at a later stage.

The purpose of P1900.4 is to improve overall composite capacity and QoS of wireless systems in a multiple RANs and technologies environment by defining an appropriate system architecture and protocols that facilitate the optimization of radio resource usage. The approach exploits information exchange between network and radio terminals, whether or not they support multiple simultaneous links and DSA.

Multimode reconfigurable devices are increasingly being adopted within the wireless industry. The choice among various supported air interfaces on a single wireless device is a reality today for second-, third-, and fourth-generation (2G, 3G, and 4G) cellular radio access technologies and IEEE 802 wireless standards. Devices and networks with DSA capabilities allowing the use of spectrum resource simultaneously among different RANs has become a reality. The P1900.4 standard addresses this need by defining the overall system architecture, functional descriptions, and information exchange between the network and devices. The overall benefit is that the composite wireless network efficiency and capacity is improved.

Benefactors and stakeholders include CR technology users (e.g., wireless device end users, regulators, operators, telecommunications companies, military users, and radio terminal manufacturers). The DSM improvements targeted in this standard will lead to the optimum exploitation of a heterogeneous wireless radio network environment for all the stakeholders. The benefits to the user community are the ability to obtain required QoS, throughput, latency at minimum cost, and assurance of access anytime, anywhere.

23.5.5 P1900.5: Policy Language for Managing CR for DSA Applications

This standard defines a policy language (or a set of policy languages or dialects) to specify interoperable, vendor-independent control of CR functionality and behavior for DSA resources and services. The standard defines a set of policy languages and their relation to policy architectures for the purpose of managing the features of cognitive radios for DSA applications. Initial work in this WG concentrated on standardizing the features necessary for a policy language to be bound to one or more policy architectures to specify and orchestrate the functionality and behavior of CR features for DSA applications. Policy generation for cognitive terminals in RANs are required to dynamically manage the spectrum via frequency band assignment context information at the radio terminal. Future work in this standard builds on this foundation to standardize how this is done in greater detail, paying special attention to interoperability concerns.

23.5.6 Additional Detail on the P1900.4 Standard

What is the system concept and architecture for DSM in a wireless heterogeneous environment that has multiple RANs? This section contains the authors' interpretation of the P1900.4 at the time of writing this chapter. The authors have taken the liberty to represent a top-level set of views in the Unified Modeling Language (UML) to describe the major classes and components in the standard. Some of the UML views were generated by members of the P1900.4 Information Model subgroup and are included here. The overall set of views is not necessarily the consensus view of the P1900.4 WG members; however, the unratified P1900.4 standard at the time of publication is presented.

Historically, network management of a RAN has been performed by single network management system owned by a single organization with some automated interfaces, but mostly manual, to other operating RANs. The P1900.4 standard is key in the suite of the SCC41 standard because of its system-level architecture and functionality descriptions.

The P1900.4 standard attempts to describe the functions and architecture for managing and controlling multiple RANs owned by a single or multiple organizations. In order to accomplish this goal, the RANs consist of DSA-enabled radios per the IEEE 802.22 WRAN standard. The P1900.4. standard also allows interoperability with legacy radio systems. As an example, in global spectrum management systems, the radio terminals and BSs can be viewed as the DSA sensors in each technology RAN. In the heterogeneous wireless environment, IEEE P1900.4 specifies a network management system comprised of a network reconfiguration manager (NRM), terminal reconfiguration manager (TRM), operator functions, and their internal and external interfaces. The purpose of this network management system is to coordinate frequency band allocation and to optimize radio resource usage, including overall capacity and quality of service among the participating RANs. Figure 23.5 shows the functional architecture of DSM in a wireless heterogeneous network per the scope of the P1900.4 standard [5, 15]. The overall composite wireless network consists of several technology RANs and legacy networks (e.g., 3G, 4G, WiFi, and WiMAX). P1900.4 assumes that the terminals in the RANs may (P1900.4 terminal) or may not be DSA enabled. Policy-based DSM is provided through the NRM and executed by the TRMs in each RAN. UML diagrams are used to describe the overview of the P1900.4 system architecture and the NRM and TRM object classes in the next sections.

The heterogeneous wireless environment in P1900.4 is called a Composite Wireless Network (CWN) and is shown as a UML deployment diagram in Figure 23.6: It is derived from the current overall system block diagram in the P1900.4 standard. A CWN may

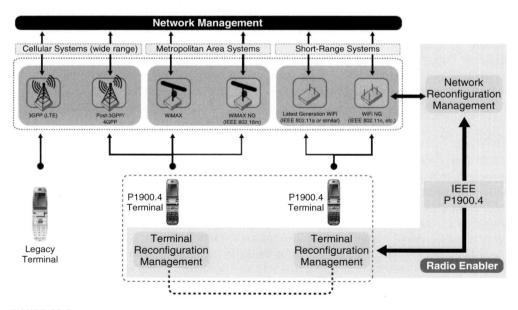

FIGURE 23.5

The P1900.4 standard describes how TRMs and NRMs interact to dynamically manage the spectrum in a wireless heterogeneous environment [5, 15].

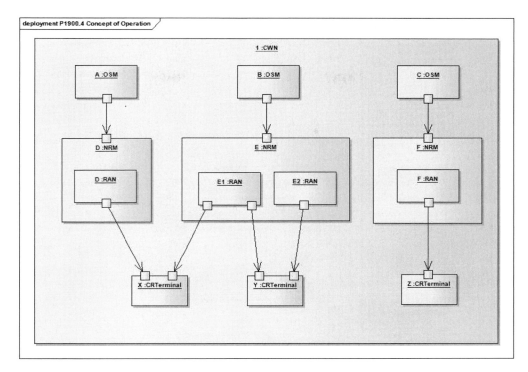

FIGURE 23.6

A heterogeneous wireless environment that forms a CWN as addressed in IEEE P1900.4.

contain multiple RAN classes, each with different network operators. The RANs are owned by a single or multiple organizations. The RANs may have different radio terminal technologies, each under the control of a single network operator. Cognitive radio terminals (called CRTerminal) are frequency reconfigurable and have a multihoming capability. The CRTerminal have the capability to connect to one or several RANs. Regional area networks are managed by the NRMs. Mulitple RANs can be controlled by one NRM. The connection mechanism of the CRTerminal classes to the RANs determines three use cases of the P1900.4 standard discussed in the following sections. The P1900.4 standard describes the architecture and functionality for management of DSA across a CWN that includes several RANs with different radio technologies.

The P1900.4 system architecture is hierarchical in nature. It defines several object classes within the CWN package. Each RAN contains the following object classes: (1) NRM, (2) TRM, and (3) operator spectrum manager (OSM). The RAN operator communicates with the OSM and the OSM communicates with the NRM and TRM (through the NRM). Each RAN may contain multiple instantiations of the BS and CRTerminal classes. These relationships are shown in a system-level class diagram in Figure 23.7.

Within the figure the NRM is the management object class that supports the RAN with terminal functionality initialization, spectrum policy and rule sets, initial software downloading, software updating, RAN state monitoring, fault monitoring, RAN

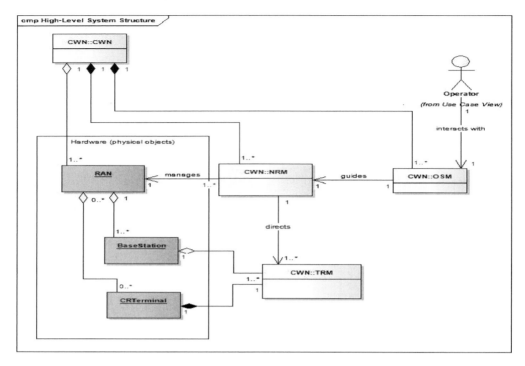

FIGURE 23.7

System-level class diagram for P1900.4 showing the main object classes that manage dynamic spectrum in a CWN.

reconfiguration, frequency band selection, and collaboration interface with other RANs. The NRM communicates with the TRM object class to provide terminal initialization, context information exchange, frequency band use, policy-based operation mode, and status query. Each RAN consists of BaseStation and CRTerminal object classes that represent the CPE. The TRM has appropriate instantiations in the BaseStation and CRTerminal objects. The OSM object class, in the case of multiple RANs existing per operator, has the role of providing levels of collaboration between the NRMs in the RANs. In the case of one RAN per operator, the OSM provides the operator the interface regarding possible changes to the spectrum assignment. In either case, the OSM manages each RAN with one operator.

At the time of this writing, a preliminary draft of the NRM and TRM functions and their interfaces were available. The NRM is further decomposed into the RAN reconfiguration controller (RRC) and the RAN measurement collector (RMC) objects. The TRM contains the terminal reconfiguration controller (TRC) and the terminal measurement collector (TMC) objects. These objects are responsible for reconfiguration and measurement services to manage the BaseStation and CRTerminal objects in each RAN. Eventually, P1900.4 will provide protocols for these managed objects in future versions of the standard. Figure 23.8 shows the preliminary NRM and TRM object class functions. Their interfaces are found in the current P1900.4 standard document.

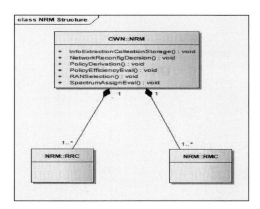
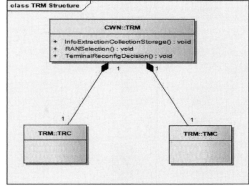

FIGURE 23.8

Functions of the NRM and TRM object classes in P1900.4.

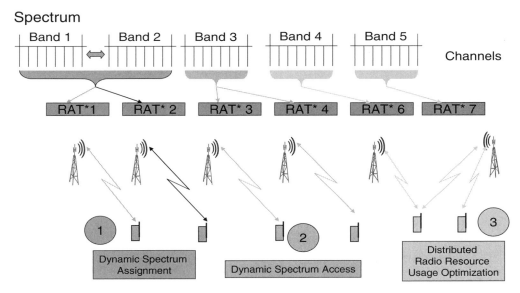

* Radio Access Technology (in commercial domain);
analogous to waveform in military context.

FIGURE 23.9

P1900.4 describes three reference use cases in the standard [7].

The P1900.4 standard defines three reference use cases as shown graphically in Figure 23.9. The terminology used for the CPE in this figure is radio access terminals (RATs). The use cases are: (1) dynamic spectrum assignment, (2) DSA, and (3) distributed radio resource usage optimization.

The three uses cases are shown in Figure 23.10 in context with the CWN, `BaseStation`, `CRTerminal`, and `Operator` classes. The component interfaces to the use

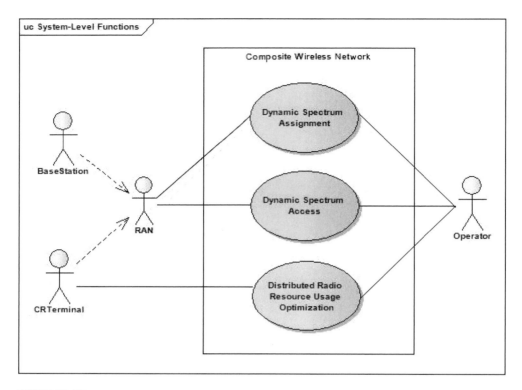

FIGURE 23.10

The P1900.4 CWM operation is defined by three reference use cases.

case functions go through the RAN and its operator. The BaseStation and CRTerminal interact with the RAN through the network and terminal resource management functions that are defined in detail next.

Dynamic Spectrum Assignment

The dynamic spectrum assignment use case describes how the OSM generates spectrum assignment policies that enforce policies and tacit operation such as changes to a RAN spectrum assignment. The OSMs provide spectrum assignment policies to the NRMs. The NRMs in each RAN analyze the spectrum assignment and dynamically make spectrum assignment decisions to make better use of the spectrum. For instance, this use case may cover situations where:

- A new carrier is added for 3G or 4G access.
- A frequency band previously used for 3G is assigned to mobile broadband wireless access, for example, 802.16e.
- The network switches the usage of a spectrum band from mobile broadband wireless access (802.16e) to wireless LAN (IEEE802.11n). This scenario may occur when a large number of wireless LAN terminals are suddenly close to the wireless LAN access point.

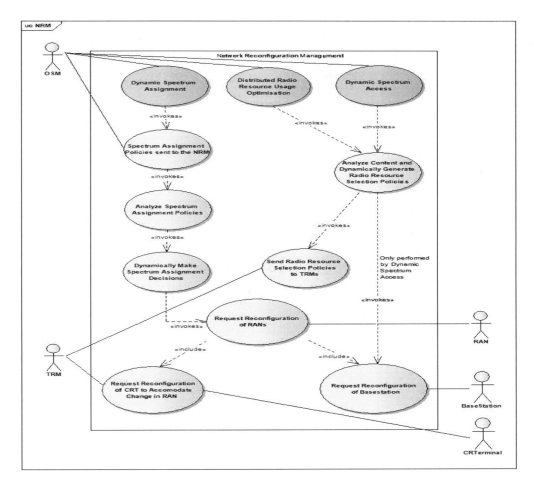

FIGURE 23.11

Three use case relationships in P1900.4.

Figure 23.11 shows the DSM use-case interaction with the other use cases. In this example, the OSM, NRM, `BaseStation`, and `CRTerminal` are used as actors to depict the interaction with the functions in the dynamic spectrum assignment scenario.

Dynamic Spectrum Access

The DSA use case describes how frequency bands assigned to RANs are used. A frequency band can be shared by several RANs. NRMs analyze the information from the DSA algorithms and dynamically make spectrum access decisions according to policies. NRMs request reconfiguration information from their RANs and dynamically generate terminal feature selection policies and send them to their TRMs. These radio feature selection policies guide these TRMs in their spectrum access decisions. Dynamic spectrum access may be done with or without negotiation. Figure 23.12 shows the DSA use

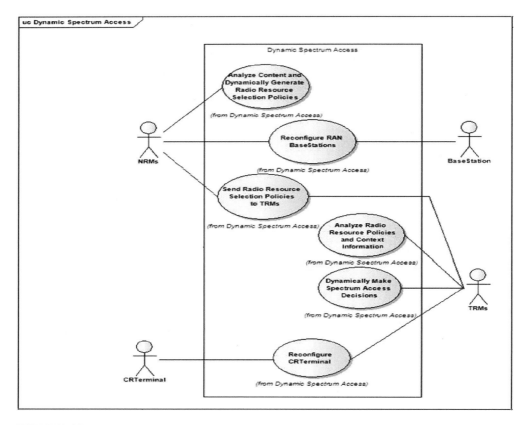

FIGURE 23.12

DSA use case in P1900.4.

case with reconfiguration to the `BaseStation` and `CRTerminal` classes. Examples pertinent to this use case are: unlicensed secondary systems (e.g., IEEE802.22) accessing licensed, but unused, TV spectrum bands, and unlicensed wireless LANs (e.g., IEEE802.11n) accessing licensed, but unused, TV spectrum bands.

Distributed Radio Resource Usage Optimization

The distributed radio resource usage optimization use case in Figure 23.13 demonstrates how the IEEE P1900.4 system can be applied to legacy RANs to make better use of the spectrum. Frequency bands assigned to RANs are fixed, and the BSs cannot change the bands. This use case assumes CR terminals have a multimode and/or multihoming capability. Multimode CR terminals can have only one active connection; however, they can dynamically reconfigure the connection by accessing different RANs with different radio interfaces. Multihoming terminals can have multiple simultaneous connections with different RANs and can dynamically reconfigure these connections. Reconfiguration decisions are made by the TRM with policy-based information from the NRM.

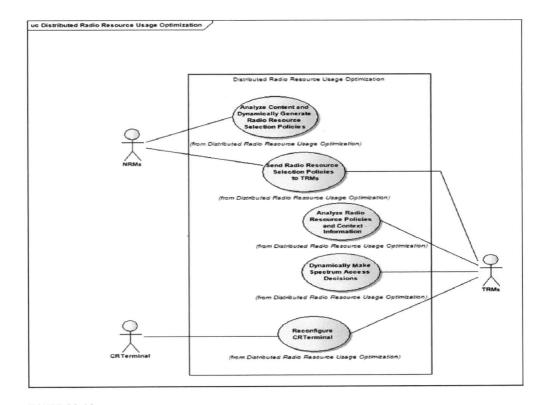

FIGURE 23.13

Distributed radio resource usage optimization use case in P1900.4.

Finally, the P1900.4 standard includes a section on an information model that describes the information representation used in the CWN (including the RAN, OSM, NRM, and TRM) and CRTerminal functional descriptions. The CWN and CRTerminal are considered managed classes in P1900.4. The CRTerminal classes abstract the users, applications, devices, and radio resource selection policy, as well as different terminal profiles, capabilities, and measurements related to the CRTerminal. The CWN classes abstract the CWN functionality, capturing the operator and RANs, spectrum access and assignment policies, and the RAN basestations and terminals. The information model abstraction is aligned with the IEEE P1900.4 scope. For example, the application class in the hierarchy of the CRTerminal class does not incorporate generic application attributes but only those that have been identified within the standard's scope. In addition, policy information abstracted by the policy classes is a fundamental part of this standard, ensuring the communication of policies to NRMs and TRMs that govern their operation and constraints. IEEE P1900.4 therefore presents a concrete representation of events that trigger policy activations and executions, the conditions within which policies must act, and the precise actions that must be undertaken should, for example, a CRTerminal be found to violate a policy. The information model classes are

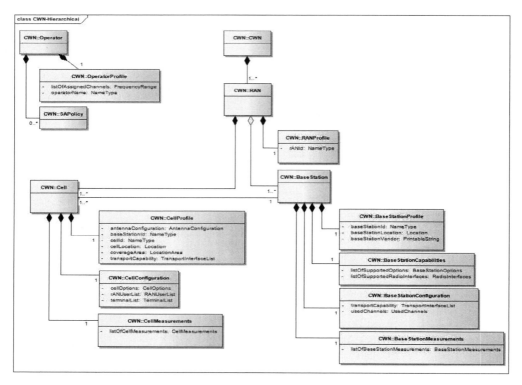

FIGURE 23.14

The CWN contains an information model representation of the data classes in P1900.4.

grouped into these categories: CRTerminal-related classes, CWN-related classes, and policy classes.

Figure 23.14 shows the information model representation for the CWN class. The consistency of the information model with the functional description of the P1900.4 has been a standardization challenge in light of the evolving standard and its aggressive schedule. Figure 23.14 also introduces the concept of a "cell."

23.6 POTENTIAL FOR NEW PRODUCTS AND SYSTEMS

IEEE standards are well known to enable worldwide markets for interface and communications products that enhance interoperability between different vendor technology implementations. The IEEE 802 and SCC41 standards are expected to make a major impact in future CR products and networks. This suite of standards is the first instance in IEEE of activities that provide adaptive, intelligent, and agile capabilities that make the radio terminal cognitive for the purpose of optimizing spectrum environment use. The standards define both radio terminal technology and DSM architectures that allow for managing several RANs that belong to one or several organizations. Radio terminals in the RANs will be DSA enabled via the 802.22 WRAN standard defining the SSF, spectrum management function, and the GLF.

The P1900.4 standard manages the radio terminals and RANs, P1900.2 defines the interference environment via measurements and analysis methods, and P1900.3 allows the conformance evaluation techniques. When completed, P1900.5 will enable the definition of several policy languages that enable RANs to interoperate under multiple policies and regulations that govern cognitive RAN operations. Radio terminal and network management products are expected to evolve in the 3G, 4G, WiFi, and WiMAX markets by 2009 and 2011. The next sections discuss potential new products and systems.

23.6.1 **Commercial**

Spectrum availability is a key enabler for NG wireless systems and their applications. For example, use of the 802.22 WRAN standard will revolutionize the use of TV frequency bands between 54 MHz and 862 MHz. Incumbent frequency band users will be able to operate without interference from white space users in the same frequency band due to 802.22 technology. 3G and 4G RANs with cognitive terminals will enable additional frequency band use for video streaming, voice communications, and text messaging. Eventually, use of the 802.22 standard will migrate to other frequency bands such as those used by WiFi and WiMAX. WiFi users will have uninterrupted services while coexisting with wireless home phones and microwaves in the same band. WiFi hotspots will become more efficient in their use of immediate spectrum.

Application areas that will benefit from the 802.22 WRAN and SCC41 standards will be public safety, intelligent transportation systems, medical systems, and wireless communications services. The FCC has provided frequency bands for applications and services in several areas and has enabled spawning of several markets in wireless communications services [16]. One can envision systems, applications, and devices using the CR and DSA management standards reviewed here. The following represent potential areas for applications of the standards:

- Wireless handset games
- Frequency use in municipalities
- Internet wireless connectivity
- Cyber attack detection and prevention
- Automobile and transportation applications
- Commercial signal detection and classification
- Wireless security systems
- Telecommunication networks frequency optimization
- Enterprise wireless frequency management

23.6.2 **US Department of Defense**

The US DoD was a research and development leader of DSA and spectrum management in the early 2000s. Military forces in countries of operations now face unique spectrum access issues. They deal with spectrum usage constraints by coalition forces, enemy use, and local civilian and government users. These constraints are codified in spectrum usage rules, regulations, policy, and tacit information, and they must be accounted for in complex mission-operation plans including spectrum utilization during operations. Spectrum management is a mechanism for achieving information superiority on the battlefield and thus a challenging priority problem for military forces.

The DARPA NeXt Generation (XG) communications program started in 2002 and developed both theoretical bases and demonstrations for DSA and management. The XG program developed dynamic control of the spectrum, the technologies and subsystems that enable reallocation of the spectrum, and the system prototypes that demonstrate coexistence between legacy and future DoD RF emitters. The XG program conducted several field demonstrations showing frequency sensing and white space discovery in the presence of several frequency users based on spectrum policy regulations [14, 15]. The XG Dynamic Spectrum Access software identifies what frequency assignments can be used at a single transceiver without causing interference to other legacy users. It does not determine which frequencies (consistent with topology and other constraints) should be used. The coordination of DSA-enabled radios must be managed on a networkwide basis. The XG program is the basis for DSA and future spectrum management systems for military networks.

23.7 SUMMARY

This chapter has presented current activities in standards for commercial CR technology. These activities include IEEE 802.16, 802.22, and the WGs of IEEE SCC41 (P1900.1, P1900.2, P1900.3, P1900.4, and P1900.5). All of these WGs are works in progress, and the authors have presented their interpretation of the features and benefits of the standards. Architectures and functional specifics are described in sufficient detail in these developments that the commercial telecommunications industry can build on for internationally harmonized processes for next-generation equipment deployment that take full advantage of cognitive principles. The IEEE SCC41 and 802.22 standards [10] are expected to evolve and meet the dynamic spectrum utilization and management in future 3G, 4G, and wireless information networks.

REFERENCES

[1] Mitola III, J., Cognitive Radio: An Integrated Agent Architecture for Software Defined Radio, PhD Dissertation, Royal Institute of Technology, June 2000.
[2] Facilitating Opportunities for Flexible, Efficient, and Reliable Spectrum Use Employing Cognitive Radio Technologies, ET Docket No. 03-108, December 30, 2003.
[3] Hoffmeyer, J., et al., Standard Definitions and Concepts for Dynamic Spectrum Access: Terminology Relating to Emerging Wireless Networks, System Functionality, and Spectrum Management, *IEEE P1900.1/D4.00*, March 2008.
[4] Guenin, J., IEEE Standards Coordinating Committee 41 on Dynamic Spectrum Access, Motorola, ITU-R WP5A SDR/CR Seminar, February 2008.

[5]   Sherman, M., A. Mody, R. Martinez, and C. Rodriguez, IEEE Standards Supporting Cognitive Radio and Networks, Dynamic Spectrum Access, and Coexistence, *IEEE Communications Magazine*, July(46):72–79, 2008.

[6]   IEEE P1900.4/D1.6 Draft Standard for Architectural Building Blocks Enabling Network-Device Distributed Decision Making for Optimized Radio Resource Usage in Heterogeneous Wireless Access Networks, SCC41 P1900.4 Working Group, June 2008.

[7]   Draft Standard for Wireless Regional Area Networks Part 22: Cognitive Wireless RAN Medium Access Control (MAC) and Physical Layer (PHY) Specifications: Policies and Procedures for Operation in the TV Bands, IEEE 802.22 Working Group, version 1.0, 2007.

[8]   Sherman, M., A. Mody, R. Martinez, R. Reddy, and T. Kieman, A Survey of IEEE Standards Supporting Cognitive Radio and Dyanmic Spectrum Access, *MILCOM*, November 2008.

[9]   Functional Requirements for the 802.22 WRAN Standard, 802.22 Working Group document: IEEE 802.22-05/0007r46, December 2008.

[10] Stevensen, C., G. Chouinard, and W. Caldwell, Tutorial on the P802.22 PAR for: Recommended Practice for the Installation and Deployment of IEEE 802.22 Systems; available at *http://grouper.ieee.org/groups/802/802_tutorials/july06/Rec-Practice_802.22_Tutorial.ppt#2*.

[11] Mody, A., R. Reddy, M. Sherman, T. Kieman, and D. Cavalanti, Protocol Reference Model Enhancements in 802.22; available at *http://mentor.ieee.org/802.22-08-0121-07-0000-text-on-protocol-reference-model-enhancements-in-802-22.doc*.

[12] Akylidiz, I., W. Lee, M. Vuran, and S. Mohanty, Next Generation Dyanmic Spectrum Access Cognitive Wireless Networks: A Survey, *Computer Networks*, 50(15):2127–2159, 2006.

[13] WRAN Reference Model, IEEE 802.22 Working Group on Wireless Regional Area Networks, document 18, IEEE 802.22-04/0002r15, 2004.

[14] Prasad, R. V., P. Pawtczak, J. A. Hoffmeyer, and H. S. Berger, Cognitive Functionality in Next Generation Wireless Networks: Standardization Efforts, *IEEE Communications Magazine*, 46(4):2–78, 2008.

[15] Buljore, S. S., and P. Martigne, Proposed System Concept, P1900.4 Working Group, document P1900.4-07-04-2007, Dublin, April 2007.

[16] Wireless Services (WCS), Federal Communications Commission, July 2008; available at *http://wireless.fcc.gov/services/index.htm?job=wtb_services_home*.

The Really Hard Problems

24

Bruce A. Fette
General Dynamics C4 Systems
Scottsdale, Arizona

"The telecommunications market is nearly 1 trillion dollars per year. With a market of that size, even a few percent is large enough to be an interesting business" [1].

24.1 INTRODUCTION

This second edition includes substantial new material providing great depth of implementation detail regarding smart antennas, reasoning for system optimization, spectrum analysis and primary user detection, accelerated rendezvous, spectrum access rights and performance prediction, waveform adaption and link optimization, cognitive networking, and the standards work of the IEEE on cognitive radio networks.

The ability to improve spectral efficiency, enhance network efficiency, and serve the telecommunication user—the purview of cognitive radio (CR)—is an application that adds significant value to the telecommunication market. The chapters in this book have provided a detailed review of the major technologies that are well understood and stand ready to enable CR. This chapter recaps those technologies, and reviews the current state of the art. It stitches together the pieces, showing how they intermesh to build CR systems, and then presents remaining problems that must be solved.

24.2 DISCUSSION AND SUMMARY OF CR TECHNOLOGIES

As stated in Chapter 1, we are taking the perspective that the CR is a radio that is sufficiently intelligent to: (1) aid spectrum efficiency, (2) aid the radio networks and network infrastructures, and (3) aid the user. At this time, many "cognitive" radios are in development or early production, prototyping in several laboratories is an ongoing activity, and several international testbeds have begun. It is important to recognize that managing spectrum efficiently by itself is completely sufficient justification for CR. At $200 million per megahertz cost of spectrum in the United States, and similar costs in other parts of the world, the ability to achieve greater utilization, and to support more users per megahertz, creates huge economic opportunity. We also envision substantial network services for the user, but as of this moment, the concierge services market in the United States seems to be about $100 million per year. We expect this to be a substantial growth market in the immediate future.

Chapter 2 presented the regulatory perspective regarding cognitive radio. The regulatory climate is positive and is encouraging the demonstration of CR technologies. US research and development (R&D) organizations are providing support for such demonstrations. In the United States, 10 MHz has been provided for experimental evaluation testbeds of cognitive technologies, and in Australia, 25 MHz has just been allocated to CR. Similarly, regulators in Ireland, in conjunction with Trinity College of Dublin, have allocated spectrum and testbed facilities for evaluation of CR techniques. Several testbed programs are under way, funded by the National Science Foundation (NSF) and the Defense Advanced Research Projects Agency (DARPA), as well as by certain corporations; and experiments are under way at several leading universities.

Chapter 3 reviewed the hardware and software architectures of software-defined radio (SDR). These architectures are real, and products are in production. Furthermore, it is clear that the cognitive applications can be added to these SDR architectures as one or more additional applications. Some of the SDRs that are in production already have sufficient general-purpose processor (GPP) computational power as well as enough memory to support the additional tasks of a cognitive engine (CE) running alongside the waveform and protocol processing.

Chapter 4 presented the technologies required to build a cognitive radio. Assuming an SDR as a platform, it showed that the CR must have an ability to sense its environment (e.g., spectrum activity, locally available networks, position, orientation, time, biometrics); a reflection capability to assess its own radio link performance behaviors; a policy engine; and a library of applicable policies, protocols, and waveforms that are useful in its current context. The chapter demonstrated that SDRs exist and are in production.

In fact, they have been integrated with SDR radios in General Dynamics Laboratories, and in at least one university research setting, and have been demonstrated at the IEEE Milcom 2005 conference [2]. Furthermore, General Dynamics's digital modular radios (DMRs) have also demonstrated that by simply adding the proper software applications, SDR radios were able to perform a spectral sensing function and geopositional awareness. The Air Force Rome Labs (AFRL) has demonstrated that an SDR could perform biometric analysis and integrate the biometric reports into the radio's waveform [3].

Chapter 5 presented the details of making efficient use of spectrum. Spectrum efficiency is not simply a matter of choosing waveforms with many bits per symbol. Rather, it is a combination of understanding the interaction of a waveform within the context of legacy communications systems and cognitive systems, as well as understanding the antenna beam management, adaptive power control, and member network topology of the existing users of the spectrum. The DARPA NeXt Generation (XG) program has demonstrated that radios can network together, and they can share sufficient information to use space–time and frequency opportunities. Chapter 5 also significantly analyzed the details of cosite interaction of radios with other nearby radios, explaining how the CR can find frequencies that will reduce cosite desense[1] of the receivers, and

---

[1] *Desense* is a condition where the radio sensitivity is reduced because the automatic gain control (AGC) has reduced the gain of the radio frequency (RF) front-end circuits to accommodate a strong signal, which is not the desired signal.

thereby optimize overall network performance. The XG program is now evolving protocols and etiquettes to standardize spectral-sharing techniques.

Chapter 6 discussed the design properties of a policy engine. The policy engine includes a current environmental status component, a policy decision/analysis component, and a policy enforcement component. The policy enforcement component acts as a final check that proposed waveforms, frequencies, or other actions to aid the network or the user, and are compliant with the radio architecture capabilities, network capabilities, and local regulatory policy. Chapter 6 also provides an extensive discussion of research in policy languages.

Chapter 7 explored how the genetic algorithm (GA) can be used to evaluate the performance of many waveforms and assess which waveform properties lead to a waveform that works well for not just the current link conditions of one network member but for the spectral efficiency of the whole network. Other optimization criteria (e.g., DC power drain on the batteries of the local radio, or even network overhead traffic) may also be included. The GA algorithm may be performed in real time on the current link by the radios currently closing the link, or non–real time by a computer server that has been provided with a channel and environment model. By moving the GA to a non-real-time server, the analysis results can be one of many choices made available to other network members when similar conditions are encountered. In addition, while the analysis is being performed, the actual spectrum and radio traffic is not encumbered by the GA experiments. The currently active radios can be updated to use the most appropriate waveforms by a network server as soon as an acceptable solution is found, whether that acceptable solution comes from over-the-air GA experiments, from prior experience discovered on a server, or from GA analysis performed on a remote computer server. Although the GA algorithm is well known and demonstrated, it is not available at this time as a network service, or to offload link and network optimization, even though there is little or no economic barrier to preclude developing and offering such a service. The immediate next step to accomplishing this is developing and publishing a standardized API for such an interface and the corresponding standardized API for the radio to download the optimized waveform recommendation, and install it.

Chapter 8 reviewed many of the techniques available and in use to perform global, regional, or local position analysis, as well as time awareness. It covered the US global positioning system (GPS), the European Galileo program, and the Russian GLObal NAvigaton Satellite System (GLONASS), as well as the concepts of very high frequency (VHF) omnidirectional ranging (VOR), LOng RAnge Navigation (LORAN), time of arrival (ToA), time difference of arrival (TDoA), and the TV ghost canceling reference (GCR) signal used with regional signal databases. The chapter also lightly touched on the topic of precision time and its availability from GPS, LORAN, TV, and WWV sources. GPS functionality is quite prevalent and is now available in single-chip implementations. Therefore, it is not necessary to use an SDR channel for this function. In addition, precision time is available from high-precision, low-power watch electronics and precision oscillators. Time, with sufficient precision for human users, includes WWV, Loran, and TV, where available. In locations where GPS may be unavailable (e.g., urban areas and the interiors of buildings), GPS may be augmented by inertial navigation techniques. Thus, these technologies are cost effective and available.

Chapter 9 discussed how wireless network protocols must be different from wired network protocols, and how a cognitive system can identify protocols that improve network performance under the current conditions. Such protocol improvements can also be made available from a server to network members, much as the waveform protocols discussed in Chapter 7. We note that considerable research has been performed in wireless ad hoc networking, and that these systems frequently define their own networking protocols. However, wireless networks often terminate into wired networks. At the gateway between the wireless and wired world, a need generally exists for protocol conversion, and it is precisely at these points where intelligent protocols provide critical functionality. It seems likely that gateway wireless servers will keep a collection of protocol applets and apply the most appropriate applets as a function of best choice for the applications under prevailing conditions. They may also serve the corresponding applets to SDRs, to a radio network, or eventually just call them out by number. Gateways have the opportunity to become a very interesting business.

In the United States, DARPA programs have been designed to demonstrate the substantial value of the components of a CR. The XG program has demonstrated improvements in spectral efficiency, and the Situation-Aware Protocols in Edge Network Technologies (SAPIENT) program has demonstrated improvements in radio networking protocols. The Adaptive Cognition-Enhanced Radio Teams (ACERT) and Brood of Spectrum Supremacy (BOSS) programs have the objectives of demonstrating the value of CR networking to the user. In addition, the DARPA Phraselator program has demonstrated the value of having language translation immediately available to the user.

Chapter 10 explored speech input and output. Because speech is analyzed with a signal analysis process that extracts vocal tract shape and excitation properties, these properties can be encoded for voice communication or can be statistically analyzed to extract speech vocal tract behaviors that map into word recognition, language recognition, or speaker identification. The radio's ability to be aware of its user is very significant to providing intelligent services that the user can appreciate and value. These technologies are commercially real. Voice coders (vocoders) are available as software products. Speech (word) recognition is commercially available in productized format. Language recognition has been demonstrated in very useful forms, but it is not generally available as a product. General Dynamics and others have demonstrated such systems. All of these technologies are readily integrated into smart radio systems and into commercial infrastructure to support CR to the user.

Chapter 11 discussed how services can be provided to a radio user from an infrastructure server. This chapter specifically focused on providing a server—called a Radio environment map (REM)—that tracks local radio activity and makes that information available to other radios. As mentioned previously, in addition to radio communication activity, other types of relevant information may also be provided as services to wireless users. Chapter 11 adds significant detail to the database structure to prepare a detailed application programming interface (API) and an analysis of the overhead involved in providing REM radio service.

Chapter 12 described knowledge representation, reasoning, and learning algorithms, ranging from neural networks to case-based reasoning (CBP). Reasoning and learning are essential parts of exhibiting cognitive behavior. If the user is to expect the radio to provide intelligent help, it must exhibit the ability to reason and learn. Humans reason and learn without being taught how they do so, but this remains a daunting computational task for a radio, regardless of whether it has large computational power or the very small computational power of a handheld radio. This task is further daunting since the human performs such tasks very quickly. The radio may take longer than a human to reach the same conclusions, depending on the complexity of the reasoning and the learning algorithms employed. There are, however, other things that the human will not learn quickly—or maybe not learn at all—that may be more readily learned by the radio. The timeliness of the cognitive processes will be quite important to the user.

Chapter 12 amplifies reasoning about time and schedules to provide the basics of reasoning about the time to plan and update a network reflecting recommended waveform and frequency changes, and to thereby determine whether the time and signaling cost is warranted when compared with the predicted improvement in network performance. The means for tying together these timeliness goals and the complexity of reasoning required is still very much a developing field, with current exploration in the artificial intelligence (AI) community.

There is other information the radio may be able to assess, such as the choices of local radio networks and their corresponding performance (e.g., data rates), and their cost effectiveness. Analysis of radio network selection can be expected to learn from the user's behaviors and choices and eventually become a fully automatic activity by the radio. We expect this type of behavior in next-generation personal digital assistant (PDA)-type wireless devices.

Chapter 13 described ontology (knowledge representation) in great detail and gave some simple examples of knowledge about radio protocols in the Web Ontology Language (OWL). For a PDA radio to deliver the level of intelligence of a "Radar O'Reilly," the radio must learn a great deal about many specialties. It is probably impractical for a radio to learn everything to help its user on its own. Rather, specialized areas of expertise may be methodically captured and transcribed, making the radio a "savant" about one topic at a time. Gradually, other economically interesting areas will be added to the domain of radio expertise. Consider, as an example, that making an airline reservation has been automated to the point of making it possible for a computer to capture the necessary keywords from speech and then perform the computerized transactions. (One DARPA contractor may actually be using this as a standard computerized service for its employees.)

In a similar fashion, we can expect other important areas of specialization in which the learning process can be presented with adequate training (including nonroutine conditions), monitored and optimized for the purpose, and then rolled out to the larger set of radios and radio users. The list of references for Chapter 13 shows that this area is an incredibly active field of AI, but commercial deployment is still limited to a few unique applications.

Chapter 14 presented the software architectures of adaptive, aware, and CRs, with the major focus of attention on the Radio Knowledge Representation Language (RKRL) and AI functions. Considerable attention is paid to "computability," to ensure that the reasoning engines can analyze and reach a conclusion in a practical finite time and not be stuck in an endless loop of contingencies. This chapter defines five perspectives of CR architecture and the basic APIs to interface the various CR functions. It relates these architectures to the process of Observe, Orient, Decide, Plan, Act, Learn (OOPDAL). In addition, it addresses how to integrate the CR architecture into the software communications architecture (SCA). Finally, the chapter provides examples of what a user might expect his radio to learn, setting goals for the industry.

Chapter 15 provides a deep mathematical treatment of game theory with application to spectrum efficiency and stability analysis. As CRs adapt their waveforms, they must avoid interference with each other. It is important that the result is not a continuous chain of frequency and waveform changes throughout all the networks (in which each radio's change of behavior causes a change to yet another radio, resulting in unbounded cycles of adaptation and readaptation). This chapter provides an analysis of the stability criteria and fairness issues related to this problem.

Chapters 16 through 23 are new in this second edition. In these chapters, we walk through the entire radio architecture from antenna to systems and standards to discuss specific implementation considerations arising from the systems now in development.

In Chapter 16, Drs. Kim and Choi and their students provided a detailed discussion about smart antennas and the use of smart antenna techniques at both the transmitter and receiver. Beginning with multiantenna beamforming, they lead the reader through decision criteria about how to choose optimal modes given the measured properties of the channel. The chapter offers an architecture for the physical (PHY) layer and cognitive engine (CE), as well as a basic software architecture, leading to a "Cognitive Engine in Multiple Antenna Systems," including the required mathematical principles. The smart antenna principles can provide very significant radio network performance enhancement, using beamforming at the transmitter, and interference nulling at the receiver. However, the price of this significant performance is increased hardware, software, and system complexity to measure the channel, choose the correct modes, and signal mode information among network members. The architecture discussed has been the subject of a standardized Smart Antenna API within the SDR Forum, and WiBro/WiMAX systems deployment.

Chapter 17 covered implementation of the system reasoner using declarative language. The chapter delves deeply into the implementation issues for a CR including permissive and restrictive policy, constraints, underspecified requests, stateless policy, policy language, subjects of policy reasoning (e.g., frequency time, location, node identity, role), and ontology. It provides implementation examples and discusses the Maude language, encoding policies in Maude, and entering and validating a policy. This is a particularly powerful example of a declarative reasoner because it can evaluate incompletely specified requests, and can return a list of additional requirements that have to be met to enable a transmission request.

Chapter 18 covered implementation of the spectral-sensing functions necessary for secondary operation in white space bands where that is allowed by the spectrum regulators. Spooner and Nicholls explain how time-domain and frequency-domain analyses allow the sensor to dig deeply into the noise in order to sense the possible presence of signals from an otherwise hidden node. By classifying the properties of detected signals, a sensor is able to determine whether a signal is a primary user (PU) or a secondary user (SU), and thus what impact the presence of the signal has to the SU. The chapter provides details for matched filtering, energy detection, and multiple types of cyclostationary analysis for ATSC TV, CDMA, GSM, and OFDM examples, as well as for nonstationary signal types. The chapter also goes into detail about efficient computational implementation techniques. In addition, it provides the level of sophistication in spectrum sensing expected of next-generation white space (cognitive) radios that will be required to see PU signals more than 15 dB below the usable communications noise floor.

Cognitive radio systems must be designed to be able to change frequency immediately should a PU begin to use the channel. When this happens they must be able to find other members of the network at new frequencies, a process called rendezvous for the first transmission, or re-rendezvous when nodes have previously been communicating. Chapter 19 provided detailed implementation strategies for aided and unaided rendezvous with important techniques necessary to reduce the search time and increase the probability of acquisition for other net members. Factors greater than 3:1 in reduced search time are demonstrated.

Chapter 20 explored a different approach to modeling and managing the dynamic spectrum, called location-based spectrum rights (LBSR). These techniques provide a means for a PU to specify location, transmit power, antenna patterns, noise, and interference based on communication performance prediction methods, and therefore can be used as a license for PUs. These models are then a constraint to secondary spectrum users. Performance prediction modeling topics include: spectral mask, underlay mask, transmit power modeling, propagation models, power density, protocol and policy, time, interference margins, impact of mobility, impact of multichannel networks, and impact of secondary spectrum market techniques on the LBSR concept. Use of these compact analytic performance prediction techniques is essential to a practical CR system.

Most CRs claim the ability to adapt the PHY layer to current channel conditions. Chapter 21 provided an implementation example of adapting the modulation and forward error correction (FEC) coding to accommodate a 21 dB range of link performance while providing best-known goodput matched to these conditions by using a library of 11 modulation and FEC types. Methods for system initialization and power-adjustment protocols are provided, as well as analysis of the time required to converge, analysis of the impact of time-varying dynamic channels on protocol design, and finally the impact of interference on protocol design. The net result of adapting to a 21 dB dynamic range of propagation loss without changing transmit power, and thus maintaining high levels of link reliability, show that CR techniques can make a profound difference in reliability and network capacity, changing wireless links from brittle to robust.

Other chapters of this book have dealt with the CR, and consequently with optimizing the communication link once that capability is in place. Once CR is in place, it is then possible to optimize the network. In optimizing the network, we coordinate all the nodes at all layers of the protocol stack to optimize for composite network performance, not just the properties of one link. Chapter 22 discussed how to create a merged layer design, and how to cope with cross-layer optimization in which the system optimization interacts with the individual parameter optimization of traditional PHY–Link–MAC design optimization. Three example architectures are discussed, and the cognitive steps of Observe–Orient–Decide–Act (OODA) are discussed with regard to one of the architectures. There is much to be done in the field of cross-layer wireless network optimization, and much performance improvement to be expected.

Chapter 23 is a detailed walk-through of the IEEE efforts to standardize selected implementation considerations of cognitive and white space WLAN and WMAN architectures, showing that CR and CN techniques are now moving into mainstream implementation. The chapter touched on 802.11h, y, 802.15.2, 4, 802.16a, 2, h, m, and 802.19, 802.22, as well as SCC41 Dynamic Spectrum Access and Management Standards. It discusses the reference architecture assumed by these standards and provided the necessary explanation to understand the essential components of the architectures and how those components operate in wireless network systems. Unified Modeling Language (UML) class diagrams are provided for a rapid system implementation startup. While much of this work is still a work in progress at IEEE, the chapter provided an early chance to see the concepts now in the development and standardization cycles, and therefore a chance to get connected to the standard implementation strategies of next-generation commercial products.

24.3 SERVICES OFFERED TO WIRELESS NETWORKS THROUGH INFRASTRUCTURE

In this final chapter, we specifically recognize that there are several classes of radios. To begin with, there is the radio designed and built for defense applications. When used in peacekeeping missions, such radios will be expected to comply with the regulatory requirements of the regions where they are used. They will need to have details of the radio networks of other allies and coalition partners, including waveforms, protocols, frequencies, and what conditions should be communicated with the partners. These radios will be the gateway for many local users to reach many global networks and have access to numerous computer servers and database servers through those gateways. Most such radios will be designed to have considerable built-in intelligence; in addition, the access to intelligent pull and push services through their networks will be very significant. The information available to such a radio is limited by the bandwidth of the most restrictive wireless links providing the access, by the battery power available, and by the total time the mission must be operative on that battery power. Consequently, distributed caching throughout the network, so that information can be found nearby, will become standard in such networked radios.

Although similar to the defense radio in that much of its capability must be embedded, the public safety and emergency response radio will have significantly different applications than the defense radio. Three-dimensional positioning inside buildings is critical, as is the ability to bridge and gateway among multiple networks, depending on which support organizations arrive to the emergency and what their radio communications capabilities are. In this application, the cognitive radio's ability to track which responders need what information is essential, and we will expect CRs to network the proper information to the correct emergency responders. The ability to synthesize multiple waveforms in multiple frequency bands, sometimes in the absence of infrastructure support, is a major issue requiring both SDR and CR functionality.

The second class of radio involves the cellular telephone subscriber unit, the corresponding basestation (BS), and its corresponding infrastructure. Because of the very high volume of subscriber units manufactured, the manufacturers go to great effort to push any complexity from the subscriber unit over to the BS. Two main drivers for this are that the power dissipation and complexity of the basestation and the infrastructure behind the BS are less limiting than for the subscriber units, and costs can be amortized across thousands or even millions of subscribers.

The third class of radios are those embedded into computing devices. This includes laptop computers, PDAs, and similar devices, for which the primary access is a wireless personal area network (WPAN) such as Bluetooth or ZigBee; wireless local area network (WLAN), usually 802.11 waveform; or wireless metropolitan area network (WMAN), which will probably be 802.16 WiMAX. These devices begin with the assumption that there is ample bandwidth in the network and that there are gateway devices providing connectivity to the Internet.

We also predict a fourth class of radio: automobiles that will soon have an add-on business of transmitting and receiving useful services to the driver and passengers by using wireless regional area network (WRAN) services. The automotive industry refers to this as *telematics*. General Motors has been successfully marketing its OnStar™ product in luxury vehicles for nearly a decade. This is the forerunner of very intelligent services provided to the driver. We can easily imagine that many of the following services would be useful and can be enabled by broadband wireless service: directions to specified locations; specific types of locally available services (e.g., gas stations, restaurants, banks, hotels, businesses); specific types of locally available products (e.g., tickets to a play, a copy of this book); interaction with dispatch services (e.g., nearest taxi, pickup and delivery services); traffic avoidance (i.e., only practical as a radio network service); entertainment (e.g., personally selected music or movies for the kids); and drive-through shopping and pickup. As these specialized services finely tune their services to the users, they will become cognitive to specialize and optimize their services for each particular user.

24.3.1 Stand-alone Radios with Cognition

SDR radios of human-pack size and larger are reasonably capable of built-in cognitive capabilities. This can include reasoning and learning about spectral activity, node locations, local networks, network protocols, data caches, standard and nonstandard activities, and functions that local users expect the radio to perform automatically and routinely. Such a radio should also be sufficiently sophisticated to assess how its own communication affects the static and dynamic performance of the network and spectral utilization density, the behavior of local legacy systems, and so forth.

Advanced versions of defense radios will likely organize and maintain themselves by using ad hoc protocols, and gateways will provide connectivity to a wide variety of wireless networks of varying data rate, varying range, and network membership. Such a collection of networks will offer robustness of message forwarding, although optimization is truly a high dimensionality problem. Defense radios can also expect to have access to wideband data networks, which can provide higher cognitive functions for authentication, nonrepudiation, speech understanding, translation, schedule keeping, and prioritized objectives. Much of this functionality is feasible through system integration of existing functionality.

24.3.2 Cellular Infrastructure Support to Cognition

The economic model of cellular telephony strongly encourages two principles: (1) shift as much complexity as possible to the BS and (2) keep the subscriber locked to his or her service provider.

Sophisticated cognitive functionality in handheld devices will exist because it is provided by the BSs and the infrastructure made available by them. Subscriber devices may become more versatile and more fully capable, but it is likely that their intelligence will be activated or downloaded only as necessary to enable selected features. The actual reasoning, spectral evaluation, and network management will be managed by the BS. The basestation will convert that logic into controls and software applets to control the subscriber units.

Higher application-level cognitive functions will first be integrated by cellular service providers. By putting highly polished special-purpose applications somewhere in the data infrastructure, time scheduling, location tracking, event awareness, location-based opportunity awareness, and business-specific knowledge niches can readily be served with incremental investment and nearly negligible cost to the position-aware subscriber unit. Furthermore, each new capability can be developed, polished, test-marketed, and rolled out to the customer base, resulting in the perception of gradually increasing cognitive capability in the subscriber unit.

To enable this, the subscriber units will need to be SDRs. The ability to shift waveform and channel protocol may be limited to the current properties of the telecommunication network, but sufficient computational resources must be provided to serve the application and the application interface to the user, and to download the application servelets on request. Thus, we will expect to see some maturing of the user interface, the memory, and computational support in subscriber devices.

However, we expect to see more significant change to the BSs and infrastructure, as cognitive services begin to be integrated. We can reasonably expect heavier data services to bring significant change to the BS. Voice over Internet Protocol (VoIP) will become a more common voice-coding choice, especially if IP is already being provided for data services. The ability to authenticate a subscriber, download a servelet to the subscriber, authenticate his or her service requests, and then perform the corresponding network transactions suggests that BSs will have significantly more server functionality and wider bandwidth data networks.

Telephony services experience high variability of traffic demand, from very few calls late at night, to incredible demand when airplanes land or when the baseball game is over. Servicing high-demand variability has a significant cost impact. The network neither wants to deny calls at peak time, nor have idle infrastructure when there is far less demand. There are significant economic opportunities associated with spectrum sharing. If there are other services willing to briefly sublease their spectrum to serve these conditions, the radio technology will need to be able to shift to new frequencies under control channel handover. This implies additional changes to the handsets and to the BSs to accommodate this degree of flexibility, and it remains to be seen if the industry considers this to be an economically cost-effective move.

24.3.3 Data Radios

Unlike telephony devices, PDAs and portable computers are first a computer with data network capabilities, possibly including WPAN (e.g., Bluetooth, ZigBee, or ultra-wideband (UWB)), WLAN (e.g., 802.11), WMAN (e.g., WiMAX), and cellular telecommunication services as additional features. Therefore, the PDA software can, in principle, make economic trade-offs on behalf of the user as to which functions ought to be performed through which networks by considering cost effectiveness, timeliness, and specialized advanced capabilities of various networks.

The choice between VoIP delivered through the WLAN or the WMAN networks or voice telephony through cellular could easily be made as a function of the user's or subscriber's habits (e.g., walking a short distance or driving a long distance). Whichever network will experience the fewest disruptions over the predicted performance of the next five minutes may be the preferred initial access, and thereby minimize disruption associated with a service handover.

It is also likely that more memory and computational resources are available than currently in a standard cellular subscriber device. Therefore, it is likely that such devices will choose which cognitive functions are performed locally and which functions are performed in the infrastructure. Because these devices are primarily data devices, they may inherently offer a broader portfolio of cognitive capabilities, matched appropriately to their form factor. For example, a PDA may act on scheduled activities and timeliness issues, operating from the user's pocket, whereas a laptop may operate on longer range time windows with more complex tasks. Both may find it convenient to draw on database servers and distributed computational servers to perform cognitive selections, trade-offs, and priorities. Although these data services may be similar to the telecommunication services, the economic model supporting deployment is likely to be quite different. Perhaps we will see academic prototypes rapidly spin out as cognitive server support businesses (CSSBs) based on drag-along advertising rather than the pay-per-month or pay-per-user-minute models of telecommunications infrastructure cost.

24.3.4 Cognitive Services Offered through Infrastructure

Chapter 11 introduced the topic of cognitive services offered through infrastructure, which is reviewed here to more fully list its relevant opportunities. Populating these opportunities will rapidly proliferate CR functionality.

At the PHY layer, the radio environment map is a powerful way to keep track of node location, waveforms, networks, timeslots, traffic volume, motion prediction, the local interference noise level, hidden nodes, silent emergency services, telecom, FM, TV, and locally relevant spectral considerations. Access to the local server in support of PDA or telecommunication BSs seeking access to extra spectrum is quite feasible. Furthermore, it is likely that the REM can be readily expanded to perform the spectrum access requests between primary and secondary spectrum users, to carry out any required financial transactions, and to manage access and priority override.

In addition, the same infrastructure server can readily serve the local regulatory policy within the local regulatory boundary. This can be handled in a fashion similar to a cellular handover between BSs, with the infrastructure recognizing a local boundary crossing and performing a handover message to the appropriate adjacent server. If financial transactions are required for spectrum access, the policy server and the spectrum server will likely serve as the banker for all parties.

The schedule keeper and the task priority manager can use the node position information transacted with the REM to keep track of how closely the user is following his or her activity plan. Significant deviations from the plan can be used to notify subsequent interaction spots of progress or schedule updates, and even schedule "meet me" locations and times.

Perhaps one of the most significant servers we all want is the extension of the GPS-based tour guide that tells you how to get from location to location, but adds the current traffic flow into consideration. If we all knew about roads with excessive congestion, accidents, or road maintenance activities, and the best current alternate choices, perhaps we would all get to work and get home more expeditiously. This function can be derived from reports to the local REM database from mobile PDA devices. By observing velocity and location reports, the average velocity and traffic light wait times along various road choices leading to the destination can be compared, and recommended alternate paths can be identified to the motorist.

Likely, there will be a server for local service advertisements. For food or services of special interest to the user, the location, the schedule pressure, the agenda of activities, and the experience of the user's common choices provide all the information necessary for local service advertisements to selectively provide welcome suggestions and avoid the nuisance factor of excessive advertisement.

Such services do not require breakthroughs of science or technology. They can be implemented from a rigorous functionality specification. The industry must prepare the functionality specifications, the interface message requirements, and the transaction protocol sequences. This work can be brought forward by a standards organization or by a de facto standard product that experiences widespread adoption.

The primary issue is the business model. How will the cost of the equipment and its maintenance be supported and be profitable? And how many subscribers must already exist before it makes sense to deploy the infrastructure? These questions are not unlike those for the rollout of cellular infrastructure. Small infrastructure designed to serve large regions is gradually reduced in radius as subscriber density grows and the business grows in cost effectiveness. Perhaps the most compelling business-case arguments are that gasoline consumption and pollution can be lowered and communities can become more desirable places to live if the cognitive services are made available as local services. It is therefore likely that funding for some cognitive services startups will be progressively municipal investments.

Other services will be invented to serve the users. Gradually many niches will be populated, providing the aggregated learning experience of academy, industry, and regulator communities; the creativity of the user community; and further validation of new marketing opportunities. New and clever services that can be provided by adding a new database server or a new computational server to the existing infrastructure can appear to make the CR grow smarter, but with no hardware change and minimal software change.

Even so, there will always be specialized services that do not exist, or are not directly supported by infrastructure servers. Some of these needs may be met as a script of many basic services woven together to meet a specific objective. But in the limit, the radio must learn to implement cognitive functions not supported by well-tuned infrastructure-based servers. These functions will truly follow the ontology-based reasoning and learning now in development in AI laboratories, as described by Bostian, Kovarik, Kokar, and Mitola in earlier chapters of this book.

24.3.5 The Remaining Difficult Problems

Many of the problems to bring the vision of CRs forward to products are technical implementation issues and business case issues. However, the truly difficult problems are going to be the regulatory ones. These nontechnical problems are frequently decided on criteria that cannot be anticipated by technologists.

Each country has a body of regulators who specify how citizens and noncitizens are allowed to use spectrum. In the United States, the FCC specifies this policy, and currently, WPAN, WLAN, and WMAN (or WRAN) devices must meet manufacturers' requirements to operate in a certain band (say 2.4 GHz) and at a certain effective isotropic radiated power (EIRP), with a specified spectral shape (spectral mask) on the out-of-band transmissions. Most countries have similar organizations, each with different rules. The rules can be provided to the CR, which can enforce its compliance to them if they are effectively transcribed. Rules will include necessary extensions to accommodate spectral cognition (e.g., etiquettes or protocols to request access to the spectrum from primary owners) and to perform the corresponding transactions. The regulators will need assurance that the protocols are robust under all network-loading conditions, that they cannot be tampered with (hacked), that radios are guaranteed to follow the policy in a regulatory region, and that the networks of the primary owners will remain stable under a wide range of CR-loading requests.

The specification and analysis of the protocols and etiquettes is currently a very difficult problem at the boundary between the technical community and the regulatory community. The regulatory organization needs to be able to specify the local rules. They need to be assured that those rules are flawlessly translated. They need to be assured that the primary spectrum owners will retain performance and network stability under all conceivable (and inconceivable) conditions. However, flawless translation into machine-readable form and protocol analysis under a wide regimen of usage conditions are difficult to ensure for the regulator. System performance analysis can be done offline by computer simulation. Formal methods have some promise to allow us to make assertions about the machine-readable policies. This appears to be a new domain, with hard problems, requiring new science.

This book should create sufficient interest in the research and development community to find solutions to these remaining problems.

REFERENCES

[1] Cook, P., and S. Hope, Technology and Application Considerations for 3G Profitability, SDR Forum Technology Conference, November 2004.

[2] Cohlman, D., Demonstration of DMR Networking and Policy Engine, *Proceedings IEEE Milcom Conference*, October 2005.

[3] Kleider, J., S. Gifford, S. Chuprun, and B. Fette, Radio Frequency Watermarking for OFDM Wireless Networks, *IEEE ICASSP Conference,* pp. 397–400, May 2004.

2G	second generation	AOA	angle of arrival
3-D	three-dimensional	AOP	aspect-oriented programming
3G	third generation	AP	access point
3GPP	Third Generation Partnership Program	APCO	Association of Public Safety Communications Officials
A/D	analog-to-digital	API	application programming interface
A3V	advanced amphibious assault vehicle	AR	access router
AA	aware-adaptive (radio)	ARIB	Association of Radio Industries and Businesses (Japan)
AACR	aware, adaptive, and cognitive radio	ARM	available resource map
AAS	Adaptive Antenna System	ARPU	average revenue per user
ABCS	Army Battle Command System	ARQ	automatic repeat request
ACERT	Adaptive Cognition-Enhanced Radio Teams	ASIC	application-specific integrated circuit
ACIP	Architectures for Cognitive Information Processing	ASIP	Advanced System Improvement Program
ACL	access control list	ASK	amplitude shift keying
ADC	analog-to-digital converter	ATC	air traffic control
AES	Advanced Encryption Standard	ATF	Alcohol, Tobacco & Firearms (US Department of Justice)
AFRL	Air Force Rome Labs	ATIS	Alliance for Telecommunications Industry Solutions (US)
AGC	automatic gain control		
AI	artificial intelligence		
ALE	automatic link establishment (protocol)	ATM	asynchronous transfer mode
AM	amplitude modulation	ATSC	Advanced Television Systems Committee
AMC	artificial magnetic conductor; Adaptive Modulation and Coding	AWGN	additive white Gaussian noise
AME	amplitude modulation equivalent	BAA	Broad Area Announcement
AML	automated machine learning	BCV	battle-command vehicle
		BER	bit error rate
AMPS	Advanced Mobile Phone System	BIST	built-in self-test
		BIT	built-in test

BOI	band of interest
BOK	bioorthogonal keying
bps	bits per second
BPSK	binary phase shift keying
BREW	Binary Runtime Environment for Wireless
BS	basestation
BWA	broadband wireless access
C2V	command-and-control vehicle
CA	Certification Authority
CAPI	cognition API
CASA	computational auditory scene analysis
CASE	computer-aided software engineering
CASIG	Cognitive Applications Special Interest Group
CB	Citizens' Band
CBDT	case-based decision theory
CBP	component-based programming
CBR	case-based reasoning
CCK	complementary code keying
CCM	configurable computing machines; custom computing machine
CCSA	China Communications Standards Association
CDD	Capabilities Deployment Document (JTRS)
CDMA	code division multiple access
CDPD	Cellular Digital Packet Data
CE	cognitive engine
CEP	circular error probability
CF	cyclic frequency
CISC	complex instruction set computer
CLDC	Connected Limited Device Configuration
CLI	command-line interface
CMRS	commercial mobile radio service

CN	cognitive network
COMREG	Commission for Communications Regulation (Ireland)
COMSEC	communications security
CONUS	continental United States
COPS	Common Open Policy Service
CoRaL	cognitive radio language
CORBA	Common Object Request Broker Architecture
COTS	commercial off-the-shelf
CPC	cognitive pilot channel
CPDA	cognitive PDA
CPE	consumer premise equipment
CPU	central processing unit
CR	cognitive radio
CRA	cognitive radio architecture
CRC	cyclic redundancy check
CRM	cognitive resource manager
CR-MAS	cognitive radio–multiantenna system(s)
CRN	cognitive radio network
CRT	cognitive radio terminal
CRWG	Cognitive Radio Working Group
CS	cyclostationary; convergence sublayer
CSA	client–server architecture
CSEL	Combat Survivor/Evader Locator
CSF	channel selection function
CSI	channel state information
CSIS	Center for Strategic and International Studies
CSL	Cognitive Specification Language
CSM	cognitive system module; collaborative spatial multiplexing
CSMA/CA	Carrier Sense Multiple Access with Collision Avoidance
CSSB	cognitive server support business

CTC	convolutional Turbo Code
CTS	clear-to-send
CTVR	Centre for Telecommunications Value–Chain Research (Trinity College, Dublin)
CVSD	continuously variable slope delta (modulation)
CW	continuous wave
CWN	cognitive wireless network; composite wireless network
CWPDA	cognitive wireless personal digital assistant
CYC	artificial intelligence project (also Cyc)
D/A	digital-to-analog
DAC	digital-to-analog converter
DAI	distributed artificial intelligence
DAMA	Demand Assigned Multiple Access (NATO)
DAML	DARPA Agent Markup Language
DARPA	Defense Advanced Research Projects Agency
dB	decibel
dBm	decibels above 1 milliwatt
dBW	decibels above 1 watt
DC	direct current
DCD	Device Configuration Descriptor; data carrier detect
DCS	dynamic channel selection
DEA	Drug Enforcement Administration (US)
DECT	digital European cordless telephone
DES	Digital Encryption Standard
DFS	dynamic frequency selection
DH3	Data High Rate 3 Frames
DL	data link (layer)
DM	delay-and-multiply
DM1	Data Medium Rate 1 Frame

DMD	Domain Manager Configuration Descriptor
DME	distance-measuring equipment
DMR	digital modular radio
DNF	disjunctive normal form
DOA	direction of arrival
DoC	Department of Commerce (US)
DoD	Department of Defense (US)
DOT	Department of Transportation (US)
DPC	dirty paper coding
DPD	Device Package Descriptor
DPSK	differential phase-shift keying
DSA	dynamic spectrum access
DSB	Defense Science Board (US)
DSL	digital subscriber link
DSM	dynamic spectrum management
DSP	digital signal processor; digital signal processing
DSR	distributed speech recognition
DSSS	direct sequence spread spectrum
DTE	digital terminal equipment
DTN	Delay Tolerant Networking; Disruption Tolerant Networking
DTV	digital television
DX	long-distance transmission and reception (ham radio)
DYSPAN	IEEE Dynamic Spectrum Access Conference
E2R	End-to-End Reconfigurability (a European research project)
EA	evolutionary algorithm
EARS	Effective Affordable Reusable Speech Recognition (DARPA program)

E_b/N_o	energy per bit over noise
ECA	event-condition-action (rules)
ECCM	electronic counter-countermeasure
ECF	Earth-centered fixed
ED	energy detection
EDGE	Enhanced Data rates for GSM Evolution
EEN	enriched experience networks
EER	equal-error rate
EGG	electroglottogram
EIRP	effective isotropic radiated power
EMC	electromagnetic compatibility
EMS	emergency medical service(s)
EPG	exact potential game
EPLRS	Enhanced Position Location Reporting System
ES	equipower surface
ETSI	European Telecommunications Standards Institute
FA	false alarm
FAA	Federal Aviation Agency (US)
FBI	Federal Bureau of Investigation (US)
FCC	Federal Communications Commission (US)
FDMA	frequency division multiple access
FDoA	frequency difference of arrival
FE	front end
FEC	forward error correction
FEMA	Federal Emergency Management Agency (US)
FER	frame error rate
FFT	fast Fourier transform
FHSS	frequency-hopping spread spectrum

FIFO	first in, first out
FIP	finite improvement path
FIR	finite impulse response
FM	frequency modulation
FOL	first-order logic
FOPC	first-order predicate calculus
FPGA	field-programmable gate array
FR	first responders
FRS	Family Radio Service
FSK	frequency shift keying
FSM	frequency smoothing method
FTP	File Transfer Protocol
GA	genetic algorithm
GAO	Government Accountability Office (US; before 2004, General Accounting Office)
GCR	ghost canceling reference
GDOP	geometric dilution of precision
GIG	Global Information Grid
GIS	geographical information system(s)
GL	geolocation
GLF	geolocation function
GLONASS	GLObal NAvigation Satellite System (Russian)
GMM	Gaussian mixture models
GMSK	Gaussian minimum shift keying
GOPG	generalized ordinal potential game
GOS	grade of service
GPP	general-purpose processor
GPRS	General Packet Radio Services
GPS	global positioning system(s)
GSC	Global Standards Collaboration (ITU)
GSM	Global System for Mobile Communications
GUI	graphical user interface

GϵPG	generalized ϵ potential game
HAL	hardware abstraction layer
HANC	Hybrid Agent for Network Control
HCI	human–computer interface
HDLC	High-Level Data Link Control (protocol)
HF	high frequency
HMI	human–machine interface
HMM	hidden Markov model
HTML	HyperText Markup Language
HTTP	HyperText Transfer Protocol
HW	hardware
Hz	hertz
I/O	input/output
IAN	Internet area network
IBL	instance-based learning
IBS	Integrated Broadcast Service
ICNIA	Integrated Control Navigation Identification Architecture
iCR	ideal cognitive radio
ICT	Information and Communication Technologies (Europe)
ICTAP	Interoperable Communications Technology Assistance Program
ID	identification
IDIS	Intra-Device Interface Standard
IDL	Interface Definition Language
IEEE	Institute of Electrical and Electronics Engineers
IETF	Internet Engineering Task Force
IF	intermediate frequency
IIP2	Input Intercept Point, Second Order
IIP3	Input Intercept Point, Third Order
IIR	infinite impulse response
IM	instant message; instant messaging
IMD	inter-modulation distortion
INFOSEC	information systems security
INS	inertial navigating system; Immigration and Naturalization Service (now US Immigration and Customs Enforcement)
IP	intellectual property, Internet Protocol
IQB	inphase-quadrature biorthogonal (modulation)
IRC	Internet relay chat
IrDA	Infrared Data Association
IRG	Internet Research Group
ISAPI	information services API
ISI	Inter-Symbol Interference
ISM	industrial, scientific, and medical
ISO	International Organization for Standardization
ISP	Internet service provider
ISSI	Inter-RF Subsystem Interface
IST	Information Society Technologies (EU)
IT	information technology
ITU	International Telecommunication Union
J2EE	Java 2 Enterprise Edition
J2ME	Java 2 Micro Edition
J2SE	Java 2 Standard Edition
JAN	jurisdictional area network
Java RMI	Java Remote Method Invocation
JPO	Joint Program Office (US JTRS)
JTP	Java Theorem Prover
JTRS	Joint Tactical Radio System (US)

JTT	Joint Tactical Terminal	MANET	Mobile Ad hoc Networking
JVM	Java Virtual Machine	MAS	multiantenna system(s); multiagent system(s)
KDD	knowledge discovery in databases	MBITR	Multiband Intra/Inter-Team Radio
KNN	K nearest neighbor	MBMMR	Multiband Multimission Radio
KQML	Knowledge Query and Manipulation Language; Knowledge Query Markup Language	MBOA	Multi-Band OFDM Alliance
		MBWA	mobile broadband wireless access
KS	knowledge source	MDS	multipoint distribution system
LAN	local area network	MELP	mixed excitation linear prediction
LBSR	location-based spectrum right(s)	MEM	microelectromechanical
LCD	liquid crystal display	MESA	Mobility Emergency Safety Applications
LDAP	Lightweight Directory Access Protocol	MF	matched filter
LDPC	low-density parity check	MIB	management information base
LE	logical elements	MIDS	Multifunction Information Distribution System
LF	low frequency		
LFSR	linear feedback shift register	MIMO	multiple input, multiple output
LID	language identification	MIPS	million instructions per second
LISP	list processing (language)		
LLC	logical link control	MIRS	Metropolitan Interoperability Radio System (Washington, DC)
LLR	log-likelihood ratio		
LM	logic modules		
LMR	land mobile radio	ML	machine learning; maximum likelihood
LNA	low-noise amplifier		
LORAN	LOng RAnge Navigation	MLE	maximum likelihood estimation
LOS	line-of-sight		
LP	linear programming	MLME	MAC layer management entity
LPD	low probability of detection		
LPI	low probability of intercept	MMDS	multichannel, multipoint distribution system
LTM	long-term memory		
LTP	local tangent plane	MMSE	minimum mean square error
M.O.S.	mean opinion score	MMU	memory management unit
M3	multiband, multimode, multimedia (radio)	MODM	multi-objective decision making
MA	multiple access; multiple antenna	mops	mathematical operations per second
MAC	medium access control; multiple access control	MoU	Memorandum of Understanding
MAN	metropolitan area network		

MRC	maximum ratio combining	NTIA	National Telecommunications and Information Administration (US)
Msps	million samples per second		
MT	machine translation		
MTBF	mean time between failure(s)		
MU	multiuser	OBR	ontology-based radio
		ODP	open distributed processing
NADC	North American digital cellular	OEM	original equipment manufacturer
NATO	North Atlantic Treaty Organization	OFCOM	Office of Communications (UK)
NC3	NATO Consultation, Command and Control	OFDM(A)	orthogonal frequency-division multiplexing (access)
NCIC	National Crime Information Center	OIL	Ontology Inference Layer
NCMS	Network Control and Management System	OIP3	output intercept point, third order
NCRs	networks of cognitive radios	OMG	Object Management Group
NE	Nash equilibrium	OODA	observe, orient, decide, act (Boyd's loop)
NET	network layer	OOK	on-off keying
NGI	Next Generation Internet	OOP	object-oriented programming
NIJ	National Institute of Justice	OOPDAL	observe, orient, plan, decide, act & learn (loop)
NIST	National Institute of Standards and Technology (US)	OPEX	operating expense
		OPG	ordinal potential game
NKRL	National Knowledge Representation Language	OPS	operations per second
		OS	operating system
NL	natural language	OSA	opportunistic spectrum access
NLECTC	National Law Enforcement and Corrections Training Center	OSI	Open Systems Interconnection
NLP	natural language processing	OSIC	ordered successive interference cancellation
NM	network manager	OSM	operator spectrum manager
NMEA	National Maritime Electronics Association	OTAR	over-the-air rekeying
		OTH	over-the-horizon
NRAO	National Radio Astronomy Observatory	OWL	Web Ontology Language
NRE	non-recurring expense	OWL-QL	OWL Query Language
NRM	network reconfiguration manager	P25	Project 25 (US public safety communications program)
NSA	National Security Agency		
NSF	National Science Foundation (US)	PA	power amplifier; power adjustment
NTDR	Near-Term Digital Radio (system)	PAM	passband amplitude modulation

PAN	personal area network
PBNM	policy-based network management
PC	personal computer; portable computer
PCIM	Policy Core Information Model
P-code	precision code (GPS)
PCS	personal communication system; personal communication service(s)
PDA	personal digital assistant
PDF	probability density function
PDL	Policy Definition Language
PDP	policy decision point
PDR	programmable digital radio
PE	policy engine
PEP	policy enforcement point; performance-enhancing proxy
PER	packet error rate
PHY	physical layer
PIM	Platform Independent Model
PKI	public key infrastructure
PKM	public key management
PL	policy language
PLL	phase locked loop
PLME	PHY layer management entity
PLMR	private land mobile radio
PLRS	Position Location Reporting System
PN	pseudorandom
PNSI	perfect next-state information
POEMA	Policy-Enabled Mobile Applications
PolySurv	Policy-Based Survivable (DARPA communications program)
POSIX	Portable Operating System Interface
PPDR	Public Protection and Disaster Relief
PPS	precise positioning service; pulse per second

PPSI	perfect previous-state information
PRF	Properties Descriptor File
PS	public safety
PSD	power spectral density
PSF	pulse shape filter
PSK	phase shift keying
PSM	Platform Specific Model
PSPP	Public Safety Partnership Project
PSTN	public switched telephone network
PTT	push-to-talk
PU	primary user
PVM	Parallel Virtual Machine
Q	quality factor (generally used to describe bandwidth of filters)
QAM	quadrature amplitude (and phase) modulation
QASK	quadrature amplitude shift keying
QoI	quality of information
QoS	quality of service
QPSK	quadriphase shift keying
R&D	research and development
R&O	Report and Order (FCC)
R&TTE	Radio & Telecommunications Terminal Equipment (ITU directive)
RA	regulatory authorities
RAM	random access memory
RAN	radio access network; regional area network
RAT	radio access technology; radio access terminal
RBAC	role-based access control
RDF	Resource Description Framework
RDFS	RDF Schema (language)
RDQL	RDF Query Language
REAL	real-world reasoning
REM	radio environment map
RF	radio frequency

RFC	Request for Comment	SATCOM	satellite communication
RFFE	radio frequency front-end	SBTC	Shorted Block Turbo Code
RFI	Request for Information	SCA	Software Communications Architecture
RFIC	radio frequency integrated circuit	SCC	Standards Coordination Committee
RFID	radio frequency identification	SCD	Software Component Descriptor; spectral correlation detector(s)
RFOPL	restricted first-order predicate logic	SCF	spectral correlation function
RFSA	radio frequency situational awareness	SCR	synchronous collision resolution
RISC	reduced instruction set computer	SDCR	software-defined cognitive radio
RKRL	Radio Knowledge Representation Language	SDMA	spatial domain multiple access
RL	reinforcement learning	SDR	software-defined radio
RMC	RAN measurement collector	serModel	model of stimulus-experience-response
ROI	return on investment	SeW	Semantic Web
RPC	remote procedure call	SFDR	spur-free dynamic range
RRC	RAN reconfiguration controller	SIC	successive interference cancellation
RSS	received signal strength; rich site summary	SIG	Special Interest Group
RSSI	received signal strength indicator	SINCGARS	Single-Channel Ground and Airborne Radio System
RSVP	Resource ReSerVation Protocol	SINR	signal-to-interference and noise ratio
RT	real time	SIP	System Improvement Program
RTD	Research and Technology Development (EU)	SIR	signal-to-interference ratio
RTS	request-to-send	SIS	self-informing system
RTT	round-trip time	SISO	single-input, single-output
RX	receiver(s); receipt(s); reception	SLF	security sublayer function
RXML	Radio XML	SM	spectrum management; spectrum manager; spatial multiplexing
SA	situation awareness; spectrum automation	SME	small-to-medium enterprise; station management entity
SAD	Software Assembly Descriptor	SMF	spectrum manager function
SAK	Swiss army knife	SMR	specialized mobile radio
SAN	software-adaptable network	SMS	short message service
SAP	service access point	SNMP	Simple Network Management Protocol
SAPIENT	Situation Aware Protocols In Edge Network Technologies	SNR	signal-to-noise ratio

SoC	system-on-chip
SOF	Special Operations Forces
SONET	Synchronous Optical Network
SoR	Statement of Requirements
SP	sensory perception
SPD	software package descriptor
SPS	standard positioning service
SPTF	Spectrum Policy Task Force (US)
SQL	Structured Query Language
SR	software radio
SRA	Software Radio Architecture
SRW	Search/Retrieve Web (service)
SS	skill set(s)
SSB	single sideband
SSCA	strip spectral correlation analyzer
SSE	system strategy engine (also known as system strategy reasoner)
SSF	spectrum-sensing function
SSL	security sublayer
SSP	subset-sum problem
SSR	system strategy reasoner
STBC	space–time block code
STC	space–time coding
STM	short-term memory
STT	speech-to-text
SU	secondary user
SUO-SAS	Small Unit Operations–Situational Awareness Systems (US)
SVM	support vector machines
SW	software
SWR	software radio
SWRL	Semantic Web Rule Language
TC	Turing-capable
TCP	Transmission Control Protocol
TCP/IP	Transmission Control Protocol with Internet Protocol

TCXO	temperature-compensated crystal oscillator
TDD	time division duplex
TDMA	time division multiple access
TDoA	time difference of arrival
TETRA	TErrestrial Trunked RAdio
TI	Tactical Internet
TIA	Telecommunications Industry Association
TICP	Tactical Interoperable Communications Plan
TMC	terminal measurement controller
ToA	time of arrival
TOC	Tactical Operations Center
TPC	transmit power control; Turbo Product Code
TRANSEC	transmission security
TRC	terminal reconfiguration controller
TRM	terminal reconfiguration manager
TTA	Telecommunications Technology Association (South Korea)
TTC	Telecommunication Technology Committee (Japan)
TTR	time-to-rendezvous
TTS	text-to-speech
TV	television
TX	transmitter(s); transmit(s); transmission(s)
UAS	unmanned aerial system
UASI	Urban Areas Security Initiative
UAV	unmanned aerial vehicle
UDP	User Datagram Protocol
UHF	ultra-high frequency
ULCS	uplink channel sounding
UML	Unified Modeling Language
UPL	Universal Policy Logic; Universal Policy Language
URI	Universal Resource Indicator

URL	Universal Resource Locator
USRP	universal software radio peripheral
UWAN	unlicensed wide area network
UWB	ultra-wideband
VCM	virtual channel management
VHDL	VHSIC Hardware Design Language (DARPA VHSIC program language)
VHF	very high frequency
VHSIC	very high-speed integrated circuit
VLSI	very large-scale integration
VM	virtual machine
VoCoder	voice coder
VoIP	Voice over Internet Protocol
VOR	VHF omnidirectional ranging
VSB	vestigial sideband
VSWR	voltage standing wave radio
VT	Virginia Polytechnic Institute and State University; Virginia Tech
VT-CWT	Virginia Tech–Center for Wireless Telecommunications
W	watt
W3C	World Wide Web Consortium
WAN	wide area network
WB	wideband
WCDMA	wideband code division multiple access
WF	waveform
WG	working group
WGN	white Gaussian noise
WGS	World Geodetic System

WGS-84	World Geodetic System (1984)
WiBro	wireless broadband
Wi-Fi®	wireless fidelity
WiMAX	Worldwide Interoperability for Microwave Access
WIN-T	Warfighter's Internet
WLAN	wireless local area network
WMAN	wireless metropolitan area network
WNaN	Wireless Network after Next (DARPA program)
WPAN	wireless personal area network
WPG	weighted potential game
WRAN	wireless regional area network
WRC	World Radiocommunication Conference
WSGA	Wireless System Genetic Algorithm
WWRF	Wireless World Research Forum
WWV	NIST short-wave radio station that broadcasts time signals
WWVB	NIST long-wave (60 kHz) signal
WWVH	NIST short-wave radio station (Hawaii)
WWW	World Wide Web
XG	NeXt-Generation (DARPA radio communications program)
XML	eXtensible Markup Language
XTM	XML Topic Maps
YIG	yttrium-iron-garnet
YPR	yaw, pitch, and roll
ZF	zero forcing